# 5th International Conference on Computing and Solutions in Manufacturing Engineering (CoSME'20)



IOP Conference Series: Materials Science and Engineering
Volume 1009

Brasov, Romania
7 – 10 October 2020

**Editors:**

**Gheorghe Oancea**
**Mircea Viorel Dragoi**
**Cristin Olimpiu Morariu**

ISBN: 978-1-7138-4545-4
ISSN: 1757-8981

**Printed from e-media with permission by:**

Curran Associates, Inc.
57 Morehouse Lane
Red Hook, NY 12571

**Some format issues inherent in the e-media version may also appear in this print version.**



Institute of Physics
Dirac House, Temple Back
Bristol BS1 6BE UK

Phone: 44 1 17 929 7481
Fax: 44 1 17 920 0979

techtracking@iop.org

**Additional copies of this publication are available from:**

Curran Associates, Inc.
57 Morehouse Lane
Red Hook, NY 12571 USA
Phone: 845-758-0400
Fax: 845-758-2633
Email: curran@proceedings.com
Web: www.proceedings.com

# TABLE OF CONTENTS

# The 5th International Conference on Computing and Solutions in Manufacturing Engineering – CoSME'20

**7-10, October, Brasov, Romania**

Selected, peer reviewed papers from the 5th International Conference on Computing and Solutions in Manufacturing Engineering - CoSME'20

Edited by:

Gheorghe **OANCEA**
*Transilvania University of Brasov, Romania*

Mircea Viorel **DRAGOI**
*Transilvania University of Brasov, Romania*

Cristin Olimpiu **MORARIU**
*Transilvania University of Brasov, Romania*

IOP Conf. Series: Materials Science and Engineering 1009 (2021) 011001 doi:10.1088/1757-899X/1009/1/011001

# Preface

The 5th **International Conference on Computing and Solutions in Manufacturing Engineering – CoSME'20** was jointly organised by *Transilvania University of Brasov, Faculty of Technological Engineering and Industrial Management, Department of Manufacturing Engineering, Advanced Technologies and Manufacturing Systems Research Center* and *Shenyang Jianzhu University, School of Mechanical Engineering, Joint International Research Laboratory of Modern Construction Engineering Equipment and Technology*, during October 7-10, 2020. The organisers of this edition are located in Brasov, Romania and Shenyang, China.

Due to nowadays exceptional conditions imposed by SARS-CoV-2 pandemic, and because of the large number of new infections, CoSME'20 Conference was organised in online mode.

The objective of CoSME'20 Conference is "to bring together professionals from universities and industry to present and discuss recent advances in manufacturing engineering, to stimulate technical and scientific innovations on sustainable development for designing and manufacturing products, to provide a focus for the exchange of latest ideas on products engineering".

The papers submitted to CoSME'20 Conference were single-blind peer-reviewed, at least by two independent referees selected from the Reviewing Committee, which mainly consists of Members of International Scientific Committee. Assigning papers to referees was done according to their expertise in the domain of the papers. After a rigorous peer reviewing process 72 papers elaborated by authors/co-authors from 15 countries and 39 universities/research institutes and industry, have been accepted to be presented in the conference.

A brief overview on the CoSME'20 Conference reveals the following: 72 presented papers in the conference, 4 invited keynote speeches, 1 plenary presentation, 7 conference sections in which the papers were distributed, 151 participants from 13 countries (Bulgaria, China, Germany, Greece, Iraq, Moldova, Poland, Romania, Serbia, Slovakia, Taiwan, United Arab Emirates, and United Kingdom). The invited keynote speakers have been professors, outstanding specialists in the field of manufacturing engineering from United Kingdom (2), China (1) and Greece (1). For each keynote speech and plenary presentation were allocated 20 minutes for presentation and 10 minutes for comments, questions and answers, and for each paper the total amount of time was 15 minutes (presentation and Q&A). Based on the technical criteria, quality criteria and presentation criteria were evaluated the papers and selected the best papers in the conference sections. In this CoSME edition, seven Best Paper Awards were offered.

The software platform that was used to deliver the conference belongs to the Transilvania University of Brasov, Romania and it is based on the BigBlueButton technology. All the presenters of the papers and chairmen shared the video camera and the presentation. Using the online platform, the participants in the conference made comments and asked questions about the paper presented. All the works and activities of CoSME'20 Conference were successfully through this platform.

We express our gratitude to all members in the International Scientific Committee, invited keynote speakers, referees, participants in the conference and members in the Organising Committee for their valuable contribution to this successful edition of CoSME Conference.

We also address thanks to all those involved in organising of this edition of the conference, as well as to our partners Shenyang Jianzhu University, School of Mechanical Engineering for the fruitful cooperation and to IOP Conference Series: Materials Science and Engineering a well-known publisher.

Professor Gheorghe OANCEA, PhD,
Chairman of CoSME'20 Conference,
Dean of Technological Engineering and Industrial Management Faculty,
Transilvania University of Brasov,

Braşov, October 2020

IOP Conf. Series: Materials Science and Engineering 1009 (2021) 011001 doi:10.1088/1757-899X/1009/1/011001

# The 5th International Conference on Computing and Solutions in Manufacturing Engineering – CoSME'20

## International Scientific Committee Members

# Chair Persons

Gheorghe OANCEA, Chairman

Yuhou WU, Co-Chairman

Mircea Viorel DRĂGOI, Co-Chairman Cristin-Olimpiu MORARIU, Co-Chairman

Nicolae-Valentin IVAN, Honorary Founder Chairman

# Organizing Committee

| | |
|---|---|
| Papers Submission Coordinator: | Nicolae EFTIMIE |
| Website Administrator: | Camil LANCEA |
| Financial Administrator: | Luminița Aurica PÂRV |

# Members

| | | |
|---|---|---|
| M. BLAJ, RO | S. S. DUICU, RO | C. O. MORARIU, RO |
| G. R. BUICAN, RO | A. E. DUMITRAȘCU, RO | M. D. NĂSULEA, RO |
| H. BULEA, RO | N. EFTIMIE, RO | I. NEAGOE, RO |
| L. A. CHICOȘ, RO | L. FAN, CN | A. NEDELCU, RO |
| M. R. CLINCIU, RO | F. FECHETE, RO | G. OANCEA, RO |
| R. CLINCIU, RO | A. C. FILIP, RO | A. L. PÂRV, RO |
| F. CORCIOVA, RO | M. FOLEA, RO | H. SHI, CN |
| R. DAICU, RO | C. LANCEA, RO | R. UDROIU, RO |
| B. DEAKY, RO | B. LEPĂDĂTESCU, RO | M. URDEA, RO |
| V. DIȚU, RO | N. B. LUPULESCU, RO | H. WANG, CN |
| M. V. DRĂGOI, RO | L. A. MIHAIL, RO | S. M. ZAHARIA, RO |

# Invited Keynote Speakers

**Dragoș AXINTE**, *Professor, The University of Nottingham, UK*
**New models for time-dependent processes enable accurate generation of freeforms**

**Felician CÂMPEAN**, *Professor, University of Bradford, UK*
**Reliability with big data: towards intelligent personalized systems healthcare**

**Panagiotis KYRATSIS**, *Assoc. Prof., University of Western Macedonia, Greece*
**Computational design and applications**

**Shongua LI**, Professor, *Shenyang Jianzhu University, China*
**Design, fabrication and service of all-ceramic ball bearings for extreme conditions applications**

# Plenary Speaker

**Gheorghe OANCEA**, *Professor, Transilvania University of Brașov, Romania*
**Faculty of Technological Engineering and Industrial Management - more than 55 years of evolution**

IOP Conf. Series: Materials Science and Engineering 1009 (2021) 011002 doi:10.1088/1757-899X/1009/1/011002

# Peer review declaration

All papers published in this volume of *IOP Conference Series: Materials Science and Engineering* have been peer reviewed through processes administered by the Editors. Reviews were conducted by expert referees to the professional and scientific standards expected of a proceedings journal published by IOP Publishing.

- **Type of peer review: Single-blind**
- **Conference submission management system: EasyChair**
- **Number of submissions received: 84**
- **Number of submissions sent for review: 84**
- **Number of submissions accepted:72**
- **Acceptance Rate (Number of Submissions Accepted / Number of Submissions Received X 100): 85.7%;**
- **Average number of reviews per paper: 2.18**
- **Total number of reviewers involved:56**
- **Any additional info on review process: The papers were evaluated based on the technical criteria, quality criteria and presentation criteria. The referees argued detailed their reviews and also provided an overall evaluation (strong accept, accept, weak accept, borderline paper, weak reject, reject, strong reject), specifying their confidence (expert, high, medium, low, none) regarding qualifications to review the allocated paper. For each paper the similarity index was determined by the organisers using Turnitin software at Transilvania University of Brasov.**
- **Contact person for queries: Prof. Gheorghe Oancea; gh.oancea@unitbv.ro**

# The forming tool, primary start position and toolpath angle movement effect on the geometry accuracy "springback", part C

**K I Abass**[1*]

[1] Mechanical Engineering Department, College of Engineering, Mustansiriya University, Baghdad, Iraq

[*]E-mail: k.i.abass@uomustansiriyah.edu.iq

**Abstract**. Incremental Forming process, IF is a technology for producing 3D-dimensional complex sheet profiles. The advantages of the process involve simple set tools, fixture and forming tool, shorter time for producing the pattern, and flexibility. While the main limitation of the process is the springback that represent geometry accuracy, the part profile precision limits the difficulty of forming toolpaths, including the wall angle. This work presents the forming toolpaths effect in the IF process. The supporting space and the product base dimension including wall angle is used to study the springback effect through the incremental forming process. Sheet metals Aluminium-A1025 are used for manufacturing the products. The research goals are to comprehend the springback defect behaviour which is affected by the supporting space-toolpath angle and the supporting space-product base ratios. The strain distributions and the thinning percentage of the final product section have been considered and analyzed in detail.

## 1. Introduction

Incremental Forming, IF is a new innovative, flexible and combination process of machine with computer technology [1]. IF is a high potential forming process for rapid prototype and for small amounts production. The process it does not need a specialized dies (just type of a fixture). Computer Aided Manufacturing, CAM software is used for programing the tool path, editing and submitted to the NC milling machine for accomplishment [2]. IF process assists for producing the complex forms without expensive die sets. Therefore, leads to reduce the time for marketing. The applications of the process, adding of the prototyping are producing the parts for automobiles, biomedical and aerospace industries [3]. The main problems in IF process are the low accuracy in geometrical concave features which involve the irregular thinning distribution, the limited wall angle, and the surface quality.

The forming parameters considered in IF technology are: geometrical shape (flat and curvatures surfaces), material type of the die- tool set, tool geometry, tool path parameters (step size, wall angle and forming speeds (rotation and feed rate), lubrication, sheet thickness before and after forming process. Each parameter is affecting to overall forming and their influences on detected effects.

Forming toolpath: first step the models created by Computer Aided Design CAD for incremental forming, and using commercial software to generate forming toolpath, as a second step. Transition method (step or ramp) are using to create a tool contours and connect it [4]. Each step down and each contour is caused a surface roughness due to the physical indentation on the surface blank [5].

Forming angle: is the angle that resulted from the horizontal xy-plane and the side walls of a part. The range of this angle mainly depends on sheet thickness and material properties. On the other hand, IF profiles are determined by the highest forming angle, Ψ. Martins et al. calculated the highest forming angle referring of the forming parameters and material properties as in equation (1) [6].

$$\Psi = \pi/2 - e^{\epsilon_t} \quad (1)$$

The equation (1) represents the beginning of fracture, combines the concepts of both the maximum forming wall angle and the fracture forming limit in the principle strain space at the onset of crack. Where, Єt is the thickness strain and t is representing the thickness at fracture at the limit of formability. In this equation can be calculated using the through-thickness fracture limit strain Єt, as resolute of equi-biaxial stretch test or a plane strain.

Depending on the sine law shown in equation (2), impossible to make the right angle walls 90° because of the thickness distribution of the part wall would be equal zero.

$$h = h_o \sin\left(\frac{\pi}{2} - \alpha\right) = h_o \sin(\Psi) \quad (2)$$

where $h_0$ is the initial thickness, h is the final thickness, and α is the spinning angle.

The toolpath represents the main variables that are select by the designer depending upon, basically the product profile. Therefore, this variable can be modified and controlled to improve the part accuracy. Tool path is also the reason of the surface roughness and springback, and because of the wide range of the forming tool paths, many researches are trying to find relevant strategies to control this variable.

Predict the translation is developed analytically and experimentally validated of the rigid body in both in-to-out (IO) and out-to-in (OI) toolpaths [7]. The incremental forming process proposed a spiral toolpath method with constant scallop height and based on triangular mesh model [8].

A new parameterized forming strategy for the tool path optimization in SPIF was developed [9], in order to reduce the manufacturing time and homogenize thickness distribution of an asymmetric part. An effective tool path generation method, by generating Cutter-Location (CL) data directly from corresponding Cutter-Contact (CC) data, has been developed [10].

FE model of the truncated cone product was performed for a single and double pass forming process and analyzed. In addition, 3D toolpath as a basis of coordinates for building up the boundary condition (displacement and moving tool) was loaded [11]. Incremental process is upgraded by a coupling methodology for correcting the deviations during the toolpath movement [12]. As well, for predicting the forming forces a systematic and a robust method is improved by derive the elastic model and FE simulation.

A feature-based toolpath generation method has developed [13]. The surface quality, geometric accuracy and thickness distribution properties of the traditional incremental toolpath are compared with feature-based toolpath generation method by using three studies including a car fender, a nonsymmetrical cone and a truncated cone with double bottoms.

The conventional trajectory (Single Slope) 'from out to in and from up to down' can improved in order to compensate the irregular thinning. Several methods were proposed experimentally to solve this defect. Decremental Slope is a suggested method based on to apply over-slop, where the blank is typically less stretched. Next, a lower angle by the forming tool is imposed and the restricted over-thinning appears. Depending on this concept, the offset of + or - units was applied to the forming tool movement [14]. A novel tool feed path generation method was developed [15]. The idea is depending on FE analyses of the multi-passes incremental forming by hydraulic bulging. The results of hydraulic bulging CAE analysis offered a series of approximating surfaces.

The influence of the processing tool paths on the accuracy of the parts and on strains and thickness reduction has been shown. Some approaches of implementing SPIF process using as technological equipment either CNC milling machines or serial industrial robots have been unrolled [16]. Both simulation and experimental data are presented.

This work present the effects, primary start position (5-15 mm from die edge center with step 2.5 mm, plus 0.98 mm) and toolpath angle movement (that is a reflect of the start position, $\theta$ = 38.09°, 39.51°, 40.77°, 42.11°, 43.52°, 45.00°) of the forming tool on the springback in incremental forming. Aluminum alloy sheets (A1025) with thickness, t=0.9 mm are used to deform the product, truncate cone geometry. The study goals to build a ratio between the wall angle and primary start position of the tool forming. Adding, study of the step size effect on the strain distribution for the deformed part section and the elements are leading the final parts to fracture have been evaluated and discussed.

## 2. Development of FE Model

The incremental forming is represented by FE simulation model. The first stage includes physical profile of SPIF model, boundary conditions, and the process components (tables 1 and figure 1). The mechanical properties of Aluminum AA1050 material that used in this work are shown in table 2. The 2D- axis symmetric model components of the blank are used. The second stage reflects the shape evaluation and FEM analysis (success or failure product), in order to estimate the final product geometry. The third stage covers the calculation and conclusions for the final product.

**Table 1**. Initial IF process data.

| Variable | Value |
|---|---|
| Die radius, $R_d$ | 2.5 mm |
| Forming tool, $D_P$ | 5 mm |
| Blank diameter, $D_b$ | 226 mm |
| Blank thickness, t | 0.9 mm |
| x and y step size | 0.1 mm |
| Coefficient of friction, μ | 0.05 |

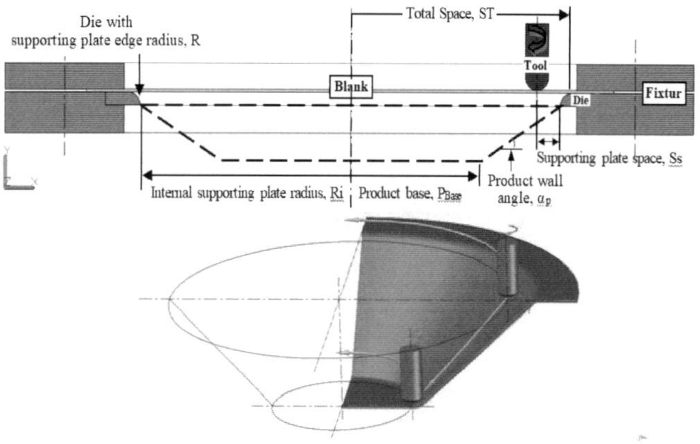


**Figure 1**. Components of the IF process.

**Table 2**. Al (AA1050), material properties of the sheet.

| Tensile Strength | Density, $\rho$ | Poisson's ratio, ν | Young's modulus, E | Tangent modules, Eτ |
|---|---|---|---|---|
| 105 - 145 MPa | 2700 g/m$^3$ | 0.3 | 75 GPa | 0.5 GPa |

The elements that have been chosen to represent the workpiece and the tool set illustrated in the previous researches [17, 18]. As well, the boundary conditions consider friction coefficient, non-linear analysis, and convergence criterion.

## 3. Results and Discussion

CAD profile and FE model are simulated of the cone shape with depth equal to 47 mm (figure 2). The boundary conditions are concerning the contact status-displacement, and the tool set motion-loading with specific profile of the constant speed as represented in figure 2-(1). The FE models develop the effect of the primary start position - toolpath angle was studied with constant values (dimension and materials) of fixture (die and blank holder) and blank.

The methodology covers two series (figure 2-(1), a and b), and both have the constant radius values of the die edge and forming tool, R=5 mm:

- Constant product base, $P_{Base}$ radius, 33 mm, with six values of the supporting space Ss, variance 2.5 mm for each model, and 0.98 mm Ss for calculating the failure statuses is used;

- Constant toolpath angle, 45°, with primary product base, $P_{Base}$ radius, 33 mm, and six values of the product base $P_{Base}$ (with variance Ss=2.5 mm for each model), and 0.35 mm Ss for calculating the failure statuses is used.

Successive stages using the toolpath angle, 45° to create the IF part profile are displayed in figure 2-(2). The wall angle is a part of the deforming path effects on the stresses values. So, the highest stress distribution values appear to be a good signal to estimate the success or failure of the deformed model.

The evaluation of stress distributions during the incremental steps 1, 10, 20, 30, 40, and 47 mm in depth are indicated the local effect of the forming tool. Increasing the forming depth is a reason for increasing the supporting space, Ss which create elastic states towered the die circumference is called the springback. Increasing the step size 'forming depth' is caused increasing the contact area and the friction region that causes failure. As a result, increasing the supporting space causes increasing the springback. In other word, increasing the forming depth increase the stresses locally (under forming tool) and along the deformed part section. The supporting space area does not suffer plastic stress during deformation steps highly, therefore the thickness distribution is close the original.

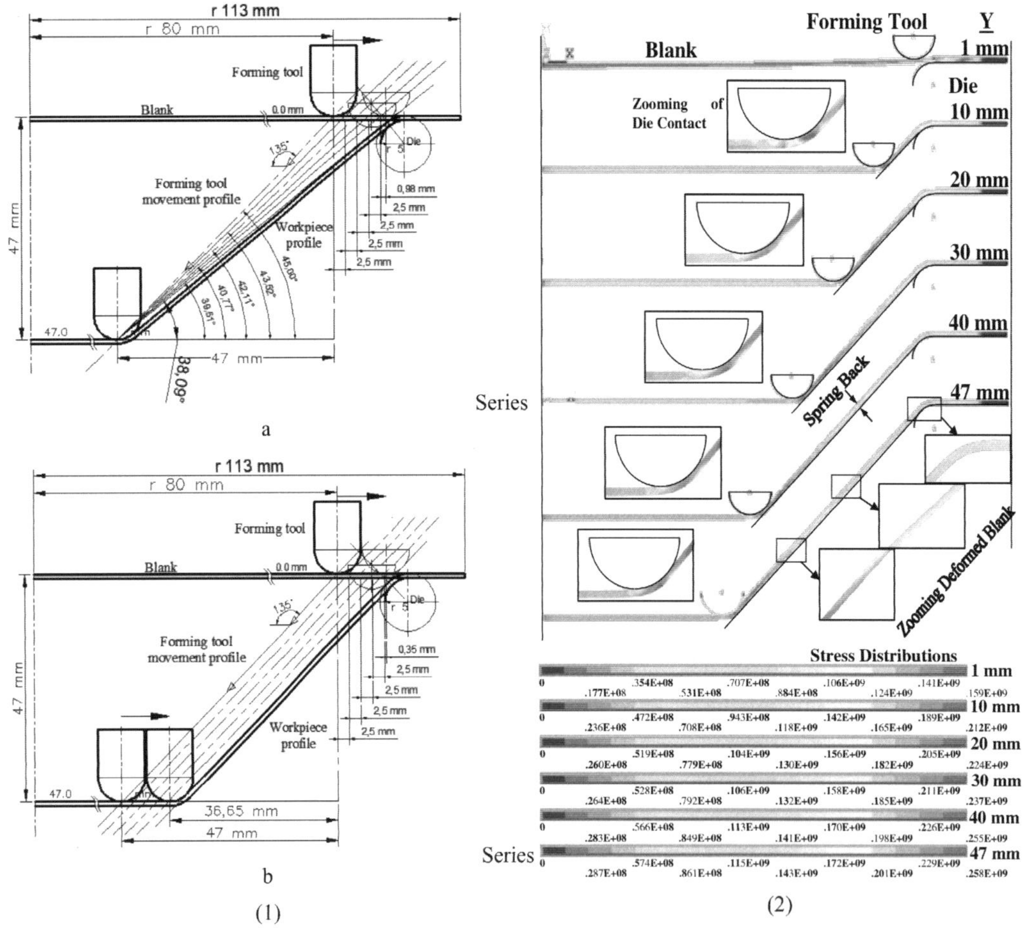


**Figure 2.** Methodology details of the primary start position - toolpath angle effect.

### 3.1. *Effect of space*

The springback and its increase with the space is appearance in both series a and b (figure 3 and 4), although fixed or changed the forming tool movement angle respectively. As well, changing the product base radius "reducing", it does not prevent increasing the springback with the space.

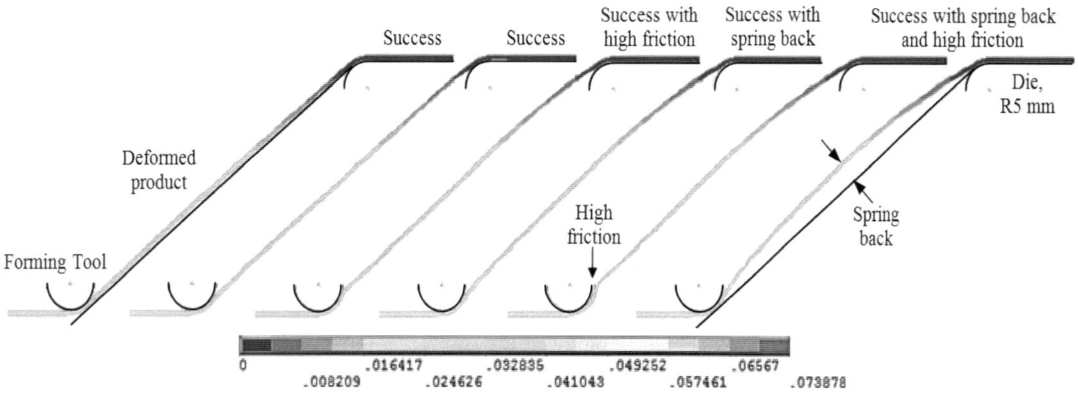


**Figure 3**. The products of SPIF process (displacements), series a.

### 3.2. *Effect of path*

The highest spring value can be recognize at the middle wall of the product, 2.4 mm and 2.41 mm of the series a and b respectively. The nature of failure that resulted in the product changed with the space value as showed in figure 4.

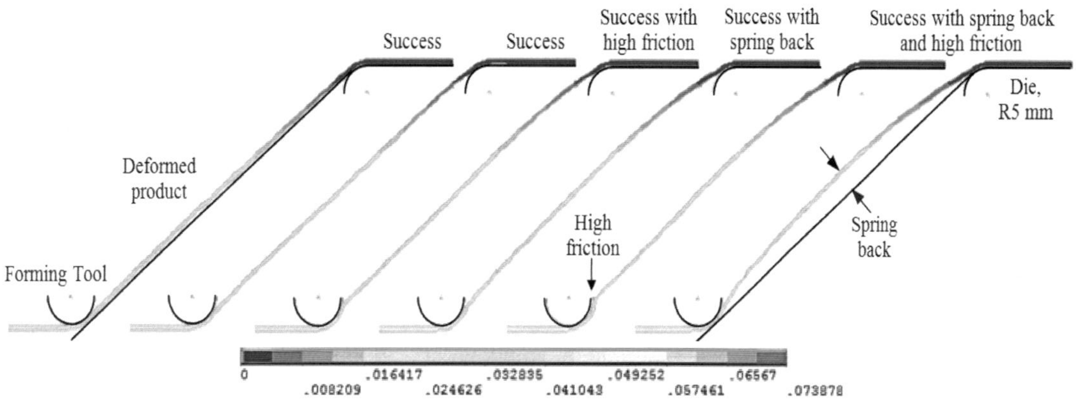


**Figure 4**. The products of SPIF process (displacements), series b.

The brief of the results are determined in the tables 3 and 4 of both series a and b as follows:

- Fixed or decrease the percentage $P_{Base}/ST$ (fixed or increase the forming tool movement angle), it does not have effect on the springback.
- Fixed or increase the percentage $\alpha°/\alpha_p°$ (fixed or increase the forming tool movement angle), it does not have effect on the springback.
- While, increasing the percentage $Ss/P_{Base}$, caused to increase the springback (fixed or increase the forming tool movement angle), because of the relation is involved the space variable too.
- The percentage relations $P_{Base}/ST$, Ri/ST and $\alpha°/\alpha_p°$, are detected between increasing, decreasing and fixing with the space-angle ratio Ss $\rightarrow$ $\alpha°$, which is do not give the clear indicate of the space-angle-springback ration. While the relation $Ss/P_{Base}$, is clear in both series, increasing the forming tool movement angle and the space causes increasing. As well, this relation involves the effect of angle on the springback.

**Table 3.** Series a, effect of toolpath angle, α° =38.09 - 45.00 - With product base fixed, $P_{Base}$ = 33.00 mm.

| Ss → α° | $P_{Base}$ | Ri, mm | ST, mm | $P_{Base}$/ST | Ri/ST | α°/$α_p$° | Ss/$P_{Base}$ | SB, mm |
|---|---|---|---|---|---|---|---|---|
| 4.02 → 38.09 | 33 | 90 | 95 | 0.35 | 0.95 | 1.00 | 0.1218 | 0.4375 |
| 5.00 → 39.51 | 33 | 90 | 95 | 0.35 | 0.95 | 1.04 | 0.1515 | 0.5491 |
| 7.50 → 40.77 | 33 | 90 | 95 | 0.35 | 0.95 | 1.07 | 0.2273 | 1.5958 |
| 10.0 → 42.11 | 33 | 90 | 95 | 0.35 | 0.95 | 1.11 | 0.303 | 2.3968 |
| 12.5 → 43.52 | 33 | 90 | 95 | 0.35 | 0.95 | 1.14 | 0.3788 | 3.0281 |
| 15.0 → 45.00 | 33 | 90 | 95 | 0.35 | 0.95 | 1.18 | 0.4545 | 3.63 |

**Table 4.** Series b, effect of product base, $P_{Base}$=43.35 - 33.00 mm - With product angle fixed, $α_p$° = 45.00°.

| Ss → α° | $P_{Base}$ | Ri, mm | ST, mm | $P_{Base}$ /ST | Ri/ST | α°/$α_p$° | Ss/ $P_{Base}$ | SB, mm |
|---|---|---|---|---|---|---|---|---|
| 4.65 → 45.0 | 43.35 | 90 | 95 | 0.456 | 0.95 | 1.00 | 0.1073 | 0.4299 |
| 5.00 → 45.0 | 43.00 | 90 | 95 | 0.453 | 0.95 | 1.00 | 0.1163 | 0.7877 |
| 7.50 → 45.0 | 40.50 | 90 | 95 | 0.426 | 0.95 | 1.00 | 0.1852 | 1.7017 |
| 10.0 → 45.0 | 38.00 | 90 | 95 | 0.400 | 0.95 | 1.00 | 0.2632 | 2.3513 |
| 12.5 → 45.0 | 35.50 | 90 | 95 | 0.374 | 0.95 | 1.00 | 0.3521 | 3.0854 |
| 15.0 → 45.0 | 33.00 | 90 | 95 | 0.347 | 0.95 | 1.00 | 0.4545 | 3.63 |

### 3.3. *Toolpath effect*

Toolpath is reflecting of the angles that are component of the product geometry. Therefore, the relation of wall angle-springback is a brief of the toolpath-springback ratio. The both series a and b are showed the proportional ration of the supporting space, Ss and forming tool movement angle, α° with the springback "Ss/ $P_{Base}$, - SB ratio", table 3, 4 and figure 6.

The aspect of series a, constant value of the supporting edge radius equal to R 5 mm. Forming tool center, with range of angle, α° from 25° to 45°, as well as, the product base is constant r = 33 mm, increase the space from 2.5 mm to 15 mm. adding S =0.98 mm to be close the edge radius and the gab equal to the thickness of blank.

The aspect of series b, supporting edge radius is constant R 5 mm, forming tool movement angle 45° and supporting spaces start 2.5 mm - with incremental level 2.5 mm until 15 mm, and with edge radius R mm respectively. Adding, the space 0.35 mm to be close to the blank thickness 0.9 mm, and squeeze the blank between the supporting edge radius and the forming tool.

In other words, testing plan is considered changing supporting space, Ss with constant edge radius of the supporting plate and fixed forming tool movement angle and changes it.

The primary fixture are more tough for vibration effect, when decreasing Ss, 15 to 2.5 mm. Increasing, Ss value make the contact area that under contact and deforming effect increase. As well as, the forming tool contact effect involves the largest surface area, and the largest deforming reign of the metal.

The deference in the thickness (thinning %) does not offer an indicator of increasing or decreasing the springback, as shown in figure 5.

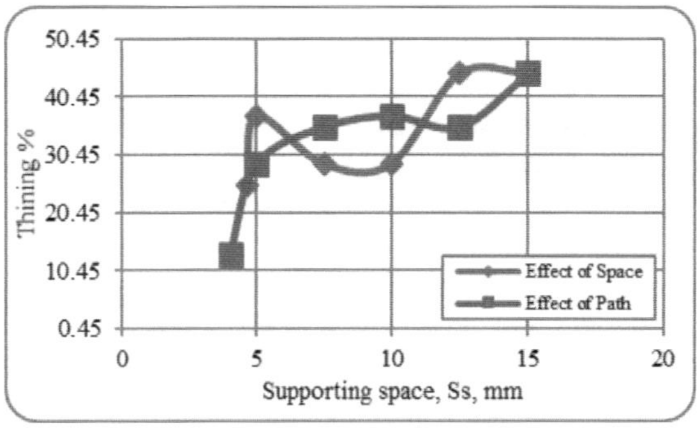


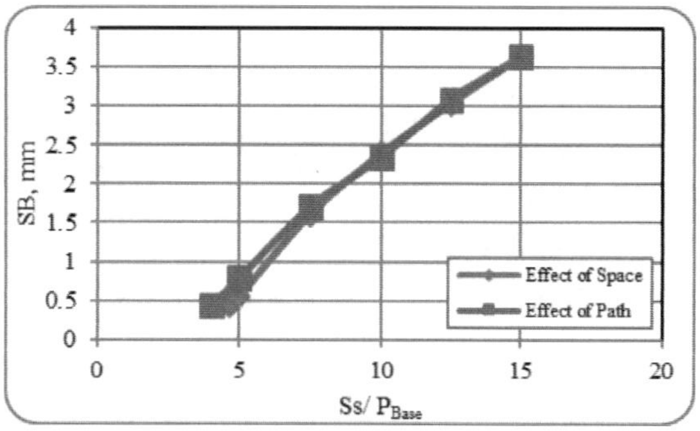


**Figure 5**. Maximum thinning % of Series a and b. **Figure 6**. Sprinkback indication of Series a and b.

The strain distributions divided into four regions form the tool center, 0-30 mm (series a), low levels of increasing appeared when compared to series b, (figure 7). The defenses are resulting due to the product dimensions, where the product base, $P_{Base}$ ranged from 43.35 to 33.00 mm. These dimensions in series b caused decreasing the strains with decreasing of supporting space Ss, while do not have real effect in Ss-$\alpha°$ ratio. Reflecting of the decreasing in the strains (region, 0-30 mm), the strains increased at the region 30-50 mm from the tool center (figure 8). The region from 50-70 mm from tool center at the series a, have been a level of strain distribution higher than at the series b. The region from 70 to the end edge of the blank have been the same behavior profile of both series a and b. The strain distributions at local point of the forming tool are increasing with increasing Ss. Remind the strains values are constant of the both series a and b, without effect due to the effect of fixture.

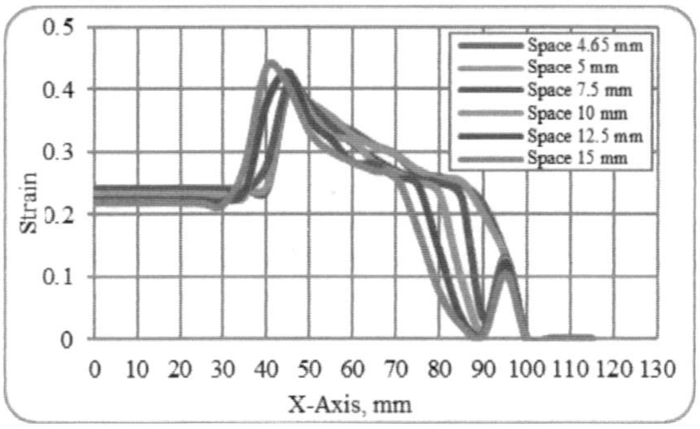


**Figure 7**. Series a, $\alpha°$ =38.09 - 45.00, equivalent strain distributions.

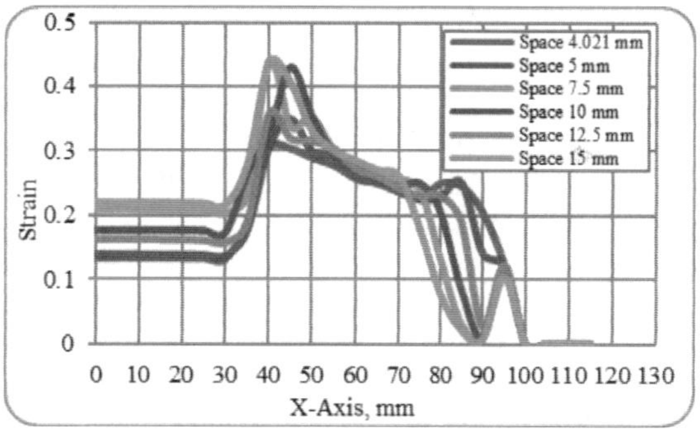


**Figure 8**. Series b, $P_{Base}$=43.35 - 33.00 mm, equivalent strain distributions.

## 4. Conclusions

The study is presenting the elements and influences of the supporting space - the angle toolpath and product base diameter ratios. Both series are respecting the correction formability in the incremental forming.

Two types of failure indicated in the resulted product, the deviation in the wall and wrap the blank part "under contact" around the forming tool. Therefore, mentioned as a nature of the failure.

The results show the supporting space with respect to the product profile ratios gives good results for reducing the springback effect. The important factor that caused increasing the springback is supporting space, which shared between both series.

The variations in strain distribution do not represent as indicator of the springback. The shared ratio, supporting space / product base, Ss/ $P_{Base}$, represent a better indicator of the springback defect.

**References**

[1] Centeno G, Silva M B, Cristino V A M and Vallellano C 2012 Hole-flanging by incremental sheet forming. *J. of Machine Tools & Manufacture* **59** 46
[2] Benmessaoud R 2014 A Tool path generation method for the multi-pass incremental forming process investigation *Int. J. Adv. Res. Computer Sci. and Software Eng.* **4** 1035
[3] Dittrich M A, Gutowski T G, Cao J and Roth J T 2012 Exergy analysis of incremental sheet forming *Prod. Eng. Res. Devel.* **6** 169
[4] Suresh K, Khan A and Regalla S P 2013 Tool path definition for numerical simulation of single point incremental forming *Int. Conf on Design and Manuf. IConDM 2013*, Procedia Engineering, **64** 536
[5] Wu S H, Reis A, Andrade Pires F M, Santos A D and da Rocha A B 2012 Study of tool trajectory in incremental forming *Adv. Mat. Res.* **472-475** 1586
[6] Silva M B, Skjoedt M, Atkins A G, Bay N, and Martins P A F 2008 Single- point incremental forming and formability–failure diagrams *J. of Strain Analysis for Eng. Design* **43(1)** 15
[7] Malhotra R, Bhattacharya A, Kumar A and Reddy N V 2011 A new methodology for multi-pass single point incremental forming with mixed toolpaths *CIRP Annals Manuf. Tech.* **60**(1) 323
[8] Zhu H, Liu Z and Fu J 2011 Spiral tool-path generation with constant scallop height for sheet metal CNC incremental forming *Int. J. Adv. Manuf. and Tech.* **54(9)** 911
[9] Azaouzi M and Lebaal N 2012 Tool path optimization for single point incremental sheet forming using response surface metho, *Simulation Modelling Practice and Theory* **24** 49
[10] Li M, Li-Chao Z, Jian-Hua and Lu Y 2012 Tool-path generation for sheet metal incremental forming based on STL model with defects *Int. J. Adv. Manuf. and Tech.* **63(5)** 535
[11] Li J, Hu J, Pan J and Pei G 2012 Thickness distribution and design of a multi-stage process for sheet metal incremental forming *Int. J. Adv. Manuf. and Tech.* **62(9)** 981
[12] Belchior J, Guillo M, Courteille E and Maurine P 2013 off-line compensation of the tool path deviations on robotic machining: Application to incremental sheet forming *Robotics and Computer-Integrated Manufacturing*, **29(4)** 58
[13] Lu B, Chen J, Ou H and Cao J 2013 Feature-based tool path generation approach for incremental sheet forming process *J. of Materials Processing Technology* **213(7)** 1221
[14] Filice L, Ambrogio G and Gaudioso M 2013 Optimised tool-path design to reduce thinning in incremental sheet forming process *Int. J. Material Forming* **6(1)** 173
[15] Zhang C, Xiao H and Yu D 2013 Negative Incremental Forming of Sheet Based on Interval Model of Hydraulic Bulging Cae Analysis *Rev. Adv. Mater. Sci.* **33** 360
[16] Bologa O 2014 Incremental Forming-An Alternative To Traditional Manufacturing Methods *Proc. in Manufacturing Systems* **9(3)** 131
[17] Abass K I 2018 Evaluation of the supporting plate edge radius and position effects on springback in incremental forming, part A, MATEC Web of Conferences **178** 02007
[18] Abass K I 2018 Supporting plate edge radius and internal hole diameter relation effect on the springback in incremental forming, part B, MATEC Web of Conferences **178** 02012

# Research on the study of the dimensional precision of the parts obtained by Additive Manufacturing

**D C Anghel[1*] and A D Rizea[1]**

[1] University of Pitesti, Manufacturing and Industrial Management Department, Târgul din Vale Street No.1, Romania

* E-mail: daniel.anghel@upit.ro

**Abstract**. This article presents research concerning the dimensional accuracy of parts obtained by additive manufacturing, 3D printing. The challenges of additive technology are exposed and also the need to have a vision on the influence of certain technological and geometric parameters on the precision of geometric surfaces. First, a virtual prototype was produced with Catia V5. Then, the virtual prototype was imported by the 3D machine software to obtain the prototyping codes for nine prototypes. The parameters of these prototypes were chosen with the Taguchi method. On the basis of the code obtained with the printer software, the parts have been prototyped, using the Z-ULTRAT material with a ZORTRAX M200 printer. All parts were measured and a database was created. The results were interpreted in order to establish the influence on the dimensional precision for the parts obtained by Additive Manufacturing.

## 1. Introduction

Today, the products have become more and more personalized, due to the high demands of the users. Additive Manufacturing is a new technology that allows the production of the parts with complex shapes and structures. Additive Manufacturing is defined as a process of joining materials to realize objects based on 3D model data, usually layer upon layer, as opposed to sub-tractive manufacturing methodologies, ISO/ASTM 52915 [1].

The development of the low-cost 3D printers was opened new perspectives for the designers and for the students at the engineering's faculty. Thus, they can design and prototype structures at low costs, rapidly, in their laboratory and the interaction with a real prototype being much more accessible than in previous years.

Most of the low-cost printer's make the pieces from ABS (acrylonitrile butadiene styrene) or PLA (polylactic acid). There are, of course, some materials developed by the printer manufacturers. This article presents a study performed on parts obtained from the Z-ULTRAT material developed by the printer manufacturer ZORTRAX. The parts obtained by Additive Manufacturing have a resistance of 80% of that of the original material.

This technology is a very complex one, based on thermomechanical processes, on high speed and precise displacements [2, 3, 4, 6]. Because of this complexity of technology an important number of problems that are still vague, like: clogged extruder, grinding filament, material stringing, overheating, layer shifting, weak infill, inconsistent extrusion and so on. This technological issues causes some inconvenient for the parts realized: dimensional instability, structural anisotropy, inhomogeneous roughness and so on [8, 10].

In order to be able to use the prototypes made for the applications they develop, the specialists need to know the influences of the technological parameters for a certain category of printers and for a certain material, on the dimensional precision of the parts obtained in the laboratory [9, 11].

An experiment was made at University of Pitești, in the laboratory of Product design and development, in order to determine the dimensional accuracy of the pieces with the thin walls obtained by Additive Manufacturing process.

## 2. Material, samples and measuring equipment

The material used for this experiment is Z-ULTRAT. This material is a thermoplastic one, characterized by good impact resistance, which gives a uniform texture to the surface of the parts. By this material we can print high durability parts, also finished parts, which can retain their original shape after a long use. With Z-ULTRAT, it is possible to produce pieces with the properties comparable to those manufactured by conventional technologies (for example the injection moulding), including functional prototypes and some mechanical parts.

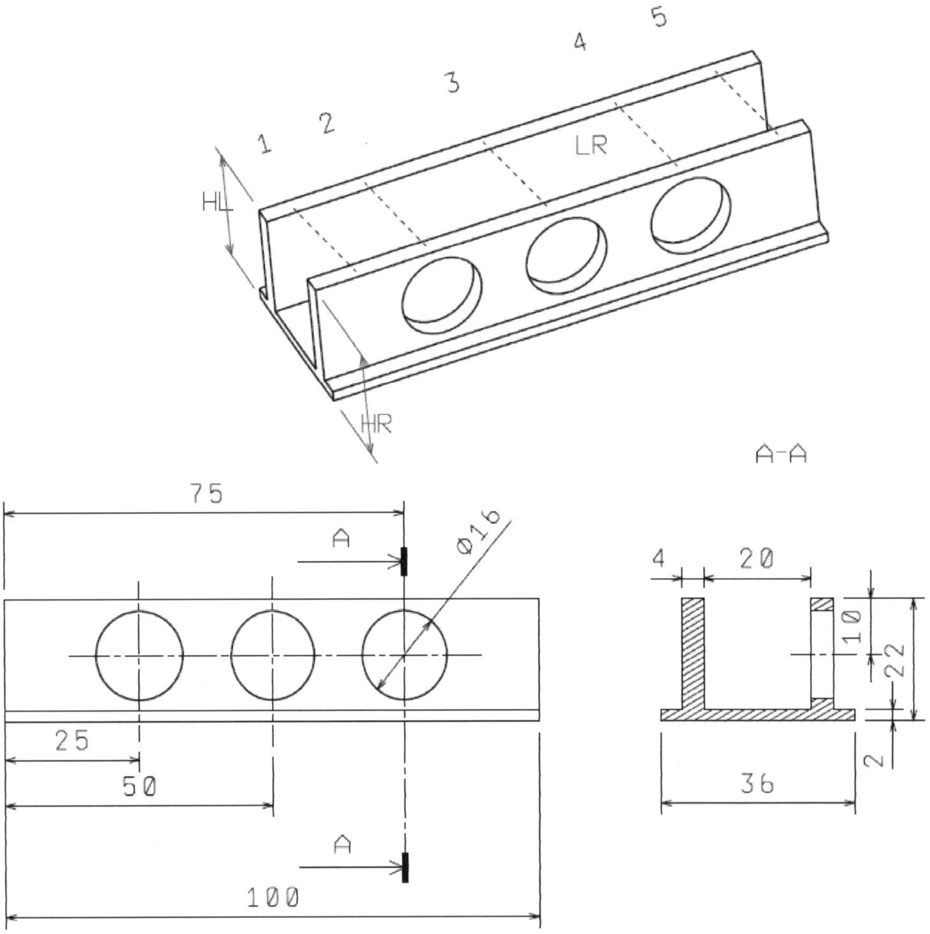


**Figure 1.** Geometry of specimen and measurement scheme.

The 3D model of the specimen is realized with Catia V5 software. The 3D model is imported in Z-Suite software in order to generate the code for the 3D Printer ZORTRAX M200, figure 2.

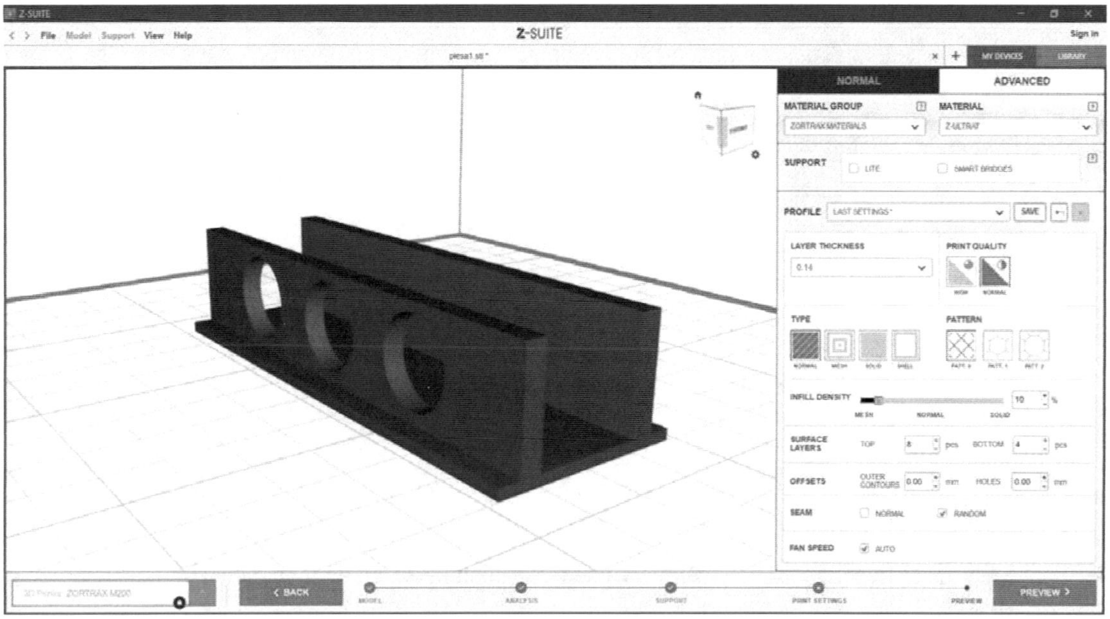


**Figure 2.** The specimen imported in Z-Suite software.

For measuring the dimensions on the samples, we used a caliper with a range of measurement between 0 to 150 mm and an accuracy of 0,01 mm.

The Minitab software was used to create a Taguchi plan for experimentation. We chose a plan with 3 factors and 3 levels for each factor. A number of 9 specimens was realized. The following parameters were chosen as inputs: layer thickness, infill density and orientation of the specimen on the table of the printer (0° - parallel with X axis, 90° - perpendicular on X axis of the printer table). The output parameters are: HR – height right and HL – height left, table 1.

**Table 1.** The dataset for the experiment.

| Exp. no. | Layer thickness (mm) | Infill density (%) | Orientation (degrees) | HR1 | HR2 | HR3 | HR4 | HR5 | HL1 | HL2 | HL3 | HL4 | HL5 |
|---|---|---|---|---|---|---|---|---|---|---|---|---|---|
| 1 | 0,14 | 10 | 0 | 22,05 | 22,20 | 22,24 | 22,18 | 21,95 | 22,05 | 22,14 | 22,24 | 22,16 | 21,94 |
| 2 | 0,14 | 50 | 45 | 22,03 | 22,20 | 22,26 | 22,17 | 22,03 | 22,01 | 22,19 | 22,23 | 22,14 | 22,04 |
| 3 | 0,14 | 90 | 90 | 21,98 | 22,22 | 22,23 | 22,22 | 21,98 | 21,96 | 22,16 | 22,23 | 22,14 | 21,95 |
| 4 | 0,19 | 10 | 45 | 22,05 | 22,15 | 22,17 | 22,18 | 22,10 | 21,95 | 22,15 | 22,16 | 22,15 | 22,04 |
| 5 | 0,19 | 50 | 90 | 21,99 | 22,15 | 22,15 | 22,14 | 21,99 | 21,85 | 22,05 | 22,15 | 22,16 | 21,91 |
| 6 | 0,19 | 90 | 0 | 22,06 | 22,21 | 22,26 | 22,19 | 22,11 | 21,90 | 22,18 | 22,25 | 22,18 | 21,97 |
| 7 | 0,29 | 10 | 90 | 21,76 | 22,00 | 22,03 | 21,95 | 21,83 | 21,78 | 21,94 | 21,95 | 21,92 | 21,90 |
| 8 | 0,29 | 50 | 0 | 21,92 | 21,97 | 22,06 | 21,98 | 21,92 | 21,76 | 21,96 | 22,00 | 21,94 | 21,87 |
| 9 | 0,29 | 90 | 45 | 21,83 | 21,89 | 22,00 | 21,98 | 21,85 | 21,76 | 21,94 | 22,03 | 22,01 | 21,84 |

## 3. Findings

The values obtained in all of the 5 measuring zones, for both parts of the specimens (on the side with holes - HR and on the side without holes - HL) have approximately the same tendency of variation, figure 3.

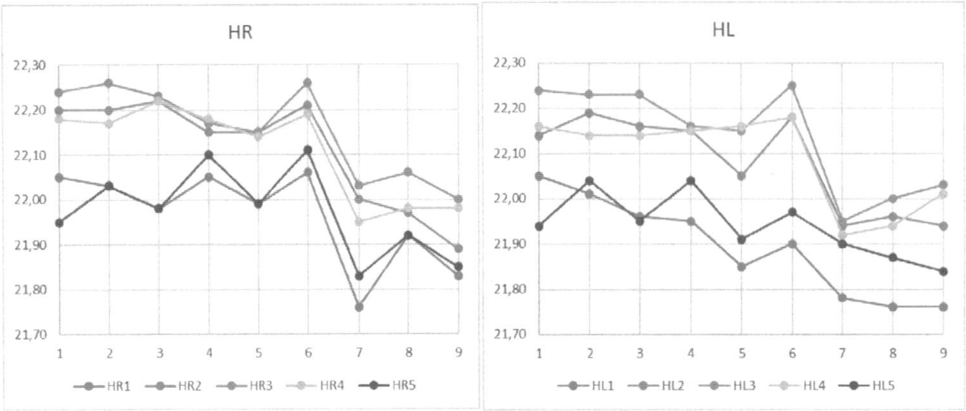


(a) (b)

**Figure 3.** The variation of the HR (a) and HD (b) for each specimen, for all of the 5 measuring points.

For data processing with regression analysis and with Artificial Neural Networks, the mean values for each side of the specimens (HR and HL) and the overall mean H, between HR and HL, were calculated, table 2 and figure 4.

**Table 2.** The average values of the datasets.

| Exp. no. | Layer thickness (mm) | Infill density (%) | Orientation (degrees) | HR | HL | H |
|---|---|---|---|---|---|---|
| 1 | 0,14 | 10 | 0 | 22,12 | 22,11 | 22,115 |
| 2 | 0,14 | 50 | 45 | 22,14 | 22,12 | 22,130 |
| 3 | 0,14 | 90 | 90 | 22,13 | 22,09 | 22,107 |
| 4 | 0,19 | 10 | 45 | 22,13 | 22,09 | 22,110 |
| 5 | 0,19 | 50 | 90 | 22,08 | 22,02 | 22,054 |
| 6 | 0,19 | 90 | 0 | 22,17 | 22,10 | 22,131 |
| 7 | 0,29 | 10 | 90 | 21,91 | 21,90 | 21,906 |
| 8 | 0,29 | 50 | 0 | 21,97 | 21,91 | 21,938 |
| 9 | 0,29 | 90 | 45 | 21,91 | 21,92 | 21,913 |

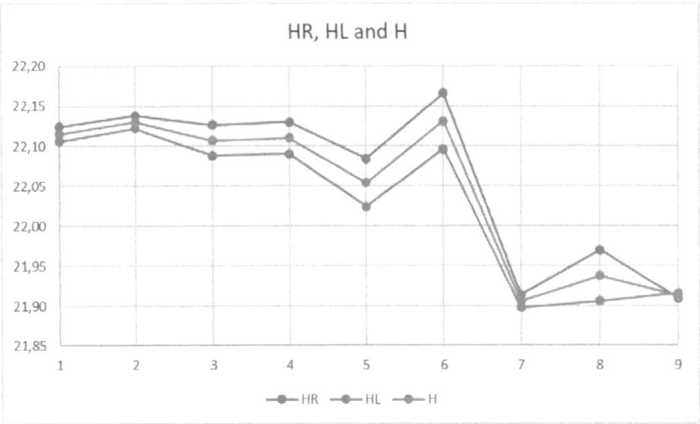


**Figure 4.** The averages values for HR and HL on each specimen and H (the average value between HR and HL).

With the Minitab software, a regression analysis was realized: H versus Layer thickness, Infill density and Orientation of the specimen, figure 5.

The regression equation is:

**H = 22,3 - 1,39 Layer thicknes + 0,000083 Infill density - 0,000433 Orientation**

| Predictor | Coef | SE Coef | T | P |
|---|---|---|---|---|
| Constant | 22,3474 | 0,0457 | 488,95 | 0,000 |
| Layer thicknes | -1,3895 | 0,1842 | -7,54 | 0,001 |
| Infill density | 0,0000833 | 0,0003518 | 0,24 | 0,822 |
| Orientation | -0,0004333 | 0,0003127 | -1,39 | 0,224 |

S = 0,0344661 R-Sq = 92,2% R-Sq(adj) = 87,5%

Analysis of Variance

| Source | DF | SS | MS | F | P |
|---|---|---|---|---|---|
| Regression | 3 | 0,069925 | 0,023308 | 19,62 | 0,003 |
| Residual Error | 5 | 0,005940 | 0,001188 | | |
| Total | 8 | 0,075865 | | | |

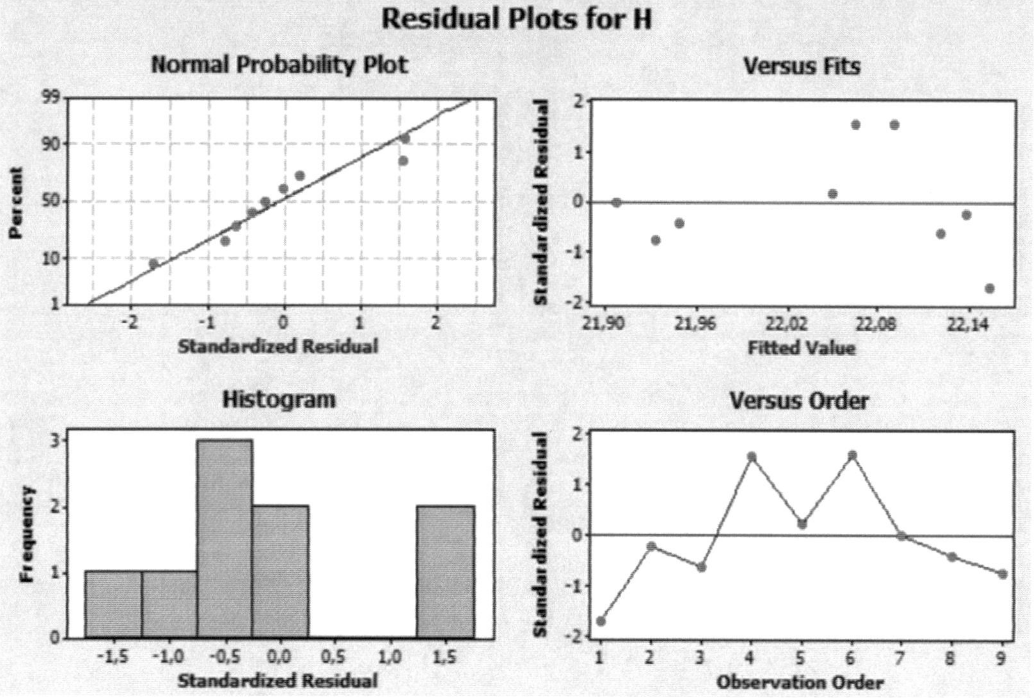


**Figure 5.** Regression rapport made by Minitab software.

## 4. Analysis with Artificial Neural Network

In order to realize this kind of analyze, the data was imported into the MATLAB Artificial Neural Network Toolbox and The Neural Fitting application was chosen. The Neural Fitting application give us the opportunity to select the data, to create and train a network, and to evaluate its performance using mean square error and regression analysis. Also, this tool has the possibility to randomly distribute the data sets in the three categories.

For the inputs of the network we created a matrix of 3x18 (Layer thickness, Infill density, Orientation), and for the output of the network a matrix (vector) of 1x18 (Height of the wall) were created.

All of the 18 datasets was divided into 3 categories: 70% for training, 15% for testing and 15% for validation.

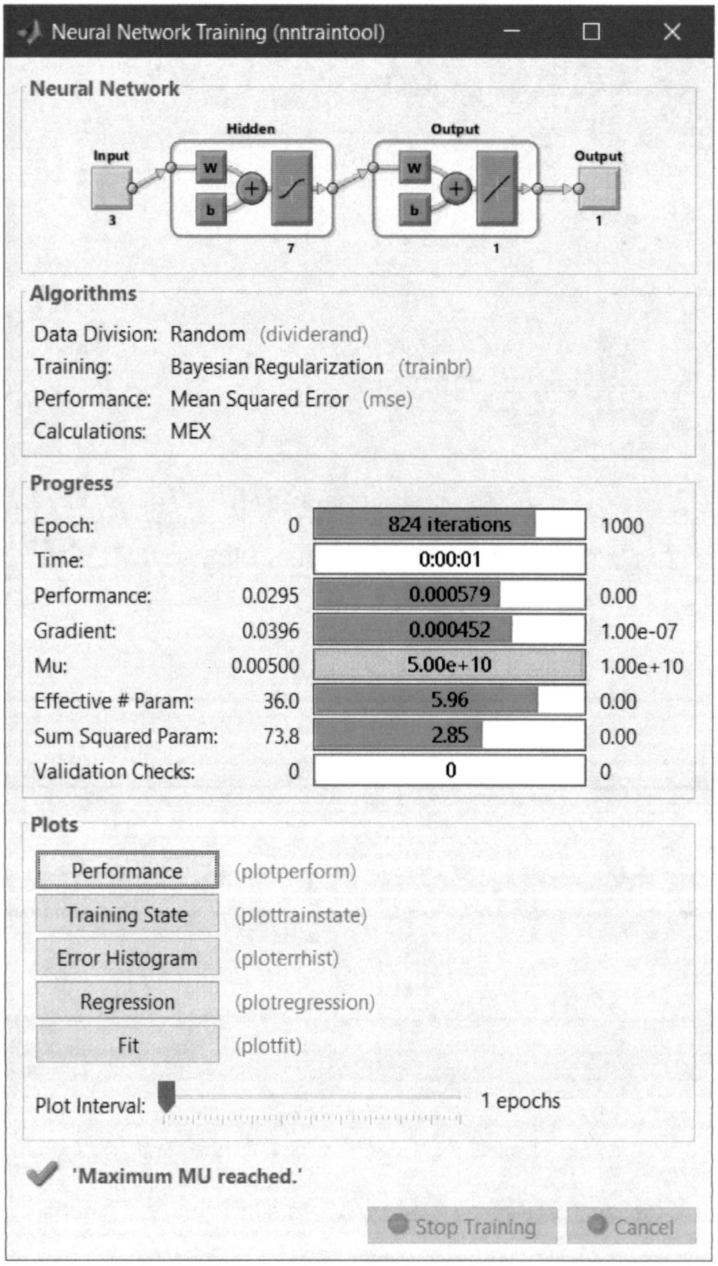


**Figure 6.** The training process of the neural network.

Using the “Bayesian-Regularization” backpropagation algorithm, the network was trained [5, 7], figure 6. The training process is an iterative one. For this case, to train the network a number of 824 iterations was needed.

When the training process is considered done, the results are displayed in the Train Network window. The values for R and the mean square error are displayed too.

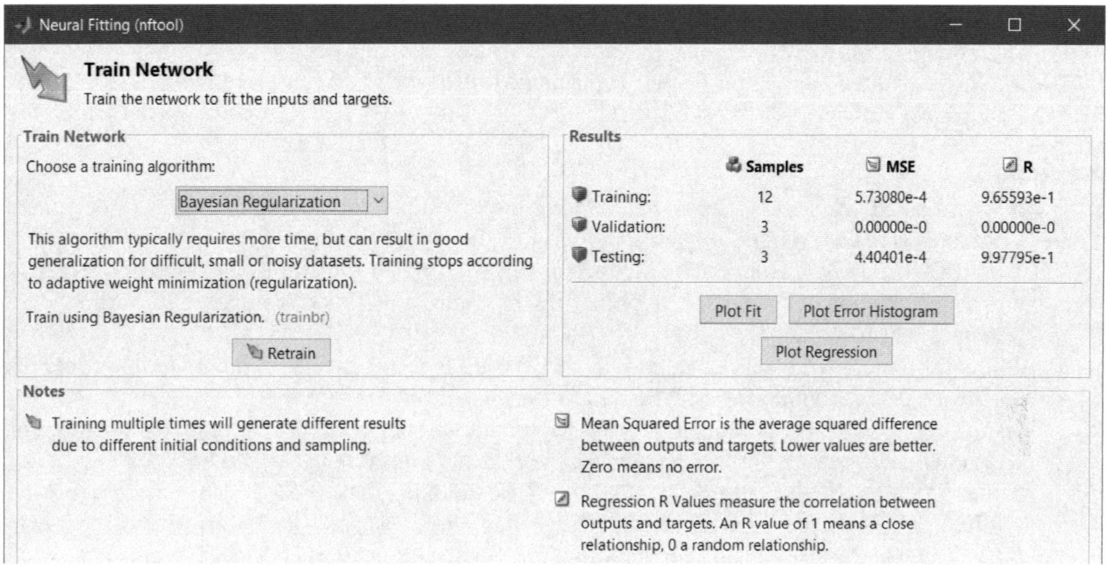


**Figure 7.** The results of network training, testing and validation.

After the training process, the network can be used to predict the value of the output parameter in function of the new values of the input parameters.

In table 3 an example of simulation by the trained network in presented.

**Table 3.** New parameters simulated with RNA.

| INPUT | |
|---|---|
| Layer thickness (mm) | 0,14 |
| Infill density (%) | 10 |
| Orientation (degrees) | 0 |
| OUTPUT | |
| H (mm) | 22,1315 |

## 5. Conclusions

The height of the thin walls obtained by Additive Manufacturing process has a strong dependency of the layer thickness and a weak dependency of infill density and orientation of the specimen on the printing table.

A regression equation was established after regression analysis: Height = f (parameters).

Artificial Neural Network Toolbox from MATLAB was used to realize an analyze for small datasets.

The network was trained, tested and validated using the results of measurements.

After the training, the network was able to "simulate" new parameters. Thus, we will introduce a new set of input data, corresponding to a new situation and thus we will be able to find the answer given by the network.

The research presented in this paper can give useful additional scientific data to understand how to improve the quality of the parts obtained by Additive Manufacturing.

In the future, it is desired to perform a series of tests on several specimens, in order to ensure a better assessment of dimensional variations. Thus, the designers can find important information regarding the influence of technological and design parameters on the dimensional accuracy of the parts obtained by Additive Manufacturing.

**Acknowledgments**
This work was supported by a grant of the Romanian Ministry of Research and Innovation, CCCDI-UEFISCDI, project number PN-III-P1-1.2-PCCDI-2017-0224/77 PCDI/2018 within PNCDI III.

**References**

[1] ISO/ASTM 52921 2013, Standard Terminology for Additive Manufacturing-Coordinate Systems and Test Methodologies

[2] Es-Said O S, Foyos J, Noorani R, Mendelson M, Marloth R and Pregger B A 2000 Effect of layer orientation on mechanical properties of rapid prototyped samples *Mat. and Manuf. Proc.* **15(1)** 107-122

[3] Ahn S H, Montero M, Odell D, Roundy S and Wright P K 2002 Anisotropic material properties of fused deposition modelling ABS *Rapid Prototyp J* **8(4)** 248–57

[4] Khan Z A, Lee B H, Abdullah J 2005 Optimization of rapid prototyping parameters for production of flexible ABS object. *J Mater Process Technol* **169** 54–61

[5] Anghel D C, Rizea A D, Iordache D M and Beşliu-Gherghescu M L 2019 Studies on the influence of design parameters on the behaviour at shock of 3D-printed components fabricated by fused deposition modelling *Mat. Sci. and Eng.* **564(1)** 012014

[6] Anghel D C, Beşliu-Gherghescu M L and Stănescu N D 2018 *Proc. Int. Congress of Automotive and Transport Eng.* (Cham: Springer) p 446

[7] Anghel D C, Ene A and Belu N 2014 A matlab neural network application for the study of working conditions *Adv. Mat. Res.* **837** 310-315

[8] Sicoe G M and Anghel D C 2019 Researches regarding the study of the dimensional precision of the geometrical elements of type shaft-hole obtained by fused deposition modelling ICMS 2019& COMEC 2019

[9] Mohamed O A, Masood S H, Bhowmik J L 2016 Optimization of fused deposition modeling process parameters for dimensional accuracy using I-optimality criterion *Measurement* **81** pp 174-196

[10] Galantucci L M et al. 2015 Analysis of dimensional performance for a 3D open-source printer based on fused deposition modeling technique *Procedia CIRP* **28** 82-87

[11] Lanzotti A, Del Giudice D M, Lepore A, Staiano G and Martorelli M 2015 On the geometric accuracy of RepRap open-source three-dimensional printer *J. of Mech. Design* **137(10)**

IOP Conf. Series: Materials Science and Engineering 1009 (2021) 012003 doi:10.1088/1757-899X/1009/1/012003

# Study of the stator geometry for a Moineau pump

**N Baroiu[1], G A Costin[1]*, G R Frumuşanu[1], V G Teodor[1] and N Oancea[1]**

[1]Department of Manufacturing Engineering, Dunărea de Jos University of Galaţi, Romania, Domnească street, No. 111, 800201, Galaţi, Romania

*E-mail: alexandra.costin@ugal.ro

**Abstract.** Moineau pumps are volumetric pumps with progressive cavities and have an eccentric rotor towards the rotation axis. The eccentric rotor performs a rotational movement towards a stator, usually made of an elastomer moulded in a cylindrical tube. Producing the pump involves knowing the cross-section of the helical rotor model, as well as knowledge of the stator shape as a complex surface, and of the enwrapping of the rotor lobes in their relative movement, in relation to the stator. An analysis for the trilobed helical pump case is proposed, in which, by defining the rotor shape, the stator model can be formed. The two objects, the rotor and the stator are in reciprocal enwrapping movement, as a result of the revolving without sliding movement (rolling) of two circular centrodes, defined in relation to their front sections. The helical surfaces of the rotor and stator are cylindrical helical surfaces.

## 1. Introduction

Current fluid extraction technologies involve multiple pumping technologies, the most common being: pumping with piston pumps, centrifugal pumping, pumping with hydraulic pumps with jet, respectively helical pumping [1 - 3].

Helical pumps with a progressive cavity, whose operation is based on Moineau's principle, involve the formation of cavities by inserting a rotor with a simple helical outer surface inside a stator with a double helical inner surface. At the beginning of the rotor movement, the cavities move from the suction to the compression, thus leading to a continuous flow [4].

Usually, helical pumps have a stator with (z+1) teeth, so with (z+1) beginnings, and a rotor with z helical flute. In this case, the ratio between the length of the stator pitch and the length of the rotor pitch became (z+1)/z [5 - 7].

The study of the active surfaces of Moineau pumps [8 - 10] can be done based on the fundamental theorems of the enwrapping surfaces, Olivier or Gohman [11,19] as well as on the basis of the complementary theorems: „minimum distance method" [12, 13] or „the method of the substitute circles family" [12, 14, 15, 20].

The problem proposed for the study can be solved based on the fundamental theorems of enveloping surfaces: Olivier or Gohman. Also, complementary theorems can be used: the „minimum distance" method [16], the method of generated plane trajectories [17].

This paper proposes approaching the problem of stator profiling based on the „family of the substitute circles" theorem [18], applied to conjugate surfaces associated with the rolling centrodes.

The method of the „substituting circles family" theory, is based on a definition of the substitute circles family of a plane profile.

*Definition:* a profile associated with a centrode, can be associated with a circles family with the centers on the associated centrode and also tangent to the profile.

The „substitute circles family" theorem is defined as follows: „the enveloping of a profile associated with a couple of rolling centrodes is defined by the transpose, in the rolling movement of the centrodes, of the substituting circles family of the given profile" [12].

The trilobed profile of the rotor - cylindrical worm, associated with a cylindrical axode, with the radius $R_{rp_1}$ - the rolling circle radius in the cross-section of the rotor, is presented in figure 1.

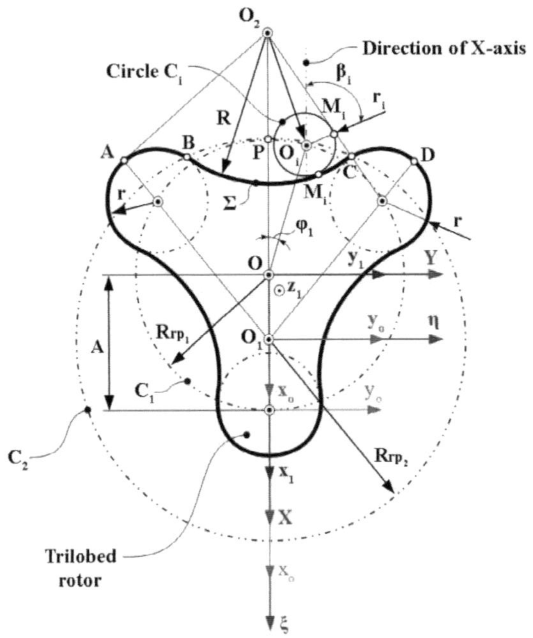


**Figure 1.** Rotor profile of the trilobed pump $\left(AB, BC, CD\right)$.

In figure 1, the rolling centrodes are shown:

- $C_1$ - centrode associated with the cross-section of the rotor;
- $C_2$ - centrode associated with the cross profile of the pump stator.

## 2. The method of „the substitute circles family" [18]

The kinematics process of generating the pump stator profile (in fact, the die in which the future stator of the pump is injected), is based on the rolling of the two centrodes: $C_1$ - associated with the trilobed rotor; $C_2$ - centrode associated with the pump stator. Constructively, the pump rotor executes a planetary movement, around the $O_1$ axis, figure 1, thus, physically, the pump stator is fixed.

The reference systems are defined:

- $x_1y_1$ - fixed system, initially superimposed on the rotor axis; in the figure, the axis of the rotor passes through the $O$-point;
- $x_oy_o$ - fixed system, superimposed on the rotor axis, the axis passing through $O_1$-point;
- $XY$ - mobile reference system, associated with the rotor in a rotational movement around its own axis, $C_1$ centrode;
- $\xi\eta$ - mobile system, in the planetary movement of the rotor (the apparent rotational movement of the $C_2$ centrode).

Noted with $R_{rp_1}$, respectively, $R_{rp_2}$ - are the radii of the two circular centrodes.

The rolling movement of the $C_1$ centrode, which apparently rotates without sliding on the $C_2$ centrode (of the stator) imposes the condition:

$$R_{rp_1} \cdot \varphi_1 = R_{rp_2} \cdot \varphi_2, \tag{1}$$

with $\varphi_1$ şi $\varphi_2$ the size of the rotation angles in the movement of the rotor ($\varphi_1$) and „apparently" of the stator, in reality the planetary movement of the rotor, of angle $\varphi_2$.

The stator has $(z+1)$ teeth (specifically, 4 teeth). From the condition of the circular pitch equality of the „teeth" on the two centrodes, the equality results:

$$R_{rp_1} \cdot \frac{2\pi}{3} = \frac{2\pi}{4} \cdot R_{rp_2}, \tag{2}$$

so, it results:

$$R_{rp_1} = \frac{3}{4} \cdot R_{rp_2}. \tag{3}$$

For the rotor profile, profile composed of arc:

- $AB$ - arc with the center in $C_1$ (with radius „$r$");
- $BC$ - arc with the center in $O_2$ (with radius „$R$");
- $CD$ - arc with the center in the $C_1$ centrode (with radius „$r$"), principled, the parametric expression is accepted:

$$\Sigma: \begin{vmatrix} X = X(v); \\ Y = Y(v), \end{vmatrix} \tag{4}$$

with $v$ - variable parameter.

The family of the „substituting circles" of the $BC$ arc is defined by the current circle of radius $r_i$, in contact with the $\Sigma$ profile, in the $M_i$ point, figure 1.

In this way, the parametric equations of the circle, with the origin in $M_i$, on the $C_1$ centrode, of radius $R_{rp_1}$, has the equations, in the $XY$-system:

$$C_i: \begin{vmatrix} X = -R_{rp_1} \cdot cos\,\varphi_1 - r_i \cdot cos\,\beta_i; \\ Y = R_{rp_1} \cdot sin\,\varphi_1 + r_i \cdot sin\,\beta_i, \end{vmatrix} \tag{5}$$

for the coordinates of the current point $M_i$ on the circle of radius $r_i$ - circle of the substitute circles family.

At this stage, the size of the parameters $\beta_i$ and $r_i$ have not been defined yet.

The condition that the $C_i$ circle (5) is tangent to the $\Sigma$ profile(4), involves the accomplishment of the conditions:

- The common point condition, meaning from (4) and (5), results:

$$\begin{aligned} X(v) &= -R_{rp_1} \cdot cos\,\varphi_1 - r_i \cdot cos\,\beta_i; \\ Y(v) &= R_{rp_1} \cdot sin\,\varphi_1 + r_i \cdot sin\,\beta_i. \end{aligned} \tag{6}$$

- The tangency condition of the two curves (4) and (6), with $\varphi_1$ - director parameter:

$$\dot{X}_v = \frac{d}{d\beta_i}\cdot\left[-R_{rp_1}\cdot cos\,\varphi_1 - r_i\cdot cos\,\beta_i\right];$$
$$\dot{Y}_v = \frac{d}{d\beta_i}\cdot\left[R_{rp_1}\cdot sin\,\varphi_1 + r_i\cdot sin\,\beta_i\right], \qquad (7)$$

or

$$\dot{X}_v = r_i\cdot sin\,\beta_i;$$
$$\dot{Y}_v = r_i\cdot cos\,\beta_i. \qquad (8)$$

The equations assembly (7) and (8) allows the determination of the enwrapping condition specific to the method, in principle, a dependence between the parameters $\varphi_1$ and $v$, as well as definitions for ri, respectively $\beta_i$, depending on the $M_i$-point position on the rolling circle, $C_1$.

From equations (8), by division, it results:

$$tg\,\beta_i = \frac{\dot{X}_v}{\dot{Y}_v}, \qquad (9)$$

where $\dot{X}_v$ and $\dot{Y}_v$ are the partial derivatives of the parametric equations of the $\Sigma$ curve (4).

The equations assembly that allows the identification of the substitute circles family and the specific enveloping condition is:

$$X(v) = -R_{rp_1}\cdot cos\,\varphi_1 - r_i\cdot cos\,\beta_i;$$
$$Y(v) = R_{rp_1}\cdot sin\,\varphi_1 + r_i\cdot sin\,\beta_i;$$
$$\dot{X}_v = r_i\cdot sin\,\beta_i;$$
$$\dot{Y}_v = r_i\cdot cos\,\beta_i. \qquad (10)$$

The enwrapping condition is deduced from the equality:

$$\frac{\dot{X}_v}{\dot{Y}_v} = \frac{Y(v) - R_{rp_1}\cdot sin\,\varphi_1}{-X(v) + R_{rp_1}\cdot cos\,\varphi_1} \qquad (11)$$

or

$$X(v)\cdot\dot{X}_v + Y(v)\cdot\dot{Y}_v = \left(\dot{X}_v\cdot cos\,\varphi_1 - \dot{Y}_v\cdot sin\,\varphi_1\right)\cdot R_{rp_1};$$
$$\dot{X}_v\cdot cos\,\varphi_1 - \dot{Y}_v\cdot sin\,\varphi_1 = \frac{X(v)\cdot\dot{X}_v + Y(v)\cdot\dot{Y}_v}{R_{rp_1}}, \qquad (12)$$

representing a dependence between the $v$ parameter of the generated profile and $\varphi_1$ - the movement parameter, in essence, the specific enveloping condition.

The circle radius of the substituting circles family, from (10) is:

$$r_i = \sqrt{\left[X(v) + R_{rp_1}\cdot cos\,\varphi_1\right]^2 + \left[Y(v) - R_{rp_1}\cdot sin\,\varphi_1\right]^2}. \qquad (13)$$

It is observed that the stator profiles corresponding to the $AB$ and $CD$ arcs are identical with the rotor profile (radius circle $r$), Figure 2, in the points on the stator centrode corresponding to the $P_a$, respectively $P_b$ arcs:

$$\text{for } \varphi_1 = \frac{\pi}{6} \text{ results } \varphi_2 = \frac{\pi}{6} \cdot \frac{R_{rp_1}}{R_{rp_2}}, \text{ see (1) and (2).} \tag{14}$$

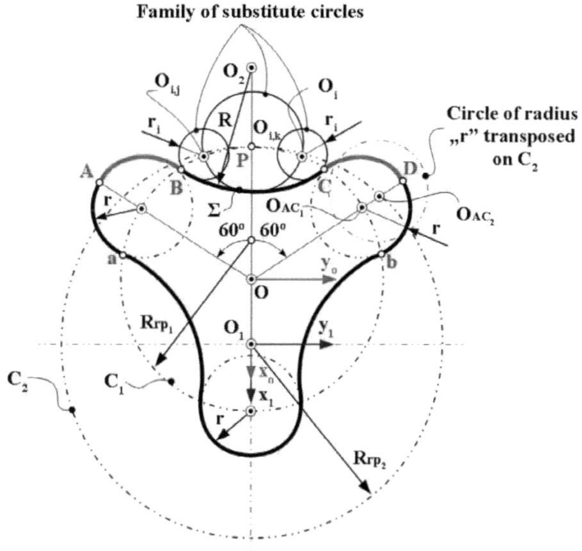


**Figure 2.** The substitute circles family of the generated profile transpose on the $C_2$ centrode.

According to the „substituting circles family" theorem, the circles family transposed on the $C_2$ centrode (of the stator), in the movements assembly:

- The rotor rotation around the $z_1$-axis, of the fixed system $x_1y_1z_1$:

$$x_1 = \omega_3^T(\varphi_1) \cdot X, \tag{15}$$

where, the $X$-matrix is:

$$X = \begin{pmatrix} X(v) \\ Y(v) \\ 0 \end{pmatrix}. \tag{16}$$

- The mobile system rotation associated with the stator (planetary movement):

$$x_o = \omega_3^T(\varphi_2) \cdot \xi; \tag{17}$$

- The relative position of the system $x_1y_1z_1$ and $x_oy_oz_o$:

$$x_o = x_1 + A, \text{ with the matrix } A = \begin{pmatrix} A \\ 0 \\ 0 \end{pmatrix}. \tag{18}$$

So the transposition movement of the substituting circles family on the $C_2$ centrode is:

$$\xi = \omega_3(\varphi_2)\cdot\left[\omega_3^T(\varphi_1)\cdot\begin{pmatrix}X\\Y\\0\end{pmatrix}+\begin{pmatrix}A\\0\\0\end{pmatrix}\right]. \tag{19}$$

In the movement (18), the substituting circles family of the $BC$ profile of the rotor, transposes on the $C_2$ centrode:

$$\begin{pmatrix}\xi\\ \eta\\ \zeta\end{pmatrix}=\begin{pmatrix}cos\,\varphi_2 & sin\,\varphi_2 & 0\\ -sin\,\varphi_2 & cos\,\varphi_2 & 0\\ 0 & 0 & 1\end{pmatrix}\cdot\left[\begin{pmatrix}cos\,\varphi_1 & -sin\,\varphi_1 & 0\\ sin\,\varphi_1 & cos\,\varphi_1 & 0\\ 0 & 0 & 1\end{pmatrix}\cdot\begin{pmatrix}-R_{rp_1}\cdot cos\,\varphi_1 - r_1\cdot cos\,\beta_i\\ R_{rp_1}\cdot sin\,\varphi_1 + r_1\cdot sin\,\beta_i\\ 0\end{pmatrix}+\begin{pmatrix}A\\0\\0\end{pmatrix}\right]. \tag{20}$$

If the matrix form is developed (19), it is reached at the analytical forms:

$$\begin{aligned}\xi &= \left[-R_{rp_1} - r_i\cdot cos(\varphi_1-\beta_i)+A\right]\cdot cos\,\varphi_2 - r_i\cdot sin(\varphi_1-\beta_i)\cdot sin\,\varphi_2;\\ \eta &= -\left[-R_{rp_1} - r_i\cdot cos(\varphi_1-\beta_i)+A\right]\cdot sin\,\varphi_2 - r_i\cdot sin(\varphi_1-\beta_i)\cdot cos\,\varphi_2,\end{aligned} \tag{21}$$

or, finally, the substituting circles family of the rotor profile $BC$ on the $C_2$ centrode, has the main form:

$$\left|\begin{aligned}\xi &= -R_{rp_1}\cdot cos\,\varphi_2 - r_i\cdot cos(\varphi_1-\beta_i-\varphi_2)+A\cdot cos\,\varphi_2;\\ \eta &= R_{rp_1}\cdot sin\,\varphi_2 - r_i\cdot sin(\varphi_1-\beta_i-\varphi_2) - A\cdot sin\,\varphi_2,\end{aligned}\right. \tag{22}$$

with definitions from (9) and (13):

$$\begin{aligned}\varphi_2 &= \frac{R_{rp_1}}{R_{rp_2}}\cdot\varphi_1;\\ tg\,\beta_i &= \frac{\dot{X}_v}{\dot{Y}_v};\\ r_i &= \sqrt{\left[X(v)+R_{rp_1}\cdot cos\,\varphi_1\right]^2+\left[Y(v)-R_{rp_1}\cdot sin\,\varphi_1\right]^2},\end{aligned} \tag{23}$$

and the enwrapping condition:

$$\dot{X}_v\cdot cos\,\varphi_1 - \dot{Y}_v\cdot sin\,\varphi_1 = \frac{X(v)\cdot\dot{X}_v + Y(v)\cdot\dot{Y}_v}{R_{rp_1}},\ \text{see (12).}$$

## 3. Applications for a trilobed profile circle arc

Figure 3 shows the substituting circles family associated with the rotor profile, circles with centers on centrode $C_1$ and also the circles family transposed on the centrode $C_2$, the stator, in the planetary movement of the pump rotor. The stator profiling for the $BC$ arc involves customizing the profile shape (3). Thus, in correlation with figure 4, the profile equations are defined:

$$\begin{aligned}X &= -e + R\cdot cos\,v;\\ Y &= R\cdot sin\,v,\end{aligned} \tag{24}$$

with $e$ and $R$ constructive sizes and $v$ variable parameter between the limits:

$$-v_{max} \le v \le v_{max}, \text{with } v_{max} = arcsin\left[\frac{R_{rp_1} \cdot sin 60^{o}}{R + r}\right]. \tag{25}$$

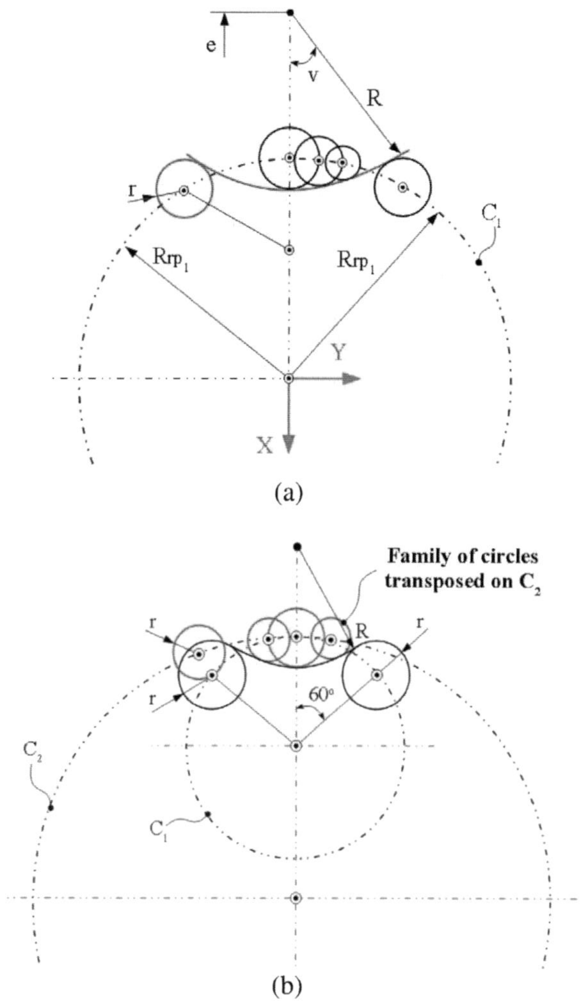


**Figure 3.** The circles families on the $C_1$ centrode of the pump rotor and transposed on $C_2$.

Equations (10) become:

$$\begin{aligned} -e + R \cdot cos\, v &= -R_{rp_1} \cdot cos\, \varphi_1 - r_i \cdot cos\, \beta_i; \\ R \cdot sin\, v &= -R_{rp_1} \cdot sin\, \varphi_1 + r_i \cdot sin\, \beta_i. \end{aligned} \tag{26}$$

The enwrapping condition (12) can be expressed, in this case, by:

$$\begin{aligned} \dot{X}_v &= -R \cdot sin\, v; \\ \dot{Y}_v &= R \cdot cos\, v. \end{aligned} \tag{27}$$

The enveloping condition (12) become:

$$-R \cdot sin v \cdot cos \varphi_1 - R \cdot cos v \cdot sin \varphi_1 = \frac{-\left(-e + R \cdot cos v\right) \cdot sin v + R \cdot sin v \cdot cos v}{R_{rp_1}}. \quad (28)$$

The condition (28) is available for values $v = 0...v_{max}$, for profile`s zone, Figure 4, $PC$ and $v_{max}$ from (25).

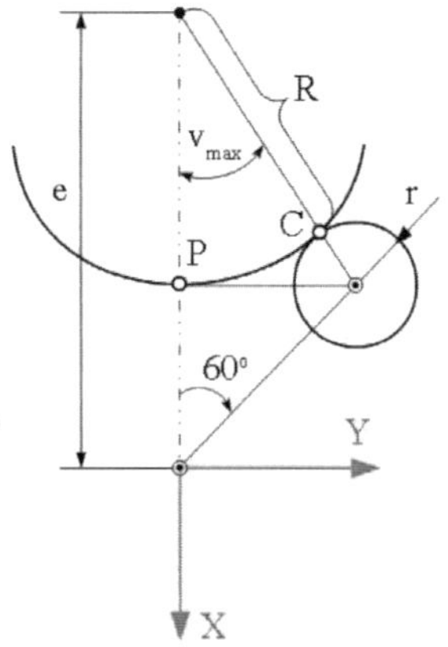


**Figure 4.** Profile`s zone *PC*.

## 4. Numerical application

Table 1 shows the stator profile coordinates, conjugated to the rotor profile with the elements:

- $R_{rp_1} = \frac{3}{4} \cdot R_{rp_2}$; $R_{rp_1} = 30$ mm;
- $r = 8.5934$ mm; $R = 35$ mm;
- $tg\beta_i = \frac{\dot{X}_v}{\dot{Y}_v} = \frac{-R \cdot sin v}{R \cdot cos v} = -tgv$;
- $r_i$ is calculated from (13) for the concrete case, the $PC$ profile`s zone, Figure 4.

**Table 1.** The stator profile coordinates conjugated to the rotor profile.

| v [°] | φ [°] | $r_i$ [mm] | X [mm] | Y [mm] | ξ [mm] | η [mm] |
|---|---|---|---|---|---|---|
| -0.638485 | -0.82 | 1.682743 | -21.895 | -20.8593 | -17.4858 | -10.24 |
| -0.510788 | -0.44 | 8.775366 | -19.4674 | -17.1103 | -26.0494 | -1.39569 |
| -0.383091 | -0.29 | 12.09141 | -17.537 | -13.0826 | -30.3894 | 1.010818 |
| -0.255394 | -0.18 | 13.83301 | -16.1353 | -8.84193 | -33.034 | 1.410554 |
| -0.127697 | -0.09 | 14.72357 | -15.285 | -4.45726 | -34.5181 | 0.897423 |
| 0 | 0.00 | 15 | -15 | 0 | -35 | 0 |
| 0.1276969 | 0.09 | 14.72357 | -15.285 | 4.457255 | -34.5181 | -0.89742 |
| 0.2553938 | 0.18 | 13.83301 | -16.1353 | 8.841927 | -33.034 | -1.41055 |
| 0.3830908 | 0.29 | 12.09141 | -17.537 | 13.08261 | -30.3894 | -1.01082 |
| 0.5107877 | 0.44 | 8.775366 | -19.4674 | 17.11026 | -26.0494 | 1.395691 |
| 0.6384846 | 0.82 | 1.682743 | -21.895 | 20.85927 | -17.4858 | 10.24002 |

Figure 5 shows the front profile of the rotor pump stator with multiple cavities.

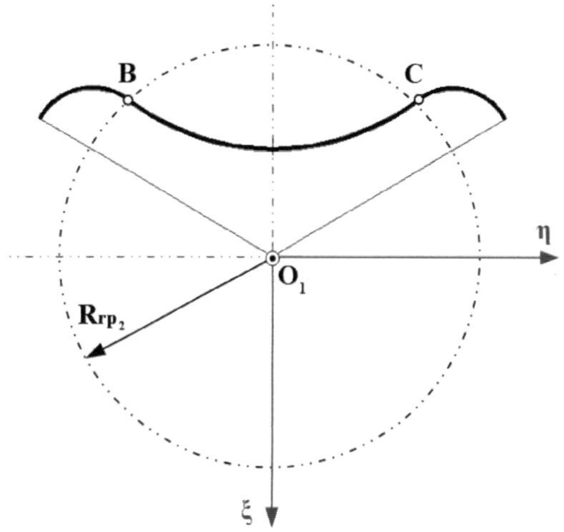


**Figure 5.** Front profile of the stator *z=4 lobes*.

## 5. Conclusions

In order to helical pumps to have increased efficiency in exploitation, they are required to have a specific geometric profile for the rotors, which allows a flow section as large as possible, but also a gearing line as short as possible, to obtain an increased flow and a constant velocity. This leads to a high volumetric efficiency, which also leads to an increase in adiabatic efficiency, because the power losses are smaller during the compression process, through internal recirculation.

In this context, constant pitch helical surfaces generating becomes a problem with multiple technical applications, the helical surfaces being usually generated by milling with the help of rotary tools bordered by revolution peripheral surfaces.

Thus, in the present paper, the theorem of the substituting circles family and the specific enwrapping condition were used to determine the envelopes of some curves belonging to the helical surface by a repetitive solution. A set of „parallel circles" of the primary peripheral surface of the future side mill tools generating of the helical surface was defined, starting from the stator profiling in circle arc.

The method of the „substituting circles family" allows a rigorous approach to the determination of the pump stator enveloping in the relative movement of the trilobed rotor (planetary movement).

The stages necessary to be completed, such as an analysis for the type of considered pump, were successively passed.

The numerical application presented proves the ability of the method to approach such a technical problem.

The presented methodology can be an exhortation for other similar applications in the field, too.

## Acknowledgments

This work was supported by a grant of the Romanian National Authority for Scientific Research and Innovation, CNCS-UEFISCDI, project number PN-III-P1-1.2-PCCDI-2017-0446 - TFI PMAIAA, within PNCDI III.

## References

[1] Hös C, Bazsó C and Haraszti P 2017 *Volumetric Pumps and Compressors* - lecture notes, (Budapest University of Technology and Economics (BME) Dept. of Hydrodynamic Systems (HDS) Hungary)

[2] Baroiu N, Popa C L, Teodor V G, Berbinschi S and Susac F 2017 *Pompe şi compresoare*

*elicoidale – profilări CAD şi analitice ale sculelor generatoare (Helical pumps and compressors - CAD and analytical profiling of generating tools)* (Ed. Academica ISBN 978-606-606-004-2)

[3] Chaudhary P 2016 Review On Optimization of PC-Pumps Installation in CBM Reservoirs and Troubleshooting Problems Faced By Them during Dewatering Imperial *Journal of Interdisciplinary Research* **2 (6)** 1150

[4] Nguyen T, Tu H, Al-Safran E and Saasen A 2016 Simulation of single-phase liquid flow in progressing cavity pump *J. Pet. Sci. Eng.* **147** 617

[5] Rotors, Rubber Stators, Solid Stators - Daurex, Schermbeck, Germany, https://daurex.com/daurex_en/production.php

[6] Chen J, Liu H, Wang F, Shi G, Cao G and Wu H 2013 Numerical prediction on volumetric efficiency of progressive cavity pump with fluid–solid interaction model *J. Pet. Sci. Eng.* **109** 12

[7] Progressive cavity pumps - S.C. Confind S.R.L., Câmpina, România, http://www.confind.ro/pompe_cavitati_progresive.html

[8] Gravesen J 2008 The geometry of the Moineau pump *Computer Aided Geometric Design* **25 (9)** 792

[9] Ceballos J B and Vivas Ó A 2019 Mathematical model of controllers for progressive cavity pumps *Revista UIS Ingenierías* **18 (2)** 17

[10] Baroiu N, Beznea E F and Vasile Șt 2018 Aspects regarding the analysys of a crude oil extraction system with helical progressive cavity pumps *Proceedings of the 24rd International Conference on Hydraulics and Pneumatics - HERVEX* 2018 Băile Govora Romania Nov. 7-9 ISSN 1454-8003 pp 66-76

[11] Litvin F L 1989 Theory of gearing NASA Reference Publications 1212 *AVSCOM Technical Report* 88 C 135

[12] Oancea N 2004 *Generarea suprafețelor prin înfăşurare vol II - Teoreme complementare (Generation of surfaces by winding vol II - Complementary theorems)* (Ed. Fundației Universitare „Dunărea de Jos" Galați ISBN 973-627-170-6)

[13] Oancea N 1996 Méthode numerique pour l'etude des surfaces envelopes *Mechanism and Machine Theory* **31** (7) 957

[14] Oancea N 1998 Metoda cercurilor de substituire, o nouă exprimare a teoriei suprafețelor în înfăşurare. Teorema fundamental (The method of substitution circles, a new expression of the theory of winding surfaces. Fundamental theorem) *Constructia de Masini* **50 (2)** 11

[15] Teodor V G, Baroiu N, Berbinschi S and Oancea N 2015 A graphical expression for the method of substitutive circles family applied for profiling side mill designed to generate sealing worms of cycloid pumps *Appl. Mech. Mater.* **809** 998

[16] Frumuşanu G R, Berbinschi S and Oancea N 2012 Profiling of the planing tool for cycloidal worms machining - CAD method versus analitycal method *Proceedings in Manufacturing Systems* **7(1)** 63

[17] Baroiu N, Teodor V G and Oancea N 2015 A new form of in plane trajectories theorem. Generation with rotary cutters *Bulletin of the Polytechnic Institute of Iaşi* **3** 29

[18] Oancea N *et al.* 1998 Metoda cercurilor de substituire. Algoritm de profilare a sculei melc (Substitution circle method. Worm tool profiling algorithm) *Construcția de Mașini* **50 (2)** 6

[19] Baroiu N, Berbinschi S, Teodor V G, Susac F and Oancea N 2017 The complementary graphical method used for profiling side mill for generation of helical surface *IOP Conf. Series: Materials Science and Engineering* **227**

[20] Berbinschi S, Teodor V G, Baroiu N and Oancea N 2013 The substitutive circles family method – graphical aproach in CATIA design environment *The Annals of "Dunărea de Jos" University of Galati Fascicle V* **53**

IOP Conf. Series: Materials Science and Engineering 1009 (2021) 012004 doi:10.1088/1757-899X/1009/1/012004

# Robot-based incremental sheet forming – the tool path planning

**A Bârsan[1]*, M O Popp*, G P Rusu[1] and A I Maroșan[1]**

[1]Lucian Blaga University of Sibiu, Engineering Faculty, 4 Emil Cioran, 550025 Sibiu, Romania

*E-mail: alexandru.barsan@ulbsibiu.ro, mihai.popp@ulbsibiu.ro

**Abstract.** To achieve a 3D shape with certain degree of complexity from metal sheets that are flat in the beginning, especially for prototype development or small series production, that is also cost efficient by avoiding high-end dies or expensive tooling, the incremental sheet forming process can represent the ideal choice due to its flexibility, reduced forming forces and increased formability. The process is performed with a round tool that follows a predefined path to deform the flat metal sheet into its final shape. The proper selection of the toolpaths affects the productivity and the accuracy of each part. The process having to focus on forming each single point in increments. For cost-effective production, the incremental sheet forming process is performed by an industrial robot. This paper aims to showcase the superior kinematic potential of the robot for complex tool paths by means of simulation. Singularity issues and robot joints limitation is checked through DELMIA software platform. After the design, simulation, and generation of the tool path, the software code can be generated and also used to steer the robot.

## 1. Introduction

Recent developments in manufacturing technology as well as the development of machining tools, especially in mechanical machining operations, reflect the flexibility requirements for adapting to changes on the market, in society and global economic environment. To handle global competition, market requirements suggest the use of small series and large variety production, even for a large series production industry.

The results of the latest developments are quantified by the high-precision machine tools [1]. However, an alternative that can meet the needs of today's and tomorrow's industry, at a low cost and ensuring a high degree of flexibility, is given by an industrial robot. In a wide range of technological processes, the industrial robots have been implemented to supplant the human operator involved in repetitive tasks. There are already robotic cells, such as the one shown in the figure 1, introduced in assembly operation, surface coatings, or mechanical machining processes.

In the previous research [2]an overview of current technology was provided based on the technical borders of industrial robots concerning robotic machining. Unfortunately, due to the serial structure, the articulated robots have a lower stiffness compared with a CNC machine. In any case, the industrial robot characteristics must fully correspond to tasks requirements [3, 4, 5, 6].

One of the most important control aspects is given by the tool path planning. The robots mechanical structure and the servo control system are less influenced by a well-planned trajectory. This paper presents the steps taken for a proper selection of the tool paths for the parts manufactured with six degrees of freedom anthropomorphic robot during single point incremental forming (SPIF) process. Using DELMIA software platform, in addition to check the singularity issues and robot joints limitation, the program code is generated to control the robot based on a proper toolpath.

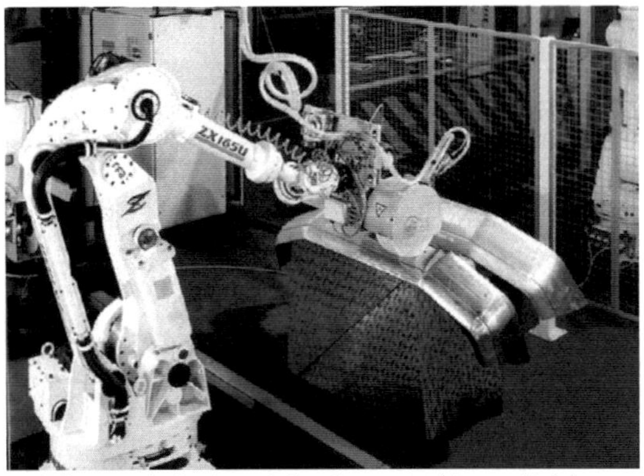

**Figure 1.** Industrial robot used in polishing a vehicle bumper [7].

## 2. Single point incremental forming

In the forming procedure of metal sheets, conventional processes, such as bending, shearing, or stamping process, are not able to meet the growing demand for shortening product development cycles due to their need to develop process-dependent tools.

Single point incremental forming is a technological manufacturing process of a flat metal sheet into a complex 3D shape, through a series of point to point small deformations, with a high cost efficiency achieved by avoiding the use of expensive dies or other forms specialized tools [8, 9, 10].In most cases, a simple hemispherical tipped tool, with a diameter between 5 to 20 mm, incrementally formed the sheet metal. The process can be performed on specialized forming machines, NC machines, or industrial robots [11, 12].

The metal sheet (2) and the fixing system, composed by support plate (1) and retaining frame (3), are fixed, while the punch (4), usually, executes a series of combined movements vertically and horizontally. For the unfolding process, a wide range of tool paths may be generated by combining these two movements. The most used processing tool paths, considered by the literature [11, 13, 14], are contour curves and spatial spirals. Figure 2 presents the SPIF process principle.

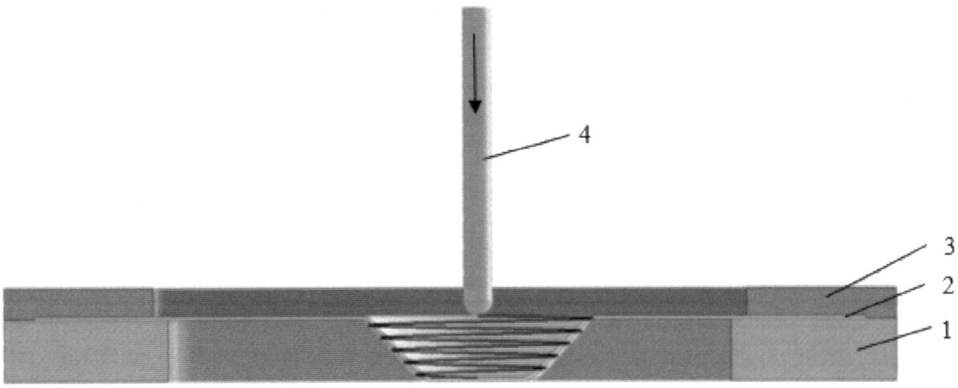


**Figure 2.** A simplified SPIF process representation.

## 3. Robot-forming

Taking advantage of their high flexibility and distribution, but at the same time losing the stiffness of CNC machines, industrial robots can be used in incremental sheet metal, process known as "roboforming". Considering the SPIF process flexibility, different studies regarding roboforming process can be found.

The manufacturing of parts to use as prototypes or in small numbers can become very expensive. Meier [15] proposed the roboforming concept to keep the cost of production as low as possible. This concept is based on a flexible modeling achieved by easy programming of a synchronous movement of the end effect or of two industrial robots with 6 degrees of freedom that have workpiece-independent forming tools. The resulting end formis made by an incremental ingoing movement of the forming tool in direct depth, and a motion made alongside the geometric contour in the lateral axis, resulting in parallel layers or a helical path.

Schäefer et al. [16] developed a forming method, without using any special die plate, in which an industrial robot is moving a hammering tool on a well-defined bearing over a sheet of metal held in a rigid frame of metal. The first time, using this forming principle, was in producing a sheet metal part of 300 x 300 mm, at a punching frequency of the hammer toolof about 100 hits / s.

The Metal Forming Research Centre from Lucian Blaga University of Sibiu has been carrying out different studies concerning the roboforming process. Chera et al. [17] presents a different way for manufacturing by using a industrial robot to form parts out of sheets of metal through forming of increments via an asymmetric approach. Oleksik et al. [18] had reported a first paper regarding the study of the strains and thinness at the end of a SPIF process. Following the experiments carried out it was observed that the vertical incremental step and the tool diameter are the most important influencing factors. In a second study [19], using a KUKA KR 6-2 industrial robots over CNC milling, showed the possibility of online measurement of the strains and relative thinness, through the Aramis optical measurement system from GOM. Crenganis and Csiszar [20] studied if the 6-axisarticulated robot, model KUKA KR6-2, could be used for SPIF manufacturing with a high degree of depend ability. After running the dynamic simulations, the study results showed that the KUKA KR6 robot is well suited for SPIF processes as well as for metallic parts that required greater forming forces, such as Ti6Al4V alloy.

However, the proper selection of tool paths influences the accuracy and productivity of the parts manufactured via SPIF process. In our future research in the Metal Forming Research Centre from Lucian Blaga University of Sibiu, further research will follow with more complex tool paths for unfolding SPIF. A preliminary step for an effective tool path strategy is to design, simulate, and generate the tool path.

## 4. Tool path planning – simulation of the robotic system

Since the incremental forming process was designed to overlap local deformations along the tool trajectory, the best way to control the final geometry is by directly influencing the main factor, namely the tool path. Designing a successful tool path depends on many variables, such as material parameters, tool diameter, step depth, tool path shape [11, 12, 21].

Since the production time of each part depends on the length of the distortion line, the punch speed and the available power of the equipment used, a preliminary step is to simulate the process.

### *4.1. Tool path strategies*

As conceptually illustrated in figure 3, greatest practiced contouring tool path layouts are Z-level and spiral tool paths. Cutting the part with equidistant planes equally distanced on Z-axis, lead to obtaining of the Z-level tool paths. A proposed tool path for a single feature part represented by dashed lines is presented in figure 3.a, as a section view of a truncated cone with a wall angle α. The vertical increment, ΔZ, has been exaggerated only for visualization purposes.

Usually, spiral tool paths deliver a better surface finish than Z-level paths due to the continuous vertical transition between following contours. A proposed spiral tool path represented by dashed lines is presented in figure 3.b, as a section view of a truncated cone. The vertical increment, ΔZ, has been exaggerated only for visualization purposes.

However, complex geometries can prove difficult with the implementation of spiral tool paths. Keeping a 90-degree angle between the surface that is in processing and the tool is a simple solution that not only favors the forming process but also improves the accuracy of the part being made.

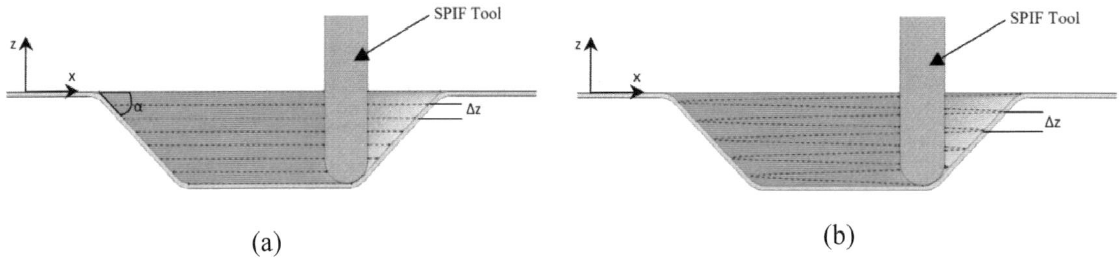


**Figure 3.** Contouring tool paths layouts: (a) Z-level tool path, (b) spiral tool path.

*4.2. Simulation of the robot-forming process*

Firstly, to simulate the tool path for incremental forming experiments, a tridimensional model (3D) of the desired part must be designed in a CAD program. The next step is the path planning. The process chain of the robot-forming process is presented in figure 4.

The usage of a CNC machine usually is dependent on 3D CAD models that help the software generate tool path strategies. On the industrial robots side, there are only a few simple yet viable solutions but none with the same potential of a classical CAD/CAM interface.

DELMIA® software from Dassault Systèmes is one of the most powerful, flexible, and easy solution for defining robot movements. CATIA is a CAD software integrated as part of DELMIA which helps not only with the production of parts but also with the modification of them if it is necessary.

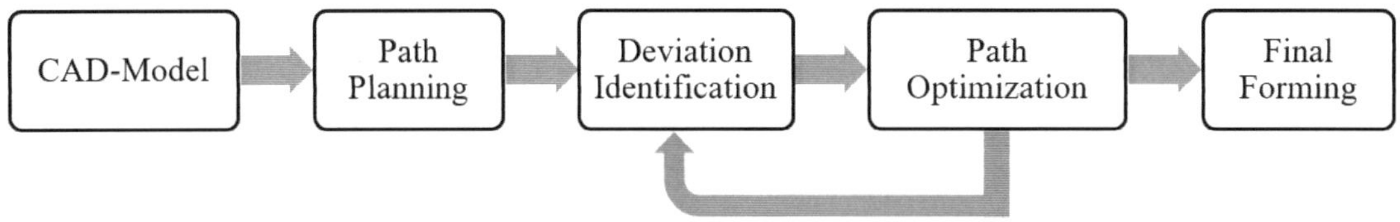


**Figure 4.** Robot-forming chain process.

As shown in figure 4, the first step is to design and assemble the components for the robotic system in CATIA V5. Then, from the DELMIA program library, the three-dimensional model of KUKA KR 6-2 robot is imported into the Device Task Definition module.

This module is a powerful 3D simulation tool for working cells, providing robot learning capabilities to evolve and practice robot tasks. Through this workbench tool, in an environment where product data and manufacturing resources can be integrated into a single 3D model, is possible to develop and program robot tasks. Each device is individually programmed with tasks that are sequenced and simulated to eliminate any interference. The analysis tools from Device Task Definition provide basic measurements, or collision checking. The simulation reproduces both simple and complex robotic manufacturing processes exactly as they are defined.

After all components were inserted in the Device Task Definition module and the punch is mounted on the robot through “Set Tool” function, the components position must be adjusted to optimize the simulation process.

Studies were carried out on a truncated pyramid and a truncated cone. The proposed tool path for a truncated pyramid represented by white lines is presented in figure 5, with a wall angle 30º, formed using Z-level tool path. The vertical increment, ΔZ, has been exaggerated only for visualization purposes.

**Figure 5.** Proposed truncated pyramid tool path.

The 3D model of the truncated cone tool path, presented in figure 7, which was previous created in CATIA V5 is now imported in Device Task Definition module. The proposed spiral tool path for manufacturing a truncated cone is shown in figure 6. From the “Sequence” toolbar, to the robot is assigned a new task. The waypoints are generated automatically on the defined continuous tool path. Because the orientation of the punch is fixed, in order to manually set the angle of the punch, a secondary path must be created via "Create new robot task" function on the primary path. Now, the waypoints appear in the form of "tags" and the orientation of each tag can be changed as defined by the user. The "Interpolate" function must be used to set the desired orientation on all tags. After setting all the components of the robotic systems, using the "Teach a device" function, the robot must be teached in which configuration to run.

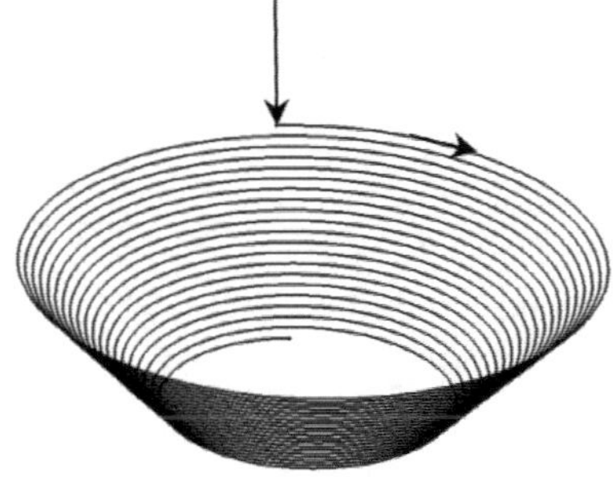

**Figure 6.** Proposed spiral tool path.

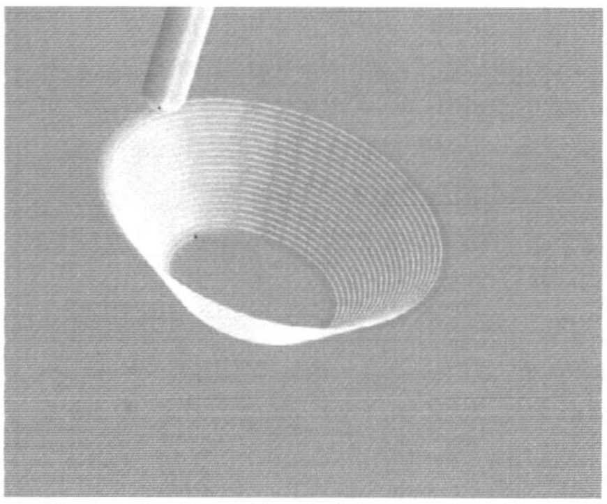

**Figure 7.** Detail of the working area.

The last step before simulating the movement of the robot according to the imposed tool path is to check the singularity problems and limit the joints of the robot. Once you have simulation with no collisions, applying the "Create Program" function and the programming system will generate robot programs. The two created files are uploaded to the KUKA controllers and based on them the forming begins for the sheet of metal. The tridimensional model of the robotic system is presented in figure 8.

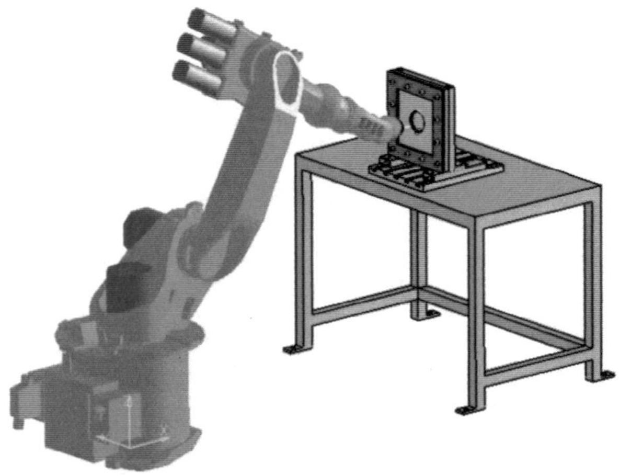

**Figure 8.** Tridimensional model of robotic system components used for the simulation.

## 5. Conclusions

To achieve a 3D shape with certain degree of complexity from metal sheets that are flat in the beginning, especially for prototype development or small series production, that is also cost efficient by avoiding high-end dies or expensive tooling, the incremental sheet forming process can represent the ideal choice. Roboforming can be a reasonable alternative to CNC machines that can meet the SPIF process requirements, at a low cost and ensuring a high degree of flexibility, especially for industry-oriented prototypes and small sized production.

The paper shows that simulation of the tool path planning has a huge positive impact in elevating the accuracy in roboforming. The proper selection of the tool paths affects the productivity and the accuracy of the parts manufactured through SPIF process.

The superior kinematic capabilities of the robot were highlighted by means of simulation. Using DELMIA software platform, the collisions, the singularity issues, and the robot joints limitation can be checked. After running the simulation, the program code is generated based on a proper tool path and is used afterwards to control the robot. DELMIA offers a flexible and easy solution for defining robot movements for many robot models.

However, further experimental research must be performed to demonstrate the capabilities of tool path planning, to assess the accuracy of the parts and the use of strategies with higher complexity regarding tool path programs.

**Acknowledgments**

The experimental test unfolded within this research work were supported by a grant of the Romanian Ministry of Research and Innovation CCCDI-UEFISCDI, project number PN-III - P1-1.2 - PCCDI-2017-0446 /nr. 82PCCDI/2018, within PNCDI III, project title: "Intelligent manufacturing technologies for advanced production of parts".

## References

[1] Cioca L I, Breaz R E and Racz S G 2019 *Sustainability-Basel* **11** 315
[2] Bârsan A 2019 *ACTA Universitatis Cibiniensis* **71** 9
[3] Breaz R E, Bologa O and Racz S G 2017 *Procedia computer science* **122** 346

[4] Crenganiş M, Tera M, Biriş C and Gîrjob C 2019 *Procedia computer science* **162** 298
[5] Klimchik A, Ambiehl A, Garnier S, Furet B and Pashkevich A 2017 *Robot Comput. Integr. Manuf.* **48** 12
[6] Schneider U, Drust M, Ansaloni M et al. 2016 *Int. J. Adv. Manuf. Tech.* **85** 3
[7] Pires J N and Bogue R 2009 *Industrial Robot: An International Journal*
[8] Racz G S et al 2019 *IOP Conf. Ser.: Mater. Sci. Eng.* **564** 012023
[9] Tera M, Breaz R, Racz S et al. 2019 *Int. J. Adv. Manuf. Tech.* **102** 1761
[10] Bologa O, Breaz R E and Racz S G 2018 *Procedia computer science.* **139** 408
[11] Behera A K, de Sousa R A, Ingarao G and Oleksik V 2017 *J. Manuf. Process.* **27** 37
[12] Duflou J R, Habraken A and Cao J et al. 2018 *Int. J. Mater. Form.* **11** 743
[13] Jeswiet J, Micari F, Hirt G, Bramley A, Duflou J and Allwood J 2005 *CIRP Ann. Manuf. Technol.* **54** 623
[14] Gatea S, Ou H and McCartney G 2016 *Int. J. Adv. Manuf. Tech.* **87** 479
[15] Meier H, Buff B., Laurischkat R and Smukala V 2009 *CIRP annals* **58** 233
[16] Schafer T and Schraft R D 2005 *Rapid Prototyping Journal*
[17] Chera I, Bologa O, Racz S G and Breaz R E 2013 *Trans Tech Publications Ltd.* **371** 416
[18] Oleksik V, Pascu A, Deac C et al. 2010 *Int. J. Mater. Form.* **3** 935
[19] Oleksik V 2014 *Procedia Eng.* **81** 2280
[20] Crenganis M and Csiszar A 2019 Materials Science Forum *Trans Tech Publications Ltd.* **957** P 156
[21] Oleksik V, Bologa O, Breaz R and Racz G 2008 *Int. J. Mater. Form* **1** 1187

IOP Conf. Series: Materials Science and Engineering 1009 (2021) 012005 doi:10.1088/1757-899X/1009/1/012005

# Using critical path method for a new project scheduling - the case of a new product launch in production

**F S Blaga[1*], A Pop[1], V Hule[1], A Karczis[2] and D Buzdugan[2]**

[1] University of Oradea, Faculty of Management and Technological Engineering, str.Universității nr.1, Oradea, Romania
[2] SC PLASTOR SA, Calea Clujului, nr. 175, Oradea, Romania, CP 410546, Oradea, Bihor, Romania

[*]E-mail: fblaga@uoradea.ro

**Abstract.** The assimilation process must take place in the shortest possible time and with the lowest costs. These desires are achievable through the most efficient procedures. The paper proposes a procedure for managing the assimilation process of a new product, based on the critical path method. The procedure is intended for a company producing plastic products. In the first phase there was a critical analysis of the existing procedure for assimilating a new product. In the following the study presents how the activities carried out for the assimilation of a new product were organized into a project. To implement this project it was used the Microsoft Project software. This is a tool that allows project management to use the critical path method. By applying this method, it can identify the critical activities within the project. In order to minimize the execution time of the project, the execution time of the critical activities must be reduced. The duration of an activity can be reduced by allocating additional resources. Using the Microsoft Project software in the analyzed case, provides to decision makers information that allows them to make decisions that will reduce the duration of the assimilation process of a new product.

## 1. Introduction

Assimilation of a new product is an important issue for any company producing goods and equipment. The importance of this process is given by the competitive dynamics of the market. The assimilation process must be done in the shortest possible time and with the lowest possible costs. It is also important that the result of the process, the product itself to meet the customer's requirements.

In [1] is presented a New Product Introduction (NPI) process template dedicated to business and engineering.

One of the commonly used methods in project management is the critical path method (CPM). In [2] is presented the implementation of CPM in the optimization of the manufacturing procedure of an All-Terrain Vehicle (ATV). The study concentrates on the time constraint. The study [3] presents how the critical path method is applied into the management of job production. The case study referred to a complex installation for packaging detergents. The application of the CPM method in the capital repair activity (retrofitting) is described in the paper [4].

In the papers [3] and [4] the project management was done using the Microsoft Project program. Microsoft Project (MS Project) is a strong tool used in critical path methods applications [5].

The paper [6] develops a qualitative and quantitative approach to selecting a new product launch strategy. In [7] the authors present the results of a study that examines the interplay between these product launch decisions and the new product development (NPD) performance. An overlapping process model to analyze the impact of process structure on the lead-time of a complex development project is present in [8].

## 2. The current procedure for assimilating a new product

The procedure used to assimilate a new product in Plastor S.A. it is called the Main Procedure (figure 1).

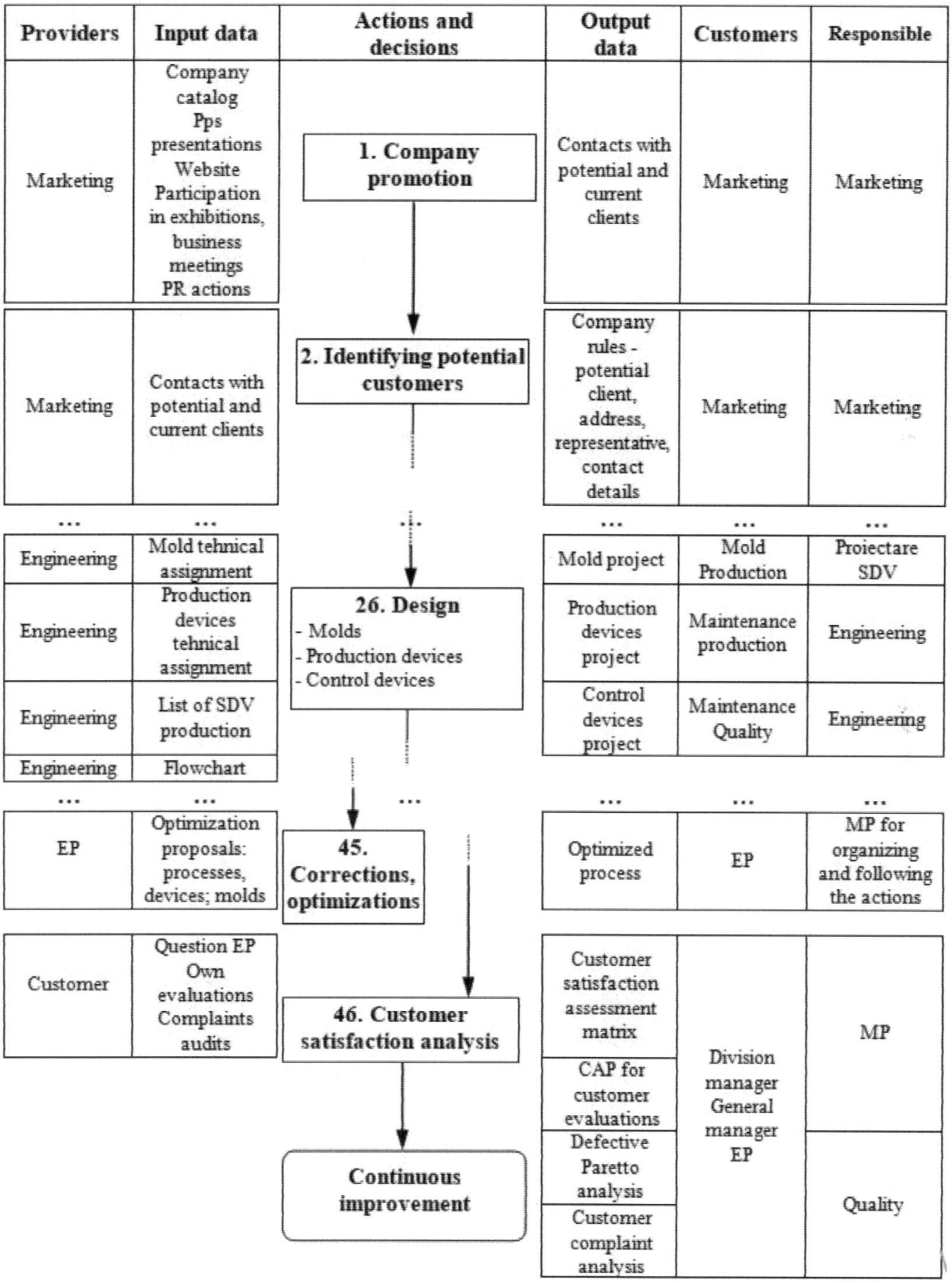


**Figure 1.** Main Procedure [9].

IOP Conf. Series: Materials Science and Engineering 1009 (2021) 012005 doi:10.1088/1757-899X/1009/1/012005

The purpose of the procedure is to specify the responsibilities and how to identify customer requirements for products and services offered by the organization, analysis of requirements and declaration of ability to meet these requirements, monitoring customer satisfaction. The procedure refers to requests for quotations, contracts and orders for the organization's products, as well as to requests for new product development (domestic and export).

For the implementation of the procedure Main Procedure Plastor S.A. use the Excel program. The file called APQP Planning (Advanced Product and Quality Planning) contains all the activities necessary to implement the project. Each activity is associated with the following information: execution time; planned duration and current duration.

An ongoing activity is colored yellow. When the activity ends, the yellow color of the cells turns green. It is also possible to highlight the proportion in which the activity was performed at a given time (in percentages). The color red is used for activities that have exceeded the deadline. To highlight the status of the project (yellow, green or red), an Excel program option called Conditional Formatting is used (figure 2).

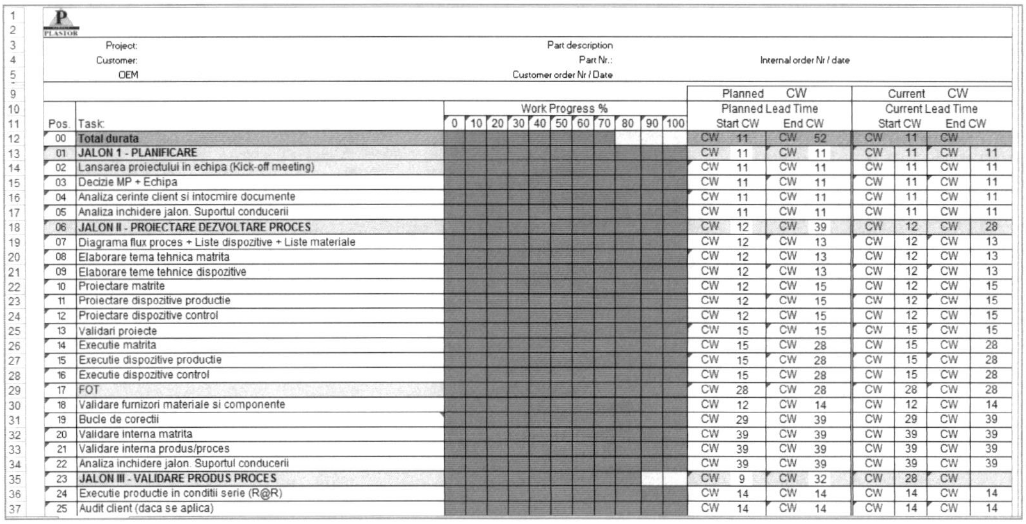


| Pos. | Task | Planned Start CW | Planned End CW | Current Start CW | Current End CW |
|---|---|---|---|---|---|
| 00 | Total durata | CW 11 | CW 52 | CW 11 | CW |
| 01 | JALON 1 - PLANIFICARE | CW 11 | CW 11 | CW 11 | CW 11 |
| 02 | Lansarea proiectului in echipa (Kick-off meeting) | CW 11 | CW 11 | CW 11 | CW 11 |
| 03 | Decizie MP + Echipa | CW 11 | CW 11 | CW 11 | CW 11 |
| 04 | Analiza cerinte client si intocmire documente | CW 11 | CW 11 | CW 11 | CW 11 |
| 05 | Analiza inchidere jalon. Suportul conducerii | CW 11 | CW 11 | CW 11 | CW 11 |
| 06 | JALON II - PROIECTARE DEZVOLTARE PROCES | CW 12 | CW 39 | CW 12 | CW 28 |
| 07 | Diagrama flux proces + Liste dispozitive + Liste materiale | CW 12 | CW 13 | CW 12 | CW 13 |
| 08 | Elaborare tema tehnica matrita | CW 12 | CW 13 | CW 12 | CW 13 |
| 09 | Elaborare teme tehnice dispozitive | CW 12 | CW 13 | CW 12 | CW 13 |
| 10 | Proiectare matrite | CW 12 | CW 15 | CW 12 | CW 15 |
| 11 | Proiectare dispozitive productie | CW 12 | CW 15 | CW 12 | CW 15 |
| 12 | Proiectare dispozitive control | CW 12 | CW 15 | CW 12 | CW 15 |
| 13 | Validari proiecte | CW 15 | CW 15 | CW 15 | CW 15 |
| 14 | Executie matrita | CW 15 | CW 28 | CW 15 | CW 28 |
| 15 | Executie dispozitive productie | CW 15 | CW 28 | CW 15 | CW 28 |
| 16 | Executie dispozitive control | CW 15 | CW 28 | CW 15 | CW 28 |
| 17 | FOT | CW 28 | CW 28 | CW 28 | CW 28 |
| 18 | Validare furnizori materiale si componente | CW 12 | CW 14 | CW 12 | CW 14 |
| 19 | Bucle de corectii | CW 29 | CW 39 | CW 29 | CW 39 |
| 20 | Validare interna matrita | CW 39 | CW 39 | CW 39 | CW 39 |
| 21 | Validare interna produs/proces | CW 39 | CW 39 | CW 39 | CW 39 |
| 22 | Analiza inchidere jalon. Suportul conducerii | CW 39 | CW 39 | CW 39 | CW 39 |
| 23 | JALON III - VALIDARE PRODUS PROCES | CW 9 | CW 32 | CW 28 | CW |
| 24 | Executie productie in conditii serie (R@R) | CW 14 | CW 14 | CW 14 | CW 14 |
| 25 | Audit client (daca se aplica) | CW 14 | CW 14 | CW 14 | CW 14 |

**Figure 2.** Main Procedure. Advanced Product and Quality Planning [9].

When completing the task execution times on the Schedule page in Excel, the cells on the Gantt Chart page are automatically colored based on the use of the Conditional Formatting option, resulting in the Gantt Chart. An illustration of the Gantt Chart can be found in figure 3.

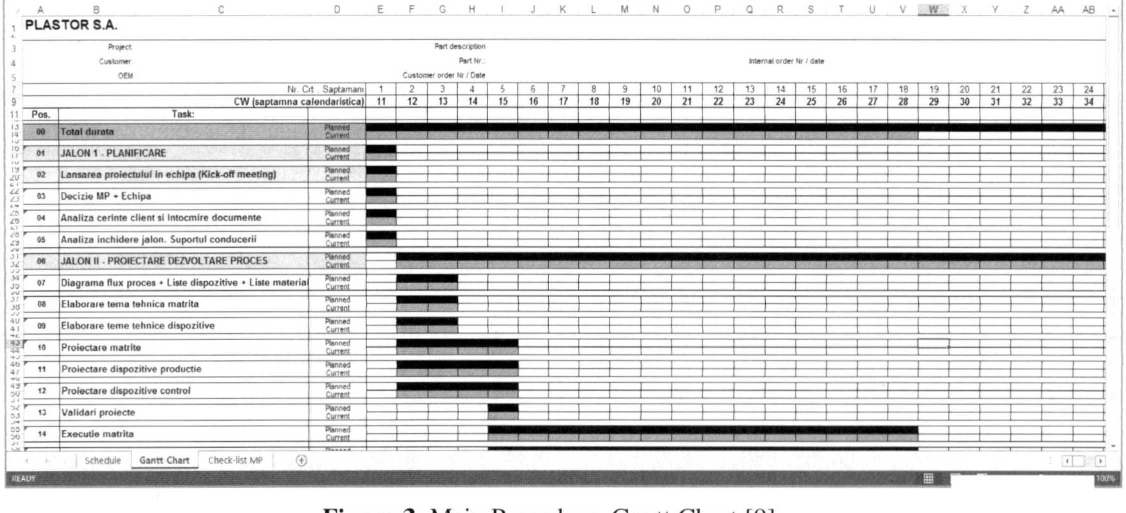


**Figure 3.** Main Procedure. Gantt Chart [9].

Critically analyzing the procedure of assimilating new products at SC PLASTOR SA, the advantages and disadvantages of its use can be identified. Among the advantages are the following: allows data entry in a simple and easy way, Excel is a program known to employees, this program is part of the Office package and does not involve high costs. Disadvantages of applying this procedure are: it does not provide a tracking of resource costs (employees, equipment, etc.); it does not provide information on those responsible for carrying out the activities, it does not highlight the critical path, respectively the critical activities. Knowing them would allow measures to be taken to reduce the duration of the project to assimilate a new product; does not provide information on the evaluation of project indicators (duration, cost).

## 3. Implementing the procedure of assimilating a new product using the critical path method

### *3.1. Product description*

The product that will be assimilated using the critical path method is called Connector (figure 4a and figure 4b). Connector 6 has an important role in the cooling circuit of the machine and is composed of: o-ring (1), sperring (2), spring (3) and injected part (5). The injected part will be manufactured in Plastor, the other components will be purchased. The material of the injected part is PA66 + 30% FS (Polyamide 66 + Fiberglass 30%), in the form of granules.

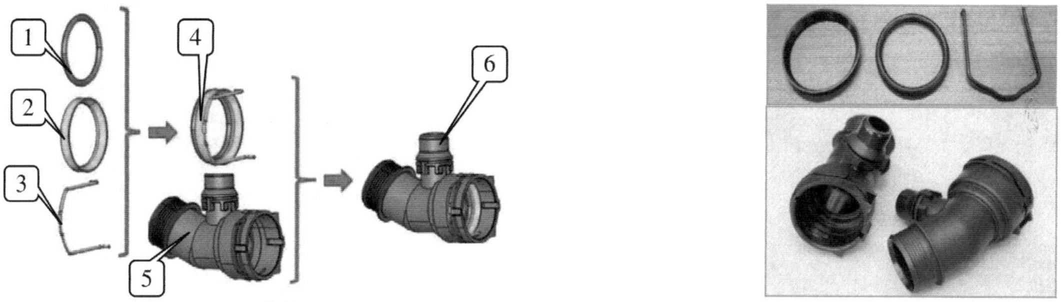


(a) (b)

**Figure 4.** Connector product [9].

### *3.2. Organizing activities for project implementation in Microsoft Project®*

The project of assimilation of the product Connector involves the realization of the activities presented in table .1. The project contains 3 main stages:

- Stage 1 - Planning;
- Stage 2 - Process design and development;
- Stage 3 - Process product validation.

Each stage includes a set of activities necessary for the assimilation of the project. Table 1 associates the following information with each activity: duration, predecessors and resources.

**Table 1.** Activities corresponding to the project of assimilation of a new product.

| Nr. Crt. | Activity name | Duration (days) | Predecessor | Resource |
|---|---|---|---|---|
| 1 | Customer input elements (product design and requirements, materials used, nomination, contract, etc.) | - | - | - |
| 2 | **STAGE 1 - Planning** | | | |
| 3 | Establishing the team and the commitment to team confidentiality | 0.1 | 1 | Project manager, Sales referent |
| 4 | Elaboration of *project objectives* | 0.05 | 1, 3 | Project manager, Sales referent |
| 5 | Project planning APQP (Advanced Product and Quality Planning) | 0.3 | 4 | Project manager, Sales referent |

| Nr. Crt. | Activity name | Duration (days) | Predecessor | Resource |
|---|---|---|---|---|
| 6 | *New Project Analysis* | 0.3 | 4 | Project manager, Sales referent |
| 7 | Establish *List of critical product / process characteristics* | 1 | 6 | Quality engineer |
| ... | ... | ... | ... | ... |
| 14 | **STAGE 2 - Process design and development** | | | |
| 15 | Realization *of Flow Chart* | 2 | 1, 7 | Technological engineer (I) |
| 16 | Elaboration *of location plan* | 0.5 | 15 | Technological engineer (I) |
| 17 | Establish *SDV manufacturing devices* | 0.2 | 15 | Technological engineer (I) |
| 18 | Establish *List of materials and parts* | 0.4 | 1 | Technological engineer (I) |
| 19 | Elaboration *of PFMEA (Process Failure Mode & Effect Analysis)* | 2 | 7, 15, 13 | Technological engineer (I) |
| 20 | Control Plan Development (PC) | 2 | 19 | Quality engineer |
| 21 | Establish SDV_control devices | 1 | 20 | Quality engineer |
| 22 | Development of a plan for MSA (Measurement System Analysis) | 0.4 | 20, 21 | Quality engineer |
| ... | ... | ... | ... | ... |
| 51 | Elaboration of *Parameter adjustment sheet* | 0.5 | 43, 44 | Technological engineer (I) |
| 52 | Technological tests | 0.1 | 15, 39, 40, 47, 51 | Technological engineer (I) |
| 53 | Internal validation of the mold | 1 | 39, 45, 52 | Technological engineer (I) |
| 54 | Validation of production devices | 1 | 40, 49, 52 | Technological engineer (I) |
| 55 | Validation of control devices and MSA (Measurement System Analysis) | 5 | 22, 41, 50 | Quality engineer |
| 56 | Internal process validation | 0.5 | 16, 52, 53, 54, 55 | Technological engineer (I) |
| 57 | **STAGE 3 - Process product validation** | | | |
| 58 | Elaboration of product file | 3 | 56 | Technological engineer (I) |
| 59 | Elaboration of folder PPAP (Production Part Approval Process) | 5 | 14, 48 | Project manager, Sales referent |
| 60 | Training of production operators | 0.4 | 49, 58 | Technological engineer (Prod) |
| 61 | Quality operator training | 0.4 | 50, 58 | Quality engineer |
| 62 | Preparation of PSW | 0.2 | 59 | Quality engineer |
| 63 | Customer report or internal pre-production audit | 2 | 58, 59, 60, 61, 62 | Project manager, Quality engineer |

The allocated resources and labor costs, in lei / hour, are presented in table 2. The number available from each resource is also specified.

**Table 2.** Allocated resources.

| Nr. crt. | Department | Resource | Available | Standard (lei/hour) | Overtime (lei/hour) |
|---|---|---|---|---|---|
| 1 | Marketing | Project manager | 1 | 30 | 60 |
| 2 | Sales | Sales referent | 1 | 21 | 42 |
| 3 | Purchases | Purchasing referent | 1 | 21 | 42 |
| 4 | Engineering | Technological engineer(I) | 1 | 25 | 50 |
| 5 | Quality | Quality Engineer | 1 | 25 | 50 |
| 6 | Design | Design Engineer | 1 | 26 | 52 |
| 7 | Tool workshop | Mold technician | 1 | 260 | 520 |
| 8 | Production | Technological engineers (Prod) | 1 | 25 | 50 |
| 9 | Equipment maintenance | Toolmaker | 1 | 22 | 44 |

*3.3 Using the Microsoft Project software for project analysis*

The program used in this case study is Microsoft Project. With the help of this program it can be follow various aspects related to project management: defining activities, determining the critical path, allocating resources, analyzing costs etc.

For a correct analysis of the activities we will consider the following: working days from Monday to Friday, working hours from $8^{00}$ to $17^{00}$ with a break of one hour between $12^{00}$-$13^{00}$.

*3.3.1. Version 1.* Considering the information in table 1 and table 2, the project was implemented in Microsoft Project obtaining version 1 (figure 4). The program automatically determines the critical path. In order to be able to view more easily, the critical activities are highlighted with red. Figure 5 shows the Gantt chart for version 1. It can be seen in the case of some resources that the available is not at the required level. More precisely, the deficit of resources appears at: project manager, purchasing referent, technological engineer, quality engineer, mold technicians and toolmaker.

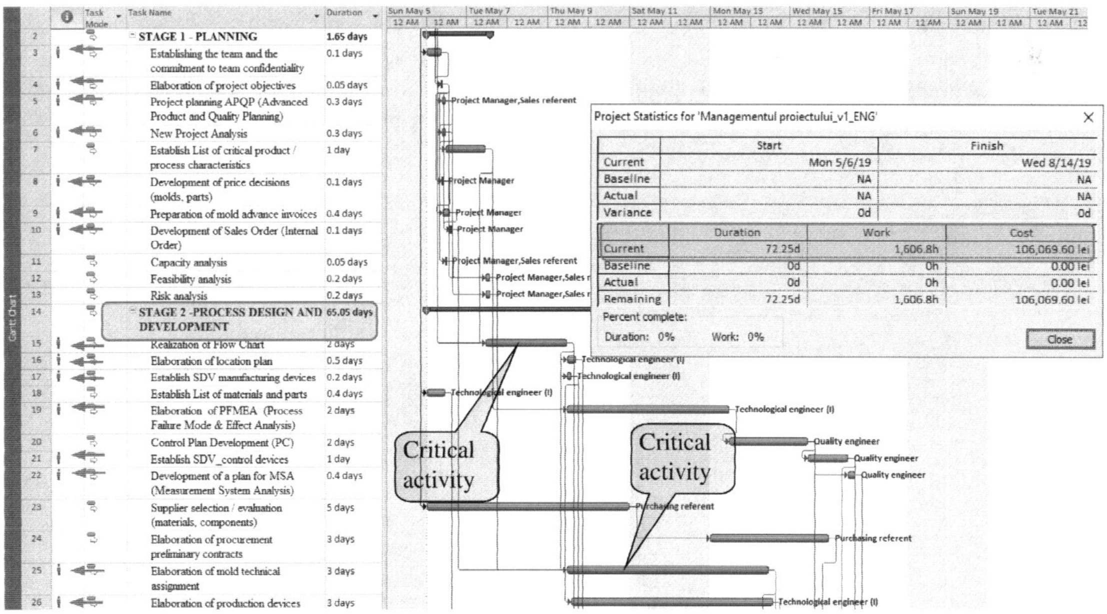


**Figure 5.** Gantt graph. Version 1.

Figure 5 also presents some summary information regarding the first variant of implementation using the Microsoft Project software. This information is:

- The project duration: 72.25 days (75 days and 2 hours);
- Work time: 1600.8 h;
- The work costs: 106,069.60 lei.

The shortage of human resources, of a certain category, can be highlighted by the *Resource Graph* option. Thus, details can be obtained regarding the date when the necessary resources are below the level of those available and what is the difference between necessary and available.

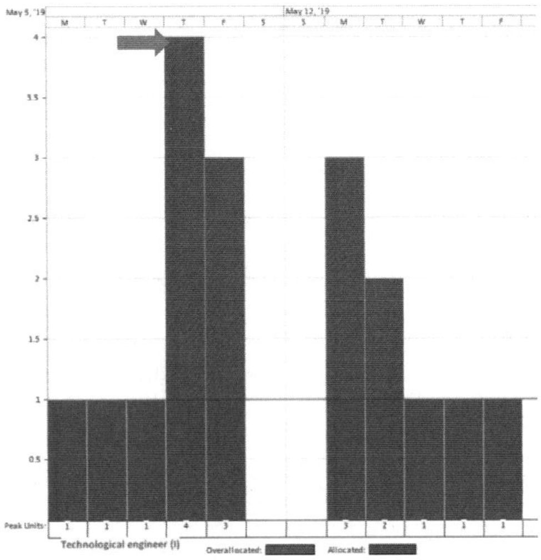


**Figure 6.** *Technological engineering (I)* resource deficit.

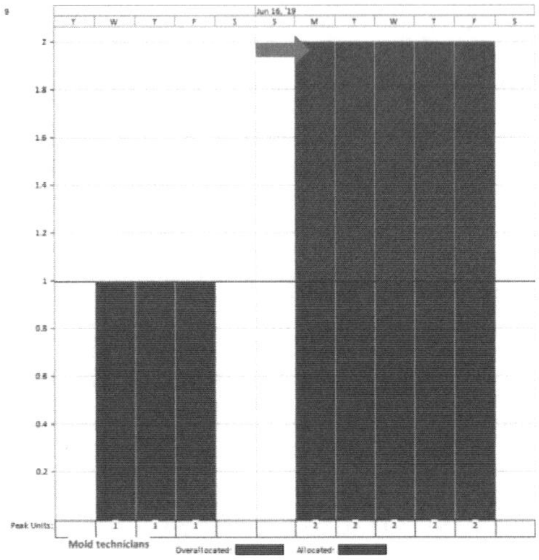


**Figure 7.** Deficit in *Mold technician* resource.

In figure 6 it can be seen that the deficit in the Technological engineering resource (I) is on 5/9/19 of 3 units. Also, in figure 7 it is highlighted the fact that the deficit in the Mold technician resource is 1 unit for 5 consecutive days.

The problem of resource scarcity can be solved in two ways.

*3.3.2. Version 2.* In the case of this version, a first solution to solve the resource deficit is applied. Specifically, it uses a Level All feature of the Microsoft Project® program. This facility allows the scheduling of activities so that the available resources exceed the required resources. Thus, if two activities that initially take place at the same time advertise the same resource, the beginning of one of them will be delayed, after the end of the other.

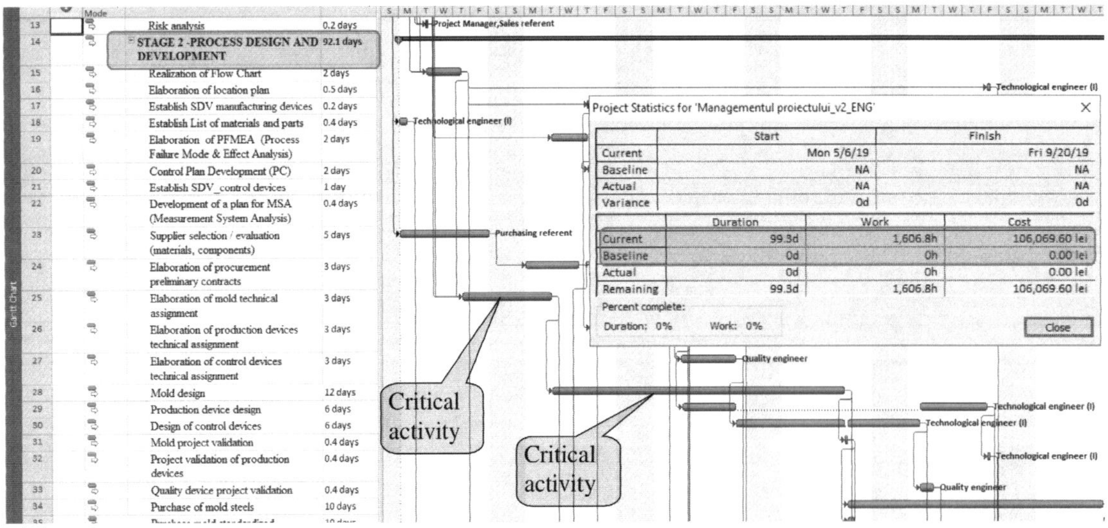


**Figure 8.** Gantt graph. Version 2.

The Gantt chart corresponding to version 2 is presented in figure 8. It is found that the program no longer signals the existence of a deficit of resources. Another shortcoming appears: the duration of execution of the main activities has increased. For example, the duration of the main activity STAGE 2- PROCESS DESING DEVELOPMENT increased from 65.05 days (figure 5) to 92.1 days (figure 8). The duration of the assimilation project of the connector product increased from 72.25 days to 99.3 days. In fact, given the available resources (table 2), this is the real duration of the project completion.

*3.3.3. Version 3.* In version 3 of the implementation of the project of assimilation of the connector product, the second solution for solving the resource deficit was applied; more precisely the number of allocated resources covers the necessary. If this solution were applied, the project could be realized, in a realistic way, in version 1. In order to minimize even more the execution time, another feature of the Microsoft Project program will be used the identification of the critical path.

The critical path defines the minimum duration to complete the project. The critical path is composed of critical activities. In order to minimize the duration of the critical and implicit path of the project, the execution times of the critical activities must be shortened. This is possible by allocating additional resources to critical activities. Two critical activities were identified: Mold design with the duration of 12 days and Mold execution lasting 26 days. For the Mold design resource, the Design Engineer resource was supplemented from one unit to two units and for the Mold execution activity the Mold Technicians resource was supplemented from one unit to three units (figure 9).

| ID | Resource Name | Type | Max. Units |
|---|---|---|---|
| 1 | Project Manager | Work | 2 |
| 2 | Sales referent | Work | 1 |
| 3 | Purchasing referent | Work | 4 |
| 4 | Technological engineer (I) | Work | 3 |
| 5 | Quality engineer | Work | 2 |
| 6 | Design engineer | Work | 2 |
| 7 | Mold technicians | Work | 3 |
| 8 | Technological engineer (Prod) | Work | 1 |
| 9 | Toolmaker | Work | 2 |

**Figure 9.** Additional resources.

| | Start | Finish |
|---|---|---|
| Current | Mon 5/6/19 | Fri 7/26/19 |
| Baseline | NA | NA |
| Actual | NA | NA |
| Variance | 0d | 0d |

| | Duration | Work | Cost |
|---|---|---|---|
| Current | 59.22d | 1,606.8h | 106,069.60 lei |
| Baseline | 0d | 0h | 0.00 lei |
| Actual | 0d | 0h | 0.00 lei |
| Remaining | 59.22d | 1,606.8h | 106,069.60 lei |

Percent complete:
Duration: 0% Work: 0%
Close

**Figure 10.** Features Version 3.

Under these conditions, the duration of the project execution decreases significantly, respectively to 59.22 days (figure 10). Compared to version 2, the realistic version in the conditions of the initial resources, the execution duration of the project of assimilation of the connector product in version 3 is approximately 40% shorter.

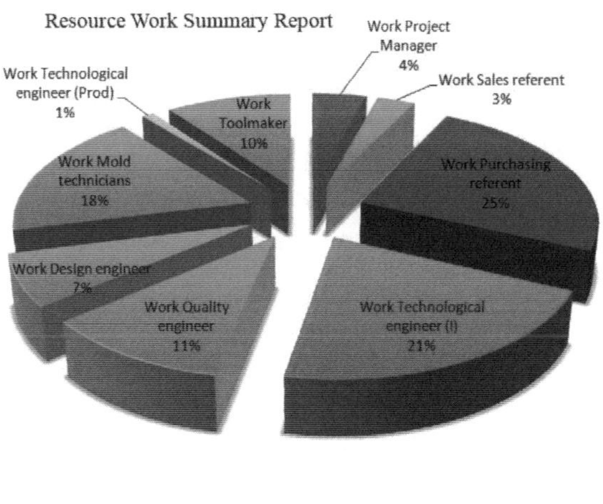


**Figure 11.** Working time corresponding to each resource.

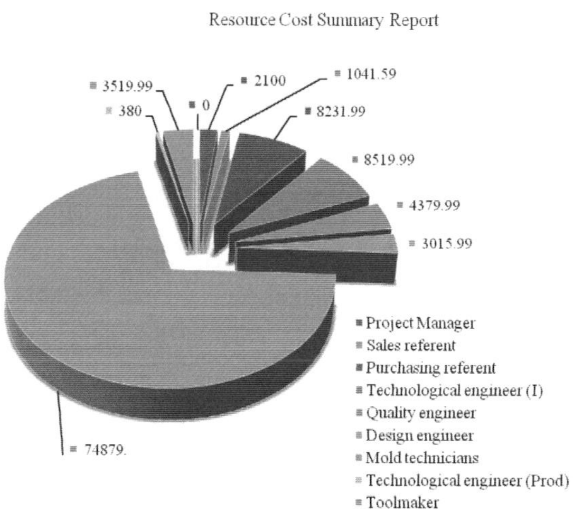


**Figure 12.** Resource costs.

The Microsoft Project® program provides information on many aspects of project management. Thus, figure 11 shows the share of working time corresponding to each resource in total time. Figure 12 shows the synthesis of the distribution of costs on project resources.

## 4. Conclusion

The procedure for assimilating a new product, the Main Procedure, used within SC Plastor SA rigorously describes the steps to be followed. It also defines the responsibilities and results of each activity. In the case of this procedure, there is an IT tool for tracking the way in which the activities of the Advanced Product and Quality Planning program are carried out over time. However, this procedure does not offer solutions to improve the assimilation process in terms of duration and costs.

The paper proposes the use of the critical path method in the management of the assimilation project of a new product. Thus, the activities corresponding to the Main Procedure were implemented in the MS Project software considering their durations, the predecessor / successor dependencies between activities, the human resources necessary for the execution of each activity, the costs associated with these resources. The result is a complex picture of how the assimilation process can take place. The estimated duration and costs were highlighted. The program also provides information on how the allocated resources meet the necessary execution of the project activities. A very important advantage is that the program highlights the critical path and, implicitly, the critical activities. If more resources are allocated for the execution of critical activities, their duration decreases, decreasing the duration of the project execution. Several versions can be analyzed, the decision makers (project manager) can choose the best option.

For the analyzed case, the assimilation of the Connector product to SC Plastor SA, by applying the Main Procedure, the duration of the assimilation process was 282 days. The paper showed that by using the critical path method, the assimilation time of the product can be reduced to about 59 days, given that additional resources are allocated for the execution of critical activities.

## Acknowledgments

We thank the company SC Plastor SA Oradea for the support provided in carrying out this research.

## References

[1] https://online.visual-paradigm.com/diagrams/templates/process-map/new-product-introduction-npi/ accessed on February 2020
[2] Razdan S, Pirgal M, Hanchate A, Sardar V and Rajhans N R 2017 *International Conference on Manufacturing Excellence* ( K.K. WAGH Institute of engineering research and education Nashik India) 5
[3] Blaga F, Stănăşel I, Hule V and Crăciun D 2015 *A. U. O. Fasc. Man. Tech. En.* **24** 13
[4] Blaga F, Buidoş T, Stănăşel I and Pop A 2009 *ModTech International Conference - New face of TMCR, Modern Technologies, Quality and Innovation* (Iasi Romania) 63
[5] Wolfgang L 2007 *Project management* (Bucureşti: Ed. ALL)
[6] Liao C. N., Lin C. H. and Fu Y. K. 2016 *Technol. and Ec. D of Ec.* **22 (5)** 715
[7] Hultink E. J., Griffin A., Hart S. and Robben H. S. 1997 *J of Prod. Inn. M.* **14 (4)** 243
[8] Wang J. and Lin Y. I. 2009 *Com. & Ind. Eng.* **57 (2)** 460
[9] ***SC Plastor SA Oradea Romania

# Fused deposition modelling process: a literature review

**M Blaj[1*] and G Oancea[1]**

[1] Department of Manufacturing Engineering, Transilvania University of Brasov, 29 Eroilor, Brasov 500036, Romania

*E-mail: madalina.blaj@unitbv.ro

**Abstract.** It is well-known that the Additive Manufacturing is a one of the key technology of the fourth industrial revolution, and Fused Deposition Modelling process is one of the most used processes with a large applicability in many domains. Based on a literature review, the paper aims to approach the topic of the main process parameters definitions which ensure the management over the printing process. The paper is also focused on the current optimized manufacturing process, presenting the latest approach from the design phase to the post processing step. Another subject is referring to the used materials, with an accent over the composite carbon filaments in the context of Fused Deposition Modelling process limitations. In order to sum up, based on the main directions and purposes of the further research, this paper aims to be a binder between the current knowledge and what will be studied regarding Fused Deposition Modelling process.

## 1. Introduction

Currently, additive technologies used in the design and manufacturing of the industrial products are the subject of an extensive process of re-evaluation and restructuring of the traditional design and manufacturing processes. In contrast to the traditional manufacturing technologies, additive technologies aim to obtain a product by creating a model, a pattern for its cross sections, until to the physical realization and integration of all cross sections which are forming the desired object, based on geometry from a CAD model [1, 2].The diagram of the realization of a product is presented in table 1.

**Table 1.** Diagram of the additive manufacturing process [3].

| Additive Manufacturing Process | | | | |
|---|---|---|---|---|
| Virtual Model | | Machine with additive manufacturing technology | | |
| 3D CAD Model | Pre-processing software program for additive manufacturing | Front-End Software | Additive manufacturing process | Post processing |

The main characteristics of the additive manufacturing are stated below:

- The models obtained by using the additive manufacturing processes are obtained after the unification of the same thickness layers. The layers are the cross-sections of the part in the XY plane of the printer and the height is provided by the thickness of the adjacent layers [3];
- Due to the each layer thickness, "the stair effect" appears - it can be reduced by decreasing the thickness of the layers [3];

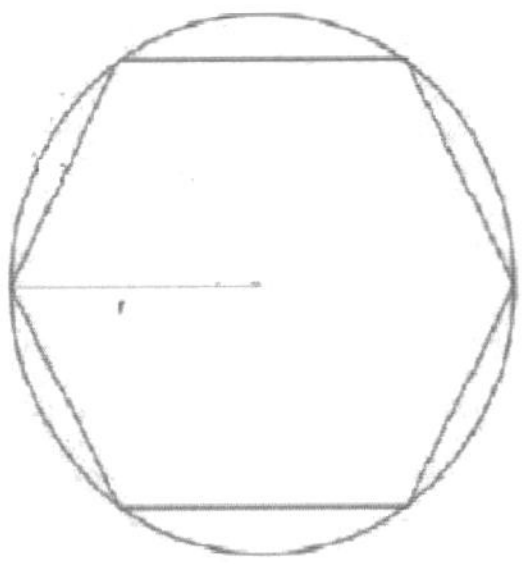

**Figure 1.** Approximation in * .stl format of a circle and its deviations from the original geometry [6].

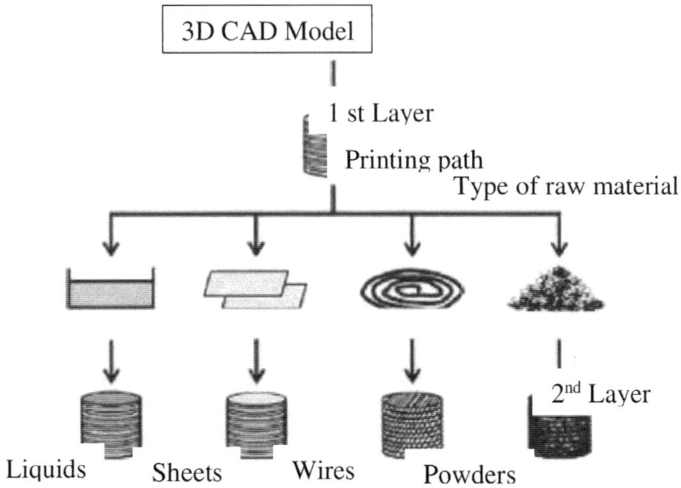


**Figure 2.** Types of raw materials used in additive manufacturing [4].

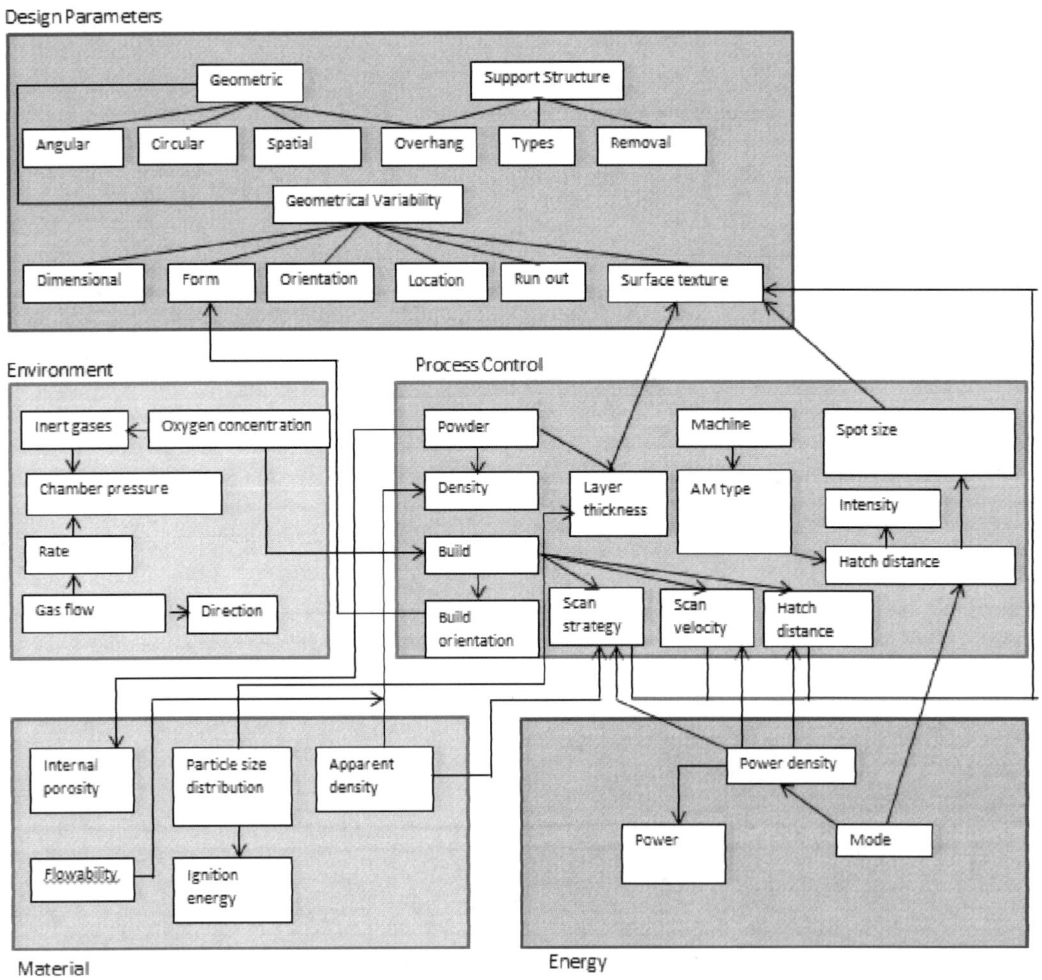


**Figure 3.** General parameters with an influence over the printing process [6].

- The output data from CAD software programs have a different accuracy in representing the geometry, with a certain level of errors in the process of meshing of each surface, for example the accuracy of approximating a curve (figure 1). Formats suitable for additive manufacturing are *.STL, * .3DMF or * .AMF [6];
- The layers of material that are deposited are generated from the virtual model of the part. Depending on the technology used, the base material can be found in various forms: liquid, sheets, wires or powder (figure 2);
- No excessive handling or constant change of work tools required, the parts being manufactured in a short time and with low stocks due to the flexibility of the process[4, 5];
- The material properties of the parts are influenced by the manufacturing process, having the possibility to control and optimize them [3, 4];
- In order to optimize the manufacturing process, the concept of "3D Nesting" can be applied - grouping parts on the printer table to maximize productivity and cost efficiency [8];
- Offers the possibility to create products with a lower level of carbon emissions, as well as the possibility to create parts with lower material consumption, at lower total costs compared by making the same part through conventional technologies [4];
- To maximize the quality of the finished part, a number of parameters are taken into account, having a certain impact on the manufacturing process (figure 3).

### *1.1. Advantages and disadvantages of additive manufacturing*

Due to the advantages of additive technologies, there is a tendency to increase their use in various fields. The main advantages are presented hereafter.

- The possibility of using a wide range of materials compared to the conventional technologies. These materials can be metallic, polymers, composites or biological materials [4, 6];
- Manufacture of parts with complex, customized or modular geometry, with the possibility of avoiding creating junctions, without additional resources that could not be achieved by the available processes offered by the conventional technologies. Moreover, depending on the use, on a part the mechanical properties can be varied, having flexible and rigid areas [2, 4, 9, 10];
- In the most cases the tool is the head of the printer that generates cross sections of the part and it is not changed from one part to another [3];
- The parts are manufactured with increased quality of the geometry approximation to nominal dimensions of the CAD models, at a lower production and energy costs compared to the traditional technologies and also at a lower level of the waste of materials [2, 6, 10];
- The obtained mechanical properties can be controlled by the printing mode [12]. According to [13], the usage of additive manufacturing offers the possibility of increasing the strength of the parts by up to 4000%, elongation to failure by up to 2000% and stiffness by up to 200% [14];
- Without the intervention of distributors and manufacturers of parts, maintenance, repair and overhaul (MRO) operations can be performed faster [14];
- Regarding the level of pollution, carbon emissions are reduced compared to traditional technologies [2].

Disadvantages currently include the following:

- Additive manufacturing technologies are in a relatively early stage of development, requiring individual studies for each technology process and type of material to determine the manufacturing process of the part, as well as the resulting mechanical properties in order to understand each process [2];
- The dimensions of the parts that can be manufactured are limited by the table sizes of the available machines, but also by the increased production time for large parts [4];
- The surfaces of the parts obtained thru various additive manufacturing processes may require mechanical processing or heat treatments for strain relief [4];
- The accuracy in the XY plane is increased, but on Z direction the “step” effect appears, affecting the quality of the surfaces and mechanical properties in Z direction [3].

### *1.2. Fused deposition modelling process*

Fused Deposition Modelling (FDM) or Fused Filament Fabrication (FFF) is an additive manufacturing process which is based on the principle of extrusion of the material. Thus, the material from the filament roll is melted and evenly distributed by the print head along the established path. If the speed of the nozzle and the flow rate of the material are constant, theoretically the circular cross section of the deposited material is also constant. After deposition, the extruded material will unify with the rest of the material in order to form the printed layer and also the whole part – see figure 4 for working principle. The basic principles of the operation of any printing system by extruding the material are based on 7 key points (figure 5).

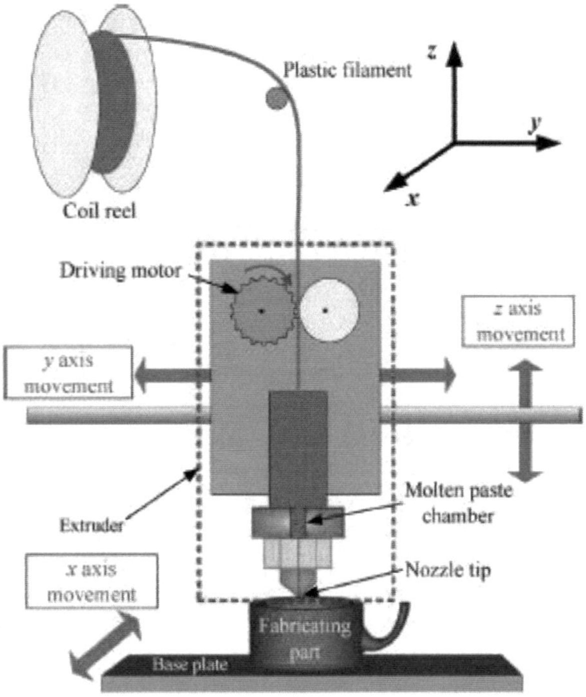


**Figure 4.** FDM working principle [16].

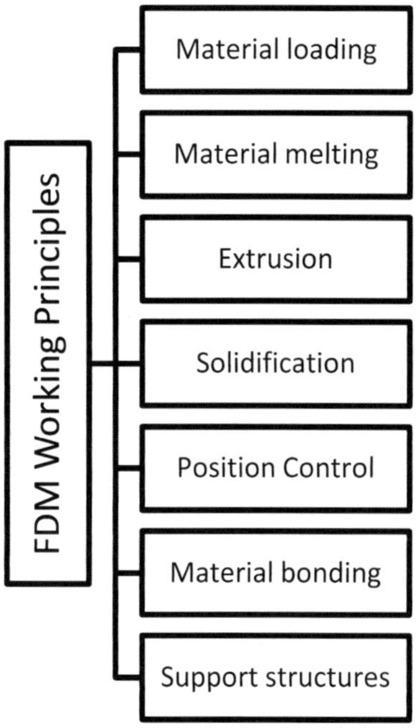


**Figure 5.** FDM – 7 working key points [15].

The FDM working principle combines the 7 key points presented above as following: if necessary, the nozzle is constantly supplied with material, where it is melted and maintained to a constant minimum temperature in this state until the extrusion in order to avoid the material burning. It flows through the tip of the nozzle over the layers of material already existing on the printer table, in the XY plane. The nozzle is moved in the XY plane and is directed to create the cross sections of the part.

After extrusion, the ideal material deposition is to remain in the same shape and dimensions, but gravity and surface stress will cause shape changes, the dimensions variations depending on cooling and drying effects, which are not linear. In order to minimize these effects, it is recommended that the temperature differences between the temperature at the building platform and the environment temperature to be minimal or manufactured in a controlled environment.

The bonding between the rows of material and also between the layers defines the quality of the part, so after the material has been extruded, it must solidify and form connections with the material in adjacent regions. The material regions can stick together, but not completely, the edges of each one being well defined. This will affect the mechanical properties of the part, as the two regions can be easily separated. It is also possible to have regions with too much deposed material, in this case the result being that the part will not be clearly defined.

Due to the complex parts which can be obtained on a printer with FDM technology, in certain situations it is necessary to make auxiliary supports that are made of either the same material as the part

IOP Conf. Series: Materials Science and Engineering 1009 (2021) 012006 doi:10.1088/1757-899X/1009/1/012006

or another material. They must be added to the part, taking into account the need for removal after manufacturing. The most effective method of making supports is to make them from special materials for this purpose [15, 16].

## 2. Main manufacturing parameters for FDM

Considering all the parameters presented in figure 3, for FDM it can be obtained a higher control over the manufacturing process, in order to obtain parts with mechanical properties and geometric accuracy as high as possible, in a shorter time and lower costs. In the printing process, certain factors have a higher influence on the final result, compared to others and the errors generated by each have an irreversible impact, the manufacturing process being unidirectional [17]. They are presented hereafter.

### *2.1. The quality of the filament*

The quality of the filament influences the manufacturing process, but also the properties of the resulting manufactured parts, considering the variations of the filament nominal diameter value – see figure 6 for the impact over the deposited material.

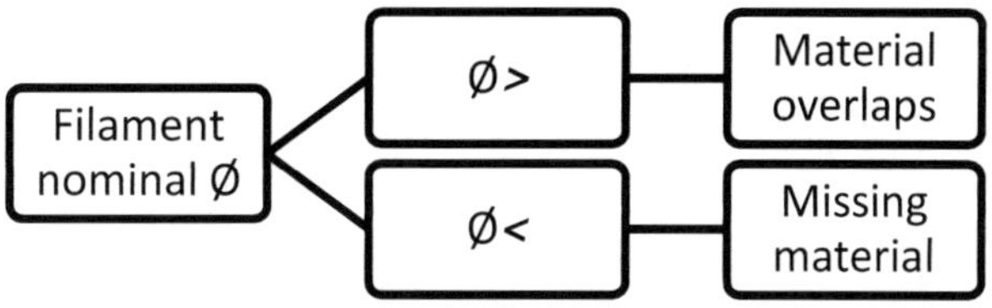


**Figure 6.** Variation of the filament diameter compared to the nominal value [18].

### *2.2. Part positioning on the building platform*

The part positioning on the building platform is an important parameter that mainly influences the mechanical properties, appearance, part "slicing", printing time, material consumption and costs. The highest mechanical properties are obtained when the part is printed in the direction of tensile stress during use, thus avoiding the delamination effect, because in the end the materials have an orthotropic behaviour similar to the composite materials [20, 21].

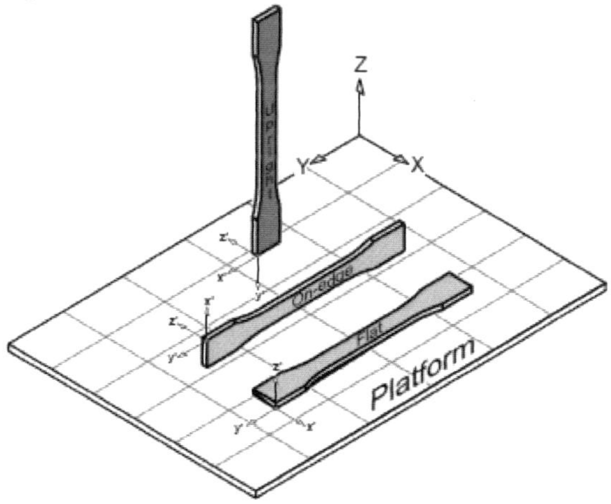


**Figure 7.** Part positioning on the building platform [20].

According to [22], from 55 design rules for additive manufacturing for FDM, 70% of them are directly or indirectly dependent on the orientation of the part on the building platform, implicitly on the direction of material deposition to form the part. Due to the importance of this parameter, work steps were established to define the orientation of the parts based on their usage, from the early stages of the projects, a work flow being presented in figure 8.

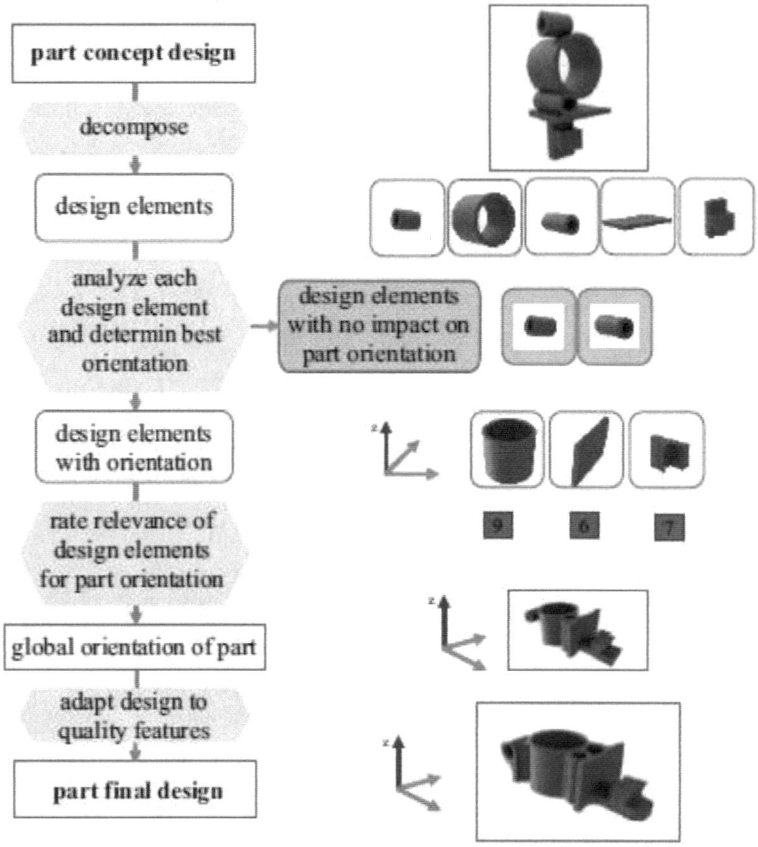


**Figure 8.** Defining the orientation of the part from early design stages [22].

*2.3. Material direction – raster angle*

The angle of deposition of the material with respect to the XY plane in the orthogonal system of the machine influences the mechanical properties of the resulting part, which can be higher or lower than the specifications of the basic filament offered by the manufacturer [23].

As a comparison to observe the importance of the material orientation, the results on stiffness presented in table 2 highlights the best result in the loading direction of tensile. The specimens are made of short carbon fibers composite filaments and TPU matrices, which also contain a different amount of fibers [24].

**Table 2.** Comparison based on the material orientation for tensile loading [24].

| Carbon fiber quantity | E [MPa] at 45° | E [MPa] at 0° |
|---|---|---|
| 0% | 239.91 | 242.37 |
| 1% | 300.23 | 313.85 |
| 2% | 331.22 | 387.89 |
| 4% | 479.04 | 642.26 |

The optimization regarding the orientation on the building platform can be performed considering the geometric parameters of the part: consider the dimensions (width, depth), the slenderness, the degree of filling with material, areas with discontinuities, but also the appearance of each layer. The aim is to obtain as large, continuous surfaces as possible and to minimize the irregularities of the material layers by analysing each layer of the part [19].

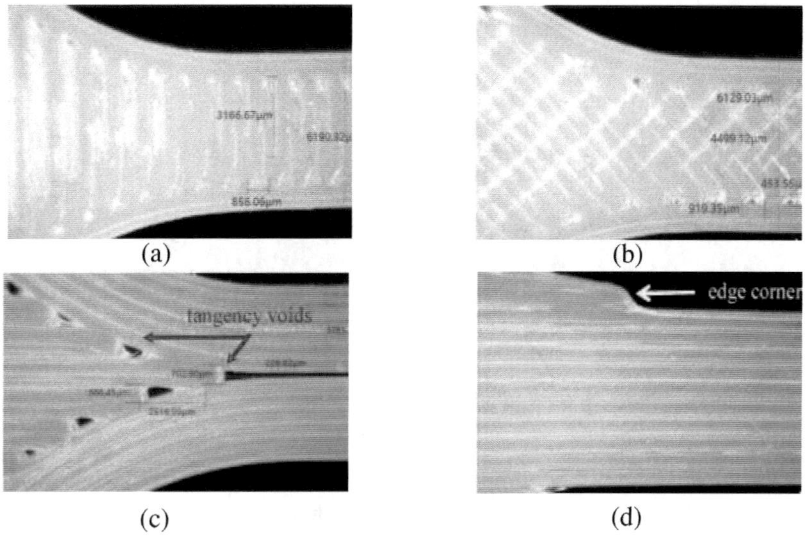


**Figure 9.** Top view of 4 specimens (a) orientation at 90°; (b) orientation at ±45°; (c) and (d) orientation at 0° [25].

Figure 9 shows four specimens made with a different material deposition angle, the manufacturing defects that appear being visible: edge corners or material gaps.

In order to increase the precision of the part, it is recommended to initially make its contour, the limits between which the material will be deposited being defined by the defined edges [19, 25].

The infill can be 100% with material or to have a different structure, as is presented in section.

The rows of material deposited and the layers between them must achieve a continuity of the material to ensure the load transfer which implicitly increases the strength of the part. If material gaps are provided, then stress peaks can appear in those areas which will cause the local failure of the material. In the printing process, initially the deposited rows of material come into contact as in area (a) of the figure 10. A heat transfer and diffusion of the material is initiated which generates a neck as in point (b), but which is influenced by the viscosity of the material. Once the temperature in that area decreases, the neck and diffusion effect decreases. In (c) and (d) after cooling, the contact and joint area stabilizes [23, 26].

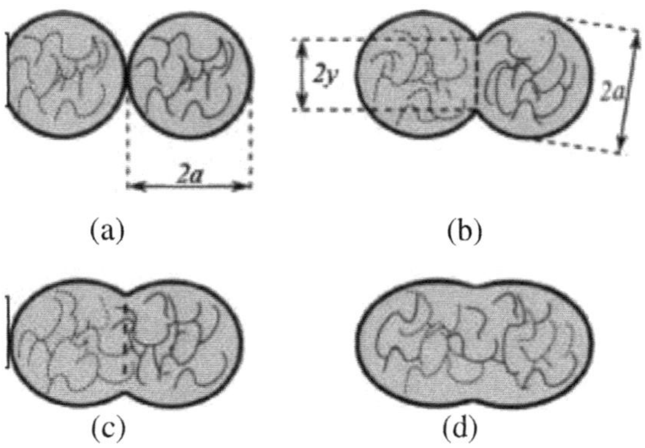


**Figure 10.** Sintering steps of polymers by FDM process [23].

Also, the approach of the raster angle is an important factor for the mechanical properties of the part. If the material rows overlap exactly, then the mechanical properties of the final part will be lower than if the layers are interspersed and the stress states are evenly distributed in all regions of the part. However, each interspersed model determines a discontinuity which materializes in weak points for the created part [15]. In figure 11 are presented two variants of material orientation in order to obtain areas without material as reduced as possible [23].

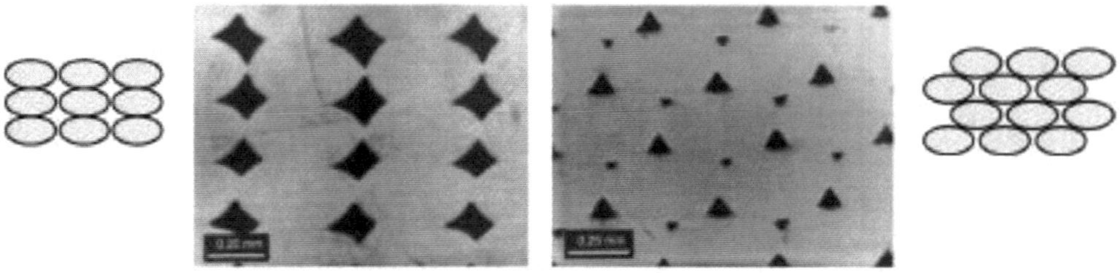

(a) (b)

**Figure 11.** Layer layout models (a) rectangular model; (b) triangular model [23].

*2.4. Temperature*

Another important parameter is temperature. An ideal situation is to maintain a constant temperature in the nozzle, reducing variations that influence the speed and flow rate of the material. As a compensatory action, the inlet pressure can be increased to maintain the same flow rate. This parameter is the decisive factor in forming the bonds between the layers [27]. Depending on the size, the cooling time of the parts varies, having an effect on the control of the FDM machine, but also on the environment [15].

Temperature measurements cannot be performed conventionally (at several points) because this stage will interfere with the manufacturing process and the measured values are only from the surface of the part or near the surface area. Areas of interest - areas from contact between the deposited layers - are not captured by classical measurements. However, to validate the simulation, surface measurements are performed to compare the obtained values with the simulation results, where the temperatures for the whole part are obtained [27].

The cooling process is important for creating the bindings between the rows and the layers in order to create the final part. As long as the cooling process is taking longer, the obtained mechanical properties are higher due to a better linkage of the material. The cooling process is influenced by the size of the part, the material, the FDM machine, but also by the cooling method. If the temperature is not evenly distributed between the layers, residual stresses can occur, causing delamination of the layers, because these stresses cannot be transmitted between the layers [28].

*2.5. Thickness of the printed layer*

An important feature is the thickness of a manufactured layer – the influence of this parameter over the quality of the final part is presented in figure 12.

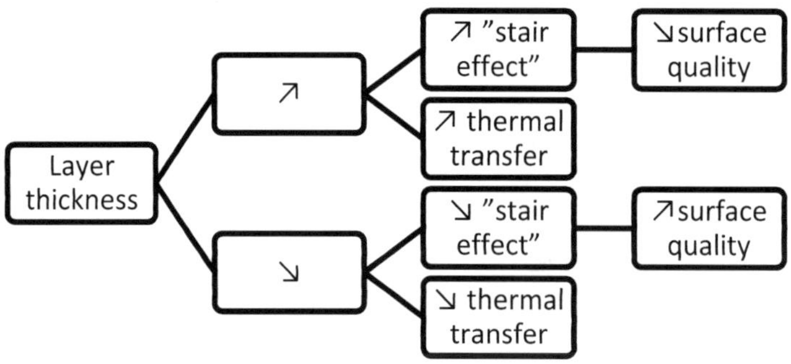


**Figure 12.** Influence of the layer thickness over the quality of the final part [16, 29].

Statistically, the dimensional accuracy is mainly influenced by the density of the inner structure (infill) and the dimensions of the layers of the material, the software program having also an influence on the quality and planning of the manufacturing with the settings of the manufacturing parameters [30].

### 2.6. Work environment

Although the working environment for the FDM manufacturing process is usually the normal environment, there are studies that aim the influence of a specific working environment on the characteristics of the parts after the manufacturing process. Thus, for a Nylon polymer the manufacturing in a nitrogen environment had an increase of the strength up to 30% and the oxidation process was diminished due to the lack of oxygen, causing an improvement during the adhesion process of the layers.

For manufacturing ABS, PLA and PA6 samples in a low pressure environment; compared to specimens of the same materials printed in normal environmental conditions, there are more qualitative parts surfaces, the traces of the manufacturing route being less visible (figure 13) and a lower roughness value. Also, there is an increase on the weight, density and volume of the part due to the longer cooling process and more deposited material due to a better bonding. There is also a decrease in the porosity of the part due to the high adhesion of the material layers [32].

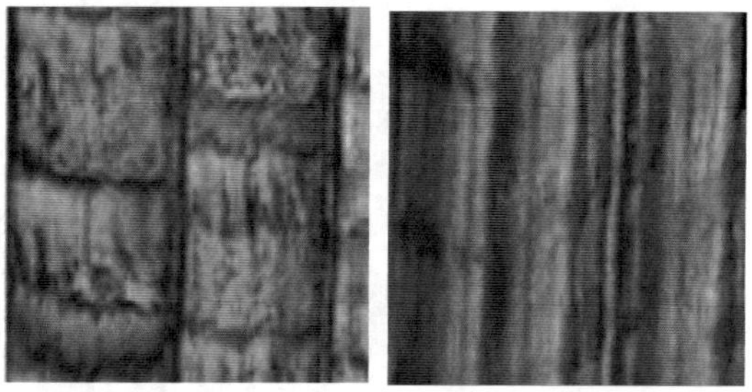

(a) (b)

**Figure 13.** Manufacturing ABS specimens (a) in normal environment;
(b) in controlled environment [32].

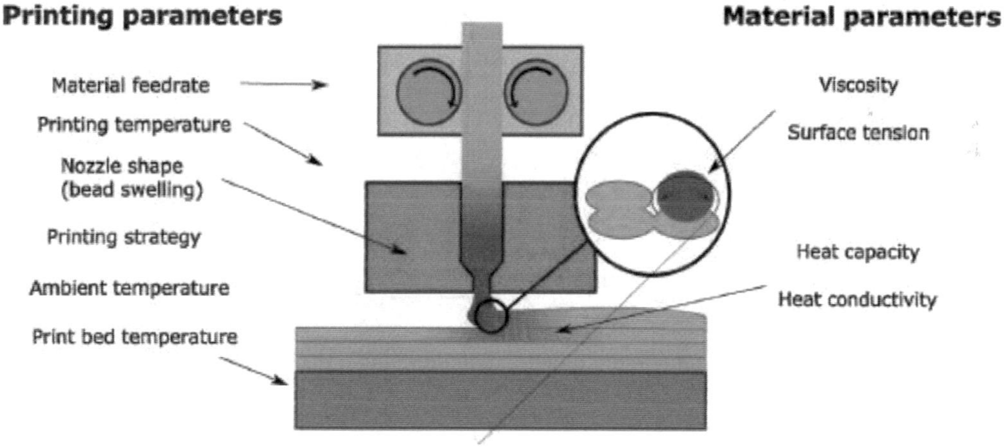


**Figure 14.** The main parameters that influence the FDM printing process [23].

### 2.7. Other parameters

In the FDM manufacturing process other manufacturing parameters are involved, which through their interdependencies, determine the user to create and use an optimal variant of them.

- Nozzle diameter - this value is a constant, but there are manufacturing systems which allow the user to change it to the desired accuracy;
- Material properties - the behaviour of the material is difficult to predict, but in most cases it is anisotropic;
- Gravity - due to gravity the molten material may flow, even if no inlet pressure is applied;

- Deposition speed – as the manufacturing speed is higher, the slower the printing time is and the time for material cooling during the manufacturing process is reduced [20, 27].
- Type of infill – presented in the next section.

In conclusion, the main parameters that influence the manufacturing process are described in figure 14 and the physical phenomena that occur during the deposition of the material are described in figure 15.

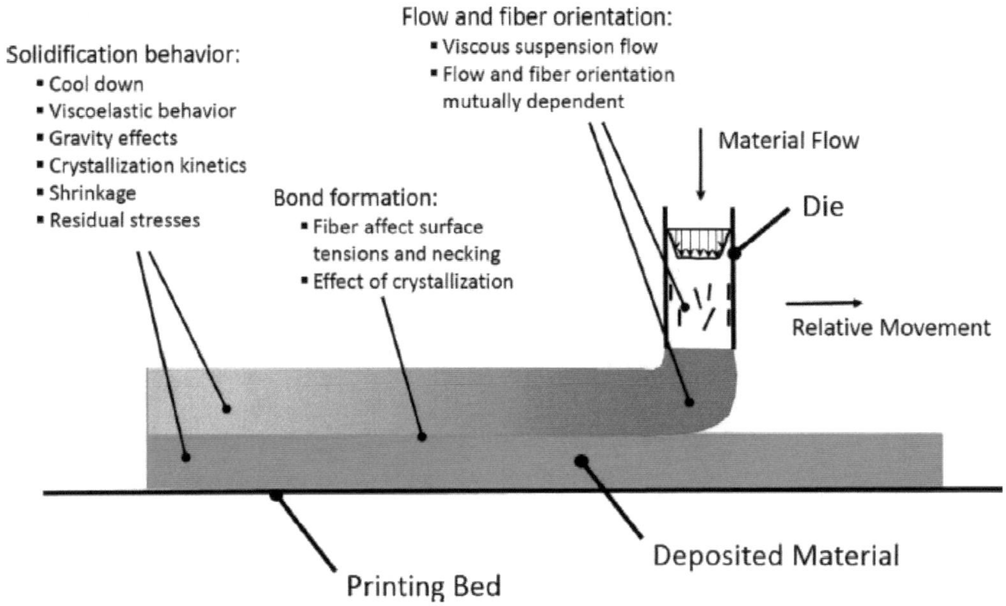


**Figure 15.** Physical phenomena in the FDM manufacturing process [31].

## 3. FDM – optimization

The optimization process for parts obtained thru FDM process occurs on each step of a product development – from design phase to its post processing after manufacturing and testing – the steps are presented below.

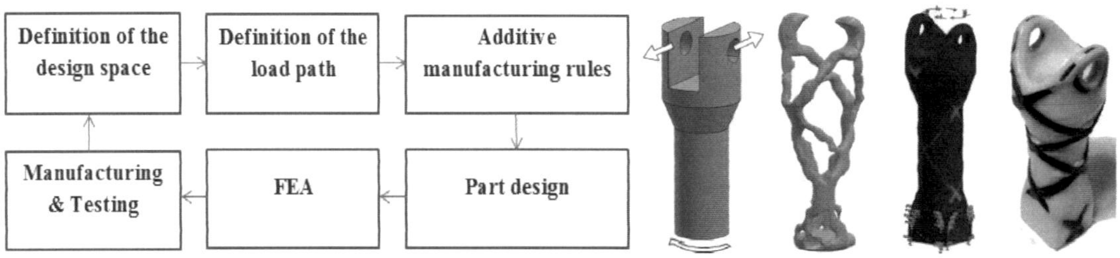


**Figure 16.** Part design & optimization process [13].

In the design phase, based on the experience obtained in Additive Manufacturing for simple polymer filaments or with reinforcement, there are several adapted rules (figure 16) which can be followed in order to obtain an improved part, taking in consideration the anisotropy of the material and lower strength values in Z direction compared to other directions [13].

- The reinforcement path must cover all areas where there are von Mises stresses, but must be aligned according to the stresses generated by tension-compression (definition of normal);
- The reinforcement route/ the deposited material row is recommended to be continuous;

- In areas of high stresses, the areas of curvature must be oriented towards the inside of the part in order to avoid the possibility of having material fibers out from the matrix;
- Internal channels are an alternative to open channels - increases the connection between reinforcement and part, having the effect of delaying buckling in areas prone to compression;
- The start and end areas of reinforcements / the deposited row material must be in low-demand areas;
- Access areas to the internal channels to allow their reinforcement;
- Inflection points must be accessible [33].

Also, for the design optimization process can be taken in consideration the infill type, the internal structure being adapted based on the application, obtaining increased properties. However, certain properties, such as rigidity, strength and energy absorption capacity, are still limited in development by the random microstructure, which translates into structural defects with local peaks [36].

Now, considering the manufacturing machine, the optimization can occur on the manufacturing parameters (see section 2 for details) or the type of the printer. Due to the large spectrum of available printers on the market, printers can be created by reconverting CNC machines or industrial robots, in this way the FDM process can be developed to be used in several planes, with several nozzles, having thus the ability to use more materials (figure 17) [21].

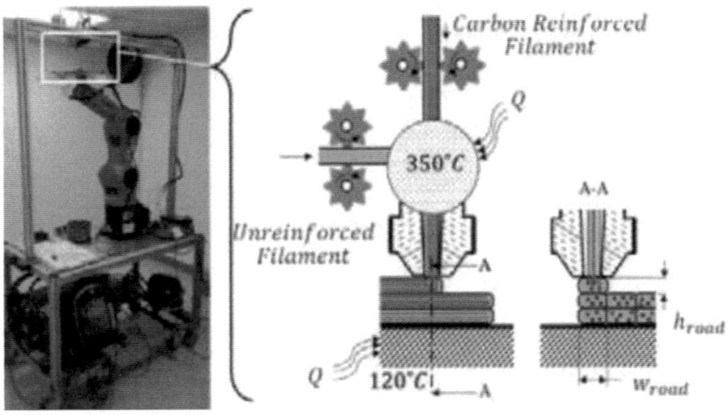


**Figure 17.** Reconversion of an industrial robot into a FDM machine [21].

(a) (b)

**Figure 18.** Part obtained (a) FDM; ( b) CLFFF [21].

Due to the weakness of the material in the Z direction for the classic FDM technology and the overlapping of the walls of the layers which may have defects in their joining, a solution is to use the improved process of FDM printing, namely curved layer printing CLFFF (Curved Layer Fused Filament Fabrication), where the print head is not static as in the case of the traditional FDM process, thus avoiding the step effect – see figure 18 for comparison between processes [34].

Another optimization process can be possibility of reducing the loss of heat. This deficiency can be reduced by infrared heating of the material already deposited to ensure a higher heat transfer on the next pass as is presented in figure 19 [35].

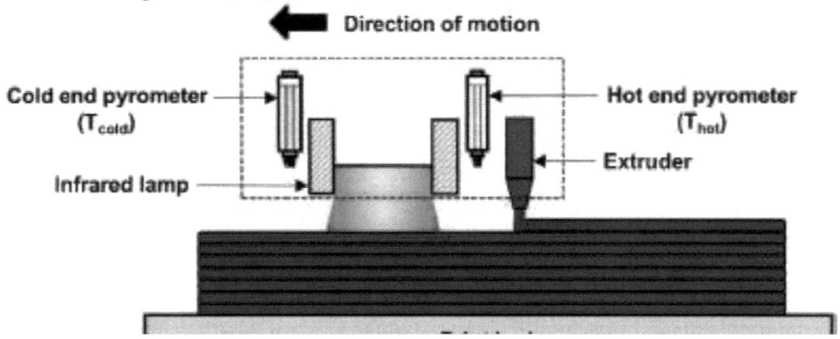


**Figure 19.** Part design & optimization process [35].

## 4. FDM – materials – carbon fiber filaments (CFF)

Due to the evolution and accessibility of the FDM manufacturing process, the field of materials used followed its trend, experiencing a rapid development of new thermoplastic materials, with properties from various categories: magnetic, conductive, flexible, soluble [36]. Research has shown that parts manufactured by the FDM process have behaviour similar to the laminated materials than that of a cast isotropic resin. [37].

If initially only polymer filaments were used, now these filaments can contain several types of materials, thus developing the concept of composite filament. By definition, a composite filament is a mixture of at least two materials which have different physical and chemical properties. After the combination of the materials, new material results with superior properties compared to the properties of each component [39].

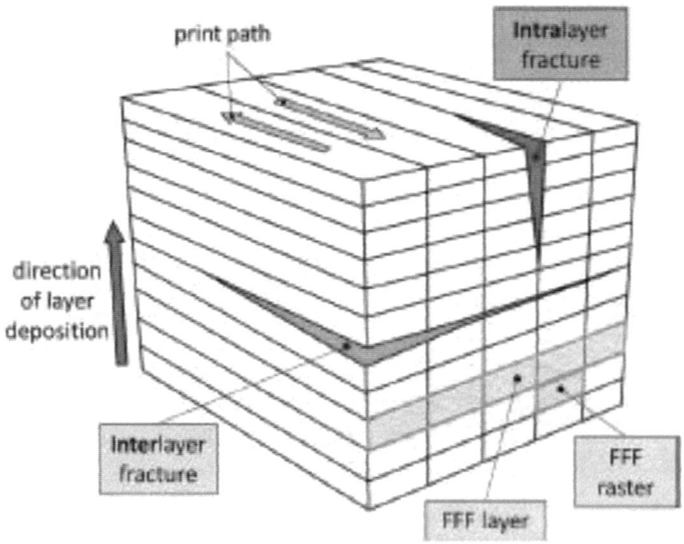


**Figure 20.** Inter-laminar area definitions and its deficiencies for CFF [40].

The topic of this section is the carbon fiber filaments – the filaments are made of carbon fibers that are embedded in a matrix of thermoplastic material in order to increase the rigidity and strength of the part in the direction of carbon fibers. In the other directions, the mechanical properties register much lower values, due to the fact that in these directions the material is dependent on the strength of the matrix and the area between the layers, called inter-laminar area (figure 20) [27].

Parts made by the FDM process with carbon fiber composite filaments have a behavior similar to conventional composite materials due to the similarity between the layers that form the manufactured part and the sheets of the composite material.

As common failure criteria observed for filament composite parts, delamination or inter-laminar failure, matrix cracks, fiber matrix detachment, fiber failure or their exit from the composite were recorded. Delamination may occur from manufacturing reasons, such as defective material deposition, cracks between layers or matrices with poor mechanical properties, or from mechanical reasons such as tensions in inter-laminar areas, compression load or a too high load or fatigue [27].

**Table 3.** Comparison between matrix polymer specimens and carbon fiber composite filament specimens [42].

| Matrix polymer vs carbon fiber composite filament (l≅100 μm) | Printing direction[°] | σ matrix polymer [MPa] | σ CFF [MPa] | σ [%] | E matrix polymer [GPa] | E CFF [GPa] | E [%] |
|---|---|---|---|---|---|---|---|
| PLA vs CFF PLA | 0 | 60.0 | 68.4 | 14 | 3.5 | 9.28 | 162.9 |
| | 45 | 49.6 | 50.7 | 2.2 | 3.4 | 5.08 | 50.7 |
| | ±45 | 52.2 | 54.6 | 4.6 | 3.3 | 5.20 | 59.5 |
| | 90 | 45.5 | 43.7 | -4 | 3.2 | 4.61 | 44.5 |
| PETG vs CFF PETG | 0 | 46.1 | 68.3 | 48.2 | 2.05 | 8.47 | 313.2 |
| | 45 | 41.9 | 45.2 | 7.9 | 1.96 | 3.60 | 83.7 |
| | ±45 | 41.3 | 50.9 | 23.2 | 1.91 | 4.23 | 121.5 |
| | 90 | 41.4 | 42.6 | 2.9 | 1.89 | 3.19 | 68.8 |
| ABS vs CFF ABS | 0 | 38.2 | 50.9 | 33.2 | 2.3 | 7.15 | 212.2 |
| | 45 | 34.8 | 30.1 | 13.5 | 2.2 | 3.14 | 44.0 |
| | ±45 | 34.7 | 29.2 | 15.9 | 2.1 | 3.09 | 45.8 |
| | 90 | 32.4 | 27.34 | -15.7 | 2.1 | 3.03 | 47.1 |
| Amphora vs CFF Amphora | 0 | 46.9 | 49.3 | 5.1 | 2.01 | 3.93 | 95.5 |
| | 45 | 43.6 | 40.9 | 6.2 | 1.91 | 2.48 | 29.8 |
| | ±45 | 44.4 | 42.0 | 5.4 | 1.96 | 2.63 | 34.2 |
| | 90 | 44.2 | 40.5 | -8.4 | 1.90 | 2.45 | 28.9 |

Composite filaments with carbon constituents may have short or long fibers or carbon fiber particles.

Short fiber filaments increase the rigidity of the part, but the increase in strength is limited by the possibility of the fibers to get off the matrix before failure. In addition to advanced mechanical characteristics, the advantages of using carbon fibers are described by their electrical conductivity, high heat transfer capacity and biocompatibility.

Studies show that short carbon fibers of about 0.1 mm that are mixed with a thermoplastic polymer have a strength and stiffness increased by 65% compared to simple polymers [41]. In table 3 is presented a comparison between matrix polymer specimens and carbon fiber composite filament specimens, considering also the printing direction.

It is emphasizing that for an orientation to 0° the strength and the stiffness have the most increased values and an orientation to 90°, the weakest values. Also, the matrix type has an increased influence over the mechanical properties, the highest strength value being recorded by PETG and the highest stiffness to PLA. Considering also the material direction, it can be observed that to a 90° orientation, the CFF can register lower strength values compared to the specimens manufactured only by the matrix polymer.

Another important topic of these filaments is the porosity of the material. It is determined as a percentage of the material gaps in the cross-section of a specimen by scanning and measuring the pixels showing the holes. The porosity is influenced by the amount of the fibers (figure 21) [39].

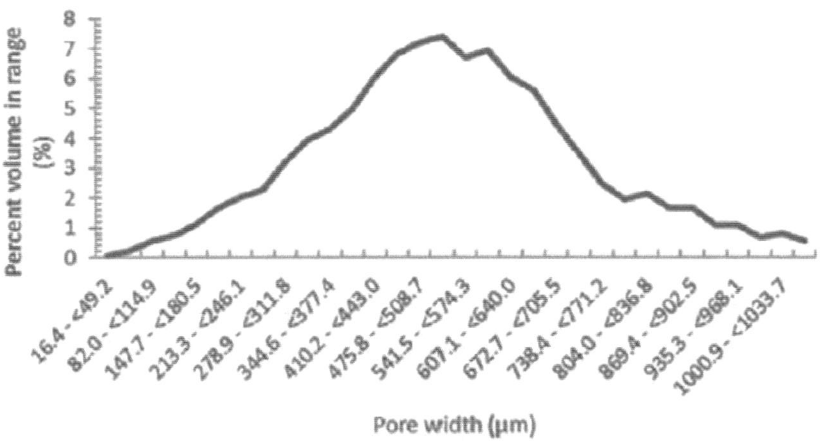


**Figure 21.** The size of the porosities depending on the amount of fibers [39].

However, material gaps can be attributed to a poor adhesion between the fibers and the matrix. A high number of fibers degenerates the heat transfer energy between the printed layers and the flow of material from the nozzle [23].

As a result of the manufacturing process, the carbon fibers undergo changes in size, which are reduced [42]. The manufacturing with continuous fibers is limited due to the possibility of arranging the rows of material. For a Nylon matrix, Mark Forged registered a maximum strength of 700 MPa and E=50GPa. As a manufacturing process, the FDM machine (figure 22) has 2 nozzles: one with not reinforced Nylon fiber and the other contains continuous Nylon filament reinforced with carbon fiber. In the outer areas of the piece, for complex and thin areas only not reinforced Nylon filament is used because the reinforced fiber does not have the capacity to manufacture these areas. Otherwise, the part is made of reinforced filament. The variation of the manufacturing parameters is limited, the material being deposited in a circular pattern (figure 23), from outside to inside - the direction of the fiber is known [23].

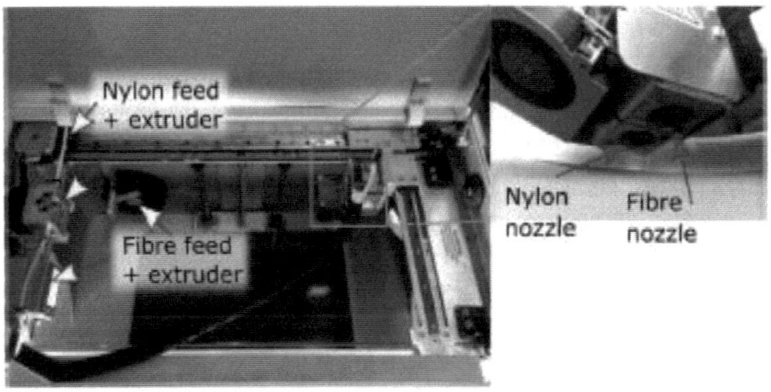


**Figure 22.** Continuous fiber printer [23].

**Figure 23.** Deposition of a continuous fiber printer [23].

Due to the carbon fiber contained by the filament, its thickness determines the thickness of the printed layer. Because continuous fiber is used, the orientation of the material is limited [23].

Studies on carbon fiber composite filaments with various matrices are presented in the literature, the results being presented in the following table.

Hereafter, in figure 24 and 25 are presented a comparison of the mechanical properties between different types of matrix, with different lengths of the fibers.

As presented in figure 24, the matrix type, the type and quantity of the fiber are the main factors which affect the stiffness of the material. It is emphasizing that the highest influence is offered by the matrix type, continuing with the type of fiber and the amount of carbon fibers. With the same type of fiber and almost the same amount of fibers, comparing the stiffness between ABS and Nylon, the ABS matrix is offering higher values of the rigidity.

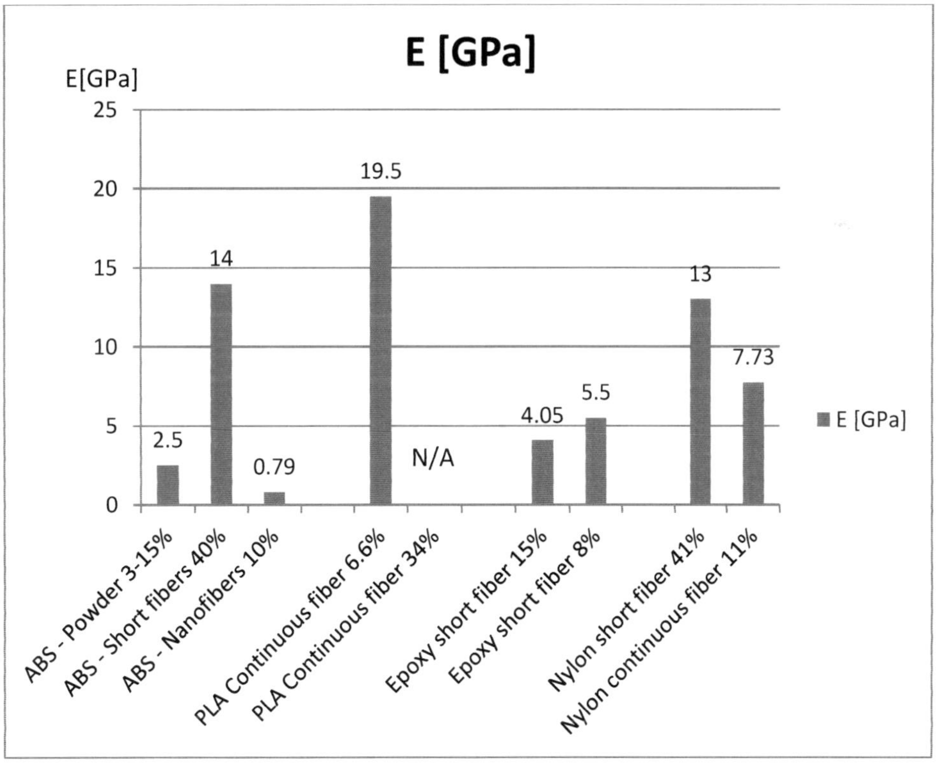


**Figure 24.** Comparison of the Young Modulus between different composite filaments with carbon fibers [23, 39, 44, 45].

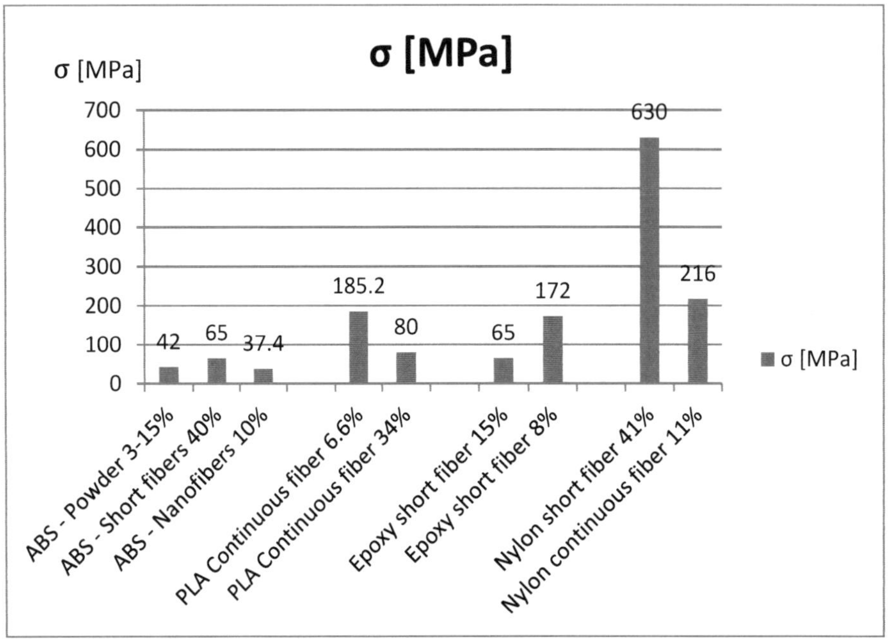


**Figure 25.** Comparison of the strength between different composite filaments with carbon fiber [23, 39, 44, 45].

As presented in figure 25, the matrix type has the highest influence over the strength of the material, the Nylon matrix having the highest strength value in comparison with ABS with the same type and amount of fibers. The second factor, after comparing the PLA and Nylon with continuous fiber is the type of the fiber and the third factor as importance is the amount of fibers.

It is emphasizing that for both types of parameters of the mechanical properties, the highest influence is offered by the type of matrix, followed by the type of fiber and by the amount of them.

For all types of carbon fiber composite filaments, different failure modes may occur as a result of stress, for example: fractures of the carbon fibers, cracks in the matrix, must be taken into account for the process of optimizing the manufacturing parameters polymer or delamination phenomenon may occur.

## 5. Limitations of the FDM process

As every manufacturing process, also the FDM process has several limitations. The main of them are listed below.

- FDM technology has a limitation in terms of the maximum temperature that can be used, this feature being directly related to the types of materials that are suitable for use;
- Parts with complex geometry can be made only by creating supports that support geometric accuracy;
- For some types of material, such as continuous carbon fiber composite filaments, the manufacturing technique does not currently allow all manufacturing parameters to be changed [23];
- The fatigue resistance of the parts is most influenced by the density of the material deposited inside the contour, followed by the size of the nozzle and the height of the layer;
- The diameter of the nozzle is recommended to be at least 1.5 times the value of the material layer height to ensure the integrity of the part;
- Printing parameters and part sizes are limited to the technical capabilities of the printer.

## 6. Conclusion

The research directions identified and detailed in the previous sections of the paper are outlined in figure 21.

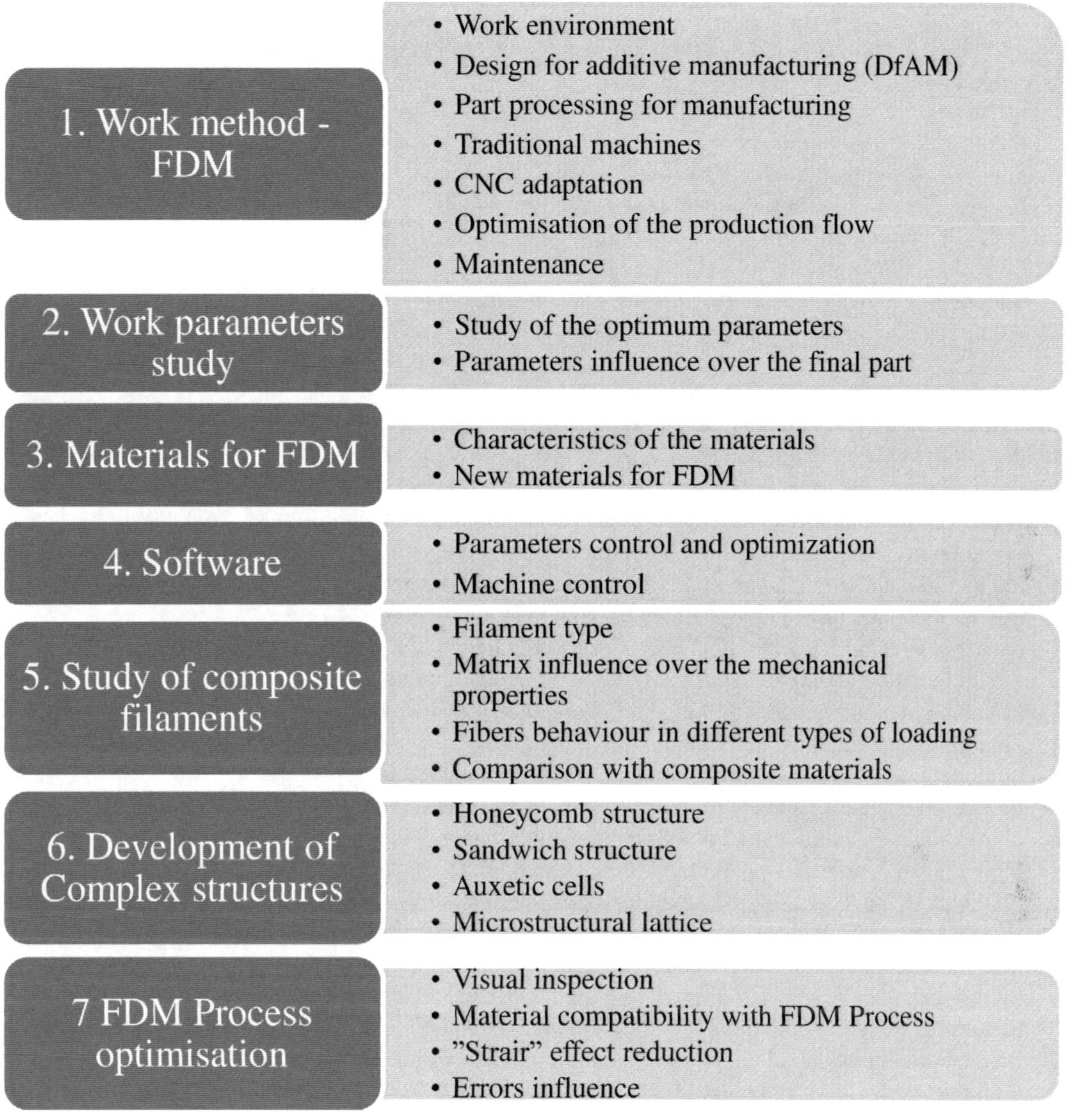


**Figure 26.** Research directions of FDM process.

As a conclusion, it is found that the manufacturing speed, the temperature of the working environment, but also the method of the deposition of the material, the infill, the thickness of the layer, the orientation of the part on the building platform are the most important parameters that lead to the improvement of the mechanical properties.

In addition to the manufacturing parameters, the material used has an important influence in improving the mechanical properties. As manufacturing with continuous carbon fiber filaments, additive manufacturing is not possible on conventional FDM machines, the use of short fiber composite filaments is being studied, in addition to the mechanical properties and failure criteria of such materials, such as the failure of the fibers, crushing and breaking them and delamination of the material.

Another possibility of obtaining increased mechanical properties is to create different complex structures, as lattice, honeycomb or auxetic cells based on the loading type.

## References

[1] Bernier S N and Bertier Luyt T R 2015 *Make: Design for 3D Printing* (Maker Media)
[2] Badiru A B, Valencia V V et al *Additive Manufacturing Handbook. Product Development for the Defense Industry* (CRC Press)
[3] Gebhardt A and Hötter J-S 2016 *Additive Manufacturing – 3D Printing for Prototyping and Manufacturing* (Hanser)
[4] Srivatsan T S and Sudarshan T S 2016 Additive Manufacturing – Innovations, Advances, and Applications (CRC Press)
[5] Spalt P and Bauemhansl T 2016 *Proc. CIRP CMS* **57** 716
[6] Hällgren S, Pejryd L and Bauemhansl T 2016 *Proc. CIRP Design Conference* **50** 518
[7] Qi Q, Pagani L et al 2018 *Proc. CIRP* **75** 87
[8] Lutters E, Ten Dam D et al 2012 *Proc. CIRP* **3** 26
[9] Cuellar J S, Smit G et al 2018 *Additive manufacturing (Elsevier)* **21** 150
[10] Kritzinger W, Steinwender A et al 2018 *Proc. CIRP Conference on Manufacturing Systems* **72** 1518
[11] Würtz G, Lasi H et al 2015 *Proc. CIRP* **33** 394
[12] Kaur M, Yun T G et al 2017 *Materials and Design (Elsevier)* **134** 272
[13] Brooks H, Molony S 2016 *Materials and Design (Elsevier)* **90** 276
[14] Wits W W, Reyese Garcia J R et al 2016 *Proc. CIRP* **40** 693
[15] Gibson I, Rosen D et al 2015 Additive Manufacturing Technologies – 3D Printing , Rapid Prototyping and Direct Digital Manufacturing (Springer)
[16] Jin Y, Li H et al 2015 *Additive Manufacturing (Elsevier)* **8** 142
[17] Galantucci L M, Bodi I et al 2015 *Proc. CIRP* **28** 82
[18] Greeff G P, Schilling M 2017 *Additive Manufacturing (Elsevier)* **14** 31
[19] Ahsan N, Khoda B 2016 *Additive Manufacturing (Elsevier)* **11** 85
[20] Chacón J M, Caminero M A et al 2017 *Materials and Design (Elsevier)* **124** 143
[21] Zhang Y, De Backer W et al 2016 *Proc. CIRP Design Conference* **50** 414
[22] Leutenecker-Twelsiek B, Klahn C et al 2016 *Proc. CIRP Design Conference* **50** 408
[23] Blok L G, Longana M L et al 2018 *Additive Manufacturing (Elsevier)* **22** 176
[24] Anwer A, Naguib H E et al 2018 *Additive Manufacturing (Elsevier)* **22** 360
[25] Plott J, Tian X et al 2018 *Additive Manufacturing (Elsevier)* **22** 606
[26] Prajapati H, Ravoori D et al 2018 *Additive Manufacturing (Elsevier)* **21** 84
[27] Damodaran V, Castellanos et al 2018 *Materials and Design (Elsevier)* **157** 60
[28] Bähr F, Westkämper E 2018 *Proc. CIRP* **72** 1214
[29] Ravoori D, Alba L et al 2018 *Additive Manufacturing (Elsevier)* **23** 132
[30] Santana L, Alves J L et al 2017 *Materials and Design (Elsevier)* **135** 159
[31] Brenken B, Barocio E et al 2018 *Additive Manufacturing (Elsevier)* **21** 1
[32] O'Connor H J, Dowling D P et al 2018 *Additive Manufacturing (Elsevier)* **21** 404
[33] Asadollahi-Yazdi E, Gardan J et al 2017 *Proc. CIRP Design* **60** 217
[34] Robert Allen J A, Trask R S et al 2015 *Additive Manufacturing (Elsevier)* **8** 78
[35] Kishore V, Ajinjeru C et al 2017 *Additive Manufacturing (Elsevier)* **14** 7
[36] Chen Y, Li T et al 2018 *Materials and Design (Elsevier)* **137** 226
[37] Christ J F, Aliheidari N et al 2017 *Materials and Design (Elsevier)* **131** 394
[38] Zaldivar R J, Witkin D B et al 2017 *Additive Manufacturing (Elsevier)* **13** 71
[39] Goh G D, Dikshit V et al 2018 *Materials and Design (Elsevier)* **137** 79
[40] Young D, Wetmore N et al 2018 *Additive Manufacturing (Elsevier)* **22** 508
[41] Vaneker T H J 2017 *Proc. CIRP Conference on Composite Materials Parts Manufacturing* **66** 317
[42] Jiang D, Smith D E 2017 *Additive Manufacturing (Elsevier)* **18** 84
[43] Dickson A N, Barry J N et al 2017 *Additive Manufacturing (Elsevier)* **16** 146
[44] Ueda M, Tae-Kun- J 2016 *Scientific Reports* **6** 23058
[45] Li N; Li Y et al 2016 *Journal of Materials Processing Technology* **238** 218
[46] Tian X; Liu T et al 2016 *Composites Part A: Applied Science and Manufacturing* **88** 198

# The modelling of detached chip area at square shafts machining with rack tools

**M Bordeanu**[1], **G Frumuşanu**[1*] **and N Oancea**[1]

[1] "Dunărea de Jos" University, Manufacturing Engineering Department, Domnească Street 111, 800201 – Galaţi, Romania

*E-mail: gabriel.frumusanu@ugal.ro

**Abstract.** The paper deals with analyzing, by 3D modeling, the shape and the area of the chips detached during the machining of shafts with square cross section, by slotting with a rack tool. The proposed modeling requires at first to determine the tool profile. This step is accomplished by finding the equations of the surfaces family described by the tool relative to workpiece, and by using the enwrapping condition, written according to Minimum distance theorem. The generating process is modeled then, on the base of CATIA environment facilities, aiming to find, in graphical manner, the variation law of the detached chip. Solving this problem could be a primary level of analysis needed for designing the hobbing tool used for machining the same type of part. A numerical and graphical application of the proposed modeling algorithm is also included. The final goal of the research is to deliver the necessary information for smoothing the addressed machining process by modifying the kinematics of the generating process.

## 1. Introduction

The detached chips modelling in the existing literature concerns almost exclusively the detaching mechanism, e.g. [1, 2]. Only few researches approached the modelling of the cutting schemes specific to cutting tools which generate by enwrapping surfaces delimited by polyhedral peripheric surfaces (shafts with square or hexagonal section, gears with involute or arc of circle profile etc.), following to reveal the unevenness of the machining process, no matter of the type of tool that is used (rack tool, pinion cutter, hob mill) [3]. This modelling can be used for finding the laws governing the addressed machining process from energetic point of view, further enabling the development of technical solutions aiming to the smoothening of the process.

Existing solutions for smoothening the machining process concern the constructive modification of the cutting tool (be it rack tool, pinion cutter or hob mill), such as the area of the chips detached by successive teeth of the tool, during the process of generating by rolling, becomes more uniform. Thus, the cutting force magnitude also becomes more uniform, leading to process smoothening from energetical point of view. Examples of such constructive modification for the rack tool are suggested in [4], while [5] presents a new type of hob mill.

The development of graphical soft packages (e.g. CATIA) are underpinning a new approach regarding the problems in connection to the graphical analysis of the cutting schemes associated to the process of generating by rolling polyhedral surfaces, like the shafts with square or hexagonal section, with tools having finite number of teeth (rack tool case).

This paper addresses the problem of identifying the cutting scheme and the variation of the detached chips area, in the mentioned case (square shaft / rack tool) in order to enable the modification of the speed of circular feed motion, seen as a new type of approach in solving the problem of machining process smoothening from energetic point of view. Problem solution was found by using the graphical modelling of the machining process, in CATIA environment, according to the specific of the rolling process when generating with the rack tool, expressed through both the fundamental theorems [6] and the complementary theorems [7] of enwrapped surfaces.

In what concerns paper structure, next section presents the finding of the analytical model of the rack tool used for generating square shafts, and the third section deals with the graphical model of the tool. The fourth section is dedicated to the modelling of the detached chips area, while the last is for conclusion.

## 2. The tool tooth analytical model

Figure 1 presents the cross section of the square shaft, the rolling centrodes, associated to the workpiece and to the cutting tool (whose profile is needed to be found) and the involved reference systems:

- $xyz$, meaning a global system, immobile, having its $z$ axis coincident to workpiece rotation axis;
- $XYZ$ – local system, mobile, attached to workpiece centrode, $C_1$ (circle of $R_r$ radius), initially overlapped to $xyz$ system;
- $\xi\eta\zeta$ – local system, mobile, associated to rack-tool centrode, $C_2$ (straight line) and having $\eta$ axis along $C_2$.

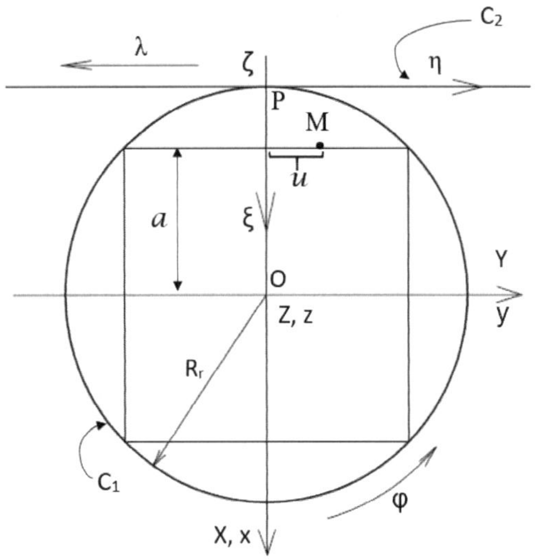


**Figure 1.** Square shaft section, centrodes and reference systems.

The rolling motions, executed by the two centrodes, are:

- $I$, meaning the rotation motion of $C_1$ centrode around $z$-axis, described by the angular parameter φ;
- $II$ – the translation motion of $C_2$, described by the linear parameter $\lambda$.

The rolling condition expresses the dependence between the parameters from above:

$$\lambda = R_r \cdot \varphi. \tag{1}$$

Note: The value of rolling radius $R_r$ must be chosen such as the normal in any point of the generated profile intersects the centrode $C_1$ [6]. Hereby, there is a minimum value of $R_r$. In the addressed problem, the selected value of $R_r$ is the radius of the circle circumscribed to the square, while the minimum value is the radius of the circle inscribed in the square.

The parametric equations of the generated profile, meaning the coordinates of the generic point belonging to square profile, written into $XYZ$ system, are:

$$\Sigma \begin{vmatrix} X = -a; \\ Y = u; \\ Z = v. \end{vmatrix} \tag{2}$$

In equations (2), $a$ means half of the square side length, while $u$ and $v$ are two independent parameters. Due to the specific of the profiling problem, it can be further considered as a plane problem. In this case, only the first two equations from (2) are needed, with $u \in [0, a]$.

The expressing of the generating process kinematics means to find the equations of the relative motion between the rack-tool and the workpiece. In this purpose, the equations of the absolute motions must be written, at first:

$$x = \omega_3{}^T(\varphi) \cdot X, \tag{3}$$

meaning the equation of workpiece rotation, and

$$x = \xi + A, \text{ with } A = \begin{pmatrix} -R_r \\ -\lambda \end{pmatrix}, \tag{4}$$

meaning the equation of rack-tool translation. In equation (3), $\omega_3$ is the well-known matrix of coordinates transform at rotation around $z$-axis.

The equations of the relative motion can be found then, by reducing $x$ between equations (3) and (4):

$$\xi = \omega_3{}^T(\varphi) \cdot X - A. \tag{5}$$

The motion (5) represents the motion of the current point from $XYZ$ space relative to the rack-tool system, $\xi\eta\zeta$. If here $X$ means the parametric equations (2) of the generated profile, then after developing and calculus we obtain the equations of the family of profiles $\Sigma$ referred to $\xi\eta\zeta$ system:

$$(\Sigma)_\varphi \begin{vmatrix} \xi = -a \cdot \cos\varphi - u \cdot \sin\varphi + R_r; \\ \eta = -a \cdot \sin\varphi + u\cos\varphi + R_r \cdot \varphi. \end{vmatrix} \tag{6}$$

The rack-tool profile results as envelop of $(\Sigma)_\varphi$ family, by associating to equations (6) the enwrapping condition, here determined according to Minimum distance method [7]. In order to find it, the gearing pole (the point of tangency between $C_1$ and $C_2$ centrodes, see figure 1) is identified at first as:

$$P \begin{vmatrix} \xi_P = 0; \\ \eta_P = R_r \cdot \varphi, \end{vmatrix} \tag{7}$$

corresponding to a generic position in the rolling process.

Then, the distance between the gearing pole and the current point of the profiles family from $\xi\eta$ space, is calculated as:

$$d = \sqrt{(\xi - \xi_P)^2 + (\eta - \eta_P)^2}. \tag{8}$$

Finally, according to minimum distance theorem, the condition of minimum is imposed to distance $d$ by annulling the derivative of its expression against $\varphi_1$ leads to the relation:

$$(\xi - \xi_P) \cdot \xi'_u + (\eta - \eta_P) \cdot \eta'_u = 0. \tag{9}$$

In the last relation, $\xi'_u$ and $\eta'_u$ represent the derivatives of $\xi = \xi(u, \varphi)$ and $\eta = \eta(u, \varphi)$ functions, calculated with (6), against $u$:

$$\xi'_u = -\sin\varphi \text{ and } \eta'_u = \cos\varphi. \tag{10}$$

After replacing (6), (7) and (10) in (9) and calculus, the final form of the enveloping condition results:

$$\varphi = \sin^{-1}\left({}^{u}/_{R_r}\right). \tag{11}$$

The ensemble formed by the equations (6) and the condition (11) determine the profile of the rack tool tooth, or, in other words, the analytical model of tool tooth.

## 3. The rack tool graphical model

The finding by graphical modelling in CATIA of variation law for detached chip area, during the machining of a square shaft by slotting with a rack tool, firstly requires to build the graphical 3-D model of the rack tool. For a better understanding, the algorithm developed in this purpose is explained when applied for solving a given problem, which addresses the case of machining a square shaft having $a = 20$ mm, starting from a cylindrical piece of raw material with radius $R = R_r = a \cdot \sqrt{2}$ mm.

The coordinates for a number of points $N_p = 101$, defining the rack tool profile necessary for generating half of square side, were calculated with relations (6) and (11) from previous section, by discretizing $u$ variation interval (which is $[0, a]$) in 100 sub-intervals. Some of these coordinates are sampled in table 1.

**Table 1.** Coordinates of profile points (excerpt).

| Point crt. no. | u [mm] | φ [rad] | ξ [mm] | η [mm] |
|---|---|---|---|---|
| 1 | 0 | 0 | 8.2842 | 0 |
| 2 | 0.2000 | 0.0070 | 8.2833 | 0.2585 |
| 3 | 0.4000 | 0.0141 | 8.2806 | 0.5171 |
| 4 | 0.6000 | 0.0212 | 8.2760 | 0.7756 |
| 5 | 0.8000 | 0.0282 | 8.2696 | 1.0341 |
| 6 | 1 | 0.0353 | 8.2614 | 1.2924 |
| 7 | 1.2000 | 0.0424 | 8.2513 | 1.5507 |
| 8 | 1.4000 | 0.0495 | 8.2394 | 1.8089 |
| 9 | 1.6000 | 0.0565 | 8.2257 | 2.0669 |
| 10 | 1.8000 | 0.0636 | 8.2102 | 2.3247 |
| ... | ... | ... | ... | ... |
| 46 | 9 | 0.3238 | 6.4600 | 11.3275 |
| 47 | 9.2000 | 0.3312 | 6.3793 | 11.5648 |
| 48 | 9.4000 | 0.3387 | 6.2970 | 11.8011 |
| 49 | 9.6000 | 0.3462 | 6.2131 | 12.0364 |
| 50 | 9.8000 | 0.3538 | 6.1276 | 12.2708 |

| Point crt. no. | u [mm] | φ [rad] | ξ [mm] | η [mm] |
|---|---|---|---|---|
| 51 | 10 | 0.3613 | 6.0404 | 12.5040 |
| 52 | 10.2000 | 0.3689 | 5.9516 | 12.7362 |
| 53 | 10.4000 | 0.3765 | 5.8613 | 12.9673 |
| 54 | 10.6000 | 0.3841 | 5.7693 | 13.1973 |
| 55 | 10.8000 | 0.3917 | 5.6758 | 13.4262 |
| ... | ... | ... | ... | ... |
| 92 | 18.2000 | 0.6990 | 1.2636 | 20.8335 |
| 93 | 18.4000 | 0.7082 | 1.1248 | 20.9970 |
| 94 | 18.6000 | 0.7176 | 0.9855 | 21.1582 |
| 95 | 18.8000 | 0.7270 | 0.8457 | 21.3169 |
| 96 | 19 | 0.7365 | 0.7054 | 21.4730 |
| 97 | 19.2000 | 0.7461 | 0.5648 | 21.6266 |
| 98 | 19.4000 | 0.7558 | 0.4238 | 21.7776 |
| 99 | 19.6000 | 0.7655 | 0.2827 | 21.9259 |
| 100 | 19.8000 | 0.7754 | 0.1414 | 22.0715 |
| 101 | 20 | 0.7853 | 0 | 22.2144 |

The points are represented by importing their coordinates in *Sketcher* module and joined then into a spline curve, hence resulting half of the 2-D profile of a tool tooth, represented in figure 2.

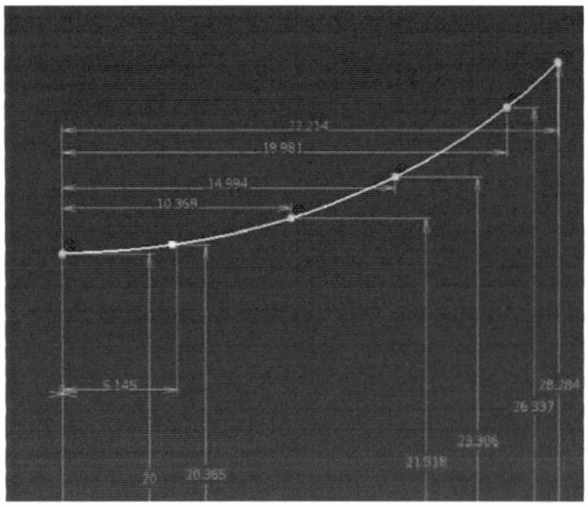


**Figure 2.** Tool tooth semi-profile.

The tool tooth semi-profile is mirrored to its left extremity, the mirror-curve being joined to the initial one, hence resulting the full tooth profile.

The rack tool is designed as having three teeth, so its profile can be obtained as an array with three items, identical to tooth profile above drawn (figure 3).

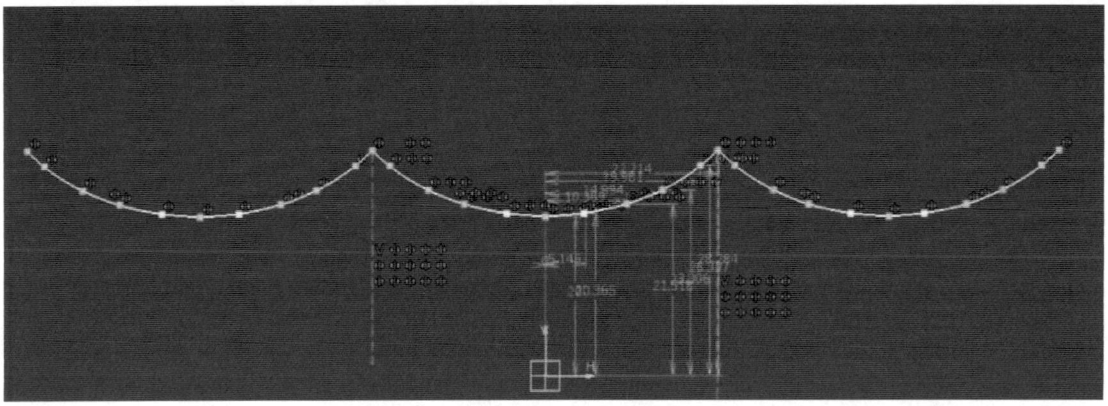

**Figure 3.** Profile of the modelled rack-tool.

A close profile is created then, starting from the active profile of rack tool teeth, in order to enable the extruding of tool 3-D model, with the help of *Pad* tool, from *Part design* module. The resulted model is presented in figure 4.

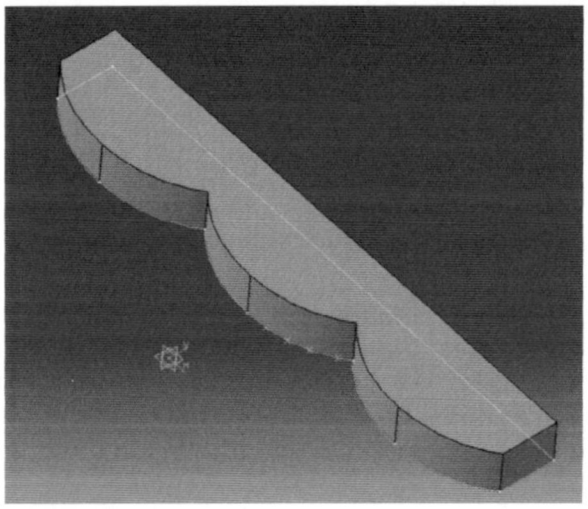

**Figure 4.** Rack tool 3-D model.

## 4. The modelling of the detached chips area

Because both entities involved in the generating process have to execute specific motions, the modelling of the detached chip area in the graphical environment CATIA must be performed by concomitantly running two sessions, dedicated to create one, the tool model and the other – the workpiece model.

The tool model building was already presented in previous section. In tool session, an incremental translation motion of $\Delta\lambda$ parameter is applied to the tool model, along $C_2$ centrode.

The workpiece model has the shape of a disc, whose diameter is equal to the diagonal of the square section of the shaft. In workpiece session, the workpiece executes an incremental rotation motion of $\Delta\varphi$ angular parameter, around the centre of $C_1$ centrode.

The two models are positioned such as the two centrodes are tangent. The rolling condition (1) is constantly imposed between the two motion increments. More specific, a workpiece rotation is sequenced in 120 steps, so $\Delta\varphi = \pi/60$ rad, while $\Delta\lambda = \pi \cdot \sqrt{2}/3 = 1.481$ mm. At each tool / workpiece repositioning, the tool model is imported and overlapped to the workpiece model, the common surface giving the detached chip shape and area (figure 5). The option *Remove* of the modelling tool *Operation* enables the removing form workpiece model of the common surface (the chip). If the workpiece area is measured before and after this action (with *Measure item* tool from *Measure* toolbar, which works very accurately), the area of the detached chip results by difference. The values of the successively detached chips can be calculated by repetitively applying the presented procedure (figure 6). At the present stage of the research, the iterations are performed manually. The future implementation of the modelling method will require a piece of software in order to automate its application.

**Figure 5.** Finding of the detached chip.

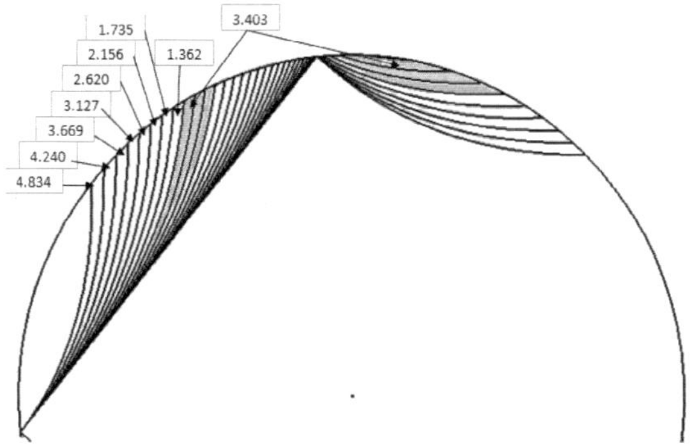


**Figure 6.** Successively detached chips.

After applying the presented procedure for measuring the detached chip area at each of the 120-tool double-strokes needed for entirely generating the square shaft, the values for all detached chips areas (some of them sampled in table 2) resulted. The variation of detached chips area $A_c$ during the machining process, presented in figure 7, has been drawn on this base.

**Table 2.** Detached chips area (excerpt).

| Dbl. stroke crt. no. | φ [deg] | $A_c$ [mm$^2$] | Dbl. stroke crt. no. | φ [deg] | $A_c$ [mm$^2$] | Dbl. stroke crt. no. | φ [deg] | $A_c$ [mm$^2$] |
|---|---|---|---|---|---|---|---|---|
| 1 | 3 | 4.834 | 51 | 153 | 10.986 | 101 | 303 | 3.111 |
| 2 | 6 | 4.240 | 52 | 156 | 10.496 | 102 | 306 | 3.046 |
| 3 | 9 | 3.669 | 53 | 159 | 9.951 | 103 | 309 | 2.847 |
| 4 | 12 | 3.127 | 54 | 162 | 9.361 | 104 | 312 | 2.591 |
| 5 | 15 | 2.620 | 55 | 165 | 8.740 | 105 | 315 | 2.329 |
| 6 | 18 | 2.156 | 56 | 168 | 8.097 | 106 | 318 | 2.062 |
| 7 | 21 | 1.735 | 57 | 171 | 7.440 | 107 | 321 | 1.814 |
| 8 | 24 | 1.362 | 58 | 174 | 6.777 | 108 | 324 | 1.579 |
| 9 | 27 | 3.403 | 59 | 177 | 6.116 | 109 | 327 | 1.359 |
| 10 | 30 | 4.126 | 60 | 180 | 5.461 | 110 | 330 | 1.151 |
| 11 | 33 | 8.718 | 61 | 183 | 4.833 | 111 | 333 | 0.958 |
| 12 | 36 | 10.163 | 62 | 186 | 4.239 | 112 | 336 | 0.777 |
| 13 | 39 | 11.077 | 63 | 189 | 3.669 | 113 | 339 | 0.609 |
| 14 | 42 | 11.659 | 64 | 192 | 3.126 | 114 | 342 | 0.456 |
| 15 | 45 | 12.006 | 65 | 195 | 2.621 | 115 | 345 | 0.322 |
| 16 | 48 | 12.176 | 66 | 198 | 2.155 | 116 | 348 | 0.212 |
| 17 | 51 | 12.192 | 67 | 201 | 1.735 | 117 | 351 | 0.126 |
| 18 | 54 | 12.037 | 68 | 204 | 1.360 | 118 | 354 | 0.062 |
| 19 | 57 | 11.765 | 69 | 207 | 3.403 | 119 | 357 | 0.021 |
| 20 | 60 | 11.414 | 70 | 210 | 4.126 | 120 | 360 | 0.000 |

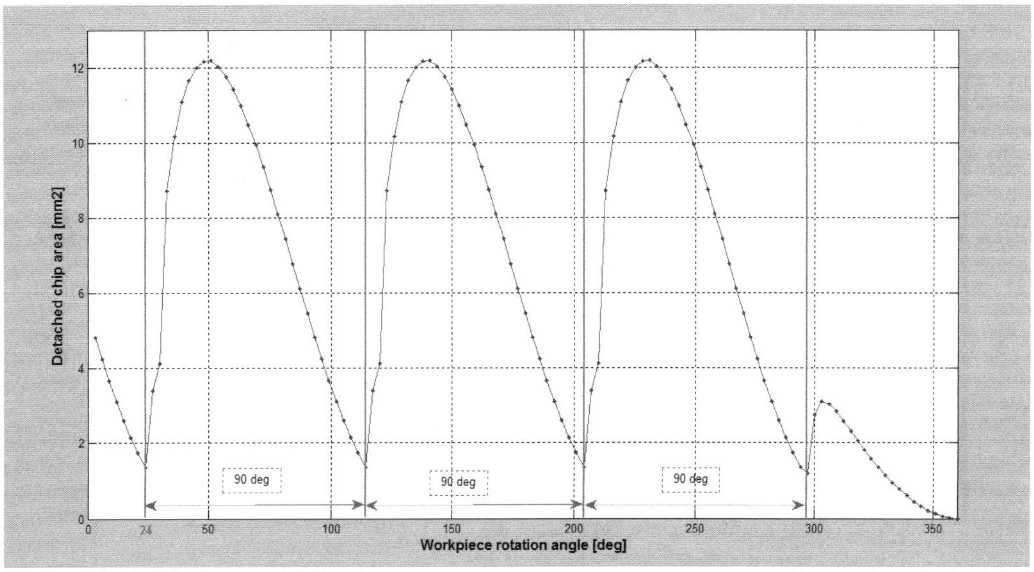


**Figure 7.** Variation of detached chips area.

## 5. Conclusion

The subject addressed in this paper is the modelling of the chips detached chip area, during the machining of a square shaft with the rack tool. In this purpose, the analytical model of tool tooth has been found at first. Then, on the base of the deducted equations for tool active profile, the tool 3-D model has been built in CATIA graphical environment. Finally, by using CATIA facilities, the detached chips areas have been found in the case of machining a square shaft having the side length of 40 mm.

The results obtained within the performed simulation enable to draw some conclusions:

- The CATIA graphical application is feasible in finding the shape and the area of detached chip;
- The detached chip area has a periodic variation, exception making the zones corresponding to rack tool engagement and disengagement;
- The detached chip area shows a significant variation (comprised between 1.36 and 12.19 $mm^2$), which sustains the opportunity of implementing a solution for cutting process smoothing.

The target of future researches is to design an experimental stand for actually validating the proposed solution for the diminishing the cutting force unevenness (the use of a variable circular feed), whose first step is the modelling of the detached chip area.

## References

[1] Mabrouki T *et al.* 2016 Some insights on the modelling of chip formation and its morphology during metal cutting operations *Comptes rendus mecanique* **344** (4-5) 138-41

[2] Rodriguez J M, Carbonell J M and Jonsen P 2018 Numerical methods for the modelling of chip formation *Archives of Computational Methods in Engineering* **27** 387-412

[3] Dima M, Oancea N and Teodor V G 2007 *Modelarea schemelor de aşchiere la danturare* (Iaşi: CERMI Publishing House Iaşi)

[4] Dima M and Oancea N 2004 Constructive modification for the energetically improvement of the toothing process *Romanian J. of Tech. Sci., Appl. Mechanics* Tome 49 237-41

[5] Georgescu V, Oancea N and Constantin E 1982 Fraise vis-module du type FREDAROM, *Buletinul Universitatii din Galaţi* **V** 33-38

[6] Litvin F L 1989 *Theory of gearing* NASA, Scientific and Technical Information Division (Washington DC)

[7] Oancea N 2004 *Surface Generation by Enwrapping, Vol. II, Complementary theorems* (Galaţi: „Dunărea de Jos" Publishing House)

# Some aspects regarding planetary precessional transmissions dynamics

**I Bostan[1], V Dulgheru[1] and I Malcoci[1*]**

[1]Technical University of Moldova, Department Fundamentals of Machinery Design, Chisinau, Republic of Moldova

*E-mail: iulian.malcoci@bpm.utm.md

**Abstract**. A mechanical system theoretically should not generate any vibrations and noise to be considered dynamically an ideal system, because vibrations mean a loss of energy. In planetary precessional reducers vibration can appear at bearings, gears, misaligned shafts, unbalanced rotors, couplings. Changes in dynamic process affects not only normal behaviors of mechanical systems but also the forces that act here. Regarding this aspects, regular vibration measurements during the operating speed provide information about any necessary repairs or maintenance. Here vibration research was made by using GUNT PT500 Machinery Diagnostic System and vibration signals was evaluated by using GUNT PT500.04 software which allow to perform correct Fast Fourier Transformation analysis. Data acquisition was made by using two piezoelectric accelerations sensor type IMI603C01 and one reference photoelectric sensor to record the shaft speed. Research concluded in this paper is made on precessional gearbox 2K-H Type. In first phase the crankshaft was installed with self-alignment on 2 radial bearings with cylindrical roller and second research phase was determination of the resonance domain and performing the spectral analysis by recording several vibrogram.

## 1. Introduction

In planetary precessional transmissions, vibration can appear at bearings, gears, misaligned shafts, unbalanced rotors, couplings. Changes in dynamic process affects not only normal behaviors of mechanical systems but also the forces that act here [1, 2].

Generation sources of vibration and its behaviors can conduct to dimensional distortions of sliding or rolling parts in direct contact, dimensional distortions at machine elements that performs rolling touching, dynamic or static imbalance and assembly errors because precessional reducer is a dynamical device where mechanical vibration can appear [2, 3].

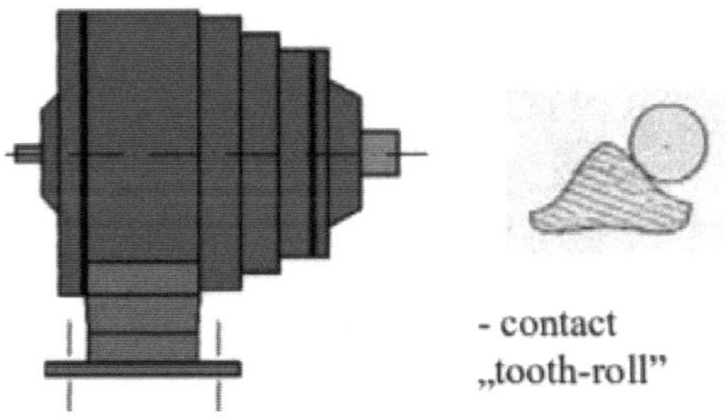


**Figure 1**. Researched precessional gearbox 2K-H Type.

Vibration analysis was carried out on power precessional reducer type 2K-H (figure 1) with the transmission ratio $i$=-10.5, nominal torque $T$=40 Nm and operational speed $n$=800 min$^{-1}$.

## 2. Constructive solution for minimizing assembly vibration errors in researched gearbox

Assembling process may occur some assembly errors (excessive clearance or too high grip) even if "satellite block and crank shaft" components node were balanced and executed perfectly. One of the most met assembly faults is the non-coincidence of the satellite precession center (point O in figure 2) with the intersection point of the O1-O1 axis of the crank (inclined part) and O-O of the drive shaft. An effective solution to compensate this error is to install the satellite and crank shaft, on radial bearings with cylindrical roller, which in turn is an additional source of vibration [2, 5].

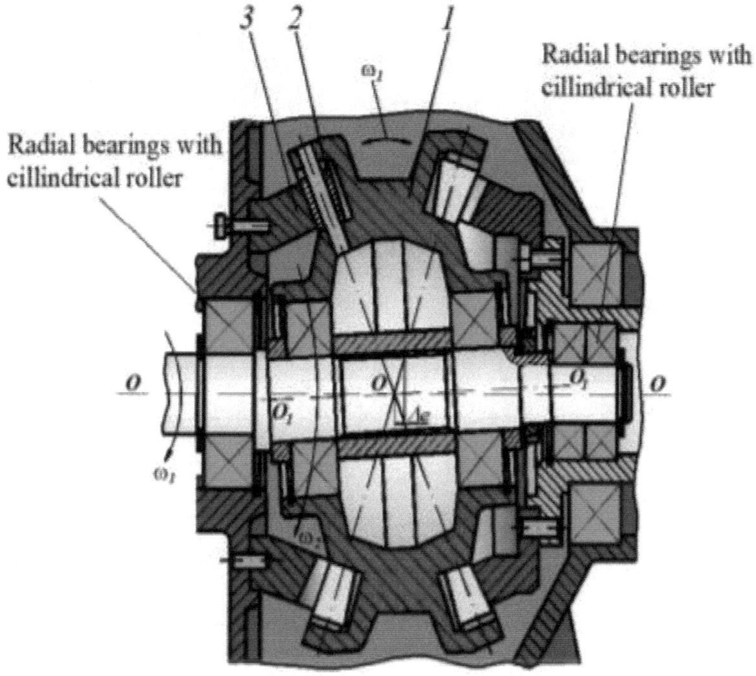


**Figure 2.** Constructive solutions.

In figure 2 it is shown one of the various constructive solutions regarding precessional transmissions where auto-aligning operation of satellite block 1 in this case can be described beneath. The constant component of the primary errors in the precessional planetary transmission elements, produce some deviation Δe among precession axis of the satellite block 1, which does not depend on the reciprocal position of the transmission parts. After the new position of the satellite block 1, the precession movement with the frequency rotation $\omega 1$ (figure 2) and the frequency rotation $\omega_2$, trajectory is determined by the first harmonic of the actual error in the satellite block gearing (radial and frontal contact of the crowns): $\omega_2=\omega_1/z_2$ – where $z_2$ represents the number of rollers of the satellite crown, that contact with the fixed central toothed wheel 3. The possibility of self-alignment, which will allow to obtain compensation of the manufacturing and installation errors of the contacting parts, us result we obtain a reduction of the vibration and noise level, by installation of the satellite block 1 on radial bearings, with cylindrical roller [2].

To minimize static and dynamic imbalance was developed a new technical solution which will conduct to increase the reliability of the gearbox by reducing the dynamic load, in particular, on the bearings and gear elements. This solution is in the realization and patenting stage and for this reason will not be considered in this paper [4, 5].

## 3. Research on the vibration behavior of the precessional planetary 2K-H reducer

The experiments were performed on the test stand (figure 3) in the laboratory "Fine Mechanics" of the "Fundamentals of Machinery Design" department in a closed room with rigid floor. The experimental stand consists from the test table GUNT PT500.01 made of aluminum profile provided with T-shaped channels, in which is fixed electric drive with power $P = 0.36$ kW, precessional planetary gearbox 2K-H with transmission ratio $i = -10.5$ and the maximum torque $T_2 = 40$ Nm, the electromagnetic brake type ПТ-6 nominal torque $T$=40 Nm.

**Figure 3.** Researched planetary precessional gearbox 2K-H Type.

To compensate the execution and assembly errors, the crankshaft was installed with self-alignment on 2 radial bearings with cylindrical roller (figure 4).

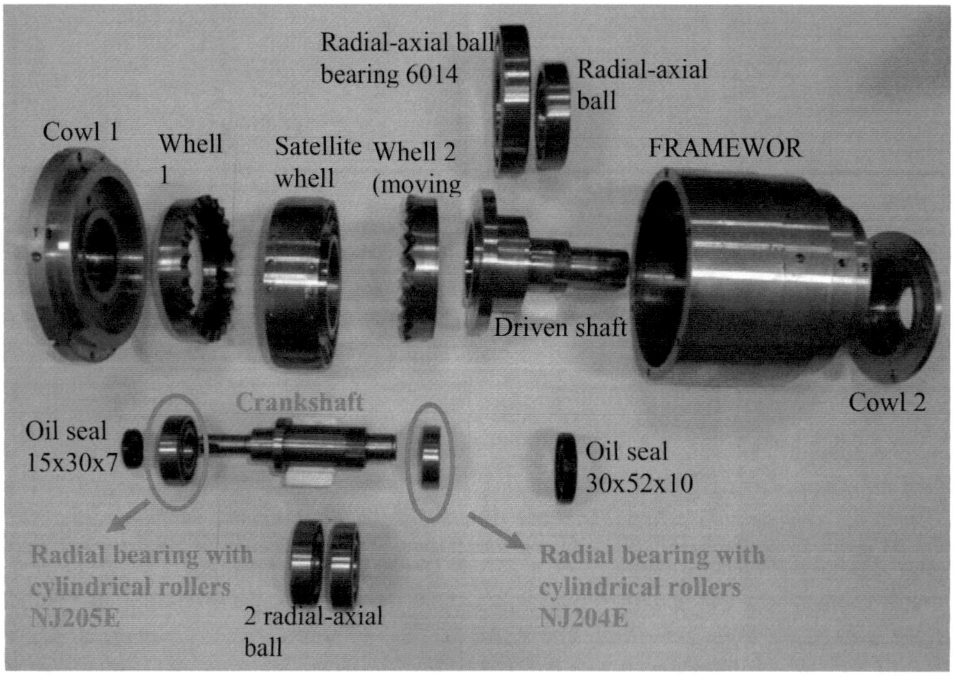


**Figure 4.** Component parts of the 2K-H precessional researched reducer.

## 4. Determination of the resonance domain

A trend in modern technology is to increase the speed of machines. This is justified by reducing the size of the machines to the same transmitting power. From the relation of the power P = F · v, it results that, at the same power, the force decreases with the increase of the speed and, consequently, the machine parts can have smaller dimensions.

However, the increase in speed is limited by the inertial forces that occur during operation and which cause additional vibrations and stresses in the machine parts. It is known that any elastic body taken out of equilibrium oscillates with its own frequency, until the deformation energy is damped. The effect of centrifugal forces and vibrations is an important factor, which must be taken into account when we are designing gearboxes. The inertia forces vary periodically, resulting in vibrations in both the reducer and the foundation. If at some point, the frequency of forced oscillations caused by inertial forces (disturbing force) coincides with the frequency of the oscillations of an element, mechanism or machine, the phenomenon of mechanical resonance occurs, which can lead to crash of elements in the reducer or destruction of connections. among the elements of the mechanism. The vibration check in the case of reducers consists in determining the natural frequency (resonance range) and therefore the corresponding speeds which are called critical range. The purpose is that the operating speed does not coincide with any of the speeds that are found in the resonance range or is not too close to it. The coincidence of the critical speed with the operating speed is accompanied by the appearance of the resonance phenomenon, situation in which the amplitude of the own vibration can increase until the occurrence of damages or accidents [6].

The resonance range for the 2K-H precession reducer was determined on the 3 basic directions (vertical, horizontal and axial) by mounting the accelerometers at the respective points on the rolling bearing. In order to determine the resonance range, the diagnostic system of the PT 500 machine parts and the PT500.04 software "Traking Analysis" module of the German company GUNT were used. After starting the electric drive, the measurement starts, slowly adjusting the speed by accelerating from 0 to 1300 rpm, the image with the diagram is recorded from computer screen with the "Print Screen" key and saved in a file on computer, the measurement is continued, slowly adjusting the speed by braking from 1300 to 0 rpm, and on the obtained diagrams the resonance range and the critical speed are identified. Next three figures are shown the diagrams obtained in vertical direction (figure 5), in horizontal direction (figure 6) and, respectively, in axial direction (figure 7).

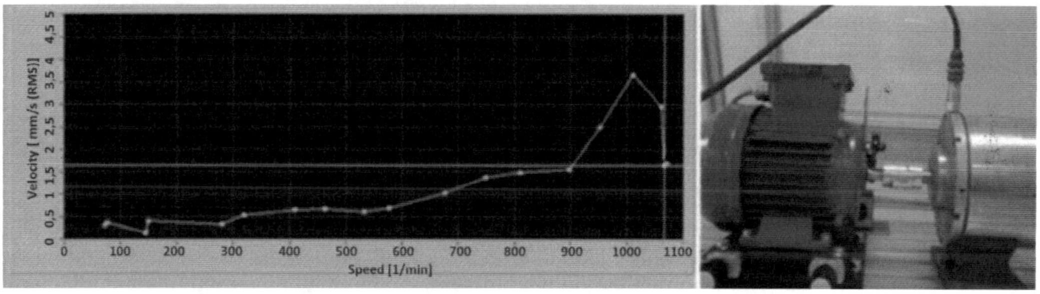


(a) (b)

**Figure 5.** Diagram in vertical direction: (a) - critical range 900 - 1200 $min^{-1}$; critical speed 1000 $min^{-1}$; (b) - vertical direction.

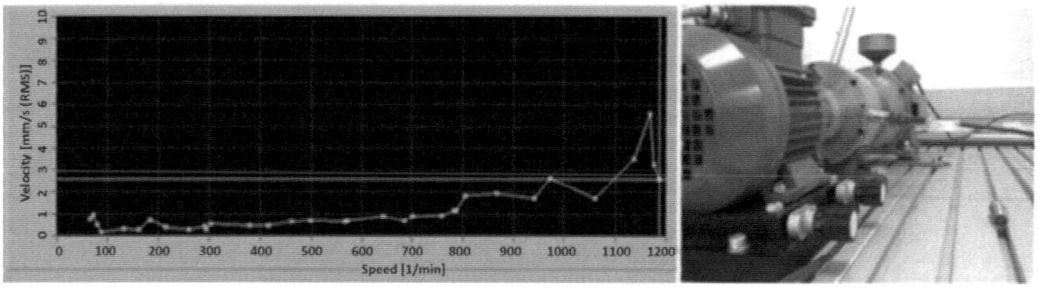


(a) (b)

**Figure 6.** Diagram in horizontal direction: (a) - critical range 900 - 1200 $min^{-1}$; critical speed 1100 $min^{-1}$; (b) - horizontal direction.

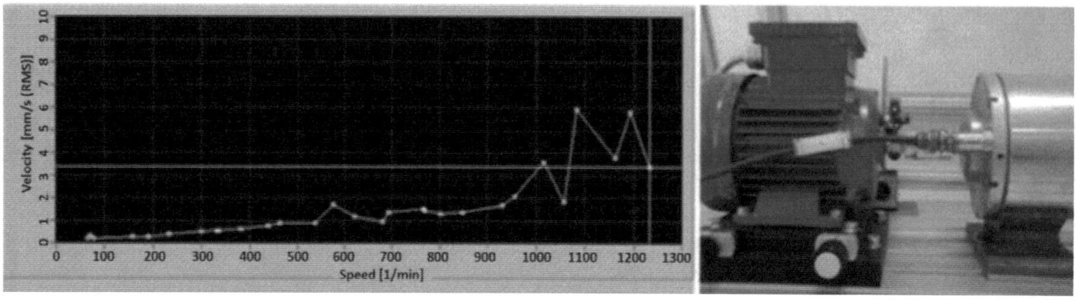


(a) (b)

**Figure 7.** Diagram in axial direction: (a) - critical range 900 - 1200 min$^{-1}$; critical speed 1100 min$^{-1}$; (b) - axial direction.

From the analysis of the resonance range, it was decided to choose the nominal speed at the regime phase $n_{nom}$ = 800 min$^{-1}$, and the loading moment of the reducer was chosen maximal $Tn$ = 25 Nm after brake calibration (figure 8), as much as 0.36 kW electric drive allowed.

**Figure 8.** Brake calibration.

## 5. Vibrodiagnosis (spectral analysis)

Spectral analysis is one of several powerfull techniques needed for analyzing and post-precesing researched and analyzed data. Analyzed data are acquisition data that have been taken in one, two, or three-dimensional space, and/or time. Spectral analysis is one of the most essential used methods for data research and analysis in oceanography, geophysics, astronomy, atmospheric science, metallography and last but not least in engineering too. In this paper vibration research, the spectral content of measured data gives information on the defects, wear and other information of mechanical parts under study [5].

The next step in the experimental research of the precessional reducer 2K-H was the spectral analysis of the reducer at the operating speed regime ($n_{nom}$ = 800 min$^{-1}$ and $Tnom$ = 25 Nm).

The PT500.04 GUNT software was used to draw up the vibrograms in the vertical (figure 10), horizontal (figure 11) and axial direction (figure 12), using the "Frequency Spectrum" module. To determine the degree of severity (damage index) of the vibrations, which must not exceed the value 4.5 according to the norms for evaluating the vibration level ISO 10816-1, the "Envelope Analysis" module was used (figure 9).

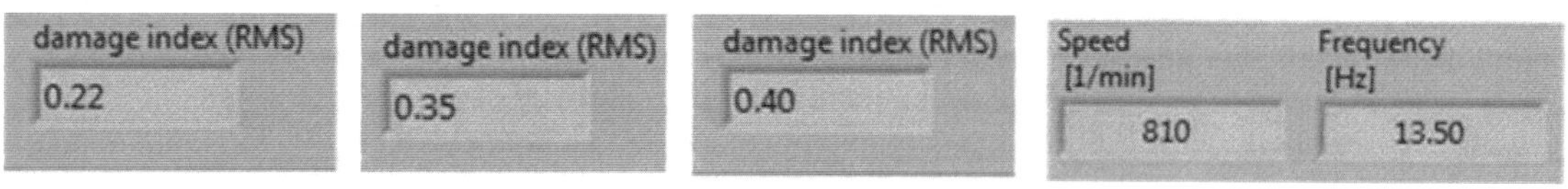


(a) (b) (c) (d)

**Figure 9.** Damage index: (a) vertical; (b) horizontal; (c) axial direction; (d) operating speed.

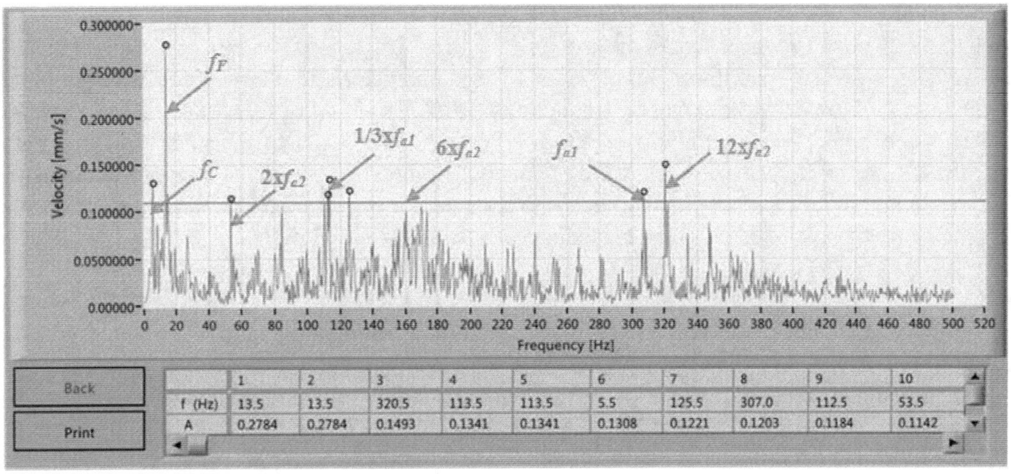


**Figure 10**. Vibrogram recorded in the vertical direction at $n_n = 800$ min$^{-1}$ and $T_n = 25$ Nm.

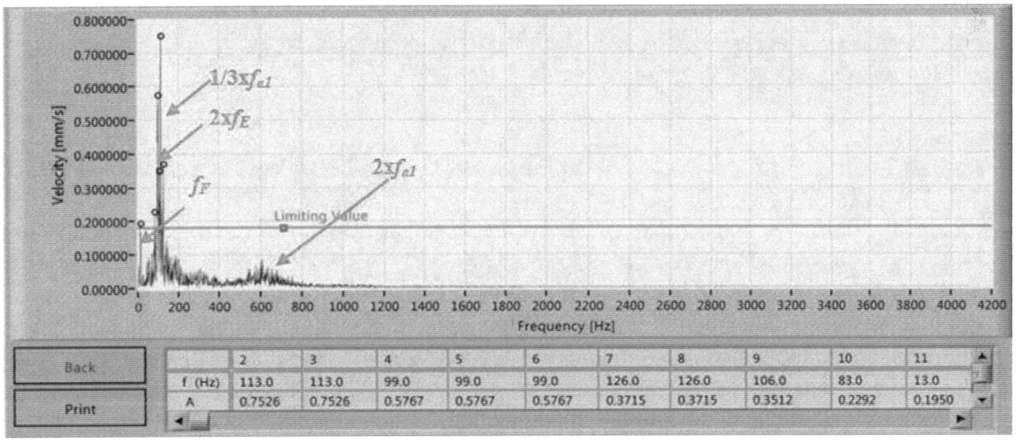


**Figure 11**. Vibrogram recorded in the horizontal direction at $n_n = 800$ min$^{-1}$ and $T_n = 25$ Nm.

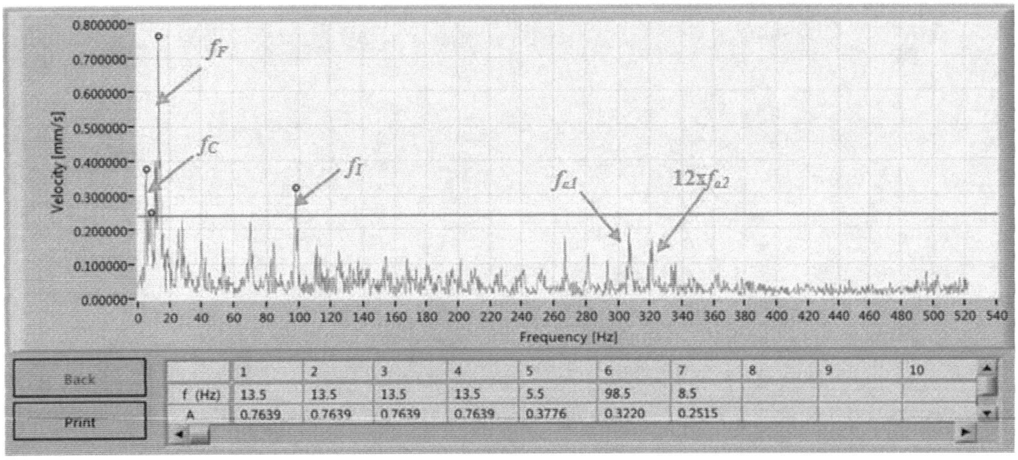


**Figure 12**. Vibrogram recorded in the axial direction at $n_n = 800$ min$^{-1}$ and $T_n = 25$ Nm.

## 6. Conclusions and future research

Analyzing the vibration spectrum obtained on the 3 measuring directions, regarding the magnitude of the vibration amplitude, which did not exceed the value of 0.8 mm/s we can say with certainty that the researched 2K-H power reducer works optimally.

As we expected, the maximum amplitudes of the characteristic frequencies were obtained at approximately ***13.5 Hz*** - the fundamental frequency $f_F$; ***307 Hz*** - gearing frequency $f_{a1}$, between the the fixed precessional wheel with number of teeth $z_1 = 23$ and the crown 1 of the satellite with number of rollers $z_2 = 22$; ***26, 75 Hz*** - gearing frequency $f_{a2}$, between the crown 2 of the satellite with number of rollers $z_3 = 22$ and mobile precessional wheel with the number of teeth $z_4 = 21$. The specific harmonics of the gearing frequency detected in the frequency spectrum 2x$f_{a2}$ = ***53.5 Hz***, 3x$f_{a2}$ = *80.5 Hz*, 4x$f_2$ = ***107 Hz***, 12x$f_{a2}$ = ***320 Hz***.

Also, in the frequency spectrum were detected the frequencies $f_C$ = *5.5 Hz* specific to the oscillations of the bearing's cages and $f_I$ = ***99 Hz*** specific to the oscillations of the inner rings of the bearings NJ205E and NJ204E on which the crank shaft is mounted with self-alignment.

Further research will be carried out on the same power 2K-H precessional gear box (figure 1) with the transmission ratio $i$=-10.5, nominal torque $T$=20 Nm and operation speed $n$=1000 min$^{-1}$ regarding acoustical recording of noise curves using Brüel & Kjær Sound Level Meter Type 2250 Light.

**Acknowledgments**

This paper was elaborated under Moldova State Project 160-PS from 03.01.2020 with title *Increasing the competitiveness of precessional transmission by developing and capitalizing „conform" toothed gear and expanding their range of application.*

**References**

[1] Malcoci Iu 2015 *Vibro acoustic research of planetary precessional transmissions* PhD Thesis UTM CZU: 621:534.6(043.2) p 135
[2] Malcoci Iu, Trifan N and Ciobanu R 2020 Precessional gear-box research regarding vibration activity behaviour *IOP Conf. Ser-Mat. Sci.* **749**
[3] Dulgheru V and Malcoci Iu 2018 Dynamic model elaboration for the planetary precessional gear reducer against vibration *MATEC Web Conf. 22nd Int. Conf. on Innovative Manufacturing Engineering and Energy* **178**
[4] Malcoci Iu and Bodnariuc I 2015 Vibroacoustical diagnosis of planetary precessional kinematical transmission (Part 1) *AMM* **809–810** 593–97
[5] Bostan I, Dulgheru V and Malcoci Iu 2015 *Precessional planetary transmission* MD-Patent MD-4354
[6] Abraham D, Boxhamer J and Mittash P 2011 *Instruction manual GUNT PT500* Gerätebeu GmbH (Barsbütel Germany)

# Estimating the warranty returns and proving root causes using statistical analysis of archived parameters measurements for an automotive mechatronic device

**I C Braga[1*], R Udroiu[1] and A Nedelcu[1]**

[1] Transilvania University of Brașov, Department of Manufacturing Engineering, 29 Eroilor, Brasov 500036, Romania

*E-mail: jbc69mc@gmail.com

**Abstract.** In the mass production of the automotive industry, there are many cases in which a problem is not detected in the production processes of the supplier, but after delivery, in the customer's assembly line. As a result, it is necessary to know both root causes to prevent them, but especially, if it cannot be detected even in the customer's processes, to estimate the probability of occurrence in the warranty period. The paper's aim is to present the analysis of the measurements of the mechatronic devices indirect parameters linked with the effect of the failure and by statistical correlation to prove the root causes. A study case will be used to understand the failure first occurrence and the duration of its effect, as the symptom description as well, in that way it will support the estimation of the warranty return number of parts, eventually, the accounting department can establish the provisions based on the output data.

## 1. Introduction

Nowadays, the automobile contains a lot of more electronic components and systems and those are present in almost all the parts from a car: the body uses systems such as power door, power window, climate control and mirror wiper, the chassis uses suspension, differential and drive shaft, the power train uses systems such as engine control and transmission, the safety systems such as EPS, ABS, airbags, traction control and tire pressure monitor and the examples can go on.

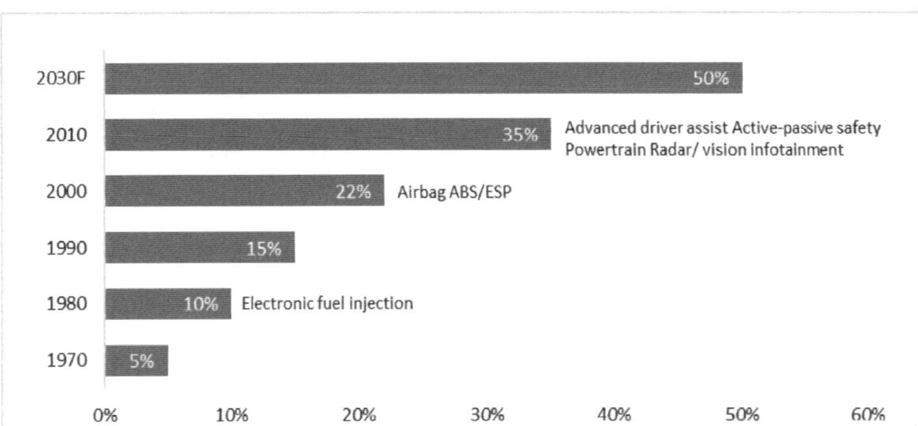


**Figure 1.** Electronic systems as % of total car cost.

A study regarding the future of the electronic systems in the cars shows that currently, the percentage of the total cost coming from these electronics is bigger than 35% and this will increase in the next ten years to 50% from the total cost of the car, as is presented in figure 1 [1].

One system present in all the new cars is the Infotainment or well known as the official term IVI, In Vehicle – Infotainment. This is a term that describes the set of systems used for entertainment and information management available in a car. IVI devices can provide entertainment services such as radio, TV, games, but also navigation and various services that are connected to mobile devices through technologies, for instance the phones via Bluetooth [2].

Now, the suppliers of these products are not out of the warranty liability [3], as the recalls campaign as well, the more so as the product is linked with the safety systems in the car. In the 2019 report analysis of Stout based on the department of transportation from NHTSA, National Highway Traffic Safety Administration and many other international recall data bases can be seen in figure 2 the trend from the ten years monitoring (starting to the 2009 year) of the electronic components which contain the software embedded [4].

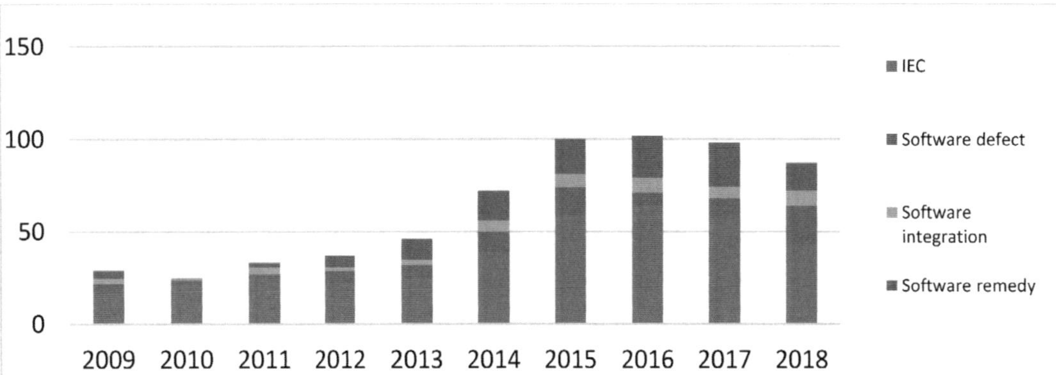


**Figure 2.** Unique campaigns recalls of electronic components by year.

From the previous report cannot be seen the last period, 2019-2020, but taken into consideration the number of total recalls from 2019 and the total number of the vehicles plus the equipment recalled, as is presented in figure 3, can be understood that the curve was switched in the wrong direction and the number increased [5].

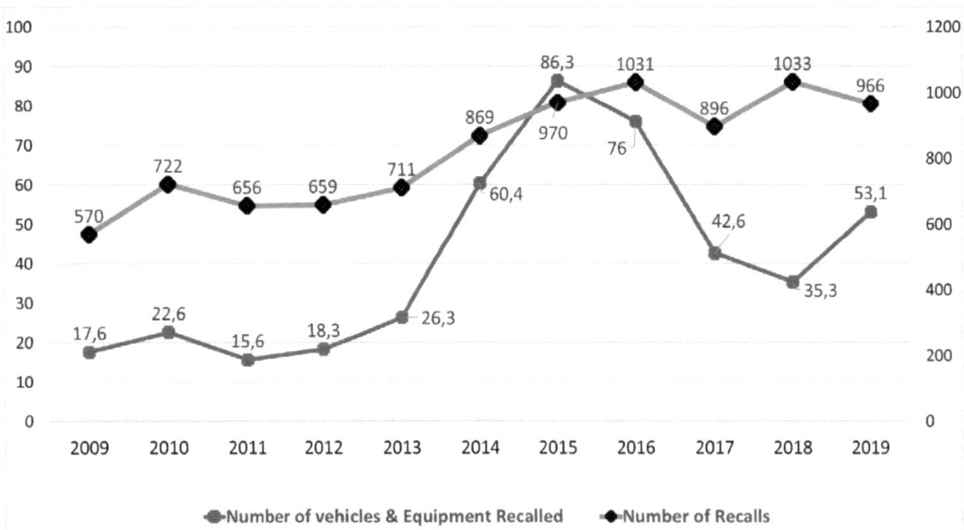


**Figure 3.** Motor Vehicle and Equipment Recalls – Source: NHTSA, 2019 Annual recall report.

Therefore, the suppliers of these mechatronic devices in the mass-production must take seriously the analysis of the risks in the development phases [6], as in the serial production as well, as long as the cost with the warranty returns and the analysis of those parts is increasing daily [7]. Moreover, the new standard IATF 16949 applicable in the automotive industry [8] is focused on the same topic: analysis, risk mitigation and reduction of the warranty return.

## 2. Study case in an automotive mechatronic device manufacturer

The study case is based on the analysis of a mechatronic devices which is part of the IVI system, a device from the category HMI, Human Machine Interface, in fact, the device that support the driver to handle with the infotainment panel and to navigate in menu or to connect the mobile phone and find the contact in the agenda.

This device consist in an assembly from the plastic injection moulded components together with a PCBA, Printed Circuit Board Assembled, containing the microprocessors and the software embedded, as the LEDs, Light Emitting Diodes, for the night vision as well, a rubber which will give the tactile feedback from the normal pushing buttons made from plastic injection moulded components, screwed together, and a joystick which will be used to navigate in the direction of north-south or in the direction of east-west.

The joystick itself is assembled from the metallic parts with the glued magnets and a stop element from rubber material which is pressed into the plastic axe will define the way of the travel on those two directions – see figure 4.

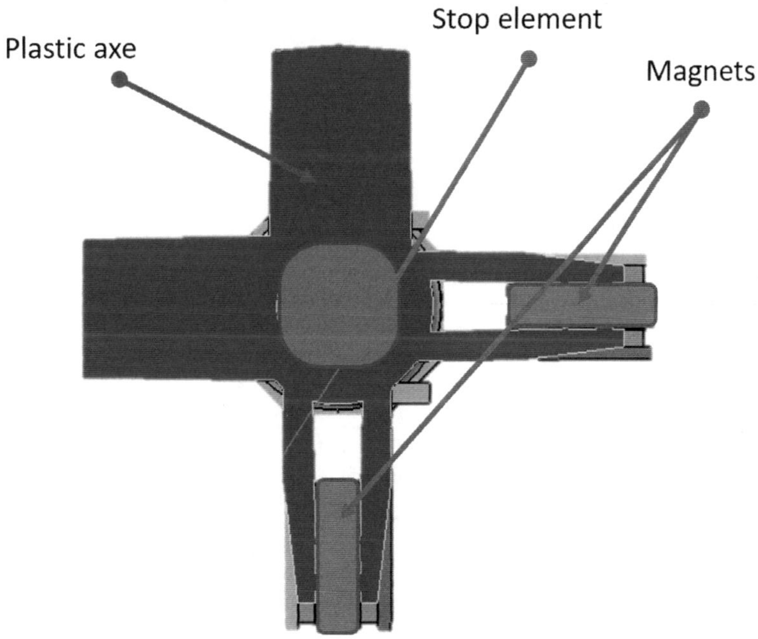


**Figure 4.** Joystick components.

As the customer tactile feedback recognition for the parts, are defined the following characteristics: a curve generated by the travel ways, S, and the forces, F, as can be seen in figure 5. Also, the report between the F1 and F2, named Snap, calculated as the difference between F1 and F2, split to F2, with a result in percentages.

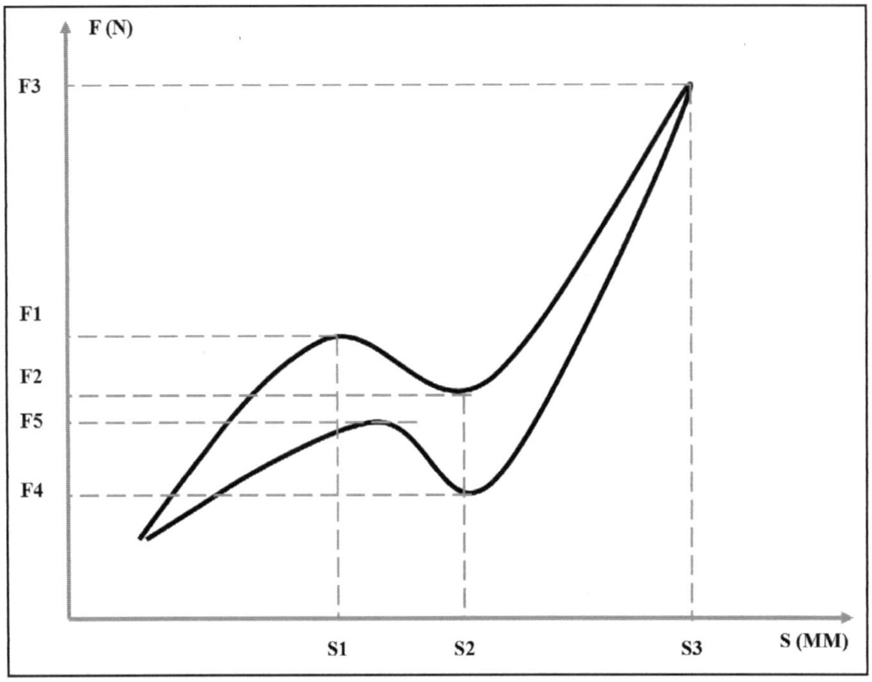


**Figure 5.** Travel ways and forces curves to describe the correspondent feeling of the user.

Those characteristics are tested and measured in an end-of-line test bench by using a dynamometer EDM20 050EL handled by a robotic arm and the auto-calibrating cycle is started when the pusher of the dynamometer starts to push over the device's joystick, in that way, the system will reach the travel start-end, in order to measure within the expected range. The robotic arm will return the travel way and, eventually, by using a LabView interface the program will present on display the curves, as the data are archived in the comma-shifted-value format in the machine hard-drive.

### *2.1. Problem description*

The issue started with a complaint from the customer and was claimed the fact that the travel way in direction of west is not according to the specification. As result of investigation, the stop element was detected in a moved position after the disassembly of the device. This stop element is inserted by the operator in the axe, then a semi-automatic machine will press it until the end position. The correct position of the stop element into the axle is ensured at this workstation, but a dedicated checking or measurement of the position is not performed in the next operation or the end of line testing machine. Moreover, if the rubber stop element will change its position in time, this effect cannot be detected as the testing step procedure is running almost immediately after the assembly (perhaps a small inventory level could be in some situations, still not enough to let the rubber to go out) and the vibration from the transport of the parts with the truck to the final facility will augment the possibility of the occurrence.

### *2.2. Investigation, analysis of the data and the result*

On the affected claimed part was assembled a stop element which has a burr on the split line from the injected tool, as can be seen in figure 6 by microscopy analysis. Due to this, the stop element didn't arrived in the end position into the axe, therefore the rubber components was kept in the position as should be and the final test at the end-of-line bench can't detect such a failure in travel way.

**Figure 6.** Burr over the split line of the rubber stop element.

The stop element is delivered by a supplier out of a toll with one hundred cavities, therefore, an investigation regarding the possibility of the affected batches and cavities was completed. A batch from a certain incoming date in week two was detected as suspect, but there was not at all the possibility to see when and at which line (out of three production lines of the joystick) were delivered these affected parts with burr because of the logistic supply method which require to split the same batch on all lines at the time.

Consequently, an analysis of the parameters from the archived data out of end-of-line hard drive was performed and the snap for the joystick in the west direction was deeply investigated. The result can be seen in figure 7 and figure 8, conclusion leading to a certain period of the suspected parts escaped and susceptible of the warranty return.

The target line is the minimum value of the snap which is 20% (in the plot chart being 20.000 as converted from the archived csv files) and the period of the suspected escaped parts is within a time frame of twelve weeks.

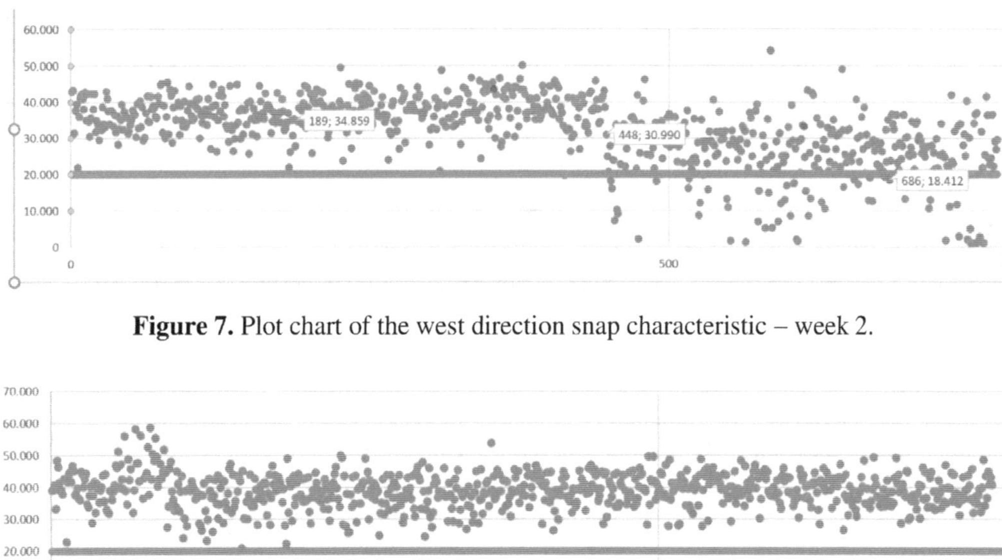


**Figure 7.** Plot chart of the west direction snap characteristic – week 2.

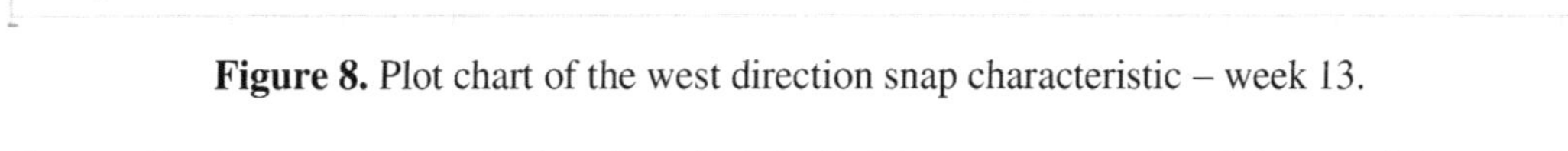


**Figure 8.** Plot chart of the west direction snap characteristic – week 13.

Eventually, the statistical analysis using Minitab 19 [9] was performed and the correlation analysis between three samples was used to prove the hypothesis of snap influence and possible detection in the end-of-line test. The first test performed was between two series of three hundred measurements, C1 and C2, and the result shows a correlation between – see figure 9.

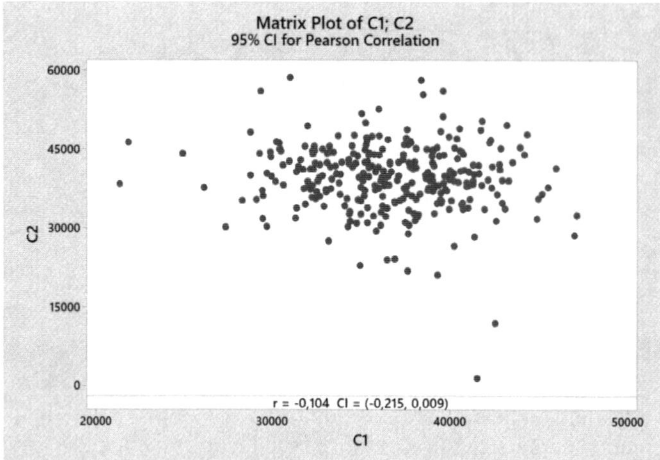


**Figure 9.** Matrix plot correlation two series three hundred measurements, C1, C2.

The second test performed was between two series of three hundred measurements as well, C1 and C3, C3 being the period of detected snap variation due to burr on the stop element, and the result shows not a correlation between – see figure 10. As a result, the collected and archived data are referring to certain characteristics and their related measurements, but not direct data regarding the insertion of the stop element or position of it, because the test bench specification was established according to the customer requirements and the risk assessments from the development phases.

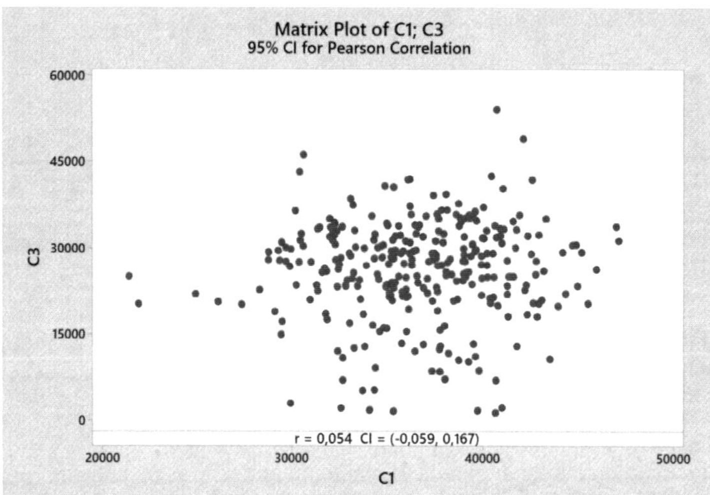


**Figure 10.** Matrix plot correlation two series three hundred measurements, C1, C3.

Due to all these facts, the paper study case proposed the possibility of using the analysis of the data concerning indirect measured characteristics, therefore, the necessity of validation from the statistical point of view was requested. Based on the information presented above, primarily, the statistical significance was proved and secondarily, the factors linked with the root causes as the period of the suspected escaped parts as well have been identified.

## 3. Conclusion

The study was initiated due to concern of the affected parts in field and the risk to overpass the provision established at the initial budget, as the image in front of the customer as well. As the effect was generated by a characteristic of the single part out of specification, the rubber stop element, and as the multi-cavities tool and the receiving inspection method based on the validation of the supplier through process audit capability and not an AQL level acceptance, was not possible to detect such a failure earlier, therefore, a batch of timeframe of twelve weeks was susceptible to be delivered with a certain percentage of the filed return risks. Based on the method selected, the root cause was proven by statistical analysis because the reproduction of the bad parts it is not quite possible at this level of the mass production.

The future research will follow, and the statistical analysis will be performed to allow a better estimation of the parts in field to be returned and the provision calculation to be accurate.

**References**

[1] Chou W, Shao J, Chung R, Chen L, Chen A and Zhou L 2019 *Semiconductors – the Next Wave Opportunities and winning strategies for semiconductor* (UK: https://www2.deloitte.com) p 14

[2] Choi D-K, Jung J-H, Koh S-J, Park J and Kim J-I 2019 *In-Vehicle Infotainment Management System in Internet-of-Things Networks* (Kuala Lumpur: Kuala Lumpur)

[3] Cohena D, Darroughb M, Huangc R and Zachd T 2011 *Warranty Reserve: Contingent Liability, Strategic Signal, or Earnings* (Florida: The Accounting Review)

[4] Steinkamp N, Levine R and Roth R 2019 *Motor Vehicle and Equipment Recalls* (Michigan: Stout)

[5] Canis B 2020 *Motor Vehicle Safety: Issues for Congress* (Washington: Congressional Research Service) Report R46398

[6] Hijazi H, Alqrainy S, Muaidi H and Khdour T 2014 *Risk factors in software development* (Almeria: European Scientific Institute, ESI January Edition) **10/3**

[7] Arnum E 2020 *Worldwide Automotive Warranty Expenses* (New York: Warranty Week)

[8] IATF 2020 *International Automotive Task Force* (Southfield Michigan: https://www.iatfglobaloversight.org/iatf-publications)

[9] Minitab 2020 *Motor Vehicle Safety: Issues for Congress* (Pennsylvania: http://www.minitab.com)

IOP Conf. Series: Materials Science and Engineering 1009 (2021) 012010 doi:10.1088/1757-899X/1009/1/012010

# Study on the application of CAM techniques on CNC lathes with Y axis and driven tools

**R E Breaz[1]*, S G Racz[1], C E Girjob[1] and M Tera[1]**

[1]Lucian Blaga University of Sibiu, Faculty of Engineering, Victoriei, 10, Sibiu, 550024, Romania

*E-mail: radu.breaz@ulbsibiu.ro

**Abstract**. CNC lathes with Y axes and driven tools are machine-tools with complex kinematic. Using CAM techniques to program these machines is a cumbersome process and could easily lead to collisions. Collisions may occur not only between tools and workpiece but also between the mobile elements of the CNC lathe. While CAM software package can identify the first type of collisions, for the second type supplementary measures must be taken. This work presents the build of a kinematic model of such a CNC lathe and how can the CAM techniques be improved by using this model.

## 1. Introduction

Nowadays machining processes require more and more complex CNC machine-tools. 3-axis milling machine centers are gradually replaced by 5-axis ones [1, 2].

5-axis machining can be performed on 3-axis CNC machine-tools by adding supplementary modules, such as rotary-tilting tables [3, 4] or even additional linear axis [5]. Adding supplementary modules could increase the machining errors, by means of additional deviations, situation which has addressed in [6].

Reconfigurable machine-tools have been seen as a solution for increasing both the machining capabilities and the flexibility of technological equipment [7, 8] but there quite few industrial implementations of them.

Increasing the number of machine axes by adding supplementary modules (during or post-fabrication) lead to a more complicated kinematic of the machine and consequently increases the probability of collisions occurrence. Avoiding the collisions, not only between tools and workpiece, but also between moving machine elements and modules was targeted in the literature [9].

Modern CAM software packages are able to identify such collisions for multi-axes machine tools but require kinematic models of the CNC machines to fulfill this task. Few kinematic models of the most common CNC machine on the market are available as standard in the libraries of the software, but most of them have to be built by the users [10].

CNC lathes, normally fitted with X and Z axes, were also considered for increasing their machining capabilities. Modern structures of lathe-milling machines, obtained by fitting normal CNC lathes with Y axis and driven tools [11, 12] are now offered as standard by most of major CNC machine-tools manufacturers.

This work presents some aspects of building a kinematic model of such lathe-milling machine and how this model was used for applying CAM techniques in machining.

## 2. GENOS L-300E-MY lathe-milling machine

GENOS L-300E-MY lathe-milling machine is a multitasking CNC lathe, equipped with driven tools and fitted with a supplementary Y -axis (which is normally not present in normal CNC lathes). Driven tools enable the machine to perform also milling operations, besides the normal turning operations. The machine is built by GENOS, a subsidiary of the well-known OKUMA company.

The machine is marketed by its manufacturer as a “simple multitasking machine with superior cost and performance”. Indeed, it allows the user to perform a variety of machining operations, by combining turning and milling capabilities.

## 3. Kinematic model

As stated before, the complex kinematic of the machine-tool requires the use of a kinematic model to be used in conjunction with CAM software. Because the machine is not very common on the market, it is not present in the standard libraries of the CAM software packages.

The first step in building the kinematic model was to build the 3D geometric model of the machine-tool, as an assembly. The disk was rather difficult because neither GENOS, not OKUMA provide no information about such model. There are companies on the market (HAAS) which provide such kind of 3D model for each manufactured machine-tool.

Consequently, the information for building the 3D geometric model has to be taken on-site, from the physical machine-tool (figure 1 and figure 2).

**Figure 1.** Overall view.

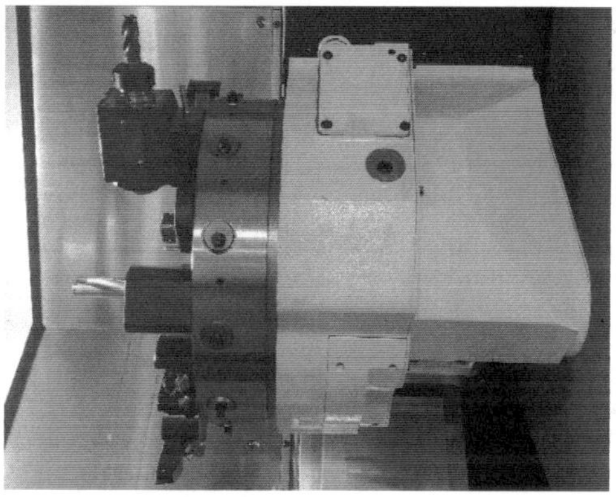

**Figure 2.** Turret with a driven tool.

After retrieving all geometric and functional information from the machine, the 3D geometric model presented in figure 3 and 4 was built and saved in .igs format.

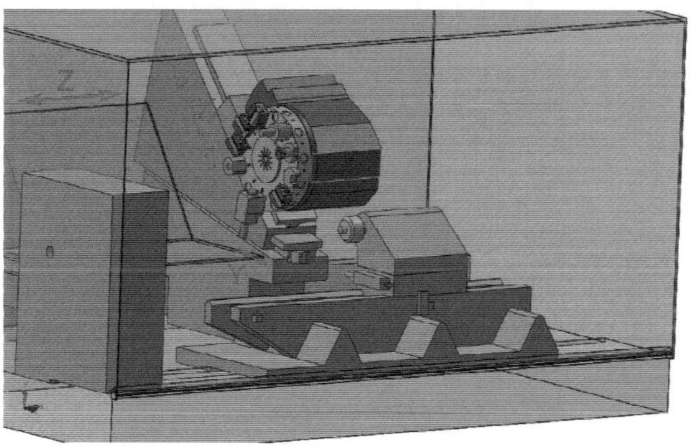


**Figure 3.** 3D model of the machine with X, Y, Z axes.

IOP Conf. Series: Materials Science and Engineering 1009 (2021) 012010 doi:10.1088/1757-899X/1009/1/012010

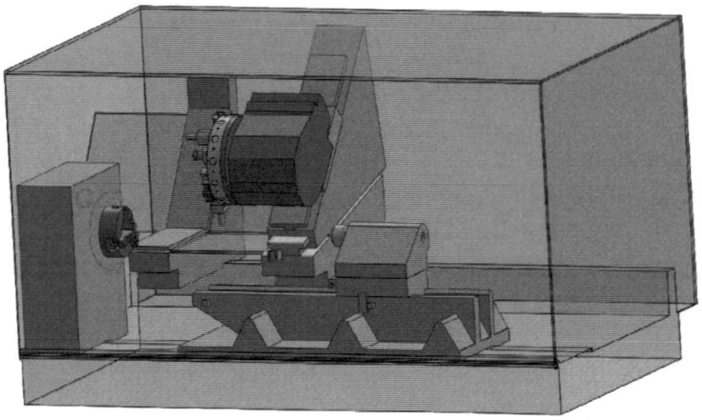

**Figure 4.** 3D model of the machine with C axis.

To build the kinematic model, the 3D geometric model was divided into the following blocks:

- Machine frame / X, Y, Z and C axes (each one separately) / turret axis / tailstock /pinole center / jaws of the holding device form C axis.

After this division, a special software was used to build the kinematic model, by defining degrees of freedoms and kinematic dependencies (parent-child type) for each above mentioned block.

Finally, the kinematic model was saved as a .xml file.

The flowchart of building the kinematic model of the machine is presented in figure 5.

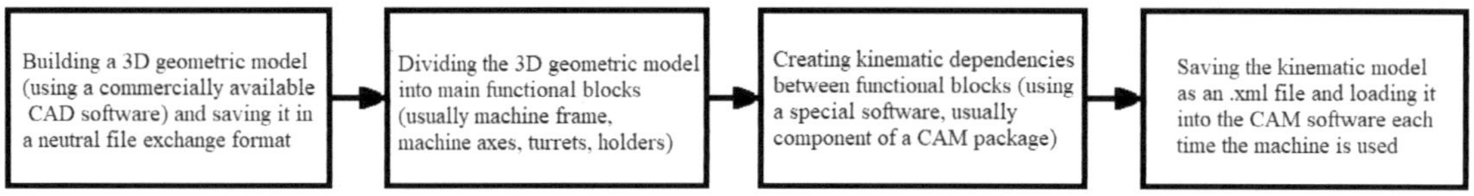


**Figure 5.** Flowchart of building the kinematic model of the machine-tool.

## 4. Machining a test part

To test the proposed kinematic model, a complex part (a screw from the structure of a hydraulic pump) was machined.

The machining process was performed at Sonnek Engineering company, form Șura Mică, Sibiu county. The programming and machining processes were supervised by engineers from the company. Figure 6 presents some machined parts, while figures 7 and 8 presents screenshots from the simulated machining proces (rough milling operation).

**Figure 6.** Machined parts.

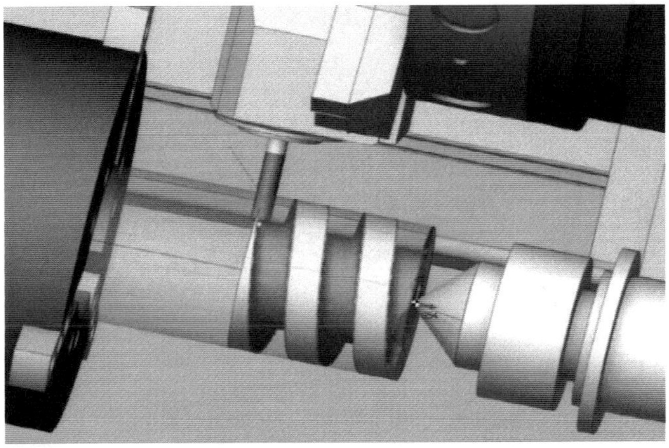

**Figure 7.** Screenshot from the machining simulation.

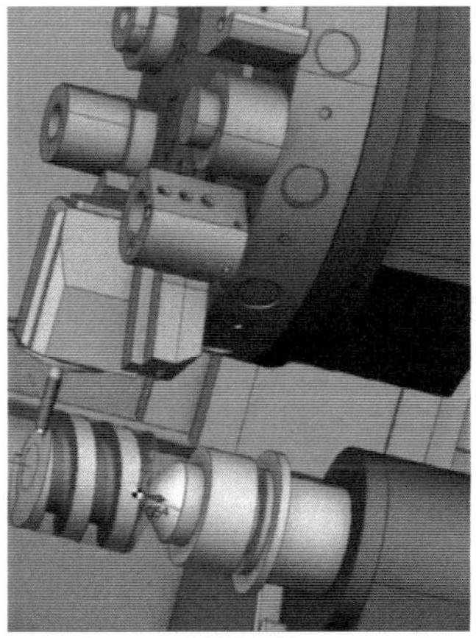

**Figure 8.** Screenshot from the machining simulation – showing the turret with the tool-holders.

## 5. Conclusion

Multitasking CNC machine-tools, which provide the user increased flexibility and machining capabilities are used increasingly in modern machining processes. Of course, modern CAD/CAM techniques have to be used to program these machines in order to obtain error-free CNC programs.

Using CAM software allows the user to simulate the relative movements between tools and machined parts in order to avoid collisions. However, when machine-tools with complex kinematic and structure are used, the collisions may also occur between moving machine elements (linear slides and/or rotary tables), a scenario which CAM software packages cannot take into consideration without supplementary information. This information may be provided by building a kinematic model of the CNC machine-tool used during the machining process.

This work presented the process of building a kinematic model for a GENOS L-300E-MY lathe-milling machine, which was further successfully tested by machining a complex part.

**Acknowledgments**

The experimental test unfolded within this research work were supported by a grant of the Romanian Ministry of Research and Innovation CCCDI-UEFISCDI, project number PN-III - P1-1.2 - PCCDI-2017-0446 /nr. 82PCCDI/2018, within PNCDI III, project title: "Intelligent manufacturing technologies for advanced production of parts".

**References**

[1] Bologa O, Breaz R E, Racz S G and Crenganiş M 2016 *Procedia computer science* **91** 184
[2] Bologa O, Breaz R E, Racz S G and Crenganiş M 2016 *Procedia computer science* **91** 683
[3] Suh S-H and Lee J-J 1998 *J. Manuf. Sci. Eng.* **120** 120
[4] Chen Z C, Dong Z and Vickers G W 2003 *Comput. Ind.* **50** 319
[5] Suh S-H, Lee J-J and Kim S-K 1998 *Int. J. Adv. Manuf. Tech.* **14** 865
[6] Tsutsumi M and Saito A 2004 *Int. J. Mach. Tool. Manu.* **44** 1333
[7] Abele E, Wörn A, Fleischer J *et al.* 2007 *Prod. Eng. Res. Devel.* **1** 421
[8] Lorenzer T, Weikert S, Bossoni S and Wegener K 2007 *J. Manuf. Syst.* **26** 167
[9] Ding S, Mannan M A and Poo A N 2004 *Comput .Aided Des.* **36** 1281
[10] Chicea A L, Breaz R E and Bologa O 2015 *Appl. Mech. Mater.* **809** 1004
[11] Martinov G M, Nezhmetdinov R A and Kuliev A U 2016 *Russ. Aeronaut.* **59** 293
[12] Aguilar A, Roman-Flores A and Huegel J C 2013 *J. Manuf. Syst.* **32** 364

# Study upon the kinematic simulation of the incremental forming carried-on using a serial industrial robot

**R E Breaz[1], S G Racz[1], C E Girjob[1] and M Tera[1]***

[1]Lucian Blaga University of Sibiu, Faculty of Engineering, Victoriei, 10, Sibiu, 550024, Romania

*E-mail: melania.tera@ulbsibiu.ro

**Abstract**. Incremental forming with contact between tool-sheet metal workpiece in a single point is an unconventional process which allow the user to manufacture parts by combining the vertical movement of the tool along Z-axis and the horizontal movements of sheet metal workpiece, which are performed in XY plane. As technological equipment able to control coordinated movements on three axes, CNC machining centers and industrial serial robots with six degrees of freedom are the most suitable. The paper introduces a study regarding the use of robots as technological equipment for the process, by means of simulation. Supplementary degrees of freedom and singularities avoidance are also considered.

## 1. Introduction

At present the industry requires an enormous variety of parts in all fields. Thus, it can be observed that if we refer to the parts made of metal sheets, one of the most important technological processing processes are the plastic forming processes.Conventional forming operations, which are specifically designed to meet high productivity lack in terms of flexibility, required were small numbers of parts with different shapes and dimensions have to be manufactured by means of plastic deformation [1, 2]

Due to the growth of sheet metal parts simultaneously with requirements of reducing processing times, incremental forming with contact between tool-sheet metal workpiece in a single point is beginning to be an increasingly studied process, although not yet accepted fully by the industry.

Serial industrial robots have usually large workspace and advanced kinematic, thus allowing the incremental forming of larger parts, with complex shapes [3]. To this is added the great flexibility in terms of obtaining parts of different types and sizes or parts in small batch or even unique parts which made incremental forming process carried on industrial robots an important topic for research.

The tools used for single point incremental forming are very simple ones as one can notice in figure 1 (a) its compound of a punch with a spherical end 1, active plate 4, holding plate 2 and sheet metal 3 [4, 5].

The process is unfolded by coordinating the vertical movement of the tool (punch) I, along Z-axis, according to figure 1 (b) with the horizontal movement of the sheet metal workpiece II, which is usually carried-on in the XY plane. Usually, the vertical movement is a discontinuous one, performed in incremental steps, while the movement in XY plane is a continuous one. The horizontal movement is performed by the blank,in case of using a CNC machine where X and Y machine slides are moving. Or, when using an industrial robot, the horizontal movement can be carried-on by the punch. In figure 1 (c)one can notice the movement on the X axis marked with II. These steps are continued until the final form of the parts is achieved, in this case in figure 1 (c) is presented a conical part. As we can see the SPIF involve three basic movements of the punch and can be achieved on CNC machines tools or industrial robots.

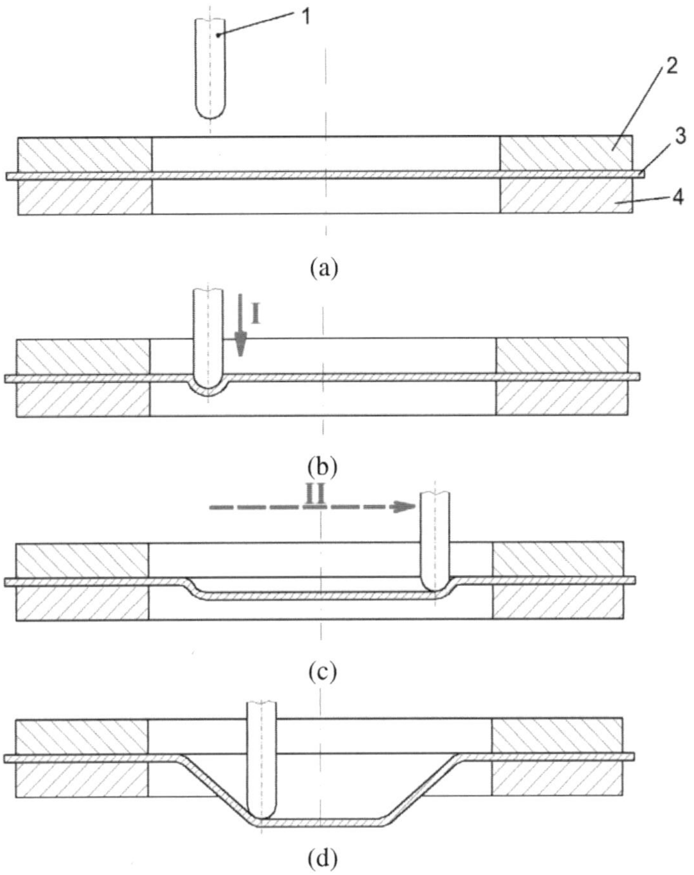


**Figure 1.** Single point incremental forming process.

Studies in this field reveal the use of industrial robots for SPIF but in general for parts obtained by simple trajectories without any difference from the technological process presented above, and which can be used identically on a CNC machine tool. So due to the fact the industrial robots has superior kinematics which allow coordinated multi-axes movements, incremental forming can be carried-on on robot in order to manufacture a conical part, as presented in figure 2 (a) [6-9].

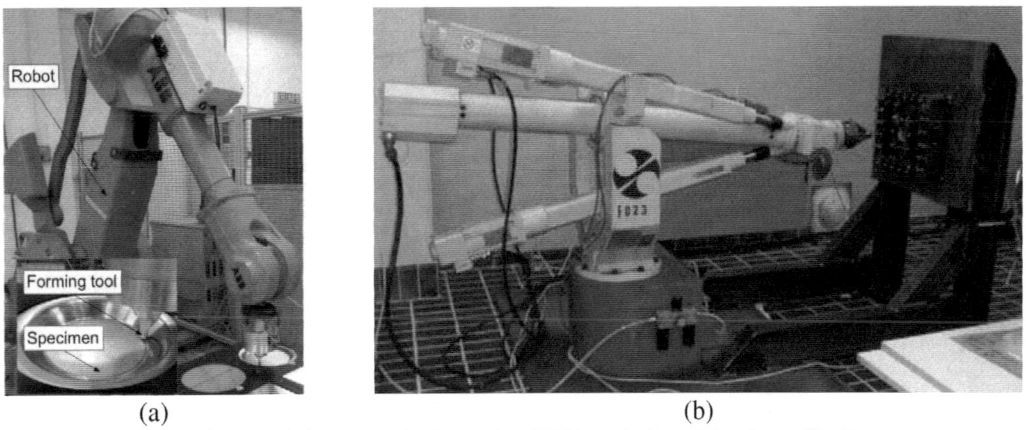


**Figure 2.** Obtaining a conical part by SPIF on industrial robots [3, 6].

## 2. Simulation of the process carried-on using a serial industrial robot

In order to simulate the machining process, a Kuka KR 210 R2700 Extra robot was used. The software used for simulation was Sprut CAM Robot. The simulation process requires a kinematic model of the robot, which is based upon the 3D geometric model provided on KUKA website. After saving every rotary axis of the 3D model as separate module, a new kinematic model is built, using the facilities of Sprut CAM Robot software. The kinematic model of the robot is based upon a specific template for KUKA robots (provided by Sprut CAM) and it is saved in .xml format.

Aside of using the standard 3D geometric model of the robot for building the kinematic model, the user has to model the end-effector of the robot, which in this case consists of the spindle motor (provided as standard for Extra class of KUKA robots and the forming tool – punch, which has to be modeled by the user).

The kinematic capabilities of the six degrees of freedom robot can be significantly enhanced by adding a supplementary unit, as working table. Thus, a supplementary Kuka DKP 400 rotary table was also used. KUKA company has specifically developed this unit to provide two supplementary degrees of freedom for the robots which are specifically designed for manufacturing tasks.

Sprut CAM Robot is a CAM software intended to be used as solution for manufacturing operations, both subtractive (milling, cutting) and additive (cladding). However, incremental forming is not supported by the software, so specific measures have to be taken:

- The punch was emulated by means of a spherical mill;
- The workpiece was emulated by taking into consideration the final form of the part and adding material to it (using the "casting workpiece" facility provided by the software);
- Forming processes involve the redistribution of the material, while cutting/cladding processes involve material removal/addition. Thus, the machining simulation was not able to represent the real incremental forming process.

A screenshot within the simulation process are presented in figures 3.Figure 4 present a detailed view of the part during SPIF process.

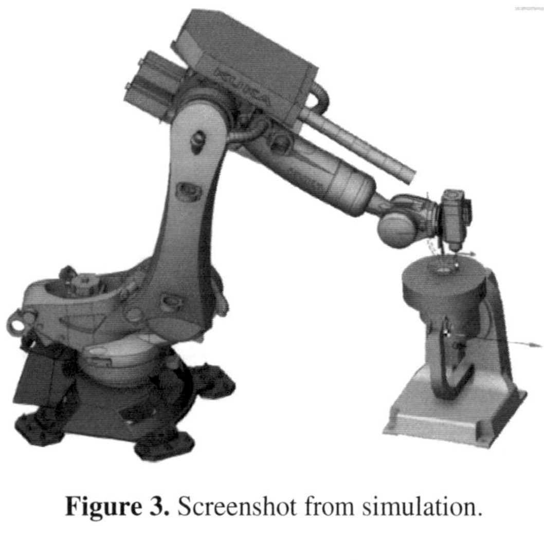

**Figure 3.** Screenshot from simulation.

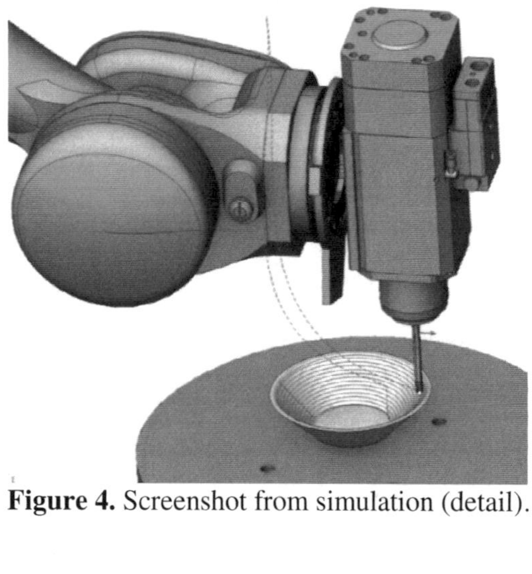

**Figure 4.** Screenshot from simulation (detail).

## 3. Trajectories used for incremental forming

The simulated machining layout allowed the use of several trajectories:

A. Circular trajectories where each circle was generated by intersecting the solid model of the workpiece using planes parallel with XY plane, at different values on Z-axis (increments). The tool axis keeps its orientation (Z axis) – figure 5 (a);

B. Circular trajectories where each circle was generated by intersecting the solid model of the workpiece by planes parallel with XY plane, at different values on Z-axis (increments). The tool axis remains always perpendicular on the workpiece surface figure 5 (b);
C. Spiral continuous trajectory, where the tool axes remain fixed - figure 5 (c);
D. 3D spiral continuous trajectory, where the tool axis remains always perpendicular on the workpiece surface - figure 5 (d).

The first two trajectories are discontinuous ones, because the tool has to performed lead-in and lead-out movements for each circle, while the last two trajectories are continuous ones. Only one lead-in and one lead-out movements are performed for the continuous trajectories.

Move-in and move-out movements on the sheet metal workpiece surface create stress concentrators on the surface of the sheet metal workpiece, which could finally lead to crack, thus continuous trajectories, which much less lead-ins and lead-outs are considered more advantageous.

On the other hand, keeping the tools axis perpendicular on the workpiece surface is believed to reduce the thickness reduction of the final part.

It is here to mention the fact that some of the elements of the proposed trajectories are generated automatically by the software, some elements have to be defined manually by the user (lead-ins, lead-outs for example), elements which can dramatically influence the plastic behavior of the processed parts. At this stage, the main purpose of the simulation was to test if the simulated experimental layout (robot + end effector with punch + working table) is able to generate and perform the above presented trajectories. The occurrence and removal of singularities and workspace limitation were also targeted by this work. Singularities can be identified by the CAM program, while workspace limitations can be avoided/removed by changing the initial position of the working table (related to the robot) and changing the initial position of the workpiece (related to the working table).

A comprehensive experimental program is now in progress to validate these assumptions.

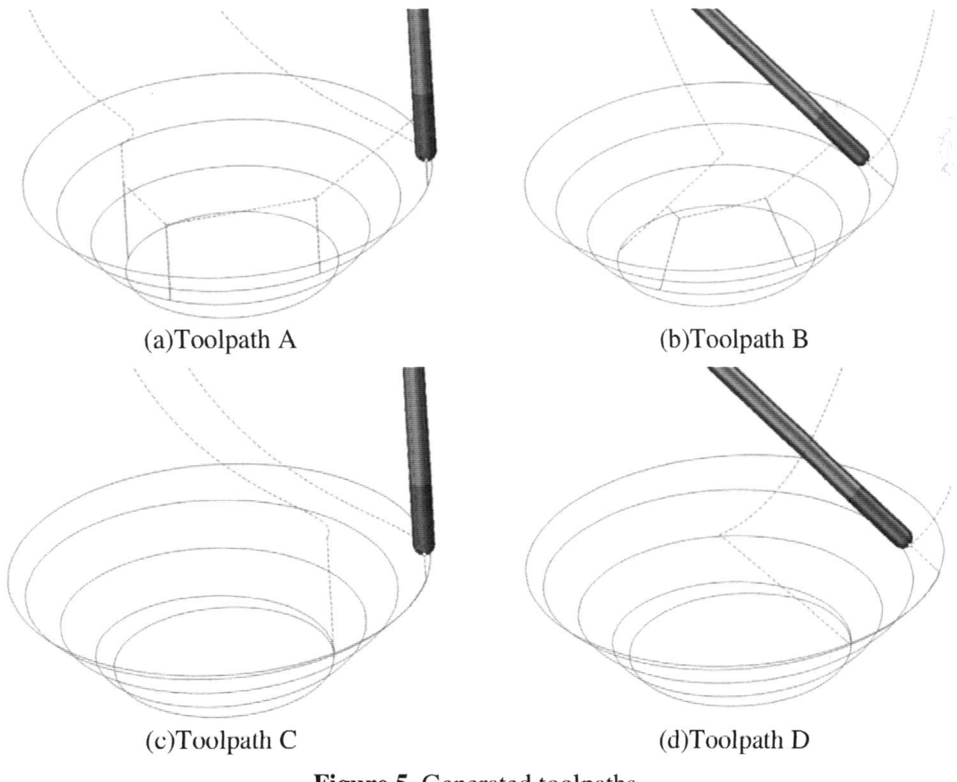

(a)Toolpath A (b)Toolpath B

(c)Toolpath C (d)Toolpath D

**Figure 5.** Generated toolpaths.

## 4. Conclusions

In order to simulate single point incremental forming machining process, a Kuka KR 210 R2700 Extra industrial robot was used and a Kuka DKP 400 rotary table.

Several types of trajectories were chosen for study in order to obtain a conical part. So, first of all a two discontinues trajectories was proposed, and after that, in order to avoid stress concentrators on the surface of the sheet metal blank, it was designed two continues trajectories.

The study also suggested using for SPIF two methods for the position of the punch related to the sheet metal workpiece. Thus, the first is the classic method in which the punch is vertical, and then the second method in which the punch is perpendicular on the sheet metal was used.

Even if the simulation process had certain drawbacks, presented in the paper, because a CAM software which does not offer support for incremental forming was used, there are also certain advantages for the user. Among them, the fact that the limits of the working space of the robot could be tested, the singularities could be identified and removed and the possibilities of achieving specific trajectories were tested are the most significant.

Further researches will be performed in order to experimentally test how will affect the proposed tool paths the plastic behavior of the parts, mainly the thickness reduction and the strain and stresses. Also, the effects upon the shape and dimensional accuracy will be studied.

**Acknowledgments**

The work presented in this paper was funded by means of a research project funded by the Romanian Ministry of Research and Innovation CCCDI-UEFISCDI, project number PN-III - P1-1.2 - PCCDI-2017-0446 /nr. 82PCCDI/2018, within PNCDI III, project title: “Intelligent manufacturing technologies for advanced production of parts”.

## References

[1] Callegari M, Amodio D, Ceretti E and Giardini C 2006 *Industrial Robotics: Programming, Simulation and Application* (Edited by: Low Kin Huat ARS/plV Germany) p 493
[2] Oleksik V, Bologa, O, Breaz R and Racz G 2008 *Int. J. Mater. Form.* **1** 1187
[3] Ismail N A, Ismail M I S, Mohd Radzman M A, Mohd Ariffin M K A and As'arry A 2019 *I. J.I. E.* **11** 217
[4] Oleksik V, Pascu, A, Mara D, Bologa O Racz G and Breaz R 2010 *Steel Research Int.* **81** 930
[5] Bologa O, Breaz R E and Racz S G 2018 *Procedia computer science* **139** 408
[6] Lamminen L 2005 Incremental Sheet Forming with an Industrial Robot – Forming Limits and Their Effect on Component *Design Proc. Shemet Erlangen Germany* 457
[7] Störkle D D, Möllensiep D, Thyssen L and Kuhlenkötter B 2018 *Procedia Manuf.* **15** 1164
[8] Crenganis M and Csiszar A 2019 *A Mater. Sci. Forum* **957** 156
[9] Schafer T and Dieter Schraft R 2005 *Rapid Prototyp J.* **11** 278

# On-line linear guideway monitoring using laser mouse sensor

**W N Cheng[1], C C Cheng[1*], C M Tsai[1], B C Chang[1], W R Chen[1] and T F Chiang[1]**

[1]Advanced Institute of Manufacturing with High-tech Innovations and Department of Mechanical Engineering, National Chung Cheng University, No.168, Sec. 1, University Rd., 621, Min-Hsiung Township, Chia-yi County, Taiwan (R.O.C.)

*E-mail: imeccc@ccu.edu.tw

**Abstract**. A low-cost, on-line linear guideway straightness monitoring system for machine tools is presented. The monitoring system which includes an embedded optical sensor from the optical mouse, an image display module, a displacement calculation module, an image enhancement module and a data visualization module are integrated into an edge computer. This proposed sensing system can measure the displacement in two orthogonal directions simultaneously, i.e. a two-dimensional displacement sensor based on digital image correlation, DIC. With image enhancement technique and a custom-made speckle patterns printed using a standard low-cost monochrome laser printer, results show that this two-dimensional displacement sensing system has a displacement resolution less than 5 μm in both orthogonal directions. The measuring range is 1mm for static measurement but is effectively infinite while moving at a low speed less than 0.05m/s. However, the accumulation error increases with the measuring range. A novel application of utilizing this displacement sensing system upon on-line monitoring the linear guideway healthy status, i.e. linear guideway straightness, of machine tools is presented. The performance and the associated advantages and disadvantages of this proposed sensor as well as the on-line monitoring system is assessed experimentally.

## 1. Introduction

Optical sensors used in displacement measurements have already proved to be reliable and accurate. However, the high cost limits their applications. A cost-effective optical displacement sensor with an acceptable accuracy which can be embedded in machines theoretically is invaluable in numerous applications. The optical mouse sensor which has a coefficient of determination of $R^2$ = 0.9998 in a linear displacement measurement is a candidate to fulfill this requirement [1]. In fact, the optical mouse sensor has been proposed in various applications, such as a displacement sensor [1-5], a rotary encoder [6, 7] and even a two-Euro counterfeit coin detector [8]. However, several drawbacks limit its industrial applications, such as very sensitive to the measurement surface [3], out of focus easily due to focus distance changes [2, 3], and poor performance while in circular trajectories [1, 5]. Most of all, both the accuracy and precision are not good enough for industrial applications. Nevertheless, with the advance both in hardware and software in recent years, the optical mouse has a great potential as a built-in sensor for machines in industrial applications.

In this study, the image from CMOS of the optical mouse is employed as an inexpensive image acquisition device. Two ways to improve the accuracy and precision in measuring the displacement. The first is to increase the resolution of the image captured by the CMOS through the image processing. The second is to optimize the non-repetitive speckles pattern in order to maximize image discrimination as well as the resulting displacement.

A custom-made speckle patterns printed using a standard low-cost monochrome laser printer was pasted on the measurement surface, which remedies the problem caused by varying sensitivity due to measurement surface. Moreover, the CMOS is mounted on a jig which not only maintains the focus distance but also shields out the noise due to environmental lights, which takes advantage of the short focal distance featured by the optical sensor. An application of utilizing optical sensor upon on-line monitoring the linear guideway healthy status, i.e. linear guideway straightness, parallelism, of machine tools is presented. The performance and the associated advantage and disadvantage of this proposed sensing system and the on-line monitoring system as well is assessed experimentally.

The remainder of this paper is structured as follows. Section 2 and 3 introduce the hardware and working principle for optical sensor measuring system, respectively. Section 4 presents the image processing techniques adopted to improve the image quality acquired from the CMOS.

The speckle pattern that enhances the resolution in displacement measurement is also illustrated. In Section 5, the performance of this displacement sensor system based on optical mouse is assessed experimentally. A novel application of utilizing this optical sensor upon on-line monitoring the linear guideway straightness of machine tools is presented. Section 6 summarized the results and contributions of this study.

## 2. Sensing principle of optical mouse

The optical mouse has a laser diode (LD) and CMOS sensor. The CMOS sensor consists of an array of pixels with a size of 30 × 30. Each pixel has a physical size around 30 μm. The black and white CMOS sensor has intensity value ranging from 0 (black) to 255 (white). When the optical mouse is moving on a surface, the lens capture images of the surface texture onto the CMOS sensor continuously. Then the mouse displacement along two orthogonal directions, i.e. x and y axes, is calculated according to the correlation between the present image of the surface texture with previous one, i.e. the digital imaging correlation (DIC) [9, 10].

To assess the performance of a typical mouse sensor in measuring the displacement, figure 1 compares the positioning accuracy from an off the shelf mouse sensor installed on the worktable of a feed drive system with the motor encoder. The mouse sensor used in this study has a resolution up to 8200 CPI (Count per inch) which is approximately 0.6 mm in terms of displacement resolution. However, figure 1 shows that the deviation in positioning accuracy as compared to that calculated from motor encoder is close to 5 mm when the work table is traveling at a speed of 2 m/min. Nevertheless, it is still far from an acceptable positioning accuracy, i.e. less than 10 μm, roughly requested by the machine tool industry.

There are two aspects to improve the resolution in measuring displacement of the optical sensor. The first is to utilize the image processing technique such as image interpolation, edge detection, etc. to improve the image quality and thus the resulting sensitivity in calculating the displacement using DIC. The second is to utilize a custom-made information patterns for DIC instead of the surface texture on the targeted object. Note that the DIC is originally adopted in measuring the strain of a deformed specimen after it is subjected to an external force. The information pattern conventionally is a speckle pattern created by paint sprayed on it, which ensures the speckle pattern is random and is strongly bonded on the specimen. The objective is to measure the speckle pattern displacement before and after the specimen is deformed based on DIC. On the contrary, the objective in this study is to measure the displacement of the optical sensor not the deformation of the targeted specimen.

Therefore, the resolution in displacement measured by the optical sensor can be increased by using a well-controlled speckle patterns pasted on the targeted object. The speckle patterns in this study are optimized aiming to increase the displacement discrimination and printed using a standard low-cost monochrome laser printer to ensure a more consistent sensing sensitivity for the optical sensor.

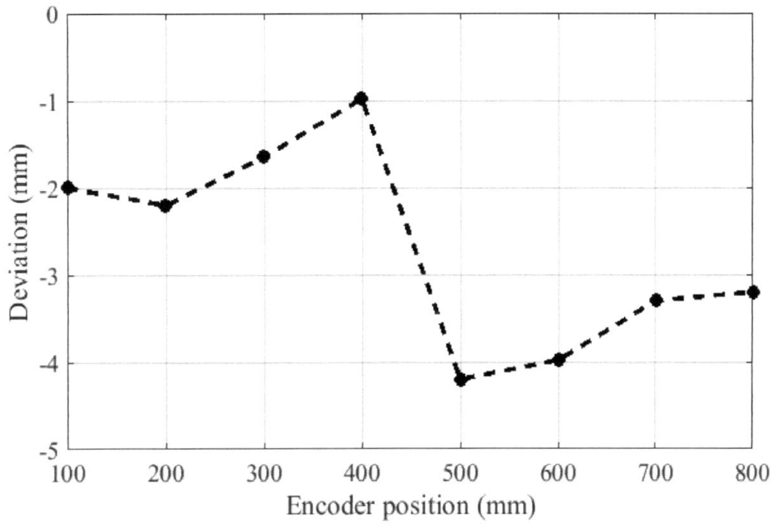


**Figure 1.** Positioning accuracy at a speed of 2 m/min.

## 3. Digital image correlation

For a typical optical mouse, the displacement is calculated from the image correlation between two consecutive surface texture images. In other words, two consecutive images captured at different locations are compared by a subset from the first image and convolute it with the second image. The similarity between two images is quantified using image correlation calculated using the summation of squared differences of the intensity value of each pixel between two images. Then the maximal similarity and then the corresponding displacement of the second image relative to the first one can be determined. The subset size is chosen to be large enough for noise reduction. In this study, a subset size of 30 x 30 pixels has been fixed in conducting the DIC. However, the pattern of the surface texture where the captured image should be non-repetitive in order to have a unique displacement of the second image relative to the first one. The accuracy of displacement measured in such a way depends not only on the pixel size of the image but also on the non-repetitive pattern of the surface texture where the image is captured by the CMOS. The pixel size from a commercial optical mouse from the market so far is more than 30 μm. Subsequently, the resulting displacement resolution is more than 60 μm, which greatly limits it applications in the machine tools.

On the other hand, a non-repetitive speckle pattern is required to cover the area of the targeted flat surface for maximizing the sensitivity in displacement measurement based on DIC algorithm. The size of the speckle as compared to that of the pixel and the associated speckle distribution as well all clearly influence the accuracy of the measured displacements. How to increase the displacement resolution through the optimal design of the non-repetitive speckles pattern is described in the next section.

## 4. Image processing and speckle pattern

In addition to the lens and the light-sensitive CMOS, the lightning condition has a great influence on the image captured by a CCD-camera. The intensity values of the monochrome image represented by the 30 × 30 pixels have a much smaller range as compared to the theoretical one ranging from 0 (black) to 255 (white). Image blurring is inevitable due to uneven lightness, the targeted speckle pattern unflatness and/or loss of focus to a certain level. Therefore, several image processing methods were adopted to improve the image quality.

The image acquired by the CMOS was resampled from 30 × 30 to 90 × 90 pixels using bicubic interpolation, i.e. upsampling, for adjusting the resolution. The intensity value for each pixel of totally

90 × 90 pixels of the resulting image is an average from 50 images captured within a second to reduce the influence caused by the time drifting or uneven lightness. The speckle pattern is designed to be chess-like and either in black or white, so an image with good contrast which has sharp differences between black and white will certainly increases the image discrimination as well as the resulting displacement calculated by the DIC. To increase the image contrast, the values of the input image intensity were manipulated pixel by pixel in order not only to saturate them at low and high intensities but also to reduce the intensity uncertainty due to the diffraction. The contrast is adjusted using multiple stages to ensure the resulting histogram to be bi-polarizing. Edge detection was also adopted to identify points in an image at which the image has discontinuities based on the abrupt changes in brightness. After through image interpolation, contrast bi-polarization and edge detection, the resulting image has higher sharpness which increases the sensitivity in calculating the displacement based on DIC.

In addition to the image processing, the second way to increase the precision in displacement calculated using DIC is to optimize the speckle pattern specially designed for the optical sensor. The numerically generated speckle pattern is characterized by the size of the speckle and its coverage over an image area with a size of 1mm×1mm occupied by the 30 × 30 pixels. Each speckle is identical throughout an entire pattern; however, needs to be distributed randomly, which is essential for a unique and correct displacement calculation based on correlation. The coverage is defined as the ratio of white/black pixels for per pattern. Different speckle size ranging from 1 pixel to 3 pixels and speckle patterns with a coverage between 10 % and 70 % as illustrated in figure 2 were examined and compared to assess the associated performance in optimal sensor displacement measurement. For a CMOS optical sensor with an array of 30 × 30, preliminary results show that the optimal speckle size is 1 pixel with a coverage ranging from 60 % to 70 %.

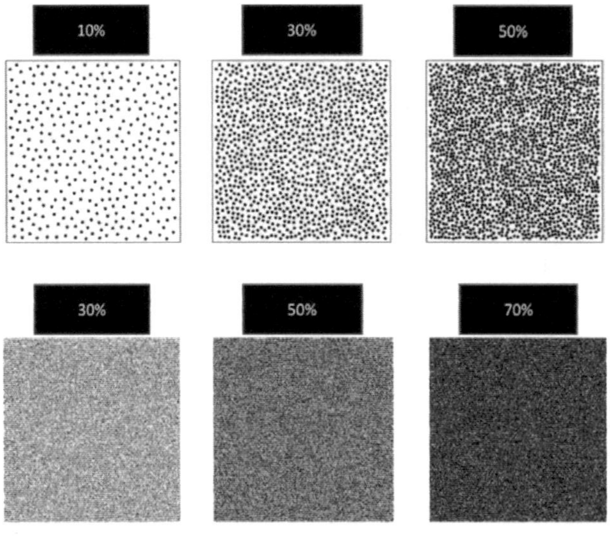


**Figure 2.** Different speckle size ranging from 1 pixel to 3 pixels and speckle patterns per pattern with a coverage between 10 % and 70 %.

## 5. Performance assessments and applications

The optical sensor was installed on the work table of a five-axis CNC machine which has a positioning accuracy less than 3 μm in both x and y axes. Figure 3 shows the work table with the optical sensor moves diagonally and stops at every 10 μm projected onto the x and y axes to measure the displacement in both x and y directions. The measurement repeats 10 times and the results are illustrated in figure 4. It shows that the accuracy in displacement and precision in percent error is less than 10 μm and 5 %, respectively, along x-direction; but has accuracy and precision less than 20 μm and 10 %, respectively along y-direction.

(a) (b)

**Figure 3.** Displacement measurement setup:(a) Five-axis CNC machine; and (b) optical sensor and planed displacement measurement.

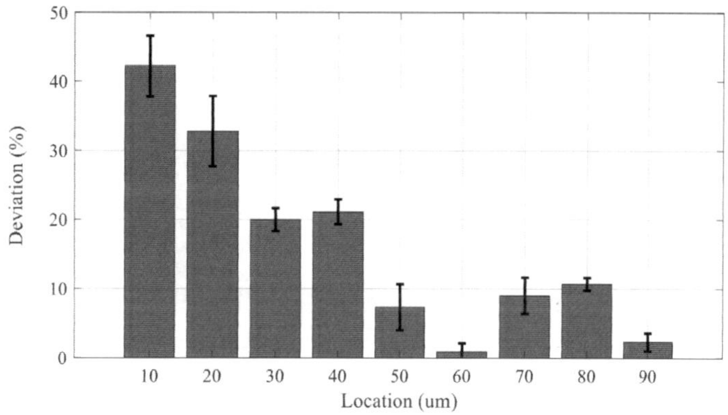


(a)

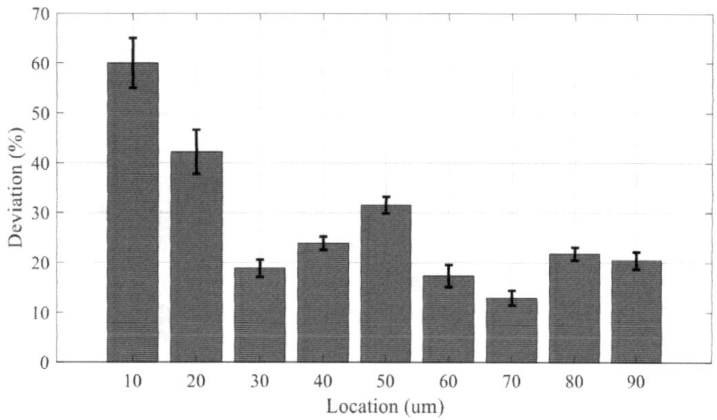


(b)

**Figure 4.** The displacement measured by optical sensors: (a) x-direction; (b) y-direction.

Monitoring the guideway straightness of a feed drive system is demonstrated as an application of the embedded optical sensor system. A custom made feed drive system which features its subsidiary guideway can be adjusted to simulate the guideway misalignment is used in this demonstration. The guideway straightness calculated by the proposed optical sensor sensing system is compared with that measured directly by the Renishaw XL-80 laser interferometer. Several speckle patterns were pasted near the guideway with a distance of 20cm between each other as illustrated in figure 5. The guideway straightness was measured while the work table with the embedded optical sensor was traveling on the guideway and stop at each speckle pattern for 1 second to capture the image of speckle patterns. The first image captured by the optical sensor at each speckle pattern is used as a reference in positioning, and the second image at the same position next time is used to determine the displacement relative to the first one according to DIC. Note that the measured displacement is two-dimensional, i.e. the displacement along y-direction is longitudinal whereas the x-direction is transverse as indicated in figure 5. The longitudinal displacement indicates the deviation in positioning, whereas the transverse one represents a sample in quantifying the rail way straightness. With several speckle patterns pasted along the y-direction, the rail way straightness can be determined with the transverse displacements, i.e. the x-direction, measured by the optical sensor at different positions of railway. The whole procedure in measuring the railway straightness is conducted automatically using a program in the CNC controller.

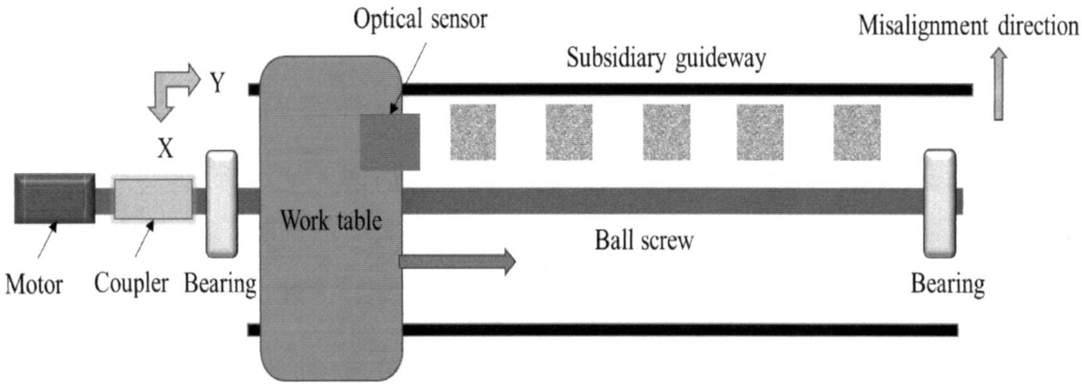


**Figure 5.** The guideway strightness monitoing system with embedded optical sensors.

To simulate the guideway misalignment, one end of the adjustable guideway of the custom-made feed drive system is pushed by a screw by 40 μm away from the bearing support away from the motor along the x- direction as shown in figure 5. The first captured image of each speckle pattern was used as reference when the work table is traveling away from the motor. Then the guideway was adjusted to deviate from 40 μm further to 80 μm and the second image of each speckle pattern was captured. Table 1 lists the linear guideway deviation along the x-direction measured respectively by the laser interferometer and the embedded optical sensor after the guideway is adjusted from 40 μm to 80 μm. Table 1 lists two measurement results for checking repeatability. It shows that the differences in both displacement measurements between the embedded optical sensor and those from laser interferometer were less than 5 μm. It is worthy of note that the carriage fixed under the work table forces the misaligned guideway back to its original position when the work table was moving. The original misalignment of 80 μm at the end of the bearing support of the guideway is forced to reduce to 4.8 μm. The experiment repeated again but the guideway is adjusted furthermore to 100 μm from 80 μm at the end of the bearing support. The displacement measured by the proposed sensing system and the laser interferometer is listed in table 2. It indicates that the differences in displacement measurements between the embedded optical sensor and those from laser interferometer were also less than 5 μm. It concludes that guideway misalignment can be monitored successfully by the proposed optical sensing system if the misalignment is larger than 5 μm.

**Table 1.** Linear guideway deviation measurements when subsidiary guideway is adjusted from 40 μm to 80 μm.

| Positioning (mm) | laser interferometer (μm) | Optical sensor (μm) | laser interferometer (μm) | Optical sensor (μm) |
|---|---|---|---|---|
| 100 | -1.6 | 0 | +0.5 | +1 |
| 200 | +1.4 | 0 | +3.0 | 0 |
| 300 | +0.9 | +1 | +2.9 | +1 |
| 400 | +3.0 | +4 | +6.7 | +4 |
| 500 | +4.8 | +6 | +6.5 | +5 |

**Table 2.** Linear guideway deviation measurements when subsidiary guideway is adjusted from 40 μm to 100 μm.

| Positioning (mm) | laser interferometer (μm) | Optical sensor (μm) | laser interferometer (μm) | Optical sensor (μm) |
|---|---|---|---|---|
| 100 | +1.7 | +1 | +1.9 | 0 |
| 200 | -0.3 | 0 | +3.0 | 0 |
| 300 | +1.2 | +1 | +4.0 | +1 |
| 400 | +8.6 | +6 | +4.9 | +5 |
| 500 | +7.4 | +9 | +8.3 | +8 |

## 6. Conclusions

A low-cost embedded optical sensor system based on optical mouse was proposed to monitor the linear guideway healthy status, i.e. linear guideway straightness. This proposed sensor system can measure the displacement in two orthogonal directions simultaneously, i.e. a two-dimensional displacement sensor based on digital image correlation. With the image enhancement technique and a custom-made speckle patterns printed using a standard low-cost monochrome laser printer, results show that this two-dimensional displacement sensor has a displacement resolution less than 5 μm. The linear guideway straightness can be on-line monitored with an accuracy within 5 μm as compared to that measured by the Renishaw XL-80 laser interferometer. With an accuracy within 5μm of this proposed embedded displacement sensing system, more industrial applications are under development.

**Acknowledgments**
This study was supported by the Ministry of Science and Technology, Taiwan, ROC (under Contract No. MOST 107-2634-F-194-001).

**References**
[1] Palacin J, Valganon I and Pernia R 2006 *Sens. Actuator A Phys.* **126** 141
[2] Ng T W 2003 *Sens. Actuator A Phys.* **107** 21
[3] Cooney J A, Xu W L and Bright G 2004 *Mechatron* **14** 623
[4] Minoni U and Signorini A 2006 *Sens. Actuator A Phys.* **128** 402
[5] Hu J, Chang Y and Hsu Y 2009 *Sens. Actuator A Phys.* **149** 74
[6] Tresanchez M, Pallejà T, Teixidó M and Palacín J 2009 *Sens. Actuator A Phys.* **155** 73
[7] Tresanchez M, Pallejà T, Teixidó M and Palacín J 2010 *Sens. Actuator A Phys.* **157** 161
[8] Tresanchez M, Pallejà T, Teixidó M and Palacín J 2009 *Sensors* **9** 7083
[9] Reu P L, Sweatt W, Miller T and Fleming D 2015 *Exp Mech* **55** 9
[10] Lecompte D, Sol H, Vantomme J and Habraken A 2006 *Proc. SPIE.* Sep. 15 2006 France

IOP Conf. Series: Materials Science and Engineering 1009 (2021) 012013 doi:10.1088/1757-899X/1009/1/012013

# Automatic generation of documents and reports for educational process

**C I Chivu[1] and C Chivu[1*]**

[1] Department of Engineering and Industrial Management, Transylvania University of Brașov, 29 Eroilor, Brasov 500036, Romania

*Email: catrina.c@unitbv.ro

**Abstract**. Management of educational process implies a large number of documents and reports that are interconnected and have high impact upon students and teachers. Coordinators of educational processes struggle with time management, correlation between implied parts and specific requirement. Present paper is focused on distance learning educational process that has specific requirement regarding students, teachers timetables and discipline management documents. Authors developed an informatic application that allows users to obtain: students' and teachers' timetables; reports regarding teachers from all study program; calendars of disciplines. Using this application, time invested in the process is reducing with almost 55%, proving the opportunity and efficiency of application.

## 1. Introduction

Management applied in standardization of different area is a real trend during these days [1]. Informatic application used in management of standardization is a real benefit in efficiency and sustainability of any process.

Management of study programs implies reports that are based on different type of document, mainly Microsoft Word documents. On national level there is no unique informatic application that may help coordinators to design these reports.

There is an international interest for managing documents, existing informatic applications designed for human resource departments in recruitment process, or for email management at organization level. These applications allow users to manage, store, protect documents, but not necessary to generate ones based on templates. Thus, there can be mentioned: OpenKM, Open DocMan, Seedddms, Kimios, GoogleDocs, Bitrix24, Nuexo, Krystal DMS, LogicalDOC, etc. [2]. All mentioned application are Open Source ones.

At universities level there is no national/ international software, agreed to be the most used. In Romania, universities, most of the time, develop their own informatic application, that include curricula development, position chart, students management, etc.

The Romanian Agency for Quality Assurance in Higher Education [3] doesn't impose specific informatic application but the format of the document as structure and aspect, not as specific type of file. The Agency doesn't have an on-line procedure to evaluate program and thus there are no restriction regarding file format of documents.

National evaluation of distance learning programs is based on the documents as:

- Curricula (specific format, standard formula implemented);
- Syllabus for each course;

IOP Conf. Series: Materials Science and Engineering 1009 (2021) 012013 doi:10.1088/1757-899X/1009/1/012013

- Timetable for each group of students per course;
- Timetable for each course;
- Timetable for each teacher.

All these documents are interconnected and dependents and thus accuracy is very important. Besides, these aspects is the copyright and license issue that restrict the applications.

## 2. Conceptual design of informatic application

In the context of minimum investment, at the universities level, an informatic application should be developed and used with more or less common software, eventually included in some packages, at low cost.

Thus, informatic application was developed using C# programming environment [4, 5]. This environment allowed authors to connect a user friendly interface to Microsoft Excel [6] input and output data. These two software are included in hardware systems acquisition. Authors choose to develop application in C# because the flexibility of it to interconnect different data. Even if input data is an *.xls file, scheduling application is not an Excel or Visual Basic application because the complexity of the conditions. Besides this, authors intention is to develop an general application, independent of Microsoft Office licence, output files being easily changed, if it is necessary.

Structural diagram of informatic application is shown in figure 1. The main idea of the application is to import data as *.xls format and to export data also as *.xls or *.xlsx format.

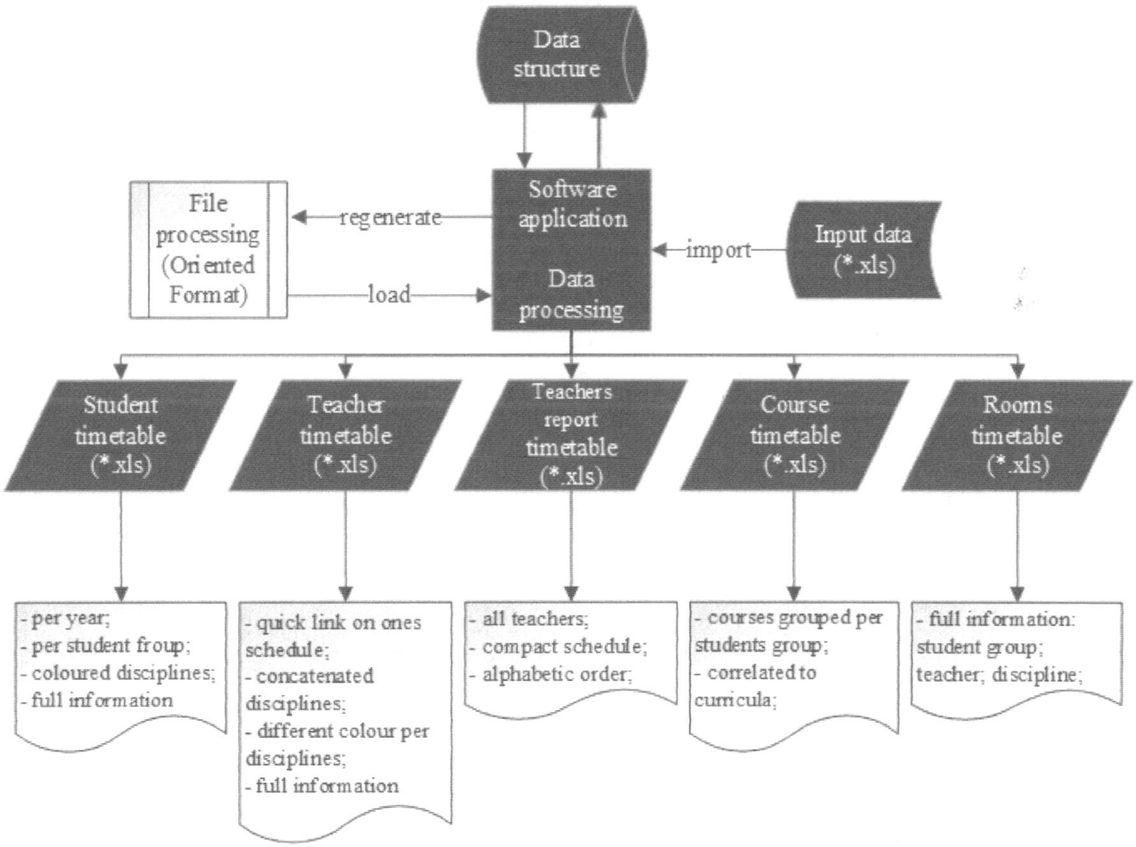


**Figure 1.** Structural diagram of informatic application.

Data structure is an executable application that concatenates data from input data base and allow interface to be active for the user. Output documents are generated in a typical Excel formats, that correspond to ARACIS standards, oriented on distance learning formats.

Input data base should be a specific formatted document, with specific conditions satisfied, as it can be seen in figure 2.

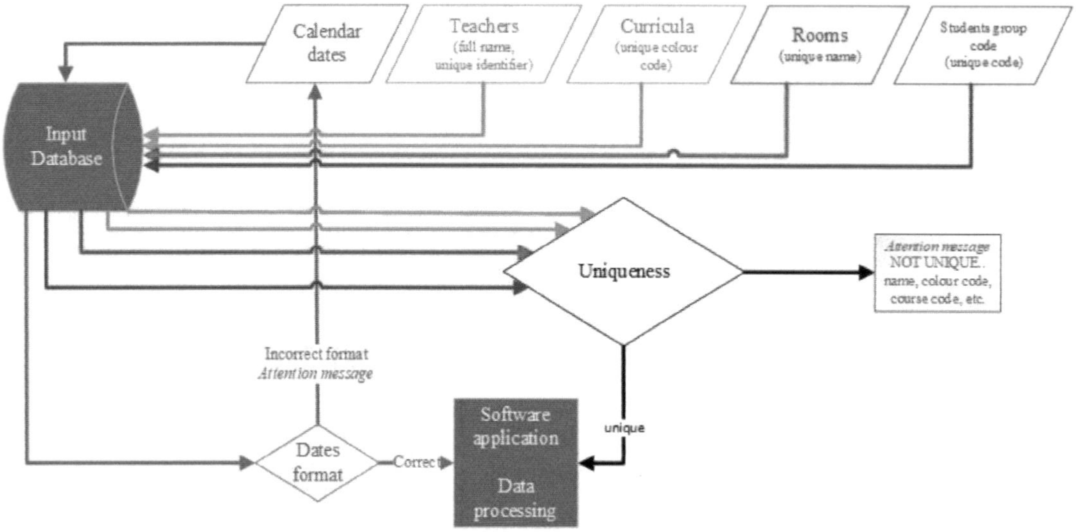


**Figure 2.** Input data base structure and conditions.

Input database has the following characteristics:

- Calendar dates: full date (time and hour, according to university laws); for distance learning, these data are weekend days; they include national holidays but mark them with different colour thus the user is well informed that should not plan any activities in those days;
- Teachers: full name, academic degree and institutional e-mail address; for each teacher is generated an unique identifier;
- Curricula: distance learning format with specific activities (individual study, tutorial activities, homework activities, applicative activities); each discipline has a unique identifier and a unique colour code (last code is to make the students and teachers timetable more visible per courses);
- Rooms: codes for available and appropriate rooms for each semester; data sheets includes coordinator department; number of places and type of room (for seminar, specific laboratory, etc.); each room has a unique identifier.
- Students groups codes: official code for each group or semigroup of students, according to secretariat; each entity has a unique identifier.

As it can be observed in the above description, performance of the application is increased by using unique identifiers for each input data. These identifiers are not visible for the user and are automatically generated using H tables.

The whole data base is design as a three dimensional matrix (figure 3).

In this matrix, students groups, hours and days work based on index. Each cell stores code for one teacher, one course (name and type of course) and one room. For each change date from data base is depth-first checked (using sub-group index) uniqueness used of room and teacher. If there is an overlap it will be marked on interface.

When it is first generated, the oriented format file stores all scheduling data, including input data. This is the only file which may restore data structure from the last saved version. This file is used to save intermediary versions or to generate all the output files.

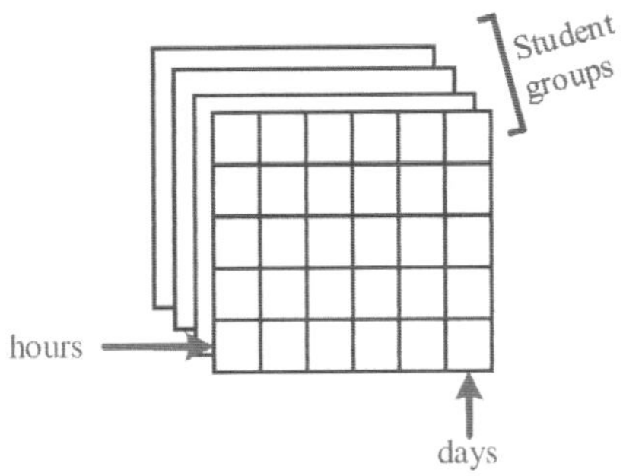


**Figure 3.** Data base: three dimensional matrix.

## 3. Functionality of informatic application

The informatic application is a user friendly application that helps to fill out timetable by marking overlaps regarding rooms, teachers, students and also include exactly the number of hours/ course from curricula.

Logic diagram of informatic application is shown in figure 4.

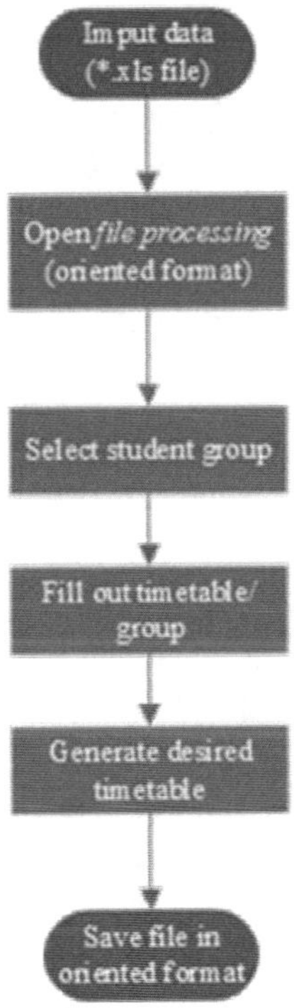


**Figure 4.** Logic diagram of informatic application.

Filling-out the schedule implies to select one student group, this one will be the active area. Interface allows user to open multiple schedules for different group of the same year or different year (figure 5). On the interface are show 4 different areas that have the following roles:

- Area 1: is the active area, on which is fill-out schedule of the current group of students (in figure group no 2LD491;
- Area 2: inactive area that allows user to see different timetables (from the same year or different year);
- Area 3: active area that allows user to choose name of the discipline, type of activity according to curricula, teacher and room; inside this area user is able to see how many hours are already allocated for one course on each activity (SI:... AT:.... TC: .... AA: ....);

- Area 4: includes schedules, with dates according to input data file; holidays or evaluation sessions are marked automatically on the schedule.

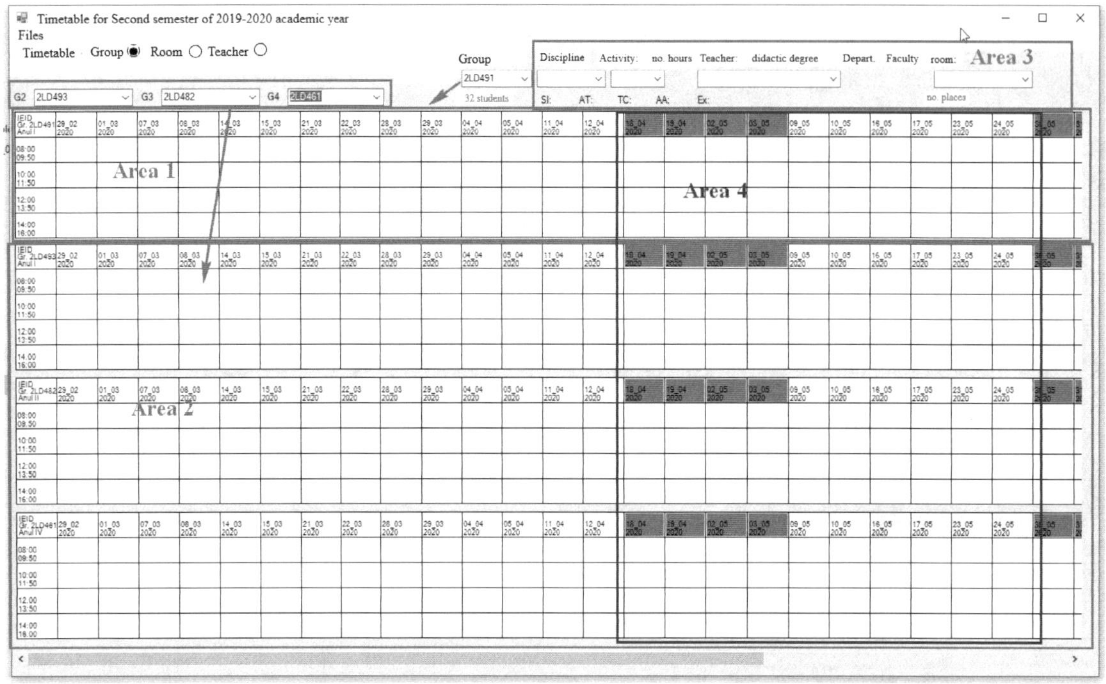


**Figure 5.** Interface of informatic application.

Students' timetable (*.xls format) is one file for all study years, a sheet for each group of students (figure 6).

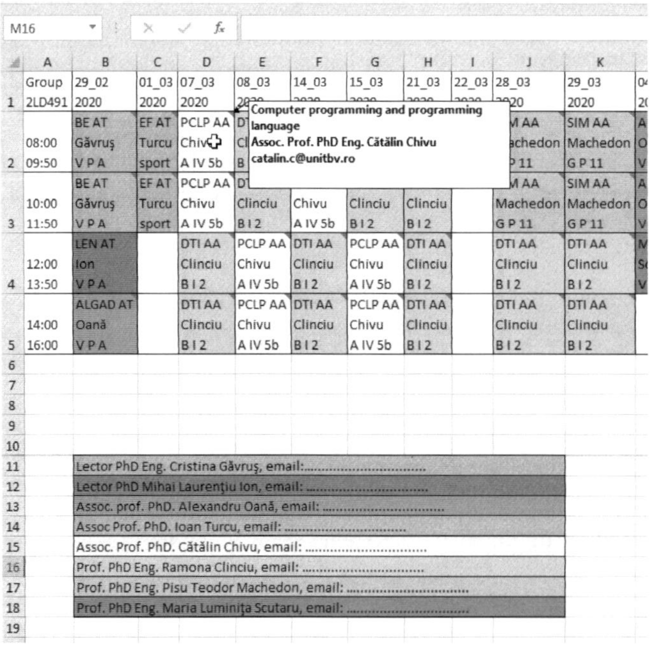


**Figure 6.** Timetable of one group of students.

IOP Conf. Series: Materials Science and Engineering 1009 (2021) 012013 doi:10.1088/1757-899X/1009/1/012013

File contains the following data:

- Each sheet is a bidimensional matrix: rows are timelines and columns are dates;
- Name of each sheet is name of the students' group;
- Schedule is coloured, each course has its own unique colour;
- Each cell (a class) contains: abbreviation of each course, type of activity, teacher name, room and a hint that contains full name of course and complete data regarding teacher;
- On each sheet, based on colour code, there are full name of teachers and their official e-mail address.

Schedule of teachers has the same structure as students schedule. A sheet for each teacher (figure 7). First sheet is a Table of contains and allows users to navigate through document.

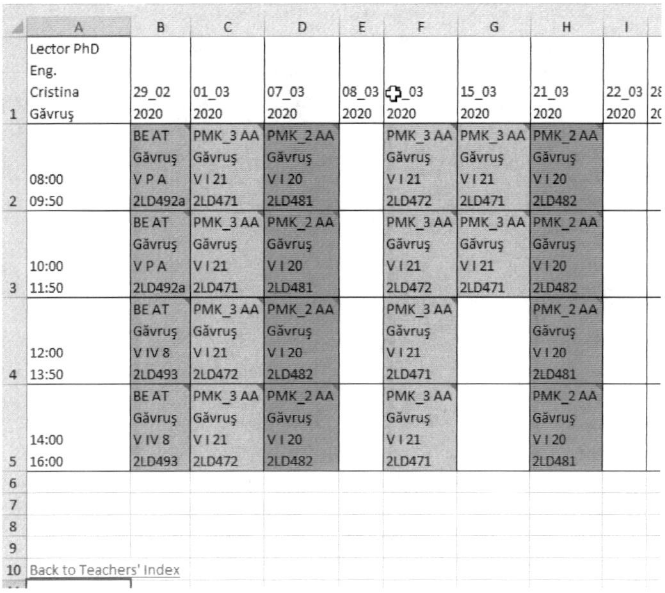

| | A | B | C | D | E | F | G | H | I | |
|---|---|---|---|---|---|---|---|---|---|---|
| 1 | Lector PhD Eng. Cristina Găvruş | 29_02 2020 | 01_03 2020 | 07_03 2020 | 08_03 2020 | _03 2020 | 15_03 2020 | 21_03 2020 | 22_03 2020 | 2 2 |
| 2 | 08:00 09:50 | BE AT Găvruş V P A 2LD492a | PMK_3 AA Găvruş V I 21 2LD471 | PMK_2 AA Găvruş V I 20 2LD481 | | PMK_3 AA Găvruş V I 21 2LD472 | PMK_3 AA Găvruş V I 21 2LD471 | PMK_2 AA Găvruş V I 20 2LD482 | | |
| 3 | 10:00 11:50 | BE AT Găvruş V P A 2LD492a | PMK_3 AA Găvruş V I 21 2LD471 | PMK_2 AA Găvruş V I 20 2LD481 | | PMK_3 AA Găvruş V I 21 2LD472 | PMK_3 AA Găvruş V I 21 2LD471 | PMK_2 AA Găvruş V I 20 2LD482 | | |
| 4 | 12:00 13:50 | BE AT Găvruş V IV 8 2LD493 | PMK_3 AA Găvruş V I 21 2LD472 | PMK_2 AA Găvruş V I 20 2LD482 | | PMK_3 AA Găvruş V I 21 2LD471 | | PMK_2 AA Găvruş V I 20 2LD481 | | |
| 5 | 14:00 16:00 | BE AT Găvruş V IV 8 2LD493 | PMK_3 AA Găvruş V I 21 2LD472 | PMK_2 AA Găvruş V I 20 2LD482 | | PMK_3 AA Găvruş V I 21 2LD471 | | PMK_2 AA Găvruş V I 20 2LD481 | | |
| 6 | | | | | | | | | | |
| 7 | | | | | | | | | | |
| 8 | | | | | | | | | | |
| 9 | | | | | | | | | | |
| 10 | Back to Teachers' Index | | | | | | | | | |

**Figure 7.** Timetable of one teacher.

File contains the following data:

- Each sheet is a bidimensional matrix: rows are timelines and columns are dates;
- Name of each sheet is name of the teacher; informatic application concatenates all courses on the same schedule, differentiating them by colour code;
- Each cell (a class) contains: abbreviation of course, type of activity, teacher name, student group index, room.

## 4. Informatic application analysis

Informatic application has a user-friendly interface, multi-management schedule and is a time-saving informatic instrument.

From type of actions point of view, application is structured as follow:

- Main timetable is done by the user, *manually*, based on input data. In this step, all checking process (not to exceed the number of hours on each type of activity, not to correlate some discipline to other teacher, not to overlap activities in the same room or for one teacher) are *automatically* done;
- Students' timetable, teachers' timetables, teachers reports timetables, courses' timetables, rooms' timetables (figure 1) are *automatically generated* by application. These timetables are generated based on the main timetable. Design of output documents are according to Romanian standards for distance learning programmes.

## 5. Conclusion

To evaluate an informatic application it is mandatory to compare two possibilities. In Transilvania University case, there is no application oriented on distance learning study-programs that may help coordinators to create schedules of any type, but there is an application designed for full-time study-programs (called here classical application). In the following there are results of time-analyse and error-analyse done by authors, obtained by simulating a schedule for one semester (four study-years and two or three groups of students/ study-year) using classical application and authors' informatic application.

Thus, regarding time-analyse there was simulated full schedules (students, teachers, courses and rooms) and all times were computed reporting to informatic application from Transilvania University for full-time study programs (figure 8). In figure 9 is compared time spent to check error and to correct those errors. In figure 8 and figure 9, in actual informatic application there is no instrument that helps teachers to generate, based on specific schedule, the course scheduling (so, it was considered Not Appropriate).

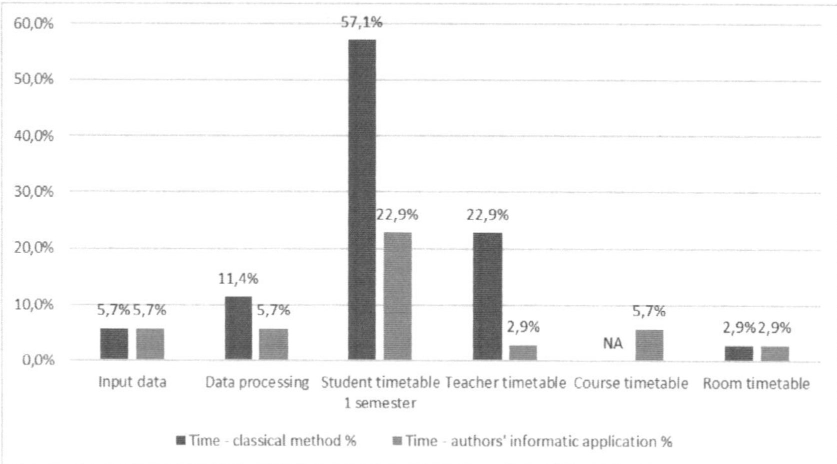


**Figure 8.** Time analysis.

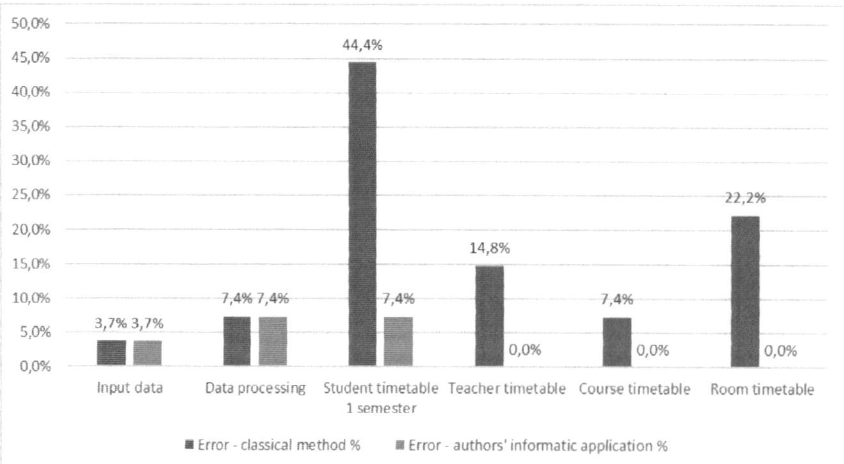


**Figure 9.** Error-time analysis.

Based on these analysis, design time, for all type of schedules (authors' information application vs. classical application) reduces with around 55%. From checking error process point of view, time dedicated for these actions (authors' information application vs. classical application) is reduced with more than 80%. From these data may be concluded that authors' informatic application is a time-saving instrument, containing a dynamic real-time error checking thus the schedules are from the beginning free from errors. This is the main advantage of the developed application. From errors point of view, in the case of informatic application, there will be no error for timetables that are automatically generated (teachers, courses, rooms). There are still errors in input data (the same error because all the process in manually done by user). Students' timetable design process is error less than classical method because the checking for error steps are dynamically done (real-time): when user select a course, automatically appears the already timetable for the course (the same happened for teacher or room).

Authors' informatic application was developed to control high complexity of interconnected documents in high-education management of study programs.

**References**

[1] Garechana G, Río-Belver R, Bildosola I et al 2017 Effects of innovation management system standardization on firms: evidence from text mining annual reports *Scientometrics* **111(3)**
[2] Singh H 2019 Top 10 Free and Open Source Document Management System *Medium* on-line Journal
[3] *** 2019 ARACIS Standards. On-line publication, available on https://www.aracis.ro/ invatamant-la-distanta-online/
[4] Troelsen A and Japikse P 2015 *C# 6.0 and the .NET 4.6 Framework* (New York: Apress)
[5] Thareja R 2018 *Data Structure using C* (London: Oxford University Press)
[6] Ostrovsky I 2014 *Reading Excel Data with C# and OpneXML* (New York: Amazon – kindle edition)

# Current perspectives on innovative methods used in teaching technical drawing

**M R Clinciu[1*] and R Clinciu[1]**

[1]Department of Manufacturing Engineering, Transilvania University of Brasov, 29 Eroilor, Brasov 500036, Romania.

[*]E-mail: clinciumr@unitbv.ro

**Abstract**. This article emphasizes some innovative methods that can be used to improve the teaching of the basic concepts of technical drawing, for students of technical faculties. These methods stimulate both the learning process and personal development, ensuring the improvement and optimization of the educational process through the exchange of ideas and experiences. The innovative methods included in the case studies presented in this paper were applied in the teaching activities, at the subject technical drawing, and demonstrated both the motivation and implication of students, as well as the fact that effective results were gained by using these methods. The results of these case studies were reflected in the improved performance of students and can also be used to improve the quality of teaching and learning technical drawing.

## 1. Introduction

In order to increase the quality of the instructive educational process in the technical university education, an important aspect is to improve the traditional methods, but also to introduce new, innovative methods.

Innovative learning methods are modern ways of stimulating learning and personal development, they are teaching tools that promote the exchange of ideas, experiences, knowledge, ensuring the improvement and optimization of the educational process. Students are not just receivers of information, they actively participate in education in the educational process, developing the ability to express opinions and assessments on the issues studied.

In university education, both classical and modern methods are used, the ability and experience of the teacher being able to make any method used to arouse the students' interest and determine them to learn.

Innovative methods involve learning through communication and collaboration, which leads to a confrontation of ideas, opinions and arguments, creates learning situations focused on the availability and willingness of students to cooperate, on their direct and active involvement, and creates social interaction of group members [1].

For creating the optimal learning environment, teacher needs to take into account certain criteria in applying modern learning methods, such as: skills to be developed by the students, their intellectual level and abilities, accessible material resources and complexity of content [2].

## 2. Innovative methods used in teaching Technical Drawing

The integration of innovative methods in the educational process of technical university education depends on a multitude of elements such as: the type of lesson (presentation of new knowledge, systematization, resuming), group level, students' willingness to work individually or in groups.

Regarding the teaching of the technical drawing subject, teacher must help the students to understand the technical concepts and to reach the highest possible level of knowledge and thinking, using analysis, synthesis, application of new knowledge or problem solving [3].

Teacher must use the knowledge and experience gained to put students in learning situations that will ensure the training and development of skills required on the labour market, but also to acquire the ability to learn throughout life.

In the case of the technical drawing subject, a multitude of modern methods can be used, accompanied by specific techniques and material resources, some of them being presented below.

### *2.1. Teaching using the "Practical examples" technique - the connection between theory and practice*

Students are interested in practical examples from the industry because it helps them understand better the theoretical notions taught in class. The use of practical examples helps to ensure the connection between technical concepts and practical applications, thus strengthening effective teaching and learning. The introduction of practical examples does not exclude theory but only helps to understand and learn theory in the classroom. It is very important to develop both the theoretical and the practical basis simultaneously [4].

The examples we are going to use in teaching must be as clear and appropriate as possible. For the effective implementation of practical examples the teacher must understand very well the given example and be able to explain it, otherwise it will cause confusion instead of helping students to understand.

Practical examples can be based on analogy, observation, demonstration and can be gathered into two classes:

- Examples which enable the explanation of the theory and of the new notions;
- Examples which present applications of the elementary principles.

### *2.2. Teaching using the "Flipping Classroom" method*

The overthrown class has become a very effective teaching method, whose popularity has grown more and more. With the help of this technique, students become active participants in the learning process, learning becoming centered on them, and teachers having the role of providing the necessary learning resources. Students have the obligation to gather all the information necessary for learning. Using various technological tools, they are encouraged to "build their knowledge", to fill in the gaps and to ask for clarifications when necessary [1, 2].

This method of learning is one of the best ways to train independent learning. To "overthrow" a class means to change the "order" in which knowledge and skills are formed. Students, instead of learning exclusively in class, learn at home by watching videos made by teachers. Discussions, practical activities and projects take place in the classroom. Students analyse and apply their knowledge by interacting with other students, but also with the teacher.

### *2.3. Collaborative learning method*

This innovative teaching method involves encouraging students to participate in various projects. Today, collaborations are a very important life skill for one's future career. Teachers can help students to develop their collaborative skills in the classroom, allowing them to study and work in groups, to achieve the various topics and objectives proposed [1, 2, 7].

Today, collaboration as a form of teaching has become a very accepted teaching tool, where each student has the responsibility of the group of students, through the role he holds in that group. Some collect information, others write it, others look for images, a leader divides the tasks and presents the whole project at the end.

Knowing each other, students know best how to share their roles in the group, for example the most casual speaks in public-in front of the class, the one who has better IT skills searches for information on the internet or writes the project, etc. This form of teaching teaches students what empathy, negotiation skills, teamwork and problem solving means.

*2.4. The method of learning with the help of virtual reality*
A virtual learning environment is a computer-simulated space where students and teachers can "meet" and interact to work together. This type of system is based on a virtual environment (virtual classroom) that provides users (students and teachers) with a series of tools that serve the teaching-learning acts; moreover, through the use of computers and the internet, these tools have the role of adding value to the learning process [5].

The tools of a virtual learning environment allow teachers to create online tests, questionnaires, but also other educational content such as presentations, lesson plans and topics, allowing curriculum creation, tracking student progress, online communication (e-mail or chat).

Students can access a variety of materials (course information, additional links, PowerPoint presentations, tests, etc.), download, open files, and solve work tasks digitally.

For example, teachers who teach technical drawing can use 3D technology to visualize bodies, parts in space, for axonometric representations.

It is possible for students to work with objects in a virtual environment, such as to obtain a better understanding. This concept is valid for cases in which students need to understand data sets, difficult formulas and abstract ideas that may be otherwise unapproachable. For some students, learning by doing seems to be less complicated than learning through listening.

Virtual reality technology offers students a valuable opportunity to learn in a certain innovative way that creates a lasting and impactful memory in their minds.

*2.5. Learning with CLOUD storage systems*
Cloud storage systems are a set of technologies and service models, based on the internet, they imply the use and transfer of computer applications, resources with storage space, information processing and loading capabilities [6].

The cloud is a recent notion in the field of computers and informatics, creating the conditions to distribute computing services, access applications and information, to store data, it is a solution for using external IT resources, according to needs

Learning with CLOUD storage systems is a method by which teachers can store vital teaching resources, such as teaching projects, notes, audio readings, videos, etc.

They can be accessed by students from the comfort of their home, whenever they need, being just a click away from accessing the necessary information. It is also useful for students who for various reasons did not attend classes, so they can recover lost classes in a timely manner.

*2.6. Stellar explosion method*
It is also known as the starbursting method ("star" = star, and "burst" = to explode) and designates a method similar to brainstorming, which is not to be confused, although it involves organizing a group of students into a group and stimulates the creation of question to question, as brainstorming develops the construction of ideas on ideas [1, 2, 7].

Starbursting is one of the most relaxing and enjoyable teaching methods; stimulates individual as well as group creativity; creates the conditions for teamwork and competition; creates the opportunity of exchange ideas; develops group spontaneity and creativity, but also teamwork skills; emphasizes the stimulation of each participant in the discussion; it does not require too much time for prior descriptions, being very accessible to all students.

Basically, write an idea or problem to be debated on a piece of paper, then line up next to the concept, as many questions related to it.

The method contains the following steps:

- Propose the issue;
- Organize students into several groups, each of which writing down the issue;
- Elaborate in every group a list with different questions on the issue;
- Communication of the results of the group activity;
- Emphasizing the most interesting questions and the teamwork.

## 3. Case studies

*3.1 Graphic organizer method - Spider diagram applied for the chapter "General rules of technical drawing".*

The graphic organizer, as an active learning method, facilitates the essentialisation of an informative material to be expressed or written, outlining the idea / ideas.

The basic premise of the spider diagram is to familiarize students with an issue at a basic level, in the case of the technical drawing, applied for the chapter "General rules for technical drawings".

Students, at first, still have a low level of knowledge in terms of technical drawing. Using the spider diagram in the first chapter of the subject, a deeper exploration is encouraged, this method allowing a multidisciplinary approach.

The spider diagram gives students a way to record and organize their ideas, by completing custom worksheets. In a spider diagram, a main concept is presented on a page and lines are used to connect the ideas, thus obtaining as many details as possible about the topic approached.

Following the application of this method, one has found that the students realized that they knew more or less than they initially thought. Also, students increased their self-confidence by synthesizing individually issued ideas. As more ideas emerged, students achieved a graphic representation of an issue that otherwise might have been difficult to understand, because, for most of them, visualization is the best way to fully understand something.

The requirement of the graphic organizer method is to organize the learned information about the technical drawing, filling in each box (figure 1 and figure 2).

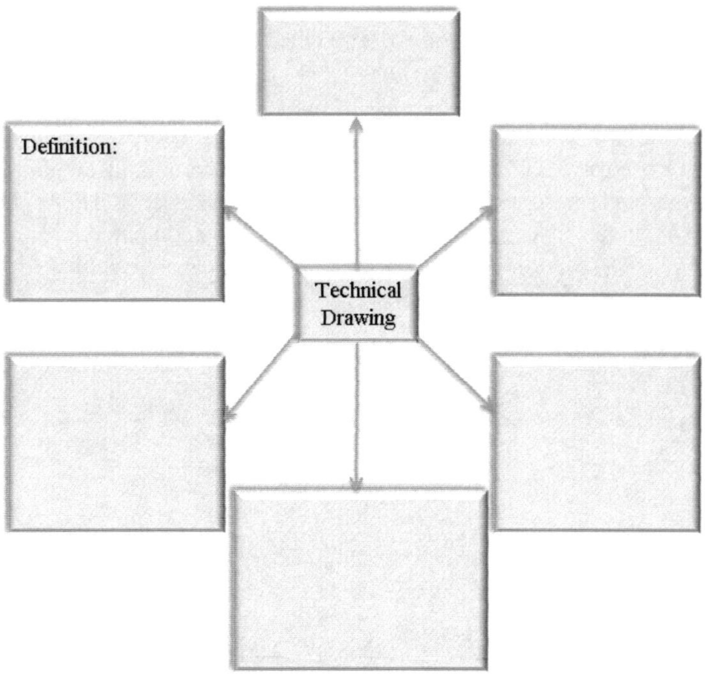


**Figure 1.** Spider Map diagram scheme

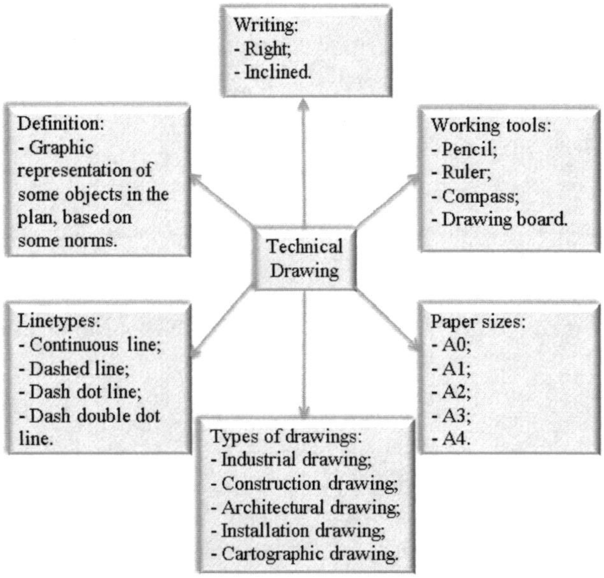


**Figure 2.** Spider Map diagram applied for the chapter General rules of technical drawing

*3.2 Graphic Pyramid method. Theme: Dimensioning in industrial drawing*

It is a combination between individual activity and group activity. It has the role of incorporating the activity of each student in an ample approach meant to solve a complex problem. This method involves organizing a structured activity in the following stages (figure 3):

- Descriptive - the teacher presents the data of the problem in question. The teacher writes on the board "Symbols used for dimensioning".
- Individual - students receive a worksheet with symbols that they solve individually in a short period of time (usually 5 minutes). Students must correlate between the symbols used in the dimension and the corresponding geometric elements.
- Pairs - groups of two students are formed, who check each other's results and try to answer the questions that were asked within the group.
- Groups of four students - formed by joining pairs two by two. They confront their results, conceive a new answer, in a formulation to which they all contribute, identifying the general conclusions in the areas of controversy resulting from each one's questions.
- The whole group - a representative of each group presents his conclusions. These can be noted on the board so that the answers between the groups can be compared.
- Decisional - the final solution is chosen and the conclusions are established on the steps taken and on the students' participation in the activity.

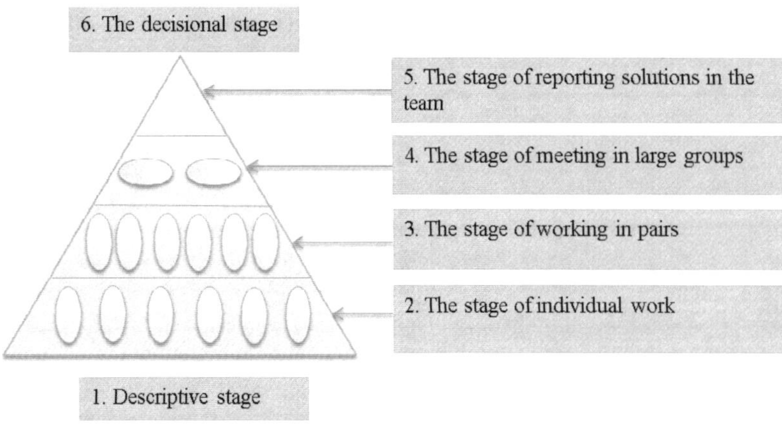


**Figure 3.** The stages of the pyramid method.

Figure 4 presents an example of using the pyramid method on the theme: Dimensioning in industrial drawing.

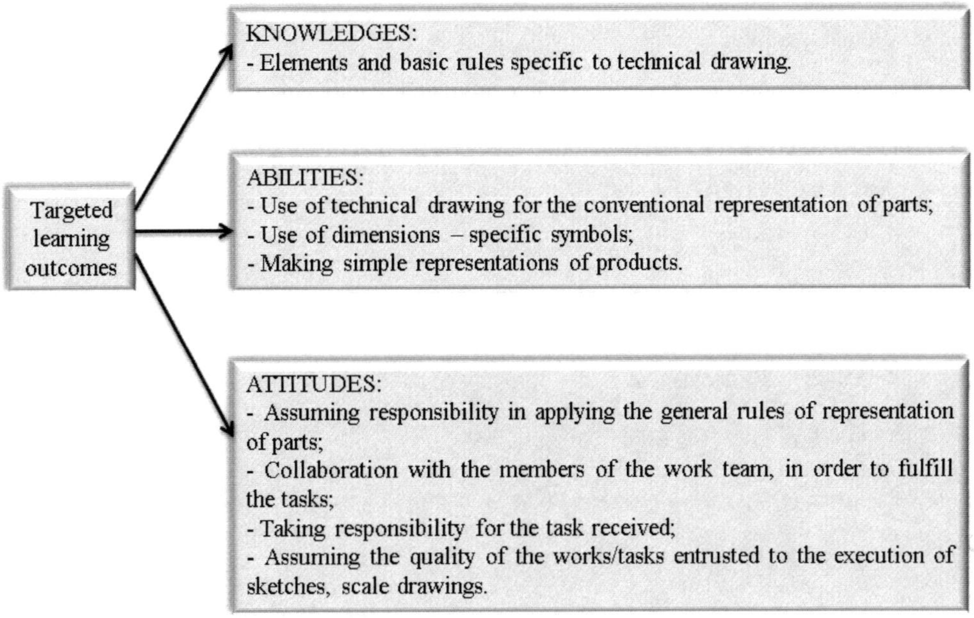


**Figure 4.** Example of using the pyramid method in dimensioning in industrial drawing.

## 4. Conclusions

This study aims to improve and diversify the teaching-learning process of technical drawing in university education, using innovative methods.

The objective of this study was to involve and motivate students to develop a greater interest in learning technical drawing, using innovative teaching methods. The methods presented in the two case studies were particularly interesting for both the teacher and the students who participated. The efficiency of using these methods is appreciated by students as having a high degree, due to the level of involvement, understanding and assimilation on the issues studied.

The results of these case studies were reflected in the improved performance of students at the final exam and can also be used to improve the quality of teaching and learning technical drawing.

## References

[1] Ionescu M and Radu I 2001 *Modern didactics* (Cluj-Napoca: Dacia Publishing House)
[2] Cerghit I 2006 *Teaching methods* (Iaşi: Polirom Publishing House)
[3] Mândru E, Borbeli L, Filip D, Gall M, Niculae A, Nemţoc M, Todoruţ D and Topoliceanu F 2010 *Interactive teaching strategies* (Bucureşti: Didactical Publishing House)
[4] Ionescu M 2003 *Instruction and education, paradigms, strategies, guidelines and models* (Cluj Napoca)
[5] Virtual reality Society, Virtual reality in the classroom, https://www.vrs.org.uk/virtual-reality-education/in-the-classroom.html
[6] Mell P and Grance T 2011 *The NIST Definition of Cloud Computing* (Gaithersburg: National Institute of Standards and Technology)
[7] Danciu E L 2004 *Collaborative learning strategies* (Timişoara: Vest Publishing House)

# A review on biomechanics of the hallux valgus pathology and its surgical treatments

**N F Cofaru[1] and A H Brănescu[1]***

[1]Faculty of Engineering "Lucian Blaga" University of Sibiu 10, Victoriei, Sibiu, 550024, Romania

*E-mail: horia.branescu@ulbsibiu.ro

**Abstract**. The present paper aims at an interdisciplinary field, regarding biomechanics aspects of the Hallux Valgus, which is a frequent deformity in our days. Owing to the importance of the study theme and the social impact, we consider it very important that the state of the art should be as detailed as possible, due to this condition being frequent in women and older people. The review will target recent research of the biomechanics of the foot, the test subjects being in normal conditions and also patients affected by Hallux valgus deformities; likewise, it will define important geometrical elements, which are required for the engineering approach of the issue. As well, the 3D modelling of the condition will be taken into consideration, beside with the surgical procedures and conservative treatments. Another important aspect covered in the paper is regarding the research and CAE simulations using the FEM, alongside with the work stands developed in this area. The bibliographic research found on the most important journals and databases shall be synthesized and structured in order to highlight the big picture of a specific research regarding a future PhD thesis.

## 1. Introduction

The following paper aims to present an interdisciplinary approach of the Hallux Valgus subject. Since the contributions brought by the Engineering are more and more useful to Medicine and that the networking between the surgeons from the Orthopedic Department and the engineers from the Sibiu Faculty of Engineering is very lucrative, we found it convenient to work together, in order to implement engineering methods more frequently. The engineering study of the Hallux Valgus deformity involves several aspects, such as: biomechanics approach, loads study, Computer aided design, surgery simulations etc.

Our main goal aimed with this article is to synthesise the information in the medicine and biomechanics fields regarding the Hallux Valgus subject, to ease the following approaches, of course, it will be taken into account the establishment of the courses for a complete research in the biomechanics domain. Accomplishing a comprehensive bibliographic study is another important purpose due to its requirement for further transposition of the data to engineering methoding, therefore we can conclude that our main objectives are:

- Shifting the medical theme of the Hallux Valgus to the engineering area by accomplishing an elaborate study of the anatomical issue;
- Study the geometrical approach of the issue for a better comprehension of the displacements;
- Research and synthesize the biomechanics of the deformity - towards the mechanical loading state;

- Elaboration of Computer aided design models alongside with Finite element analysis of the Hallux Valgus deformity and its treatments.

## 2. The foot - anatomy and biomechanics

In order to accomplish correct bioengineering studies a first important step is to comprehend the biomechanics and anatomy of the issue, only in this way, approaching the issue with engineering methods and principles is possible. This structure is very interesting from mechanical point of view, being extremely complex, although it's relatively small sized. The complexity is due its plurality of bones, muscles, ligaments and nerve endings. The 26 bones, 33 joints and all the other previously mentioned elements assure several functions such as: bipodal gait, body propulsion and shock absorption during running or jumping [1, 2].

Due to the fact that an average person walks approximately200.000 kilometers and that the foot could be affected by loadings varying between 1-4 of its bodyweight, for example in running and jumping the loadings are maximum, the foot is protected by a special architectural structure from stress related injuries [3, 4].

The main parts of the foot from anterior to posterior are: forefoot, midfoot and hindfoot, detailed in figure 1.

Talus and calcaneus are the two hindfoot bones. The talus is the bone jointing two important bones of the inferior member, Tibia and Fibula. The joint between these three bones is called ankle joint and in provides the vertical plane movement of the foot: up and down. As well, the talus joints with the calcaneus through the subtalar joint, which assures the pendulum foot motion: left-right. Further, the hindfoot is connected with the next structure, the midfoot, which is composed by 5 bones: cuboid, navicular and three cuneiforms, working together as a group. The previously mentioned bones describe the tarsal region of the foot and joint through the tarsal-meatatarsal rigid joint with five long bones, named metatarsals. Each one of the five bones joint with the phalanges; which are groups of two or three bones (two for the first finger and three for the others) [5, 6]. The metatarsal and phalanges region form together the forefoot. The positioning of the bones and joints is illustrated in figure 2.

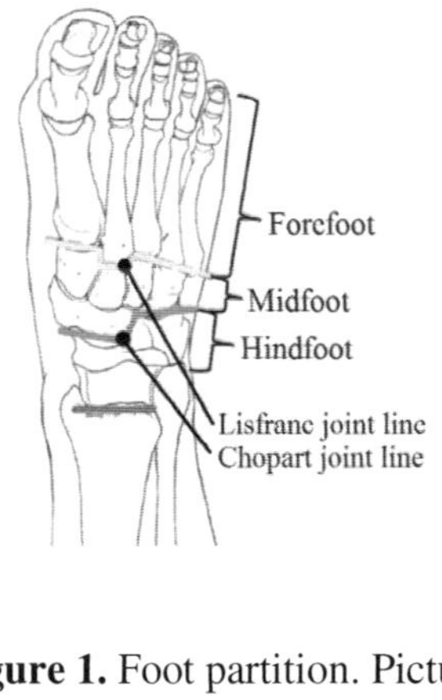


**Figure 1.** Foot partition. Picture illustrated and modified from [5].

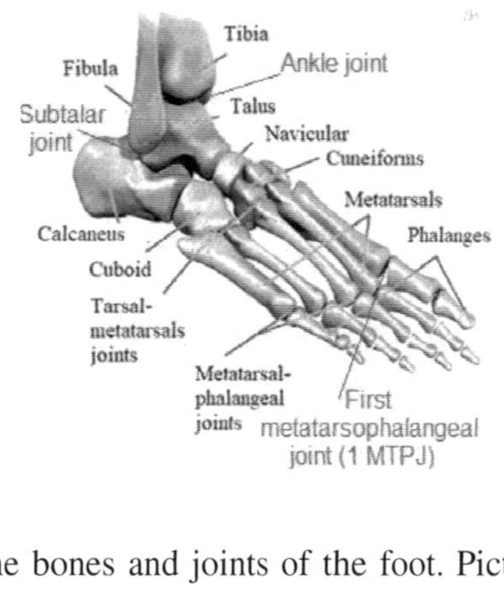


**Figure 2.** The bones and joints of the foot. Picture illustrated from [2].

The articulation of the medial and lateral sesamoid bones (located on the inferior side of the foot), the first metatarsal head (MTH) and the corresponding phalanx base, are summing up the First metatarsophalangeal joint (1 MTPJ). Because in this joint the movements have great amplitude, the sesamoid bones, first metatarsal and first phalanges are connected in the same synovial capsule [7].

The 1 MTPJ has its mechanical mobility within the traverse and sagittal planes. The vertical and traverse axis are consisting the two main axes of the joint. The metatarsophalangeal joint has a complex structure (condylar), which has repercussions regarding the rotational motions of it. Those types of movements are impossible without harming the joint.

The backward bending and contracting of the foot is called dorsiflexion, and such a movement is done by the 1 MTPJ as well, there are three main components of motion which can be separated during it: rolling motion, sliding motion and finally compression motion. During gait, the angle between metatarsal axis and phalangeal axis could have significantly values of 50-60 degrees, as shown in figure 3 [8].

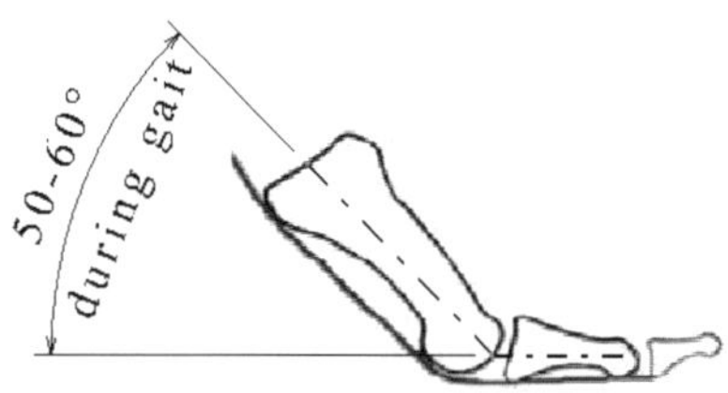


Figure 3. Dorsiflexion of the 1 MTPJ during gait.

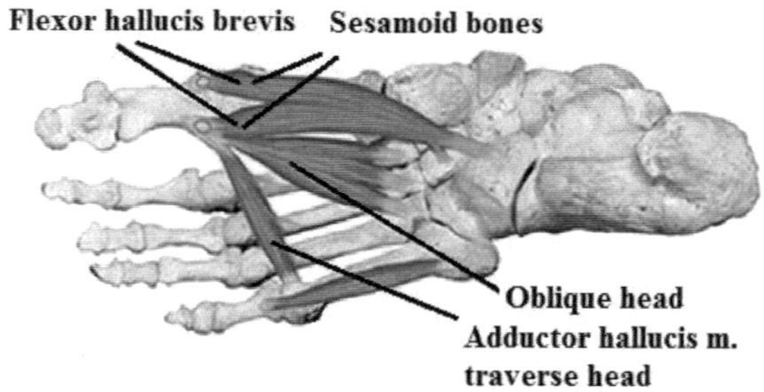


**Figure 4.** The main tendons which link the proximal phalanx. Illustrated and modified from [11].

The anatomy of the 1 MTP joint is a complex structure consisting from bones mentioned above but also from small ligaments and soft-tissue, attached to the head of the metatarsal (MTH) and phalanx. The first MTH is located into a concave cavity, surrounded by soft-tissue. The medial and lateral support is provided by the tendon attachments in order to supplement the ligamentous structures [2], [9]. The most important tendons which link the proximal phalanx are: Flexor hallucislongus (FHL), the adductor halluc is oriented in two directions (oblique and transverse) and finally the flexor hallucis brevis, which anchor the sesamoid bones to the foot [4]. The propulsion is provided with the required stability by the abductor hallucis and the flexor hallucis brevis, illustrated in figure 4 [2]. Several factors are required for the correct biomechanical function of the flexor hallux: stability and flexion of the assembly consisting of first metatarsal and first cuneiforms as a single bone structure (first ray), standard function of the sesamoid bones and adequate activity of the muscles which stabilize the 1 MTPJ [10].

## 3. Pathoetiology and development of hallux valgus

### *3.1. Epidemiology of Hallux Valgus*

The Hallux Valgus (HV) is a frequent pathology of the Hallux, with propensity towards feminine gender; it can be considered a manifestation of the flat forefoot. Based on recent studies, almost 1 out of every 10 people might be affected by this deformity with a women-men ratio of about 9:1 [12].

The leading factors of the Hallux Valgus are still unclear, but plenty of intrinsic and extrinsic aspects such as uncomfortable sharp shoes, overweight or static deformations of the foot inflict the apparition of it [13].

### *3.2. Genetic factors*

Some inherited characteristics have been presumed to predispose people to Hallux Valgus, plenty of genetic related features such as specific foot type and joint hypermobility might be relevant to HV [9]. Mom-to-child transmission has been found in over 90% of subjects with a family background, in terms of juvenile HV [14].

Racial differences have been found as well between black Africans and whites, with a slightly prevalence of the HV in whites. The ratio is estimated to 2:1 detrimental to the whites [9].

*3.3. Biomechanical factors*
The pathoetiology and development of HV is strongly related to the biomechanics of the first ray. During propulsion, it is strongly required a stable and congruent MTP joint, although there is no interference between tendon attachments and the first metatarsal. The metatarsophalangeal joint articulations have a condylar structure which occurs to being enough to keep from happening the sesamoid subluxation, in terms of joint reaction force of the metatarsosesamoid.

Subluxation is engaged by the unloading of the metatarsosesamoid joint and this may occur by raising the Metatarsal head by the laterally loaded body weight [9]. There are plenty of displacements which occur in a chain reaction in engaging the HV, which are illustrated in figure5regarding the adduction, abduction and flexion and in figure 6for the deviations and rotations.

According to Lelievre, the congenital adduction of the first Metatarsal bone (MT1), the hypermobility of the tarsometatarsal joint and the unusual length of the metatarsal bones are the main factors of the Hallux Valgus [15]. The abduction angle of the Hallux might reach 90° and the Hallux could mount over the second finger.

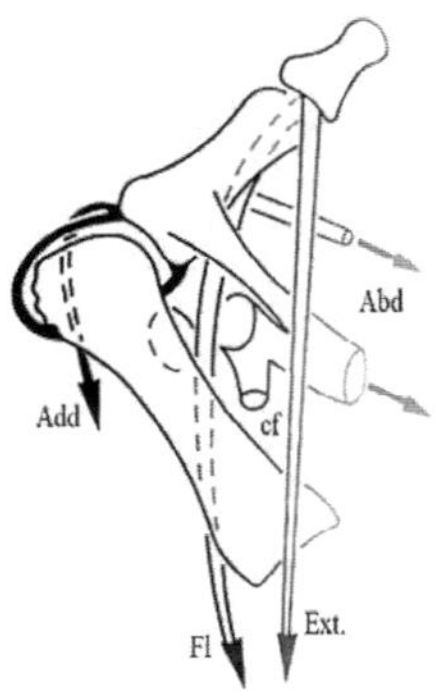


**Figure 5.** Adduction of MT1. Picture illustrated from [15].

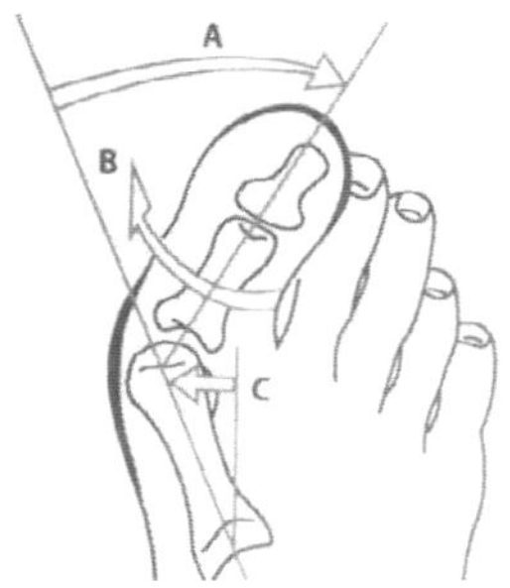


**Figure 6.** Valgus deviation (A), rotation of Hallux (B), Varus deviation of the 1st Metatars (C). Picture illustrated from [15].

*3.4. Other factors*
Factors such as trauma, joint diseases and some neurological conditions such as Parkinson's disease, which disturb the normal functioning of the foot, may occur to HV; being considered etiological elements [6]. Recent research articles claim that there is a bounding between the HV deformity and metatarsal pain by measuring and recording several parameters such as rate of loading, duration and area of contact, during dynamic gait [16].

Regarding the symptomatology of the deformity, in premature phases, patients feel a vague pain in the Metatarsophalangeal joint, but in evolved forms, the pain may expand up to the calves.

## 4. Approach of the Hallux Valgus deformity

*4.1. Hallux Valgus geometrical classification*
For a better assessment of the bunion condition, certain angles must be measured, such as: the Hallux valgus angle, the Intermetatarsophalangeal angle, the Interphalangeal angle, but there is also a series of observation items regarding the Tibialsesamoid position. The most important of them are the Hallux

Valgus Angle (HVA) and the Intermetatarsophalangeal Angle (IMA), which are determined as following:

- The Hallux Valgus Angle is defined bythe two longitudinal axes - one of the proximal phalanx and the second one of the first metatarsal. For healthy subjects the HVA is less than 15°, but for increased values, it characterizes a hallux valgus deviation.
- The Intermetatarsophalangeal Angle is between the longitudinal extension of the first and second metatarsals axis. The normal values of this angle are less than 10° [17, 18].

The mentioned angles (HVA and IMA) are presented in figure 7 alongside with their normal values.

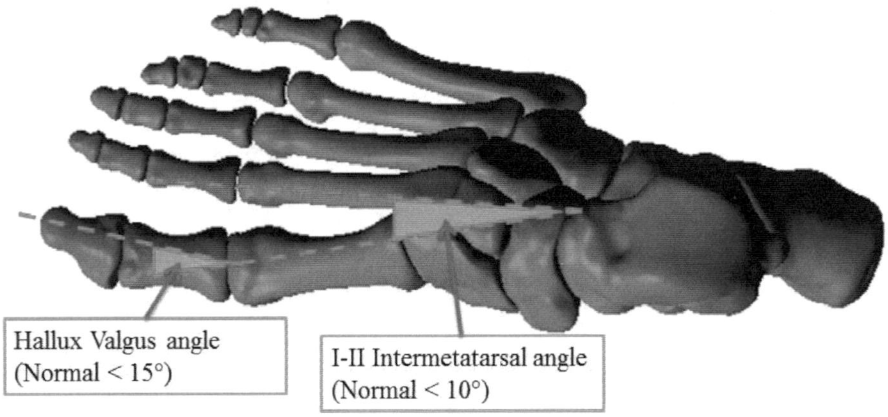


**Figure 7.** Normal values of the HVA and IMA.

The radiographic exam usually requires the basic AP (anteroposterior) and LL (lateral) x-rays, in order to state the Hallux Valgus type and for scheduling the eventual surgical correction.

In 1960, based on the congruity of the 1st MP joint, Pigott classified HV into 3 types:

- Congruous joint (1st Type)
- Deviated non congruous joint (2nd Type)
- Subluxated joint (3rdType), as shown in figure 8.

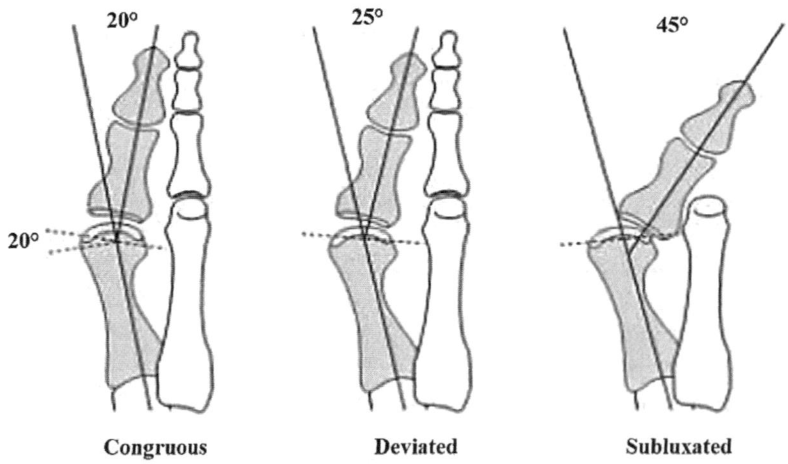


**Figure 8.** The three types of HV, based on Pigott. Picture illustrated from [19].

A second classification was made by LaPortain 1974, which was later issued in 1999. The deformity could be classified in three categories, starting from mild, with a lapse between 15 to 20 degrees, continuing with moderate, between 20 and 40 degrees and finally, severe above 40 degrees, relating to the Hallux Valgus Angle [20]. The LaPorta classification has it's benefits due to the fact that can classify the tibialsesamoid position relative to the first metatarsal bisector byassessingthedorso-plantar projection, ranging from normal to most laterally shifted [21].

In the early 2000s, researchers established the basic elements and values in hallux abducto valgus (HAV), establishing the first IMA as normal for values lower than 10°, mild for values ranging from 10° to 11°, moderate for angles between 11° to 16° and severe for values above 16° [22].

According to the latest research, the triplane method is the most accurate one relating to distinguishing the Hallux Abducto Valgus (HAV), besides the 2-dimensional algorithms, this method includes the frontal plane. This method takes into consideration the anatomic findings and the status of the MTPJ, concluding with recommendations regarding the treatment as well.

The first type (Class 1) has the following anatomical aspects: the HV and IM angles are increased but there is no evident pronation at the first metatarsal on AP X-ray or sesamoid axial X-ray, although, the sesamoids might be subluxed. Regarding the MTPJ status, the first class HV does not have any kind of medical or imagistic evidence of degenerative joint disease, and the algorithm recommend an osteotomy to the metatarsal or a correction on the Lisfranc joint.

The second class is divided into 2 subclasses (A and B), with the same MTPJ status and anatomical findings beside the metatarsal pronation and sesamoid status. Class 2A is described by evident pronation of the first metatarsal on AP X-rays but the sesamoids are not subluxed and the treatment recommendations are:triplane correction including the 1 MT inversions, with or without lateral capsulotomy.

The difference between class 2A and 2B is strictly related to the sesamoids status, which are subluxed in the second subclass. The treatment for this case is alike the previously mentioned one, but imply in addition before correction a conservative lateral capsular release.

The third class is distinguished from the others by being characterized with an increased Hallux Valgus angle and metatarsus adductus over 20°. Its main treatment recommendations consist in plane correction of the second and third metatarsal, Lisfranc correction or metatarsal osteotomy.

The final class from this classification has medicalor imagistic evidence of degenerative joint disease and increased values of Hallux Valgus and intermetatarsal angle but there is no pronation of the first metatarsal. The surgical recommendation for the 4$^{th}$ class consists in arthrodesis of the first metatarsophalangeal or even joint arthroplasty.

### *4.2. Geometrical references*

In order to make a correct biomechanical approach from dimensional and geometrical point of view, certain pieces of information regarding the axis and planes of the human body must be presented.

The most frequently used system in the body biomechanics is the triortogonal system, which consists of 3 planes, each one mutually perpendicular, as following: the sagittal plane, the transversal plane and the frontal plane, illustrated in figure 9.

The sagittal plane is a vertical plane which crosses the body, dividing it into two perfectly symmetrical sides, reason for what it is called the bilateral symmetry plane. The two mentioned parts are left and right.

The secondary plane is horizontal, and it is called transversal plane. This plane crosses the body on his height midline, dividing it into to asymmetrically parts, superior or cranial and inferior or caudal.

The third plane is vertical, positioned parallel with the forehead, and it's called frontal. It divides the body into anterior (front) and posterior (back) parts.

The intersection of these planes is defining the axis of the human body. Therefore, the intersection of the sagittal and transversal describes the sagittal axes, with two poles, one anterior and one posterior, describing the body's thickness. The transversal line is described by the intersection of the transversal and fontal plane, with two opposite planes, left and right, describing the width of the body.

The third line is the result of the sagittal and frontal plane intersection, named the longitudinal axis, having a superior and inferior pole. The position of the superior pole is on the top of the head and the inferior one is located on the sole [24, 25].

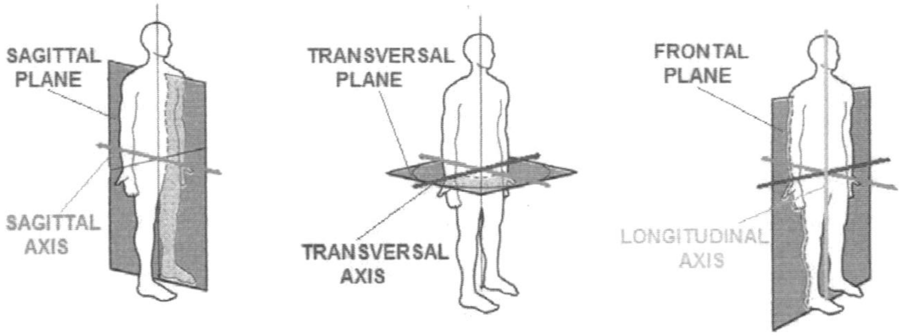


**Figure 9.** Axes and planes of the human body.

Obviously, the intersection of the three axes previously defined, describes the origin of the human body reference system, which concur with its symmetry center.

Other important terminology must be defined for a better comprehension of the problematic, this terms are found in specialized literature are very useful in geometrical describing: proximal – for elements located closer to the joints and distal for elements positioned further from them, or volar for describing hand elements and plantar for the foot ones.

Those geometrical elements have their usefulness in dimensional and geometrical defining of bones, and especially for describing their movements, translations and rotations, and even for some displacements owed by pathological conditions.

Relatively to this main reference system of the human body, it can be defined by translating and rotating movements, secondary coordinate systems, which are necessary for the required biomechanics elements studied. For example, for a foot study, it can be defined in an incipient phase a coordinate system by moving the main one to the ground surface. The origin of the new system is the projection of the main one on the sole, having its planes and axis parallel to the body's. A more comprehensive view of the system is illustrated in figure 10.

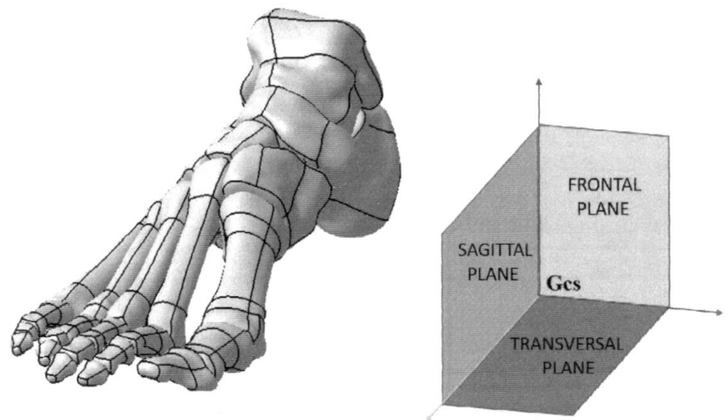


**Figure 10.** The ground reference coordinate system (Gcs) of the foot.

Starting from this approach, for a better understanding of the displacements that cause the Hallux Valgus deformity, we decided to place several coordinate systems in points which we consider important. Those coordinate systems are associated with the human reference planes previously presented.

The main reference coordinate system is located on the frontal and sagittal planes and the XoZ plane is associated with the ground system (Gcs), as presented in the previously image [24].

A first coordinate system will be placed in the talocrural joint (ankle) and it will be considered the main reference and all the movements will be associated with it. This system will be further referred to as Talocrural Coordinate System - Tcs. Due to the fact that the Hallux Valgus Angle is crucial in the decision making regarding the surgical method, we considered proper to place another system in the intersection of the longitudinal axis of the first metatarsal and the extension longitudinal axis of the second metatarsals, which we named Angular Coordinate System - Acs. The final and the most relevant to the subject in discussion, is the Displacement Coordinate System - Dcs, placed on the metatarsophalangeal joint, exactly where the whole pathological condition has its origins. The three coordinate systems are presented more specific in figure 11:

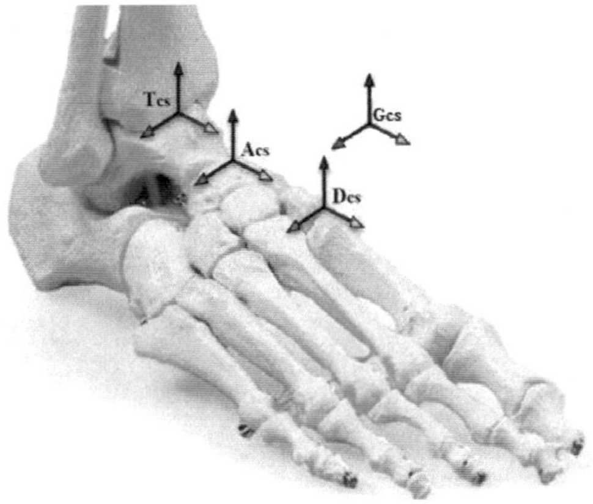


**Figure 11.** Setting up the 3 main coordinate systems:
Tcs, Acs, Dcs, relatively to the Gcs.

### *4.3. FEA considerations*

The Finite Element Analysis is an Engineering approach of the medical issue, established for the simulation of the behavior of a part or an assembly under specific conditions, in our case, this element plays a crucial role in developing and designing adequate models. Plenty of data is needed to begin such a study, such as mechanical properties of the bones, loads and tension state etc.

Due to the fact that the structure of the bone is complex, engineers had to make several researches to determine the most precise pattern that provided its properties and roles.

In terms of biomechanical research, lately was adopted in clinical applications the subject specific finite element analysis of the skeleton. Reproducing the subject specific models by processing the information from the X-ray images is one of the most popular methods. This method is based on interpreting the resulted computer tomography images' grey tones using mathematical relationship between them and their mechanical correspondent. This process will lead to obtaining important data regarding bones' mechanical properties [26, 27].

Due to the fact that the bones affected by the studied deformity are mostly trabecular bones, their properties will be presented in brief.

Considering the fact the real bone structure is not homogenous, having rough areas (cortical) and trabecular ones (with different densities) the mechanical characteristics such as Young's Modulus must be correlated with them. This specific action is necessary for performing correct analysis regarding the bone, or the bone-implant assembly [28]. Accurate determination has a high difficulty,due to the fact that the heterogeneity and because of the trabecular struts, which are very small sized. In order to establish a Young's modulus estimation, plenty of methods were used, regarding the testing of single trabeculae from mechanical point of view, nanoindentationor ultrasonic testing [29, 30].

Another important aspect which must be taken into consideration is the load status of the foot, during posture and gait. Clinical applications require a good comprehension of the functional biomechanics of the foot and ankle, information which is also vital for an accurate engineering approach [31].The procedure's workflow has 3 steps:

- Data gathering of 3D foot kinematics and kinetics during gait;
- 3Dmodelling;
- Finite Element Analysis of internal stresses and loads [32].

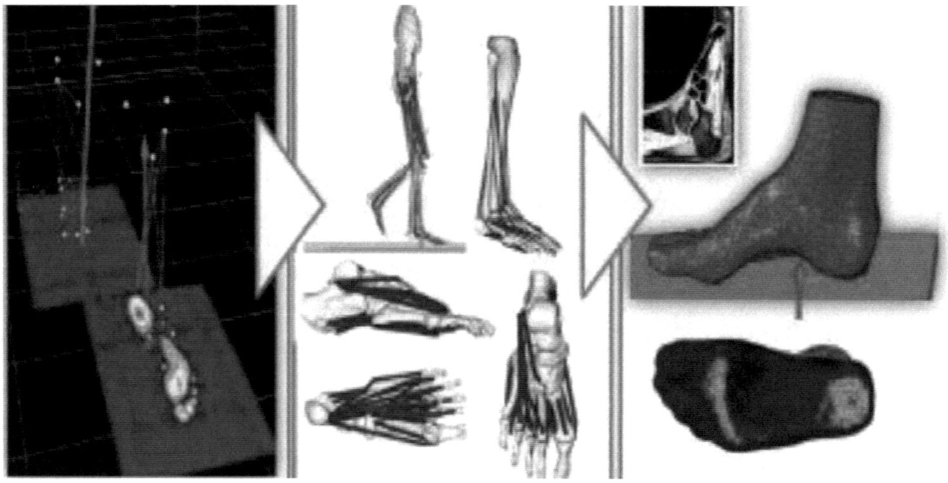

**Figure 12.** Workflow of the procedure. Picture illustrated from [32].

### 4.4. *Principles of treatment*

The suitable treatment methoding is picked according to the severity of the displacement. Mild cases are befitting for non operative treatments, the so called conservative treatments, but more severe one require surgical interventions. Therefore, a much more efficient solution for mild cases of Hallux Valgus is the conservative alternative, due to the fact that the patient can return to his activities much faster and it's less pricy [33]. Special footwear, orthotic and plenty of exercises are the main techniques involved by this method. The main exercises are for strengthening and mobilization of the affected elements, which are performed mostly in the sitting position by the patient, for a better overview, such a exercise program will be exemplified: abducting the hallux, which literally involves taking far away the hallux from the 2$^{nd}$ phalanx [34, 35].

Although these conservative methods could have effects only for the mild cases, the severe types of HV were assigned by Mann with their suitable surgery. The following paragraphs present the soft-tissue, bony and combined soft-tissue-bony Hallux Valgus surgical treatment procedures. In order to pick the suitable surgical method of treatment of the HV, Mann described a remarkable algorithm, which was recently was transmuted into a more easily comprehensive scheme, recreated in figure 13 [21, 36].

The decision making starts from the status of the 1 MTPJ, which can be congruent, incongruent or arthritic. In the first case, if the Hallux valgus interphalangeus (measured between long axis of distal phalanx and proximal phalanx) is above 10° will lead to a Akin osteotomy, if the IM angle is above 15° but the HV angle is under 45°, the suitable treatment will be an Chevron or Scarf osteotomy.

In the case of a incongruent first MTP joint, alongside with an 10-15° IM angle and the value of the HVA under 30°, the only surgical treatment shall be the Chevron osteotomy, for values over 15° of the IMA and under 45° regarding de HVA, the treatment will be strictly the scarf or chevron osteotomy. Besides of all the mentioned findings, if the patient has an arthritic first tarsometatarsal joint, it should be taken into consideration a modified lapidus procedure.

Finally, in case of an arthritic 1st MTP joint, the treatment shall be a fusion of the first MTP joint.

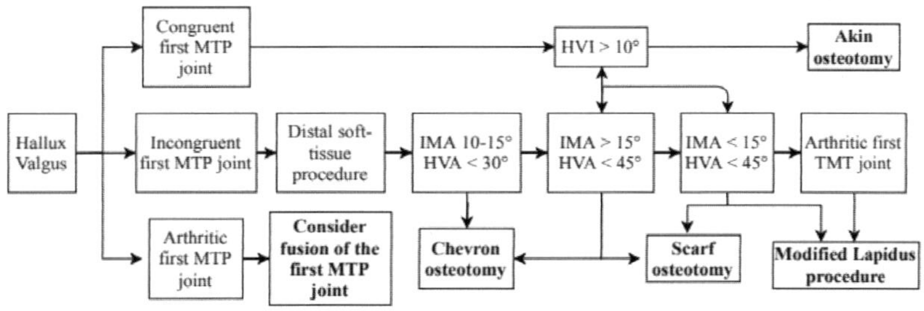


**Figure 13.** Hallux Valgus Treatment - reissuedafter Mann. [21], [36], [37].

A thorough review of the surgical techniques has been done, focusing on the osteotomy procedures since their suitability for engineering approaches. We have chosen to present the scarf osteotomy procedure to illustrate the necessary steps of such a surgical treatment, because, as illustrated in the scheme, is the only one suitable for both moderate and severe deformities, and those are the main types of HV which require surgical interventions. The mentioned treatment method is made at the diaphyseal level of the bone aided by a horizontally directed displacement osteotomy Z-shaped. The term “scarf” refers to jointing of two pieces’ends and this specific medical procedure is illustrated in figure 14 and has the following steps:

1. Osteotomy cuts;
2. Lateral displacement;
3. Repositioning the bone parts for correcting the deviation;
4. Fastening of the osteotomy with two minifragment screws [36].

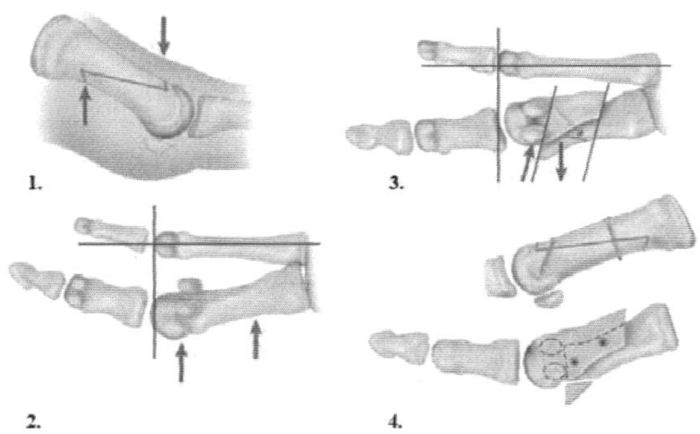


**Figure 14.** The five steps of the scarf osteotomy procedure.
Picture illustrated from [36].

Besides the scarf osteotomy there are also other procedures which are preformed, such as: the crescentic, chevron and Ludloff, osteotomies. Those are done when the patient is no suitable for the scarf osteotomy; these types of situations are less frequent. The scarf osteotomy, alongside with the other three interventions’ brought up are presented in figure 15in terms of osteotomy cuts and fastening methods.

For the proximal crescentic osteotomy (1), the cut is done parallel with the frontal plane, as for the proximal chevron osteotomy (2), the cut is L-shaped, both requiring a single screw for the fastening. The Ludloff osteotomy (3) has an angular cut and requires two screws, displaced in converse directions. Finally, the scarf osteotomy (1) demands a Z-shaped cut of the bone, the fastening being done with two minifragment screws [36].

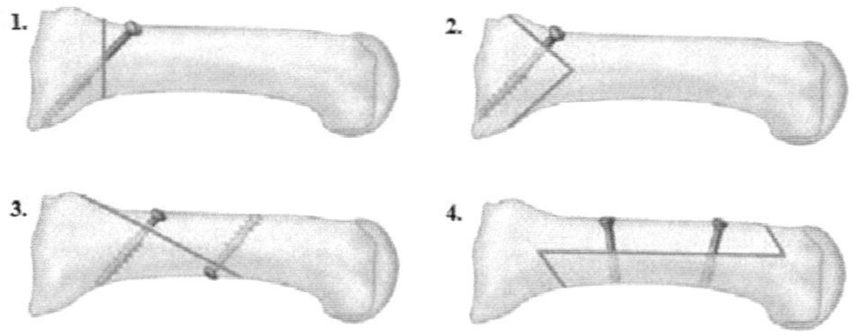


**Figure 15.** Metatarsal shaft osteotomies.
Picture illustrated from [36].

Proximal osteotomy is also necessary in most hallux valgus deformities that require a distal soft-tissue procedure but this procedure is not recommended if excessive valgus posturing, above the HVA angle 15 degrees [36, 38].

## 5. Conclusions

As noted, this article aims a detailed bibliographic research regarding a very frequent medical condition, Hallux Valgus. Studying, synthesizing and structuring the data regarding the axial deviation of the foot are very important for several reasons:

- The first reason is that the biomechanics of the deformity and the afferent treatment methods are an interdisciplinary field which involves in-depth studies. The interest area which is referred to is in relation to analytical approaches, alongside with CAD-CAE research regarding geometric, dimensions, reference systems, 3Dmodelling systems or the applied simulation technologies for the Hallux Valgus research, in the event of prosthesis and necessary surgical interventions;
- The second reason is the favourable social impact that the study will have, considering that this deformity causes unbearable pain alongside with the high frequency of it in women and older people;
- Another reason is linked to the increasing surgical accuracy by training the doctors with surgery simulations on virtual or 3D printed models;
- The final reason is the interdisciplinary research of the two fields, medical and engineering. The principle followed for this approach is that "the whole is much more than the sum of its parts", which will lead to superior results.

Analyzing the researches presented in this paper, which are the start point of an ample research, we structured a number of certain research objectives, as following:

- Designing the 3D models of the bones affected by the Hallux Valgus deformity;
- Elaboration of a generalized model with the mentioned bones, assembled with constraints, in Skeleton systems, for healthy subjects and for subjects affected by Hallux Valgus;
- Accomplishing 3D simulations that include all the geometrical and dimensional possibilities of the deformity;
- Designing by Rapid Prototyping of several physical models, starting from the considerations previously mentioned, with the purpose to submit them to mechanical testing;
- Parameterized 3D modelling of the surgical strategies, alongside with their afferent simulations;
- Finite Element Analysis of the deformity and the treatment methods as well;

- Specific experimental research.

We consider that the approached field is interesting and there is plenty of potential in accomplishing high interdisciplinary research results.

**References**

[1] Robinson A, Bhatia M, Eaton C and Bishop L 2009 Prospective comparative study of the Scarf and Ludloff osteotomies in the treatment of hallux valgus *Foot Ankle Int.* **30** 955
[2] Valmassy R L 1996 *Clinical biomechanics of the lower extremities* (St. Louis [u.a.]: Mosby) p 1
[3] Alazzawi S, Sukeik M, King D and Vemulapalli K 2017 Foot and ankle history and clinical examination: A guide to everyday practice *World J. Orthop* **8** 21
[4] Ferner H and Staubesand J 1982 *Sobotta atlas of human anatomy (10th ed.)* (In Ferner Helmut bStaubesand Jochen (Eds.) Munchen: Urban & Schwarzenberg) p 327
[5] Fraissler L, Konrads C, Hoberg M, Rudert M and Walcher M 2016 Treatment of hallux valgus Deformity *EFORT Open Reviews* **1** 295
[6] Hansen S T 2000 *Functional reconstruction of the foot and ankle* (In Hurley R, Seigafuse S L and Marino-Vasquez D (Eds.), Philadelphia u.a: Lippincott Williams & Wilkins) p 1
[7] Haines R W and McDougall A 1954 *The anatomy of hallux valgus* (Lancet) p 1308
[8] Klemola T 2018 *Flexible Hallux Valgus. Results of a new surgical technique* (University of Oulu, Faculty of Medicine; Medical Research Center Oulu; Oulu University Hospital) p 19
[9] Perera A M, Mason L and Stephens M M 2011 The pathogenesis of hallux valgus *J. Bone. Joint. Surg* **93** 1650
[10] Root M L, Orien W P and Weed J H 1977 *Forefoot deformity caused by abnormal subtalar joint pronation* (Los Angeles: Clinical Biomechanics Corporation) p 349
[11] Wesker K 2007 *Thieme Atlas of Anatomy: General Anatomy and Musculoskeletal System* (New York, Thieme, Stuttgart. Copyright (2005) by Georg Thieme Verlag)
[12] Xu H, Jin K, Fu Z, Ma M, Liu Z, An S and Jiang B 2015 Radiological Characteristics and Anatomical Risk Factors in the Evaluation of Hallux Valgus in Chinese Adults *Chin. Med. J* **128**
[13] Donatelli R 1985 Normal Biomechanics of the Foot and Ankle *J. Orthop Sports. Phys. Ther.* **7** 91
[14] Coughlin M J and Jones C P 2007 Hallux valgus: Demographics, etiology, and radiographic Assessment *Foot Ankle Int.* **28** 759
[15] Pop S P, Gergely I, Russu O M and Roman C O 2013 *Elemente de ortopedie (Elements of orthopedics)* (ed. II, ISBN 978 973-169-259-3)
[16] Cheng C, Wang Q F, Guo J C, Li D D, Fan Y B and Wen J M 2019 The biomechanical relationship between Hallux Valgus Deformity and Metatarsal Pain *Journal of Healthcare Engineering* **2020**
[17] Xu H, Jin K, Fu Z, Ma M and Liu Z 2015 Radiological Characteristics and Anatomical Risk Factors in the Evaluation of Hallux Valgus in Chinese Adults *Chinese medical journal*
[18] Bryant A, Tinley P and Singer K 2000 A comparison of radiographic measurements in normal, hallux valgus, and hallux limitus feet. *J. Foot Ankle Surg.* **39** 39
[19] Gui J C, Hou M F, Shen H Q, Wang L M, Gu X J and Ma X 2001 X-ray evaluation of the normal and hallux valgus feet and it's clinical values (in Chinese) *Chin. J. Orthop* **21** 137
[20] Laporta G, Melillo T and Olinsky D 1974 X-ray evaluation of hallux abducto valgus deformity *J. Am. Podiatr. Med. Assoc.* **64** 544
[21] Mann R A and Coughlin M J 1999 *Adult hallux valgus. in Surgery of the foot and ankle (7$^{th}$ ed.)* (St. Louis: Mosby) p 159
[22] Condon F, Kaliszer M, Conhyea D, O'Donnell T, Shaju A and Masterson E 2002 The first intermetatarsal angle in hallux valgus: an analysis of measurement reliability and the error involved *Foot Ankle Int.* **23** 717

[23] Hatch J D, Santrock D R, Smith B, Dayton P andWeil Jr. L 2018 Triplane Hallux Abducto Valgus Classification *J. Foot Ankle Surg.* **57** 972

[24] Cofaru I I 2013 *Cercetări privind biomecanica deviațiilor axiale ale membrului inferior uman și dezvoltarea unor echipamente chirurgicale aferente (Research on biomechanics axial deviations of the member human inferior and the development of some related surgical equipment) (*PhD thesis, Lucian Blaga University of Sibiu)

[25] Glasoe W, Phadke V, Pena F, Nuckley D and Ludewig P 2013 An Image-Based Gait Simulation Study of Tarsal Kinematics in Women With Hallux Valgus *Phys. Ther.* **93** 10.2522/ptj.20130025

[26] Viceconti M, Olsen S, Nolte L P and Burton K 2005 Extracting clinically relevant data from finite element simulations *Clin. Biomech.* **20** 451

[27] Kaneko T S, Bell J S, Pejcic M R, Tehranzadeh J and Keyak J H 2004 Mechanical properties, density and quantitative CT scan data of trabecular bone with and without metastases. *J. Biomech.* **37** 523

[28] Helgason B, Perilli E and Schileo E 2007 Mathematical relationships between bone density and mechanical properties: A literature review *Clin. Biomech.* **23** 135

[29] Wu D, Isaksson P, Ferguson S J and Persson C 2018 Young's modulus of trabecular bone at the tissue level: A review *Acta Biomater.* **78** *1*

[30] Suzuki T, Matsuura Y, Yamazaki T and colab. 2019 Biomechanics of callus in the bone healing process, determined by specimen-specific finite element analysis *Bone* **132** 115

[31] Horwood A 2019 The biomechanical function of the foot pump in venous return from the lower extremity during the human gait cycle: An expansion of the gait model of the foot pump *Med. Hypotheses* **129** 109

[32] Scarton A, Guiotto A, Malaquias T, Spolaor F, Sinigaglia G, Jonkers I, Cobelli C and Sawacha Z 2017 A methodological framework for detecting ulcers' risk in diabetic foot subjects by combining gait analysis, a new musculoskeletal foot model and a foot finite element model *Gait and Posture* **60** 279

[33] Ferrari J, Higgins J and Prior T D 2004 Interventions for treating hallux valgus (abductovalgus) and bunions *Cochrane Database Syst. Rev.*: CD000964

[34] Srivastava S, Chockalingam N and El Fakhri T 2010 Radiographic angles in hallux valgus: comparison between manual and computer assisted measurements *J. Foot Ankle Surg.* **49** 523

[35] Gul O K, Nilgun B and Ugur T 2015 Short-term effects of the kinesiotaping of pain and joint alignment in conservative treatment of the hallux valgus *J. Manipulative Physiol. Ther.* **38** 564

[36] Canale S T, Azar F M, Beaty J H and Campbell W C 2017 *Campbell's operative orthopaedics (Thirteenth edition*, Philadelphia, PA : Elsevier, Inc.)

[37] Fraissler L, Konrads C, Hoberg M, Rudert M and Walcher M 2016 Treatment of hallux valgus Deformity *EFORT Open Rev* 2016;1:295-302.

[38] Chapman M W 2001 *Chapman's orthopaedic surgery* (Philadelphia: Lippincott Williams Wilkins)

IOP Conf. Series: Materials Science and Engineering 1009 (2021) 012016 doi:10.1088/1757-899X/1009/1/012016

# The optimization of cutting parameters at turning, using electrical current at cutting, in order to increase the durability of the cutting tool

**R Daicu[1]* and V Dițu[1]**

[1]Manufacturing Engineering Department, Transilvania University of Brașov, Eroilor nr. 29, 500036 Brasov, Romania

*E-mail: raluca.daicu@unitbv.ro

**Abstract**. The optimization of the turning cutting parameters was a permanent concern for research but also for production. The emergence of new metal materials to be processed as well as the development of new cutting edges has always led to new optimizations of the cutting regime. The appearance of coated plates, single layer or multilayer, determined the companies producing them to indicate information about the cutting parameters, but the values are given in a certain interval. In order to reduce the necessary experiments to indicate the optimal cutting regime, the paper proposes the analysis of the electrical current at cutting and its use for optimization. The processing of the experimental data obtained as well as the optimization of the cutting parameters was performed using Design Expert software.

## 1. Introduction

In case of machining metals with cutting tools that conduct electricity, heat is released in the cutting area, which leads to a higher temperature than the cold end of the cutting tool. This causes the appearance of the electrical current at cutting, which in the literature is also known as thermocurrent.

If the temperature difference between the cutting area and the cold end of the cutting tool increases, the current voltage also increases [1-8].

In research [9] it is shown that when drilling metals with high speed steel drills, there is a dependence between the quality of the drill edge and the voltage of the occurred thermocurrent. Research [10] proposes an algorithm for acquiring drills depending on the size of the thermocurrent voltage that occurs when cutting metals.

Research [11-13] present several approaches for using the thermocurrent that occurs during turning process, involving determination of quality differences between cutting edges and an algorithm for acquiring cutting inserts.

The mentioned previous research attests the direct connection between the voltage of the thermocurrent and the imperfections of the cutting edge. The more and larger these imperfections, the higher the temperature in the cutting area, which leads to an increased voltage and finally to a low durability of the cutting edge.

Based on the presented facts, this paper proposes the use of the thermocurrent for optimization, by analyzing the influence of the cutting parameters.

## 2. The influence of the cutting regime on the thermocurrent value when turning with coated plates

To study the influence of the cutting regime on the thermocurrent value as well as to optimize its values in order to increase the durability of the tool, it was used a three-factor model, each having three levels, according to table 1, and the response was introduced as the thermocurrent voltage.

**Table 1.** Input parameters for the complete factorial model.

| Factors | Name | Unit | Levels | Level 1 | Level 2 | Level 3 |
|---|---|---|---|---|---|---|
| A | Cutting speed | m/min | 3 | 86 | 107 | 137 |
| B | Cutting feed | mm/rev | 3 | 0.151 | 0.208 | 0.25 |
| C | Cutting depth | mm | 3 | 1 | 1.5 | 2 |

Having three factors with three levels each, according to the complete factorial plan, the number of experiments is, $3^3$, which means 27 experiments.

### *2.1. Performing experiments*

For the experiments it was used the configuration presented in figure 1, defined by the next elements: 1 – lathe; 2 – part; 3 – cutting tool;4 – collector for the electrical current; 5 – multimeter; 6 – computer.

**Figure 1.** The configuration used to measure the voltage of the thermocurrent.

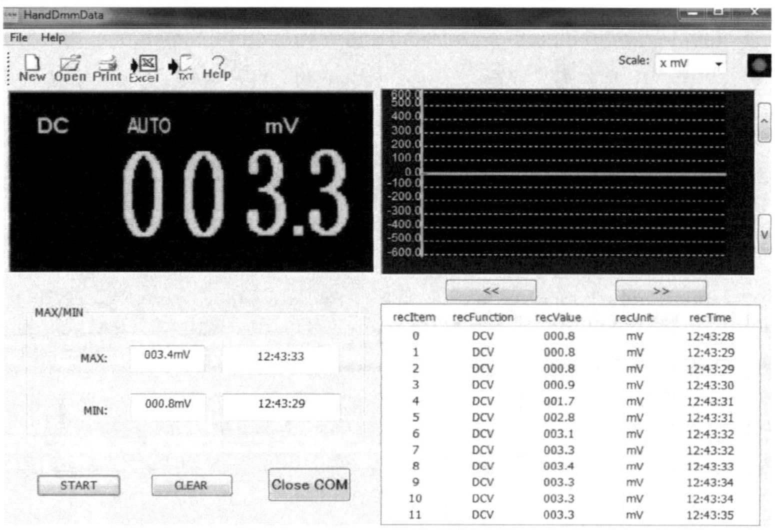


**Figure 2.** The interface of the digital multimeter.

**Table 2.** Experimental results using the complete factorial model.

| No. of experiment | Speed [m/min] | Feed [mm/rev] | Depth [mm] | Voltage [mV] |
|---|---|---|---|---|
| 1 | 86 | 0.208 | 2 | 2.6 |
| 2 | 137 | 0.151 | 2 | 3.5 |
| 3 | 137 | 0.151 | 1.5 | 3.8 |
| 4 | 107 | 0.208 | 2 | 3.4 |
| 5 | 137 | 0.25 | 2 | 5 |
| 6 | 86 | 0.208 | 1.5 | 3 |
| 7 | 107 | 0.208 | 1 | 3.2 |
| 8 | 107 | 0.151 | 2 | 3.3 |
| 9 | 137 | 0.151 | 1 | 3.9 |
| 10 | 86 | 0,25 | 1.5 | 3.2 |
| 11 | 107 | 0.151 | 1.5 | 3.4 |
| 12 | 137 | 0.25 | 1 | 3.5 |
| 13 | 137 | 0.208 | 1 | 3.2 |
| 14 | 107 | 0.25 | 1 | 3.5 |
| 15 | 86 | 0.25 | 1 | 3.1 |
| 16 | 107 | 0.208 | 1.5 | 3.2 |
| 17 | 107 | 0.25 | 2 | 3.7 |
| 18 | 107 | 0.25 | 1.5 | 5.5 |
| 19 | 107 | 0.151 | 1 | 3.3 |
| 20 | 137 | 0.208 | 2 | 4.8 |
| 21 | 86 | 0.25 | 2 | 3.5 |
| 22 | 137 | 0.25 | 1.5 | 4.6 |
| 23 | 86 | 0.208 | 1 | 2.9 |
| 24 | 86 | 0.151 | 1 | 3.3 |
| 25 | 137 | 0.208 | 1.5 | 3.9 |
| 26 | 86 | 0.151 | 1.5 | 2.9 |
| 27 | 86 | 0.151 | 2 | 2.4 |

With the help of the configuration presented in figure 1, the 27 experiments that are centralized in table 2 were performed.

The experiments were performed when turning C45 steel with titanized metal carbide plate, type WNGB080408.

The multimeter is used as an acquisition system and it has a sample rate of three values/second. The acquisitioned values can be visualized in real-time, as presented in figure 2, and they also can be saved.

*2.2. Data processing. ANOVA Analysis.*

To see how each parameter of the cutting regime influences the value of the thermocurrent voltage, an ANOVA analysis was performed using Design Expert version 11. Also, experiments no. 18 and 21 were eliminated from the analysis. They were identified as aberrant values by studying DFFIT indicator.

The final results for ANOVA are presented in table 3, and table 4 shows the resulted statistical indicators.

**Table 3.** ANOVA analysis.

| Source | Sum of squares | Freedom degrees | Mean square | F Value | P value |
|---|---|---|---|---|---|
| Regression | 8.39 | 7 | 1.20 | 24.71 | < 0.0001 |
| A-Speed | 5.02 | 1 | 5.02 | 103.44 | < 0.0001 |
| B-Feed | 0.0034 | 1 | 0.0034 | 0.0708 | 0.7933 |
| C-Depth | 0.6865 | 1 | 0.6865 | 14.15 | 0.0016 |
| AB | 0.1596 | 1 | 0.1596 | 3.29 | 0.0874 |
| AC | 1.24 | 1 | 1.24 | 25.59 | < 0.0001 |
| BC | 0.0094 | 1 | 0.0094 | 0.1934 | 0.6656 |
| ABC | 0.2382 | 1 | 0.2382 | 4.91 | 0.0407 |
| Residual | 0.8249 | 17 | 0.0485 | | |
| Total | 9.22 | 24 | | | |

In order to interpret the results, there were used the specifications provided by Design Expert software.

The row of the Regression source shows how much of the variation of the response is explained by the model, the rest of the variation left unexplained being shown in the Residual row, the total variation around the average of the observations being calculated in the Total row.

The F value represents, in case of ANOVA analysis, the value of the effect and it is calculated as the ratio between the square mean of the analysed parameter and the residual square mean. For the analysed model, the F value of 24.71 represents that it is statistically significant.

It can also be seen that the cutting speed has the greatest influence on the voltage of the electrical current, being followed by depth and feed. But to find out if the model parameters have statistical significance on the analysed response, the p value, which must be less than 0.05, is analysed. In this case, the cutting speed and the cutting depth parameters are significant while the cutting feed has no statistical significance on the electric current voltage value.

**Table 4.** Statistical indicator resulted from ANOVA analysis.

| Indicator | Value |
|---|---|
| Standard deviation | 0.2203 |
| Mean | 3.46 |
| Coefficient of variation % | 6.36 |
| R2 | 0.9105 |
| Adjusted R2 | 0.8737 |
| Predictable R2 | 0.7507 |
| Signal to noise ratio | 20.9610 |

Predictable $R^2$ measures the variation of newly introduced experimental data, which can be explained by the current model. In order for the model to be used for optimizations and predictions regarding the variation of the response to new values of the input parameters, the difference between adjusted $R^2$ and predictable $R^2$ must be less than 0.2, which in this case it is fulfilled.

The signal-to-noise ratio must be greater than 4 for the model to be used in predictions. This requirement is also fulfilled.

The resulted linear mathematical function that models the experimental data is:

$$U = -0.276 + 0.0514202 * v_c + 27.53036 * f + 1.87008 * a_p - 0.34882 * v_c * f - 0.032826 * v_c * a_p - 23.22924 * f * a_p + 0.29698 * v_c * f * a_p \quad (1)$$

where:
U [mV] – thermocurrent voltage;
$v_c$ [m/min] – cutting speed;
f [mm/rev] – cutting feed;
$a_p$ [mm] – cutting depth.

## 3. The optimization of the cutting regime, using the thermocurrent, in order to increase the durability of the cutting tool

As it was mentioned in the introduction, a direct relationship is created between the wear of the cutting edge and the value of the thermocurrent voltage. For a high value of the voltage, the wear intensity will be high, so the durability will be low.

In this way the criterion chosen to increase the durability of the cutting tool is to minimize the value of the thermocurrent voltage. The solution generated by the previously created experimental model is presented in figure 3.

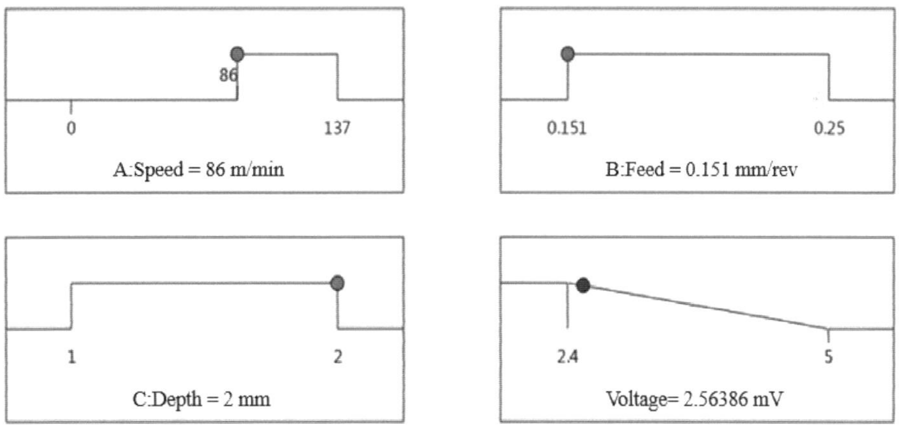


**Figure 3.** The values proposed for the cutting parameters to obtain the minimum voltage.

## 4. Conclusions

A first conclusion consists in the fact that the electric thermocurrent that appears when processing by cutting the nets with good electrically conductive edges can be used to optimize the cutting regime.

In this paper, it was determined how the cutting regime influences the thermocurrent voltage of the cutting by performing experiments according to a complete factorial plan. Using Deign Expert and using ANOVA analysis it was demonstrated that the cutting speed has the greatest influence on the voltage of the electrical current, being followed by depth and feed.

The obtained linear dependence was used to optimize the cutting regime in order to minimize the thermocurrent voltage, which leads to maximizing the durability of the cutting edge.

The values of the cutting regime that lead to the minimization of the thermocurrent voltage obtained when turning C45 steel with titanized metal carbide plate type WNMG080408 are: $v_c$ = 86 m/min, f = 0.151 mm/rev and ap = 2 mm.

**References**

[1] Daicu R and Oancea G 2015 *Appl. Mech. Mater.* **808** 40
[2] Bagchi H and Basu S K 1973 *Wear* **26** 39
[3] Plotnikov A L, Krylov E G and Frolov E M 2009 *STIN* **1** 2
[4] Murata M, Kurokawa S and Ohnishi O 2013 *CSME* **79** 2546
[5] Murata M, Kurokawa S, Ohnishi O, Uneda M and Doi T 2012 *CSME* **6** 958
[6] Hirota H, Murata M, Kohima A and Yanagihara K 1998 *JSPE* **64** 1191
[7] Dițu V 2010 *AJME* **8** 18
[8] Tanaka R, Hosokawa A, Yamada K and Ueda T 2004 *Proceedings of the Seventh International Conference on Progress of Machining Technoogy* **786**
[9] Dițu V, Oancea G and Daicu R 2017 *MATEC Web Conf.* **94** 02005
[10] Daicu R, Dițu V and Oancea G 2017 *MATEC Web Conf.* **112** 01023
[11] Nurul Amin A K M, Sarker M R A, Ahmed M, Karim A N M and Mater J 1998 *Process. Technol.* **77**
[12] Dițu V and Lepădătescu B 2013 *WSEAS* **12**
[13] Daicu R, Dițu V and Oancea G 2017 *MATEC Web Conf.* **137** 03004

# The determination of the relation between the drill diameter and the correction coefficient of the relative thickness of the drill core when calculating the axial force

**R Daicu[1]* and V Dițu[1]**

[1]Manufacturing Engineering Department, Transilvania University of Brașov, Eroilor nr. 29, 500036 Brasov, Romania

*E-mail: raluca.daicu@unitbv.ro

**Abstract**. When drilling metals, the determination of the axial force is necessary for multiple applications. In the literature the algorithm for calculating the axial force is quite well presented but there is an exception, namely, the correction coefficient of the axial force in relation to the relative thickness of the drill core where several iterations are needed. To determine the relationship between the diameter of the drill and the correction coefficient of the axial force, we start from the expression of the axial force developed based on the hypothesis of polytropic compression of the material to be processed where the correction coefficients are applied when practical situations require it. From the methodology of calculating the correction coefficient in relation to the relative thickness of the drill core it is found that there are three stages. In this paper we eliminate two steps by determining the relationship between the diameter of the drill and the correction coefficient for the relative thickness of the drill core and also improves the calculation both by developing a relationship that can be easily applied and by building a graph that helps upon rapid identification of the correction coefficient.

## 1. Introduction

The knowledge of the forces in metal cutting is particularly important due to the multitude of applications for which they are used. Since the beginning of metal cutting, researchers have paid special attention to estimating cutting forces by developing calculation relations, tables and graphs useful for practice. The sources [1-7] are just a few examples where relations for cutting forces are presented.

In paper [8] is presented a device for smooth entrance in cutting of the drill. Both for its design and for its reengineering, the axial force at drilling is needed to dimension the springs that are part of it. Research [9] and [10] are examples that show the interest for the cutting forces at drilling.

The source [1] presents, in detail, the methodology for determining the axial force at drilling and the relationship (1) was developed based on the hypothesis of polytropic compression of the material to be processed.

$$F = C_F \times D^{x_F} \times f^{y_F} \times K_F \times 10 \text{ [N]} \quad (1)$$

where:
D [mm] – drill diameter;
f [mm/rev] – cutting feed;
$K_F$–compound correction coefficient.

$$K_F = K_{a_F} \times K_{sa_F} \times K_{K_F} \times K_{\eta_F} \quad ([1]) \tag{2}$$

where:

$K_{a_F}$–sharpening coefficient;

$K_{sa_F}$ –coefficient of oversharpening (adjustment of the transverse edge);

$K_{K_F}$ –coefficient for nose angle 2K (K – angle of attack);

$K_{\eta_F}$ –coefficient for the relative thickness of the drill core (the ratio between the core diameter and the drill diameter).

From the correction coefficients presented in relation (2) the coefficient for the relative thickness of the drill core (the ratio between the diameter of the core and the diameter of the drill) is determined in three steps which is time consuming.

In this paper two steps are eliminated by determining the connection between the diameter of the drill D and the correction coefficient $K_{\eta_F}$ for the relative thickness of the drill core.

That also improves the determination both by developing a relationship that can be easily applied as well as by building a graph that helps to quickly determine the correction coefficient.

## 2. Determination of $K_{\eta F}$- correction coefficient of the relative thickness of the drill core

Relative thickness of the drill core η is the ratio between the drill core diameter $d_0$ and the drill diameter D.

To determinate $K_{\eta F}$ the next steps will be followed:

- The determination of the relation between the drill diameter D and relative thickness η of the drill core using the data from [1] and [5];
- The determination of the relation between the relative thickness η of the drill core and the correction coefficient $K_{\eta F}$ using data from [1];
- The determination of the relation between the drill diameter D and correction coefficient for the relative thickness of the drill core $K_{\eta F}$ using the relation obtained in the previous step.

### *2.1. Determination of the relation between the drill diameter D and the relative thickness η of the drill core*

In the literature [1] and [5], the relative thickness of the drill core η depending on the drill diameter is presented using intervals, as in table 1.

**Table 1.** Relative thickness η of drill core depending on the drill diameter [1].

| D – drill diameter [mm] | η – relative thickness of the drill core [mm] |
|---|---|
| 0.25 ÷ 1.25 | 0.28 ÷ 0.2 |
| 1.5 ÷ 12 | 0.19 ÷ 0.150 |
| 13 ÷ 80 | 0.145 ÷ 0.125 |

The data from table 1 is processed using Curve Expert 1.4 software (figure 1 and figure 2).

Figure 1 shows the relation between the relative thickness of the drill core and the drill diameter obtained by processing the data with the Curve Expert software.

Figure 2 shows the graphical representation of the relative thickness η depending on the drill's diameter D processed by the same software.

Relation (3) is the resulted relation between the relative thickness of the drill core and the drill diameter.

$$\eta = \frac{0{,}359+0{,}332\cdot D}{1+2{,}314\cdot D+0{,}005\cdot D^2} \tag{3}$$

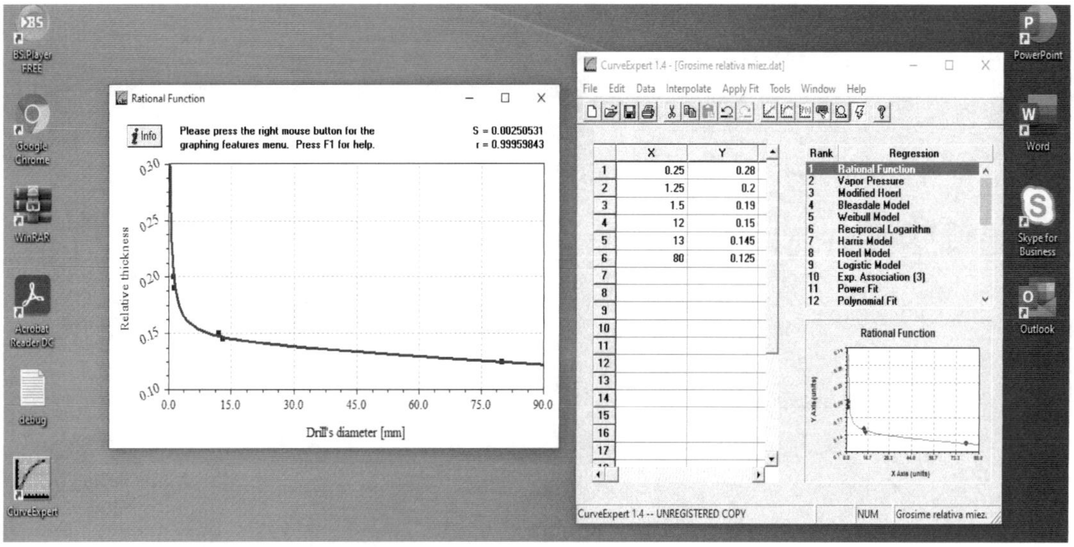


**Figure 1.** Using Curve Expert to determinate the relation between the relative thickness η and the drill diameter D.

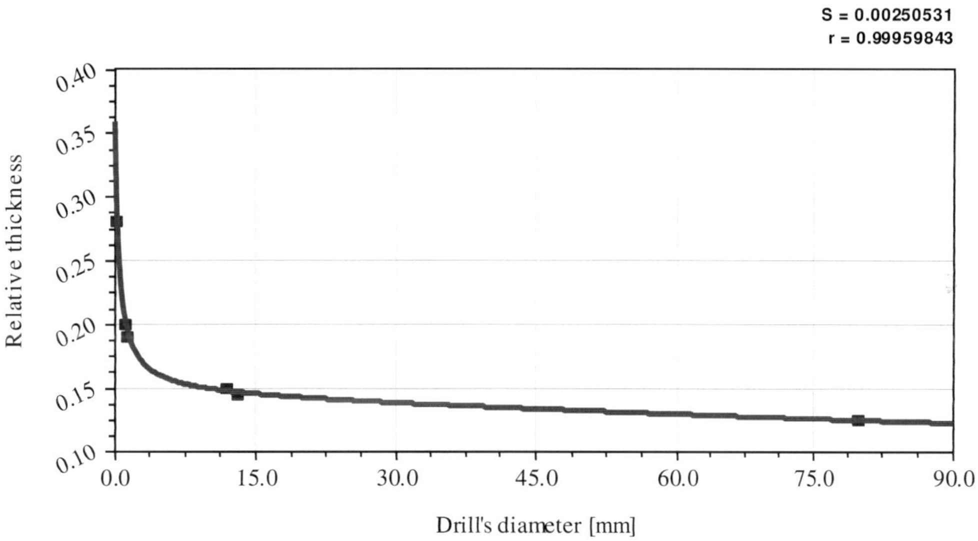


**Figure 2.** Graphical representation of the relative thickness η depending on the drill's diameter D.

*2.2. Determination of the relation between the relative thickness $\eta$ of the drill core and the correction coefficient $K_{\eta F}$*

Relation (4) is the resulted relation between the relative thickness of the drill core and the correction coefficient $K_{\eta F}$.

$$K_{\eta_F} = 0.085 + 6.115 \times \eta \qquad (4)$$

**Table 2.** Determination of relation between relative thickness η of drill core and correction coefficient $K_{\eta F}$.

| No. | η | $K_{\eta F}$ | No. | η | $K_{\eta F}$ |
|---|---|---|---|---|---|
| 1 | 0.12 | 0.82 | 8 | 0.19 | 1.25 |
| 2 | 0.13 | 0.88 | 9 | 0.20 | 1.31 |
| 3 | 0.14 | 0.94 | 10 | 0.21 | 1.37 |
| 4 | 0.15 | 1 | 11 | 0.22 | 1.43 |
| 5 | 0.16 | 1.06 | 12 | 0.23 | 1.49 |
| 6 | 0.17 | 1.12 | 13 | 0.24 | 1.55 |
| 7 | 0.18 | 1.19 | | | |

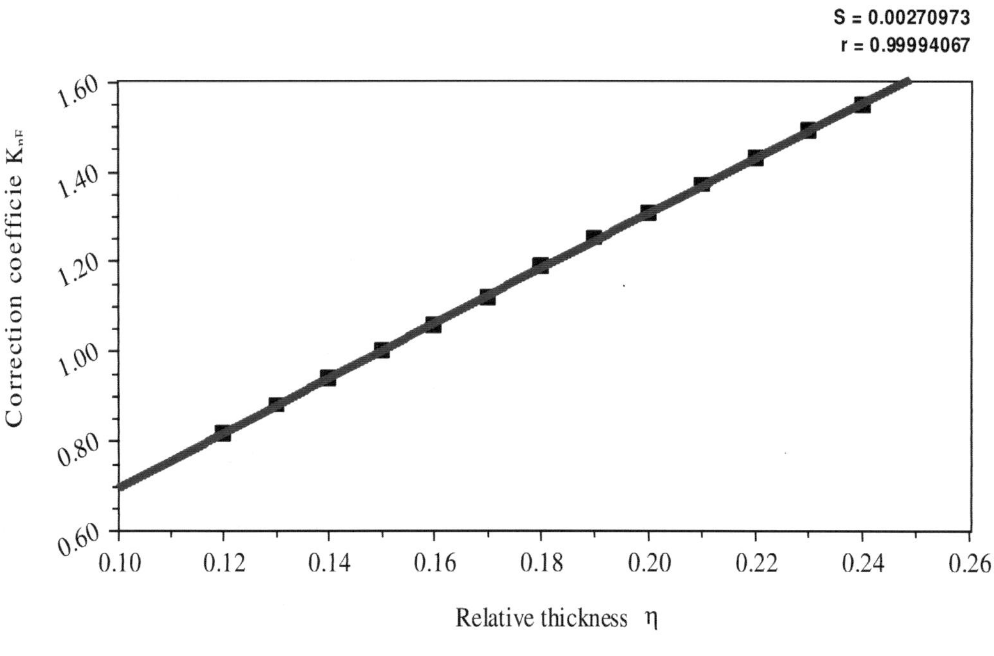


**Figure 3.** Graphical representation of the correction coefficient $K_{\eta F}$ depending on the relative thickness η.

*2.3. Determination of the relation between the drill diameter D and correction coefficient for the relative thickness of the drill core $K_{\eta F}$*

The last step in determining the relationship between the diameter of the drill, D, and the correction coefficient for the relative thickness of the drill core, $K_{\eta F}$, is to use the relationships (3) and (4) to generate the table 3.

Having created table 3 it is easy to use the Curve Expert software to see and graphically the dependency stated above.

The data presented in table 3, processed using the Curve Expert software, give both figure 4 and relation (5).

**Table 3.** Connection between drill diameter D and the correction coefficient $K_{\eta F}$.

| D [mm] | η | $K_{\eta F}$ | D[mm] | η | $K_{\eta F}$ |
|---|---|---|---|---|---|
| 1 | 0.208 | 1.357 | 9 | 0.151 | 1.008 |
| 2 | 0.181 | 1.181 | 10 | 0.150 | 1.002 |
| 3 | 0.170 | 1.124 | 11 | 0.149 | 0.996 |
| 4 | 0.164 | 1.087 | 12 | 0.148 | 0.990 |
| 5 | 0.159 | 1.057 | 13 | 0.147 | 0.984 |
| 6 | 0.156 | 1.039 | 14 | 0.146 | 0.977 |
| 7 | 0.154 | 1.026 | 15 | 0.145 | 0.971 |
| 8 | 0.152 | 1.014 | | | |

Relation (5) is the resulted relation between the correction coefficient $K_{\eta F}$ and the drill diameter.

$$K_{\eta F} = \frac{2,926+4,112\cdot D}{1+4,176\cdot D+0,012\cdot D^2} \tag{5}$$

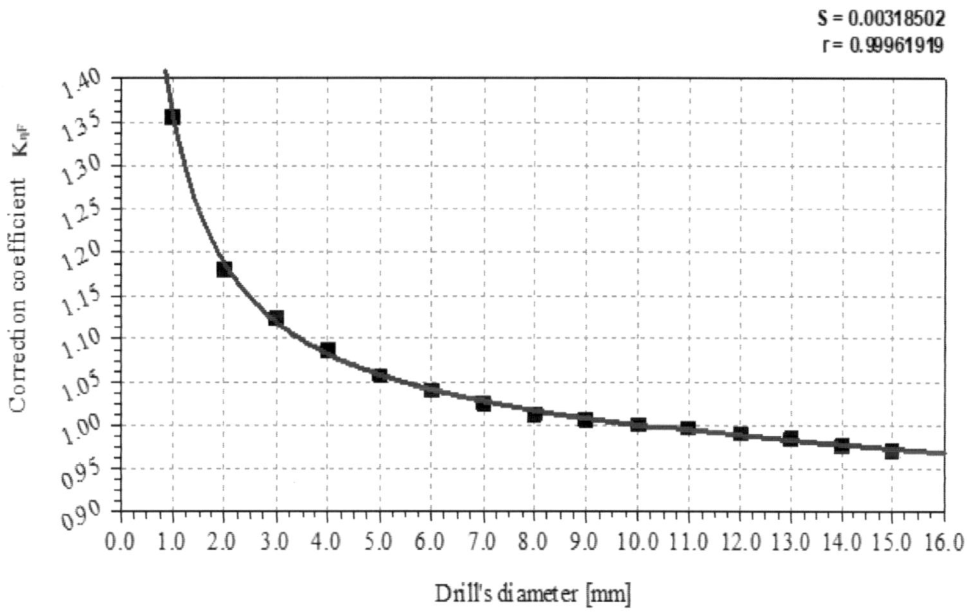


**Figure 4.** Graphical representation of the correction coefficient $K_{\eta F}$ depending on the drill diameter D.

Relation (5) determines directly the correction coefficient $K_{\eta F}$ depending on the diameter of the drill, the drill being sharpened according to the conical method and without over sharpening.

As a rule for choosing a value from the intervals shown in table 1 is not indicated, the error in determining the axial force compared to the proposed method depends on the adoption of the value in the range. Thus, for the drill diameter of 1 mm, where table 3 indicates for η the value of 0.208 which corresponds for $K_{\eta F}$ the value of 1.357 it can be taken $\eta = 0.28$ and the correction coefficient gets the value of 2.235. In this situation the error is +64.7%.If $\eta = 0.2$ is chosen from table 1, it results $K_{\eta F} = 1.24$ and the error is -8.6%.

If the drill diameter of 6 mm is chosen for the calculation of the axial force, based on the previous reasoning, it results that the error is between + 13.4% and -0.5%. For the 15 mm drill the error is in the range 0 ÷ -20.2%. It is found that the error depends on the diameter of the drill and the smaller the diameter of the drill is, the higher the error gets.

## 3. Conclusions

When determining the axial force at the drilling of metals the most commonly used relationship is determined based on the hypothesis of polytropic compression of the material to be processed.

When the concrete working conditions differ from the conditions in which the constants and exponents in the relation have been determined, correction coefficients are needed.

For one of the correction coefficients, namely the coefficient for the relative thickness of the drill core (the ratio between the core diameter and the drill diameter) the direct relation between the drill diameter and the coefficient was developed, which simplifies its determination and also builds a graph that can be used to quickly determine the analyzed correction coefficient.

The proposed method eliminates the errors that could be committed when determining, by calculation, the axial force at the drills by the classical method.

## References

[1] Picoş C *et al.* 1992 *Design of mechanical machining technologies* vol 1 and (Chișinău: Universitas)
[2] Fetecău C 2001 *Parametric relations in machining* (Bucharest: Technical)
[3] Cristian I 2015 *Generation of surfaces by cutting* (Brașov: Lux Libris)
[4] Ștețiu G et al 1994 *Theory and practice of cutting tools. Elements of metal cutting theory* vol I (Sibiu: University)
[5] Minciu C et al 1995 *Cutting tools. Design guide* vol 1 (Bucharest: Technical)
[6] Cozmîncă M et al 1995 *Cutting basis* (Iaşi: Gheorghe Asachi)
[7] Cordebois JP et al 2003 *Machining manufacturing* (Paris: Dunod)
[8] Dițu V, Oancea G and Daicu R 2017 *MATEC Web Conf*. **94** 02005
[9] Pirtini M and Lazoglu I 2005 *Int. J. Mach. Tool. Manu.* **45** 11
[10] Hamade R F, Seif C Y and Ismail F *Int. J. Mach. Tool. Manu.* **46** 3

# Selecting an alternative to solve a problem from several available alternatives

**O Dodun[1], L Slătineanu[1], G Arcidiacono[2], M Coteaţă[1], G Nagîţ[1] and A Hriţuc[1*]**

[1]"Gheorghe Asachi" Technical University of Iaşi, Department of Machine Manufacturing Technology, blvd. D. Mangeron, 39 A, 700050 Iaşi, Romania
[2]Guglielmo Marconi University, Department of Innovation and Information Engineering, Rome, Italy

*E-mail: hrituc.adelina3295@yahoo.com

**Abstract.** When for a certain design problem many alternative solutions could be susceptible to be used, a selection activity should be developed. If in the case of simple problems, the identification of an applicable alternative could be made very fast, in the case of the complex problems, the identification of the most convenient alternative could constitute itself a difficult problem. Over the years, simpler or more complex methods were found and applied to help the problem solver in finding the most convenient applicable alternative. The selection of a certain alternative when many such alternatives seem to exist could be made using the double input matrix method and the method of the pairwise comparison matrix and distinct evaluation criteria. The results of a applying the two selection methods in the case of necessity to establish a machining scheme for an electrical discharge machining process were presented in the paper.

## 1. Introduction

When there are several alternatives to solve a certain problem, it is necessary to take actions to select the most convenient of the available alternatives. If the problem is relatively simple and the number of alternatives is low, it is possible that the most convenient alternative will be quickly identified. This is the case of the so-called tactical problems, specific to the current activity. When the problem is more complex and involves the involvement of high material values and important human resources, we are dealing with a strategic problem and identifying the most convenient alternative can be difficult. In the last situation, we will have to use a method of optimal selection of the alternative to solve the problem we face.

The significance of the optimization concept takes into account the need to undertake an activity to select one or more alternatives considered as maximum convenient [1 - 4].

Optimization is a common activity in engineering activities of constructive or technological design, but also in actual practical activities. When the problem of establishing a certain machining process for obtaining a surface of a part is raised, the technology designer must resort to identifying the process alternatives usable in that case, and then select that process considered to be the most convenient in relation to criteria needed to be considered [5, 6].

Selecting a processing process from several processes that could lead to similar results is a matter of establishing a decision. To solve a problem of this nature, the so-called research field of decision analysis has been developed. In principle, such a field considers a group of methods that could be applied to select an optimal alternative from several available alternatives.

Taking into account the circumstances in which a decision must be made, it is possible to identify several groups of methods for the optimal selection of a particular alternative. Thus, in the case of decisions characterized mainly by individual applicability and in conditions of certainty, methods can be found such as: global utility method, Technique for Order the Preference by Similarity to Ideal Solution, TOPSIS, group of methods ELECTRE I, II, II, IV, IS, TRI, weighted sum model (WSM), weighted product model (WPM), analytic hierarchy process (AHP), analytic network process (ANP), multi-attribute utility theory (MAUT), dominance-based rough set approach (DRSA), aggregated indices randomization method (AIRM), nonstructural fuzzy decision support system (NSFDSS), gray relational analysis (GRA), superiority and inferiority ranking method (SIR method), potentially all pairwise rankings of all alternatives (PAPRIKA), value analysis (VA), axiomatic design method, etc. If decisions are made at risk, it can be used the method of the waited value, the method of decision tree, etc. Finally, for decisions that must be adopted in conditions of uncertainty, the optimistic method, pessimistic method, temperate optimism method, equal probability method (proportionality method), regret minimization method, etc. can be used.

Over the years, the issue of selecting the most convenient option from several available alternatives has preoccupied many researchers. Thus, some of these researchers have conducted research on the use and better understanding of the principles of the pairwise comparison matrix method [7 - 10].
The objectives of the research presented in this paper aimed at the possibilities of using two distinct methods, with different levels of complexity, for the selection of the scheme of electrical discharge machining of cylindrical test pieces.

## 2. The problem of an optimal selection

During the research activities, it appeared at a certain moment the need to separate a cylindrical test piece, of relatively small dimensions (diameter of 6 mm and length of 10-15 mm) from a part made of a material characterized by special physical and mechanical properties. These test pieces were to be subsequently subjected to mechanical tests specific to the use of the part in question. Given the mechanical properties of the part material, it was decided to detach the test pieces by electrical discharge machining using a copper tubular tool electrode and the usual working movement of the work head of the machine, along a vertical direction, from top to bottom, towards the workpiece (part) fixed on the machine tool table, in a vise type device. The results were not really good, because due to the longer maintenance in the space between the two electrodes of the metallic particles detached from them, there were spurious lateral electrical discharges and the test piece thus obtained had a lateral surface with a certain taper. In these circumstances, there was a need to identify another technological solution, which would improve the accuracy of obtaining the lateral cylindrical surface of the test piece. Some of these machining schemes are those presented in figure 1.

Each of these proposed schemes has advantages and disadvantages and it has been proposed to use a method to facilitate the selection of the most convenient alternative, by taking into account the proper conditions available.

In figure 1, the first machining scheme was relatively simple, but it generated an unacceptable taper of the machined surface. The machining was performed by immersing the working area in the dielectric liquid.

In the case of the processing scheme from figure 1, b it was necessary to resort to the aspiration of the dielectric liquid through the inside of the tubular tool electrode, other characteristics of this machining scheme being similar to those of the scheme shown in figure 1, a. Such a machining scheme could lead to a decrease in the conicity of the lateral surface, due to a faster evacuation (by aspiration) of the metallic particles suspended in the dielectric liquid.

The third machining scheme (figure 1, c) is based on the realization of the working movement by the workpiece from top to bottom, towards the tubular tool electrode fixed in a mandrel type device, located on the table of the electrical discharge machine. Since the evacuation of the detached metallic particles from the electrodes was performed under the action of gravity, it was expected that the taper of the outer surface of the test piece detached from the part would be lower.

In the case of the processing scheme from figure 1, d, compared to the scheme presented above, the aspiration of the dielectric liquid with metallic particles in suspension through the tubular tool electrode appears.

Since the machining schemes previously exposed used a non-rotating tool electrode, with the risk of copying possible deviations of the tool electrode in the outer surface of the test piece separated from the workpiece by electrical discharge machining, the fifth version of the machining scheme involved the rotational movement of the tubular tool electrode. An improvement in machining accuracy was expected, but also an increase in the complexity of the tool electrode positioning and clamping device.

The sixth machining scheme considered was based on the use of a cylindrical bar tool electrode, which performs a planetary movement around the axis of symmetry of the outer cylindrical surface of the test piece to be detached by electrical discharge machining from the plate type workpiece. In addition to the planetary motion, the tool electrode of course also performs the working movement from top to bottom specific to the normal use of the electrical discharge machine with massive electrode. The machining scheme became more complicated, the productivity decreased, but an improvement of the precision of the test piece surface obtained by electrical discharge machining was expected. In the previously mentioned aspects, only a few characteristics of each machining scheme considered were revealed very briefly. In reality, the comparison of such machining schemes makes necessary to identify and analyze other aspects, such as, for example, those related to the influence of tool electrode wear on machining accuracy, etc.

Without using methods to optimize the selection of an alternative when there are several alternatives and especially when the number of alternatives is large, it seems very difficult to select the most convenient alternative and here can be applied the various methods proposed by researchers to select the alternative, optimal or close to the optimal one. Furthermore, in order not to make the analysis too complex and to facilitate its understanding, only 5 alternatives for solving the problem and 4 criteria for their evaluation were considered.

## 3. Selecting a solution using the pairwise comparison matrix method

The pairwise comparison matrix method is also called the double entry matrix method, because both

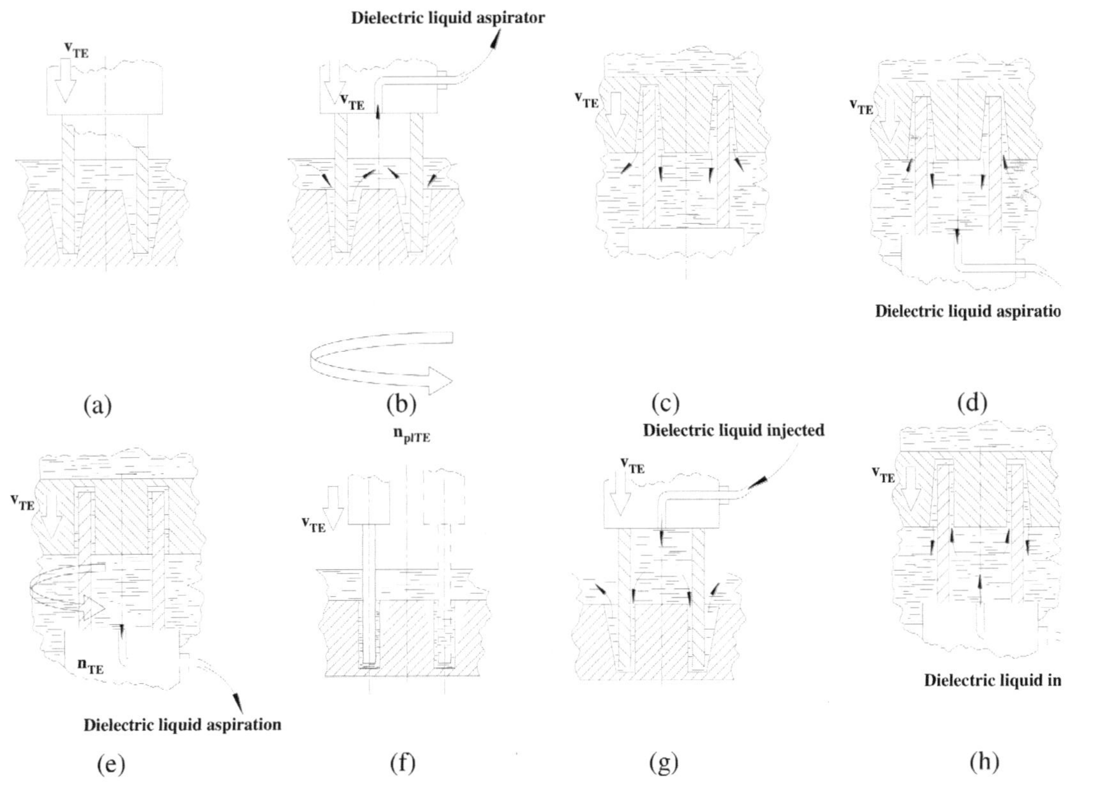


**Figure 1.** Machining schemes that could be used to detach a cylindrical test piece from a plate type workpiece.

along the first row and the first column, the identified alternatives for solving the approached problem are included. Next, in the places obtained at the intersection of a column with a line, marks resulting from the comparison of the pairs of alternatives defined by the line and the column in question are entered, granting evaluation marks of type 1, 0 and 0.5, for example. If the first alternative of the pair seems to be more advantageous, mark 1 will be entered in that column, while mark 0 will be entered in the column corresponding to the other alternative of the pair. When the two alternatives seem to be equal importantly, marks of 0.5 will be registered in both places. However, there is also the possibility of awarding marks varying between wider limits [7-10]. Along the diagonal whose position corresponds to the comparison of the same alternative, the symbol X will be inscribed.

**Table 1.** Table elaborated using the pairwise comparison matrix method when selecting a machining scheme.

| Alternative | Alternative a | Alternative c | Alternative d | Alternative e | Alternative f |
|---|---|---|---|---|---|
| Alternative a | X | 1 | 1 | 1 | 0.5 |
| Alternative c | 0 | X | 1 | 1 | 0.5 |
| Alternative d | 0 | 0 | X | 1 | 1.0 |
| Alternative e | 0 | 0 | 0 | X | 0,5 |
| Alternative f | 0.5 | 0.5 | 0 | 0.5 | X |
| Sum of marks | 0.5 | 1.5 | 2 | 3.5 | 2.5 |
| Relative sum | 0.05 | 0.15 | 0.2 | 0.35 | 0.25 |
| New order | e-f-d-c-a since 0.35>0.25>0.2>0.15>0.05 | | | | |

After completing all the places in the tables with the results of the comparisons, a line follows in which the values of the sums of the marks awarded for each alternative are entered, along the column corresponding to that alternative. These sums already provide an image of the order of importance of the alternatives, an order established by comparing two by two of all the alternatives. In the penultimate line of the table, it can be entered the values of some relative sums, obtained by dividing the sum of the marks of each alternative to the number of comparisons. This $N_c$ number of comparisons is determined by means of a relation of form:

$$N_c = \frac{n(n-1)}{2} \quad (1)$$

where $n$ is the number of alternatives considered. The last line of the table will include the new order of the alternatives, established by arranging them in descending order the relative sums of the marks obtained by each alternative.

It can be seen that the method described here does not use criteria when comparing two alternatives, which leads to a less objective evaluation, but the advantage of the method is the shorter duration of its application. Table 1 shows the results of applying the method in the case of the 5 versions of machining schemes previously identified. The most convenient alternative established by applying the method is the alternative *e*, which is assigned the highest value of the relative sum.

## 4. Selecting a solution using a matrix of comparing pairs of alternatives and distinct evaluation criteria

A method that uses a comparison of existing pairs of alternatives using distinct criteria will be addressed in this chapter. A version of this method was promoted by Professor Belous (from the "Gheorghe Asachi" Technical University of Iaşi, Romania) [11] and called the technique of the imposed decision. In principle, this method can be included in the broader group of value engineering methods [12].

To illustrate how to use this method, the information contained in table 2 was used. A first part of the table was intended to compare the selection criteria used two by two and respectively to establish weights given to each of the criteria.

**Table 2.** Table elaborated using a matrix of comparing pairs of alternatives and distinct evaluation criteria.

1. Calculation of the weighting coefficient of each criterion

| Criterion | Results of comparisons | | | | | | Sum | Weighting coefficient |
|---|---|---|---|---|---|---|---|---|
| Column 1 | 2 | 3 | 4 | 5 | 6 | 7 | 13 | 14 |
| *A* | 1 | 1 | 0.5 | | | | 2.5 | 0.41 |
| *B* | 0 | | | 0 | 0.5 | | 0.5 | 0.08 |
| *C* | | 0 | | 1 | | 1 | 2.0 | 0.33 |
| *D* | | | 0.5 | | 0.5 | 0 | 1.0 | 0.16 |

2. Calculation of the importance coefficient for each alternative using the criterion of achievable accuracy

| Alternative | Results of comparisons | | | | | | | | | | Sum | Importance coefficient |
|---|---|---|---|---|---|---|---|---|---|---|---|---|
| *a* | 0 | 0 | 0 | 0 | | | | | | | 0 | 0.0 |
| *c* | 1 | | | | 0 | 0 | 0 | | | | 1 | 0.1 |
| *d* | | 1 | | | 1 | | | 0 | 0 | | 2 | 0.2 |
| *e* | | | 1 | | | 1 | | 1 | | 0.5 | 3.5 | 0.35 |
| *f* | | | | 1 | | | 1 | | 1 | 0.5 | 3.5 | 0.35 |

3. Calculation of the importance coefficient for each alternative using the criterion of the expected rate of material removal

| Alternative | Results of comparisons | | | | | | | | | | Sum | Importance coefficient |
|---|---|---|---|---|---|---|---|---|---|---|---|---|
| *a* | 0 | 0 | 0 | 0.5 | | | | | | | 0.5 | 0.05 |
| *c* | 1 | | | | 0 | 0 | 1 | | | | 2 | 0.2 |
| *d* | | 1 | | | 1 | | | 0 | 1 | | 3 | 0.3 |
| *e* | | | 1 | | | 1 | | 1 | | 1 | 4 | 0.4 |
| *f* | | | | 0.5 | | | 0 | | 0 | 0 | 0.5 | 0.05 |

3. Calculation of the importance coefficient for each alternative using the criterion of the constructive simplicity of the devices necessary for the materialization of the alternative of the machining scheme

| Alternative | Results of comparisons | | | | | | | | | | Sum | Importance coefficient |
|---|---|---|---|---|---|---|---|---|---|---|---|---|
| *a* | 1 | 1 | 1 | 1 | | | | | | | 4 | 0.4 |
| *c* | 0 | | | | 1 | 1 | 1 | | | | 3 | 0.3 |
| *d* | | 0 | | | 0 | | | 1 | 1 | | 2 | 0.2 |
| *e* | | | 0 | | | 0 | | 0 | | 1 | 1 | 0.1 |
| *f* | | | | 0 | | | 0 | | 0 | 0 | 0 | 0.0 |

4. Calculation of the importance coefficient for each alternative using the criterion of the risk of copying the deviation of the shape of the tool electrode in the machined surface

| Alternative | Results of comparisons | | | | | | | | | | Sum | Importance coefficient |
|---|---|---|---|---|---|---|---|---|---|---|---|---|
| *a* | 0.5 | 0.5 | 0 | 0 | | | | | | | 1.0 | 0.1 |
| *c* | 0.5 | | | | 0.5 | 0 | 0 | | | | 1.0 | 0.1 |
| *d* | | 0.5 | | | 0.5 | | | 0 | 0 | | 1.0 | 0.1 |
| *e* | | | 1 | | | 1 | | 1 | | 0.5 | 3.5 | 0.35 |
| *f* | | | | 1 | | | 1 | | 1 | 0.5 | 3.5 | 0.35 |

Calculation of the value numbers associated with each alternative

| | |
|---|---|
| *a* | $N_{va}$=0.41·0 +0.08·0.05 +0.33·0.4 +0.16· 0.1=0.152 |
| *c* | $N_{vb}$=0.41· 0.1+0.08· 0.2+0.33·0.3 +0.16· 0.1 =0.172 |
| *d* | $N_{vd}$=0.41· 0.2+0.08· 0.3+0.33·0.2 +0.16· 0.1 =0.188 |
| *e* | $N_{ve}$=0.41·0.35 +0.08·0.4 +0.33· 0.1+0.16·0.35 =0.7125 |
| *f* | $N_{vf}$=0.41·0.35 +0.08· 0.05+0.33· 0+0.16· 0.35 =0.2035 |
| New order | *e-f-d-c-a* since 0.7125>0.2035>0.188>0.172>0.152 |

In the case of the five alternatives proposed for establishing the machining scheme usable for separating cylindrical test pieces by electrical discharge machining from a plate-type workpiece, it could be as differentiating criteria: the achievable accuracy, expected rate of material removal, constructive simplicity of the devices necessary for the materialization of the proposed machining scheme, the risk of copying the deviation of the shape of the tool electrode in the machined surface, the machining cost (including the level of complexity of the device), efficiency of the removal of detached metal particles from electrodes in the machining area, the risk of accumulation of gas bubbles (from the gases released as a result of the electrical discharge machining process) between the tool electrode and the workpiece (these bubbles would have a low electrical conductivity, which could change the nature of discharges electrical and a possible more pronounced deterioration of the processed surface), etc.

As can be seen, the number of criteria that could be taken into account when comparing alternatives can be quite large, but in the case of this paper only the first four criteria were used, to which the symbols A, B, C, and D were assigned.

Specifically, in the first part of table 2, the criteria for evaluating the alternatives of machining schemes were compared and received weighting coefficients. As in the case of the method pairwise comparison matrix, the criteria considered were compared two by two, assigning marks of 1-0, when the first criterion was assessed as more important, 0-1 when more important was assessed as being the second criterion and respectively 0.5-0.5 when the two compared criteria were appreciated as being of equal importance. In the penultimate column of this first part of table 1 were entered the sums of marks given to each criterion, and in the last column was entered the value of the weighting of the criterion, by relating the sum of marks to the number of comparisons made (there are four criteria, number of comparisons will be 4 x 3/2 = 6 comparisons).

Similar working principles were applied in the following four components of table 2, when the five alternatives were compared in pairs and using each of the four criteria considered. In this way it became possible to establish a coefficient of the importance of each alternative by taking into account each evaluation criterion.

In the last part of table 2 there were calculated the value numbers of each alternative, by summing the products of the importance coefficients of each alternative determined in the case of a certain evaluation criterion with the value of the weighting coefficient of the respective criterion. The calculation relation of the value numbers has the aspect:

$$N_{vj} = \sum_{i=1}^{n} C_{wi}\, C_{ji} \tag{2}$$

where $C_{wi}$ is the weighting coefficient of each criterion i, and $C_{ji}$ is the value of the importance coefficient of alternative j when using criterion i.

By applying the matrix method of comparing pairs of alternatives and distinct evaluation criteria, it can be seen that the most convenient machining scheme alternative among the five alternatives considered is the alternative e, since in this case the highest value of the value number $N_{ve}$ was obtained. As expected, the least convenient alternative is alternative a, this being in fact the one that later allowed to outline the problem of selecting a solution alternative when several such alternatives are available.

## 5. Discussion

If we compare the time required to select the most convenient alternative from the five available alternatives, we find that the use of the matrix method of comparing pairs of alternatives and distinct evaluation criteria requires a much longer duration than that of the pairwise comparison matrix method. We will note, however, that by considering several evaluation criteria, it is expected that the matrix method of comparing pairs of alternatives and distinct evaluation criteria can provide a more objective ordering of the alternatives considered and therefore, when available, long enough, it is preferable to use the latter method.

We will find further that the use of the second method (method that uses the pairwise comparison matrix and distinct evaluation criteria), through the more detailed analysis it involves, can lead to useful suggestions for further improvement including the selected alternative, even by taking into account those criteria for which it was not well rated.

## 6. Conclusions

When it is necessary to select an alternative from a low number of available alternatives, the selection is relatively easy to make. If the number of available alternatives is large, there is a need to use a method that allows the justified choice of the most convenient alternative. In the literature there are mentioned many methods that can be used in this regard. In the present paper, the problem of selecting a machining scheme that would allow the detachment by electrical discharge machining of a cylindrical test piece from a plate-type workpiece was considered, identifying several such possible schemes. It was thus found that applying the pairwise comparison matrix method is possible to select the most convenient in a relatively short time, but the use of a pairwise matrix comparison and distinct evaluation criteria allows a more objective selection of the most convenient alternative from the considered alternatives. It was also appreciated that a more detailed method may facilitate the emergence of suggestions for improving one or more of the alternatives analyzed. In the future, the identification and analysis of other methods designed to facilitate the selection of the most convenient alternative will be pursued when several available alternatives are available.

## References

[1] Arcidiacono G, Matt D T and Rauch E 2017 Axiomatic design of a framework for the comprehensive optimization of patient flows in hospitals *J Healthc Eng* **2309265**
[2] Optimization 2018. Available: http://www.dictionary.com/browse/optimization, accessed: 30.04.2018
[3] Gonçalves-Coelho A M and Mourão A J F 2007 Axiomatic design as support for decision-making in a design for manufacturing context: A case study *Int J Prod Econ* **109** p 81
[4] Slătineanu L and Duşa P 2002 *Management of technological innovation* (in Romanian) (Iaşi: Tehnopress Publishing House)
[5] Arcidiacono G and Brown C 2013 Automatic design of production systems for operational excellence *Proc. of ICAD2013 The Seventh International Conference on Axiomatic Design Worcester – June 27-28,* **ICAD-2013-29**
[6] Slătineanu L, Coteaţă M, Dulgheru V, Duşa P, Banciu F and Beşliu I 2017 Selection of a solution when using Axiomatic Design *The 11th International Conference on Axiomatic Design (ICAD 2017), MATEC Web Conf.* **127** 01019
[7] Harker P T 1987 Incomplete pairwise comparisons in the analytic hierarchy process *Mathl Modelling* **9**(11) p 837
[8] Song B and Kang S 2016 A method of assigning weights using a ranking and nonhierarchy comparison *Adv Dec Sci* Article ID **8963214**
[9] Kou G, Ergu D, Lin C and Chen Y 2016 Pairwise comparison matrix in multiple criteria decision making *Technol Econ Dev Eco* **22**(5) p 738
[10] Ramik J 2017 Ranking alternatives by pairwise comparisons matrix and priority vector *Scientific Annals of Economics and Business* **64** (SI) p 85
[11] Belous V 1992 *Inventics* (Iaşi: Asachi Publishing House)
[12] Crum L W 1976 *Value engineering* (in Romanian; translation from English language) (Bucharest: Editura Tehnică)

IOP Conf. Series: Materials Science and Engineering 1009 (2021) 012019 doi:10.1088/1757-899X/1009/1/012019

# Computer program for single sampling plans by variables with known lot standard deviation

**N Eftimie**[1*]

[1]Department of Manufacturing Engineering, Transilvania University of Brasov, 5 Mihai Viteazul Street, Brasov, Romania

*E-mail: eftimien@unitbv.ro

**Abstract**. The paper presents a computer program that can be used at design and application of single sampling plans by variables when the standard deviation of the lot is known. The developed program may be applied for quality characteristics with one single or two specification limits. When the quality characteristics have two specification limits, the program determines if two separate plans with one specification limit may be applied for making the decision about the lot acceptance or rejection, and takes into account all the cases that may appear in this situation. The computer program calculates first the plan parameters and then, using the sample measurements, makes the decision about the lot acceptance or rejection. Besides the program presentation, the paper also presents several examples that cover all cases of its utilization.

## 1. Introduction

One important aspect of statistical quality control is making the decision about the acceptance or the rejection of the lots of products. In the case when the quality characteristics are continuous, this operation may be accomplished by means of the acceptance sampling plans by variables.

In the last years, some authors studied different types of such sampling plans [1, 2, 3].

The present paper studies the single sampling plans by variables with known lot standard deviation. The plan design and application will be realized according to the *k* method [4, 5]. The application of this method is presented below.

Depending on the specification limits, there are several cases for the plan application.

If the quality characteristic has one single lower limit of specification *L*, by means of a sample of size *n* that is selected at random from the lot, it is calculated the statistic $z_L$ using the equation (1):

$$z_L = \frac{\bar{x} - L}{\sigma} \tag{1}$$

In equation (1), by $\bar{x}$ it is denoted the mean of the sample. The decision is made using the acceptance constant *k*: if $z_L \geq k$ the lot will be accepted, and if $z_L < k$ the lot will be rejected.

The second situation, for the quality characteristic that has only an upper limit of specification *U* is analogous. In this case, the statistic $z_U$ is calculated using the equation (2):

$$z_U = \frac{U - \bar{x}}{\sigma} \tag{2}$$

If $z_U \geq k$ the lot will be accepted, and if $z_U < k$ the lot will be rejected.

In the third situation, for the characteristics with two specification limits $L$ and $U$, the application of the single sampling plan by variables becomes more complicated.

When the variability is high compared to the tolerance interval, both the upper and the lower proportion of nonconforming units may have high values at the same time. In order to determine if two plans with a single limit may be applied in this case, a modified procedure from that proposed by Duncan in 1974 may be used. This modified procedure that is presented in [5], may be applied as follows:

- First, it is calculated the value of $z_p$ using the equation (3):

$$z_p = \frac{U - L}{2\sigma} \tag{3}$$

- Then, it is calculated the value of $p^*$, which is the area under the normal distribution curve situated on the right of $z_p$. This value represents half of the minimum proportion of defective products.
- The decision is made as follows:
  1) If $2p^* \leq p_1/2$ then two plans with a single limit may be applied;
  2) If $p_1/2 < 2p^* \leq p_1$ the specifications limits may be too close and nonconforming items may appear on both sides of the tolerance interval if the distribution is centered. In this case there must be determined the values for the lower and upper proportion of defective items, whose sum gives the value $p_1$ when the normal curve is displaced between the limits of specification. The maximum of these two values will will replace the value of $p_1$ at the calculation of the plan parameters $n$ and $k$;
  3) If $p_1 < 2p^* < p_2$ there must be reconsidered the plan specifications;
  4) If $2p^* \geq p_2$ the lot must be directly rejected.

The sample size can be calculated using equation (4) [6]:

$$n = \left( \frac{z_\alpha + z_\beta}{z_1 - z_2} \right)^2 \tag{4}$$

In order to calculate the acceptance constant $k$ (using equation (7)), there must be calculated first the values of $k_1$ and $k_2$, using equations (5) and (6) respectively [6]:

$$k_1 = z_1 - \frac{z_\alpha}{\sqrt{n}} \tag{5}$$

$$k_2 = z_2 + \frac{z_\beta}{\sqrt{n}} \tag{6}$$

$$k = \frac{k_1 + k_2}{2} \tag{7}$$

In equations (4) - (7) the values of $z_\alpha$, $z_\beta$, $z_1$ and $z_2$ represent the $(1 - \alpha)*100$, $(1 - \beta)*100$, $(1 - p_1)*100$ and $(1 - p_2)*100$ percentiles of the standardized normal repartition respectively.

By $\alpha$ and $\beta$ there are denoted the producer's risk value and the customer's risk value.

The acceptable quality level is denoted by $p_1$ and the rejectable quality level is denoted by $p_2$.

## 2. Objectives

Depending on the specification limits, there are three cases for the design and application of the single sampling plans by variables in the case when lot standard deviation $\sigma$ is known.

These are the following:

- The case of the quality characteristics with one single lower specification limit $L$;

- The case of the quality characteristics with one single upper specification limit $U$;
- The case of the quality characteristics with two specification limits.

Furthermore, depending on the lot variability compared to the tolerance interval, the last case has four different kinds of application.

All these variants of plan application require difficult calculations, especially the second case of the plan with two specification limits, what it requires the development of an algorithm for calculating the value that will replace the acceptable quality level at the calculation of the plan parameters.

The objective of the paper is to develop a computer program that may be used for the design and application of the single sampling plans by variables when the lot standard deviation $\sigma$ is known.

The program performs the selection of the appropriate case of plan application, calculates the values for the sample size and the constant of acceptance, and using the sample measurements that are provided by the user, will make the decision about the lot acceptance or rejection.

## 3. Methods

The computer program that may be used for the design and application of the single sampling plans by variables with known lot standard deviation was developed in Visual Basic for Applications for Excel 2019.

The functions and procedures respect the syntaxes presented in [7].

When the program is run, a dialog box will be displayed on the screen (figure 1):

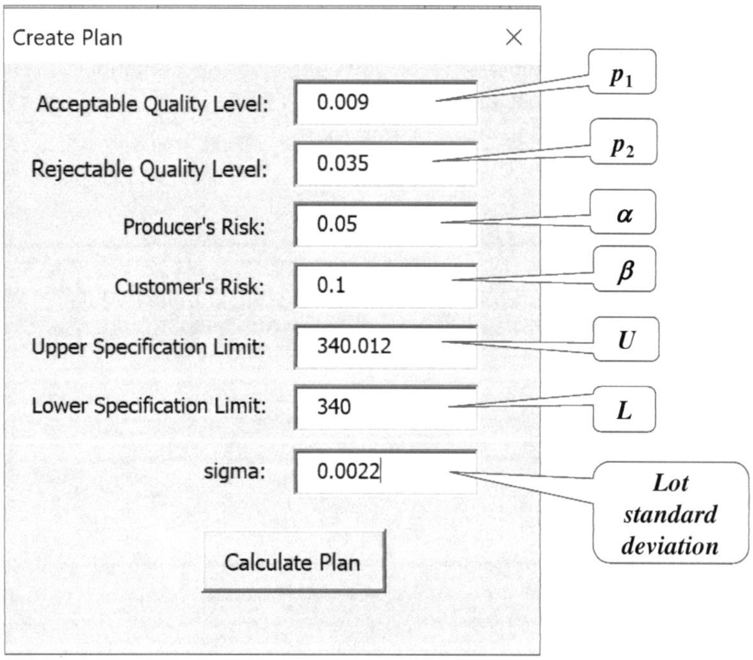


**Figure 1.** The Create Plan dialog box.

The first data that the user of program must record into this dialog box are the inspection conditions: the acceptable quality level $p_1$, the rejectable quality level $p_2$, the producer's risk $\alpha$ and the customer's risk $\beta$.

Then, the user must record at least one specification limit: the upper specification limit, or the lower specification limit, or the upper and the lower specification limits (in the presented example the measured characteristic has two specification limits).

At the end, the lot standard deviation $\sigma$ will be recorded.

In order to calculate and display the plan parameters the command button *Calculate Plan* must be clicked. This action will run the procedure *Calculate_Plan_Click(),* that will collect first the data from the dialog box *Create Plan.*

If the quality characteristic has one single specification limit, the procedure will call further another procedure that calculates the plan parameters for this case.

If the measured characteristic has two limits of specification, the procedure *Calculate_Plan_Click()* calculates first the value of $z_p$ using the equation (3). By means of $z_p$, the value of $2p^*$ will be computed. Depending on the value of $2p^*$, the procedure can call one in four procedures, each one being developed to correspond to a case defined by the modified procedure that is presented in [5].

The sample size $n$ is calculated using the equation (4) and the acceptance constant $k$ by means of equation (7). The results will be recorded into the worksheet *Plan*. The procedure will also create on the worksheet a table with $n$ rows. This table will be used by the program user for recording the sample measurements. If the user did not record $n$ measurements exactly, the program will display an error message.

The data from the dialog box in figure 1 correspond to the most difficult case that can appear at the design of this type of sampling plan. This is a plan with two specification limits that corresponds to the case when $p_1/2 < 2p^* \leq p_1$. In order to calculate the value of $p_1calc$, that will replace the acceptable quality level $p_1$ in the equations used for the plan parameters calculation, a function was developed. In figure 2 it is explained how the algorithm of this function was created.

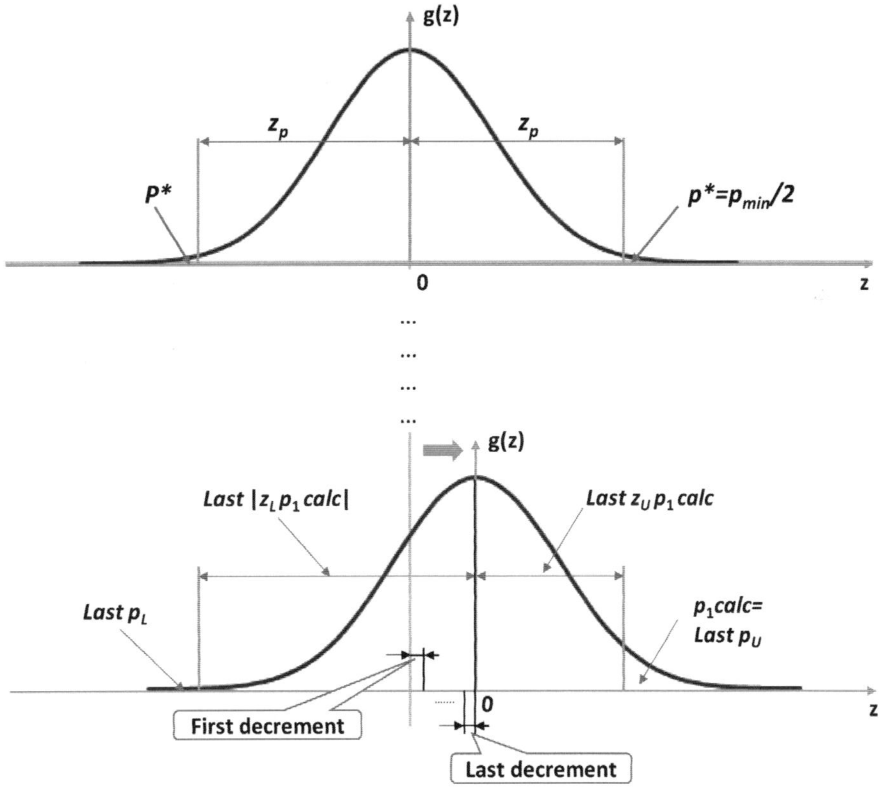


**Figure 2.** The algorithm explained.

The position of first distribution in figure 2 corresponds to the minimum proportion of nonconforming items: $2p^*$. In order to model the displacement of the distribution, the ordinate axis will be moved to the right, step by step, with the value of a decrement that equals to 0.0001. After each ordinate axis shift, the total proportion of nonconforming products $p_{tot}$ is calculated. If the closest value of $p_{tot}$, compared to acceptable quality level $p_1$, has been reached then $p_1calc$ will take the value of the last $p_U$ (there is no need to continue the displacement because $p_{tot}$ would continue to increase more above $p_1$), otherwise the shift of the ordinate axis will continue. The algorithm for the function that returns the value of $p_1calc$ is presented in figure 3:

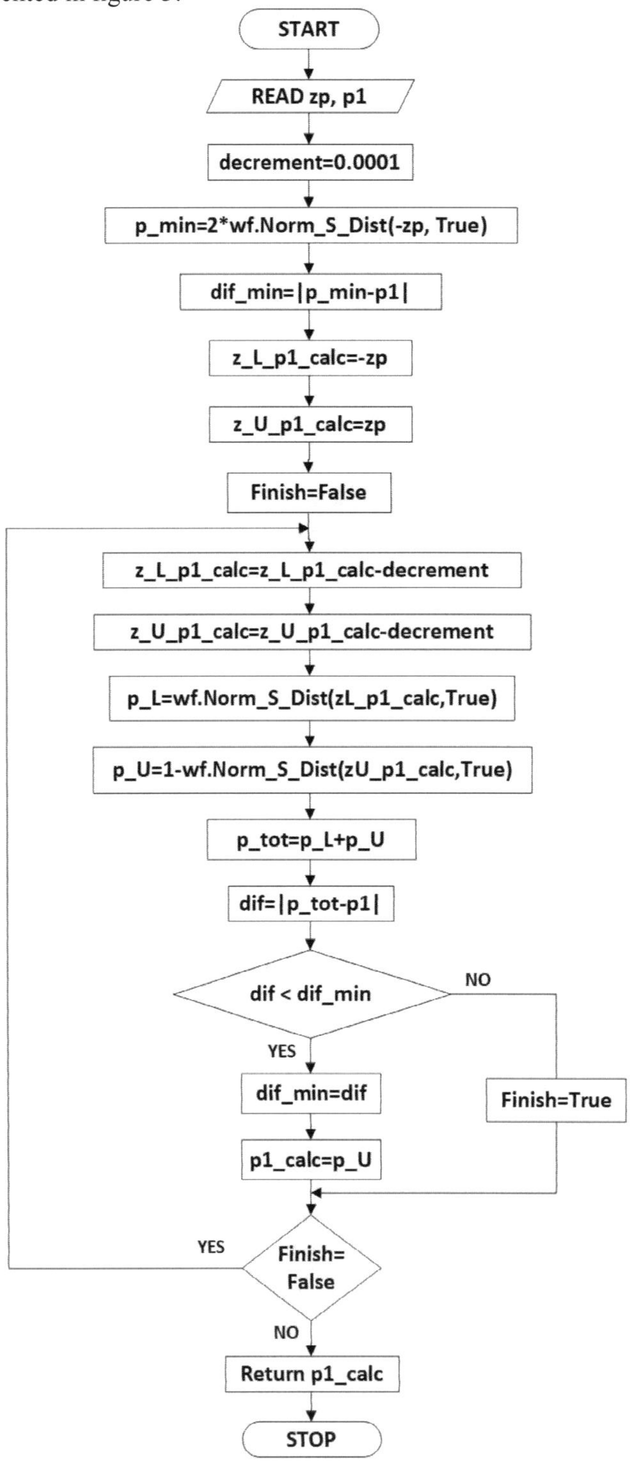


**Figure 3.** The algorithm for the function that returns the value of $p_1\_calc$.

The listing of the developed function is presented in figure 4:

```
Function p1_calc(z_p As Double) As Double
Dim Finish As Boolean, p_min As Double
Dim p_L As Double, p_U As Double, p_tot As Double
Dim decrement As Double, dif_min As Double, dif As Double
Dim zL_p1_calc As Double, zU_p1_calc As Double
  decrement = 0.0001
  p_min = 2 * wf.Norm_S_Dist(-z_p, True)
  dif_min = Abs(p_min - p1)
  zL_p1_calc = -z_p
  zU_p1_calc = z_p
  Finish = False
  Do
    zL_p1_calc = zL_p1_calc - decrement
    zU_p1_calc = zU_p1_calc - decrement
    p_L = wf.Norm_S_Dist(zL_p1_calc, True)
    p_U = 1 - wf.Norm_S_Dist(zU_p1_calc, True)
    p_tot = p_L + p_U
    dif = Abs(p_tot - p1)
    If dif < dif_min Then
      dif_min = dif
      p1_calc = p_U
    Else
      Finish = True
    End If
  Loop While Finish = False
End Function
```

**Figure 4.** The function that returns the value of $p_1\_calc$.

The result of the run of procedure that uses the function $p_1\_calc()$ is presented in figure 5. All information but the sample measurements was recorded on the worksheet by the procedure.

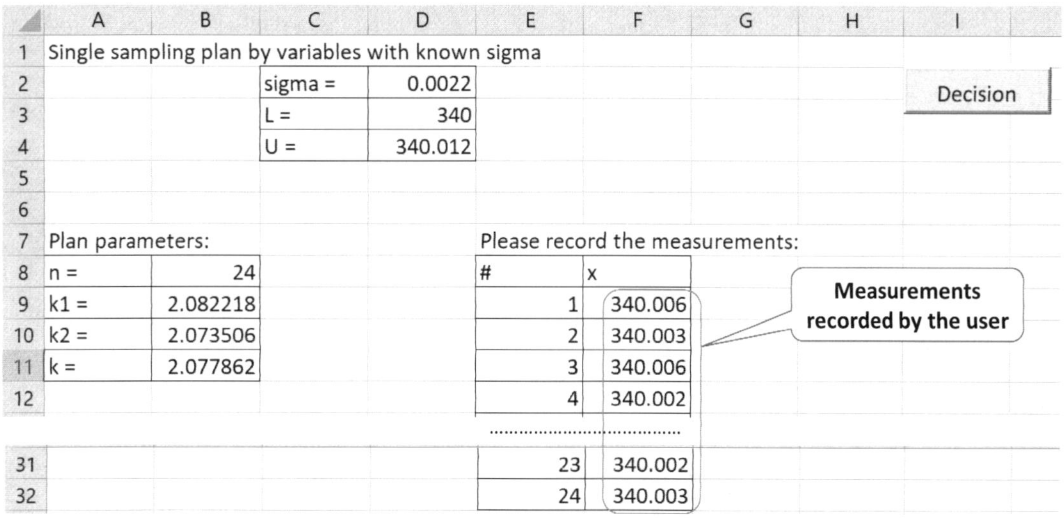


**Figure 5.** The plan parameters and the sample.

In figure 5 there are presented only the first and the last part of the recorded measurements.

After the recording of the sample measurements, the user must click the command button labelled *Decision*. This action will run a procedure that makes the decision about the lot acceptance or rejection. In figure 6 there are presented the results of the run of procedure:

| | |
|---|---|
| xbar = | 340.0039 |
| ZL = | 1.750229 |
| ZU = | 3.704317 |

The lot is rejected

**Figure 6.** The decision.

The procedure that makes the decision, depending on the specification limits of the quality characteristic, will call one of the following three procedures: a procedure for a plan with only one single upper limit of specification, a procedure for a plan with one single lower limit of specification, or a procedure for a plan with two limits of specification.

For the presented case the decision was made by the procedure for the plan with two specification limits. This procedure calculated first the sample mean and then applied two plans with one single specification limit (a plan for each specification limit). There were calculated the statistics $z_L$ and $z_U$ and these values were compared with the acceptance constant $k$. In the presented case the lot was rejected.

## 4. Results and discussions

The developed program was presented for an operating case that corresponds to a measured characteristic with two limits of specification and the lot variability compared to the tolerance interval corresponds to the case when $p_1/2 < 2p^* \leq p_1$.

The case of the measured characteristic with two limits of specification and $2p^* \leq p_1/2$ is analogous to that presented before, with the difference that in this situation there is no need to calculate a new value for $p_1$. In figure 7 there are presented the results of the program run in a situation of this type. The inspection conditions were: $p_1 = 0.01$, $p_2 = 0.025$, $\alpha = 0.05$, $\beta = 0.1$, $U = 235.015$, $L = 235$ and $\sigma = 0.002$.

In all examples presented in this section, only the first measurements of the samples are presented.

| | A | B | C | D | E | F | G | H | I | J |
|---|---|---|---|---|---|---|---|---|---|---|
| 1 | Single sampling plan by variables with known sigma | | | | | | | | | |
| 2 | | | sigma = | 0.002 | | | | | Decision | |
| 3 | | | L = | 235 | | | | | | |
| 4 | | | U = | 235.015 | | | | | | |
| 5 | | | | | | | | | | |
| 6 | | | | | | | | | | |
| 7 | Plan parameters: | | | | Please record the measurements: | | | | | |
| 8 | n = | 64 | | | # | x | | | xbar = | 235.0072 |
| 9 | k1 = | 2.120741 | | | 1 | 235.009 | | | ZL = | 3.579294 |
| 10 | k2 = | 2.120158 | | | 2 | 235.011 | | | ZU = | 3.920706 |
| 11 | k = | 2.12045 | | | 3 | 235.006 | | | | |
| 12 | | | | | 4 | 235.007 | | | The lot is accepted | |

**Figure 7.** An example for the case when the measured characteristic has two limits and $2p^* \leq p_1/2$.

In the case when the measured characteristic has two limits of specification and $p_1 < 2p^* < p_2$, when the command button *Calculate Plan* is clicked, the program will display a message that proposes the reconsideration of the plan specifications.

In figure 8 it is presented a situation of this type. The plan specifications were: $p_1$ = 0.0025, $p_2$ = 0.008, $\alpha = 0.05$, $\beta = 0.1$, $U = 520$, $L = 519.98$ and $\sigma = 0.0035$.

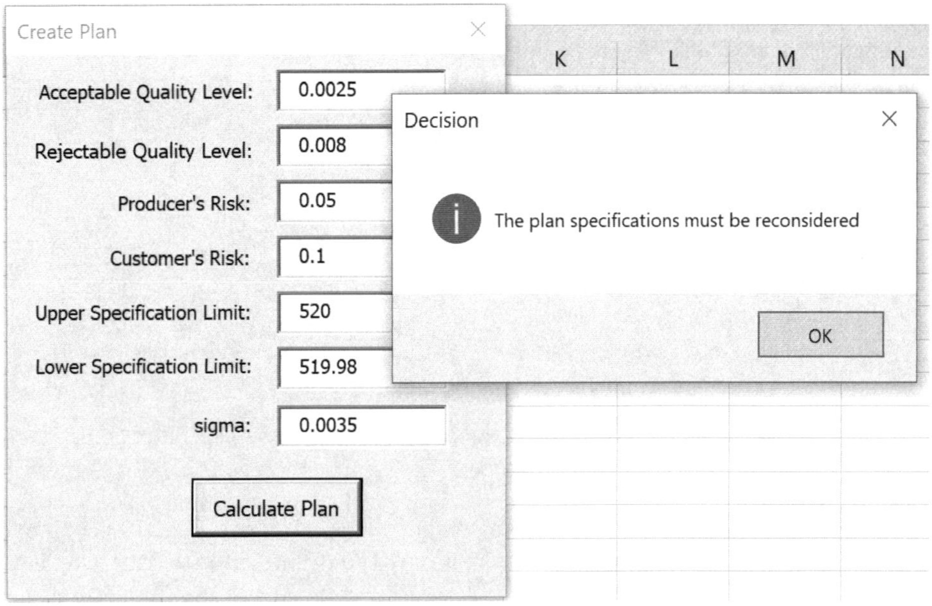


**Figure 8.** A measured characteristic with two limits and $p_1 < 2p^* < p_2$.

If the measured characteristic has two limits of specification and $2p^* \geq p_2$, when the command button *Calculate Plan* is clicked, the program will display a message that proposes the directly rejection of the lot. In figure 9 it is presented a situation of this type. The specifications of the plan were: $p_1 = 0.045$, $p_2 = 0.065$, $\alpha = 0.05$, $\beta = 0.1$, $U = 75.018$, $L = 75$ and $\sigma = 0.007$.

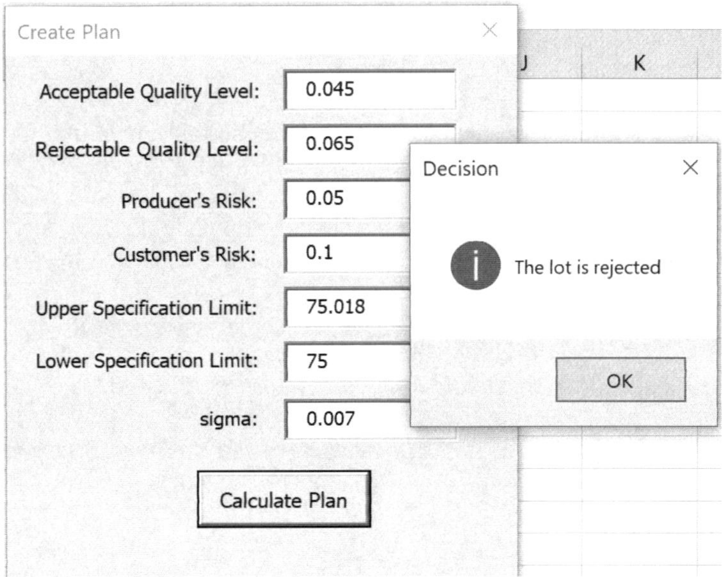


**Figure 9.** A measured characteristic with two limits and $2p^* \geq p_2$.

Next, the paper presents a case for a measured characteristic with only one single lower limit of specification (figure 10). The case when the measured characteristic has only an upper limit of specification is analogous. The plan specifications for this case are: $p_1 = 0.035$, $p_2 = 0.08$, $\alpha = 0.05$, $\beta = 0.1$, $L = 235$ and $\sigma = 0.015$.

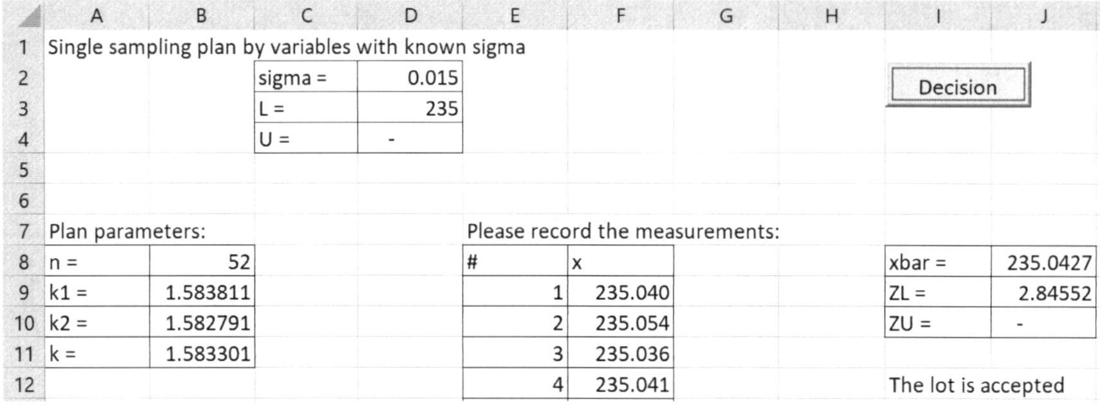


**Figure 10.** A measured characteristic with a lower limit of specification.

## 5. Conclusion

The presented computer program may be used for design and application of single sampling plans by variables when the standard deviation of the lot is known. The quality characteristic may have one or two specification limits.

The function $p_1\_calc()$ that were developed on the basis of the algorithm presented in the paper may be used to determine the new value of $p_1$ that will be used at the computing of the plan parameters in the case when the measured characteristic has two limits of specification and the lot variability compared to the tolerance interval corresponds to the case $p_1/2 < 2p^* \leq p_1$.

The program selects the appropriate case of application for the plan and, using the inspection conditions, calculates the plan parameters. By means of the calculated parameters and the sample measurements the program makes the decision about the lot acceptance or rejection.

## References

[1] Chen J Li K H and Lam Y 2007 Bayesian single and double variable sampling plans for the Weibull distribution with censoring *Eur. J. Oper. Res.* **177** 1062–73

[2] Lee A H I Wu C-W and Wang Z H 2018 The construction of a modified sampling scheme by variables inspection based on the one-sided capability index *Comput. Ind. Eng.* **122** 87-94

[3] Nadi A A Gildeh B S and Afshari R 2020 Optimal design of overall yield-based variable repetitive sampling plans for processes with multiple characteristics *Appl. Math. Model.* **81** 194-210

[4] Lieberman G J and Resnikoff G J 1955 Sampling plans for inspection by variables *J. Am. Stat. Assoc.* **50** 457–516

[5] Schilling E G and Neubauer D V 2017 *Acceptance Sampling in Quality Control*, Third Edition. (Boca Raton: CRC Press Taylor & Francis Group)

[6] Duncan A J 1974 *Quality Control and Industrial Statistics* 4th ed. (Homewood, IL.: Richard D Irwin)

[7] Alexander M and Kusleika D 2019 *Excel 2019 Power Programming with VBA* (Indianapolis: John Wiley & Sons, Inc.)

# Simulation research on pressure shock of load sensing system of large boom tower crane

**Y Fei [1], J Li [1*] and C Sun [1]**

[1]School of Mechanical Engineering, Shenyang Jianzhu University, Shengyang, 110168, Liaoning, China

*E-mail: whale_dajing@hotmail.com

**Abstract.** Large boom tower cranes have large masses, heavy loads, and large inertial dynamic loads during the conversion of working conditions. Even if there is a buffer device, hydraulic shock will still occur. The problem of hydraulic shock seriously affects the leakage of hydraulic oil and decreases system reliability. Now the closed hydraulic system of a certain type of boom tower crane is taken as the research object, and the joint modeling of machinery and hydraulic is completed in AMESim. Without changing the basic hydraulic components, the hydraulic system of the whole machine just adjusts the ratio of the pulley block or changes the hydraulic speed regulation method to study the hydraulic shock of the servo-sensitive hydraulic system. AMESim simulation results show that in this load-sensitive system, the double-rate hoisting of the pulley block is compared with the single-rate hoisting of the pulley block, the maximum dynamic load generated by the load off the ground is increased by 5.4%; Speed regulation of the variable pump and variable motor combination is compared with the speed control of only variable motor, the maximum dynamic load of the hydraulic system is reduced by 7.6%.

## 1. Introduction

Boom tower cranes have super-large lifting moments, super-high lifting heights, high lifting speeds, and small tail slewing radius, which have unparalleled advantages in the construction of high-rise buildings or narrow construction sites. When the working condition of the boom tower crane is switched, the pressure of the tower crane hydraulic system will rise temporarily. Even at a slower speed, due to the huge inertial dynamic load, the pressure in the oil return chamber of the motor will instantly rise to a high level [1], and the oil intake chamber of the motor will not absorb enough oil, which will cause a pressure shock to the system. The occurrence of hydraulic shock will seriously affect the performance and life of each key component in the system and increase the failure rate of the system [2].

In view of the hydraulic shock generated during the lifting process of a large inertia load system, there have been many researches on the hydraulic shock by adding or replacing different hydraulic components to the system. Qing Xiao et al. proposed a parallel rotary buffer valve in a two-chamber hydraulic motor, which can buffer the instantaneous high pressure after the impact pressure is higher than the set pressure of the buffer valve [3]. Jingfang Zhang et al. proposed an accumulator with a throttle valve as a buffer device for the load-sensitive hydraulic system of a coal shearer. Through analysis and optimization of the accumulator parameters, the hydraulic shock is reduced to a certain extent [4]. Linyi Gu et al. used load-sensitive modulation valves instead of throttle valves to improve the starting and braking characteristics of the load [1], and so on.

IOP Conf. Series: Materials Science and Engineering 1009 (2021) 012020 doi:10.1088/1757-899X/1009/1/012020

This paper conducts simulation research on the closed load sensitive system of large scale boom tower cranes, and explores the maximum additional dynamic load generated by the mechanical system when using different hydraulic speed regulation method and single-rate or double-rate lifting, and the resulting hydraulic system pressure overshoot and the resulting pressure shock. Combined with STL2400 full hydraulic boom tower crane produced by Fushun Yongmao Construction Machinery Co., Ltd, using AMESim software as a platform, a simulation model of the lifting mechanism including hydraulic closed system and boom and other mechanical structures is built to simulate the lifting process of the crane under maximum load. During the lifting process, study the hydraulic impact of switching work condition to the system during different hydraulic speed regulation method modes and different load ratios during lifting, thus providing a theoretical basis for the improvement of this type of tower crane hydraulic system.

## 2. Modeling of lifting mechanism

During the lifting process of the crane, the motor in the hydraulic drive circuit rotates and is transmitted to the drum through the coupling and the reducer, drum rotates to realize the retracting of the steel rope, which in turn drive the load up and down. If the pulley is added to the hook, a double-rate lifting pulley group is formed to realize the function of saving effort. The hydraulic driving device and the hoisting drum are installed on the tower crane balance arm platform, and the fixed pulley in the working device is installed on the boom jib. When the lifting mechanism is converted under different working conditions, additional inertial dynamic load will be generated. The inertial dynamic load will cause the change of the internal tension of the wire rope, causing the boom to vibrate at a high frequency and small amplitude around its hinge point with the balance arm, resulting in pressure shock inside the hydraulic system. The above factors affect each other, and in combination affect the generation and attenuation of dynamic load effects. It shows that the hydraulic shock brought by the conversion of the working condition of the hoisting mechanism is the result of a series of complex machine-hydraulic coupling effects [5]. Using mechanical and hydraulic joint modeling can better simulate the hydraulic impact and some dynamic characteristics brought by the boom crane when it is lifted.

The lifting circuit of STL2400 boom tower crane is composed of two Rexroth swash plate axial piston pumps A4VG and four Rexroth swash plate axial plunger motors A6VM connected in parallel. Four motors are used to drive the load together, and the load torque is evenly divided by the four motors. This avoids hydraulic system leakage, increased oil temperature rise, and damage to hydraulic components caused by excessively high working pressure when only one motor drives a large load torque [6].

In order to facilitate the description of the working principle of this closed lifting hydraulic circuit, after omitting part of the control oil circuit, it is simplified into a closed circuit with a single variable pump driving a single variable motor as shown in figure 1.

In the servo-variable pump, the spool acts upon receiving the displacement signal input by the electrical-mechanical converter, changes the flow area of the control side orifice and then changes the flow into the two chambers of the displacement control cylinder, and drives the piston to extend, through the calculation of the function formula, the displacement of the piston extension is converted into the displacement control signal of the main pump. The established model uses a spring damping element to replace the feedback rod in the pump to achieve the feedback function. This displacement signal can make the valve sleeve return to zero position, the control cylinder is also in a "floating" state [7], the pump displacement is fixed, and the feedback adjustment function is realized. The control piston of the servo-sensitive pump is built with a "piston element with rod cavity and a return spring", the function of the return spring is to ensure that the control piston returns automatically when there is no control signal input or the control pressure oil is cut off due to overload, to make the displacement of the main pump 0. Based on the above principles, a hydraulic lifting system model with two pumps driving four motors is built.

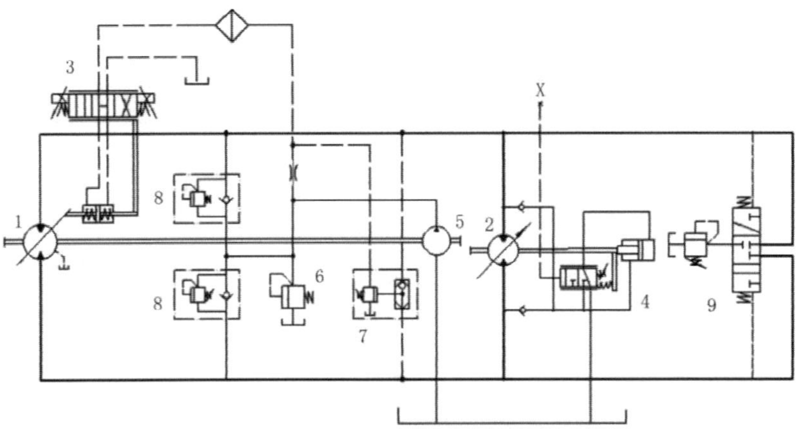


1. Variable pump
2. Variable motor
3. Pump servo variable mechanism
4. Motor servo variable mechanism
5. Charge pump
6. Charge relief valve
7. Pressure shut-off valve
8. High pressure relief valve
9. Flush valve

**Figure 1.** Simplified schematic diagram of closed hydraulic drive system.

The torque and speed output by the motor are firstly decelerated by four Brevini reducers, and then driven by the meshing transmission of the small gear of the motor output shaft and the large ring gear of the reel to achieve two-stage deceleration.

The main parameters of the system are given in table 1.

**Table 1.** The main parameters of the system.

| Name | Parameter |
|---|---|
| Control pressure of motor | 1.3～2.3MPa |
| Maximum speed of motor | 3400rot/min |
| System maximum lifting pressure | 31MPa |
| Rated pressure of pump | 40MPa |
| Geometric displacement of pump | 180mL |
| Maximum output flow of pump | 450L/min |
| Geometric displacement of motor | 200mL |
| Rated pressure of motor | 40MPa |
| Reduction ratio of Brevini reducer | 20.4 |
| Reduction ratio of motor and drum | 170/23 |

STL2400 lifting pulley group is a single ratio or double ratio pulley block composed of guide pulley, force measurement pulley, fixed pulley and movable pulley. Generally, the pulley group adopts double ratio. After the wire rope passes through the fixed pulley and the movable pulley, the rope end is fixed to the tip of the arm. In some occasions where light load and high-speed lifting are required, the pulley block can adopt a single ratio. The lifting rope is directly connected to the hook after passing through the arm-tip pulley, and a complete lifting mechanism model of the boom tower crane is built according to the above principle as shown in figure 2.

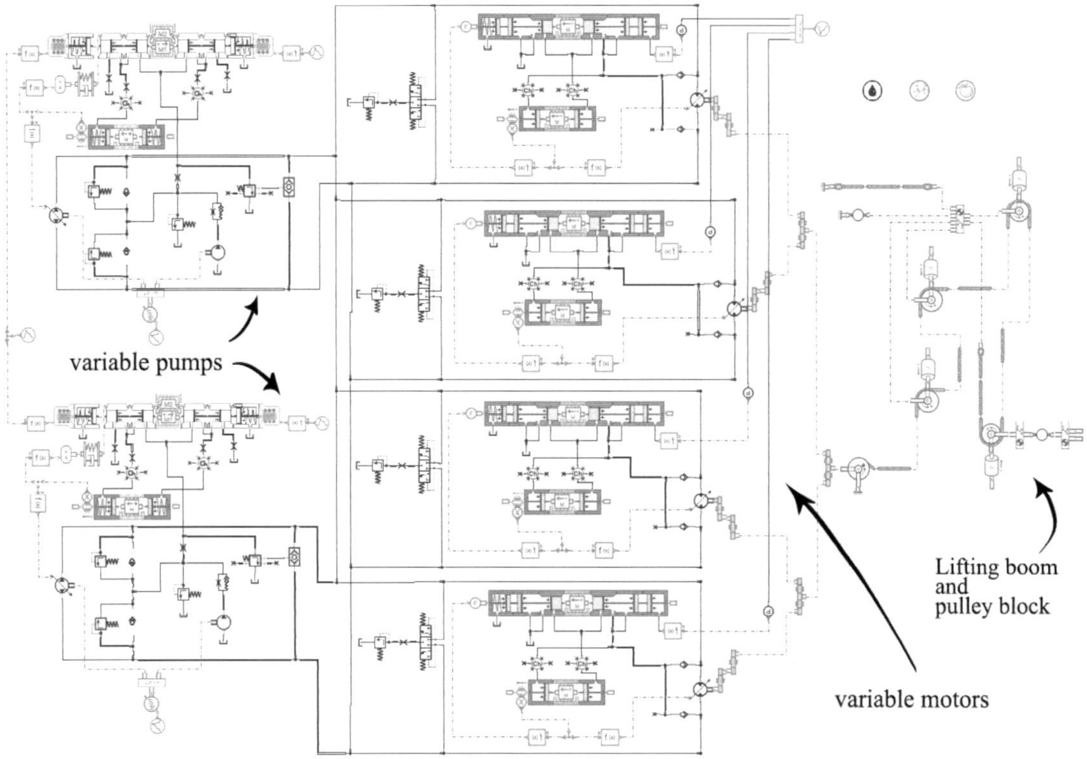


**Figure 2.** AMESim model of complete hoisting mechanism.

The simulation takes the crane lifting weight as the designed maximum lifting weight of 100t, and the boom is at the minimum swing angle of 17°. The complete simulation working conditions include six working stages: accelerated lifting, constant speed lifting, brake hovering, accelerated descending, constant speed descending, decelerating descending until it stops. The real speed regulation process is very short during lifting [8]. In order to visually observe the changes of the simulation curves of various variables, the duration of the speed regulation process in each working stage is specially set to 5s. Take the signal gain of 0.5, the pump and motor variable displacement mechanism control signal source in the model is set according to figure 3.

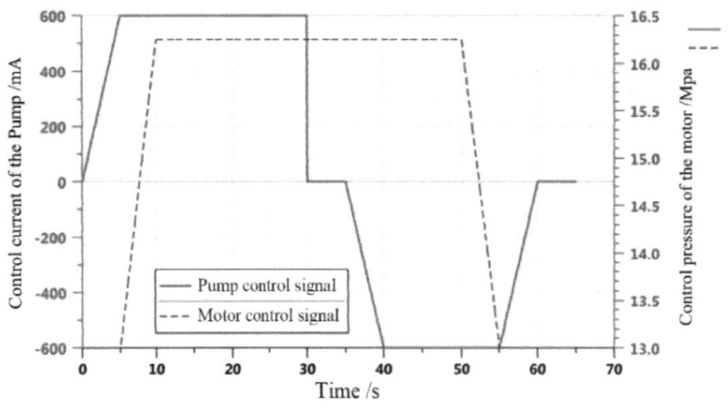


**Figure 3.** Lifting mechanism control signal.

As shown in figure 4, the model simulation verifies that the simulation curve in the startup phase has obvious fluctuations, which is caused by the additional dynamic load caused by working condition conversion in the startup phase.

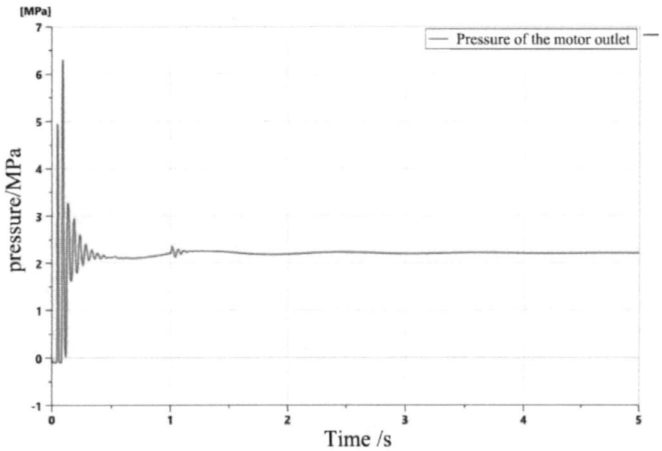


**Figure 4.** Motor oil outlet pressure.

## 3. Simulation analysis

### *3.1. Influence of single-rate or double-rate on hydraulic dynamic characteristics during lifting*

In practice, the boom tower crane changes the position of the hoisting rope to realize the conversion of the pulley block from double-rate to single-rate. Figure 5 shows the AMESim model of the mechanical part of the working device when using single-rate.

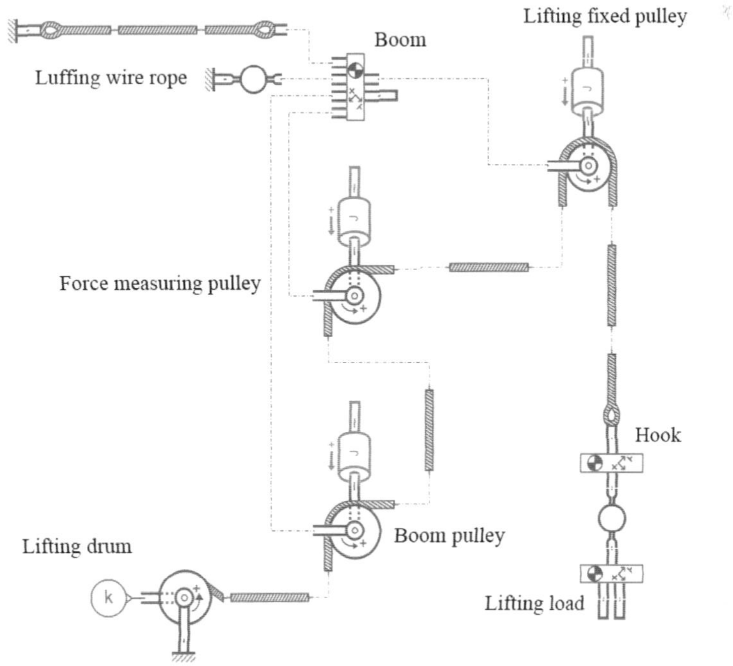


**Figure 5.** Model of single-rate lifting working device.

IOP Conf. Series: Materials Science and Engineering 1009 (2021) 012020 doi:10.1088/1757-899X/1009/1/012020

Set up a comparison simulation of single-rate and double-rate lifting, where the variable pump control current in the single-rate comparison group is half of the double-rate comparison group, as shown in figure 6(a), to ensure that the loads in the two comparison groups move at the same speed. As shown in figure 6(b).

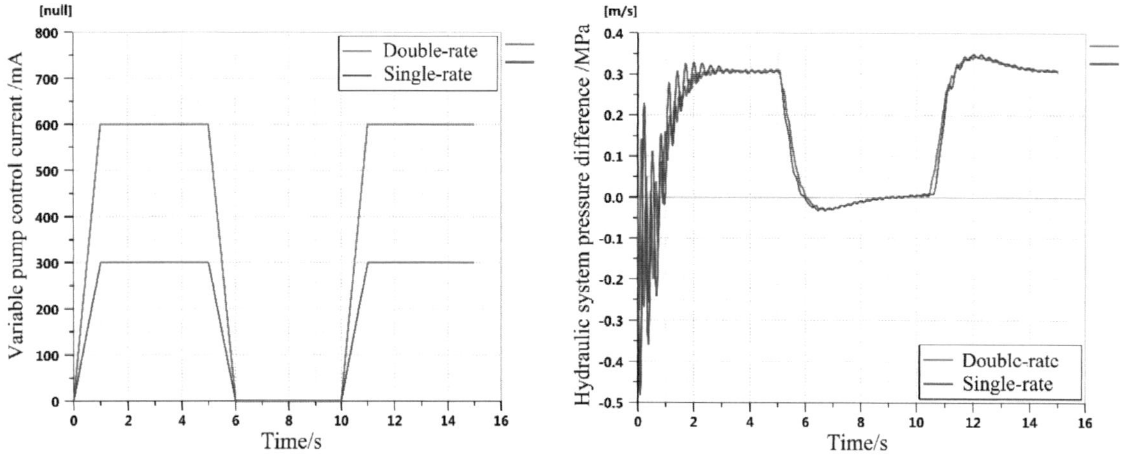


(a) (b)

**Figure 6.** Pump control current and load lifting speed at different rates: (a) variable pump control current; (b) lifting speed curve comparison.

The simulation results are shown in figure 7.

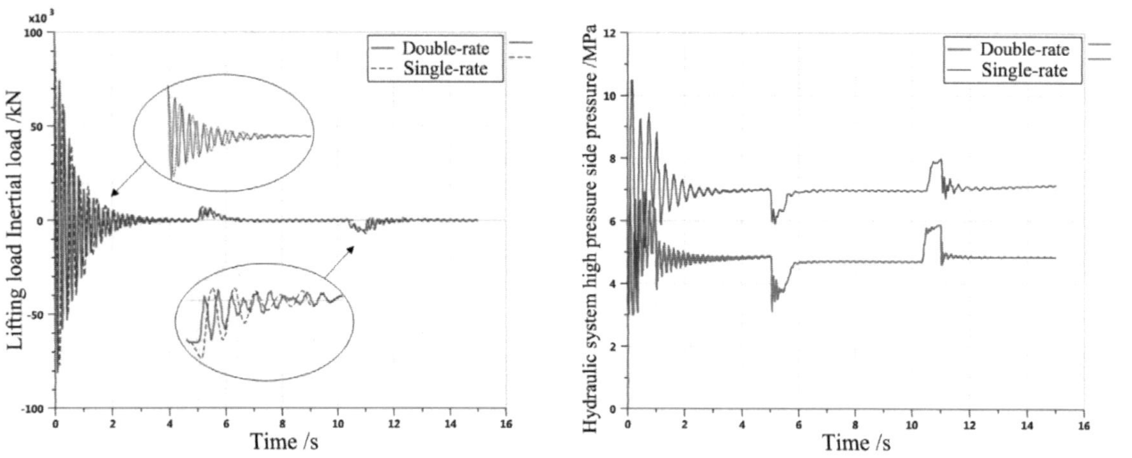


(a) (b)

**Figure 7.** Simulation comparison under different magnification conditions: (a) contrast curve of inertial load during lifting; (b) pressure curve of high-pressure side of hydraulic system.

The dynamic performance indicators of the hoisting system at different rate are shown in table 2.

**Table 2.** Dynamic performance index of lifting system under single-rate or double-rate.

| Working condition | Magnification | Maximum dynamic load (kN) | Maximum overshoot of dynamic load | Peak pressure (MPa) | Maximum overshoot of pressure | Adjusted time (s) |
|---|---|---|---|---|---|---|
| **Lift off the ground** | Single-rate | 77.0 | 78.0% | 10.5 | 50.0% | 2.27 |
| | Double-rate | 81.2 | 82.9% | 7.0 | 45.8% | 2.16 |
| **Lifts twice after hovering** | Single-rate | 5.0 | 5.1% | 8.8 | 25.7% | 1.37 |
| | Double-rate | 6.9 | 7.0% | 5.9 | 22.9% | 1.16 |

When hoisting with double-rate, the maximum dynamic load generated in the stage of the load leaving the ground and the stage of second lift after hovering are both slightly larger than the single-rate hoisting conditions, increasing by 5.4% and 3.8% respectively. The single-rate of the system under dynamic load is slightly better than the double-rate, but on the vibration waveform, the single-rate ratio lags the double-rate, and the adjustment time required to return to a stable state is also longer. As far as the system pressure is concerned, when the single-rate is raised, the system steady-state pressure is increased by 46% compared to the double-rate. The system is under high-pressure working conditions for a long time, which is not conducive to the heat dissipation of the closed system and will cause oil leakage problems.

*3.2. Influence of speed regulation method on dynamic characteristics*

In this example, the variable pump-variable motor combination speed regulation method is used to simulate the load lifting phase, and the simulation results are compared with the results of using a single variable pump to drive the load lifting. In order to ensure that the loads in the control group have the same lifting speed, the control signals of the variable pump and variable motor are set according to figure 8(a). The load speed curve obtained in this way and the load speed curve obtained using only pump variables are placed in figure 8(b).

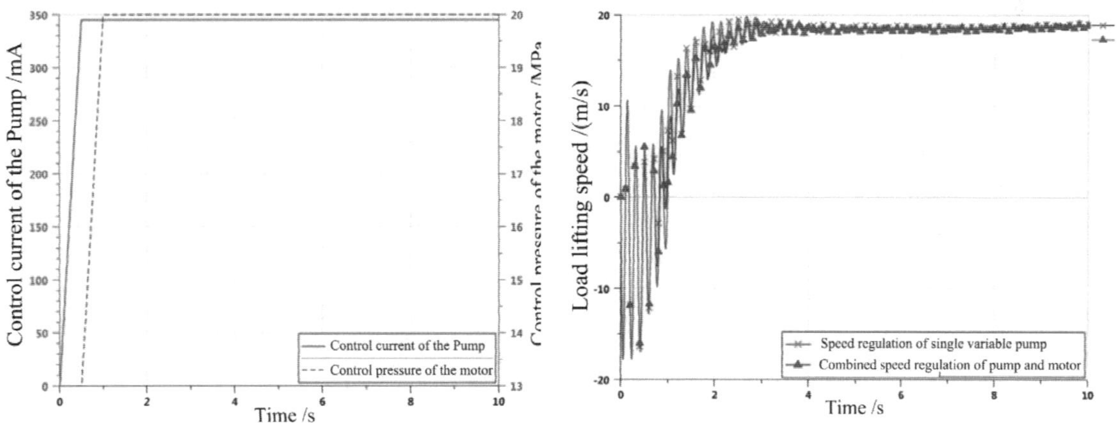


(a) (b)

**Figure 8.** Comparison of control signals and load lifting speed of different variable modes: (a) control current of the pump; (b) contrast curve of lifting speed.

Figure 9 shows the simulation comparison curves under different speed adjustment modes.

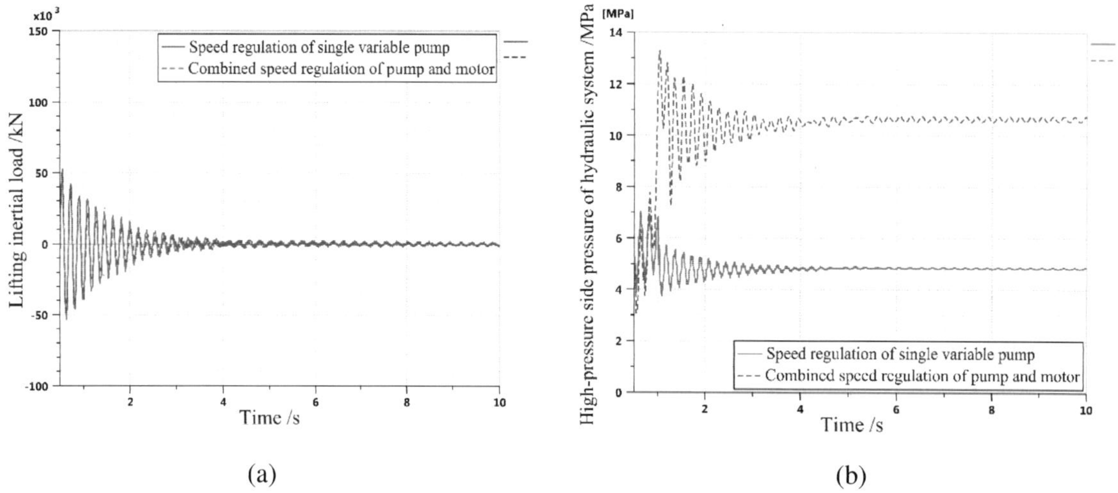


**Figure 9.** Simulation comparison under different speed adjustment modes: (a) contrast curve of inertial load during lifting; (b) pressure curve of high-pressure side of hydraulic system.

Table 3 gives the dynamic performance indexes of the lifting system under different speed control modes.

**Table 3.** Dynamic performance of hoisting system under Different speed regulation modes.

| Speed mode | Maximum dynamic load (kN) | Maximum overshoot of dynamic load | Peak pressure (MPa) | Maximum overshoot of pressure | Adjusted time (s) |
|---|---|---|---|---|---|
| **Adjust speed with pump** | 53.5 | 54.6% | 7.1 | 47.9% | 1.7 |
| **Adjust speed with motor** | 49.7 | 50.7% | 13.3 | 26.7% | 0.9 |

In this hydraulic lifting mechanism, the variable speed ratio of the variable pump-variable motor combination is only with variable pump speed control, the maximum dynamic load of the hydraulic system is reduced by 7.6%, and the adjustment time is shortened by 89%. However, the high-pressure side pressure (10.5MPa) under steady state increased by 118.8% compared with the single pump speed control group (4.8MPa), and the peak pressure increased by 87%. It can be seen that using variable pump speed regulation under the same load condition than the variable pump-variable motor combination speed regulation, the steady-state working pressure is reduced by about 50%, and some closed-system heating can be reduced. Therefore, on the premise of meeting the speed requirements, variable pump speed regulation method should be selected.

## 4. Conclusion

This article takes the full hydraulic boom tower crane lifting mechanism with a maximum lifting torque of 24000N·m as the research object, and uses AMESim simulation software to deal with the hydraulic shock caused by the inertial dynamic load during the conversion of the working condition of the lifting mechanism. Simulation is a means to analyze the influencing factors of the dynamic characteristics of the hoisting mechanism. The inertial dynamic load generated during the working condition change of the lifting mechanism will cause the pressure impact of the hydraulic system, and the pressure overshoot will reach 50.3% of the steady state working pressure.

Under the same lifting load conditions, the maximum dynamic load of single rate is reduced by 5.2% compared with double rate, which has no obvious effect on improving the force of the mechanism. However, the steady-state pressure has increased by 46%, which will bring the problem of increased system temperature and increased leakage. Therefore, when high-speed or light-load is not required, single rate is not recommended, but double rate should be used;

The hydraulic circuit of the hoisting mechanism is a closed hydraulic system that uses variable pump and variable motor speed regulation. Under the same load condition, a single variable pump speed control is compared with variable pump-variable motor combination speed control. The steady-state working pressure is reduced by nearly 50%, which can effectively reduce the closed system heating. Therefore, under the premise of meeting the speed requirements, the variable pump should be used for speed regulation first.

The above data can provide useful conclusions for the structure and hydraulic system improvement of this type of tower crane. In future research, the dynamic characteristics of the switching of working conditions during hoisting can be further optimized through comprehensive consideration of various parts in the hoisting mechanism. The load-sensitive closed hydraulic system model of the lifting mechanism of the boom tower crane in the article can provide a reference for the optimal design and research of the system in the future.

## References

[1] Linyi G, Yingjun X and Qingfeng W 2001 *J. Construction Machinery and Equipment* **2** 30-33
[2] Ning L, Yufeng Z and Jiancheng W 2007 *J. Machine Tool & Hydraulics* **4** 154-197
[3] Qing X 2013 *J. Machine Tool & Hydraulics* **15** 75-77
[4] Jingfang Z, Hao H and Shangqing H 2019 *J. Coal Technology* **038** (5) 139-41
[5] Zheng C 2012 *D. Research on Dynamic Characteristics of Hydraulic System of Hydraulic Crawler Crane Lifting Mechanism* (Central South University )
[6] Pengyan Y, Qiuxia Z and Pingxi Y 2015 *J. Chinese Hydraulics & Pneumatics* **15** (9) 52-55
[7] Jiwei W, Hongjia Z *et al. 2009 M. Hydraulic Transmission* (Beijing: China Machine Press) 153
[8] Ming'an L 2011 *J. Fluid Power Transmission & Control* **15** (6) 23-25

# Analysis of the influence on the hydraulic lifting circuit of the boom crane when the characteristics of the steel wire rope changes

**Y Fei[1], J Jiang[1*] and C Sun[1]**

[1]School of mechanical engineering, Shenyang Jian Zhu University, Shenyang, 110168, Liaoning, China

*E-mail: 18609880787@163.com

**Abstract**. In the process of lifting and landing of large hydraulic crane fully loaded, the working condition conversion makes the stress of steel wire rope change frequently and generate additional dynamic load, which makes the movement of the crane unstable and affects the working stability of the hydraulic system, and also leads to the fatigue damage of the tower structure and reduces the effective working life of the tower crane. In order to find the optimal dynamic characteristics of hydraulic system, the initial tension force, stiffness of hoisting wire rope and rigidity of variable amplitude wire rope of hoisting wire rope are studied respectively by means of machine-liquid integration simulation based on the AMESim, a kind of simulation software. The lifting mechanism of a type of closed hydraulic system is modelled and simulated. The results show that the maximum dynamic load of hoisting mechanism is reduced by 92.6% and the stabilization time is reduced by 49%. The optimized stiffness of hoisting wire rope reduces the dynamic load and peak pressure of hoisting mechanism by 5.2% and 4.3% respectively. The stiffness of the variable amplitude steel wire rope has little effect on the dynamic characteristics of lifting.

## 1. Introduction

For the lifting mechanism of the full-hydraulic tower crane with heavy-duty and full-power, a large inertial dynamic load will be generated when the operating conditions change, such as acceleration lifting, braking hover, acceleration descent, and deceleration descent. It will not only cause the vibration of the overall structure of the tower but also cause pressure fluctuations in the hydraulic system, so we need to analyze the dynamic characteristics of the lifting mechanism when the working conditions change. In paper [1], the values and curves of various performance parameters of the hoisting system will be obtained by means of integrated modeling and simulation of the machine and fluid. First, the mechanical-hydraulic system model of the hoisting mechanism will be established by using The AMESim, a kind of modeling and simulation software. Then, by analyzing the transmission devices of the hoisting mechanism, it is considered that the steel wire rope (hereinafter referred to as the rope) is the most typical elastic-damping member. The difference in initial tension and rigidity will alleviate the impact of the inertial dynamic load on the system to a certain extent. Therefore, on the basis of the model that has been built, the simulation is carried out under different conditions of the initial tension of the hoisting rope, the rigidity of the hoisting rope and the rigidity of the amplitude-changing rope. By comparing and analyzing the simulation results, the influence of various factors on the dynamic characteristics of the hoisting mechanism is obtained [2, 3].

## 2. Working principle of the lifting system

The function of the tower crane lifting mechanism is to drive and control the lifting device to complete the vertical lifting, lowering or hovering of the load at a certain speed within the lifting height range. The lifting mechanism of the boom tower crane studied in this paper is mainly composed of working device and hydraulic drive circuit [4]. Among them, the working device includes a steel wire rope, a drum, a pulley block and a hook, and the hydraulic drive circuit is a variable pump-variable motor closed hydraulic system. Figure 1 is a schematic diagram of the lifting mechanism.

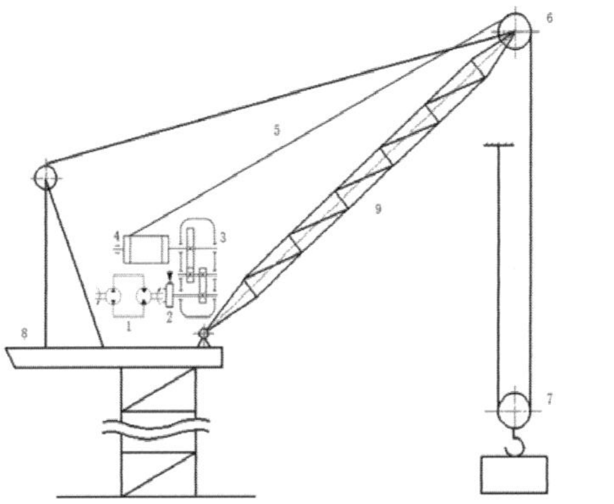

1-Hydraulic drive circuit 2-Coupling 3-Reducer 4-Reel 5-Wire rope 6-Fixed pulley 7-Moving pulley 8-Balance arm 9-Boom

**Figure 1.** Lifting mechanism diagram.

## 3. Establish the mechanical and hydraulic integration model of lifting system

### *3.1 Modelling the mechanical structure*

First of all, use SolidWorks modeling software to build a three-dimensional solid model of the cantilever and balance arm, as shown in figure 2. Taking the hinge point of the cantilever and the balance arm as the coordinate origin, the weight of the cantilever obtained by the "mass attribute" function in SoildWorks is 30365kg, and the coordinate position of the center of gravity is (43.25, 0) [5].

Then, select the six reserved ports physical module in the AMESim mechanical library. Use port 1 as the hinge point connecting the boom and the balance arm, and set the coordinate positions 2, 3, 4, 5, 6 according to the attribute characteristics shown in the three-dimensional model, each position respectively indicates the articulated point which connected the boom with crown block, force fixing pulley, boom lifting point and arm tip pulley, and the lock line position. After that, build a sketch of the hoisting mechanism model. In this example, a double-rate hoisting pulley set is used, so four sets of pulley modules, drum modules, wire rope modules and physical modules are selected in the mechanical library to build a model sketch, the built model is shown in figure 3.

**Figure 2.** Three-dimensional model of boom and balance arm.

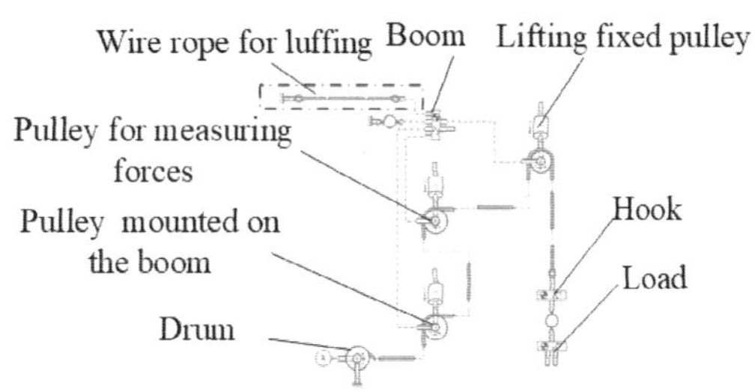


**Figure 3.** Lifting device working device model.

In particular, it is pointed out that the difficulty of wire rope modeling is the setting of its initial parameters, which is also the difficulty and focus of the entire lifting working device modeling. Its initial elongation needs to be set manually and has a great influence on the simulation results. In order to avoid the forced vibration of the wire rope due to the gravity of the boom when we run the simulation, the initial elongation of the luffing rope can be set as follows:

- Fix one end of the luffing rope at the top of the tower, and connect the other end to the boom lifting point. Run the simulation with the initial elongation of the wire rope set to zero;
- Use the elongation of the wire rope in a stable state after oscillation as the initial elongation of the wire rope, reset the parameters and run the simulation again.

Repeat the steps above repeatedly until the satisfactory initial elongation of the rope is obtained [6].

*3.2 Modelling the hydraulic system*

This simulation analysis of the hydraulic system uses a typical closed volume speed control circuit called 'variable displacement pump-variable displacement motor' circuit. However, it will cause hydraulic system leakage, increased oil temperature and damage of hydraulic components by high working pressure when a single motor drives a large load torque, so we use two variable pumps to drive four variable motor drives, the system schematic diagram is shown in figure 4.

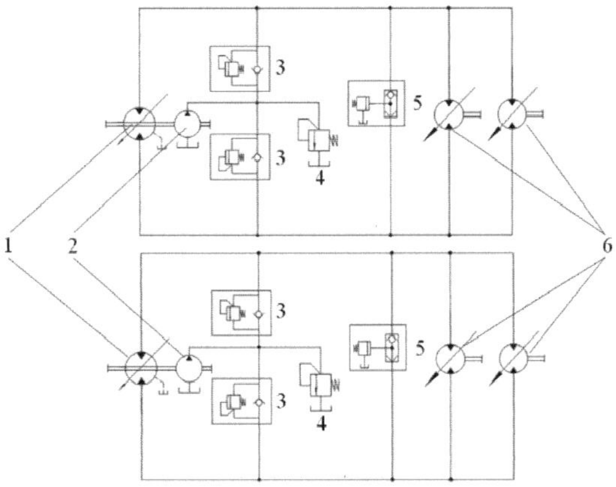


1. Hydraulic pumps 2. Hydraulic charge pumps 3. High pressure relief valve 4. Charge relief valve 5. Pressure shut-off valve 6. Hydraulic motors

**Figure 4.** Schematic diagram of hydraulic system.

Use pumps, shuttle valves, relief valve variable motors and other components in the AMESim hydraulic library, then refer to the hydraulic schematic shown in figure 3.3 to build a complete model of the hydraulic system, as shown in figure 5.

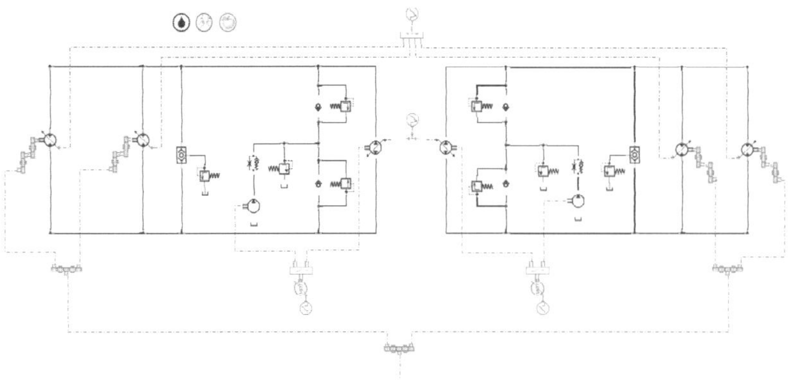

**Figure 5** Simulation model diagram of hydraulic system.

## 4. Simulation analyse the system with different wire rope characteristics

For large tower cranes, the speed of the lifting mechanism to drive the load will not be very fast, the moving speed of the wire rope is slow, and the damping force generated accounts for a small proportion of the wire rope pulling force [7, 8]. Therefore, it will not be discussed here. Analysis of the dynamic characteristics of the lifting mechanism. Set the working conditions as shown in figure 6, and analyse based on the simulation model established above.

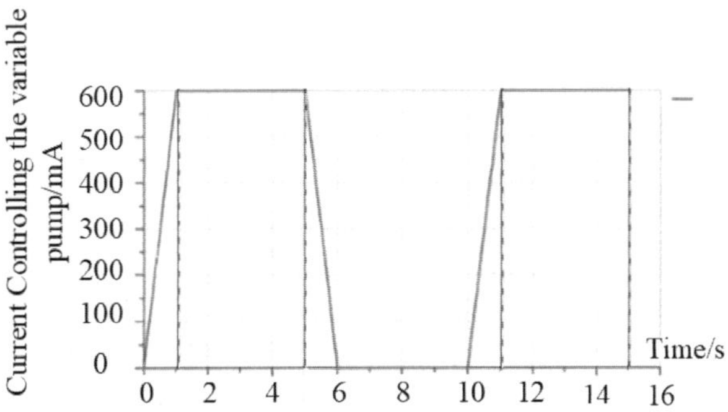


**Figure 6.** Variable pump control current and working condition division.

### *4.1 Simulation analyse the system when the initial tension of hoisting rope is different.*

Two sets of comparative simulation experiments are set up here to verify the influence of the steel wire tension on the dynamic characteristics of the mechanism. Both of the simulation experiments will use the same lifting weight of 10t, a single ratio pulley block, 17-degree boom swing angle and the acceleration of gravity is $g=9.8m/s^2$. The effect of other disturbances is not considered, When the load remains stationary or at a constant speed, the pulling force of a single rope is $F_{r1}=10000\times9.8/2=49kN$. The batch processing function is used to set the initial tension of the steel wire ropes in the two comparison groups to be 0 and 49kN, respectively, then run the simulation.

The curve of the inertial of the hoisting load and the system pressure is shown in figure 7(a) and figure 7(b) under two different initial tensions of the hoisting rope.

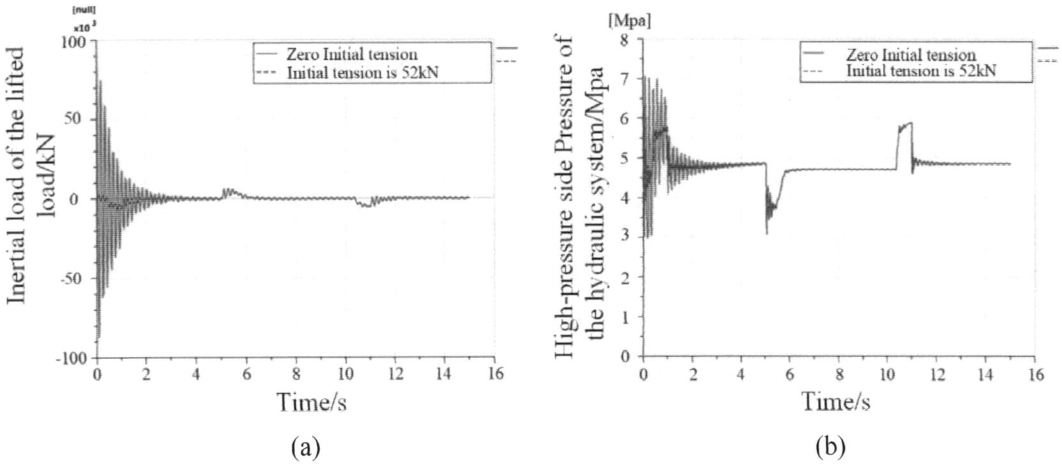


**Figure 7.** Simulation comparison of different initial tension forces: (a) load inertia load comparison curve; (b) high pressure side pressure curve of hydraulic system.

Table 1 lists the dynamic performance indicators of the hoisting system under different initial tensions, where the adjustment time error band to restore the steady state is taken as ±5%.

**Table 1.** Dynamic performance index of lifting system under different initial tension.

| Working condition | Initial tension (kN) | Maximum dynamic load (kN) | Maximum overshoot of dynamic load | Peak pressure (MPa) | Maximum overshoot of pressure | Adjustment time (s) |
|---|---|---|---|---|---|---|
| Lifting from the ground | 0 | 89.70 | 91.6% | 4.90 | 93.6% | 2.05 |
| | 52 | 6.60 | 6.8% | 4.03 | 59.0% | 1.04 |
| Secondary lifting | 0 | 5.27 | 5.1% | 3.70 | 44.0% | 1.02 |
| | 52 | 5.37 | 5.5% | 3.60 | 42.0% | 1.03 |

According to the figure 7 and table 1, it is not difficult to find that the control group with the initial tension set to 52kN has a 92.6% reduction in the maximum dynamic load under the lifting condition of the load compared to the control group with the initial tension set to 0. The adjustment time reduced by 49%. In the control group with the initial tension of 52kN, the lifting force acting on the load at the initial moment is equal to the gravity of the load, so in this case, lifting from the ground is equivalent to the second lifting condition. The dynamic load and pressure curve at the high-pressure side while lifting the load from the ground basically consistent with the second lifting condition. The data given in table 1 can also verify this point well. After going through the load off-ground stage, the tension of the steel wire ropes of the two control groups stabilized at the same value. When they experienced other working conditions, the simulation curves of the two basically coincided.

*4.2 Simulation Analyse the system when the Stiffness of lifting rope is different.*

To investigate the influence of the hoisting rope stiffness on the dynamic characteristics of the hoisting mechanism, five groups of control groups are set up here. Set the initial tension of the hoisting rope is 0, the unit rigidity of the luffing rope is $1.2\times10^8$N/m, and the unit stiffness of the hoisting rope is selected according to the sample, the range is $1.2\times10^8$~$2.8\times10^8$N/m. In the batch processing function, the step size of the stiffness change is set to $2\times10^7$N/m and the rest of the settings are the same as above example. Due to the short hoisting time, the length of the hoisting rope changes very little compared to the total length. So the rope length can be regarded as a fixed value, which is 166m in this mechanism. The total stiffness of the lifting wire rope can be calculated from this.

In order to facilitate the observation of the simulation results, only the dynamic load and high-pressure side pressure curves corresponding to the maximum and minimum stiffness are intercepted, as shown in figure 8.

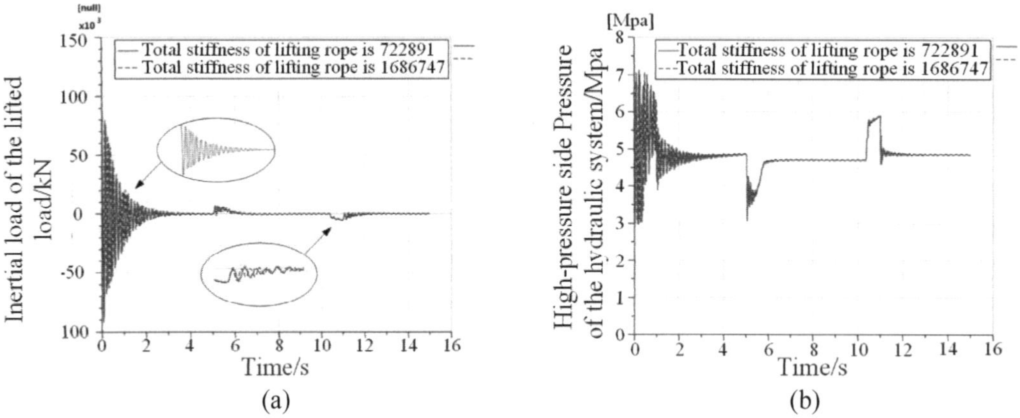


**Figure 8.** Simulation comparison of different Hoisting wire rope stiffness: (a) load inertia load comparison curve; (b) high pressure side pressure curve of hydraulic system.

The dynamic performance indexes of the hoisting system under different hoisting rope stiffness conditions are listed in table 2.

**Table 2. (a)** Dynamic performance index of hoisting system under different stiffness of rope.

| Working condition | Lifting rope stiffness (N/m) | Maximum dynamic load (kN) | Maximum overshoot of dynamic load | Peak pressure (MPa) | Maximum overshoot of pressure | Adjustment time (s) |
|---|---|---|---|---|---|---|
| Lifting from the ground | 722891 | 87.2 | 88.9% | 7.0 | 45.8% | 1.99 |
| | 963855 | 90.6 | 92.4% | 7.1 | 47.7% | 1.79 |
| | 1204819 | 88.3 | 90.1% | 7.2 | 50.0% | 1.43 |
| | 1445783 | 90.5 | 92.3% | 7.3 | 52.0% | 1.12 |
| | 1686747 | 91.8 | 93.7% | 7.3 | 47.9% | 1.68 |

**Table 2. (b)** Dynamic performance index of hoisting system under different stiffness of rope.

| Working condition | Lifting rope stiffness (N/m) | Maximum dynamic load (kN) | Maximum overshoot of dynamic load | Peak pressure (MPa) | Maximum overshoot of pressure | Adjustment time (s) |
|---|---|---|---|---|---|---|
| Secondary lifting | 722891 | 5.4 | 5.5% | 5.8 | 20.8% | 1.02 |
| | 963855 | 5.7 | 5.8% | 5.8 | 20.8% | 1.00 |
| | 1204819 | 5.5 | 5.6% | 5.8 | 20.8% | 0.89 |
| | 1445783 | 5.5 | 5.6% | 5.8 | 20.8% | 0.74 |
| | 1686747 | 5.3 | 5.4% | 5.8 | 20.8% | 0.55 |

This shows that when the total stiffness of the hoisting rope changes from 722891N/m to 1686747N/m, the maximum dynamic load is only increased from 87.2kN to 91.8kN, and the peak pressure on the high pressure side is only increased from 7MPa to 7.3MPa, that is, the ratio of the maximum stiffness of the wire rope to the minimum stiffness is 2.33, and the difference between the dynamic load and the peak pressure caused by the two is only 5.2% and 4.3%.

*4.3 Simulation Analyse the system when the Stiffness of luffing rope is different.*

Also set up five groups of control groups with a lifting weight of 10t, a boom swing angle of 17°, and an initial tension of the hoisting rope of 0. The simulation conditions follow the settings in Figure 4.1. After checking the sample, the variation range of the unit stiffness of the luffing rope is selected as: $8\times10^7$~$1.2\times10^8$N/m, and the simulation step size is set to $10^7$N/m in the AMESim batch processing function. When setting the stiffness of the wire rope in AMESim, the stiffness per unit length needs to be given. According to the equation (1) and (2). It's not hard to find that the $L$ and $k$ are automatically calculated by the software based on the wire rope tension during the simulation.

$$F = \frac{EA}{m} \tag{1}$$

$$k = \frac{EA}{L} \tag{2}$$

Since the length change caused by the tensile deformation of the luffing rope is very small compared with the total rope length, both the length of the luffing rope $L$ and the total stiffness $k$ can be regarded as fixed values. The simulation results are shown in figure 9.

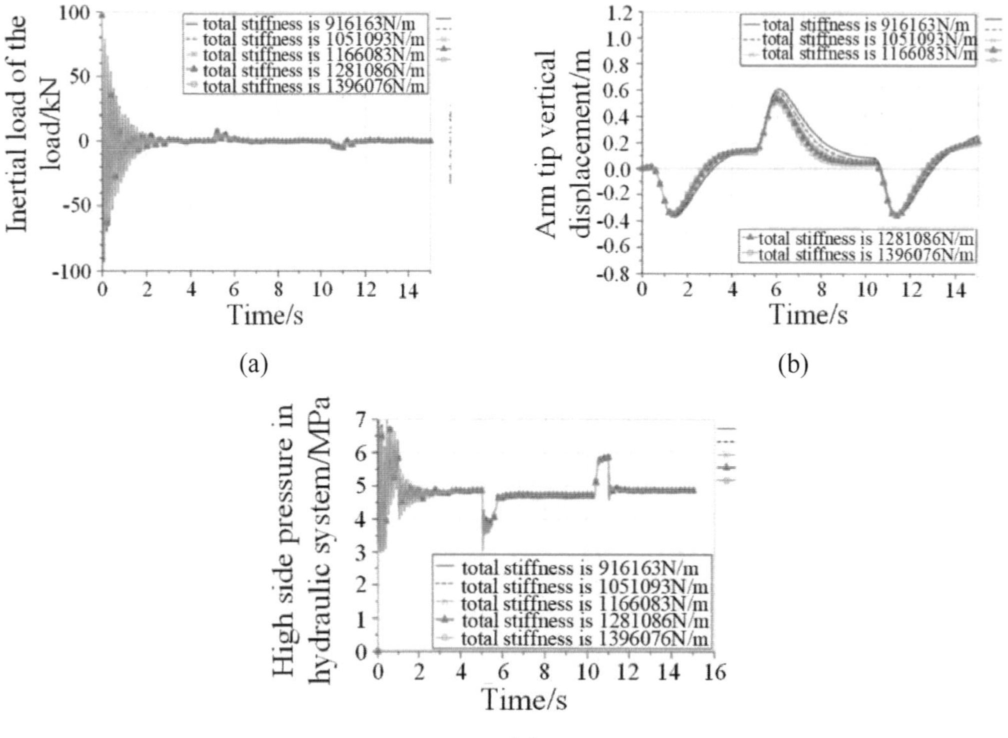


**Figure 9.** Simulation comparison of different amplitude-variable rope stiffness: (a) load inertia load comparison curve; (b) arm tip vertical displacement contrast curve; (c) high pressure side pressure curve of hydraulic system.

The total stiffness which is automatically calculated by the software and some other dynamic performance indexes of five groups different unit stiffness ropes are listed in table 3.

**Table 3.** Dynamic performance index of hoisting system under different stiffness of rope.

| Working condition | luffing rope stiffness (N/m) | Maximum dynamic load (kN) | Maximum overshoot of dynamic load | Peak pressure (MPa) | Maximum overshoot of pressure | Adjustment time (s) |
|---|---|---|---|---|---|---|
| Lifting from the ground | 916163 | 93.4 | 95.3% | 6.9 | 43.8% | 2.16 |
| | 1051093 | 90.6 | 92.4% | 6.9 | 43.8% | 2.16 |
| | 1166083 | 88.9 | 90.7% | 6.9 | 43.8% | 2.16 |
| | 1281086 | 87.4 | 89.1% | 6.9 | 43.8% | 2.16 |
| Secondary lifting | 916163 | 5.2 | 5.3% | 5.8 | 20.8% | 1.02 |
| | 1051093 | 5.2 | 5.3% | 5.8 | 20.8% | 1.02 |
| | 1166083 | 5.3 | 5.4% | 5.8 | 20.8% | 1.02 |
| | 1281086 | 5.3 | 5.4% | 5.8 | 20.8% | 1.02 |
| | 1396076 | 5.3 | 5.4% | 5.8 | 20.8% | 1.02 |

It can be seen that the variation of the stiffness of the luffing rope within the allowable range from 916163N/m to 1396076 N/m causes very small changes in load inertial load, vertical displacement of the arm tip and pressure on the high-pressure side of the system, which can be ignored. Therefore, the change of the rigidity of the luffing rope within the allowable range has little effect on the dynamic characteristics of the hoisting. It can be noticed from Figure 9 that the arm tip vibrates up and down around the initial position during the simulation, and the displacement range is between -0.4m~0.6m.

## 5. Conclusions

It can be clearly seen from the combination of figure 7 and table 1 that setting a certain initial tension is equivalent to transforming the ground-lifting condition into a secondary lifting condition. The simulation results show that when the initial tension is set on the hoisting rope, the dynamic load and high-pressure side pressure oscillations when lifting the load are much smaller than the case when the initial tension is set to zero;

The change in the rigidity of the hoisting wire rope does not significantly improve the dynamic characteristics of the hydraulic system, so when the stiffness of the hoisting wire rope is selected within the specified range, it can be selected with the principle of lowest cost;

The change of the stiffness of the luffing rope within the allowable range has little effect on the lifting dynamic characteristics.

## References

[1] Mingliang Y and Zhengyan C 2006 *J. Modern Manufacturing Engineering* **04** 104-07
[2] Hanjun P, Xiaopeng X and Liang G 2013 *J. Journal of Southwest Jiaotong University* **48** (02) 205-10
[3] Donglai L 2012 *D. Dynamic Characteristic Analysis of Flexible Attachment Boom Tower Crane* (Shenyang Jian Zhu University)
[4] Lefu C 2011 *M. Building Tower Crane* (China Environmental Science Press)
[5] Standards Press of China 2017 *GB/T13752-2017 Design rules for tower cranes* (Beijing: Standards Press of China & General Administration of Quality Supervision, Inspection and Quarantine of the People's Republic of China)
[6] Phillips J W and Costello G A 1985 *J. Journal of Applied Mechanics* **52** (03) 510
[7] Kai F, Yujun X and Fang Y 2018 *J. Coal Technology* **37** (06) 32-38
[8] Shiqing Z 2008 *J. Science and Technology Innovation Herald* **16** 65-65,71

# Study of input parameters influence on the surface quality at laser processing of Hardox 400 steel

**C C Girdu[1*], M V Dragoi[1], M Mileșan[1], L A Mihail[1], L Cirtina[2] and C Radulescu[2]**

[1] Manufacturing Engineering Department, Transilvania University of Brașov, Eroilor nr. 29, 500036 Brasov, Romania
[2] University "Constantin Brâncuși" Targu-Jiu, Str. Republicii, nr. 1, 210135 Targu-Jiu, Romania

*E-mail: girdu2003@yahoo.com

**Abstract**. The paper presents the result of the experimental research regarding the influence of the parameters of the laser cutting regime of the HARDOX 400 steel on the roughness of the processed surface. The study was carried on by means of experimental research, based on a plan developed according to the Taguchi method, a fractionated factorial one, applied for three input parameters: assistant gas pressure, laser power, and cutting speed, each having three levels. The statistical processing of the experimental data was performed by ANOM (analysis of means) and S/N indexes (signal to noise), according to Taguchi Method approach, validated by ANOVA (analysis of variation). Particularly, the paper analyses the combined influence of laser power and the pressure of the assistant gas on the roughness. The main conclusion is that among the two parameters, the most significant influence is that of the pressure of assistant gas.

## 1. Introduction

Roughness is a parameter that characterizes the quality of surfaces. When it comes to laser cutting of metal, the cutting parameters – the laser power, the assistant gas pressure, and the cutting speed - have an important influence on the roughness of the processed surfaces. The scientific literature displays some researches related to cutting parameters at laser processing of metals, and their influence on the output, as follows (selection): Adelmann B and Hellmanna R present an optimization algorithm of input parameters (laser power, cutting speed, the distance between noozle and part, the assistant gas pressure, and the of their interaction when it comes to laser cutting of Aluminum sheet having the thickness of 1 mm [1]. Lutey A et al. concludes that when cutting steel with part material having thickness of 1…4 mm. the best result in terms of surface quality is obtained when use Oxygen as assistant gas, medium laser power, and low cutting speed [2]. Ivarson A et al. describe a study regarding to the influence of alloying elements in the material on the quality of surface processed by laser [3]. In [4] Bachy B et al. claim the novelty of using Artificial Neuronal Network (ANN) to estimate the roughness of the part processed by laser, based on the values of input cutting parameters: laser power, pulse rate, and cutting speed. Parthiban A claims that the quadratic model is the best in terms of correlation to experimental data and the profile shape (linear or curved) have an important influence on kerf width [5]. Hashemzadeh M et al. presents a study which shows that the width of kerf is directly influenced by the size of laser beam, but it is independent from interaction duration of laser with part material [6]. In [7] Pocorni J et al. analyse the shape of the kerf al laser processing of stainless steel with $CO_2$ laser. The conclusion is that there is not a direct relationship between the tilt of strips on the flank of part and the angle of kerf. A different approach on engendering the strips on the flank of laser cut parts has Singh et al. [8]; they affirm that strips are caused of light reflation on the tart surface. The source of the reflection is supposed to be some optical elements locate behind the lens of laser device. Thombansen U et al. says that roughness is influenced by the wavelength of the laser beam, cutting speed, and the position of the focal point [9]. Zhang and Lei affirm in [10] that the perturbations in the flow of melt material cause defects of surface quality, as deep strips and adherent slag. In [11] Madic et al. presents an iterative search algorithm, with six variables to study laser cutting of metal sheets. Pramanik D and collaborators studied the Ti-Al-4V alloy in laser processing, applied to increase the hardness and corrosion resistance, of parts used for medical purposes. The effects of the cutting angle and the melting depth on the Ti alloy were analyzed depending on the power, operating cycle, frequency of pulse, and cutting speed. Inclination of kerf and surface roughness were measured and analysed, as well [12]. Wust et al. presents in [13] an experimental research on laser processing of general usage steel. Their conclusion was that the most relevant influence on the surface quality is that of cutting speed.

Different from the cited studies, the present paper aims to present some aspects concerning with the influence of two of the main cutting parameters on the quality of the surface at processing (cutting) by laser the Hardox 400 steel. This material was selected for study, because it is hard to be processed by chip removing technologies, so laser processing is considered to be more proper. In this context the quality of the surface is expressed by roughness. The experimental research considered the influence of pressure of assistant gas, laser power, and cutting speed. Particularly, this paper discuss only the influence of the first two mentioned parameters. For this study, cutting speed is considered to be constant, that is it does not act as influence factor on the roughness of the processed part. Thickness of the metal sheet was of 10 mm.

## 2. The experimental research methodology

The study is based on an experimental research, conducted according to a plan developed through Taguchi method, a fractionated factorial one. The plan is applied for three of the important cutting parameters: assistant gas pressure, laser power, and cutting speed, each having three levels. Processing of the experimental data was performed by ANOM (analysis of means) and S/N indexes (signal to noise), according to Taguchi Method approach, validated by ANOVA (analysis of variation).A series of nine combinations of input data level was selected. The input data are presented in table 1. The set of processings was repeated five times.

**Table 1.** Input/output data (input data set was replicated four timed).

| Part # | Laser power [w] | Gas pressure [bar] | Cutting speed [mm/min] | Roughness [µm] |
|---|---|---|---|---|
| 1 | 4100 | 0.35 | 1200 | 3.48 |
| 2 | 4100 | 0.45 | 1400 | 3.00 |
| 3 | 4100 | 0.55 | 1600 | 4.40 |
| 4 | 4200 | 0.35 | 1400 | 2.32 |
| 5 | 4200 | 0.45 | 1600 | 1.70 |
| 6 | 4200 | 0.55 | 1200 | 4.19 |
| 7 | 4300 | 0.35 | 1600 | 2.94 |
| 8 | 4300 | 0.45 | 1200 | 3.74 |
| 9 | 4300 | 0.55 | 1400 | 3.52 |

Material used was Hardox400, having the chemical composition presented in table 2.

**Table 2.** Chemical composition of Hardox400 steel [%].

| | Content of alloying elements | | | |
|---|---|---|---|---|
| **Hardox 400 steel** | C | Si | Mn | Cr |
| | 0.19 | 0.50 | 1.25 | 0.70 |

The thicknes of 10 mm. is a commonly used one in industry to produce parts that are subject of hard mechanical abrasion and wear. This thickness is sufficient to allow measuring in appropriate conditions the hardness and roughness. By means of *screening trials* have been determined the limit values of input parameters. These values were chosen such that even the extreme combinations of parameters (min, min, min, and max, max, max) to ensure appropriate conditions for processing. The values of input parameter *assistant gas pressure* were varied within the ranges of 0.35...0.55 bar, which means low pressures. The first conclusion was that bigger values of pressure lead to worse output in terms of surface quality. The melt material is eliminated from the kerf being blown by the assistant gas. The normal piercing and cut is done with a laser power ranged 4100...4300 W. When interacting with the processed material, the laser beam produces an intense flow of incandescent drops that cause strips and craters occurrence on part surface. The cut is facilitated by the low level of light reflection of the steel. The low thermal conductivity of the steel makes more difficult heat dissipation; this helps melting material in the cutting area. The burnt material caused of presence of $O_2$can affect negatively the surface quality.

The type of experimental plan, a fractionated factorial one, was selected such that to fit best to production conditions, which impose to make correct decision by means of a low time consuming experiment. The working parameters used in the research are presented in table 3.

**Table 3.**Working parameters used in processing.

| Parameter | Value |
|---|---|
| Focal lens length | 190.5 mm |
| Focal position above the part surface | -0.5 mm |
| Diameter of nozzle | 1.5 mm |
| Stand-off distance | 0.8 mm |
| Assistant gas | $O_2$ |
| Pulse regime | Continuous |

The experimental processing was performed by means of a Bystronic Bysprint Format 3015 machine, presented in figure 1.

(a)  (b)

**Figure 1.** BystronicBysprint Format 3015 machine.(a) general view; detail of working space.

## 3. The experimental research

The roughness was measured on the side flanks of the laser processed parts. The side flanks follow both straight and curved contour. Roughness measurements were performed using, a roughness meter Mitutoyo SJ 301 (figure 2).

(a) (b)
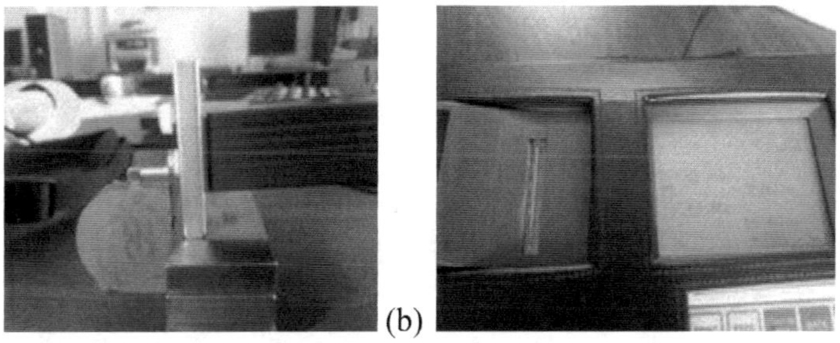

**Figure 2.** The roughnessmeter Mitutoyo SJ 301 (a) general view (b) detail of working space.

In figure 3 some processed parts are presented.

**Figure 3.** Processed parts.

The values of the measured roughness were ranged in the domain of 2.32...4.40 μm. Measurements have been performed on a planar flank of the part. For each part have been determined three parameters: $R_a$ – the average deviation of roughness, $R_z$ – the mean value of roughness measured in 10 points, and $R_q$ – the quadratic mean deviation of roughness. In the next sections all the referings will point to $R_a$.

## 4. Data processing

In order to evaluate laser processing of Hardox400 steel, and to find out the mathematical relationship between the output and input parameters an analysis with *Statistica 7.0* was performed. In the experimental research three input variable parameters were considered: laser power, assistant gas pressure, and cutting speed. In *Statistica*, cutting speed was considered constant. Analysis is based on statistical models. Two different models were used to emphasize the dependence of roughness on laser power and assistant gas pressure: the quadratic and the linear ones. From this point, for easyness of reading purpose, laser power will be named shortly *power*, and assistant gas pressure, shortly, as well *pressure*.

### *4.1. The quadric model*

For the interpretation of data the Response Surface Model – RSM was used. Based on the experimental darta a chart was built; it is shown in figure 4. Analysing this chart, one can observe:

- The lowest values of roughness occur at low values of power and pressure;

- The increase of the roughness determined by the increase of the pressure, at low values of the power, is a linear one;

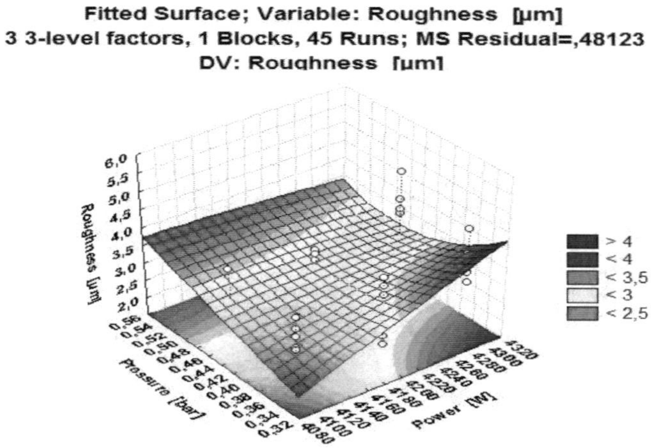


**Figure 4.** The predictive quadratic model of the combined influence of power and pressure on roughness, when cutting speed is constant.

- At high values of power, roughness varies in a quadratic way under ihe influence of gas pressure. In these conditions, roughnes has the minimum value corresponding to medium values of gas pressure, and increases at higher values of gas pressure;
- Acceptable values of roughness are obtained for medium values of power and pressure (the yellow area on the chart);
- Bad values of roughness are obtained when extem values of power and pressure are combined ($p_{max}$, $p_{min}$) and ($p_{min}$, $p_{max}$);
- On the chart, the plane surface, corresponding to roughness of 1.5 µm is the plane projection of quadric surface. This shows clearly the zone where the best values of roughness are obtained, that is the zone of the lowest pover and pressure;

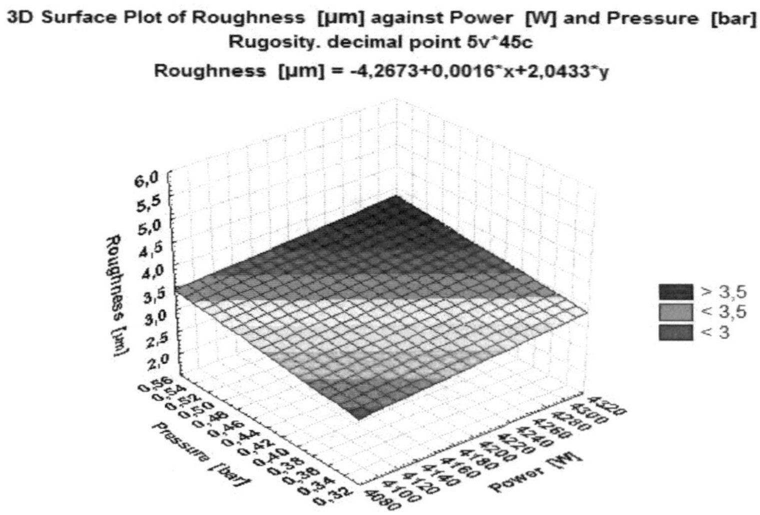


**Figure 5.** The linear mathematical model of dependence of roughness on pressure and power.

IOP Conf. Series: Materials Science and Engineering 1009 (2021) 012022 doi:10.1088/1757-899X/1009/1/012022

Based on the remarks above, some conclusions can be drawn, as follows:

- A more intense gas flow stimulates buning of material, that causes suplimentary and non uniform heating of the surface, that is oxidation is a reason of increasing roughness;
- Prediction shows that the best conditions to get a good roughness are low values of pressure and power, if cutting speed is kept constant.

$$R_a = -4.2673 + 0.0016P + 2.0433p \quad (1)$$

where $B_0$ = -4.273 is the constant of the model, $B_1$=0.0016 and $B_2$ = 2.0433 are the coefficients of the two variables, power, and pressure, respectively.

### *4.2. ANOVA linear model*

The linear model for computing the value of roughness using ANOVA variations reveals that the most influent factor is pressure of assistant gas, since it has the biggest coefficient. The tilt of planar surface confirms the dependence of roughness on the input parameters - pressure and power. It is a direct dependence, increasing the value of any input parameter causes an increasing of roughness. Combining the influence of both input parameters emphasizes the negative effect of each independent factor.The flow of the melted material on the flank of the part influences negativelly the roughness.

### *4.3. ANOVA quadratic model*

Using specific means, i. e. quadratic regresion, a relationship of form (2) is determined. The function is explicited in equation (3).

$$R_a = B_0 + B_1*X + B_2*Y + C + B_{11}*X^2 + B_{12}*X*Y + B_{22}*Y^2 \quad (2)$$

$$R_a = -71.275 + 0.0322X + 13.9533Y + 3.1667*10^{-6}*X^2 - 0.0088*X*Y + 27.8333*Y^2 \quad (3)$$

In (2) and (3) were marked down power with X, and pressure with Y. In equation (3) the magnitude of coefficients $B_2$ and $B_{22}$, which expres the quadratic and respectively linear interaction of pressure, reveals that this input parameter has the most significant influence on the roughness. The coefficient of interaction of power and pressure is the smallest; that shows that the input parameters potentiate each other in a small measure.

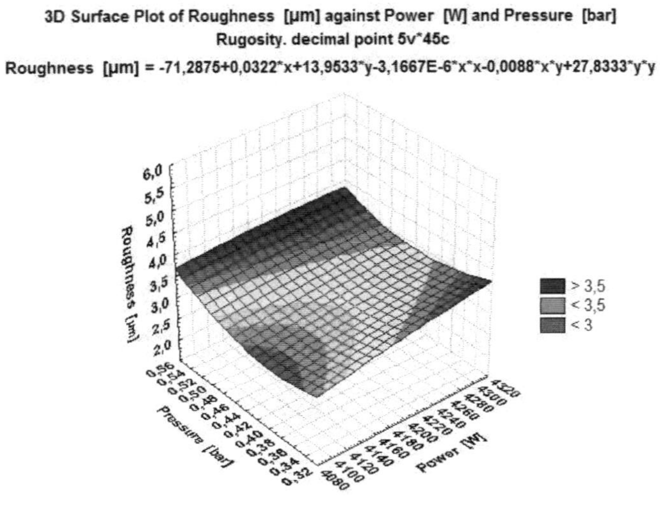


**Figure 6.** The quadratic model of influence of power and pressure on the surface roughness, obtained by ANOVA.

The 3D graph (figure 6) allows drawing some conclusions concerning with the way the roughness is influenced by the input paraeters:

- Roughness increases exponentially with the pressure at high values of power;
- Roughness increases linearly with power when pressure has low values;
- At low values of power, the best values of roughness migrate slightly to the middle of presure range compared to graph in figure 4, where clearly, the best roughness coresponds to the lowest levels of both power and pressure;
- While values of pover increase the values of roughness are distributed closer to its mean value;
- Values of roughness are simetrically distributed arround the mean value.

## 5. Results and discussion

The prediction of the roughness indicates in the case of the linear model that the low Ra roughness is obtained at low values of pressure and power. Roughness bigger than 2.5μm occur at values of pressure ranged in 0.35… 0.40 bar, respectively of power ranged in 4100 … 4150 bar. As the pressure and power increase, roughness values equal to 3 μm are reached. This trend remains constant at average pressure values and close to the average for laser power. For values of pressure higher than 0.50 bar and power above 4200 W, the roughness exceeds 3.5 μm. The increase of roughness follows the Gaussian curve from the minimum to medium values of pressure and power. The roughness graph is flattened at low values of pressure and power, and the average roughness is placed around 3 μm. Above these values the roughness deteriorates with increasing laser pressure and power. Roughness also gets values to the maximum in the situation of high pressure and low power, respectively low pressure and high power. The average value of roughness for the 45 parts is Ra = 3.0440 μm. In the power range 4150…4200 W, the roughness values fall within the confidence limits. In the power range 4250…4300 W, roughness values obtained are outside the 95% range. The good prediction of the roughness as a function of pressure to obtain the average value Ra = 3.0440 μm is in the range 0.40…0.45 bar. Between 0.45…0.55 bar the roughness prediction as a function of pressure displays a quadratic increase.

Comparing the linear graph for prediction with those for the calculation the roughness by analysing ANOVA variations results in a good compatibility, with similar results. The best values of roughness are at low values of pressure and power, but they are high at high values of influencing factors.

However, comparing the results obtained by the two methods, some differences can be observed. The differences of performance between prediction and regression are caused by the modelling techniques that take into account supplementary criteria. The regression equations are more accurate because they establish a relationship between the dependent value (roughness), and input parameters, pressure and power, while prediction takes into account the performance, concordance, and tuning, described by RSM model. Some small differences might be induced by the fact that cutting speed was considered to be constant in modelling, despite that it was slightly varied in the experiment.

## 6. Concluding remarks

It is important to emphasize again, that these conclusions are based on anexperimental plan developed according to the Taguchi method, a fractionated factorial one, applied for three input parameters, each having three levels. This is the method that allows an experiment with as less as possible number of tests, which still to provide with confident results. This type of plan was selected because of it fits the best to production conditions, when experimental research must be not a long time consuming process.

Analyzing the results of the experimental research on the influence of input parameters on roughness of parts made of Hardox400 steel by laser cutting process, some conclusins can be drawn. The main of them are as follows:

- The experimental research allowed to find the range of assistant gas pressure and laser power that provide a proper processing of Hardox400 steel in sheets having a thickness of 10 mm; these are 0.35...0.55 bar, and 4100...4300 W, respectivelly;

- The best results in terms of roughness are obtained when both pressure and power are at low levels;
- Pressure has a more significant influence on roughness than power has;
- Prediction and regression produce slightly differnt result. The difference are explicable
- The reccomendation for a proper lasser cutting process of Hardox400 steel (10 mm thickness) is the combination of assistant gas pressure of 0.35 bar and the laser power of 4100 W; according to regression results, the best combination should be 0.40 bar and 4100 W.

**Acknowledgments**
The present paper was supported by SC BYSTRONIC SA Brasov, which made possible the experimental research of Hardox400 steel laser processing.

**References**

[1] Adelmann B and Hellmann R 2011 Fast Laser Cutting Optimization Algorithm *Physics Procedia* **12** 591
[2] Lutey A H A, Ascari A, Fortunato A and Romoli L 2018 Long-pulse quasi-CW laser cutting of metals *Int. J. Adv. Manuf. Technol.* **94** 155
[3] Ivarson A, Powell J and Siltanen J 2015 Influence of alloying elements on the laser cutting process *Physics Procedia* **78** 84
[4] Bachy B and Al-Dunainawi Y 2020 Influence of the effective parameters on the quality of laser micro-cutting process: Experimental analysis, modeling and optimization *J. Laser Appl.* **32**
[5] Parthiban A, Sathish S, Chandrasekaran M and Ravikumar R Optimization of $CO_2$ laser cutting parameters on Austenitic type Stainless steel sheet *IOP Publishing materials Science and Engineering* **183**
[6] Hashemzadeh M, Suder W, Williams S, Powell J, Kaplan A F H and Voisey K T 2014 The application of specific point energy analysis to laser cutting with 1 μm laser radiation *Physics Procedia* **56** 909
[7] Pocorni J, Powell J, Frostevarg J and Kaplan A 2018 Dynamic laser piercing of thick section metals *Optics And Lasers In Engineering* **100** 82
[8] Singh S S, Baruah P K, Khare A and Joshi S N 2018 Effect of laser beam conditioning on fabrication of clean micro-channel on stainless steel 316L using second harmonic of Q switched Nd: YAG laser *Optics And Lasers In Engineering* **99** 107
[9] Thombansen U, Hermanns T and Stoyanov S 2014 Setup and Maintenance of Manufacturing Quality in $CO_2$ Laser Cutting *Procedia CIRP* **20** 98
[10] Zhang YL and Lei J H 2017 Prediction of Laser Cutting Roughness in Intelligent Manufacturing Mode Based on ANFIS *Procedia engineering* **174** 82
[11] Madic M, Mladenovic S, Gostimirovic M, Radovanovic M and Jankovic P 2020 Laser cutting optimization model with constraints: Maximization of material removal rate in CO2 laser cutting of mild steel *Journal Of Engineering Manufacture* **234**
[12] Pramanik D, Kuar AS, Sarkar S and Mitra S 2020 Enhancement of sawing strategy of multiple surface quality characteristics in low power fiber laser micro cutting process on titanium alloy sheet *Optics And Lasers In Engineering* **122**
[13] Wust P, Edelmann A and Hellmann R 2020 Areal Surface Roughness Optimization of Maraging Steel Parts Produced by Hybrid Additive Manufacturing *Materials* **13**

# Novel trend in case study of industrial robot programming and production system design using the virtual reality

**R Holubek[1*], R Ružarovský[1], K Velíšek[1] and M Janíček[1]**

[1]Slovak University of Technology in Bratislava, Materials Sciences and Technology in Trnava, Institute of Production Technologies, 91701 Trnava, Slovakia

*E-mail: radovan.holubek@stuba.sk

**Abstract.** The current trend of the production systems designing using by industrial robotics aims to reduce deployment time and minimize the value of financial investments. A large number of CAD software of various standards with little shared associativity are currently used for the design of production systems. Pre-production phases are the most demanding time and cost item for designing and programming industrial robots in terms of currently used methods (online and offline programming) at the design stage of new production devices and systems. Especial the new possibilities of using virtual reality in designing manufacturing systems, and the use of non-standard methods for programming of industrial robots in immersive virtual reality environments helps to shorten the pre-production phase of design, time-consuming programming trajectories of industrial robots, as well as tune of collision situations, debugging of robot trajectories and programs, testing and simulation of future production system.

## 1. Introduction

Virtual reality as the one of the Industry 4.0 technologies with others is a tool of the 21st century, which uses hardware and software resources in cooperation with computer support. In the last period, in addition to the term virtual reality (VR), other areas began to appear and add, which are called Augmented Reality (AR), Augmented Virtuality (AV), as well as the term Mixed Reality (MR). These mentioned terms on the figure. 1 were defined in this issue as early as 1994 in one important and cited publication [1].

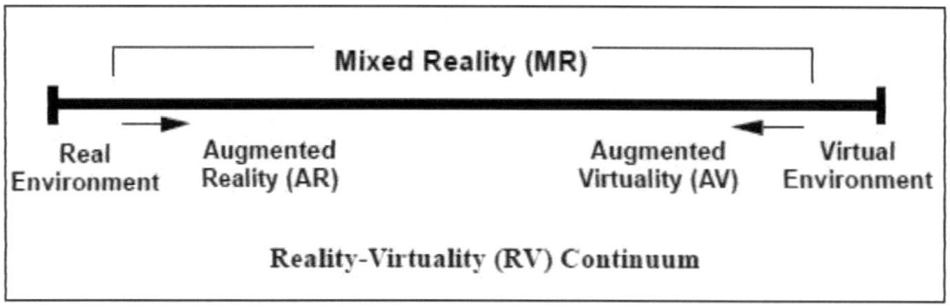


**Figure 1.** Simplified representation of a RV Continuum [1].

Based on figure 1, the authors wanted to bring closer and clarify in their publication the concepts that have been created over time and with advances in technological, hardware, software and computer innovations improved to currently used support elements such as various "headset" glasses for VR display with inseparable hardware - software. computer support, smartphones and tablets supporting various applications for augmented reality AR, as well as mixed reality MR of the two described areas [2, 3]. Based on this, but also other publications in this field of research, the original figure 1 was modified to the following form figure 2.

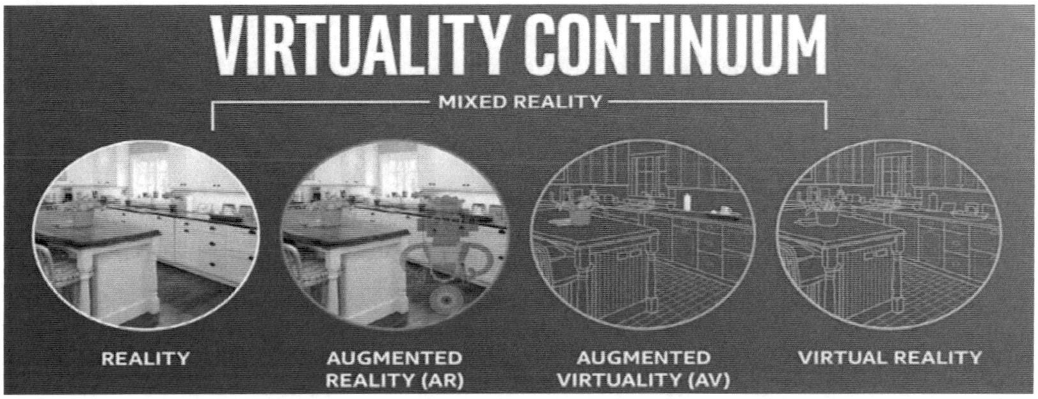


**Figure 2.** Virtuality continuum referred to as mixed reality MR [2].

Due to the constant shortening of time from the initial assignment of the research task to the final deployment of various production and robotic systems in real production, many companies are moving in addition to standard procedures for creating desktop simulations using CAD data and pre-production verification of comparison designs in a Virtual Commissioning (VC) and VR environment [4, 5]. At present, most companies are deploying and implementing the pre-production phase of designing future robotic systems with various innovative tools such as verification or creation of proposals for future production in an immersive VR environment and virtual commissioning. Just the area of virtual prototyping of the future production and new possibilities of designing and programming of industrial robots that are one of the important areas of strategies Industry 4.0 [6].

The creation, testing and verification of future production systems design using various VR headsets is subject to an iterative and concurrent design process based on the use of 3D CAD models databases and displayed VR studies in a suitable simulation environment, figure 3 [7].

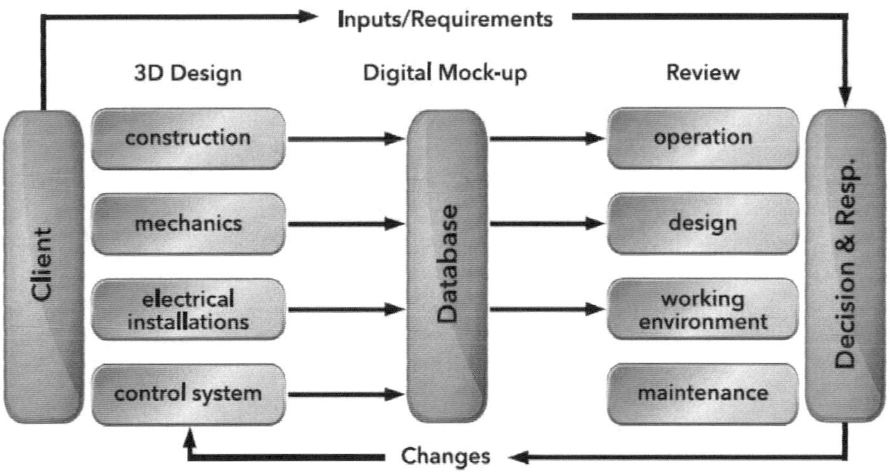


**Figure 3.** Subsequence process design with connect specified modules displayed in VR environment [7].

The finally immersive VR simulation models provide structured and easy-to-understand design information for the design teams. This is done in such a way that would not be possible using a traditional design approach based on 2D and 3D CAD drawings. The stakeholders can analyse the design from both a general and a more detailed perspective through the possibility of navigation in the models. VR models representation allows simpler to outline and discuss about the various design solutions instead of displaying through 2D drawings. This is a new perspective from the point of view of model design and we can talk about a new approach. This new approach makes it possible to detect possible errors and correct them even at the design stage [7].

## 2. Objectives

The basic goals of deploying modern tools in the form of the use of immersive imaging in a virtual reality environment are defined as the follows:

- Display of new simulation designs and "factory of the future" concepts;
- Virtual prototyping of future production;
- Shortening the deployment time of future production;
- Design of new automated and robotic systems;
- Verification of designs for future robotic production (verification of a suitably selected type of
- Robot reachability, verification of CAD data for the design of functional units, etc.);
- Testing of designs in the form of verification of simulations (elimination of collision states,
- Verification of utilization of individual machines and devices);
- Debugging of industrial robot trajectories;
- Safety simulation;
- Validation of logistic layouts;
- Analysis of ergonomics of workers.

Our assumption is that it is possible to use VR tools in creating a new robotic workplace and find out to what extent it is possible to use the scheme. Our assumption is that it is possible to use VR tools in creating a new robotic workplace and find out to what extent it is possible to use the scheme. The aim of this research article is to present the usability rate of using modern technologies VR for displaying or creating a new design for the future robotic workplace. ABB RobotStudio software will be used to validate the robotic workplace design method via a VR headset.

## 3. Methodology of research

The research consists in the possibility of creating a new robotic workplace in a new way by use of VR in order to simplify and speed up the design process compared to the standard methods. The standard method which is used is the Offline method of industrial robots programming [8]. This method is most commonly used through the software developed by the individual producers of the industrial robotics or universal robotics software. The method represents the current standards in the design and creation of simulations of various robotic systems. Our method consists in a new approach and expand these possibilities and methods from the classic creation of OLP simulation to the creation of robotic simulation in the environment of virtual reality. For these purposes, we chose software that allows offline programming of industrial robots and has the ability to connect, respectively. integration of robotic simulation imaging in VR. We have experience with the use of ABB RobotStudio software, which enables these functionalities. The simulation software RobotStudio and VR headset glasses from HTC Vive were used for the research figure 4 [9].

### *3.1. Design of layout solution in VR*

Standardly used methods of designing production as well as robotic systems are dispositional layout in appropriately selected CAD software or simulation software [10]. This method of design is currently used in the pre-production phases. The use of virtual reality as an auxiliary tool in the pre-production design phase appears to be a very effective method, as it allows not only the most realistic display of layout designs but also fast changes in layout design with immediate interaction of coordinate system change in OLP simulation study implemented in RobotStudio software between OLP and VR module [11].

**Figure 4.** Virtual reality simulation environment in RobotStudio software displayed by HTC Vive VR headset.

**Figure 5.** Design and layout of a future robotic system in VR environment.

In this case, VR imaging can be characterized as a standard online design method robot programming but fully implemented in a digital environment (creation of a digital twin of a future robotic system). It is therefore a matter of connecting online and offline methods using VR environment, figure 5.

*3.2. Verification of robot reachability in the case study VR*

Another necessary step in verifying the correct layout design in a future robotic case study is to verify the reachability of a selected industrial robot with a technological or manipulation tool. Using by the VR haptic controllers, the end effector together with the industrial robot is guided to the required targets

IOP Conf. Series: Materials Science and Engineering 1009 (2021) 012023 doi:10.1088/1757-899X/1009/1/012023

and the reachability of the robot in each desired target is verified, figure 6. In our case study, it is a robotic application of spraying a selected 3D CAD model of chassis using the industrial painting robot IRB 5400, paint tool Robobel926, positioner IRB L-ABB and CAD model of the selected part. If the layout of the robot and positioner together with the sprayed part does not suit, it is possible to change and modify this layout very easily using the haptic controllers with use the Grab and Local functions.

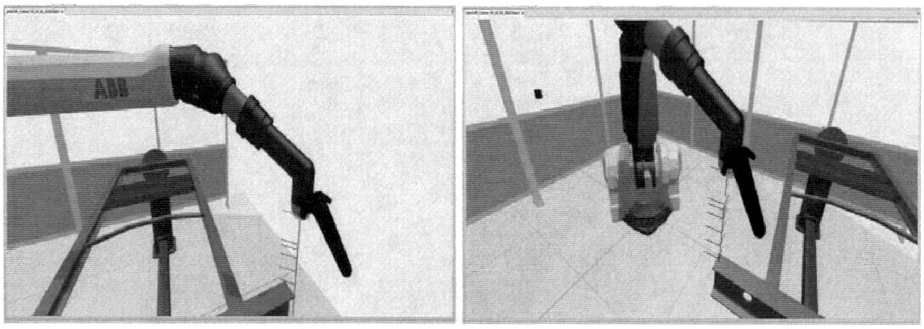


**Figure 6.** Testing of robot reachability in the VR case study.

*3.3. Creation of robotic targets and paths in the case study VR*

The programming of the targets and the subsequently created Path is created using haptic controllers by selecting the "Teach" function and guiding the tool together with the robot to the desired targets in the RobotStudio VR environment. Using the "Play" function, the future path of the spray robot together with the tool can be verified very fast.

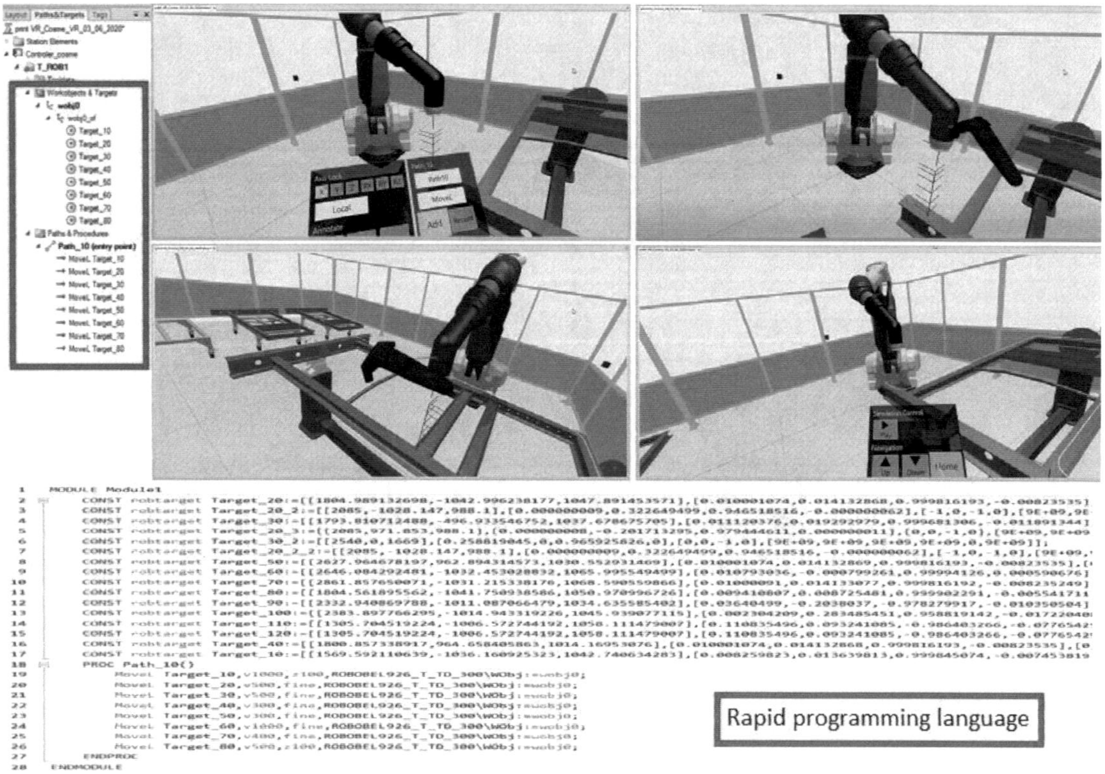


**Figure 7.** The case study displayed in VR robot programming environment and classical OLP environment and Rapid programming language.

On the figure 7 shows the possible interaction of the created targets and the path in VR conditions and in the standard OLP mode in RobotStudio software and Rapid programming language. The targets and path created in the VR environment can be used and modified as a standard PC simulation using the classic OLP method of programming industrial robots or Rapid programming language [12].

*3.4. Modification and debugging of robotic targets and paths in the case study VR*

The advantages of creating a robotic simulation in the VR environment, in addition to shortening time of the creation of future robotic simulation, lies in the ability to quickly verify created targets and paths, but also in the ability to modify targets, zone adjust, speed adjust and quick tune by the translation and reorientation of created targets in a coordinate system at VR environment. Such a modification represents the user's interaction in the virtual reality environment, as the modifications executed in this way guide the user through the displayed cube with commands on the haptic controller. This method of using the Zone, Speed, Tune targets modification seems to be intuitive and should be mastered by a less skilled user as in classical OLP simulation the implementation of these modifications is more complicated and more experience and expertise of the OLP method user is expected.

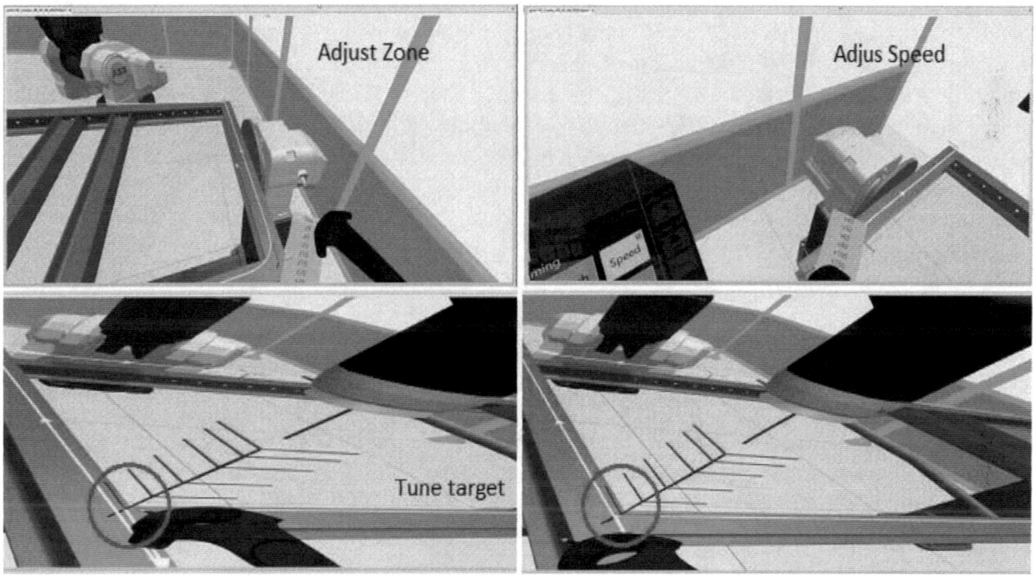


**Figure 8.** The adjust of Zone, Speed and Fine Tune targets in the VR environment by the displayed cube with commands on the haptic controller.

## 4. Conclusion

The expected scientific contribution of this research is to point out and verify in the form of the resulting simulation the possibilities of designing and programming industrial robots by a different method than are the standard online and OLP methods for programming industrial robots. VR programming appears to be the future between the connection between Online and OLP programming of industrial robots, as the VR environment allows to simulate, test, eliminate collision state and create an interactive environment similar to the real one, but with immediate interaction with the OLP environment. The simulations created in this way help to shorten the deployment time of future robotic systems with using of various production, assembly and technological operations. The authors of the scientific article verified the possibilities of using virtual reality in the field of dispositional design, planning and prototyping of future production, as well as creating a simulation of a robotic workplace designed to perform a selected technological operation with a spray tool in the VR environment. figure 9 shows a QR code which, when read via a smartphone or other suitable device recognizing QR codes, displays the resulting simulation created in the environment of the simulation software VR - RobotStudio from ABB.

**Figure 9.** The case study of the novel trend of the industrial robot programming and production system design using the virtual reality.

The authors of the scientific article see the future research in addition to developing their skills and possibilities in creating various DT-VR simulations (digital twin- virtual reality simulations) through the VR, but also through the interaction of several participants in one creating case study [13]. The advantage is to connect several participants to a study using a suitably selected VR device using the Create or Join meet support implemented in the RobotStudio VR environment and to participate interactively from anywhere in the world in creating a joint shared study either in the preproduction phase or creating innovations of existing production systems and robotic workplaces.

**Acknowledgments**
This research was supported by the project: KEGA-021STU-4/2018. Development of a laboratory for the design and maintenance of production systems supported by the use of Virtual Reality.

**References**

[1] Milgram P, Takemura H, Utsumi A and Kishino F 1995 *Augmented Reality: A Class of Displays on the Reality-Virtuality Continuum - Telemanipulator and Telepresence Technologies* vol **2351** (SPIE) pp 282–92
[2] Anon Looking to the future of mixed reality - Part I - Unity Technologies Blog
[3] Manou E, Vosniakos G C and Matsas E 2019 Understanding industrial robot programming by aid of a virtual reality environment *Int. J. Mech. Eng. Educ.* **47** 135–55
[4] Kizil M S, Joy J, Manou E, Vosniakos G C, Matsas E, Çakiroğlu Ü and Gökoğlu S 2019 A design model for using virtual reality in behavioral skills training *J. Educ. Comput. Res.* **57** 1723–44
[5] Choi S, Jung K and Noh S Do 2015 Virtual reality applications in manufacturing industries: Past research, present findings, and future directions *Concurr. Eng. Res. Appl.* **23** 40–63
[6] Damiani L, Demartini M, Guizzi G, Revetria R and Tonelli F 2018 Augmented and virtual reality applications in industrial systems: A qualitative review towards the industry 4.0 era *IFAC-PapersOnLine* **51** 624–30
[7] Rovaglio M *The Role of Virtual Reality in the Process Industry*
[8] Semjon J, Janos R, Sukop M, Tuleja P, Hajduk M, Jurus O, Marcinko P, Virgala I and Vagas M 2020 Verification of the UR5 robot's properties after a crash caused by a fall of a transferred load from a crane *Int. J. Adv. Robot. Syst.* **17** 172988142090420
[9] Ružarovský R, Holubek R, Sobrino D R D and Velíšek K 2019 A case study of robotic simulations using virtual commissioning supported by the use of virtual reality *MATEC Web Conf.* **299** 02006
[10] Almusawi A J R, Dulger L C and Kapucu S 2019 Simulation in virtual reality: Robotic training and surgical applications *Sci. Iran.* **26** 3369–74

IOP Conf. Series: Materials Science and Engineering 1009 (2021) 012023 doi:10.1088/1757-899X/1009/1/012023

[11] Sága M, Bulej V, Čuboňova N, Kuric I, Virgala I and Eberth M 2020 Case study: Performance analysis and development of robotized screwing application with integrated vision sensing system for automotive industry *Int. J. Adv. Robot. Syst.* **17** 172988142092399
[12] Crespo R, García R and Quiroz S 2015 *Virtual Reality Application for Simulation and Off-line Programming of the Mitsubishi Movemaster RV-M1 Robot Integrated with the Oculus Rift to Improve Students Training* Procedia Computer Science vol 75 (Elsevier) pp 107–12
[13] Dahl M, Albo A, Eriksson J, Pettersson J and Falkman P 2017 Virtual reality commissioning in production systems preparation *IEEE Int. Conf. on Emerging Technologies and Factory Automation, ETFA* (Institute of Electrical and Electronics Engineers Inc.) pp 1–7

# Application of fractal theory in fault diagnosis of nonlinear mechanical equipment system: a review

**X L Hou[1] and X Wang[1*]**

[1]School of Mechanical Engineering, Shenyang Jianzhu University, Shenyang 110168, Liaoning, China

*E-mail: 1360077271@qq.com

**Abstract**. Fractal theory takes the irregular and complex things as the research object, and uses the fractal dimension to describe the object. The review classifies and summarizes the application development of fractal theory in the diagnosis of common nonlinear mechanical equipment system faults such as diesel engines, gear boxes, hydraulic systems, etc. It is showed that fractal theory can extract and analyse irregular features and simplify complex research. It has certain advantages among various methods and is gradually being widely used. It provides a new reference and idea for the future research direction after fully understanding its current status and summarizing its characteristics. At present, most of the research on fault diagnosis of mechanical equipment is based on the signals under steady running state, while ignoring the transitional signals with high information content. Fractal theory has unique advantages in dealing with this kind of transitional signals such as the start and shutdown of equipment. It is a new trend to solve practical problems that study the short-time signals at transition process with strong non-linearity based on fractal theory to diagnose the fault of the mechanical equipment.

## 1. Introduction

The pace of development of modern industrial production is accelerating in the context of social economic development and technological progress. Mechanical equipment system plays a pivotal role in industry. Its characteristic advantages such as high speed, automation and continuity accelerate industrial development and bring great economic benefits. The operation of the equipment has a significant impact on the progress, achievements, and benefits of related industrial production. A failure in a certain part will affect the entire industrial production process, and then will have a different degree of negative impact on economic benefits. In severe cases, a safety accident may even occur, causing a huge harm to the personal safety of the staff and resulting in irreparable damage. Therefore, the reliability and safety of mechanical equipment are highly valued, the related research of fault diagnosis is particularly important. At the same time, the fault diagnosis technology is also put forward higher requirements.

Mechanical equipment is constantly being researched and innovated due to its importance. Its structure is more complex, the technical content is higher, and the complexity of its vibration signal is also increasing. The information content is more abundant, and most of them have nonlinear characteristics, so the single fault diagnosis method is more inadaptable to the feature recognition of increasingly complex mechanical equipment. For the fault diagnosis of nonlinear mechanical equipment systems, the fractal theory method has great application significance. So it is very necessary to introduce and apply the fractal theory to the fault diagnosis problems of nonlinear mechanical equipment systems. In this regard, scholars have conducted a lot of research on this issue, innovated on the basis of existing methods and technologies, and achieved certain results.

## 2. Fractal theory

### *2.1. Concepts and characteristics of fractal*

The fractal theory was founded by the American mathematician Mandelbrot. In the 1960s, he discussed the length of the British coastline in the American "Science" magazine, and proposed the self-similar nature of the coastline curve and other geographic boundaries. Since then, fractal thought has sprouted. Then he formally introduced the concept of fractal in 1975. Taking the natural irregular phenomena as the description object, he created the fractal geometry theory, which laid an important foundation for the analysis and research of irregular sets and complex sets in the future. And the theory has been widely used in various subject fields quickly.

Fractal theory is a subject with a high degree of activity and a wide range of applications in recent years. It takes irregular and complex things as the research object, and uses the fractional dimension to describe the object. Fractal can be reflected not only in geometric models, but also in mathematical models such as "information", "time", and "energy". If the rule set F is fractal, it must be satisfied that its part and the whole have self-similar properties [1]. This form of self-similarity can be either a strict one like the Koch curve, or a non-strict, statistically self-similar such as rolling mountains and coastlines. Through the above self-similarity differences, fractal can also be divided into regular fractal and random fractal. Fractal also has scale-free nature which means that the enlargement and reduction of the research object in the local area will not change the complexity, irregularity, shape and other characteristics of the original shape. Generally, the fractal dimension is larger than its own topological dimension.

### *2.2. Fractal dimension*

In the 1910s, the mathematician Hausdorff proposed a concept that broke the stereotyped understanding of the concept of dimension. In traditional public opinion, people think that the spatial dimensions of objects are integers, however in reality, the dimensions of objects in nature exist in the form of fractions, and we can also regard the integer dimension as a special case in fractional dimension.

Fractal dimension is an important parameter to describe fractal. The results are presented in numerical form, which can be used to identify the research object quantitatively and intuitively. The complexity of fractal dimension is relatively high, and mathematicians have been conducting a lot of research on it, from a single fractal dimension to multiple fractal dimensions [2]. And through the exploration, a variety of fractal dimension theories and their corresponding dimension calculation methods have been developed. Among them, the box dimension is widely used, and the correlation dimension is also widely applied. The common calculation method of the correlation dimension is the G-P algorithm. When using the correlation dimension, attention should be paid to the noise of the vibration signal, because the correlation dimension is more sensitive to the noise of the vibration signal. As the noise increases, the correlation dimension's ability to recognize the signal state will be weakened. Therefore, the vibration signal of the research object should be processed to eliminate the noise [3]. In addition, there are information dimension, generalized dimension, capacity dimension and other fractal dimension theories.

## 3. Application of fractal theory in fault diagnosis of nonlinear mechanical equipment system

Fractal theory is one of the important means to study nonlinear science. In the research of equipment fault diagnosis based on vibration signals, many vibration signals are likely to mutate in practice, also are unstable and irregular. When this type of signal has strong nonlinear characteristics, the fractal theory method is suitable for this kind of irregular and complex phenomena, which can effectively solve the calculation of signal dimension and construct fractal filtering [4]. For example, Ziwen Jia et al. [5] used mathematical morphological filters to analyse and modify the vibration signals of mechanical equipment in the time domain. Then used the G-P algorithm to calculate the correlation dimension of the signal, took it as the state identification characteristic index of the nonlinear system. Finally verified the feasibility of the algorithm through the simulation experiment of the Lorentz system and the actual simulation experiment on the Bently-RK4 rotor test rig.

On the basis of studying a large number of relevant literatures, this paper classifies and summarizes the application of fractal theory in the following common nonlinear mechanical equipment system fault diagnosis.

### *3.1. Application in diesel engine fault diagnosis*

Diesel engines have good economic performance and are widely used in large and medium-sized trucks, ships, engineering machinery and other manufacturing. It also plays an important role in the fields of agricultural machinery and aeroengine. Compared with other thermal engines, diesel engines also have advantages in energy saving and emission reduction. In addition, the small and high-speed diesel engines are getting new development gradually, so it is necessary to apply the advantageous fractal theory to the fault diagnosis technology. Sheng Yang et al. extracted the vibration signal characteristics of diesel engine. After signal noise reduction, the correlation dimension in fractal theory was selected as the characteristic value, and the fault signal was characterized by combining with PNN neural network. Hailong Lan et al. [6] combined the local wave decomposition method and the correlation dimension to perform fault diagnosis on the diesel engine. The signal was decomposed into the form of each basic mode component by the local wave method, and then the fractal theory was used to calculate the correlation dimension, which is used as the characteristic quantity to identify the diesel engine fault. In An Dong et al. [7] 's research on diesel engine fault diagnosis, the vibration signal of the diesel engine cylinder head was taken as the research object, and then the correlation dimension of the signal was calculated. The improved G-P algorithm is adopted in the calculation process, and the results obtained can identify the fault of the diesel engine.

### *3.2. Application in gearbox fault diagnosis*

Liwei Tang et al. [8] applied the box-counting dimension algorithm based on fractal theory to the gearbox, and studied the fractal characteristics under the conditions of gearbox bearing cracks, gear pitting, etc., and then identified the gearbox fault. Saifei Zhang et al. [9] selected the correlation dimension as the characteristic value of the fractal, used G-P algorithm to perform fault diagnosis of the engine bearing of the gearbox, and verified the effective effect of the method on fault diagnosis by comparing the fractal dimension of each working condition. Pengcheng Yan et al. [10] extracted the characteristic information of vibration signal of spur gear transmission system with EMD and SVD methods to reduce signal noise. Then used Kolmogorov entropy, multiple fractal characteristic parameters to quantitatively identify the fault of the gearbox more comprehensively. And it is verified that the method can achieve the goal of identifying the nuance of faults. Zhichuan Liu et al. [11] used Kalman filter to reduce the noise of the vibration signal in the time domain of the gearbox, and then calculated its dimension based on the fractal dimension method of mathematical morphology. This extraction method is feasible for identifying gear wear, broken teeth and other faults. Qingqing Chu et al. [12] collected vibration signals in the normal state of gears and in several fault states such as gear wear, broken teeth and circumferential joint errors. Next used the EMD method to layer the signal and perform noise reduction. Then combined fractal theory with neural network to form a new fault diagnosis method which identify the status of gear more efficiently.

### *3.3. Application in wind turbine fault diagnosis*

Wind turbines use renewable and non-polluting wind energy resources in nature, which is of great significance for alleviating energy shortages and achieving sustainable development [13].

Yang Li et al. [14] studied the fault diagnosis of the transmission system of wind turbines, analysed the multiple fractal characteristics of the vibration signals of fixed-speed bearings and variable speed bearings, and verified the effectiveness of the method for fault diagnosis of wind turbines through experiments. Jianqin Fan [15] conducted fault diagnosis research on gears and rolling bearings that are prone to failure in wind turbines, constructed a nonlinear simulation signal for equipment operation, also measured and analysed the signals of direct-drive wind turbine and doubly-fed wind turbine. In the literature, MATLAB was used as a tool to calculate the correlation dimension of each signal, Kalman filtering and wavelet scale domain filtering were combined with the correlation dimension, which further improves the fault feature extraction of the wind turbine to achieve more effective fault diagnosis. Peiming Shi et al. [16] collected the vibration signal of the gearbox bearing of the wind turbine, extracted the time-domain characteristic parameters, and used the box dimension as the fault characteristic parameter, combined with the genetic algorithm support vector machine diagnosis algorithm, improved the identification rate of the wind turbine bearing fault.

### *3.4. Application in hydraulic system fault diagnosis*

The hydraulic system has the characteristics of complexity, the occurrence of faults has greater randomness and non-linearity, the fault points are also more concealed, and there is a stronger correlation between faults. Simple condition monitoring is more limited for hydraulic systems, but the comprehensive application of fractal theory and other methods can effectively improve such problems.

Qinghua Wang et al. [17] simulated three kinds of faults such as hydraulic pump sliding shoe wear, collected and pre-processed vibration signals in the fault state and normal state of the hydraulic pump. Then extracted fractal characteristic parameters and found that the correlation dimension in different states was obviously different. The applicability of correlation dimension in hydraulic pump fault diagnosis is verified. Xiliang Liu et al. [18] measured simulated fault signal of the CB-KP63 hydraulic gear pump, used the packet decomposition method to pre-process the noise, and then calculated the box dimension of the vibration signal. The research showed that the box dimension is obviously different under different fault states which means that the box dimension is suitable for hydraulic pump state identification, and the accuracy is high. Tongguang Yang et al. [19] studied the problem of cracks in aviation hydraulic pipe. They captured the vibration signals of cracked and non-cracked hydraulic pipes through experiments. And used generalized fractal dimension to analyse whether the hydraulic pipe has cracks. This method can also effectively identify the early failure of hydraulic pipe cracks.

### *3.5. Application of fault diagnosis in other nonlinear systems*

Xianglin Hou et al. [20] defined the fractal calculation dimension based on the signal fractal dimension, combined with the neural network of the multilayer perceptron to form a fault diagnosis method for nonlinear systems. Then verified the feasibility and accuracy of the method through the example of rolling bearings. Meijun Zhang et al. [21] filtered and denoised the original engine signal, and used the correlation dimension to diagnose the fault, which improved the accuracy of the diagnosis. Then by comparison, analysed the correlation dimension of the engine start signal before and after noise reduction in normal state and in different degree of air leakage state to verify the reliability of the method. Yu Yuan et al. [22] established a new diagnosis principle for fault diagnosis of reciprocating compressor, decomposed the vibration signal in the time-frequency domain of local wave, determined the generalized dimension for multifractal, extracted the characteristic quantity of fault diagnosis, and obtained the practical conclusion of the method through case analysis. Zhongyuan Zhang et al. [23] proposed a fault diagnosis method for rolling bearings combining wavelet transform and fractal dimension. After collecting the vibration signals of rolling bearings, the wavelet transform method was used to carry out layered reconstruction. The fractal dimension calculated in the time-frequency domain is more suitable to be used as the basis for judging the failure of rolling bearing and is more accurate.

In the fault diagnosis of various nonlinear mechanical equipment systems, fractal theory has been applied more and more frequently due to the particularity of its research objects. It can effectively process nonlinear and non-stationary signals. In practical application, combining the fractal theory with other analysis methods, the fault features can be more clearly extracted to describe the fault state of the system, which makes fault diagnosis more accurate, faster and more effective.

## 4. Development status and prospect of fractal theory

The research of fractal theory started late and is a relatively new and developing discipline. This paper summarizes the successful cases of the current fractal theory in the fault diagnosis of nonlinear mechanical equipment systems, but there is still a lot of research space in the application of fault diagnosis. Fractal theory usually needs to be combined with other theories in practical applications, which is its limitation. Therefore, improving the adaptive ability of itself is a major direction worth studying in the future. At the same time, fractal theory has not been developed for a long time, so some related schemes still need to be further verified in theory and practice. It may be the case that some schemes are feasible in the theoretical stage, but new problems may appear in practice, which still need to be explored and perfected.

Based on the research of the characteristics, advantages and the application of fractal theory in this paper, it can be seen that fractal theory has unique advantages in the practical application of nonlinear systems, and can process short-term, non-stationary and non-periodic signals more effectively. It is very promising to apply this theory to the short board of fault diagnosis of mechanical equipment at this stage. At present, there is still a lack of research on fault diagnosis of mechanical equipment based on transition process signals. Most are to study the vibration signal of the equipment in a stable running state, while ignoring the research on the transition signal of the equipment starting and stopping. The information content of the signal during the transition process is high. It is an important basis for fully grasping the status of mechanical equipment. The transient process has strong signal non-linearity and non-stationarity, and the process time is short. It is a good idea to study the transition process signal based on fractal theory to realize the fault diagnosis and recognition of mechanical system. It has a broad research prospect.

In addition, it is not only effectively applied in the fault diagnosis of nonlinear systems in the mechanical field described in this paper, but also plays an important role in fields such as geography, biology, chemistry, physics and social science, with a very wide range of applications [24, 25].

## 5. Conclusion

Due to the uniqueness of its research object, fractal theory is very suitable for extracting and analysing irregular features in fault diagnosis of nonlinear mechanical equipment systems. In practical applications, combined with other analysis methods, it can accurately and effectively describe the fault state of the system, with high accuracy and wide application. Based on a large number of related literatures, this paper summarizes the source, principle and application of the theory, and classifies and summarizes the successful cases of fractal theory's application in fault diagnosis of nonlinear mechanical equipment systems. Then briefly describes the advantages and limitations of the theory in combination with the development status. Through the exploration, it can be seen that most of the current researches on fault diagnosis of nonlinear mechanical equipment are based on the signals during the smooth operation of mechanical equipment, while the signal information content of the transition process such as the start and stop of the equipment is rich, which has great research value. However, the nonlinearity and non-stationarity of the signal during the transition process are high, and the time is short, thus the research is relatively difficult. The fractal theory is suitable for dealing with such irregular and unstable research objects. Therefore, combining with the advantages of fractal theory, this paper puts forward an idea which is based on fractal theory to study the transition process signals to identify mechanical equipment faults. In general, fractal theory has a good prospect and a large development space.

## References

[1] Du B and Tang G 2007 Fractal data compression algorithm for vibration signal in fault diagnosis *Int. Conf. on Wavelet Analysis and Pattern Recognition (Beijing)* pp 809-13

[2] Yuan Y, Li B, Shang J and Yao S 2010 The application of vibration signal multifractal in fault diagnosis *2nd. Int. Conf. on Future Networks (Hainan)* pp 164-7
[3] Du B, Tang G and Wang S 2008 Application of morphological noise reduction in rotor fractal fault status recognition *Congress on Image and Signal Processing* pp 161-5
[4] Lopes R and Betrouni N 2009 Fractal and multi-fractal analysis: a review *Medical Image Analysis* **13** (4) 634-49
[5] Jia Z 2013 Research on fractal dimension algorithm in mechanical vibration fault diagnosis *North China Electric Power University (Beijing)*
[6] Lan H, Pan H and Gong M 2013 Application of local wave fractal method in diesel engine fault diagnosis *Machine Tool and Hydraulics* **41** (05) 180-3
[7] Dong A and Pan H 2014 Fractal characteristics of diesel engine vibration signal and fault diagnosis *Noise and Vibration Control* **34** (02) 144-7
[8] Tang L, Wang W, Luan J and Zheng H 2001 Application of box counting dimension in gear box fault diagnosis *Journal of Ordnance Engineering College* **13** (4) 12-4
[9] Zhang S, Gao Y and Xue B 2007 Research on gearbox fault diagnosis based on fractal theory *Journal of Artillery Launch and Control* **2** 65-8
[10] Yan P, Sun H, Mao X and Feng G 2012 Research on fractal diagnosis method of gearbox based on EMD and SVD *Journal of Electronic Measurement and Instrument* **26** (5) 404-12
[11] Liu Z, Tang L and Cao L 2014 Gearbox fault diagnosis based on Kalman filtering and mathematical morphology fractal analysis *Mechanical Transmission* **38** (07) 141-4 + 60
[12] Chu Q, Xiao H, Lv Y and Yang Z 2015 Gear fault diagnosis based on multifractal theory and neural network *Vibration and Shock* **34** (21) 15-8
[13] Liu W, Tang B, Han J, Lu X, Hu N and He Z 2014 The structure healthy condition monitoring and fault diagnosis methods in wind turbines: a review *Renewable and Sustainable Energy Reviews* **44** 466-72
[14] Li Y, Li C and Yang Y 2019 Fault diagnosis of wind turbine bearing based on multifractal detrend fluctuation analysis *Acta Energiae Solaris Sinica* **40** (11) 3235-43
[15] Fan J 2016 Fault feature extraction method of wind power transmission chain based on fractal dimension and its application *North China Electric Power University (Beijing)*
[16] Shi P, Liang K, Zhao N and An S 2018 Fault diagnosis of gearbox bearing of wind turbine based on fractal dimension and GA-SVM *Journal of Metrology* **39** (01) 61-5
[17] Wang Q, Zhang X, Zhang H and Sun Y 2004 Application of fractal theory in fault diagnosis of hydraulic pumps *Journal of Dalian Maritime University Natural Science Edition* **30** (2) 40-3
[18] Liu X, Chen X and Wang H 2009 Fault diagnosis of gear pump based on fractal box dimension *Machine Tool and Hydraulics* **37** (7) 255-7
[19] Yang T, Yu X, Wang Z, Sun J and Dou J 2019 Application research of fractal theory in fault diagnosis of aviation hydraulic pipeline crack *Machine Tool and Hydraulics* **47** (22) 169-73
[20] Hou X, Li Y, Yu H and Ji S 2001 Fractal neural network diagnosis method for nonlinear mechanical equipment system *Journal of Vibration Engineering* **2** 100-3
[21] Zhang M, Chen J and Hou B 2006 Analysis and processing of noise signals in fractal diagnosis of engine faults *Journal of PLA University of Science and Technology* **4** 380-4
[22] Yuan Y and Ma X 2007 Application of local wave time-frequency domain multifractal in fault diagnosis *Vibration and Shock* **5** 60-63 + 153
[23] Zhang Z, Wu J, Ma J and Wang X 2015 Fault diagnosis for rolling bearing based on lifting wavelet and multi-fractal dimension *Proc. In. Conf. on Advances in Mechanical Engineering and Industrial Informatics* (Paris: Atlantis Press) pp 295-300
[24] Zhang C, Sun Q, Wu L, Chen Z and Li D 2019 Coastal fractal calculation method based on vector data and its application *Geomatics spatial information technology* **42** (7) 36-7 + 45
[25] Huang D 1994 Fractal sets and fractal dimensions *Journal of Jiangxi Normal University* **1** 54-61

# Voice frequency impact for voice control of collaborative robots

**M Janíček[1*], R Holubek[1], R Ružarovský[1] and K Velíšek[1]**

[1]Slovak University of Technology in Bratislava, Materials Sciences and Technology in Trnava, Institute of Production Technologies, 91701 Trnava, Slovakia

*E-mail: matej.janicek@stuba.sk

**Abstract**. The project is focused on research of voice control of collaborative robots. Previous research of voice control and speech recognition research have given examples and evidence that machines can truly be controlled by human voice. The main problem in the voice control of collaborative robots is the lack of reliability and safety of control compared to manual control. Of the results achieved, the most and least relevant voice control problem has been clearly identified, next step is voice frequency impact for voice control. The experiments were performed in the robotics laboratory at the Faculty of Materials Science and Technology in Trnava, and this using an Arduino kit and a small robot "DOBOT Magician". In the final part of the project are summarized all experimentally found facts and confronted with the assumptions set in the design of the experiment.

## 1. Introduction

Nowadays, we try to make our work easier where possible. The same is true in industry. The robot has become part of many manufacturing processes. In the automotive industry, for example, it has become an integral part of the production process. The growing popularity of robots has brought robotics to households as well. It is no longer a problem to buy a robot for vacuuming, a food processor or a robot controlling the entire smart home. Until recently, voice control was just a dream that people dreamed of. After all, controlling things without moving your hands must be incredibly comfortable. But technology moves big strides forward every day. What was a dream yesterday may become a reality today. Voice control has become part of new cars, smart TVs, smartphones or home helpers. The question is whether voice control has already reached such a level that we can use voice control safely in industry as well. Research into voice control in the past has yielded a lot of new information, such as knowledge about a voice controlled robotic arm vehicle [1], Li-Fi based voice control robot [2], a voice controlled robot using Arduino [3], speech recognition system for a voice controlled robot [4] voice-controlled autonomous vehicle using IoT [5]. This research again focuses on the issue of voice control safety, uses the knowledge from previous research [6], but also involves the collaborative robot DOBOT Magician. The research is original of its kind, similar research has not yet been carried out.

## 2. Objectives and Methods

The aim of this research is to verify the effect of changing the frequency of a voice command on the success probability of voice control of a collaborative robot. For this purpose, the collaborative robot DOBOT Magician and its accessory for voice control, the Arduino kit, were used. DOBOT Magician is a small collaborative robot designed for teaching and simulating various processes, figure 1.

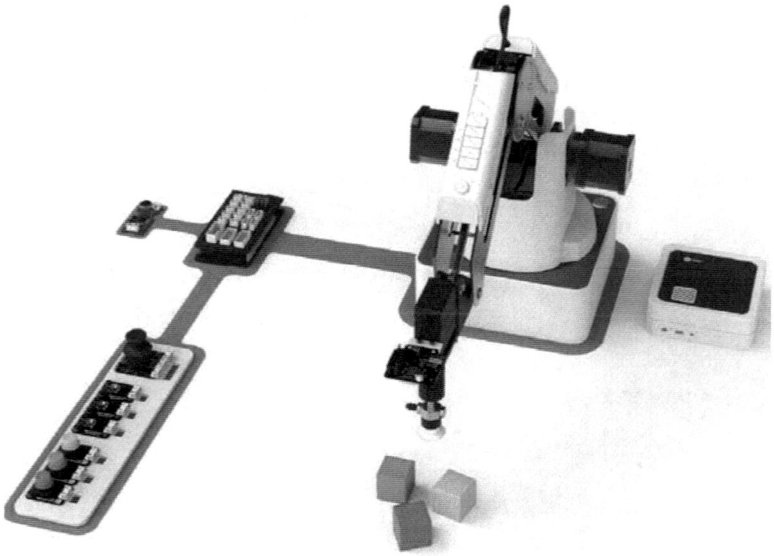

**Figure 1.** Collaborative robot DOBOT Magician and Arduino kit.

DOBOT Magician has 22 original fully programmable voice commands, for example Up, Down, Close the door. Table 1 shows all voice commands for DOBOT Magician.

**Table 1.** List of DOBOT´s voice commands.

| Voice command | Voice command |
|---|---|
| Open the door | Mode 1 |
| What's the time | Turn off the light |
| Play music | Right |
| Decrease temperature | Next |
| Increase temperature | Stop |
| Close the door | Start |
| Turn on the TV | Left |
| Turn off the TV | Pause |
| Mode 2 | Down |
| Previous | Go |
| Turn on the light | Up |

Previous experience has shown that DOBOT Magician does not have the same success probability for all its voice commands. Voice commands composed of several words (Play music, Close the door) had a significantly higher degree of success probability than voice commands composed of only one word (Up, Go). Therefore, the most reliable voice command (Open the door) was used to determine the effect of changing the frequency on voice control.

The voice command Open the door was programmed so that the robot performed a physical action = movement, which verified whether the voice command was successful.

Sound is any longitudinal mechanical wave in a material environment that is able to evoke auditory sensation in the human ear.

The frequency of this wave is approximately in the range of 20 Hz to 20 kHz, outside these limits one does not perceive sound. The human voice handles a fundamental tone of around 400 Hz. Simulations in this research were also performed around the 400Hz spectrum.

The voice command Open the door was recorded with different voices with different frequencies Voice commands were recorded via a headset microphone to a personal computer. The personal computer was equipped with Audacity audio software. The analysis of frequencies of voice commands in Audacity is shown in figure 2.

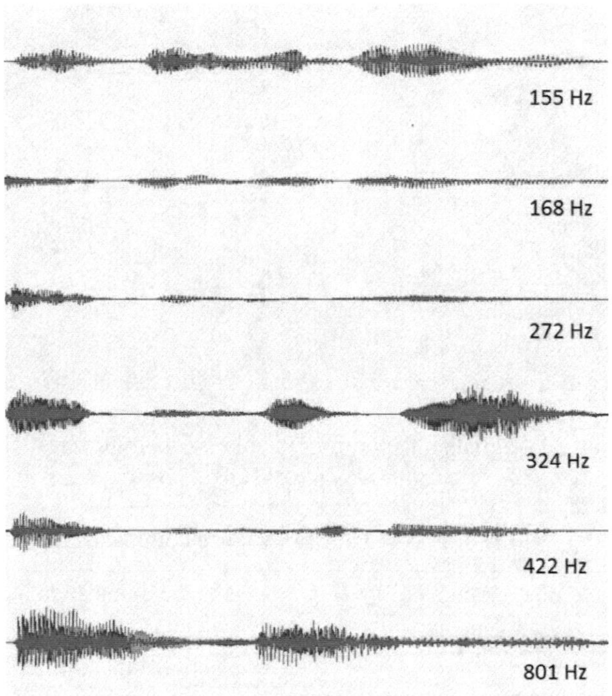


**Figure 2.** Analysis of frequencies of voice commands in Audacity.

As mentioned, the voice command was recorded in 6 different voice frequencies. In order to determine the effect of changing the voice frequency on the success probability of the voice command, a simulation with 100 identical voice commands with the original frequency was recorded. The results are shown in table 2.

**Table 2.** Success probability of original voice commands frequencies.

| Voice frequency (*Hz*) | Success probability (%) |
|---|---|
| 155 | 82 |
| 168 | 88 |
| 272 | 95 |
| 324 | 95 |
| 422 | 95 |
| 801 | 94 |

## 3. Results

Audacity audio software allows you to change the voice frequency from -99% to + 99%. Simulations in this research were performed at 5% until the DOBOT Magician stopped responding to voice commands.

Each change in frequency involved 100 simulations performed with a change in frequency and an assessment of success probability. The effect of individual frequency changes on the success probability of voice commands is shown in tables 3, 4, 5.

**Table 3.** Frequency change impact for voice commands.

| Frequency change (155 *Hz*) | Success probability (%) | Frequency change (168 *Hz*) | Success probability (%) |
|---|---|---|---|
| +5% | 94 | +5% | 61 |
| +10% | 74 | +10% | 16 |
| +15% | 32 | +15% | 0 |
| +20% | 0 | +20% | 0 |
| -5% | 39 | -5% | 90 |
| -10% | 14 | -10% | 98 |
| -15% | 2 | -15% | 73 |
| -20% | 0 | -20% | 47 |
| -25% | 0 | -25% | 8 |
| -30% | 0 | -30% | 0 |

**Table 4.** Frequency change impact for voice commands.

| Frequency change (272 *Hz*) | Success probability (%) | Frequency change (324 *Hz*) | Success probability (%) |
|---|---|---|---|
| +5% | 76 | +5% | 79 |
| +10% | 44 | +10% | 64 |
| +15% | 29 | +15% | 5 |
| +20% | 0 | +20% | 0 |
| -5% | 96 | -5% | 96 |
| -10% | 99 | -10% | 95 |
| -15% | 38 | -15% | 70 |
| -20% | 3 | -20% | 58 |
| -25% | 0 | -25% | 16 |
| -30% | 0 | -30% | 0 |

**Table 5.** Frequency change impact for voice commands.

| Frequency change (422 *Hz*) | Success probability (%) | Frequency change (801 *Hz*) | Success probability (%) |
|---|---|---|---|
| +5% | 71 | +5% | 77 |
| +10% | 37 | +10% | 53 |
| +15% | 32 | +15% | 21 |
| +20% | 0 | +20% | 0 |
| -5% | 96 | -5% | 97 |
| -10% | 93 | -10% | 91 |
| -15% | 68 | -15% | 77 |
| -20% | 29 | -20% | 62 |
| -25% | 12 | -25% | 16 |
| -30% | 0 | -30% | 0 |

It is clear from the results that no change in the frequency was able to achieve 100% success probability of voice commands. DOBOT Magician could no longer recognize any of the voice commands when the frequency changed to + 20%.

On the other end, when changing the frequency to -30%, the same applies, none of the voice commands could be recognized by the DOBOT magician. To better represent the results, a table 6 was prepared, which well illustrates the results of this research.

**Table 6.** Research results.

| Voice frequency | 155 *Hz* | 168 *Hz* | 272 *Hz* | 324 *Hz* | 422 *Hz* | 801 *Hz* |
|---|---|---|---|---|---|---|
| Frequency change | Success probability (%) | | | | | |
| +20% | 0 | 0 | 0 | 0 | 0 | 0 |
| +15% | 32 | 0 | 29 | 5 | 32 | 21 |
| +10% | 74 | 16 | 44 | 64 | 37 | 53 |
| +5% | 94 | 61 | 76 | 79 | 71 | 77 |
| **Original** | **92** | **88** | **95** | **95** | **95** | **94** |
| -5% | 39 | 90 | 96 | 96 | 96 | 97 |
| -10% | 14 | 98 | 99 | 95 | 93 | 91 |
| -15% | 2 | 73 | 38 | 70 | 68 | 77 |
| -20% | 0 | 47 | 3 | 58 | 29 | 62 |
| -25% | 0 | 8 | 0 | 16 | 12 | 16 |
| -30% | 0 | 0 | 0 | 0 | 0 | 0 |

The results showed that in all cases, changing the frequency of the voice command contributed to improving the success probability of the voice command. Only in one case, with a voice command with a frequency of 155 Hz, better results occurred after increasing the frequency, in other cases, the improvement was always achieved by decreasing the frequency of the voice command.

## 4. Conclusion

This research was to verify the effect of changing the frequency of the voice on the success probability of the voice control of a collaborative robot. Voice commands of different frequencies were selected to achieve the most objective results possible. In 5 of 6 cases, changing the voice frequency by reducing the frequency between 5% and 10% was shown to have a positive effect on the success probability of voice commands. Thus, it is expected in the future that if an input device such as a microphone changes the voice frequency, the success probability of the voice commands will be higher than at the original voice frequency. Voice control can make our work easier and the comfort of using various devices. However, its success probability is still not 100%. The authors of the scientific article still see great scope for improving the voice control of collaborative robots, which can be of great help to us in the future.

**Acknowledgments**

This research was supported by the projects: KEGA-021STU-4/2018. Development of a laboratory for the design and maintenance of production systems supported by the use of Virtual Reality.

## References

[1] Alka N U, Salihu A A, Haruna Y S and Dalyop I A 2017 A voice controlled pick and place robotic arm vehicle using android application *American Journal of Engineering Research* **6** 207-15

[2] Sawasakade S, Palkar M, Khankal R, and Kale S D 2015 Li-fi based voice control robot *Journal of Electronics and Communication Engineering* **33** 45-49

[3] Teeda V, Sujatha K and Mutukuru R 2016 Robot voice a voice controlled robot using arduino *International Journal of Engineering and Advanced Technology* **5** 60-67
[4] Memon Y A, Motan I, Akbar M A, Hameed S and Hasan M U 2016 Speech recognition system for a voice controlled robot with real time obstacle detection and avoidance *International Journal Of Electrical, Electronics And Data Communication* **4** 33-37
[5] Sachdeva S, Macwana J, Patela Ch and Doshia N 2019 Voice-controlled autonomous vehicle using IoT *3rd International Workshop on Recent Advances on Internet of Things: Technology and Application Approaches* **160** 712-717
[6] Janicek M, Ruzarovsky R, Velisek V and Holubek R 2020 Analysis of voice control of a collaborative robot *Int. Conf. on Applied Sciences* (Hunedoara, Romania)

# Performance diagnosis of the drive unit for high speed motorized spindle based on hidden state mapping model

**Z W Jia[1], L T Fan[1*], C X Zhu[1], Z P Si[1] and J Z Liu[1]**

[1]School of Mechanical Engineering, Shenyang Jianzhu University, Shenyang 110168

*E-mail: fanliting@sjzu.edu.cn

**Abstract.** As the key component of the motorized spindle system, the performance of drive unit will directly affect the processing quality of CNC machine tools. This paper proposed a performance diagnosis method of hidden state mapping model to prevent the motorized spindle from failure. The hidden state model was obtained by feature extraction of the parameter vectors identified from the dynamic responses to test input signals. The thresholds of performance degradation were determined by the hidden state to evaluate the performance of the drive unit in the motorized spindle system. The feasibility and effectiveness of the model can be verified by taking IGBT degradation failure as an example. The results showed that the model parameters extracted from different degradation degrees of IGBT devices can be mapped to hidden state characteristics and used for performance evaluation of the drive unit. The proposed hidden state mapping model can be used in the framework of monitoring the safety and reliability of the drive unit, and lay the foundation for the prognostic and health management of the motorized spindle system.

## 1. Introduction

High-end CNC machine tools provide guarantee for the manufacture of high-tech products in my country, it is a processing equipment commonly used in my country's manufacturing industry, basic equipment for manufacturing, automotive and power generation equipment manufacturing industries. The performance of the system is used to describe the ability of the entire system to maintain a certain degree of reliability and continue to achieve predetermined goals. Health monitoring is a kind of periodic detection of a certain part of the system parameters or characteristics failure. During the long-term use of the equipment, the phenomenon of aging and wear inside the equipment, the performance of the equipment is degraded. With long-term performance degradation, the device is malfunctioning. In large mission-critical systems, an advanced detection system can continuously assess the health of the system. Production equipment should be regularly evaluated for performance to avoid downtime maintenance.

There are many ways to evaluate performance, the evaluation method introduced in the reference [1] to determine the evaluation weight, this method can provide a reference for mechanical system performance evaluation and similar problems, but the scale workload is too large, it should cause disgust and confusing judgment experts. Reference [2] describes the method of fuzzy comprehensive evaluation, aiming at the problem of many uncertain factors in the diagnosis process of industrial equipment, it realizes the simplicity of the equipment diagnosis process and the accuracy of the evaluation results through the conversion of qualitative analysis to quantitative analysis, but it has strong subjective certainty to the index weight vector; this article uses the method of system identification performance evaluation, determine the parameters of the system model by performing system identification on the system in the normal state, identify systems at different stages of degradation. Study the mapping relationship between the geometric distance of the parameter vector and the performance degradation, and evaluate the performance of the current drive system [3]. Through IGBT performance degradation simulation examples to verify, the result shows: as the degree of performance degradation continues to intensify, the geometric distance of the system parameter vector gradually increases. Confirm the feasibility of this performance evaluation method.

## 2. High-speed motorized spindle drive system simulation design and system identification

### 2.1. Drive system simulation design

The high-speed motorized spindle drive system is a closed-loop system composed of a frequency converter and spindle motor. The spindle motor has synchronous motor and asynchronous motor, this design selects asynchronous motor for design analysis. The closed-loop drive system is built according to the idea of rotor orientation control, which converts the alternating current in the system into direct current for control, so as to realize the separation control of the electric spindle torque and speed, and achieve the equivalent effect of DC power control [4]. In the process of simulation design, combined with the equation expression of asynchronous motor.

$$\begin{bmatrix} u_{sd} \\ u_{sq} \\ u_{rd} \\ u_{rq} \end{bmatrix} = \begin{bmatrix} R_s + L_s P & -\omega_1 L_s & L_m P & -\omega_1 L_m \\ \omega_1 L_s & R_s + L_s P & \omega_1 L_m & L_m P \\ L_m P & -\omega_1 L_m & R_r + L_r P & -\omega_s L_r \\ \omega_s L_m & L_m P & \omega_s L_r & R_r + L_r P \end{bmatrix} \begin{bmatrix} i_{sd} \\ i_{sq} \\ i_{rd} \\ i_{rq} \end{bmatrix} \tag{1}$$

$$\begin{bmatrix} \psi_{sd} \\ \psi_{sq} \\ \psi_{rd} \\ \psi_{rq} \end{bmatrix} = \begin{bmatrix} L_s & 0 & L_m & 0 \\ 0 & L_s & 0 & L_m \\ L_m & 0 & L_r & 0 \\ 0 & L_m & 0 & L_r \end{bmatrix} \begin{bmatrix} i_{sd} \\ i_{sq} \\ i_{rd} \\ i_{rq} \end{bmatrix} \tag{2}$$

$$T_e = n_p L_m (i_{sq} i_{rd} - i_{sd} i_{rq}) \tag{3}$$

In two-phase rotating coordinate system, $\psi_{rd} = \psi_r, \psi_{rq} = 0$;

$$T_e = n_p \frac{L_m}{L_r} i_{sq} \psi_r \tag{4}$$

$$i_{sd} = \frac{1 + T_r P}{L_m} \psi_r \tag{5}$$

$$\psi_r = \frac{L_m}{1 + T_r P} i_{sd} \tag{6}$$

$$\omega_s = \frac{L_m}{T_r \psi_r} i_{sq} \tag{7}$$

In the equations: $\omega_1$ is the synchronous speed; $\omega$ is the rotor speed; $\omega_s$ is slip angular velocity; $\psi$ is flux linkage; $i$ is the current; $R$ is the resistance; $L$ is the inductance; $n_p$ is polar logarithm; $T_r$ is the rotor time constant; $P=d/dt$ is the differential factor; $s$ is stator; $r$ is the rotor; $d$ is d-axis; $q$ represents q-axis; $m$ is the mutual inductance.

*2.1.1. Transformation from 3 phases to 2 phases of stationary coordinate system.* When the asynchronous motor is connected with three-phase alternating current, it will generate a rotating magnetic field inside the motor. Under the condition of ensuring a certain power, the rotating magnetic field generated by the three-phase alternating current can be equivalent to the rotating magnetic field generated by the two phases, and the process is realized. The equivalent transformation of the static coordinate system from three phases to two terms. Under the premise that the magnetomotive force remains unchanged, determine the relationship between the voltage, current and magnetomotive force of the three-phase winding and the two-phase winding. Set the current of two symmetrical windings, current of three-phase symmetrical winding, the transformation equation between them is:

$$\begin{bmatrix} i_\alpha \\ i_\beta \\ i_0 \end{bmatrix} = \sqrt{\frac{2}{3}} \begin{bmatrix} 1 & -\frac{1}{2} & -\frac{1}{2} \\ 0 & \frac{\sqrt{3}}{2} & -\frac{\sqrt{3}}{2} \\ \frac{1}{\sqrt{2}} & \frac{1}{\sqrt{2}} & \frac{1}{\sqrt{2}} \end{bmatrix} \begin{bmatrix} i_A \\ i_B \\ i_C \end{bmatrix} = C_{3/2} \begin{bmatrix} i_A \\ i_B \\ i_c \end{bmatrix} \tag{8}$$

In equation (8), it is a one-phase zero-sequence component added for inverse transformation.

*2.1.2. Transformation between two-phase stationary coordinate system and two-phase rotating coordinate system.* The two-phase static winding is connected with a two-phase balanced alternating current to generate a rotating magnetomotive force. According to the analysis of the relative relationship between static and motion, assuming that the fixed current is direct current, the winding becomes rotating, and the angular velocity of rotation is consistent, then the two-phase windings are connected with direct current to produce space rotating magnetomotive force. This transformation process is called two-phase rotation-two-phase static transformation. The transformation equation is:

$$\begin{bmatrix} i_\alpha \\ i_\beta \end{bmatrix} = \begin{bmatrix} \cos\varphi & -\sin\varphi \\ \sin\varphi & \cos\varphi \end{bmatrix} \begin{bmatrix} i_d \\ i_q \end{bmatrix} = C_{2r/2s} \begin{bmatrix} i_d \\ i_q \end{bmatrix} \tag{9}$$

In equation (9), $\varphi$ is the angle between the d axis of the d-q coordinate system and the axis of the coordinate system, $\omega$ $(\varphi = \omega t)$ is the rotational angular velocity of the d-q rotating coordinate system. $C_{2r/2s}$ is the transformation matrix from two-phase rotation to two-phase stationary coordinate system.

$$C_{2r/2s} = \begin{bmatrix} \cos\varphi & -\sin\varphi \\ \sin\varphi & \cos\varphi \end{bmatrix} \tag{10}$$

Inverse transformation of equation (10) can get the conversion from static to rotating coordinate system.

$$C_{2s/2r} = C_{2r/2s}^{-1} = \begin{bmatrix} \cos\varphi & \sin\varphi \\ -\sin\varphi & \cos\varphi \end{bmatrix} \tag{11}$$

*2.1.3. Transformation between three-phase stationary coordinate system and two-phase rotating coordinate system.* Through equations (8)-(11) in section 2.1.2, comprehensive analysis of the formulas can derive the matrix relationship between the three-phase stationary to the two-phase rotating coordinate system. The conversion relation is as follows:

$$\begin{bmatrix} i_a \\ i_\beta \\ i_0 \end{bmatrix} = C_{2s/2r} \begin{bmatrix} i_a \\ i_\beta \\ i_0 \end{bmatrix} = C_{2s/2r} C_{3s/2s} \begin{bmatrix} i_A \\ i_B \\ i_C \end{bmatrix} = C_{3s/2r} \begin{bmatrix} i_A \\ i_B \\ i_C \end{bmatrix} \tag{12}$$

The transformation matrix from the three-phase stationary coordinate system to the two-phase arbitrary rotating coordinate system is:

$$\sqrt{\frac{2}{3}} \begin{bmatrix} \cos\varphi & \cos(\varphi-120^\circ) & \cos(\varphi+120^\circ) \\ -\sin\varphi & -\sin(\varphi-120^\circ) & -\sin(\varphi+120^\circ) \\ \frac{1}{\sqrt{2}} & \frac{1}{\sqrt{2}} & \frac{1}{\sqrt{2}} \end{bmatrix} \tag{13}$$

The transformation matrix from the two-phase arbitrary rotating coordinate system to the three-phase stationary coordinate system is:

$$C_{2r/3s} = C_{3s/2r}^{-1} = \sqrt{\frac{2}{3}} \begin{bmatrix} \cos\varphi & -\sin\varphi & \frac{1}{\sqrt{2}} \\ \cos(\varphi-120^\circ) & -\sin(\varphi-120^\circ) & \frac{1}{\sqrt{2}} \\ \cos(\varphi+120^\circ) & -\sin(\varphi+120^\circ) & \frac{1}{\sqrt{2}} \end{bmatrix} \tag{14}$$

*2.1.4. Under the d-q coordinate system, the motor model.* In the d-q coordinate system, the basic formula of asynchronous motor can be obtained.

$$Te = n_p \frac{Lm}{Lr}(i_{sq}\psi_{rd} - i_{sd}\psi_{rq}) \tag{15}$$

$$Te - T_L = \frac{J}{n_p}\frac{dw_r}{dt} \tag{16}$$

$$i_{sd} = \frac{U_{sd}}{L_{sc}s + R_s} - \psi_{rd}\frac{1}{L_r}\frac{L_m s}{L_{sc}s + R_s} \tag{17}$$

$$i_{sq} = \frac{U_{sq}}{L_{sc}s + R_s} - \psi_{rq}\frac{1}{L_r}\frac{L_m s}{L_{sc}s + R_s} \tag{18}$$

$$\psi_{rd} = \frac{1}{R_r + L_r s}(i_{sd}L_m R_r - w_r L_r \psi_{rq}) \tag{19}$$

$$\psi_{rq} = \frac{1}{R_r + L_r s}(i_{sq}L_m R_r + w_r L_r \psi_{rd}) \tag{20}$$

From (15) to (20), the following motor models can be built.

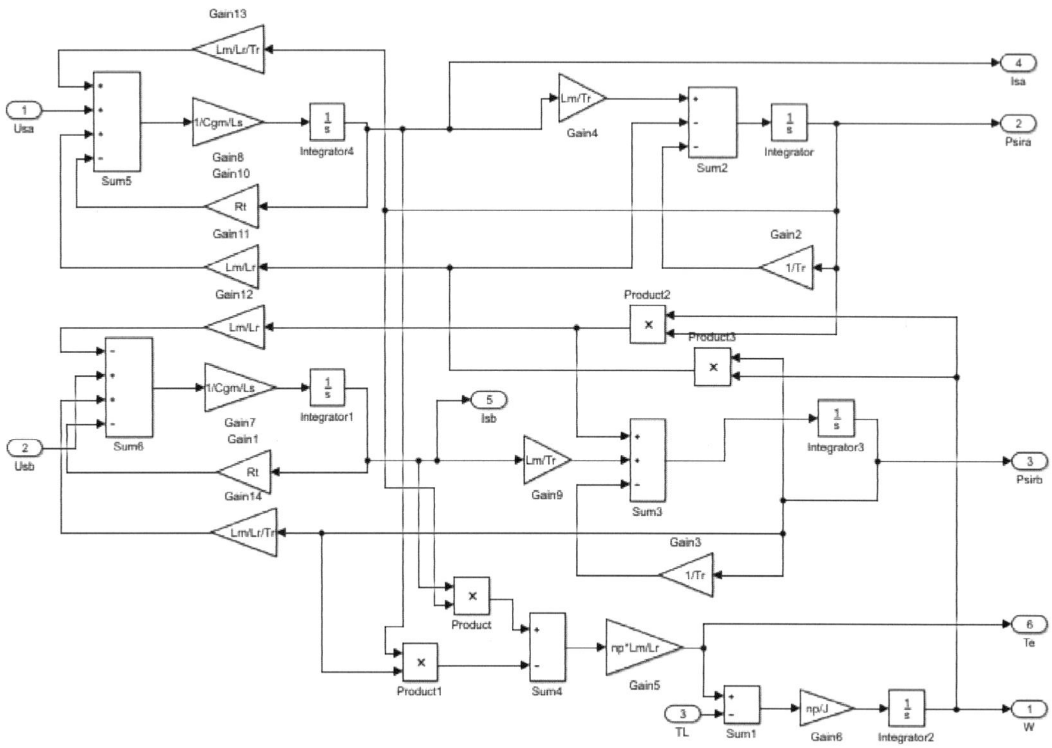


**Figure 1.** Asynchronous motor simulation model.

Figure 1 is a model simulation diagram of an asynchronous motor, the motor model under ideal conditions is established to eliminate the influence of motor performance degradation on the IGBT degradation case study.

*2.1.5. Establish the entire system simulation model.* Figure 2 is the simulation model of the motorized spindle frequency conversion speed regulation system. Use the from workspace and to workspace of the simulink module to transfer input and output data to and from the workspace, use this model for the following identification steps.

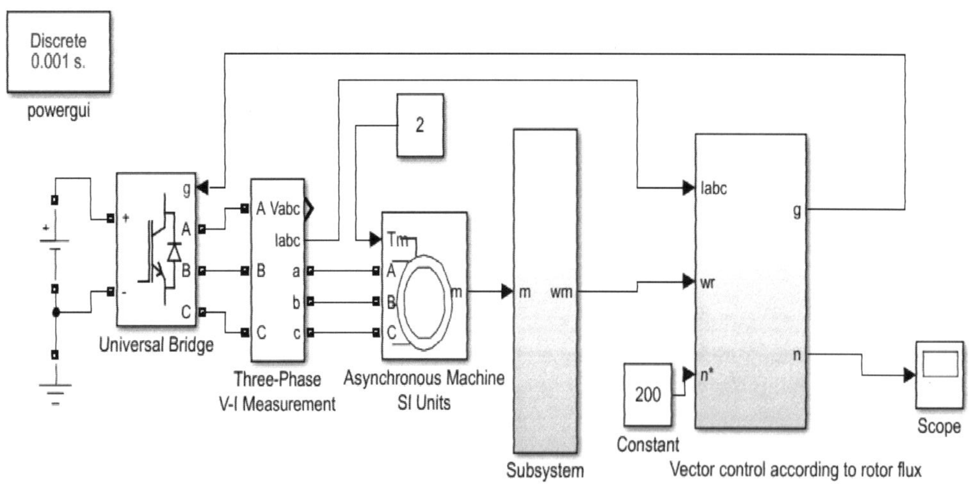


**Figure 2.** System simulation model.

### 2.2. Drive system identification

*2.2.1. Principles of system identification.* System identification is widely used in control systems. The input and output signals of the system are collected through experiments, the system model is identified using the identification toolbox, and the system model can be determined by further verification [5]. It can get a description of the system through the processing and calculation of input-output signals. For manual control in industrial processes, the process operator needs to know how the process output responds to different control behaviours. System identification is to study how to obtain the mathematical model of the system from the measured values.

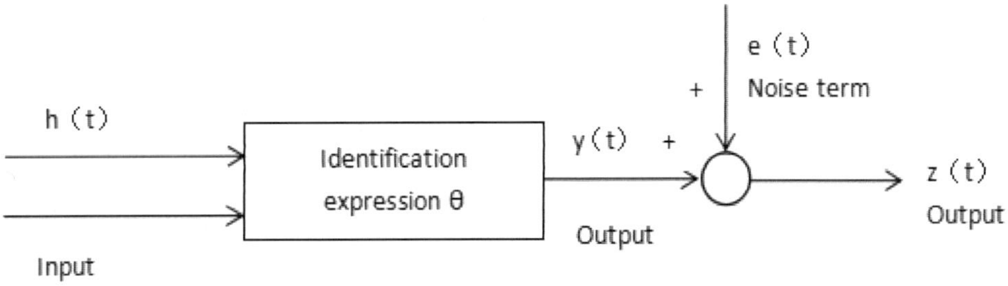


**Figure 3.** Expressions to identify problems.

The system identification method is to identify the transfer function. The process equation is:

$$\begin{aligned} & y^{n}(t)+a_1 y^{(n-1)}(t)+\cdots+a_n y\ (t) \\ & =b_0 u^{m}(t)+b_1 u^{(m-1)}(t)+\cdots+b_m u(t) \end{aligned} \tag{21}$$

Among them, *y(t), u(t)* are the output and input signal in the time domain, n, m are the system order of the model, $a_1,\ldots,a_n$, $b_0,\ldots,b_m$ are constant, the above formula is transformed by Laplace, we can get a general expression in the time domain of the transfer function [6].

$$G(s)=\frac{Y(s)}{U(s)}=\frac{b_0 s^{m}+b_0 s^{m-1}+\cdots+b_m}{s^{n}+a_1 s^{n-1}+\cdots a_n} \tag{22}$$

*Y(s), U(s)* are the output input signals. In the study, we introduce unit forward operator and unit delay operator, get a description of the discrete process of the system

$$y(t)=G(q)u(t) \tag{23}$$

Pass operator *G(q)* expression is

$$G(q)=\frac{b_1 q^{-1}+\cdots+b_m q^{-m}}{1+a_1 q^{-1}+\cdots a_n q^{-n}} \tag{24}$$

*2.2.2. System identification steps.* Generally speaking, the system identification process includes the following basic steps:

*Step 1* - identification experiment.

Reference [7] uses Chirp signal as excitation signal to study the identification of electro-hydraulic servo system. Select the generalized binary noise (GBN) signal [8] as the input signal, the selection of sampling time should refer to the response speed of the system. This article selects GBN signal as the excitation signal, the GBN signal is a randomly generated noise signal, suitable for control identification experiments related to industrial processes, the generation rule of this signal is that the value of u(t) is switched back and forth between the constant ±a, and switch at a certain conversion moment, meet the following formula with the conversion time t.

$$P[u(t)=-u(t-1)]=P_{sw} \quad (25)$$

$$P[u(t)=u(t-1)]=1-P_{sw} \quad (26)$$

$P_{sw}$ is the conversion probability that is the probability of the signal switching at each conversion time, each time is distributed independently. Therefore, the expectation of the GBN signal is zero.

*Step 2* - model order/Structure selection.

Choose the optimal model order, the finalized model can best describe the system under test. In the system identification research, different transfer function models are selected for different measured objects to carry out experimental design. Generally speaking, the models commonly used in system identification are the most commonly used ARX (auto regression model) models, ARMAX (noisy autoregressive moving average) models, GLS (generalized least squares) models, etc. These models are selected for different system structures, the ARX and ARMAX models are used in common systems. In this paper, the ARX model is selected as the system identification model.

*Step 3* - parameter estimation.

Use the optimization algorithm to get the best parameter estimate.

*Step 4* - model checking.

For different recognition situations, determine different accuracy standards, perform accuracy test on the obtained model, if the accuracy does not meet the requirements, then optimize the previous experimental steps, do the model test again, until it meets the standard. The test method adopted in this article is 0.3 precision test method, That is, 0.3 times the size of the performance degradation evaluation interval is greater than or equal to the identification parameter error.

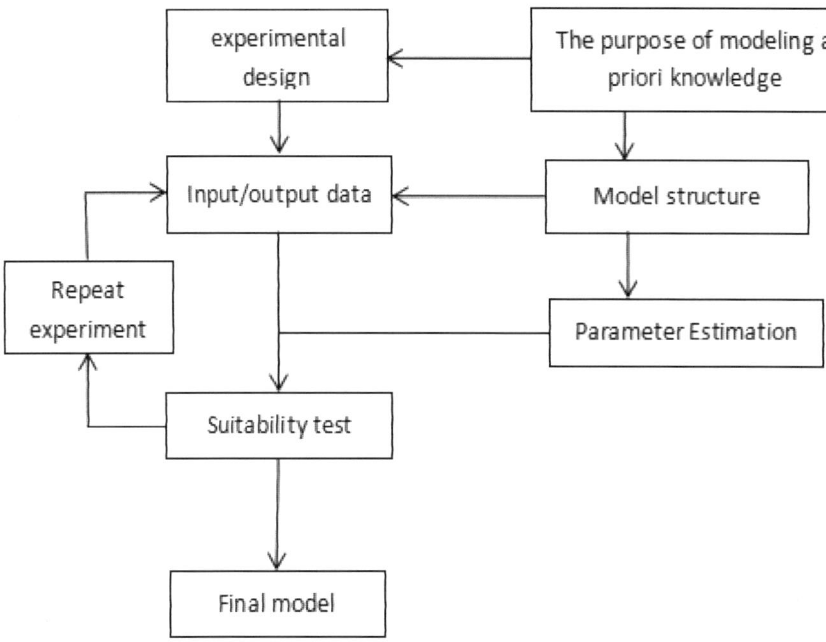


**Figure 4.** Identification process block diagram.

*2.2.3. Determination of system performance evaluation level.* According to the literature [9], we can know that the operation performance of the motorized spindle will decrease, the overall performance gradually decreases. In order to make the different results obtained by the system identification have different measurement standards, we need to rank the evaluation results in advance, determine the status of different levels.

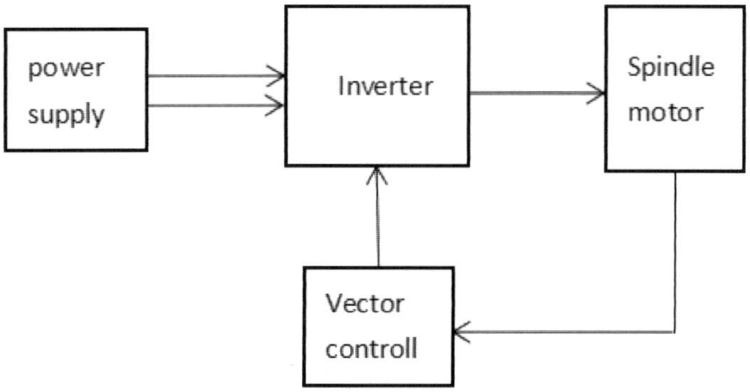


**Figure 5.** High-speed electric spindle drive system block diagram.

This paper determines the degree of performance degradation based on the geometric distance between the parameters identified by the system and the identification parameters of the normal system, the geometric distance ($g$) reflects the degree of deviation of the values [10, 11], the evaluation value is obtained by normalizing the calculated distance.

$$g = \sqrt{(a_1 - b_1)^2 + (a_2 - b_2)^2 + \cdots + (a_n - b_n)^2} = \sqrt{\sum_{i=1}^{n} (a_i - b_i)^2} \tag{27}$$

$$x = \frac{g - g_{\min}}{g_{\max} - g_{\min}} \tag{28}$$

where $a_1, a_2, \ldots, a_n$ are the transfer function parameters identified by the normal system, $b_1, b_2, \ldots, b_n$ are the transfer function parameters identified during operation, $g$ is the Euclidean distance of the parameter vector, $x$ is the normalized value of geometric distance.

Establish four performance evaluation levels between normal and faulty systems, on the basis of summarizing the current classification, this article, this paper proposes a model table 1 for the evaluation of the high-speed motorized spindle drive system [12].

**Table 1.** Evaluation grade division of drive system.

| Evaluation level | Evaluation value | Recommended measures |
|---|---|---|
| excellent | $0 \leq x < 0.25$ | Excellent performance |
| good | $0.25 \leq x < 0.5$ | Normal maintenance |
| average | $0.5 \leq x < 0.75$ | Strengthen maintenance and local repair |
| poor | $0.75 \leq x \leq 1$ | Overhaul |

## 3. IGBT degradation performance evaluation example analysis

### *3.1. IGBT performance degradation principle*

The performance of IGBT is degraded, the literature [13] analyse the degradation process from its internal chip and external package. In terms of internal chips, one is the aging of the solder layer, the solder layer is used to fix the IGBT chip and the package shell, the shell temperature cycle rises and falls, it causes electron-thermal shock to the solder layer, leads to fatigue degradation of the solder layer, reduces the heat dissipation capacity of IGBT, causes the junction temperature to rise continuously as the degree of degradation deepens, eventually fails and a short circuit or open circuit fault occurs. The second is that the bonding wire falls off, the IGBT module continuously withstands the impact of temperature fluctuations during the work process, resulting in the accumulation of plastic strain at the connection point between the bonding wire and the silicon chip, and the cracks at the edge of the connection point between the bonding wire and the chip initially appear. Under the repeated impact of temperature fluctuations, the crack gradually expands, accelerating the fall of the bonding wire, at the same time, the aluminium substrate of the emitter is restructured, which leads to the degradation of IGBT performance. Literature [14] through temperature cycle experiments, after the device failed, the phenomenon of bond wire shedding, cracking and metal reconstruction was observed through scanning electron microscope pictures. After the performance of the IGBT chip is degraded, the results of bond wire drop and metal reconstruction are shown in figure 7: (a) bond wire off, (b) bond wire cracks, (c) the surface of the aluminium substrate without reconstruction before use, (d) after the performance degradation and reconstruction, the surface of the aluminium substrate.

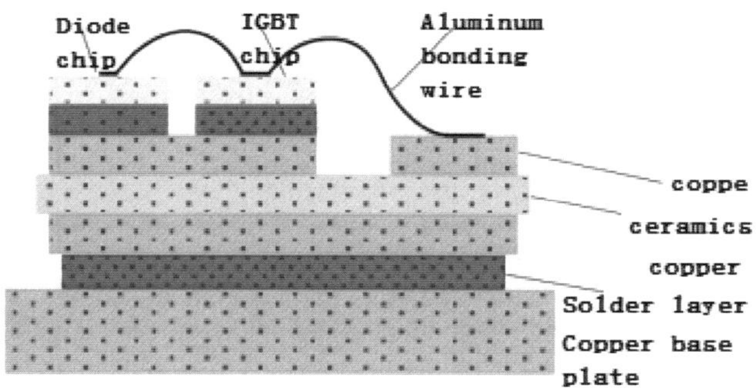


**Figure 6.** Picture of IGBT device structure degradation.

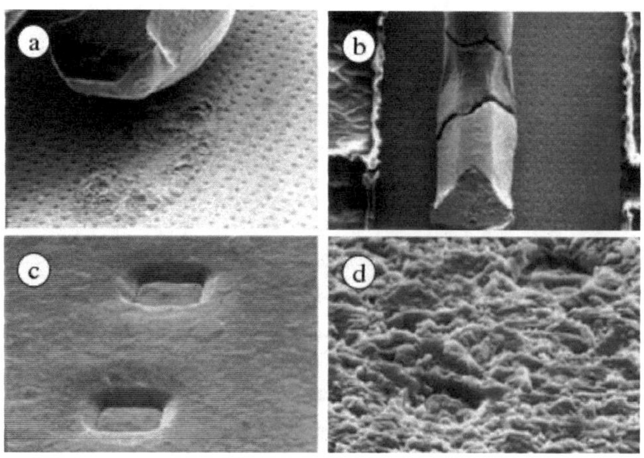


**Figure 7.** IGBT structure diagram.

In terms of the material composition of the package, there are various materials in the packaged IGBT. From table 2, the material will expand and contract due to temperature changes, under equal pressure (P-fixed) conditions, measured by the change in volume of the material caused by the change in unit temperature, this change can be called the coefficient of thermal expansion (CTE) [15].

**Table 2.** CTE coefficient of each material in IGBT package structure.

| Name | Material used | Thermal expansion coefficient (ppm/°C) |
|---|---|---|
| Aluminium bonding wire | aluminium | 22 |
| IGBT chip | silicon | 3 |
| Welding layer | solder | 28 |
| Ceramics | ceramics | 7 |
| Substrate | copper | 17.5 |

The connection between the IGBT chip and the diode chip is made of aluminium metal layer and bonding wire, moreover, the chip material silicon and the metal layer and the material of the bonding wire have a large difference in thermal expansion coefficient, resulting in the thermal cycling process, repetitive mechanical stresses caused by different degrees of expansion on the contact surface, this will lead to continuous fatigue degradation of the lead metal layer, and the bonding wire subjected to this repeated stress impact.

There are two forms of failure caused by this degradation. One form of failure is to cause reconstruction of the surface of the lead metal layer, On the other hand, the resistance of this part and the power generated is increased, and the junction temperature is further increased, On the other hand, reducing the heat dissipation capability also makes the junction temperature further increase. When the junction temperature is greater than a certain threshold, the parasitic crystal tube in the IGBT is turned on, the gate loses control, and a collector-emitter short circuit occurs; another type of failure is to cause the bonding wire to break, and an open collector-emitter circuit will occur. The breaking of the bonding wire will also cause the resistance of this part to increase. In summary, both the lead metal layer and the bonding wire will increase the resistance of this part, both resistances are part of the resistance of the IGBT collector-emitter channel. Therefore, online monitoring of the resistance of the IGBT collector-emitter channel enables monitoring of the degree of degradation of the IGBT lead metal layer and bonding wire.

### *3.2. Performance evaluation of IGBT performance degradation process*

*3.2.1. IGBT performance degradation simulation design.* According to the analysis of the degradation mechanism of IGBT, there are two types of IGBT degradation, bond wire breakage and aluminium metal layer damage. The two forms of degradation eventually lead to an increase in the internal resistance $R_{on}$, and the internal resistance of the IGBT varies from 0.105 to 0.115 [16]. The resistance change inside the IGBT can characterize the degradation degree of the aluminium metal layer and the bonding wire inside the IGBT, and not affected by changes in load current $I_c$. Through this part of the analysis, simulation experiments of performance degradation can be established, in the simulation, the value of internal resistance Ron is continuously increased to simulate the degradation process of IGBT performance. Set the Ron value to 0.110 to evaluate the performance of the current state.

*3.2.2. Performance evaluation.* The parameters of the asynchronous motor model are set as: rated power $P_N=20kW$, rated voltage $U_N=350V$, rated frequency $f_N=500Hz$, stator resistance $R_S=0.11\Omega$, rotor resistance $R_r=0.21\Omega$, stator leakage inductance $L_S=0.00030239H$, rotor inductance $L_r=0.00030876H$, stator and rotor mutual inductance $L_r=0.0086H$, number of pole pairs $p_n=2$, moment of inertia $J=0.01kg{\cdot}m^2$.

The system identification method is used to gain the transfer function, it contains the dynamic and static characteristics of many models. Under the guidance of practical industrial process experience and knowledge, by comparing and analysing some parameter information of these system transfer functions, in order to obtain some changes in key information, data information with reference value for system testing personnel, make corresponding judgments and repair work.

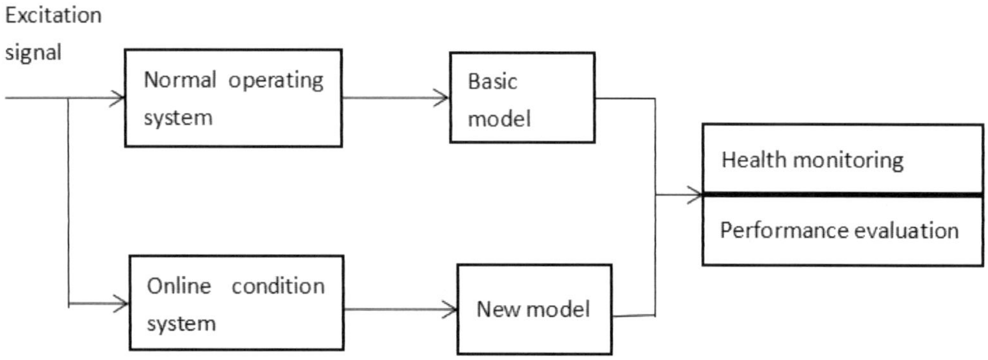


**Figure 8.** Performance evaluation method based on system identification.

The basics model in the figure is the transfer function parameter identified by the normal system $a_1,a_2,\ldots, a_n$, new model is the system transfer function parameter identified during detection. This figure describes the overall idea of using the system identification method to evaluate the performance of the system proposed in this article. From the framework, it can be seen as a research method based on model discrimination. Identify the system during the period of normal operation of the system and the period of being tested, compare the transfer function information obtained at different times, if some key variables and information in the industry have serious deviations from normal operating conditions, it is necessary to carry out troubleshooting or troubleshooting in advance.

When the internal resistance of the IGBT is set to 0.110, according to the theoretical analysis, the evaluation grade of IGBT is good. Follow the system identification steps in chapter 2 to perform identification under Ron=0.105 and Ron=0.110 states, as shown figure 9.

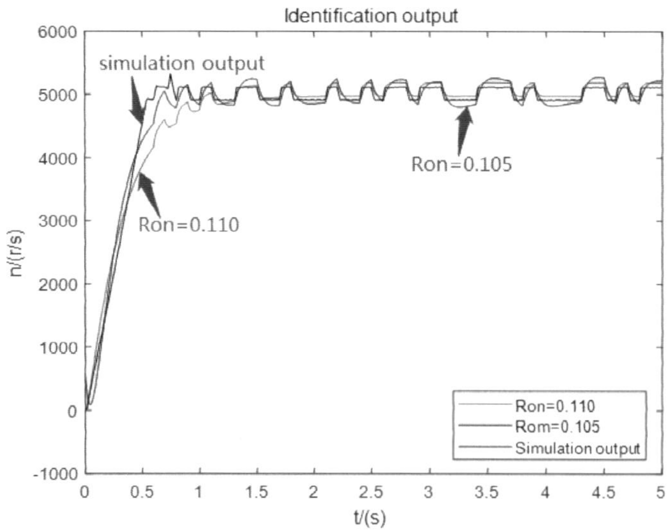


**Figure 9.** System identification output.

The transfer function can be obtained through the identification method as shown in table 3.

**Table 3.** Identification transfer function table.

| | |
|---|---|
| $G_{\text{normal}}$ | $\dfrac{-0.7136z^{-1}-0.1279z^{-2}}{1+0.3247z^{-1}+0.2312z^{-2}}$ |
| $G_{R_{on}=0.110}$ | $\dfrac{-0.7113z^{-1}-0.1051z^{-2}}{1+0.3035z^{-1}+0.2335^{-2}}$ |
| $G_{\text{fault}}$ | $\dfrac{-0.6712z^{-1}-0.0763z^{-2}}{1+0.2521z^{-1}+0.2310z^{-2}}$ |
| Performance evaluation | X=0.317, according to the performance evaluation level model table, the performance level is good. |

The above examples of IGBT degradation confirm the correctness of the established model and system identification performance evaluation method, using the system identification parameter mapping mechanism to diagnose the system performance, it provides a great help to the system performance evaluation.

## 4. Conclusion

In this paper, the geometric distance between the identification parameters of the high-speed motorized spindle drive system and the normal working system in the actual working process is obtained. Determine the degree of system performance degradation through the geometric distance, the corresponding evaluation level is given to the current system performance. Using this method, an example of IGBT device performance degradation is analysed. The result shows, the method can evaluate the performance of IGBT degradation process well, the degree of degradation is mapped through the identification parameters. There are few studies using this method to evaluate system performance. The system performance evaluation method based on system identification parameter changes and geometric distance determination lays the foundation for future system performance evaluation research.

## References

[1] Luo Y S and Chen J Z 2009 Application of AHP-FCE in mechanical system performance evaluation *Mechanical Manufacturing and Automation* **38** (03) 112-14

[2] Tian Q H and Du Y X 2004 Research on performance evaluation of mechanical products based on entropy weight fuzzy comprehensive evaluation method *China Manufacturing Informatization* pp 97-99

[3] Abolfazl S, Abolhassan R and Vladimir I S 2020 *System Identification and Control Design of a Nonlinear Continuously Stirred Tank Reactor* (Elsevier B.V) p 173

[4] Chen X A, Kang H M, He Y, Chen M, Miao Y Y and Chen W Q 2010 Analysis of dynamic performance of high-speed electric spindle under speed sensorless vector control *Journal of Mechanical Engineering* **46** (07) 96-101

[5] Chavan S L and Talange D B 2018 System identification black box approach for modeling performance of PEM fuel cell *The Journal of Energy Storage* 327-32

[6] Sari R, Joki D, Ustovi E and Juri E 2019 Experimental parametric identification in z closed-loop feedback using novel approach to the linear control systems *IFAC Papers online* **52** (27) 13-18

[7] Wang H, Zhang M R, Li J F, Du J 2016 Chirp signal excited electro-hydraulic servo system identification. Machine Tool and Hydraulics, **44** (13) 128-131
[8] Tulleken H J A F 1990 Generalized binary noise test-signal concept for improved identification-experiment design *Automatica* **26** (1) 37-49
[9] Guo L, Chen J, Zhao F G, Dong G M and Wang G W 2008 Application of geometric distance method based on support vector machine in equipment performance degradation evaluation *Journal of Shanghai Jiaotong University* no. 07 1077-80
[10] Yang B 2018 Reliability *Evaluation of Electric Spindle Based on Performance Degradation* Jilin University
[11] Lu Y and Christou A 2019 Prognostics of IGBT modules based on the approach of particle filtering *Microelectronics Reliability* **92** 96-105
[12] Wang X Q, Niu L Y, Huang Y and Jiang J C 2014 Electric and safety performance evaluation systemem of electric vehicle chargers *Power Technology* **38** (04) 734-36
[13] Du Y, Li F, Zeng Z Y and Zheng J F 2013 Research on the construction method of IGBT performance degradation model *Power Electronics Technology* **47** (01) 104-06
[14] Smet V, Forest F and Huselstein J J 2011 Ageing and failure modes of IGBT modules in high-Temperature Power Cycling *IEEE Transactions on Industrial Electronics* **58** (10) 4931-41
[15] Liu D 2016 *On-line Monitoring of IGBT Degradation Under Repeated Overcurrent Impacts* Zhejiang University
[16] Mohamed S M H, Karthik K, Tseng K J and Prasanth S 2016 Online computation of IGBT on state resistance for off-shelf three-phase two-level power converter systems *Microelectronics Reliability* **64** 379-86

# Study of milling machining roughness prediction based on cutting force

**Y C Kao[1,2*], S J Chen[1], T K Vi[2], G H Feng[2] and S Y Tsai[2]**

[1]Advanced Institute of Manufacturing with high-tech Innovations, National Chung Cheng University, No. 168, Sec. 1, University Rd., Minhsiung, Chiayi 621301, Taiwan (R.O.C.)
[2]Department of Mechanical Engineering, National Chung Cheng University, No. 168, Sec. 1, University Rd., Minhsiung, Chiayi 621301, Taiwan (R.O.C.)

*E-mail: imeyckao@ccu.edu.tw

**Abstract.** Cutting parameters can to be properly adjusted to increase machining precision, to enhance cutting efficiency, and to reduce cost. These parameters are generally chosen according to the recommended data from the tool maker's technical manual with the synergy of the operator's experiences. However, inappropriate parameters selection might not only damage cutter and the machine tool, but also worsen the surface roughness of the workpiece. Researches on improving surface roughness has been studied over the years, for example, on the optimization of cutting conditions, vibration measurement and analysis, and cutting forces measurement and analysis, to name only a few. However, most of them must have sensors installed on the machine tool for data collection before the analysis. This will increase the cost. Furthermore, the measurement of cutting force needs to have dynamometer which is not only expensive but is also not appropriate to be installed on a small-sized machine tool. This research focused on establishing the relationship between the cutting force and the surface roughness for the workpiece machining in a small-sized machine tool. In order to calculate the cutting force, cutting force coefficients were obtained based on the cutting test on the other machine tool rather than the small-sized machine. Tapping test was also conducted to obtain the frequency response function of the machine. The cutting forces were calculated based on the cutting force coefficients and the frequency response function. The calculated cutting forces were then converted through the proposed methodologies into eleven characteristic values. Finally, two commonly used machine learning algorithms, multivariate linear regression and generalized regression neural network were adopted for regression analysis to establish the relationship between the cutting forces and the surface roughness. The results have shown that the prediction by generalized regression neural network is better than that of multivariate linear regression. The major contribution of this paper is that appropriate algorithms have been studied, compared and verified so that the relationship between cutting force and surface roughness can be established. This means that surface roughness can be estimated based on the calculated cutting forces incorporating the machine tool's frequency response function.

## 1. Introduction

During the cutting process, the changes of cutting force during the machining will affect the stability of the whole machine. In severe cases, it will worsen the quality of the workpiece, incur fast tool wear, affect the reliability, shorten useful life of the machine, and directly reduce production efficiency.

Smith and Tlusty [1] had shown that the analysis of "stability lobe diagram" is more important in high-speed milling, because there is a wide and highly stable cutting area during high-speed milling. Chiou, Chung and Liang [2] used a dynamometer to verify and analyse the chatter stability diagram and found that the tool wear loss could be calculated using the Coriolis force equation to obtain a difference in feed rate, which indirectly affects the damping coefficient and the cutting force. Tlusty and Andrews [3] developed processing technology for automatic detection; they used three kinds of sensors to detect cutting force, vibration and sound wave, respectively. Experimental analysis of turning and milling results showed that the cutting force sensor is the best because the chatter is more difficult to measure than that of vibration signal. Cutting forces was able to very clearly distinguish Chatter. Taylor [4] found the cutting chatter is mainly caused by the discontinuous cutting, the periodic cutting force and the mechanical parts of the machine tool (such as a part of the table and the drive system) are caused by the same natural frequency. Smith et. al [5] discussed the stability of cutting after changing the tool length - after the tool holder holds the tool, the length of the tool extends outward. When the tool is clamped on the spindle, it is like a cantilever beam structure. When the external length of the tool is different, the rigidity of the machine structure is also different. Because the tool, workpiece and machine tool structure form a closed system during cutting. If the external length of the tool is different, it directly changes the stability of cutting, and also changes the rigidity of the cutting process. When the tool overhang is longer, the main stable processing area becomes narrower, and it has no effects on the depth of cut.

Since stable and unstable cutting will affect the processing quality, in order to achieve the required upper quality level, the processing parameters are usually adjusted that the optimized application parameters could be obtained. The most commonly used method is the trial-and-error method of empirical rules, but the use of this method will conduct actual processing cutting and testing, and therefore, it is not only time-consuming but also waste of materials. Milton [6] discussed various phenomena and interactive effects in cutting. It was mentioned that cutting was a highly nonlinear behaviour. Davim [7] provided surface detection methods, from the earliest offline surface measurement to online direct surface measurement, and to online indirect surface measurement. Dimla, Lister and Leighton [8] organized various tool monitoring methods for cutting, and proposed analysis from multi-sensors through signal feature analysis by the use of neural network to find key features. Lauro, Brandao, Baldo, Reis, Davim [9] organized the commonly used methods and theories for cutting, signal acquisition, signal processing, and signal analysis cutting. Varghese and Radhakrishnan [10] discussed the integration of multiple sensors. It used the binary value of the root mean square of capacitive, inductive, and optical distance sensors as input. The binary values of the median, root mean square, and average roughness of the surface roughness were the output. A feed forward back propagation neural network was constructed. The number of neurons was tested in three hidden layers, and the surface roughness was predicted successfully by the experiment with the error less than 20%. Hossain, Amin and Patwari [11] only used the spindle speed, feed rate, and cutting depth as the input, and the surface roughness as the output; a two-layer Log sigmoid hidden layer with genetic algorithm (GA), the feedforward back-propagation neural network model was successfully constructed for nickel-aluminium alloy end milling. The error was below 8.3%, which showed that even under the condition of insufficient information, the function approximation ability of the neural network still has reliability. Wang [12] used spindle speed, feed rate, depth of cut and workpiece acceleration value as input, and surface roughness as output. A feed-forward backward neural network with LS-SVR algorithm was constructed. The obtained average relative error (ARE) by least square vector regression neural network is about 7%. Azlan, Habibollah and Safian [13] optimized the processing parameters through GA, found the best cutting conditions could be used to obtain the smallest surface roughness value. Ilhan and Mehmet [14] used artificial neural networks (ANN) and multiple regression methods to simulate the surface roughness prediction of AISI 1040 steel. The input parameters include cutting speed, feed rate and cutting depth, and the output was the surface roughness value. It was clearly seen from the results that ANN algorithm could optimize the processing parameters and reduce the surface roughness value compared with the multiple regression model. Reddy [15] used composite materials (GFRP) for cutting, and found key factors through neural network algorithm. After analysis, it was found that cutting speed and cutting depth are important factors affecting surface roughness, which were 26.84% and 40.44%, respectively. Grynal, Srinivasa and Rashmi [16] thought that surface roughness prediction was too complicated because there were uncontrollable factors such as the impact of blade wear and machining vibration in the cutting process. Therefore, neural network algorithm (ANN), radial basis function neural network (RBFNN) and Summation Wavelet-Extreme Learning Machine (SW-ELM) were used to predict surface roughness. The research results showed that SW-ELM had a higher prediction accuracy. Gopan, Leo and Surendran et. al. [17] claimed that there were high requirements for surface roughness in the field of wear. Therefore, prediction of the surface roughness through neural network algorithm (ANN) were researched, but the ANN was easy to fall into the local solution. The optimization of ANN model through genetic algorithm (GA) could have a greater improvement in the accuracy of ANN. Titus, K., R, S., Prabhu, Sivakumar, Devaraj and Sathiskumar [18] developed neural network models for composite materials. Each input parameter is checked by using the ANOVA method, and then the material removal rate (Material Removal Rate, MRR) and surface roughness (Ra) were predicted by a neural network algorithm. According to the above literatures, the prediction of surface roughness is very important for a machined component but it seems that study on a small-sized machine has not been studied yet. This research is therefore aimed for predicting the surface roughness (Ra) by using neural network algorithm based on dynamic cutting forces according to the stable and unstable cutting through the SLD.

## 2. Dynamic cutting force operation process and method

In the machining process of a machine tool, the cutting energy of the machine is dispersed and the tool vibration is caused, which causes the unstable phenomenon of the cutting system and affects the surface of the workpiece. This phenomenon is called chatter. Generally self-excited chatter can be divided into three categories, the Regenerative Chatter, the Mode Coupling Chatter, and the Friction Chatter (Friction Chatter). Regenerative chattering is the most common case. It is due to the wavy surface left by the tool during the previous cutting, which causes unstable phenomena in the closed loop system of cutting and machine structure. On the surface of the workpiece, deeper cuts marks will be produced, resulting in a larger roughness on the surface of the workpiece, and fast tool wear. Such a vicious cycle will eventually cause damage to the tool, workpiece and even the machine. Therefore, this paper calculates the dynamic cutting force through the cutting force methodology [19]. The working process is shown in figure 1, and it includes two parts: (1) cutting experiment, and (2) machine tapping test.

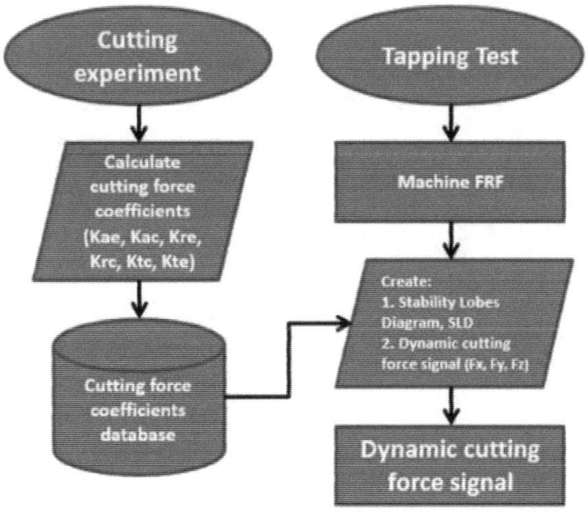


**Figure 1.** Flow chart of dynamic cutting force operation.

### *2.1. Cutting experiment*

The cutting experiment is used mainly to obtain six cutting force coefficients (CFCs) - Kae, Kac, Kre, Krc, Ktc, and Kte. These CFCs are then saved into cutting force coefficient (CFC) database. This research will establish the cutting force coefficients for Cu material. The tool type is shown in table 1.

The cutting experiment is Down-milling, the cutting depth is 0.5mm, and the spindle speed is 5120 rpm. According to the use of CuPro [19] software, five experiments are conducted, as shown in table 2, in order to obtain the average cutting force and establish a linear regression model (as figure 2). The linear regression equations are shown from equation (1) to (3).

**Table 1.** Tool type.

| | |
|---|---|
| Tool type | Solid carbide mill |
| Tool diameter | 3 mm |
| Tool material | Tungsten Carbide |
| Angle | 25° |
| Number of flutes | 2 |
| Cutter type | Flat-end |

$$\mathrm{Avg}(F_x)=a_xF_x+b \quad (1)$$

$$\mathrm{Avg}(F_y)=a_yF_y+b \quad (2)$$

$$\mathrm{Avg}(F_z)=a_zF_z+b \quad (3)$$

Avg() is the average cutting force. $F_x$ is the cutting force in the x direction. $F_y$ is the cutting force in the y direction. $F_z$ is the cutting force in the z direction. $a_x$ is the slope in the x direction. $a_y$ is the slope in the y direction. $a_z$ is the slope in the z direction. b is a constant.

**Table 2.** Cutting force experiment table.

| Test no. | Feed per flute (mm/min) | Feed rate (mm/min) |
|---|---|---|
| 1 | 0.012 | 123 |
| 2 | 0.014 | 143 |
| 3 | 0.016 | 164 |
| 4 | 0.018 | 184 |
| 5 | 0.02 | 205 |

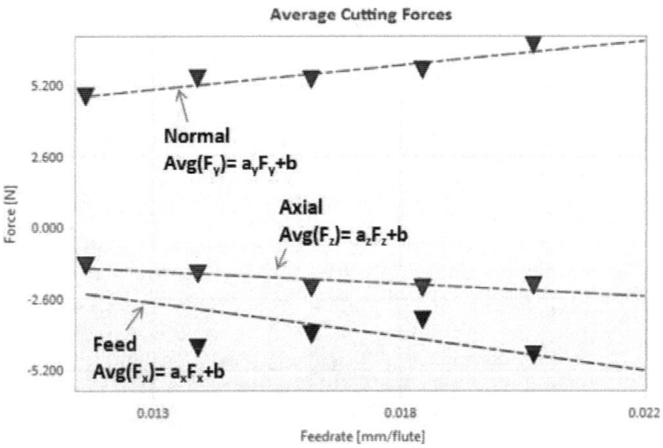


**Figure 2.** Regression model of average cutting force.

Next, the CutPro software was used to calculate the six CFCs through the average cutting force regression model, as shown in Equation (4) to Equation (9).

$$K_{tc}=\frac{4a_yF_y}{Na} \tag{4}$$

$$K_{rc}=\frac{-4a_xF_x}{Na} \tag{5}$$

$$K_{ac}=\frac{\pi a_zF_z}{Na} \tag{6}$$

$$K_{te}=\frac{\pi a_yF_y}{Na} \tag{7}$$

$$K_{re}=\frac{-\pi a_xF_x}{Na} \tag{8}$$

$$K_{ae}=\frac{2a_zF_z}{Na} \tag{9}$$

where N is Blade count, and a is Depth of cut.

It is found that these CFCs could be calculated if ax, ay, az are known. And ax, ay, az are the slopes of the linear regression equations. Therefore, as long as the slope could be obtained, the CFCs tend to be consistent, and the results of SLD and dynamic cutting force will also be similar to each other.

*2.2. Machine's Frequency Response Function (FRF)*

This research uses the impact hammer and acceleration sensor, as shown in figure 3, to obtain the vibration signal of the machine. The Frequency Response Function (FRF) could then be calculated from CutPro software.

After established the cutting force coefficient and the FRF of the machine, the stability lobe diagram (SLD) and dynamic cutting force could be calculated next. The SLD shows the relations between the depth of cut and the spindle speed with stable and unstable zones. This is important information for the machining processes, because it is necessary to avoid machining in the unstable zone. In addition, the generation of dynamic cutting forces is also important for the subsequent introduction of machine learning algorithms to calculate surface roughness.

**Figure 3.** Machine tapping test.

## 3. Space problem of small machine.

This research adopted a small turning and milling centre (TA20) manufactured by Arix Machinery Co. Ltd. The correlation between the cutting force and the surface roughness is calculated and predicted through adopted AI algorithms. The performance, specifications and dimensions of the machine are shown in table 3 and figure 4.

**Table 3.** Small turning and milling centre (TA20).

| | TA20 |
|---|---|
| Controller | Syntec 220MA-5 |
| Work space (mm) | 400*300*100 |
| X/Y/Z axis travel (mm) | 100/200/200 |
| X/Y/Z axis fast speed (m/min) | 10/10/10 |
| Max spindle speed (rpm) | 30,000 |
| Spindle motor output (kw) | 1 |

**Figure 4.** The adopted small turning and milling centre (TA20).

The dynamometer models used in this research are Kister 9265B and 9443B, the size of it is 220 mm * 100 mm * 186 mm, as shown in figure 5.

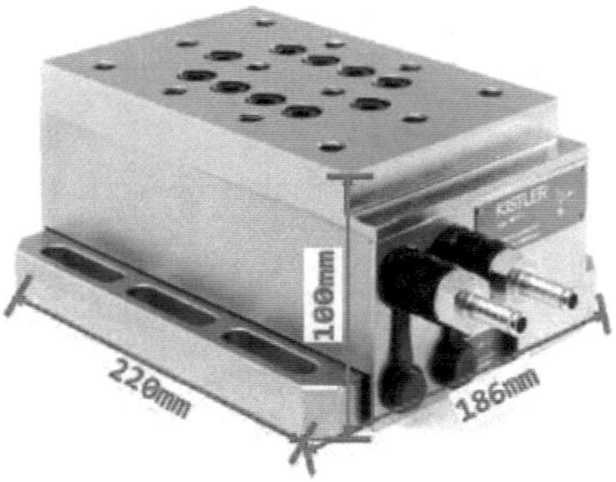


**Figure 5.** Kister dynamometer models: 9265B and 9443B.

It can be seen from table 3 and figure 4 that the Kister dynamometer (9265B and 9443B) is too large to be installed in the work space of TA 20. This will lead to the problem that the cutting force signal cannot be obtained by cutting test in TA 20. Therefore, this study used the CFCs obtained based on the cutting tests on the other industrial machine tools, and then apply it to the small machine (TA20). In this way, the dynamic cutting force signal is obtained, and the surface roughness could be estimated for the subsequent machine learning algorithm.

## 4. Surface roughness processing experiment

Since the output end of the machine learning algorithm is surface roughness (Ra) value, therefore, this research carried out cutting experiments on the TA 20 so as to measure the machined surface roughness; the experimental parameters include processing speed and cutting depth, as listed as follows:

- Cutting speed (rpm): 5000, 7000, 9000, 11000, 13000, 15000, 17000, 19000.
- Depth of cutting (mm): 0.1, 0.3, 0.5, 0.7, 0.9, 1.1, 1.3, 1.5.

The tool type is shown in table 1. The workpiece is mainly copper rod, with a diameter of 20mm and a length of 100mm. Before the cutting processing, the copper rod was milled into a slope to facilitate subsequent surface roughness measurement. The machining experiment is carried out by milling in the Y direction, and eight layers of steps were cut on the inclined plane, and the cutting depth equals to the cutting width; and the eight layers of steps corresponded to the depth of cut, as shown in figure 6.

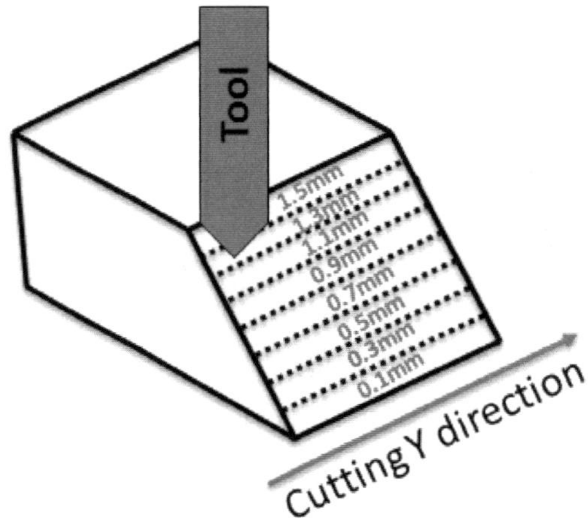


**Figure 6.** Surface roughness processing experiment.

The total cutting experiment has a total of eight processing speeds and eight cutting depths, and then sixty four processing experiment data were obtained. Next, measure the surface roughness (Ra) of the workpiece through the surface roughness instrument produced Mitutoyo, model SJ-410, as shown in table 4.

**Table 4.** Workpiece surface roughness (Ra).

| Cutting Speed (rpm) | Ap 0.1mm | Ap 0.3mm | Ap 0.5mm | Ap 0.7mm | Ap 0.9mm | Ap 1.1mm | Ap 1.3mm | Ap 1.5mm |
|---|---|---|---|---|---|---|---|---|
| S5000 | 1.21 | 1.31 | 3 | 4.55 | 4.93 | 2.96 | 6.73 | 3.63 |
| S7000 | 0.31 | 1.147 | 0.797 | 2.807 | 2.261 | 4.2 | 4.34 | 4.72 |
| S9000 | 0.26 | 0.53 | 0.6 | 1.19 | 1.94 | 3.31 | 4.07 | 4.61 |
| S11000 | 0.242 | 0.9885 | 0.564 | 0.9515 | 1.127 | 1.425 | 1.39 | 0.675 |
| S13000 | 0.222 | 0.5115 | 0.537 | 0.9643 | 1.109 | 0.794 | 1.3935 | 1.011 |
| S15000 | 0.293 | 0.289 | 0.5785 | 1.1795 | 1.8675 | 1.174 | 1.4187 | 1.3585 |
| S17000 | 0.2287 | 0.4675 | 0.9507 | 0.8737 | 1.0973 | 0.6255 | 1.0105 | 0.963 |
| S19000 | 0.639 | 0.4633 | 0.5175 | 0.6955 | 0.78 | 0.905 | 1.289 | 0.9705 |

Since the maximum cutting depth of the processing parameter is already the radius of the tool, the vibration caused by the processing will increase with the cutting depth, and the vibration will also be greater. If the blade frequency of this vibration is just at the natural frequency of the machine, resonance phenomena or chatter will be induced, and the surface quality of the cutting workpiece will deteriorate as the vibration amplitude increases. Next, this research will try to find out the stable zone and the unstable zone from the SLD, and examine the relationship between the processing parameters and the SLD.

## 5. Stability Lobe Diagram (SLD) to analysis

Because the small machine cannot have a Dynamometer installed. So, the CFCs came from the other industrial normal-sized machine. Large industrial machines include the vertical machine centre by Tongtai since the worktable area is enough to setup the Kister dynamometer. The cutting force coefficient experiment were carried out and established. Therefore, based on the dynamic cutting force operation process (figure 7), this study obtained the CFCs from the VC608 machine centre, and then transferred to the TA20 processing machine, as shown in table 5 and figure 7.

**Table 5.** VC608 Cutting force coefficient.

| Ktc | Krc | Kac | Kte | Kre | Kae |
|---|---|---|---|---|---|
| 1567.32 | 695.04 | 216.75 | 2.71 | 7.03 | -3.11 |

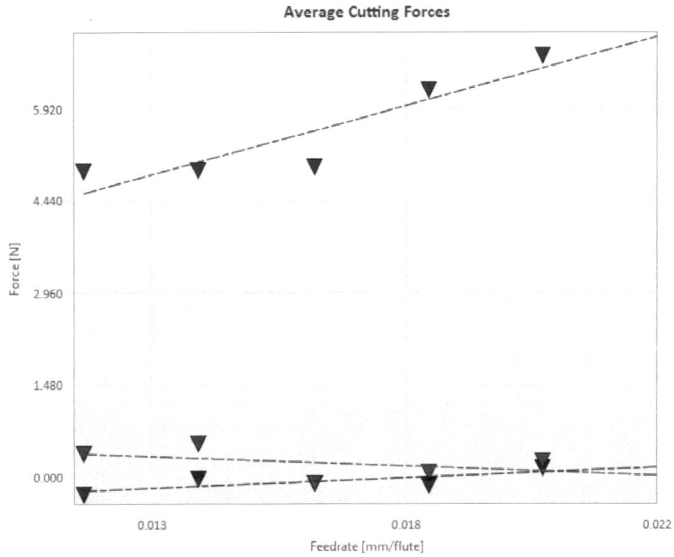


**Figure 7.** VC608 average cutting force regression model.

Next, according to the dynamic cutting force operation flow, the FRF of the TA20 machine is obtained, as shown in figure 8. If the cutting frequency is the same as or close to the natural frequency, resonance will occur. At this time, if the rigidity of the machine is not good enough, chatter will occur, resulting in an increase in poorer surface roughness. The quality of the workpiece will decline, and even cause problems such as machine damage and tool damage. So, this figure will be able to provide users' reference so as to avoid the natural frequency.

Because the surface roughness processing experiment used eight different speeds for machining, and the cutting width (Ae) is equal to the cutting depth (Ap) processing parameter. Therefore, in this study, CutPro will create a stability lobe diagram (SLD), examines the stable zone and unstable zone, and to mark them as shown in figure 9 to figure 16. Due to the influence of the machine's FRF, the maximum depth of cut and the stable depth of cut are also different due to the difference in processing speed and cutting width. For example, the processing speed is 5000rpm, the blade frequency is 166.67Hz, and the natural frequency is listed as follows:

- The 5th edge frequency 833Hz is close to the natural frequency in the y direction (825Hz).
- The 18th blade frequency 3012Hz is close to the natural frequency in the y direction (3012Hz).

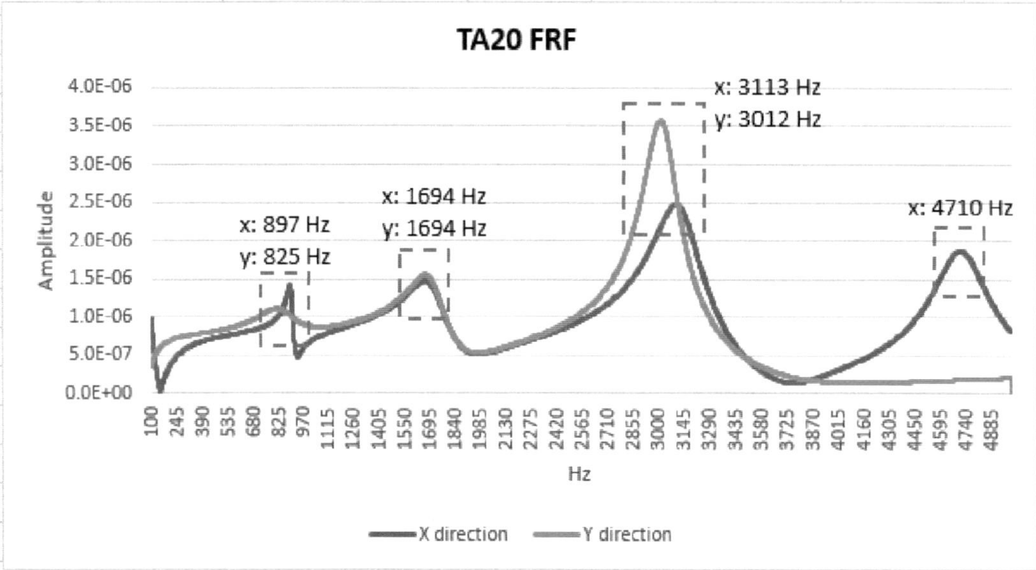


**Figure 8.** FRF of TA20 machine.

When the tool blade frequency is close to the natural frequency, the maximum allowable depth of cut will be reduced. That is, the stable zone will also be limited. Next, according to the cutting depth range of the SLD, there are a stable zone and an unstable zone at different speeds, and the arrangement is shown in table 6. It can be known from this table that at the processing speeds of 13000 rpm, 15000 rpm and 19000 rpm, due to the influence of the machine FRF, that the maximum cutting depth will be extended to 0.9 mm.

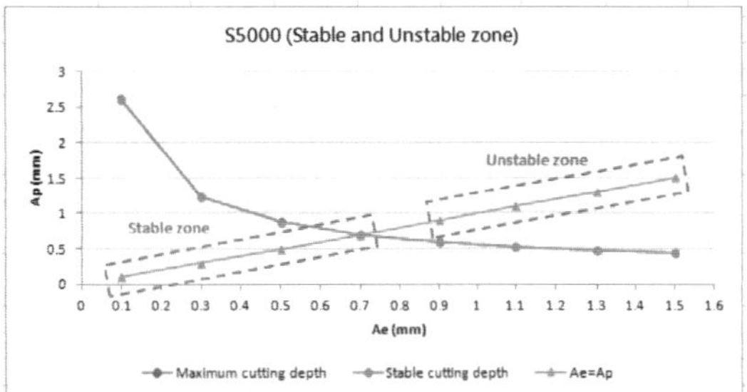


**Figure 9.** S5000 (Stable and Unstable zone).

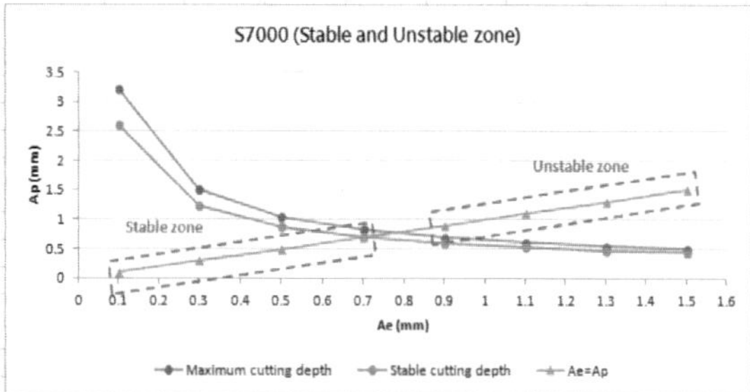


**Figure 10.** S7000 (Stable and Unstable zone).

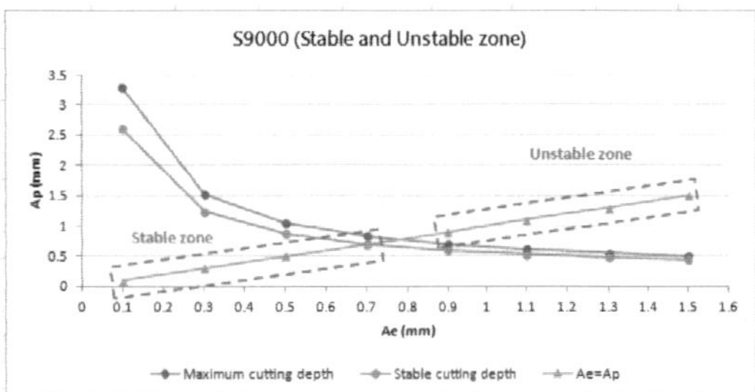


**Figure 11.** S9000 (Stable and Unstable zone).

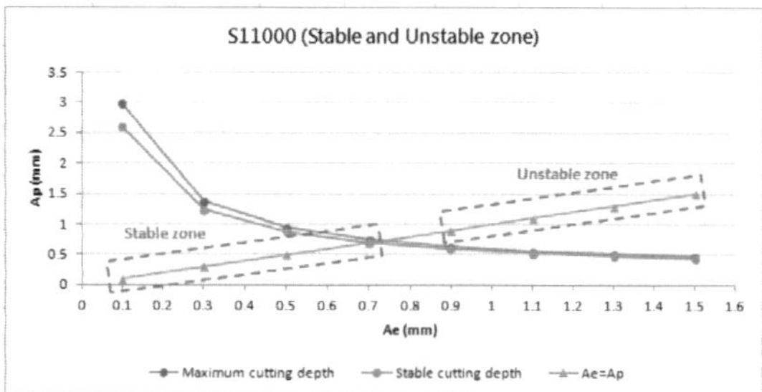


**Figure 12.** S11000 (Stable and Unstable zone).

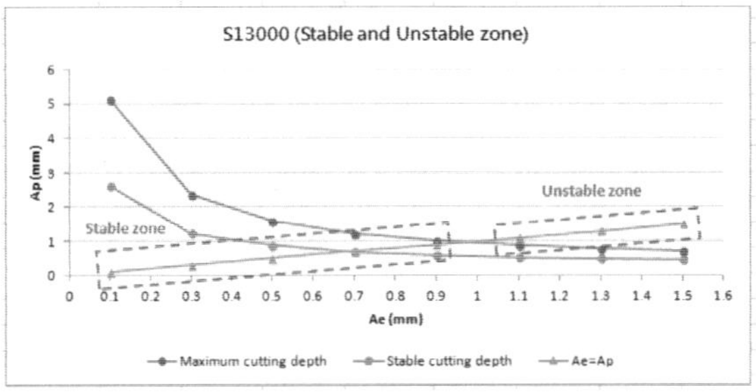


**Figure 13.** S13000 (Stable and Unstable zone).

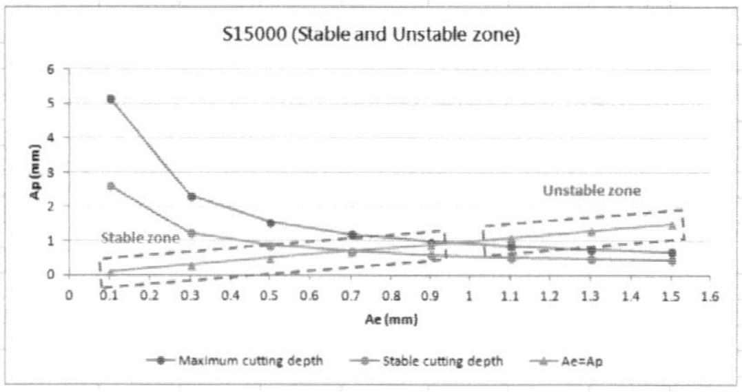


**Figure 14.** S15000 (Stable and Unstable zone).

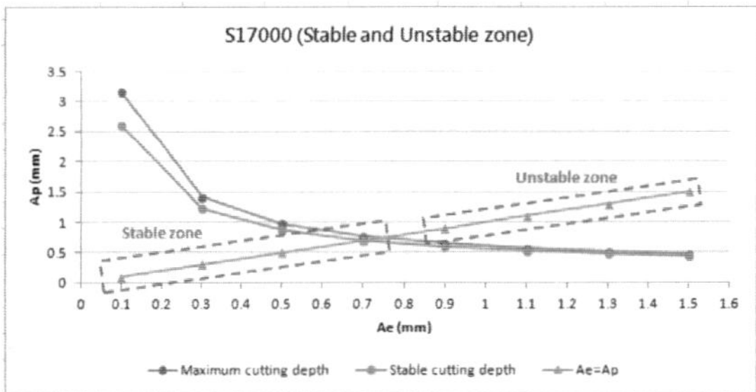


**Figure 15.** S17000 (Stable and Unstable zone).

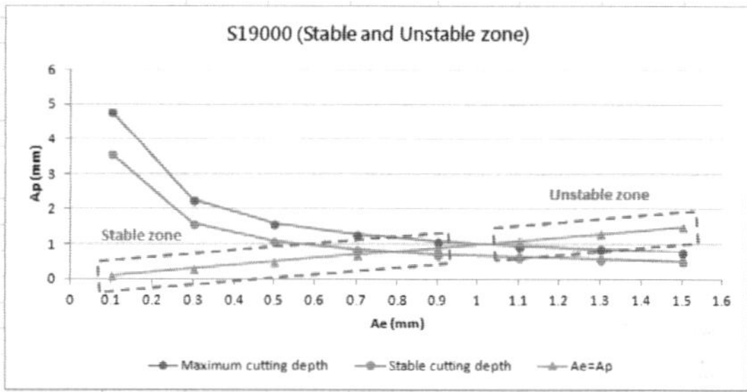


**Figure 16.** S19000 (Stable and Unstable zone).

**Table 6.** The range of stable zone and unstable zone at different speeds.

| Cutting Speed (rpm) | Stable zone Ap (mm) | Unstable zone Ap (mm) |
|---|---|---|
| S5000 | 0.1 ~ 0.7 | 0.9 ~ 1.5 |
| S7000 | 0.1 ~ 0.7 | 0.9 ~ 1.5 |
| S9000 | 0.1 ~ 0.7 | 0.9 ~ 1.5 |
| S11000 | 0.1 ~ 0.7 | 0.9 ~ 1.5 |
| S13000 | 0.1 ~ 0.9 | 1.1 ~ 1.5 |
| S15000 | 0.1 ~ 0.9 | 1.1 ~ 1.5 |
| S17000 | 0.1 ~ 0.7 | 0.9 ~ 1.5 |
| S19000 | 0.1 ~ 0.9 | 1.1 ~ 1.5 |

In this study, SLD was used to obtain stable depth of cut, maximum depth of cut, stable zone and unstable zone. Combined with the surface roughness (Ra), the trend between SLD and surface roughness (Ra) will be shown. However, due to errors in measurement, this study will use polynomial regression equations to solve. First, the surface roughness (Ra) in the stable zone was assumed to be polynomial regression (the trend of Ra in the stable zone has reproducible characteristics), as shown in figure 17. In addition, because the unstable zone contains chattering and instability, and contains randomness problems. So, it is not considered to do polynomial regression.

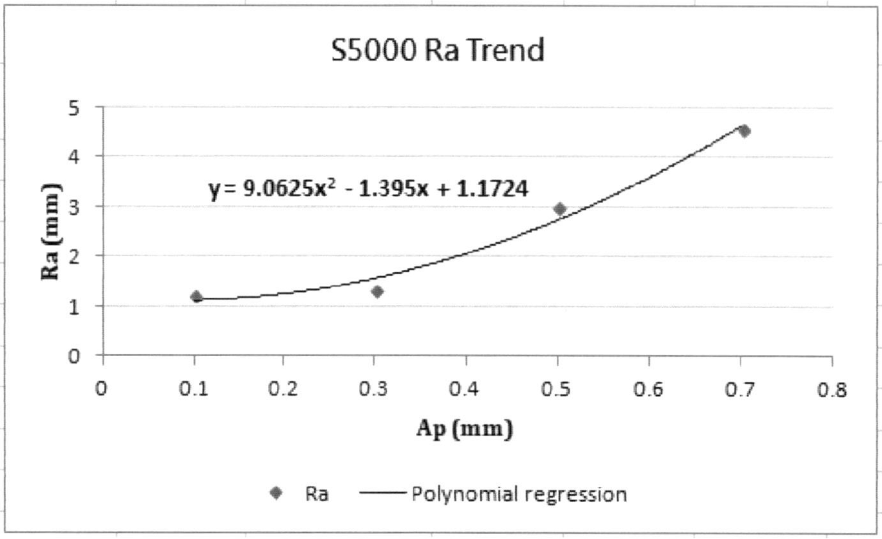


**Figure 17.** S5000 Ra polynomial regression.

This experiment used the parameter that the cutting width (Ae) is equal to the cutting depth (Ap), so the effective material removal area "Ae * Ap" was used for correlation. Then the polynomial regression equation in the stable zone is numerically normalized to the minimum value (ex: S5000, Ap 0.1mm) and maximum value (ex: S5000, Ap 0.7mm) in the stable zone of Ae*Ap. Arrange it as shown from figure 18 to figure 25.

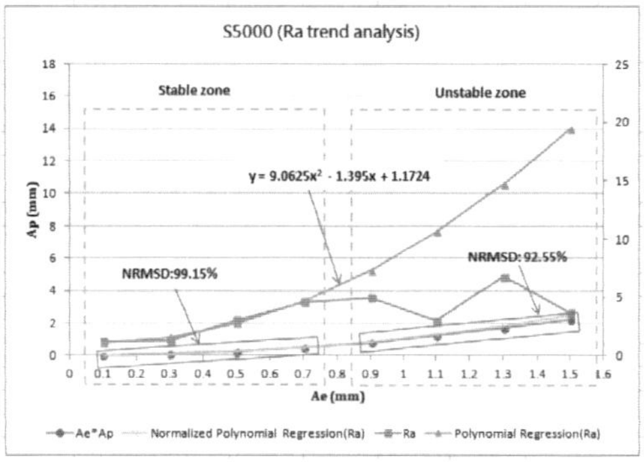


**Figure 18.** S5000 Ra tend analysis.

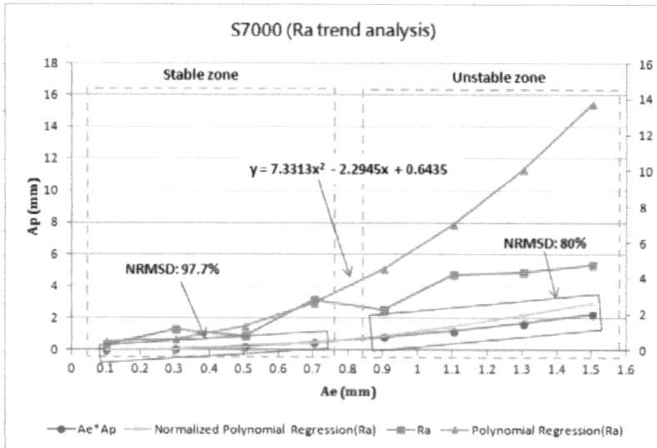


**Figure 19.** S7000 Ra tend analysis.

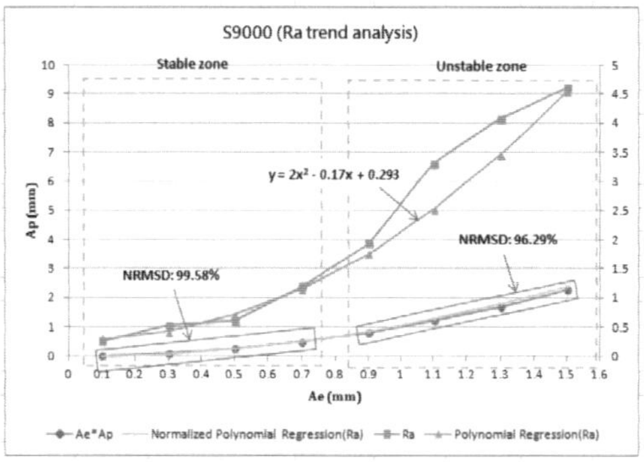


**Figure 20.** S9000 Ra tend analysis.

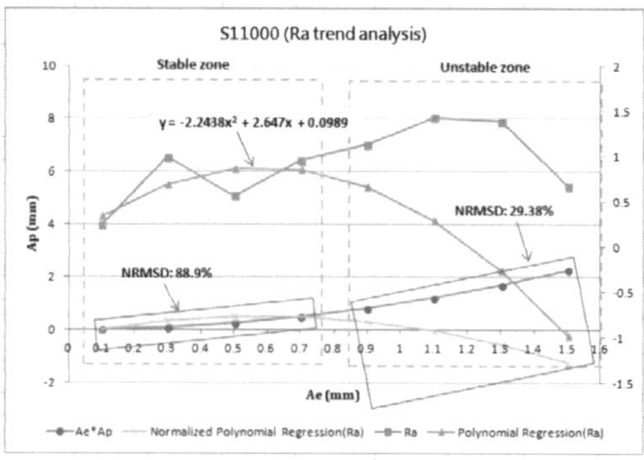


**Figure 21.** S11000 Ra tend analysis.

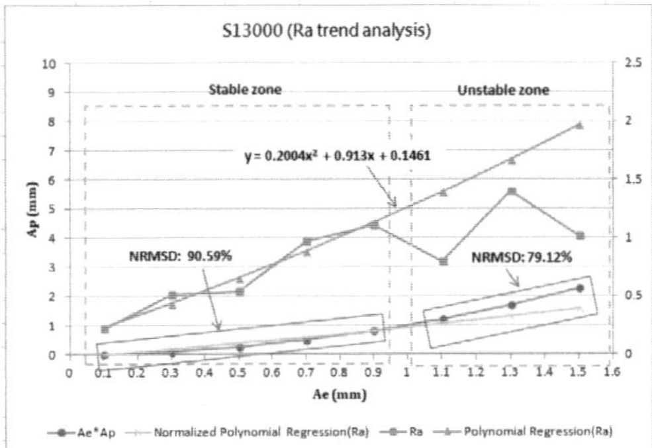


**Figure 22.** S13000 Ra tend analysis.

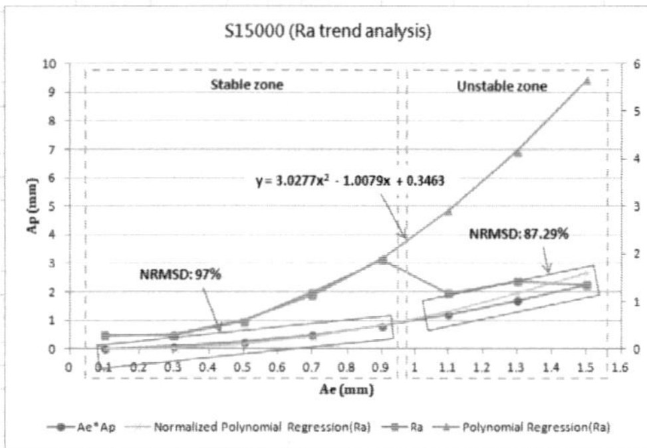


**Figure 23.** S15000 Ra tend analysis.

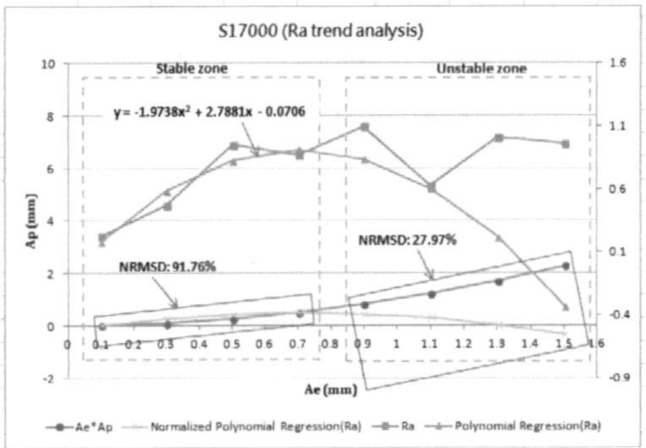


**Figure 24.** S17000 Ra tend analysis.

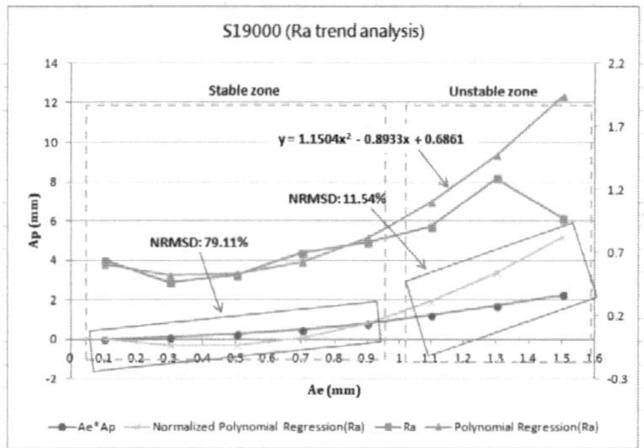


**Figure 25.** S19000 Ra tend analysis.

In this study, the Normalized Root-Mean-Square Deviation (NRMSD) formula was used to calculate the accuracy rate, as shown in Equation 10 and Equation 11. The accuracy rate calculated by the NRMSD formula were interpreted here as the confidence level.

$$\mathrm{NRMSD}(\%)=\left(1-\left(\frac{\mathrm{RMSD}}{y_{max}-y_{min}}\right)\right)*100\% \tag{10}$$

$$\mathrm{RMSD}=\sqrt{\frac{1}{n}\sum_{i}^{n}(y_i-\hat{y}_i)^2} \tag{11}$$

Where $Y_{max}$ – Maximum of measured value (ex: Ra or Ae*Ap);
$Y_{min}$ – Minimum of measured value (ex: Ra or Ae*Ap);
N - Total count;
$\hat{y}_i$ - measured value (as calculated from the polynomial regression equation).

Since the machining state has stable cutting and unstable cutting and the NRMSD can be calculated separately for both, as shown in table 7. It can be seen from the table that in the stable zone, there is an average confidence level of 92.97%, while in the unstable zone, there has only 63.02% confidence level. This is because the random phenomenon occurs in unstable cutting, which it is difficulty for SLD to achieve a higher accuracy. Next, the machine learning algorithm will be followed to establish and predict the dynamic cutting force and surface roughness (Ra).

**Table 7.** NRMSD in the stable and unstable zone.

| Cutting Speed (rpm) | Stable (%) | Unstable (%) |
|---|---|---|
| 5000 | 99.15 | 92.55 |
| 7000 | 97.7 | 80 |
| 9000 | 99.58 | 96.29 |
| 11000 | 88.9 | 29.38 |
| 13000 | 90.59 | 79.12 |
| 15000 | 97 | 87.29 |
| 17000 | 91.76 | 27.97 |
| 19000 | 79.11 | 11.54 |
| Average: | 92.97 | 63.02 |

## 6. Machine learning algorithms predict surface roughness (Ra)

According to Fayyad, Piatetsky-Shapiro and Smyth [18], the process of defining machine learning knowledge discovery includes Data Selection, Data Processing, Data Transformation, Data Mining, and Interpretation and Evaluation. There are five major steps, and gradually establish the relationship between the surface roughness (Ra) and then prediction.

### *6.1. Data selection*

According to the dynamic cutting force operation flow, through the MAL Lab's CutPro software [19]. The dynamic cutting force has eight different cutting speeds (rpm) and eight cutting depths (Ap), and can be calculated with a total of sixty four data.

### *6.2. Pre-processing*

Since these 64 data are cutting force signals (time-domain signals) and subsequent algorithm analysis cannot be performed. At this stage, the feature extraction will be performed to facilitate algorithm analysis. This study uses Equations (12) to (23) to extract dynamic cutting, but because the dynamic cutting force contains three axes, all three axes must be extracted. After feature extraction, the data matrix size is 64*33.

$$\text{Mean Value (MV)} = \sum_{i}^{N} y(i)/N \tag{12}$$

$$\text{Mean Square Error (MSE)} = \sqrt{\sum_{i}^{N} (y(i)\text{-}MV)/N\, K_{ae} = \frac{2a_z F_z}{Na}} \tag{13}$$

$$\text{Square Mean Root (SMR)} = \left( \sum_{i}^{N} |y(i)/N| \right)^2 \tag{14}$$

$$\text{Square Mean Root (SMR)} = \left(\sum_{i}^{N} |y(i)/N|\right)^2 \quad (15)$$

$$\text{Maximum Absolute Value (MA)} = \max|y(i)| \quad (16)$$

$$\text{Skewness Factor (SF)} = \frac{\sum_{i=1}^{N}(y(i)\text{-MV})^3}{(N\text{-}1)\times \text{MSE}^3} \quad (17)$$

$$\text{Kurtosis Factor (KF)} = \frac{\sum_{i=1}^{N}(y(i)\text{-MV})^4}{(N\text{-}1)\times \text{MSE}^4} \quad (18)$$

$$\text{Crest Factor (CF)} = \frac{\text{MA}}{\text{RMS}} \quad (19)$$

$$\text{Margin Factor (MF)} = \frac{\text{MA}}{\text{SMR}} \quad (20)$$

$$\text{FC} = \frac{1}{N}\sum_{f=1}^{N} S(f) \quad (21)$$

$$\text{Root Mean Square Frequency (RMSF)} = \sqrt{\frac{1}{N}\sum_{f=1}^{N} S(f)^2} \quad (22)$$

$$\text{Root Variance Frequency (RVF)} = \sqrt{\frac{1}{N}\sum_{f=1}^{N} (S(f)\text{-FC})^2} \quad (23)$$

*6.3. Data transformation*

Since thirty-three features are not conducive to analysis, this study uses PCA (Principal Component Analysis) for dimensional reduction, as shown in table 8. In the nineteenth principal component (PC No.), the cumulative contribution ratio can reach 100%, so the data is reduced to a matrix size of 64*19.

Dimension reduction of data was through PCA, and then machine learning algorithms was used; and the algorithm chose linear and nonlinear algorithms such as the multivariate linear regression (MLR) and generalized regression neural network (GRNN). However, because the GRNN algorithm uses the Gaussian function as the basis for prediction, it is easily affected by other value ranges (such as other cutting speeds or cutting depths), and then it decreases accuracy. Therefore, this study used linear interpolation, that is, at the same processing speed that between the previous cutting depth and the next cutting depth, it used linear interpolation and insert eight data. So, it extends to 568 data, that the matrix size is 568*19. This approach will reduce the learning bias of the GRNN algorithm.

*6.4. Data mining*

In the research of pattern recognition in the machine learning, the data set is often divided into two subsets: a training set and a validation set. The former is used to build a model, and the latter is used to evaluate the accuracy of the model in predicting unknown samples. K-fold Cross Validation is a more commonly used verification method. The method is to randomly divide the data into k sets, and then use a certain set as a validation, and the remaining k-1 sets as training, so repeat until each set is regarded as the verification information. The final result (Predication results) is compared with the ground truth (Performance Comparison), as shown in figure 26. In this study, 10-Fold Cross Validation were used to establish 10 sets of training data sets and validation data sets.

**Table 8.** Principal Component Analysis.

| PC No. | Variation | Variation (%) | Cumulative contribution ratio (%) |
|---|---|---|---|
| 1 | 24.6121 | 74.5822 | 74.58 |
| 2 | 3.9817 | 12.0657 | 86.65 |
| 3 | 2.8461 | 8.6244 | 95.27 |
| 4 | 0.6927 | 2.0991 | 97.37 |
| 5 | 0.3029 | 0.9178 | 98.29 |
| 6 | 0.2020 | 0.6120 | 98.9 |
| 7 | 0.1591 | 0.4821 | 99.38 |
| 8 | 0.0953 | 0.2887 | 99.67 |
| 9 | 0.0418 | 0.1265 | 99.8 |
| 10 | 0.0271 | 0.0820 | 99.88 |
| 11 | 0.0171 | 0.0518 | 99.93 |
| 12 | 0.0071 | 0.0214 | 99.95 |
| 13 | 0.0044 | 0.0133 | 99.97 |
| 14 | 0.0030 | 0.0090 | 99.98 |
| 15 | 0.0021 | 0.0064 | 99.98 |
| 16 | 0.0015 | 0.0046 | 99.99 |
| 17 | 0.0013 | 0.0040 | 99.99 |
| 18 | 0.0011 | 0.0034 | 99.99 |
| 19 | 0.0006 | 0.0017 | 100 |
| … | … | … | … |
| 33 | 0.0004 | 0.0011 | 100 |

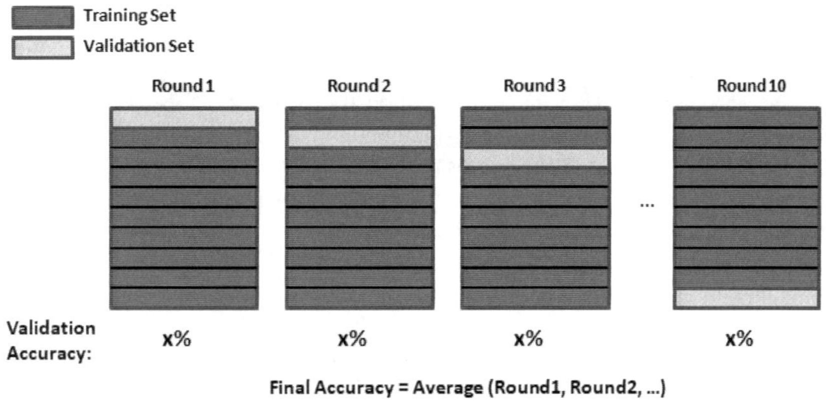


**Figure 26**. k-Fold Cross Validation.

In this research, according to 10 sets of data sets and machine learning algorithms, the relationship between dynamic cutting force and surface roughness (Ra) was established. This research chose multivariate linear regression (MLR) and generalized regression neural network (GRNN) algorithms to learn training data sets and predict validation data sets. The prediction results of these algorithms are shown in figure 27 and figure 28.

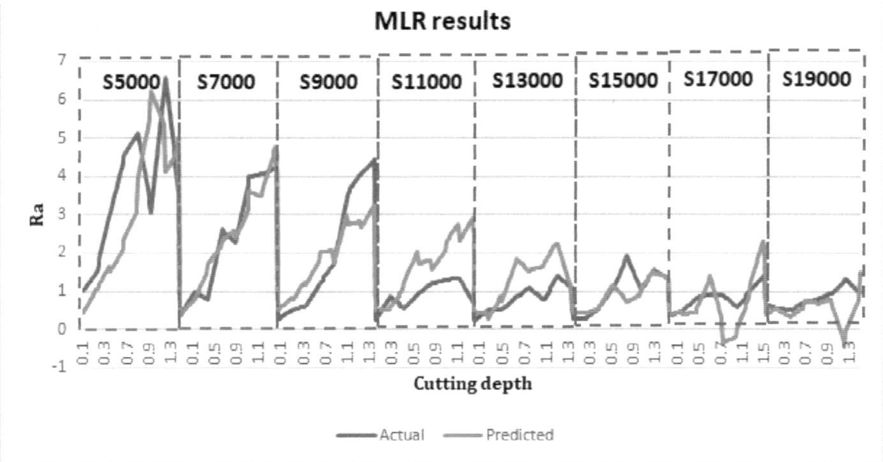


**Figure 27.** MLR results.

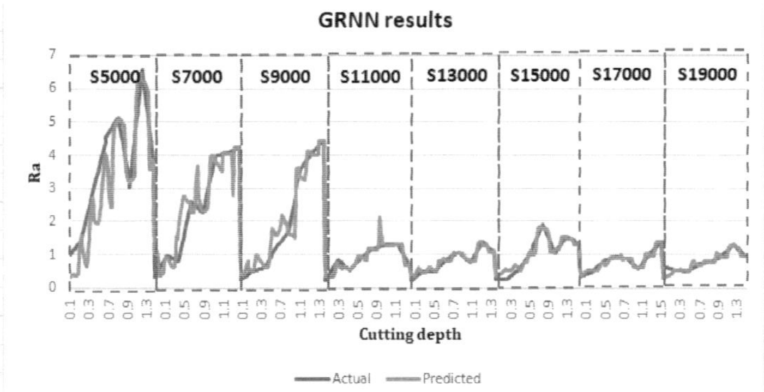


**Figure 28.** GRNN results.

After the machine learning algorithm targets 10 sets of verification data sets were collected, the NRMSD was used to calculate, as shown in Equation (10). The average accuracy of the MLR algorithm is 58.66%, while the average accuracy of the GRNN algorithm is 88.69%. Obviously, it could be known that the data is not linear, but it is a non-linear data type. So, the GRNN algorithm has a great contribution on the prediction of non-linearity. In addition, on the results of the GRNN algorithm, it is also clear that the performance of the TA20 machine is a low torque and high speed. So, when the processing speed is 11000rpm, the prediction accuracy of the surface roughness (Ra) can be above 85% and the maximum error is below 0.893mm. And in the unstable zone, it can also have a higher accuracy than the stability lobe diagram (SLD).

### *6.5. Interpretation and evaluation*

The respective execution results of the MLR algorithm and the GRNN algorithm must estimate indicators to obtain the accuracy rate of the two algorithms for the data set. The evaluation indicators were aimed at inputting the same data into different algorithm models and giving quantitative indicators of the results. In evaluation indicators, most of them can only partially reflect the model performance. If the evaluation indicators could not be used reasonably, the problems could not be found with the model itself, and the wrong conclusions might be concluded. Therefore, this research calculated the accuracy of MAPE (Mean Absolute Percentage Error) through Equation (24).

$$\text{MAPE}=100\%-\left(\frac{1}{n}\sum_{i=1}^{n}\left|\frac{\text{actual(i)-predicted(i)}}{\text{actual(i)}}\right|\times 100\%\right) \tag{24}$$

The prediction results of the two algorithms (MLR and GRNN) were calculated by Equation (10) to calculate NRMSD and Equation (24) to calculate MAPE, as shown in figure 29 and figure 30. Combining the analysis of the two evaluation indicators (NRMSD and MAPE), the GRNN algorithm is better than the MLR algorithm, and can improve the accuracy of the SLD in the unstable zone.

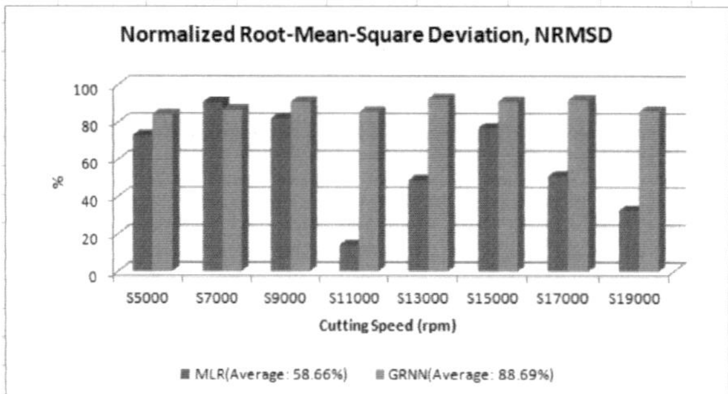


**Figure 29.** The accuracy of the prediction of the each cutting speed in NRMSD.

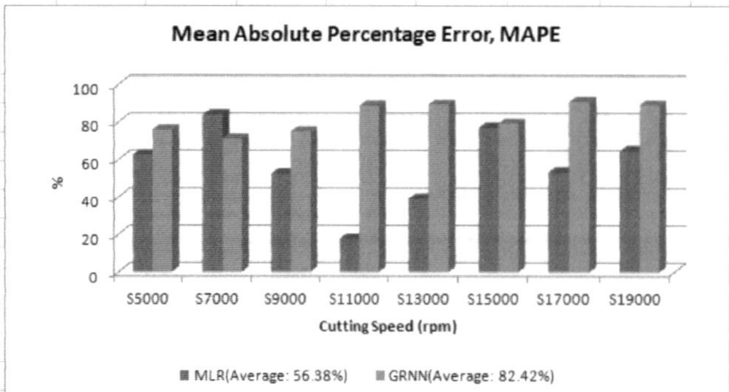


**Figure 30.** The accuracy of the prediction of the each cutting speed in MAPE.

## 7. Conclusion

Since it is generally not able to install dynamometer on a small machine such as the TA20, the cutting force could not be measured during the machining. Therefore, this research has successfully generated dynamic cutting force signals of the small machine according to cutting force coefficients obtained from the cutting test on the other industrial machine tool, and also the small machine's FRF. The calculated cutting force have been found to be able to approximate the cutting force signal of the physical dynamometer on a small machine (TA20). In this research, the actual machining was conducted on the small machine and the machined surface Ra was measured, and the roughness prediction algorithm has been established by using multivariate linear regression (MLR) and generalized regression neural network (GRNN). It is concluded that the GRNN (non-linear) algorithm is better than MLR (linear) in the predicted accuracy. From another perspective, the dynamic cutting force was adopted to analyse this study, and some use machine learning algorithms such as GRNN were used for prediction. The accuracy rate of the surface roughness prediction could be reached to that of more than 80% (NRMSD and MAPE), which means that the proposed methodology is useful for surface roughness prediction on small machines or machines that cannot have a dynamometer installed.

## Acknowledgments

The authors thank the support from the Ministry of Science and Technology (MOST), Taiwan, under the grant MOST 108-2218-E-194-006. Appreciation extends to the Advanced Institute of Manufacturing with High-tech Innovations (AIM-HI), National Chung Cheng University (CCU), Taiwan.

## References

[1] Smith S and Tlusty J 1990 Update on High-Speed Milling Dynamics *ASME J. of Eng for Industry* **112(2)** 142
[2] Chiou Y S, Chung E S and Liang S Y 1995 Analysis of Tool Wear Effect on Chatter Stability in Turning *Int. J. of Mechanical Sciences* **37(4)** 391
[3] Tlusty J and Andrews G C 1983 A Critical Review of Sensors for Unmanned Machining *CIRP Annals Manufacturing Technology* **32(2)** 563
[4] Taylor F W 1907 On the Art of Cutting Metals (American society of mechanical engineers)
[5] Smith S, Winfough W and Halley J 1998 The Effect of Tool Length on Stable Metal Removal Rate in High Speed Milling *Annals of the CIRP* **47(1)** 307
[6] Shaw M C 2005 *Metal Cutting Principles - Second Edition* (New York: Oxford University)
[7] Davim J P 2008 *Machining* (University of Aveiro)
[8] Dimla D E Jr Lister P M and Leighton N J 1997 Neural network solutions to the tool condition monitoring problem in metal cutting - A critical review of methods *Int. J. Mach. Tools Manufact* **37(9)** 1219
[9] Lauro C H, Brandao L C, Baldo D, Reis R A and Davim J P 2014 Monitoring and processing signal applied in machining process-A review *Measurement* **58** 73
[10] Varghese S and Radhakrishnan V 1994 A multi sensor approach to in-process monitoring of surface roughness *J. Mater. Process* **44(3-4)** 353
[11] Hossain M I, Nurul Amin A K M, Patwari A U 2008 Development of an artificial neural network algorithm for predicting the surface roughness in end milling of Inconel 718 alloy *Int. Conf. on Computer and Communication Engineering* (Kuala Lumpur:IEEE) 10130030
[12] Wang X 2009 Intelligent Modeling and Predicting Surface Roughness in End Milling *Fifth Int. Conf. on Natural Computation* (Tianjin:IEEE) 11024377
[13] Zain A M, Haron H, Sharif S 2010 Application of GA to optimize cutting conditions for minimizing surface roughness in end milling machining process *Expert Systems with Applications* **37(6)** 4650

[14] Asiltürk I, Çunkaş M 2011 Modeling and prediction of surface roughness in turning operations using artificial neural network and multiple regression method *Expert Systems with Applications* **38(5)** 5826
[15] Sreenivasulu R 2013 Optimization of Surface Roughness and Delamination Damage of GFRP Composite Material in End Milling Using Taguchi Design Method and Artificial Neural Network *Procedia Engineering* **64** 785
[16] D'Mello G, Pai P S and Shetty R P 2017 Surface roughness modeling in high speed turning of Ti-6Al-4V – Artificial Neural Network approach *Materials today: Proceedings* **4(8)** 7654
[17] Gopan V, Wins K L D and Surendran A 2018 Integrated ANN-GA Approach for Predictive Modeling and Optimization of Grinding Parameters with Surface Roughness as the Response *Materials Today* **5(5)** 12133
[18] Thankachan T, Prakash K S, Malini R, Ramu S, Sundararaj P, Rajandran S, Rammasamy S, Jothi S 2019 Prediction of surface roughness and material removal rate in wire electrical discharge machining on aluminium based alloys/composites using Taguchi coupled Grey Relational Analysis and Artificial Neural Networks *App. Surface Science* **472** 22
[19] Manufacturing Automation Laboratories Inc. Available: http://www.malinc.com/. Accessed 2020/06/15

IOP Conf. Series: Materials Science and Engineering 1009 (2021) 012028 doi:10.1088/1757-899X/1009/1/012028

# Prediction of radial clearance based on bearing vibration using artificial neural network

**I Knežević[1*], A Živković[1], M Rackov[1], Ž Kanović[1], A Buljević[1], M Bojanić Šejat[1] and S Navalušić[1]**

[1]Faculty of Technical Sciences, University of Novi Sad, 21000 Novi Sad, Serbia

*E-mail: ivanknezevic@uns.ac.rs

**Abstract**. Artificial neural networks are used for diagnosing of ball bearings condition. They are capable to recognize bearing fault from information extracted from signal of vibration. The aim of the work is to investigate the relationship between vibration velocity in the characteristic band frequencies and radial clearance. Changing the radial clearance during operation may indicate changes in the operation of the rotating system. Experimental research is based on deep groove ball bearings and measuring the vibration of the bearings. The vibration velocities for each bearing were measured under laboratory conditions, and the processed signals were used to train the artificial neural network. An artificial neural network trained in this way can predict the size of the radial clearance based on the vibration of the bearing.

## 1. Introduction

Deep groove ball bearings (DGBB) are widely used in rotating machines. There are almost no machines, appliances or devices that do not contain a larger or smaller number of ball bearings in their construction. In a large number of cases, they are considered as critical elements in the assembly, on whose working abilities the functionality, reliability and operational properties of the entire assembly in which they are installed largely depend. The quality of the bearings plays a significant role in all rotating machines. The quality of bearings and the parts of which it is assembled is checked during and after production, and the condition of the bearings in operation has been successfully monitored for a long time using advanced techniques. Condition monitoring provides large amounts of bearing vibration data. The analysis of the recorded vibrations in the frequency spectrum provides insight into the cause of the dominant amplitudes. For these purposes, artificial neural networks (ANN) have been successfully used for the last 20 years. Internal radial clearance is one of the influential parameters of the ball bearing and it is introduced to provide thermal dilatations and is a significant source of vibration and nonlinear dynamic behaviour of the bearing. Tiwari et al. [1] determine the stiffness of ball bearings and their dynamic behaviour due to the nonlinearity of the Hertz contact and the radial clearance. Higher order Poincare maps and interpolation techniques were used to determine the stability of a balanced rotor supported by ball bearings. The results show that instability and chaotic dynamic response occur when the system speed changes. The influence on the increase of nonlinearity due to the increase of the radial gap has been experimentally proven. In [2], Harsha provides a model that investigates the structural vibrations of a ball bearing depending on the influence of the radial clearance. It is shown that for certain combinations of rotational speed and radial clearance, the nonlinearity of the system turns into chaotic behaviour. Reducing the radial clearance affects the increase in system linearity. Analysis of the dynamic behaviour of ball bearings without failure to rolling surfaces using a mathematical model that describes the stiffness of the bearings by Hertz's contact theory was performed by Ghafari et al. [3]. They found that as the radial clearance increases, the bearing changes from stable behaviour to the area of unstable dynamic behaviour. Analysing bearings with larger radial clearances, a map of chaotic behaviour has been developed, which can help to determine for which combinations of number of revolutions and radial clearance the bearing passes into a state of chaotic (unstable) behaviour. Bai et al. [4] investigated the stability and dynamic behaviour of a shaft supported in bearings with radial clearance. Theoretically, they determined that there are three ways for the system to transition from a stable state to a chaotic state, and concluded that there are combinations of the number of revolutions and the size of the radial gap that introduce the system into chaotic behaviour. Meier et al. [5] propose an artificial neural network that can predict the size of the radial clearance depending on the energy spectrum of vibrations in two-row ball bearings. Artificial neural networks have been used to determine the remaining service life [6], the size of the bearing stiffness [7], the influence of the amount of lubricant on the vibrations generated by the bearing [8]. Fathiah Waziralilah et al. [9] presented review of applied Convolutional Neural Networks (CNN) for detection of ball bearing failure. Deep learning neural networks are used widely in ball bearing diagnosis, especially DCNN (Deep Convolutional Neural Networks). Results presented in [9-12] shows a high precision of DCNN in ball bearing diagnosis.

## 2. Experimental setup

The experimental research was performed on a sample of 30 ball bearings, marked 6006. The first part of the experimental research is based on the measurement of internal radial clearance ($G_r$). The size of the internal radial clearance is in the range from 6 to 51 µm. The second part of the experimental research is the measurement of bearing vibrations. The vibration measurement and control system is installed on a measuring instrument for testing, analysing and diagnosing ball bearings and is based on vibration measurement of ball bearings using an electrodynamics speed encoder (pickup). The bearing is mounted over the inner ring to the spindle, which is supported by hydrodynamic bearings. The spindle speed is constant during the measurement ($n$ = 1800 RPM). The outer ring is stationary and loaded with axial force ($F$) over the pneumatic cylinder, figure 1. Axial force is variable from 200 to 1000 N in 100 N increment. The measurement was performed by measuring the vibration velocity, and the measured vibration signals were processed using Fast Fourier Transformation (*FFT*). The measurement results are Root Mean Square (*RMS*) amplitude of vibration velocity. For smaller dimensions ball bearings vibrations are measured in three frequency bands, according to ISO 15242-4. It prescribe low (50 - 300 Hz), medium (300 - 1800 Hz) and high frequency (1800 - 10000 Hz) band.

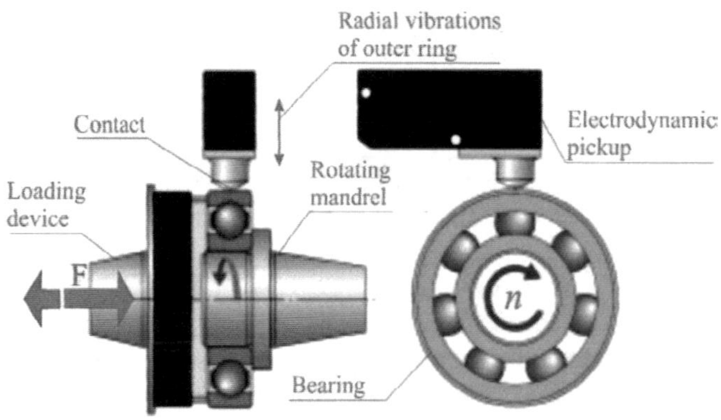


**Figure 1.** Schematic view of measuring principle of electrodynamics pickup.

## 3. Design of ANN

The problem that is tried to be solved within this paper is to predict the size of the internal radial clearance of the ball bearing based on the *RMS* value of the vibration velocities in the characteristic frequency ranges. To solve this problem, MultiLayer Perceptron (*MLP*), Feedforward Shallow neural network with Backpropagation algorithm was chosen. The input parameters to the neural network are *RMS* vibration velocities for low, medium and high band frequencies, and the output from the network is the size of the internal radial clearance ($G_r$), figure 2.

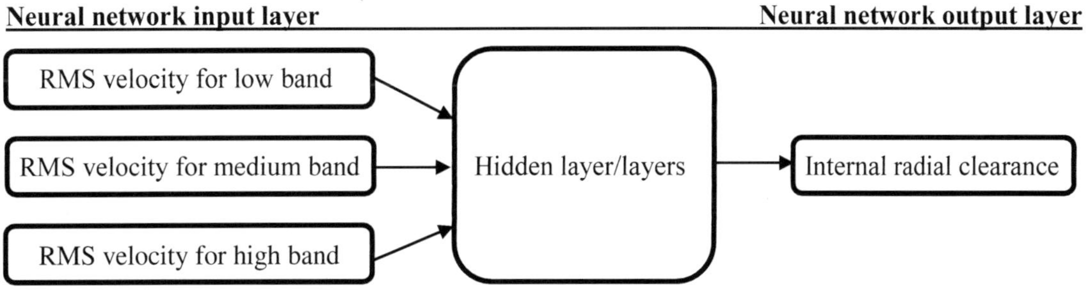


**Figure 2.** Structure of input and output layers of ANN.

As there is no recommendation for choosing the architecture of an artificial neural network for a specific problem, a large number of networks have been trained and tested. Networks with one, two and three hidden layers were tested. One neuron was selected in each hidden layer in the initial training, after which the number of neurons was increased up to 20. The Levenberg-Marquardt training algorithm was applied, and the activation function for neurons in the hidden layers is a sigmoid. The neurons in the output layer have a PureLinear activation function. All artificial neural network architectures were trained 100 times and a prediction assessment was performed for each. The results for the neural network that have the best prediction is presented.

## 4. Results and discussion

The architecture of the Artificial Neural Network was adopted based on a prediction quality assessment. The quality of the prediction was assessed based on the Pearson correlation coefficient (R), the coefficient of determination ($R^2$) and the mean value of the prediction error expressed as a percentage. The neural network with the best performance by all criteria is a network that has 3 hidden layers and 15 neurons in each layer, figure 3.

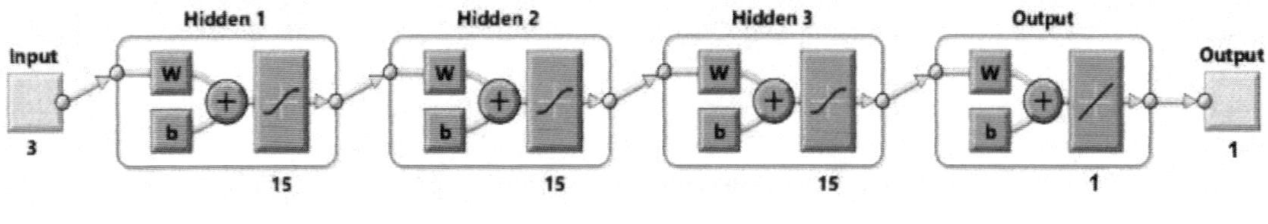


**Figure 3.** Architecture of ANN.

Table 1 shows the results for the chosen neural network. The Pearson correlation coefficient shows that there is a strong relationship between the values of the internal radial clearance obtained by experimental measurement and the values predicted by the artificial neural network. The coefficient of determination indicates that the model predicts the value of the internal radial clearance with 89.8% accuracy.

**Table 1.** Results of quality assessment of prediction.

| | |
|---|---|
| Pearson correlation coefficient (R) | 0.947 |
| Coefficient of determination ($R^2$) | 0.898 |
| Average prediction error, % | 16.5 |

The average prediction error indicates that the adopted model predicts with an accuracy of 83.5%. According to the diagram in figure 4, it is shown that for the bearing with the internal radial clearance of 30 μm, the network predicts 50 μm, which has the effect of reducing the accuracy of the prediction. Other predicted values are close to real.

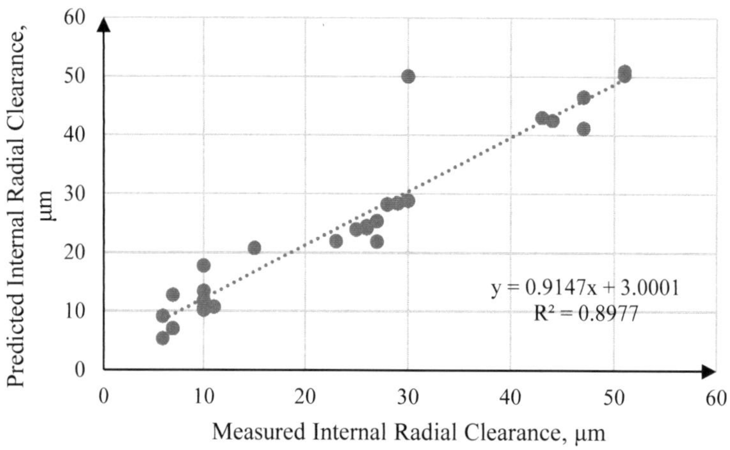


**Figure 4.** Measured and predicted values of internal radial clearance.

## 5. Conclusion

The paper shows that by using artificial neural networks on experimental data, it is possible to obtain a model that is able to predict the size of the internal radial clearance depending on the *RMS* velocity of vibrations in characteristic bands. It is shown that only from *RMS* velocity extracted from vibration signal it is possible to predict radial clearance of ball bearing. The Pearson correlation coefficient shows that there is a strong relationship between the values of the internal radial clearance obtained by experimental measurement and the values predicted by the artificial neural network. The coefficient of determination indicates that the model predicts the value of the internal radial clearance with 89.8% accuracy. The internal geometry of the bearing also has a significant influence on the *RMS* of vibration velocity in all band frequencies, which is not considered in this paper and it can affect the accuracy of the prediction. *MLP*, Feedforward Shallow neural network with Backpropagation algorithm was applied. The results show that accuracy is acceptable, but the application of Deep Learning Neural Networks could improve accuracy, which is one of the directions for further research. This paper presents the results of a preliminary study of the application of artificial neural networks for the analysis of the dynamic behaviour of deep groove ball bearings.

## Acknowledgment

This research has been supported by the Ministry of Education, Science and Technological Development through the project no. 451-03-68/2020-14/200156: "Innovative scientific and artistic research from the FTS (activity) domain".

**References**

[1] Tiwari M, Gupta K and Prakash O 2000 Dynamic response of an unbalanced rotor supported on ball bearings *J. of S. and Vib.* **238** 757-79
[2] Harsha S 2006 Nonlinear dynamic response of a balanced rotor supported by rolling element bearings due to radial internal clearance effect *M. and M. Theory* **41** 688-706
[3] Ghafari S, Abdel-Rahman E, Golnaraghi F and Ismail F 2010 Vibrations of balanced fault-free ball bearings *J. of S. and Vib.* **329** 1332-47
[4] Bai C Q, Xu Q Y and Zhang X L 2006 Nonlinear stability of balanced rotor due to effect of ball bearing internal clearance *A. M. and Mechanics* **27** 175-86
[5] Meier N, Biyani Y and Georgiadis A 2018 *IEEE SENSORS: IEEE* (New Delhi, India) pp 1-4
[6] Hong S and Yin J 2018 *Proc. of the 2nd Int. Conf. on Big Data and Internet of Things* pp 175-9
[7] Pavlenko I, Simonovskiy V, Ivanov V, Zajac J and Pitel J 2018 *Advances in Design, Simulation, Manufacturing - Chapter: Application of Artificial Neural Network for Identification of Bearing Stiffness Characteristics in Rotor Dynamics Analysis* (Springer International Publishing) pp 325-35
[8] Knezevic I, Živkovic A, Rackov M, Kanovic Ž and Sejat Bojanic M 2020 Analysis of the impact of lubrication on the dynamic behavior of ball bearings using artificial neural networks *Romanian Journal of Acoustics Vibration* **16** 178-83
[9] Waziralilah N F, Abu A, Lim M, Quen L K and Elfakharany A 2019 A review on convolutional neural network in bearing fault diagnosis *MATEC Web of Conferences: EDP Sciences* **255** p 06002
[10] Afrasiabi S, Afrasiabi M, Parang B and Mohammadi M 2019 Real-time bearing fault diagnosis of induction motors with accelerated deep learning approach *10th Int. Power Electronics, Drive Systems and Technologies Conf. (PEDSTC): IEEE* pp 155-9
[11] An Z, Li S, Wang J, Xin Y and Xu K 2019 Generalization of deep neural network for bearing fault diagnosis under different working conditions using multiple kernel method *Neurocomputing* **352** 42-53
[12] Zhao X and Jia M 2019 A new local-global deep neural network and its application in rotating machinery fault diagnosis *Neurocomputing* **366** 215-33

# The heat treatment influence on microstructure and mechanical properties of TiAl6V4 parts manufactured by SLM

**C Lancea[1]*, S M Zaharia[1], M A Pop[2] and G R Buican[1]**

[1] Department of Manufacturing Engineering, Transilvania University of Brasov, Eroilor nr. 29, Brasov 500036, Romania
[2] Department of Materials Science, Transilvania University of Brasov, Eroilor nr. 29, Brasov 500036, Romania

*E-mail: camil@unitbv.ro

**Abstract**. Nowadays titanium alloys are widely used for building various parts for the aeronautical or aerospace industries, in the energy field, petroleum industry, in medicine and recently in the automotive industry, the sporting goods industry, etc. In this paper is analysed the influence of the heat treatment on the microstructure and compressive strength of the TiAl6V4 alloy samples, having a Lattice structure. The analyses results indicate differences between the as-built samples and the heat-treated samples. In this regard, heat-treated samples have a much uniform and stabile structure, a micro-structure with fewer pores and better mechanical resistance to compression stress. These results confirm that the heat treatment favourably influences the microstructure of the samples, reduces the internal stress and their mechanical characteristics too.

## 1. Introduction

Additive manufacturing allows the fabrication of metal parts with complex shapes and high density that cannot be obtained through traditional manufacturing methods [1]. These structures allow obtaining of both strong and lightweight parts [2] that are highly demanded mostly in astro- and aeronautical industries (production of satellite parts, cabin accessories, jet engine components for Boeing and Airbus etc.) [3].

Lately, the interest for these kinds of parts increases in many other industries such as: automotive [4] –starting with high-performance Formula 1 racing cars and continued with the production vehicles [5] (Daimler AG, Ferrari, Mitsubishi Motors Inc. etc.); medicine (implantology, dentistry, surgery etc.) [6]; marine engineering (parts with a complex shape such as: engine blades, turbine impellers etc.) [7]; nuclear-power engineering (SLM parts for nuclear fusion reactors) [8]; chemical engineering etc.

In this regard, the lattice structures [9], which are three-dimensional closed or open celled structures, generated as a periodic arrangement of one or more unit cells, offer the best way to obtain lightweight structures with very good mechanical properties. The best material that meets these requirements has proven to be titanium, or to be more precise, titanium alloys [10]. That's why nowadays, the most used material in SLM industry (also for building these kinds of structures) is TiAl6V4. This alloy is as much used (more than 50% of the market) because of its excellent properties in obtaining light, complex shapes having very good mechanical properties [11]. Besides this, another advantage of the TiAl6V4 alloys refers to a very good corrosion resistance and ahighly biocompatibility[12].

The manufacturing of TiAl6V4 complex parts, through selective laser melting (SLM), decreases significantly the production costs by shortening the machining processes, no need of tools or devices and a very good material utilization [13]. Besides these advantages, the SLM process gives rise to some unwanted problems too. During the SLM manufacturing process the part is subjected to a high thermal stresses because of the high energy release in a very short time, followed by a rapid solidification and fast cooling [14]. These conditions can lead to some problems as segregation phenomena, a higher surface roughness, big internal porosity or the appearance of non-equilibrium phases [15].

Given this situation, after the SLM manufacturing a heat treatment is very necessary to relieve the cumulated stress, to stabilize the microstructure and finally to improve the mechanical properties of the part [16].

## 2. Material preparation and methods

A titanium Grade 23 powder (also known as TiAl6V4 ELI) was chosen to produce, through the SLM process, six specimens. All specimens have identical shapes, and were built as a Lattice structure. This structure was obtained as a three-dimensional pattern generated along three rectangular directions (figure 1-b) of a 3D cell (figure 1-a).

For determining heat treatment influence on microstructure and mechanical properties of the samples, three of them were not subjected to any heat treatment (figure 1-b), and the other three were subjected to a homogenization heat treatment (figure 1-c).

### *2.1. Samples material*

TiAl6V4 ELI is the high purity version of TiAl6V4 (Grade 5), the world most used titanium-based alloy. The chemical composition of this titanium-based alloy, expressed as a percentage, is given in table 1:

**Table 1.** TiAl6V4 (Grade 5) chemical composition.

| Ti | Al | V | C | O | N | Fe | H | other |
|---|---|---|---|---|---|---|---|---|
| balance | 5.5–6.5% | 3.5–4.5% | 0.08% | 0.13% | 0.03% | 0.25% | 0.0125% | 0.5% |

### *2.2. Geometry of the samples*

The cell is composed of three rectangular isosceles triangles (figure 1-a), placed in three rectangular planes, having base and high of 25mm. The cylinder has a 5mm diameter. The cell was three times patterned along rectangular directions so that the sample dimensions become 75mm x 75mm x75mm (figure 1-b).

After manufacturing the dimensions of the physical samples were compared with the dimensions of their CAD model and the results showed a difference of about 0.03 mm along all the three rectangular axes.

(a) (b) (c)

**Figure 1.** Sample geometry: the cell (a), as-built samples (b) and heat treated samples (c).

### *2.3. Fabrication of samples*
Samples were produced at SLM Solutions Group AG headquarters, in Lübeck, Germany on a SLM125 3D printer machine with a single (1x 400 W) IPG fibre laser. The scanning speed was set to 2 m/s, with a total laser power of 250 W, 140 µm hatch spacing and a 40 µm layer thickness. In order to obtain fully dense and good quality samples [15], the layers were scanned using a zigzag pattern, which was rotated with 90°after exposing each layer. The laser spot size was 80 µm and the building process was performed in an inert atmosphere (argon). The building platform was preheated and maintained at 35° C during the entire building process.

### *2.4. Microstructure and surface feature analysis*
For analyzing microstructure, two sets of samples: heat-treated and as-builtwere cut with a diamond cutting disk, followed by grinding. The samples were cold-mounted in Dentacril resin and polished with 60, 30 and 15 µm grit magnetic discs on a Buehler Phoenix Beta Grinder (figure 2-a).

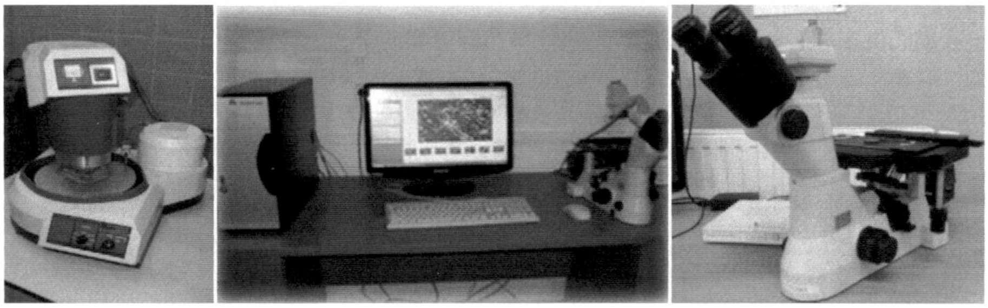

(a) (b)

**Figure 2.** Buehler Phoenix Beta Grinder (a) and the Nikon Eclipse MA 100 optical microscope (b).

After polishing on felt and 0.05µm alumina paste a mixture of 30 ml HCl + 10 ml HNO3 was used for etching the specimens.

The microstructural analysis was conducted using a Nikon Eclipse MA 100 optical microscope (figure 2-b).

### *2.5. Influence of temperature and tensile properties*
To homogenize structure and to increase the mechanical properties [17], three of the specimens were heat-treated after the following schedule: homogenization treatment at 850°C, for 2 hours, followed by furnace cooling in a Nabertherm N 41/H/P300 furnace.

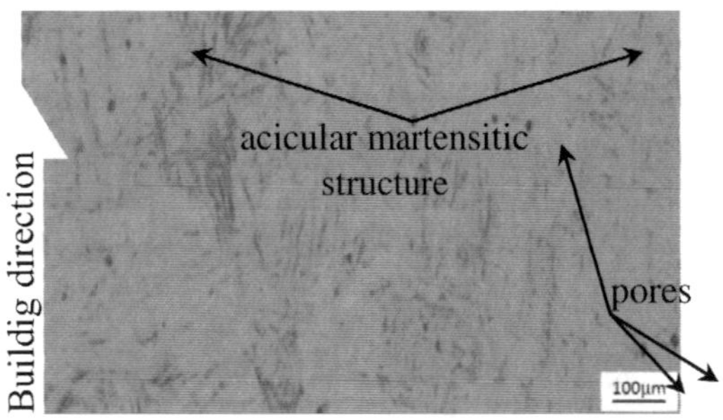


**Figure 3.** Microstructure of the as-built samples (magnification 200X).

**Figure 4.** Microstructure of the heat treated samples (magnification 200X).

A metallographic analysis was performed to study the microstructural changes that appears after the heating treatment. The as-built samples (figure 3) has a fine acicular martensitic structure (α' phase); as a result of the high cooling rate, given by the SLM process [18]. The small spherical pores (figure 3) indicate a good scanning speed within the SLM process. The heat treated samples have an improved structure, much uniform where the α' martensite was transformed into more stable α+β phases with white regions (given by the α phase) and dark areas (given by the β phase) (figure 4).

The compression properties were analysed on as-built and stress relieved samples. For determining the compression resistance the fixture bearing blocks and the ends of the samples were cleaned with acetone and subjected to a quasi-static uniaxial compression.

The tests were made on a WDW-150S universal testing machine with a crosshead displacement speed of 2 mm/min, according to ASTM standards [19]. The obtained strain-stress curve is shown in figure 5.

A comparative study regarding the compression properties of the samples were made also using the ANSYS Workbench 19 R3finite element analysis software, (figure 6). For meshing the model were used TET10 type elements of 5 mm. The mesh was composed of 1244206 nodes and 616852 elements.

By comparing the results it is shown that FEA results are very similar to the experimental results (figure 5 and figure 6).

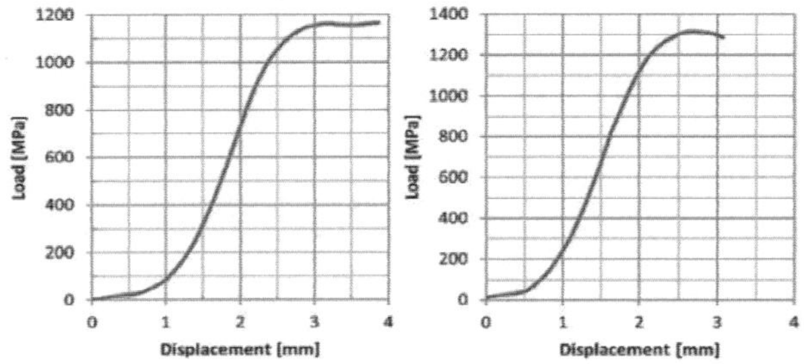


(a) (b)

**Figure 5.** Compression testing results for: as-built samples (a) and heat treated samples (b).

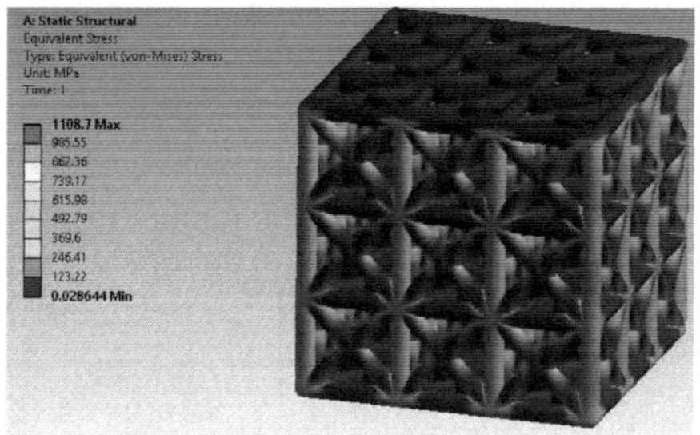


**Figure 6.** Compression testing results after FEA.

## 3. Conclusions

After analyzing the differences between the as-built samples and the heat treated samples the following conclusions can be drawn:

- The heat treatment lead to a structure where the α' martensite was transformed into more stable α+β phases;
- Microstructure of the heat treated samples is much uniform;
- The heat treated samples have less pores;
- Compression resistance of the heat-treated samples was improved.

In this regard was concluded that heat treatment improves the microstructure of the samples by reducing the internal stress that occurs during the building process. By decreasing the internal stress a better mechanical stress resistance was obtained too.

**Acknowledgments**

The authors are grateful for the financial support provided by a national research-development grant of the Romanian Ministry of Research and Innovation, CNCS-UEFISCDI, project number PN-III-P1-1.1-MC-2017-0391.

The authors would like to acknowledge the structural funds project PRO-DD (POS-CCE, O.2.2.1, ID 123, SMIS 2637, ctr. No 11/2009) too, for providing the infrastructure used in this work. The authors acknowledge also the technical support of the staff from the PRO-DD Research and Development Institute of the *Transilvania* University of Brasov.

**References**

[1] Hao L and Raymond D 2011 Design and additive manufacturing of cellular lattice structures, *The International Conference on Advanced Research in Virtual and Rapid Prototyping* (Bartolo, P J. et al. Leiria, Portugal Taylor & Francis Group) 249254

[2] Zadpoor A A 2019 Mechanical performance of additively manufactured meta-biomaterials *Acta Biomater.* **85** 4159

[3] Zhou H, Zhang X, Zeng H, Yang H, Lei H, Li X and Wang Y 2019 Lightweight structure of a phase-change thermal controller based on lattice cells manufactured by SLM *Chinese J.* Aeronaut. 32 1727

[4] Ming W, Chen J, An Q and Chen M 2019 Dynamic mechanical properties and machinability characteristics of SLM and forged Ti6Al4V *J. Mater. Process. Technol.* **271** 284

[5] Baudana G, Biamino S, Ugues D, Lombardi M, Fino P, Pavese M and Badini C 2016 Titanium aluminides for aerospace and automotive applications processed by Electron Beam Melting: Contribution of Politecnico di Torino Met. Powder Rep. 71 193

[6] Gao P et. all 2020 Biofunctional magnesium coated Ti6Al4V scaffold enhances osteogenesis and angiogenesis in vitro and in vivo for orthopedic application *Bioact. Mater.* **5** 680

[7] Elanchezhian C, Vijaya Ramnath B, Ramakrishnan G, Sripada Raghavendra K N, Muralidharan M and Kishore V 2018 Review on metal matrix composites for marine applications *Materials Today: Proceedings* **5** 1211

[8] Gong Y, Ma F Q, Xue Y, Jiao C S and Yang Z G 2019 Failure analysis on leaked titanium tubes of seawater heat exchangers in recirculating cooling water system of coastal nuclear power plant *Eng. Fail. Anal.* **101** 172

[9] Maconachie T, Leary M, Lozanovski B, Zhang X, Qian M, Faruque O and Brandt M 2019 SLM lattice structures: Properties, performance, applications and challenges *Mater. Des.* **183** 108

[10] Ma R, Liu Z, Wang W, Xu G and Wang W 2020 Microstructures and mechanical properties of Ti6Al4V-Ti48Al2Cr2Nb alloys fabricated by laser melting deposition of powder mixtures *Mater. Charact.* **164** p 110

[11] Castellanos S D, Lino J A and Neto R J 2017 A comparative study of manufacturing processes of complex surface parts in Titanium Ti6Al4V *Ciência Tecnol. dos Mater.* **29** 73

[12] Froes F H, 2015 Titanium Physical Metallurgy Processing and Applications (Ohio: ASM International 1st ed.)

[13] Liang Z, Sun Z, Zhang W, Wu S and Chang H, 2019 The effect of heat treatment on microstructure evolution and tensile properties of selective laser melted Ti6Al4V alloy *J. Alloys Compd.* **782** 1041

[14] Ali H, Ma L, Ghadbeigi H and Mumtaz K 2017 In-situ residual stress reduction, martensitic decomposition and mechanical properties enhancement through high temperature powder bed pre-heating of Selective Laser Melted Ti6Al4V *Mater. Sci. Eng.* **695** 211

[15] Huang Q, Liu X, Yang X, Zhang R, Shen Z and Feng Q 2015 Specific heat treatment of selective laser melted Ti–6Al–4V for biomedical applications *Front. Mater. Sci.* **9** 373

[16] Chicos L A, Zaharia S M, Lancea C, Pop M A, Canadas I, Rodriguez J and Galindo J 2018 Concentrated solar energy used for heat treatment of Ti6Al4V alloy manufactured by selective laser melting *Sol. Energy* **173** 76

[17] Vrancken B, Thijs L, Kruth J P and Van Humbeeck J V 2012 Heat treatment of Ti6Al4V produced by Selective Laser Melting: Microstructure and mechanical properties *J. Alloys Compd.* **541** 177

[18] Wauthle R, Vrancken B, Beynaerts B, Jorissen K, Schrooten J, Kruth J P and Van Humbeeck J, 2015 Effects of build orientation and heat treatment on the microstructure and mechanical properties of selective laser melted Ti6Al4V lattice structures *Addit. Manuf.* **5** 77

[19] ASTM, E9-09 Standard Test Method Compression Testing of Metallic Materials at Room Temperature, ASTM International, West Conshohocken, Pa, 2009

# Parameters setting of selective laser melting machine

**D T Le[1] and S L Chen[1,2*]**

[1] Department of Mechanical Engineering, National Chung Cheng University, Taiwan
[2] Advanced Institute of Manufacturing with High-tech Innovations (AIM-HI), National Chung Cheng University, Taiwan

[*] Email: imeslc@ccu.edu.tw

**Abstract**. Additive Manufacturing or Selective Laser Melting, in particular, has been used in industries in a vast number of applications since the 1980s. It has become common in various manufacturing processes, ranging from car engines, ships, and building constructions. In Additive Manufacturing applications, getting better accuracy and surface roughness as the main criteria with time efficient are challenging tasks that requires understanding of the fundamental Selective Laser Melting process. This study will investigate the relationship of process parameters and system parameters in Scanner Head Control, experimentally. Following the given result of this work, the machine operators will be able to understand the root of systems dynamics, the effects of process parameters, and speed up the initial setting with time efficiency.

## 1. Introduction

Additive manufacturing (AM) has become a trend in manufacturing technology in recent decades. It has developed for decades as 3D Printing, Rapid Prototyping (RP), Direct Digital Manufacturing (DDM), Selective Laser Melting, and Selective Laser Sintering (SLS) in many industrial applications [1, 2]. In the case of SLM, it is an Additive Manufacturing (AM) technique, in which laser power will melt powder alloys on a platform, layer by layer. This manufacturing technology has proved the potential of an existing procedure to fabricate frameworks for complex objects with flexibility, mass customization, and low material waste. This type of manufacturing also can be applied to various materials such as Fe, Ni, and Cu [3-5]. Although this technique is with many advantages, encountered problems in SLM are surface quality and shape accuracy that is related to process and motion control in terms of laser motion and system dynamics. In order to tackle these problems, some researches tried to optimize these parameters that influence processes such as scanning speed local geometry, oxygen content, and powder properties [6, 7]. During manufacturing process, the laser spot characteristics, such as spatial distribution and colour temperature, were monitored by camera to improve fabricated products [8, 9]. However, there is a gap in finding the experimental solutions for setting the procedure of process parameters based on the galvanometers systems' behaviour which is a major issue for tuning parameters for the laser motion controller called Scanner Head Control (SHC) [10].

This study aims to find out the relationship between system dynamics and motion of galvanometer systems experimentally. It also presents an overview of SLM process and points out these compensated parameters on final fabricated commercial products. The following objectives were presented:

- Understand about SLM controller system, Laser control;

- Analyse the effects of Delay parameters;
- Provide a procedure for setting Delay parameters.

## 2. Experimental framework

In the following experiments, the motion of mirrors was identified by conducting measurements on the fabricated workpiece's marked line.

### *2.1. Apparatus*

Since the machine with any inconsistency setting may give poor accuracy and validity of the experimental results, all hardware and software of the machine need to calibrate to ensure that the apparatus used throughout the experiments were in an optimal state and remained consistent throughout the study. Several unlisted parameters, such as air pressure, ambient temperature, and humidity, were provided by the user technical manual and kept constant during all experiments. All of the measurements were acquired using the Optical Measurement Machine (OMM) [11]. The material Titanium Alloy-Ti6Al4V samples were fabricated using an SLM-AM250 machine equipped with a fiber laser power-500W with a spot size of 50um and scanning speed of 900mm/s. The laser spot on platform was controlled by two mirrors in the galvanometer system of SHC [10].

### *2.2. Experimental details*

Movement of laser beam and delay times of laser directly indicate on melted Ti6Al4V plate material. When laser beam moves, the heat-affected zone of the laser within a half linewidth in the radius is homogeneous. The transient and end zone, at the time laser beam powered on and off is a semicircle, as shown in figure 1. By the assumption that the linewidth keeps constant along the path, if the boundaries of starting and end zone can be identified, it implies that the center of laser spot size should be identified.

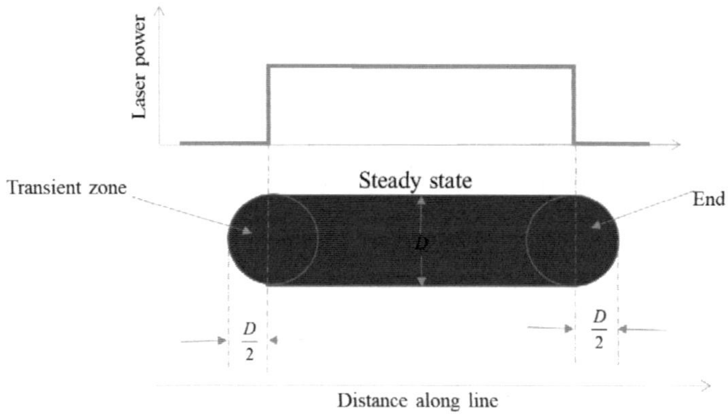


**Figure 1**. Terminal geometry of a melted vector (D is the laser beam diameter).

There are two parts in SHC changes when executing commands, including the mirrors - physical part and laser power - an electronic part. Since the bandwidth of electronic components is far higher than in physical parts, two kinds of delay parameters are added into SHC to compensate for the lagging in between. This is to keep laser intensity in constant along the designed path and reduce the processing time. The delay time includes Laser Delays (On/Off) and Scanner Delays (Jump/Mark/Polygon).

The designed experiments were conducted to find proper parameters based on effect zones of delay parameters. The designed path is fundamental in the construction of complex shapes. The beginning and end of each line in the designed path was involved at least one or two Delay parameter. For example, at a starting point of Mark command, the mirrors execute a Mark command right after the Jump command. It means that at starting points of Mark command involve the Jump Delay effects.

Moreover, at the transient zone, the laser power is switched from Off to On state. Therefore, the transient zone must involve Laser On Delay.

On the contrary, the motion of mirrors changes from executing a Mark command to stop in the end zones. Therefore, end zone must involve Mark Delay. Similarly, the laser power is switched from On to Off state at the end zone, which means the end zone must involve the Laser Off Delay. If a further Mark command follows a Mark command, Polygon Delay will be inserted instead of Mark Delay. Thus, the effects of Mark Delay on the end zone must change to the effects of Polygon Delay, instead.

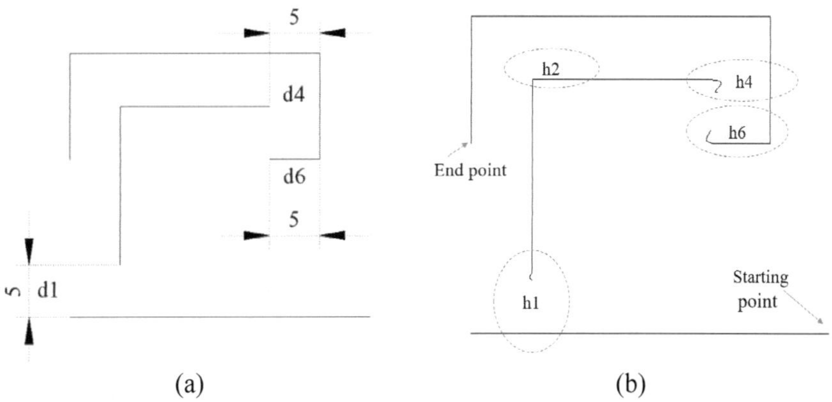


**Figure 2**. Designed path; (a) dimensions (unit: mm); (b) denote zone.

The input setting of Laser and Scanner Delays were set in a large range from 20us to 590us (given in Appendix). The experimenting process was considered to show the effect of single parameters by fixing either Laser Delays or Scanner Delays. Thus, the effect of Laser and Scanner Delays were visible in different areas. The details of its effects are shown in table 3, following denotes in figure 2.

**Table 1**. Effective areas of Delay parameters.

| Zone | Effects of Delay parameters | |
|---|---|---|
| | Specimen 1~9 | Specimen 10~18 |
| h1 | Laser On Delay | Jump Delay |
| h2 | Polygon Delay | Polygon Delay |
| h4 | Laser Off Delay | Mark Delay |
| h6 | Laser On Delay | Jump Delay |

All of the 1~9 experiments, by using fixed Scanner Delays, including Jump and Mark Delays, the Laser Delays caused the topology at each zone at the beginning of each Mark command (zone h1, 4, and 6). These experiments, from No.10~18, were accompanied by using vary Scanner Delays (Jump, Mark and Polygon), and fixed Laser Delays. The measurements were conducted at zone h2, h4 and h6, following denote in figure 2(b).

## 3. Results and Discussions

### *3.1. Effects of Laser Delays*

The differences in the obtained distances value of d1 and d4 are clearly visible and result from Laser Delay values' varied settings. The value of d1 was increased with a longer value of Laser On Delays. The value of d4 decreased with a longer value of Laser Off Delays. Figure 3 shows the relationship between Laser On Delay and distance d1. Since the designed distance of d1 is 5mm, the Laser On Delay needs to set equal to 173us. Similarly, the Laser Off Delay requires to be 450us.

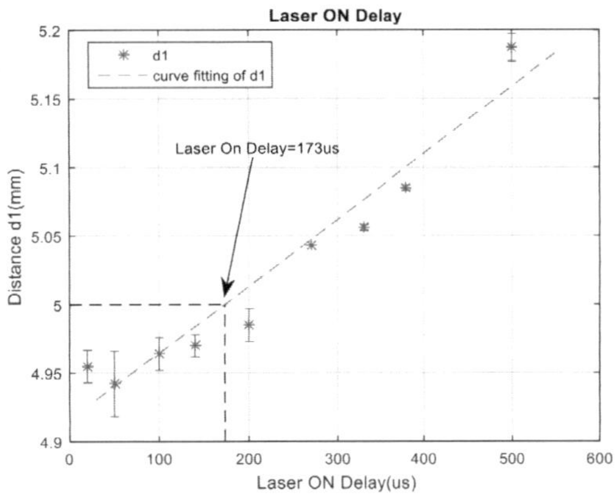


**Figure 3**. Laser On Delays and its effects on the distance d1.

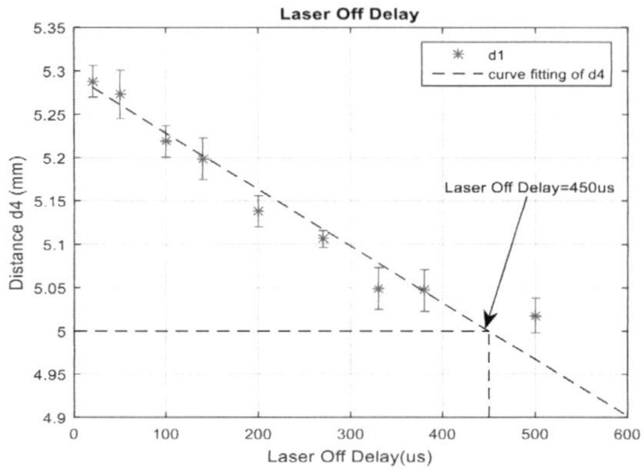


**Figure 4**. Laser Off Delays and its effects on the distance d4.

### *3.2. Effects of Scanner Delays*

The effect of Scanner Delays was obtained from concept number 9 to 18, changing its values and using fixed Laser Delays in figure 2. The Mark Delay was defined as the distance d4. The Jump Delay, the effect zones were divided into horizon and vertical. In figure 5 and 6, there are significant changes in both vertical and horizontal direction in the zone h1. With the Jump Delay smaller than 100us, the errors are about 0.25mm in both directions, making a large contour error in total. The errors tend to be in steady states when Jump Delay set larger than 200us. This can explain that when Jump Delays are longer than the systems' settling time, the error caused by this Delay parameter becomes a single value.

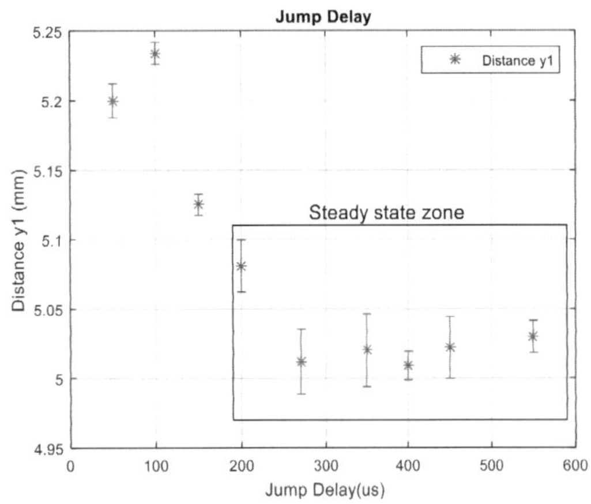


**Figure 5**. Vertical effects of Jump Delays.

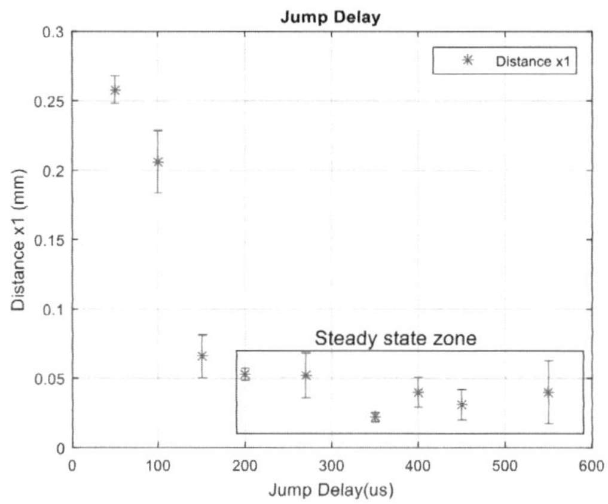


**Figure 6**. Horizontal effects of Jump Delays.

The Mark Delay effects were visible at the zone h4, and the distance d4 was used to evaluate the Mark Delay because in this zone is at the end of Mark commands. Following the indicating trend in figure 7, when the value set larger than 150us, the distance d4 tends to keep constant values. This means that the motion part needs more than 150us to be settling after executing Mark commands.

The Polygon Delay was validated in the corner errors because, at the corner zone h2, there are two consecutive Mark commands. Figure 9 shows the un-sharps and burn-in effects at the corner. The corner errors tend to a steady-state value with the setting of Polygon Delay of 160us in figure 8.

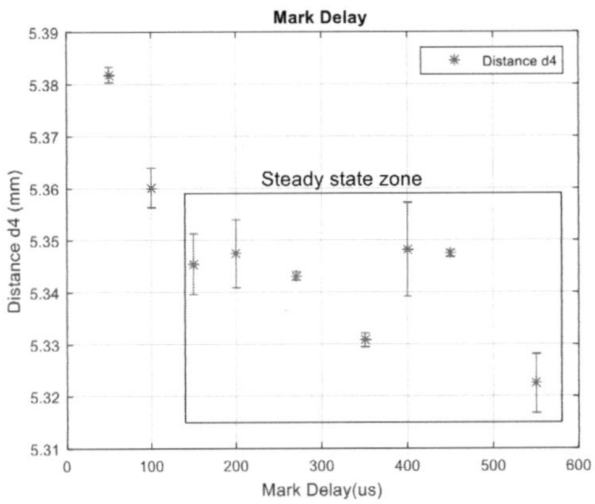


**Figure 7**. Effects of Mark Delays at zone h4.

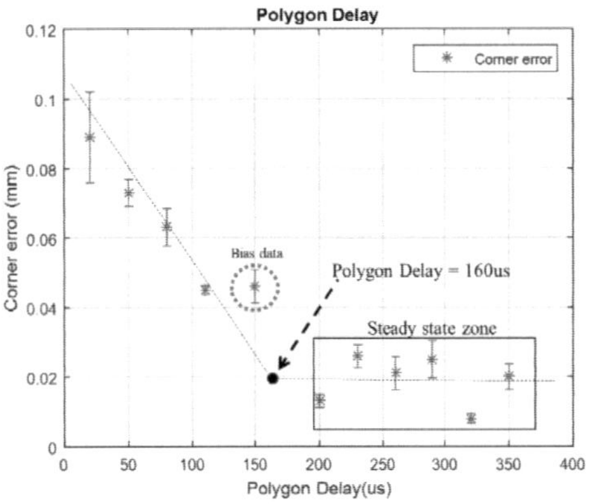


**Figure 8**. Corner errors with variant Polygon Delay values at zone h2.

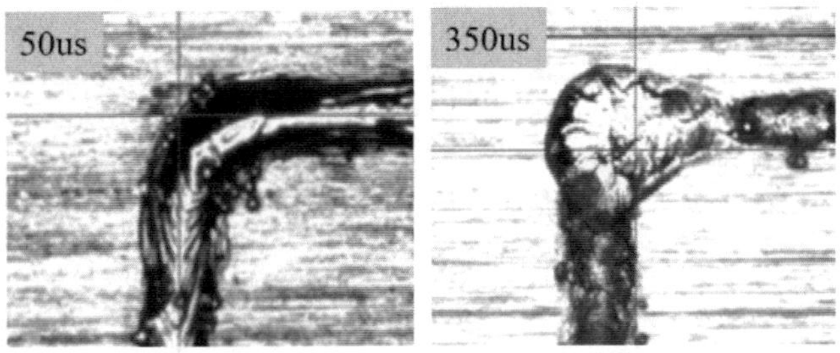


(a) (b)

**Figure 9**. Polygon Delay; (a) un-sharp with low value; and (b) burn-in effects at the h2 zone with high value.

## 4. Conclusions

The purpose of these analysing methods experimentally is to understand the SLM system and the effects of the Delay process parameters. In these experiments, the movement of mirrors was identified basing on melted vectors on materials. However, the uncertainty of determining boundaries of melted vectors caused uncertainty in measurement results. Also, experimenting with Ti6Al4V, some disturbances caused unreasonable and bias data resulting from Laser Delays. The reason could be listed:

- The experiments were focused on laser spot size and compensated parameters;
- The engineer might input wrong or mark the sample cause to singular data;
- The particle materials, and other environment conditions were not suitable.

**Table 2**: Obtained parameter thought out experiments.

| Parameters | Obtained value(us) |
|---|---|
| Laser On Delay | 168~173 |
| Laser Off Delay | 450~460 |
| Jump Delay | 200 |
| Mark Delay | 150 |
| Polygon Delay | 100~160 |

By applying in a varied setting, obtained results demonstrate the relationship between the effects of Delay parameters and changes in related measurement as expected utility theory. Thus, finding an experimental procedure based on the fabricated prototype is a feasible scheme as following steps:
Step 1: Provide a well-designed path which should include:

- A Jump command to Mark command to involve Laser On and Jump Delay;
- An endpoint of Mark command to involve Laser Off and Mark Delay;
- Two consecutive Mark command to involve Polygon Delay.

Step 2: Setup the machine and manufacture one by one with these parameters in the ranging of Predicted value +/- n*(step change). (n=0…4; step change = 20us).
Step 3: Measure these fabricated products.
Step 4: Use curve fitting for Laser Delays and plot these values of Scanner Delays.
Step 5: Evaluate the measurement results.

## 5. Future work
Although this study presented the galvanometer's motion, other factors such as absorption and absorption rate of material, inert gas, and humility of environment and particle size need to be taken into consideration to get a better performance in SLM machine. Further research on observation for transient response in the starting and end zones will be required, specifically in evaluating the lagging in motion and laser control. Thus, setting up peripheral equipment such as High-Speed Camera for supporting to do systems identification are needed for fundamental research.

## Appendix: Range of setting Delay parameters
In the Scanner Head Control, there are two specifications dynamic range in concern which are Step response time and tracking error (unit: time). Since the *Step command* can be understood as a Jump command with a big change, Jump/Mark Delay must be smaller than Step response time. If Jump/Mark Delay set equal to Step response time, these effects of Jump/Mark Delay will not visible on fabricated products. Range of setting for Laser Delay was according to the recommendation from technical manual of SHC. In the model intelliSCAN® III 20, Step response time and Tracking error are 700us and 320us, respectively. From the definition of the manufacturer [10]:

- Step response time is the time interval required to execute a jump over a specified distance;
- Tracking error is the time interval between the set and actual positions when traversing a vector.

## Acknowledgements
This work was supported in part by the Advanced Institute of Manufacturing with High-tech Innovations (AIM-HI) from The Featured Areas Research Center Program within the framework of the Higher Education Sprout Project by the Ministry of Education (MOE) in Taiwan and was also supported in part by the Ministry of Science and Technology, Taiwan, ROC, under Grants MOST 109-2634-F-194-004, MOST 108-2218-E-194-007, and MOST 109-2634-F-194-003.

## References
[1] Olakanmi E O, Cochrane R and Dalgarno K 2015 A review on selective laser sintering/melting (SLS/SLM) of aluminium alloy powders: Processing, microstructure, and properties *Prog. Mater. Sci.* **74** 401
[2] Murr L E, Gaytan S M, Ramirez D A, Martinez E, Hernandez J, Amato K N, Shindo P W, Medina F R and Wicker R B 2012 Metal Fabrication by Additive Manufacturing Using Laser and Electron Beam Melting Technologies *J. Mater. Sci. Technol.* **28** 1
[3] Prashanth K G 2020 Selective Laser Melting: Materials and Applications *J. Manuf. Mater. Process.* **4** 13

[4] Herzog D, Seyda V, Wycisk E and Emmelmann C 2016 Additive manufacturing of metals *Acta Mater.* **117** 371
[5] Yap C Y, Chua C K, Dong Z L, Liu Z H, Zhang D Q, Loh L E and Sing S L 2015 Review of selective laser melting: Materials and applications *Appl. Phys. Rev.* **2** 041101
[6] Raus A A, Wahab M S, Shayfull Z, Kamarudin K and Ibrahim M 2016 The Influence of Selective Laser Melting Parameters on Density and Mechanical Properties of AlSi10Mg *MATEC Web Conf.* **78** 01078
[7] Gorsse S, Hutchinson C, Gouné M and Banerjee R 2017 Additive manufacturing of metals: a brief review of the characteristic microstructures and properties of steels, Ti-6Al-4V and high-entropy alloys *Sci. Technol. Adv. Mater.* **18** 584
[8] Chivel Y and Smurov I 2010 On-line temperature monitoring in selective laser sintering/melting *Phys. Procedia* **5** 515
[9] Lott P, Schleifenbaum H, Meiners W, Wissenbach K, Hinke C and Bültmann J 2011 Design of an Optical system for the In Situ Process Monitoring of Selective Laser Melting (SLM) *Phys. Procedia* **12** 683
[10] SLM Solutions GmbH 2015 intelliSCAN III 20.
[11] MitutoyoCorporation 2016 CNC Vision Measuring System-Quick Vision Series No.E14007(2).

# Experimental research on dynamic thermal performance of silicon nitride all-ceramic ball bearings

**S H Li[1,2], Y H Wang[1*], C Wei[1], Z N Wang[1] and Z X Xia[1]**

[1] Mechanical Engineering College, Shenyang Jianzhu University, Shenyang, Liaoning 110168, China
[2] National-Local Joint Engineering Laboratory of NC Machining Equipment and Technology of High-Grade Stone, Shenyang, Liaoning 110168, China

*E-mail: yonghua0514@163.com

**Abstract**. In order to verify that the silicon nitride all-ceramic ball bearings have more excellent high-speed performance, thermal rise and vibration are the two main factors that restrict the bearing to move at high speed. In this manuscript, based on the establishment of the mechanical model of bearings, the structural model of all ceramic motorized spindle-bearings is established. The temperature field of the motorized spindle-bearing system is simulated through analysis of bearing heat generation. The thermal rise of silicon nitride all-ceramic ball bearings was tested by setting up all ceramic motorized spindle-bearing test platform under the operating speeds of 6,000, 9,000, 12,000, 15,000 and 18,000 rpm respectively, the results are compared with the simulation results. Vibration characteristics of the all-ceramic ball bearings were compared with those of similar steel bearings. The test results show that the thermal rise of all-ceramic ball bearings gradually tends to be stable at different rotational speeds after the running time exceeds 15 minutes, no-load thermal rise is less than 10℃, the maximum temperature of ceramic bearing is nearly 10 ℃ lower than the maximum temperature of the simulated metal bearing . With the increase of rotating speed, the maximum amplitude of metal bearing without loading is 2.38 times of that of ceramic bearing. The conclusion shows that ceramic bearings have better dynamic characteristics and stable reliability than metal bearings, so they have better high-speed performance, providing reference value for the future design of the same type of all-ceramic bearings.

## 1. Introduction

Silicon nitride, as an important engineering structural ceramic material, has excellent properties as bearing material, such as high strength, high hardness, thermal shock resistance, good wear resistance, stable chemical properties, good insulation, good self-lubricating properties, etc. In addition, the ceramic material has low density, strong bearing capacity and low failure rate in high-speed environment, thus the working life is greatly increased [1, 2].

With the development of science and technology, new materials are widely used and gradually applied in the spindle-bearing system of CNC machine tools [3, 4]. Among them, the outer ring, inner ring and ceramic rolling elements of the all-ceramic bearing components are all made of ceramic materials, which are more used in spindle-bearing system. Ceramic ball bearings have excellent performance and good performance under high-temperature and high-speed environment, which can guarantee the normal operation of high-speed equipment [5, 6]. The thermal rise and vibration characteristics of ceramic ball bearings will have a great impact on the accuracy and vibration characteristics of the equipment.

Bearing heat generation and vibration have great influence on spindle thermal rise, thermal displacement, rotation accuracy and vibration characteristics. Harris [7] calculated the heat generation of the contact unit inside the bearing and tested the bearing temperature according to the test, which verified that the local calculation of the heat generation of the bearing is relatively close to the actual working condition. Bossmans [8] established the thermal model of high-speed motorized spindle, deeply explored the factors such as thermal conductivity and thermal resistance, and analyzed the thermal rise characteristics of motorized spindle at different rotational speeds. Chen Xiao-an et al. [9] established a thermal-mechanical coupling dynamic model of angular contact ball bearings, and considered the influence of thermal displacement and pre-tightening mode in this process, and analyzed the friction loss and dynamic bearing stiffness of the spindle-bearing system in running state. Marhomy [10] put forward an analytical method to determine the stability of elastic spindle-bearing system by using nonlinear parameters. Wardle [11] established a model of dynamic characteristics of spindle-bearing system, which simplified the preload constant. Chen [12] established a model to determine the response of the spindle bearing system under high speed operation. Jones model was adopted for the bearing in the model, and only axial pretension was considered. Bert R. Jorgensen [13] established the dynamic model of the "spining-bearing" unit of motorized spindle based on the centralized parameter method in Timoshenko Beam theory, and studied its dynamic characteristics by numerical method. Altintas and Cao [14, 15] proposed how to establish the dynamic analysis model of spindle-bearing system by using finite element method under the influence of centrifugal force and bearing gyro torque. It is proved that the bearing dynamic model has good accuracy in predicting the stiffness, mode of vibration and natural frequency of contact force bearing.

The contact load and contact angle of high-speed rolling bearing are the basis of analyzing and calculating thermal characteristics, stiffness characteristics and dynamic characteristics of the spindle - bearing system [16]. In this manuscript, considering the influence of the gyroscopic moment and centrifugal force to balance problems, based on the Hertz elastic contact theory of rolling bearing, all-ceramic ball bearing structure analysis model is established, through constructing the all-ceramic motorized spindle - bearing test platform, the rise of temperature of the silicon nitride all-ceramic ball bearing is tested, and compared with the vibration characteristics of similar steel bearing through the contrast test. It provides reference value for the future design of the same type all-ceramic bearing.

## 2. Establishment of bearing mechanical model

### 2.1. Geometrical relation of bearing internal deformation

As shown in figure 1, the relative positions of the $j$ ball center and the center of curvature of the inner and outer ring of the raceway before and after the rolling bearing bears load at the Angle position $\varphi_j$. Assuming that the central position of the curvature of the outer raceway remains unchanged and the central position of the curvature of the inner raceway changes, $\delta_a$ is the axial relative displacement, $\delta_r$ is the radial displacement, and $\theta$ is the angular deformation.

When no load is borne, the distance between the curvature centers of the inner and outer ring channels is:

$$A=(f_i+f_e-1)d_w \tag{1}$$

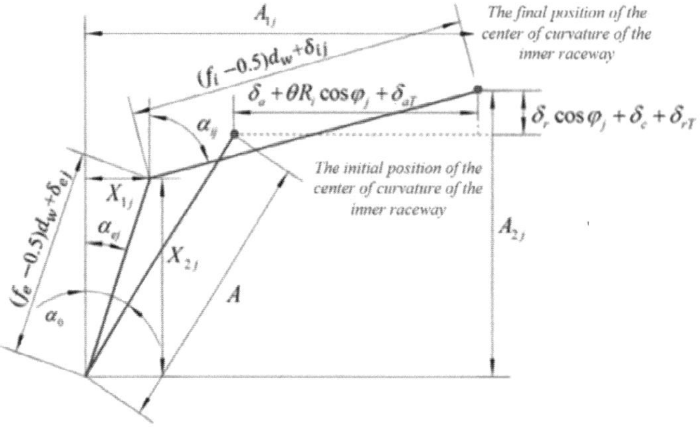


**Figure 1.** Relative position of the center of curvature.

Considering the effect of thermal displacement of the bearing, the axial distance and radial distance between the final positions of the curvature centers of the inner and outer raceways can be obtained from equations (2) and (3) at the ball $j$ during constant pressure preload [17].

$$A_{1j} = A\sin\alpha_e + \delta_a + \theta R_i \cos\varphi_j + \delta_{aT} \tag{2}$$

$$A_{2j} = A\cos\alpha_e + \delta_r \cos\varphi_j + \delta_{rT} \tag{3}$$

At the position of ball $j$ during positioning preloaded, Considering the influence of bearing thermal displacement and centrifugal radial expansion, the axial distance and radial distance between the final positions of curvature centers of inner and outer raceways can be obtained by equations (4) and (5) [18].

$$A_{1j} = A\sin\alpha_e + \delta_a + \theta R_i \cos\varphi_j + \delta_{aT} \tag{4}$$

$$A_{2j} = A\cos\alpha_e + \delta_r \cos\varphi_j + \delta_c + \delta_{rT} \tag{5}$$

In the formula, $\alpha_o$ is the contact angle before being loaded; $\delta_{aT}$ and $\delta_{rT}$ are the axial thermal displacement and radial thermal displacement respectively; $\delta_c$ is the radial expansion of the bearing inner ring caused by centrifugal force under high-speed operation. The calculation method that affects it is shown in formula (6).

$$\delta_c = \frac{\rho_i \omega^2}{16E_i}\left[(3+\upsilon)d^2 + (1-\upsilon)(2R_i)\right]^2 2R_i \tag{6}$$

In the formula $\rho_i$ is the inner ring material density; $d$ is the inner diameter of the inner ring; $R_i$ is the diameter of inner ring channel; $\upsilon$ is the inner circle material Poisson's ratio; $E_i$ is the elastic modulus of inner ring material, and $\omega$ is the revolving speed of inner ring.

By introducing new variables $V_{1j}$ and $V_{2j}$, the geometric compatibility equation of ball and raceway deformation is:

$$(A_{1j} - V_{1j})^2 + (A_{2j} - V_{2j})^2 - [(f_i - 0.5)d_w + \delta_{ij}]^2 = 0 \tag{7}$$

$$V_{1j}{}^2 + V_{2j}{}^2 - [(f_e - 0.5)d_w + \delta_{ej}]^2 = 0 \tag{8}$$

According to figure 1, the contact Angle between the ball and the inner and outer raceway can be expressed as:

$$\cos\alpha_{ej} = \frac{V_{2j}}{(f_e - 0.5)d_w + \delta_{ej}}$$
$$\sin\alpha_{ej} = \frac{V_{1j}}{(f_e - 0.5)d_w + \delta_{ej}}$$
$$\cos\alpha_{ij} = \frac{A_{2j} - V_{2j}}{(f_i - 0.5)d_w + \delta_{ij}} \quad (9)$$
$$\sin\alpha_{ij} = \frac{A_{1j} - V_{1j}}{(f_i - 0.5)d_w + \delta_{ij}}$$

*2.2. Ball force and balance equation*

When the bearing is running at high speed, the ball is subjected to centrifugal force, gyro torque, and t he contact load between the ball and the raceway. The force of the *jth* ball is shown in figure 2. $F_{cj}$ is t he centrifugal force, $M_{gj}$ is the gyroscopic moment, $Q_{ij}$ and $Q_{ej}$are the contact loads between the ball r olling body and the inner and outer raceways [17, 18, 19].

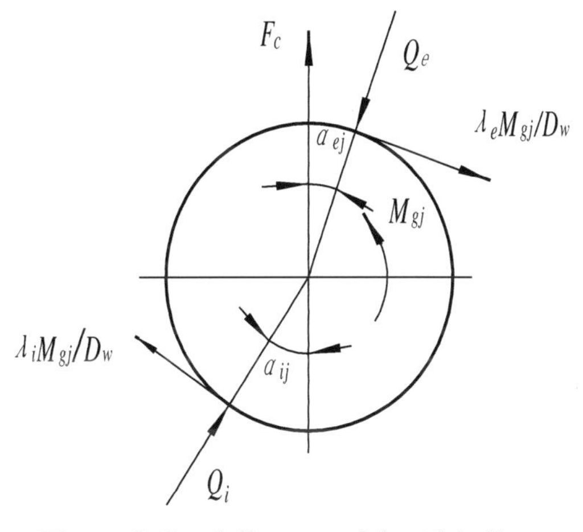


**Figure 2.** Load diagram of the *jth* ball.

The centrifugal force of the jth ball is calculated by the following equation:

$$F_{cj} = \frac{1}{2} d_m \omega^2 (\frac{\omega_m}{\omega})^2 \quad (10)$$

The gyroscopic moment of the j-th ball is calculated by the following equation:

$$M_{gj} = J(\frac{\omega_b}{\omega})_j (\frac{\omega_m}{\omega})_j \omega^2 \sin\beta \quad (11)$$

$$J = \frac{1}{60} \rho_m \pi d_w \quad (12)$$

In the equation, $\rho_m$ is the material density.

The contact loads of the ball and the inner raceway are:

$$Q_{ij} = K_{ij}\delta_{ij}^{3/2} \tag{13}$$

$$Q_{ej} = K_{ej}\delta_{ej}^{3/2} \tag{14}$$

In the equation, $K_{ij}$ and $K_{ej}$ are the load deformation constants between the rolling element and the inner and outer raceways [20].

According to the outer race control theory, it is believed that the frictional force only occurs on the outer race, taking $\lambda_{ij}$=0 and $\lambda_{ej}$ =2.

The balance equation of the ball is:

$$Q_{ij}\sin\alpha_{ij} - Q_{ej}\sin\alpha_{ej} - \frac{M_{gj}}{d_w}(\lambda_{ij}\cos\alpha_{ij} - \lambda_{ej}\cos\alpha_{ej}) = 0 \tag{15}$$

$$Q_{ij}\cos\alpha_{ij} - Q_{ej}\cos\alpha_{ej} + \frac{M_{gj}}{d_w}(\lambda_{ij}\sin\alpha_{ij} - \lambda_{ej}\cos\alpha_{ej}) + F_{cj} = 0 \tag{16}$$

Assuming that the outer raceway of the bearing is fixed and the external load received by the bearing is the axial force Fa and the radial force Fr, the balance equation of the inner raceway of the bearing is:

$$F_a - \sum_{j=1}^{Z}(Q_{ij}\sin\alpha_{ij} - \frac{M_{gj}}{d_w}\lambda_{ij}\cos\alpha_{ij}) = 0 \tag{17}$$

$$F_r - \sum_{j=1}^{Z}(Q_{ij}\cos\alpha_{ij} + \frac{M_{gj}}{d_w}\lambda_{ij}\sin\alpha_{ij})\cos\varphi_j = 0 \tag{18}$$

During positioning preloading, the initial preloading force of the bearing has been changed due to the radial expansion amount $\delta_c$ generated by centrifugal force, the axial thermal deformation amount $\delta_{aT}$ at and the radial thermal deformation amount $\delta_{rT}$ caused by the themal rise of each component. The changed preloading force value needs to be calculated by equation (19).

$$F_a = \sum_{j}^{Z}\sin\alpha K_n\delta_n^{3/2} \tag{19}$$

In the formula, δn is the total deformation of the ball after loading, which can be obtained by equation (20); α is the contact angle after loading, and kn is the load deformation constant obtained by equation (18) [21].

$$\delta_n = \left[(A\sin\alpha + \delta_a + \delta_{aT} + R_i\theta\cos\varphi)^2 + (A\cos\alpha + \delta_r + \delta_c + \delta_{rT})^2\right]^{1/2} - A \tag{20}$$

$$\sin\alpha = \frac{A\sin\alpha_0 + \delta_a + \delta_{aT} + R_i\theta\cos\varphi}{\left[(A\sin\alpha_0 + \delta_a + \delta_{aT} + R_i\theta\cos\varphi)^2 + (A\cos\alpha_0 + \delta_r\cos\varphi + \delta_c + \delta_{rT})^2\right]^{1/2}} \tag{21}$$

## 3. Motorized spindle-bearing system model

### *3.1. All ceramic motorized spindle-bearing structural model*

The research in this manuscript takes the angular contact ball bearing 7009C as an example, and the performance analysis is carried out for different materials. The structural parameters of bearings with different materials are shown in table 1.

**Table 1.** Structural parameters of bearings with different materials.

| Bearing material | Inner diametr (mm) | Outer diameter (mm) | Initial contact angle (°C) | Density /(g/cm3) | Diameter of balls (mm) | Elasticity modulus (GPa) | Poisson ratio |
|---|---|---|---|---|---|---|---|
| $Si_3N_4$ | 45 | 75 | 15 | 3.24 | 8.731 | 320 | 0.26 |
| Steel | 45 | 75 | 15 | 7.85 | 8.731 | 206 | 0.3 |

This manuscript uses the ceramic spindle as the prototype to study the dynamic characteristics of silicon nitride all-ceramic ball bearing, as shown in figure 3. The front (7009C) and rear (7008C) ceramic bearings are angular contact ball bearings, the two bearings are paired and installed back to back. The bearing and shaft adopt interference fit. The rotor is installed on the shaft by hot pressing, and the air gap between the rotor and the stator is 0.3 mm. Spindle cooling uses a water chiller to adjust the temperature and flow rate generated at high speeds. The cooling water jacket is hot-pressed in the main shaft housing and has a rectangular cross section. In the running process of the high-speed ceramic motorized spindle, the dynamic characteristic of bearing is the main factor affecting the running precision of spindle [22]. The spindle bearing system adopted in this experiment is based on the high-speed motorized spindle model of 170-30/15 and all-ceramic ball bearing as support. The bearing material selected is silicon nitride and the cage is PEEK composite material.It has higher wear resistance, because the density is less than the steel bearing, the inertia force and centrifugal force produced is smaller, because the friction force is smaller the heat produced is lower, has a better development prospect.

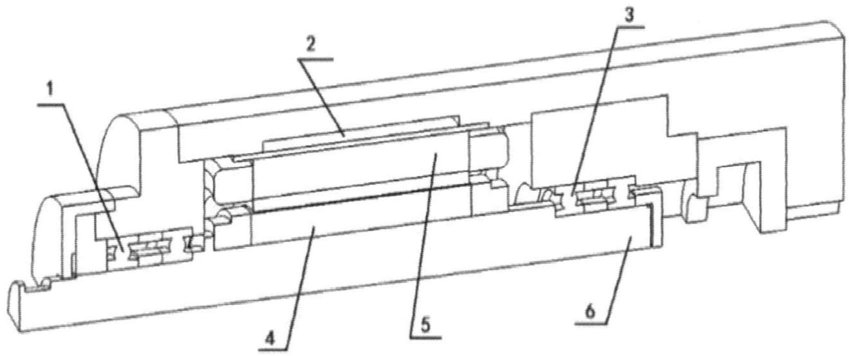


**Figure 3.** The schematic diagram of spindle-bearing system structure:
1-Front bearing 7009C; 2-Cooling Jacket; 3-Rear bearing 7008C;
4-Rotor; 5-Stator; 6-Shaft.

### *3.2. The themal field of the motorized spindle-bearing system*

Based on the boundary conditions of the spindle parts, the simulation results through Comsol software show that the distribution of the thermal field of the motorized spindle-bearing system in figure 4.

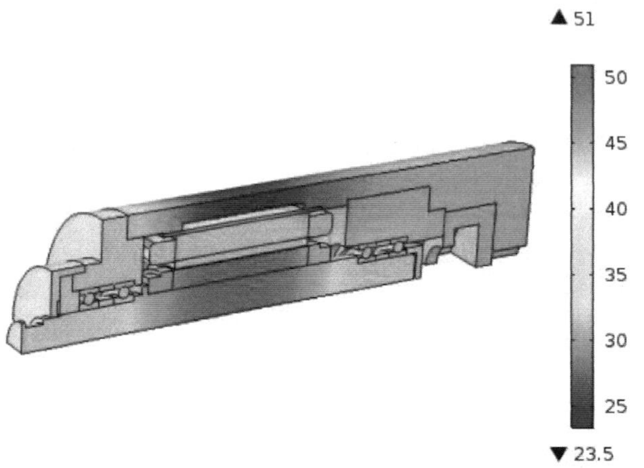


**Figure 4.** The temperature field of the motorized spindle-bearing system.

The steady-state temperature field cloud diagram of 170-HT-30 through simulation calculation is shown in figure 4, the simulation results show that the spindle thermal reaches steady state, the internal thermal distribution is very obvious, and the temperature at the rotor is the highest, up to 51 ℃. The temperature difference between the front and rear bearings is also obvious. The temperature of the front bearing is slightly higher than that of the rear bearing, up to 47 ℃. The main reason is that the front and rear bearing models are different. Under the same load, the internal contact state of the bearing is not the same. As a heat source, the heat output of the front bearing is greater than that of the rear bearing. Secondly, from the structure of the electric spindle, the front bearing is closer to the built-in motor, and the heat will also be transferred from the rotor to the front bearing, so that the themal rise of the front bearing is slightly higher than that of the rear bearing.

## 4. Experimental test and analysis of dynamic performance of all ceramic motorized spindle-bearing system

### *4.1. Temperature Vibration test*

The test conditions are shown in table 2, in which the bearing speed starts from 6000 rpm and increases by 1000 rpm every time until 18000 rpm.

**Table 2.** Basic parameters of spindle.

| Lubrication mode of bearings | Oil intake (ml/h) | Pressure (Mpa) | Cooling water temperature (℃) | Cooling water flow (L/min) | Room temperature (℃) | Bearing preload (N) |
|---|---|---|---|---|---|---|
| Oil-air | 5 | 0.25 | 18 | 10 | 25 | 400 |

The equipment used during the test is shown in figure 5.

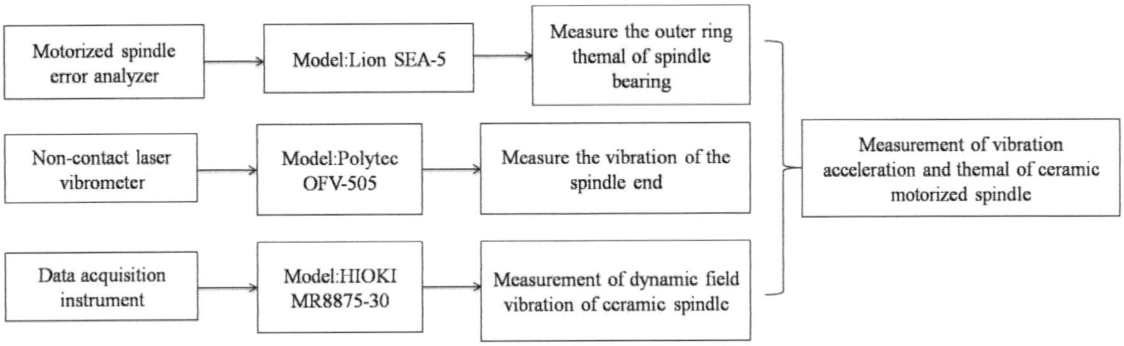


**Figure 5.** Test equipment.

The measurement of the vibration acceleration and themal of the ceramic motorized spindle is shown in figure 6. The basic parameters of ceramic motorized spindle are shown in table 3.

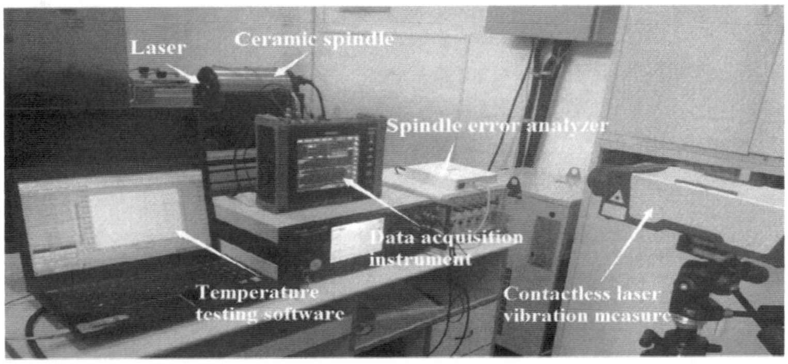


**Figure 6.** The measurement of the vibration acceleration and themal of the ceramic motorized spindle.

**Table 3.** Basic parameters of spindle.

| Basic parameters of spindle | 170-30/15 ceramic spindle | 170HT30 metal spindle |
|---|---|---|
| Rated speed (rpm) | 30000 | 30000 |
| Rated voltage(V) | 350 | 350 |
| Rated frequency(Hz) | 1000 | 1000 |
| Pole logarithm | 4 | 4 |
| Shaft material | $Si_3N_4$ ceramic | 40Cr steel |
| Bearing inner and outer ring material | $Si_3N_4$ ceramic | GCr15 steel |
| Bearing ball material | $Si_3N_4$ ceramic | GCr15 steel |

### *4.2. Experimental results and analysis*

The experiment on the bearing outer ring thermal as shown in figure 6. At the operating speed of 6000~12000 rpm, the curve of the themal of the outer ring of the ceramic bearing changes with time, as shown in figure 7. Before the running time of 15min, it can be seen that the themal rise at different speeds has an obvious upward trend. Friction heat generation is the main reason for the heat generation of high-speed electric spindle bearings. In addition to the differential friction heat generation between the ball and the track, there is also the ball gyro Rotating friction heat generation and sliding friction heat generation between the ball and the cage [23]. When the motorized spindle started to operate, the ceramic bearings were still in the adaptation stage, and the coolant circulation was not smooth enough. Oil and gas lubrication does not play a certain role in bearings. After the running time exceeds 15min, it can be seen that the themal rise of the ceramic bearing gradually stabilizes at different speeds, and load themal rise is less than 10°C. At this time, as the rotation time is longer, the high temperature resistance and self-lubricity of the ceramic bearing are more reflected , The performance of the spindle at high speed tends to be stable, and the life can be extended.

Compared with the simulation results in figure 4, it can be seen that the maximum temperature of ceramic bearing is 36.5°C, which is nearly 10°C lower than the maximum temperature of the simulated metal bearing. It can be seen that ceramic bearings have better high-speed performance.

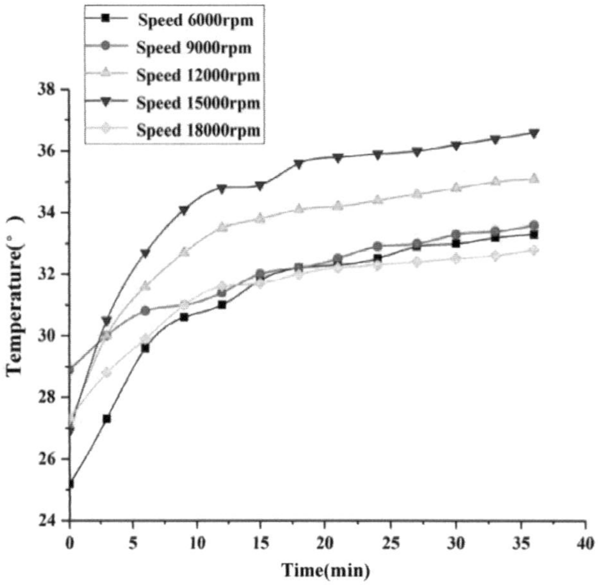


**Figure 7.** The temperature of the outer ring of the bearing changes with time at different speed.

The ceramic bearing and the rotating shaft of the ceramic motorized spindle adopt interference fit, and the vibration of the spindle can reflect the dynamic characteristics of the inner ring of the ceramic bearing. The acceleration amplitude is used to measure the intensity of the vibration, and the data collected in the motorized spindle operation by the Non-contact laser vibrometer are exported and sorted out. Under the working conditions of 6000, 9000, 12000, 15000, 18000 rpm, ceramic motorized spindle and metal motorized spindle respectively take 2000 sets of data for spectrum analysis, and obtain the vibration frequency domain of ceramic motorized spindle at 6000~18000 rpm as shown in figure 8. As can be seen from figure 8, under the working condition of 6000 rpm, the fast Fourier transform (FFT) transform shows that the first-order frequency is 99.06Hz, and the amplitude is 0.0158m/s$^2$. The second-order frequency is 199.63 Hz, and the amplitude is 0.0258 m/s$^2$. The third-order frequency is 299.20Hz, and the amplitude is 0.0234 m/s$^2$. Under the working condition of 9000 rpm, the first-order frequency is 149.72Hz, and the amplitude is 0.0623m/s$^2$. The second-order frequency is 299.94 Hz, and the amplitude is 0.0037 m/s$^2$. The third-order frequency is 450.68 Hz, and the amplitude is 0.0266 m/s$^2$. Under the working condition of 12000rpm, the first-order frequency is 200.63Hz, and the amplitude is 0.082m/s$^2$. The second-order frequency is 399.77Hz, and the amplitude is 0.0229 m/s$^2$. The third-order frequency is 601.16 Hz, and the amplitude is 0.025 m/s$^2$. Under the working condition of 15000 rpm, the first-order frequency is 249.29Hz, and the amplitude is 0.1121m/s$^2$, the second-order frequency is 500.34Hz, and the amplitude is 0.0291 m/s$^2$. the third-order frequency is 750.38 Hz, and the amplitude is 0.0284 m/s$^2$. Under the working condition of 18000 rpm, the first-order frequency is 300.20Hz, the amplitude is 0.157m/s$^2$. The second-order frequency is 599.90Hz, the amplitude is 0.0238 m/s$^2$. The third-order frequency is 899.33 Hz, and the amplitude is 0.0394 m/s$^2$. When the bearing vibration signal increases with the number of revolutions, the double frequency amplitude is larger in the low frequency region than in the high frequency region. The vibration amplitude of each frequency level tends to change steadily while the number of revolutions of the motorized spindle increases, no-load vibration is less than 0.16m/s$^2$. Because of the excellent characteristics of silicon nitride material, such as high temperature resistance, high rigidity, good precision retention, good self-lubricating performance, etc., the themal rise of the motorized spindle makes the thermal deformation of the internal ceramic bearing small, thus affecting its own natural frequency small.

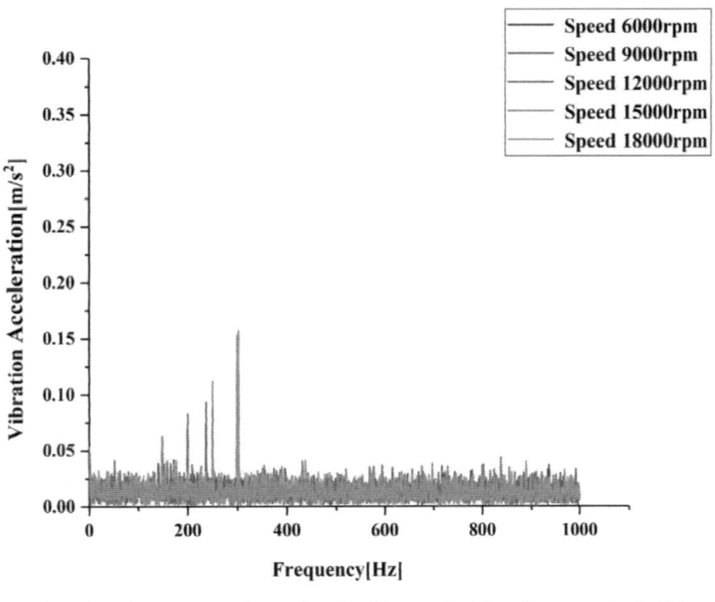


**Figure 8.** Vibration frequency domain of silicon nitride all-ceramic ball bearings.

The vibration frequency domain of metal motorized spindle at 6000~18000 rpm is shown in figure 9. As can be seen from figure 5, under the working condition of 6000 rpm, the fast Fourier transform (FFT) transform shows that the first-order frequency is 99.73Hz, and the amplitude is 0.0122m/s$^2$. The second-order frequency is 199.72 Hz, and the amplitude is 0.005 m/s$^2$. The third-order frequency is 300.81Hz, and the amplitude is 0.0063 m/s$^2$. Under the working condition of 9000 rpm, the first-order frequency is 150.39Hz, the amplitude is 0.0712m/s$^2$. The second-order frequency is 300.26 Hz, the amplitude is 0.0661 m/s$^2$. The third-order frequency is 449.68Hz, and the amplitude is 0.0097 m/s$^2$. Under the working condition of 12000 rpm, the first-order frequency is 199.63Hz, the amplitude is 0.1161m/s$^2$. The second-order frequency is 399.77 Hz, the amplitude is 0.0458 m/s$^2$. The third-order frequency is 601.16Hz, and the amplitude is 0.0168 m/s$^2$. Under the working condition of 15000 rpm, the first-order frequency is 250.26Hz, the amplitude is 0.1808m/s$^2$. The second-order frequency is 500.34 Hz, the amplitude is 0.0732 m/s$^2$. The third-order frequency is 749.38Hz, and the amplitude is 0.0321 m/s$^2$. Under the working condition of 18000 rpm, the first-order frequency is 282.14Hz, the amplitude is 0.3743m/s$^2$. The second-order frequency is 601.16 Hz, the amplitude is 0.0072 m/s$^2$. The third-order frequency is 939.99Hz, and the amplitude is 0.0483 m/s$^2$. The vibration amplitude of each frequency of the motorized spindle increases with the increase of the number of revolutions. On the one hand, the thermal rise causes the thermal deformation of the internal parts of the motorized spindle, that is, the shape of each part changes, which may affect its own natural frequency. On the other hand, it should be the result of the increase of electromagnetic force after the shape change, no-load vibration is less than 0.40m/s$^2$.

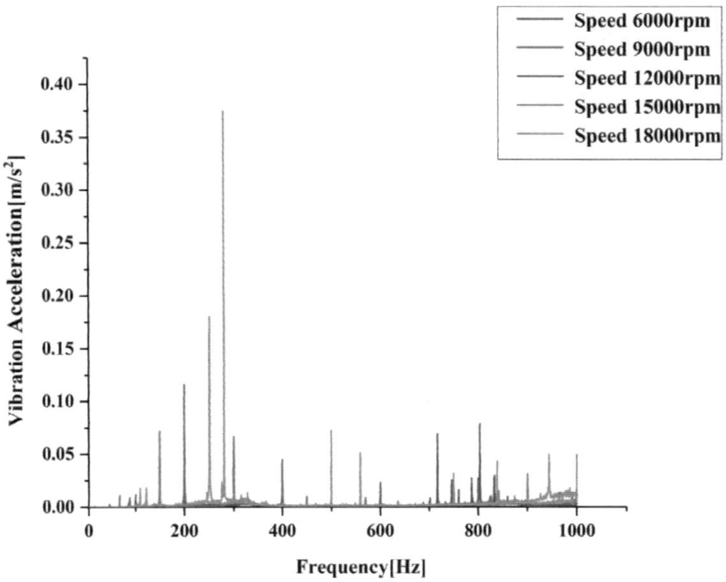


**Figure 9.** Vibration frequency domain of metal bearings.

It can be seen from the experimental data that as the speed increase, the amplitude fluctuation of ceramic bearings is much smaller than that of metal bearings. The amplitude fluctuation corresponding to each order of frequency at each speed is also smaller, and the maximum amplitude of metal bearing without loading is 2.38 times of that of ceramic bearing. It can be seen that ceramic bearings have better dynamic characteristics and stable reliability than metal bearings.

## 5. Conclusion

In this manuscript , through the establishment of all-ceramic motorized spindle bearing test platform, the themal rise of silicon nitride all-ceramic ball bearing was tested, and compared with the same type of metal bearing temperature simulation. The vibration characteristics of the bearings were compared with those of similar type steel bearings. The conclusions are as follows:

- Based on the rotation time of the motorized spindle is longer, the themal rise of the ceramic bearing gradually stabilizes at different speeds, the maximum temperature of ceramic bearing is nearly 10 °C lower than the maximum temperature of the simulated metal bearing, indicating that the high temperature resistance and self-lubricity of the ceramic bearing play a role, so that the performance of the spindle at high speed tends to be stable and the life is extended;
- With the increase of spindle speed, the vibration amplitude of each order frequency of ceramic spindle changes little and tends to be stable. The vibration amplitude of each frequency of the metal motorized spindle increases with the increase of revolution, which indicates that the

ceramic bearing has obvious frequency-time response characteristics, good dynamic characteristics and stable reliability;

- As the rotation speed of the electric spindle increases, the vibration amplitude of the ceramic motorized spindle at each order frequency changes less and tends to stabilize. The vibration amplitude of each frequency of the metal motorized spindle increases with the rotation speed. The maximum amplitude of the metal bearing is 2.38 times that of the ceramic bearing under no load, indicating that the ceramic bearing has better dynamic characteristics and reliability than the metal bearing.

**Acknowledgments**
This work was supported by NSFC Grant Numbers 51975388, Liaoning Natural Science Foundation 2019-MS-266, Shenyang Science and Technology Project RC170216. The Project is also sponsored by "Liaoning BaiQianWan Talents Program".

**References**
[1] Moazami G M, Jenabali S A and Nazarboland A 2009 *Mater Des.* **6** 2283-95
[2] Bai X T, Wu Y H, Zhang K 2017 *Journal of Sound and Vibration* **410** 35-48
[3] Umehara N, Kirtane T, Gerlick R, Jain V K and Komanduri R 2006 *Int. J. Mach. Tool. Manu.* **46** 151-95
[4] Golabczak A and Koziarski T 2005 Assessment method of cutting ability of CBN grinding wheels *Int. J. Mach. Tool. Manu.* **45** 1256-60
[5] Ohta H and Kobayashi K 1996 *Journal of sound and vibration* **192** 481-93
[6] Engel T, Lechler A and Verl A 2016 *CIRP.ANN.* **65** 353-56
[7] Harris T A 2001 *Rolling Bearaig Analysis* (New York: John Wiley & Sons Ina)
[8] Bossmanns B and Tu J F 1999 *Int. J. Mach. Tool. Manu.* **39** 1345-66
[9] Chen X A, Liu J F and He Y 2013 *Chinese Journal of Mechanical Engineering* **11** 135-42
[10] El-Marhomy A A 1999 *Heat and Mass Transfer* **35** 334-44
[11] Wardle F P, Lacey S J and Poon S Y 1983 *Precision Engineering* **5** 175-83
[12] Chen C H, Wang K W and Shin Y C 1994 *Int. J. Vib. Acoust.* **4** 506-22
[13] Bert R J and Yung C S 1998 *J. Manuf. Sci. Eng.* **120** 387-94
[14] Cao Y and Altintas Y 2007 *Int. J. Mach. Tool. Manu.* **47** 1342-50
[15] Altintas Y and Cao Y 2005 *Manufacturing Technology* **54** 379-82
[16] Houpert L 2010 CAGEDYN: a contribution to roller bearing dynamics calculations part II: description of the numerical tool and its output *Tribology Transaction* **1** 10-21
[17] Li H, He H and Liu L 2012 *Bearings* **9** 37-39
[18] Guo X and Wang Y 2013 *Bearings* **4** 30-33
[19] Than V T and Huang J H 2016 *Tribology International* **96** 361-72
[20] Wang W, Hu L, Zhang S *et al.* 2014 *Mechanism and Machine Theory* **82** 154-72
[21] Ma C, Yang J, Zhao L *et al.* 2015 *Applied Thermal Engineering* **86** 251-68
[22] Zhang K, Wang Z *et al.* 2020 *Adv. Mech. Eng.* **12** 1-12
[23] Chen G, Wang L and Gu L 2007 *Journal of Aerospace Power* **1** 163-68

# Fabrication and service of all-ceramic ball bearings for extreme conditions applications

**S Li[1], C Wei[1*] and Y Wang[1]**

[1]Mechanical Engineering College, Shenyang Jianzhu University, Shenyang 110168, China

*E-mail: panda_way@126.com

**Abstract.** Hot isostatic pressing silicon nitride material has properties that most metal materials do not have, such as low specific gravity, high strength, high stiffness, wear resistance, corrosion resistance, extreme temperature resistance, electrical insulation, no magnetic permeability, and excellent dimensional stability Performance, self-lubricating performance, etc. It is one of the ideal alternative materials for the manufacturing of bearings applied to various extreme conditions. This manuscript focuses on the practical needs of bearings for environmental equipment such as extreme temperature and marine corrosion. These bearings are used to improve the ultimate speed and load capacity of the whole steel bearing design system in ultra-high speed applications (such as machine tool spindle and gas turbine engine); to eliminate corrosion and reduce friction under extreme temperature or corrosion conditions, so as to improve system performance. In this paper, from the preparation of ceramic bearing materials to the ultra-precision machining process of ceramic bearing components, the development status of ceramic bearings in recent years is introduced, and the future development and application prospect of ceramic bearings are prospected.

## 1. Introduction

Bearing is the key basic component in the field of industrial machinery, known as "joint of equipment" [1]. The quality of bearing performance directly affects and determines the key indicators of high-end equipment, such as accuracy, life, ultimate speed, bearing capacity, temperature resistance, stability, reliability and dynamic performance. With the development of high-end technology, higher requirements are put forward for the limit of equipment, especially the performance of bearings of rotating parts (such as large-scale cryogenic wind tunnel and liquid rocket turbine pump) under extremely complex working conditions is particularly important. With the diversification of the use environment of bearings, the performance requirements of bearings are also improved [2, 3]. At present, some bearing steel can't meet or fully meet the requirements of the main engine for bearings. Due to the limitation of materials and the lack of high temperature and corrosion resistance, the application limitations of steel bearings in high-speed machine tools are gradually expanded. A series of new strengthening technologies for bearing parts, including ion implantation, enhanced ion-plating, nano-coating, functional gradient coating and laser surface treatment, have not been widely used in bearing research and manufacturing [4, 5]. The first set of ceramic bearings in the world was made by NASA in 1972, major industrial countries in the world are competing to develop a new generation of high-performance ceramic bearings, which have achieved good application results in high-speed, high-temperature, corrosion, insulation, vacuum and other industries [6, 7]. For example, in the spindle unit of high-speed machine tool, the *Dn* value of ceramic bearing is more than 3 million, in the high-temperature field, ceramic bearing can still maintain the material stability under the harsh conditions higher than 1273K, it can work stably in some high corrosive media such as strong acid, strong alkali and salt for a long time. Engineering ceramic materials, including silicon nitride (Si3N4), silicon carbide (SiC), zirconia (ZrO2), alumina (Al2O3), etc., have the excellent characteristics of low density, high strength, high hardness, wear resistance, high temperature resistance, corrosion resistance, good self-lubrication, etc. among them, hot isostatic pressure silicon nitride (HIPSN) is currently considered to be the most suitable for manufacturing high-performance rolling shaft ceramic materials for bearing [8, 9, 10].

## 2. Service behaviour of bearing in extreme temperature environment

Extreme temperature environment includes ultra-high temperature, ultra-low temperature and wide temperature range environment which alternates between high temperature and low temperature. The long-term stable service of bearing in extreme temperature environment is a new research hotspot with the continuous expansion of bearing application range. When the bearing is operated in the environment of 73K~1273K, new requirements are put forward for bearing materials, lubrication, structural parameters and many other aspects [11].

Low temperature and ultra-low temperature environment generally refers to the service environment of bearings in low temperature fluid medium, including liquid hydrogen, liquid oxygen, liquid nitrogen, etc. For example, the internal core bearing parts of large-scale low-temperature wind tunnel and liquid rocket turbine pump built for large-scale aircraft simulation experiment need to operate in the nitrogen atmosphere of ultra-low temperature and wide temperature range (110K-323K) and normal temperature air atmosphere for a long time. In the extreme temperature environment, the complexity of the operation conditions and the high requirements of the service performance of the bearing are far different from the traditional ordinary bearing [12, 13]. The traditional metal bearing cannot work in the ultra-low temperature environment, and the conventional lubricating oil and grease cannot meet the use requirements.

Taking 9Cr18 bearing steel as an example, cold welding occurs between the rolling body and raceway due to the rupture of the metal oxide film, which results in bearing failure. Through the surface modification of bearing materials, such as vapour deposition, ion implantation, plasma spraying and other methods, researchers can obtain a modified layer of bearing rolling surface [14, 15]. Among them, tungsten disulfide (WS2) and molybdenum disulfide (MoS2) are the most important solid anti cold welding lubrication molds in the liquid oxygen environment. They are widely used in the field of low temperature bearing lubrication. The performance comparison is shown in table 1.

**Table 1.** Comparison of basic properties of $WS_2$ and $MoS_2$ films.

| Project | $WS_2$ | $MoS_2$ |
|---|---|---|
| **Colour** | Silver grey | Blue silver grey |
| **Density（kg.m$^{-3}$）** | 7500 | 5060 |
| **Melting point（K）** | 1523-1533 (decompose) | 1458 (decompose) |
| **Thermal stability in air** | Friction factor<0.1(867K) | Friction factor<0.1(587K), Friction factor = 0.1(867) |
| **Operating temperature range（K）** | 0~923 | 88~623 |
| **Coating thickness（μm）** | 0.5-1 | 0.2-2 |
| **Plated material** | Iron, steel, aluminum, copper and other alloys, plastics and synthetic solids, etc. | |

The ideal solid lubricating film can be obtained by modifying the surface of metal materials, but it is difficult to form a wide range of applications because of its complex manufacturing process and high equipment cost. At present, solid lubricating materials have developed from single micro powder, adhesive film, etc. to composite materials with complex components [16].

Polymer based self-lubricating materials have good wear resistance and are used in the design and manufacture of cage at low temperature. Based on this, the researchers start with the solid lubrication of materials and transfer the film produced by the friction between the polymer material cage and the rolling body to the surface of the rolling body to form the polymer transfer film. Under the condition of ultra-low temperature and lack of oil lubrication, they can keep the basic lubrication behaviour of the bearing, such as PTFE, PEEK, NYLON and other materials [17, 18].

High performance HIPSN all ceramic ball bearing has the excellent performance that traditional material bearing does not have, such as wear resistance, high and low temperature resistance, corrosion resistance, high stiffness, good accuracy retention, long service life, electrical insulation, non-magnetic, light weight, good self-lubricating performance. It has a wide application prospect in aerospace, national defence and military industry, energy and chemical industry, equipment manufacturing and other fields. In particular, for applications requiring service in extreme temperature environment (extremely high temperature, extremely low temperature and extremely wide temperature range), all ceramic ball bearing will become an indispensable core and key basic component [19, 20]. For example, in a large ultra-low temperature wind tunnel, the ambient temperature for bearing operation is below 93K, in a liquid rocket engine turbine pump, the ambient temperature for bearing operation in liquid hydrogen or liquid oxygen and other media is below 20K, as shown in figure 1. All kinds of spacecraft bearings used in space must not only bear the environment of large temperature change of 73K~323K, but also face extreme space environment such as vacuum and electromagnetic radiation. Bearings used in aircraft engines and accessory casings are required to work stably in a wide temperature range of 223K~673K, as shown in figure 2. For bearings used in these fields, first of all, it is required that bearing materials have excellent adaptability to extreme temperature environment, which can realize long-term stable operation in extremely low temperature and high temperature without brittle fracture, softening or extreme wide temperature environment without rapid wear, scratch, adhesion, bite and other phenomena.

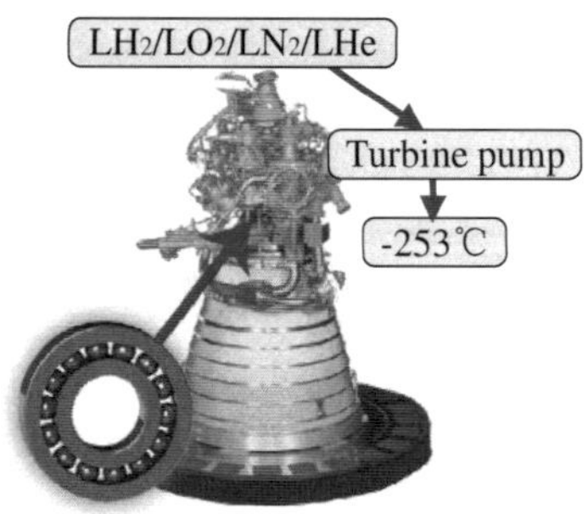


**Figure 1.** Application of low temperature bearing in liquid rocket turbine pump.

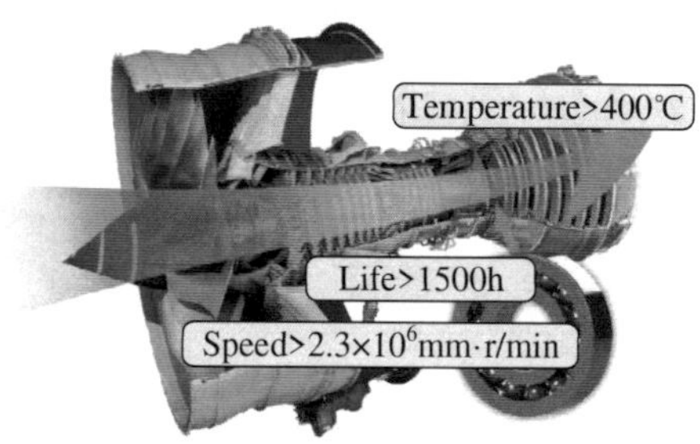


**Figure 2.** Application of high temperature bearing in aero engine.

The traditional bearing steel, even the new steel processed by strengthening the material properties, has already failed [21, 22, 23]. Secondly, conventional oil lubrication or grease lubrication is difficult to apply, new solid self-lubrication or oil-free lubrication will be the main form, which requires the bearing material to have better self-lubricating performance and wear resistance; in addition, it also requires the accuracy retention and long-term operation reliability of the bearing in extreme temperature environment. Because the thermal deformation coefficient of HIPSN ceramic material is only 1/4~1/5 of that of bearing steel, all ceramic ball bearing shows good thermal shock resistance, stable service performance and long service life under the working conditions of high temperature, low temperature and large temperature difference. Moreover, the ceramic material has small density, strong bearing capacity, good wear resistance and low failure rate under high-speed environment, so its working life increases greatly. At present, the limit working temperature of all ceramic bearing has been able to exceed 1273K, and the continuous working time can be more than 100h, and it has self-lubricating characteristics, which can still ensure the working accuracy and service life under the condition of lack of oil lubrication. In conclusion, ceramic bearing may be one of the best choice [24, 25].

## 3. Development and application of ceramic materials in bearing field

Common bearing materials include GCr15, 9Cr18, etc. however, with the requirements for ultra-high and low temperature resistance, corrosion resistance, ultra-high speed, high reliability, etc. in aerospace, marine shipping, military industry and other fields, bearing materials are also developed from traditional metal materials to non-metal materials with better performance. A large number of experimental studies show that the engineering ceramic materials, including silicon nitride (Si3N4), silicon carbide (SiC), zirconia (ZrO2), alumina (Al2O3), etc., have the excellent characteristics of low density, high strength, high hardness, wear resistance, high temperature resistance, corrosion resistance, good self-lubricating as bearing materials. Figure 3 shows the comparison of common performance parameters between Si3N4 ceramic shaft material and common bearing material 9Cr18 [26, 27, 28].

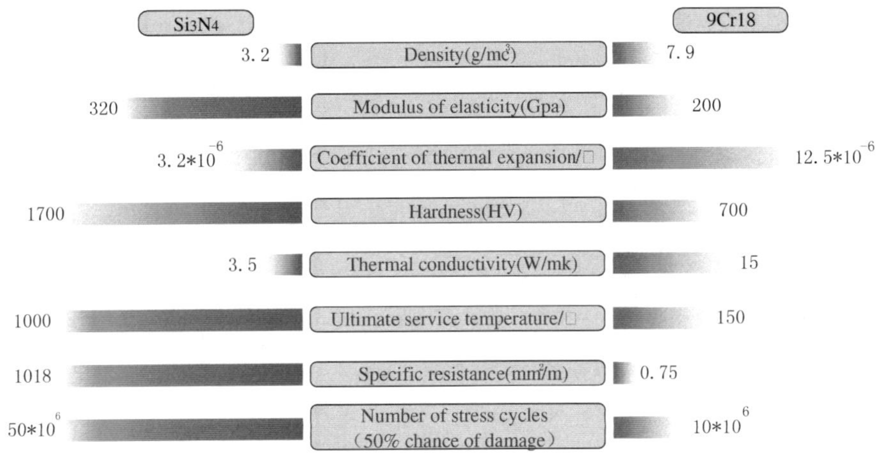


**Figure 3.** Comparison of basic parameters of $Si_3N_4$ and 9Cr18 bearing materials.

The synthesis of silicon nitride (Si3N4) material was first completed by German scientists in 1896 by carbon reduction synthetic method, and it was found that it has the advantages of high hardness, wear resistance, corrosion resistance and low high temperature creep. By the 1950s, silicon nitride has been widely used in the field of high temperature refractories. In the 1960s and 1970s, the application of silicon nitride materials in the field of high temperature has been fully and in-depth studied. Scientists focus on replacing metal materials with silicon nitride materials in higher precision equipment parts in order to achieve better reliability. In this period, ceramic materials began to be applied in the field of bearings. From the 1970s to the 1980s, with the United States, Japan, Germany and other countries as the representatives, research began in the field of hybrid ceramic bearings. In the late 1980s, silicon nitride hybrid ceramic bearings have formed large-scale production and application. Since the 1990s, with the improvement of the requirements of high-end bearings in different fields, and the rapid development of research on extreme working conditions and even intelligent bearings, higher requirements have been put forward for the performance of silicon nitride materials themselves. Through continuous efforts, researchers have carried out in-depth research in the fields of basic unit structure transformation of silicon nitride, toughening mechanism of brittle materials, sintering process and processing technology [29, 30, 31]. The sintering process of silicon nitride materials is complex, because the internal silicon nitride molecules are easily transformed in different sintering processes. A silicon nitride molecule consists of a silicon nitride molecule with four N atoms of a Si nucleus combined in the form of a covalent bond to form a Si-N4, tetrahedral structure unit. Each silicon nitride molecule consists of three Si-N4 tetrahedrons sharing the same N atom plane to form a Si3N4 continuous and solid network structure, as shown in figure 4 [32].

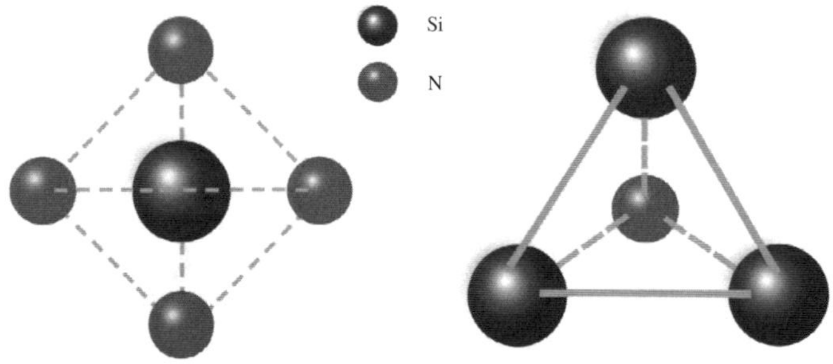


**Figure 4.** Basic structural unit of silicon nitride.

Silicon nitride (Si3N4) has three crystal forms, two crystal forms are common. One is $\alpha$ - Si3N4, acicular crystal, white or gray white, the other is $\beta$ - Si3N4, dark color, dense granular polyhedron or short prism. Both of them are hexagonal. The longer stacking sequence results in higher hardness of α phase than $\beta$ phase. However, compared with $\beta$ phase, α phase is chemically unstable. Therefore, at high temperature of liquid phase, $\alpha$ phase always changes into $\beta$ phase. Therefore, $\beta$ - Si3N4 is the main form used in silicon nitride ceramics [33].

Silicon powder is one of the basic raw materials for the preparation of silicon nitride ceramic materials. The shrinkage of high-quality silicon nitride powder is much smaller than that of ordinary silicon nitride powder, which helps to control the size of precision ceramic parts in the sintering process, and the preparation of high-quality silicon nitride powder has a strict process flow. Reaction sintering silicon nitride process is usually divided into two steps, including silicon powder nitriding and silicon nitride powder densification. Because the reaction temperature of silicon powder nitriding is about 1400℃, close to the melting temperature of silicon powder, the difficulty of silicon powder nitriding is to prevent silicon powder from melting before nitriding. In the nitriding process of silicon powder, in order to avoid the melting of silicon powder, many researchers add different kinds of catalysts to the nitriding reaction to promote the nitriding of silicon powder at low temperature, such as Cr, CA, Cu, Au, Zr and other metal oxides [34].

Silicon nitride materials are easy to crack and break because of its brittleness, so it is difficult to process. As a bearing material, it has limited its further development to a higher precision level. Therefore, in order to improve the processing performance of silicon nitride materials and improve the brittleness defects in the process of processing, through microstructure engineering, from the perspective of improving the toughness of materials, by controlling the growth and transformation of silicon nitride grains, the cracks deflect according to the different shapes and directions of grains, thus forming the mechanism of controlling the direction of cracks.

## 4. Development status of ceramic bearing technology

Ceramic ball bearing has the excellent performance that traditional material bearing does not have, such as wear resistance, high and low temperature resistance, corrosion resistance, high stiffness, good precision retention, long service life, electrical insulation, non-magnetic conduction, light weight, good self-lubricating performance. It has a wide application prospect in the fields of aerospace, national defense and military industry, energy and chemical industry, equipment manufacturing, etc. Full ceramic ball bearing will become an indispensable core and key basic component in the application field of service in extreme temperature environment (extremely high temperature, extremely low temperature and extremely wide temperature range). Although ceramic bearing is a kind of basic bearing spare parts, it is also a kind of particularly important mechanical spare parts in the process of mechanical operation. When compare ceramic bearing with bearing steel, it is not difficult to find that the characteristics of ceramic bearing itself provide a considerable role for the development of ceramic bearing, and at the same time promote the ceramic bearing to mature [35, 36].

Ceramic bearing manufacturing process is mainly divided into raw material proportioning, molding and sintering, component processing, assembly testing and other parts, in which component processing involves ultra-precision processing technology of hard and brittle materials, which is the technical difficulty of the whole bearing manufacturing technology, and also the technical bottleneck limiting the development of ceramic bearing to P4 or P2 level. The ultra-precision processing technology of ceramic bearing components includes the processing and manufacturing of the rolling body and inner and outer rings. There is a big gap between the processing method and the processing difficulty. Therefore, the initial ceramic bearing product is a hybrid ceramic bearing, the rolling body is made of silicon nitride, and the inner and outer rings are made of metal, as shown in figure 5.

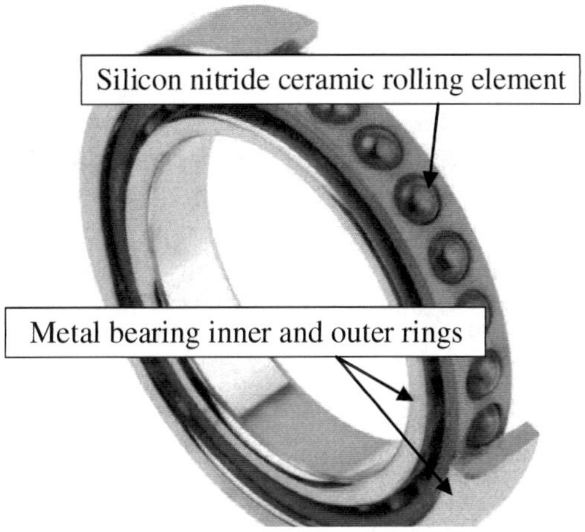


**Figure 5.** Hybrid ceramic ball bearing.

The inner and outer raceways of metal bearings are relatively simple in processing technology compared with silicon nitride materials, with good accuracy retention. With the development of ultra-precision machining technology, researchers design all ceramic bearings. Both the rolling body and the inner and outer ferrules are ceramic materials, and the rolling body is silicon nitride ceramic. The inner and outer ferrules can be made of silicon nitride or zirconia materials, which greatly improves the performance of ceramic bearings, as shown in figure 6 [37, 38].

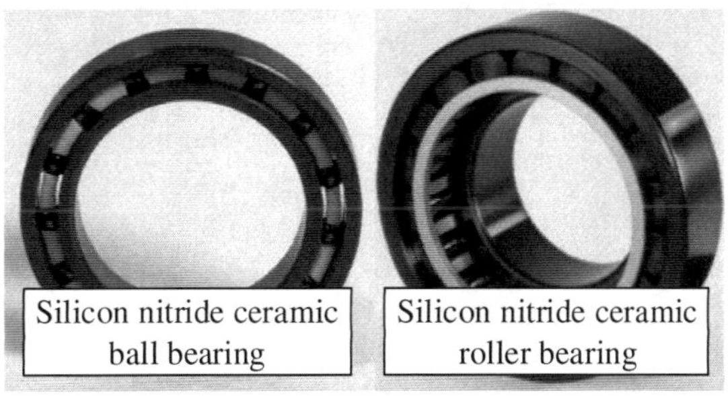


**Figure 6.** All ceramic bearing.

*4.1. Manufacturing technology of silicon nitride rolling element*

The rolling body of ceramic bearing includes ceramic ball and ceramic roller. The processing and manufacturing of ceramic ball has relatively mature theoretical basis and processing methods in the world. The classification of processing equipment mainly includes single disc grinding equipment and multi disc grinding equipment, among which single disc grinding equipment is the most widely used, mainly including traditional V-shaped grinding method and conical grinding method. Compared with the V-shaped grinding method, the rotation angle of the ceramic ball cannot be changed, the conical grinding method increases the rotation angle, the grinding trajectory can achieve the entire envelope of the whole spherical surface, and the efficiency of the conical grinding is improved by 30%, and the grinding quality is improved by 20% [39, 40].

The machining method of ceramic roller is similar to that of metallic roller, and the centreless grinding method is generally adopted. Ceramic roller itself is hard and brittle material, and it is easy to produce micro cracks, scratches and other uniform defects in the process of processing, so the development of ceramic roller bearing lags behind that of ceramic ball bearing [41, 42].

*4.2. Manufacturing technology of silicon nitride ceramic inner and outer rings*

Compared with metallic bearing, the manufacturing of ceramic inner and outer rings is very difficult. At present, diamond wheel grinding is the most commonly used and effective processing method for ceramic bearing rings, accounting for 80% of all processing technologies, and its grinding cost accounts for about 90% of the production cost of the whole bearing. The key technologies involved include high-speed and ultra-high-speed grinding technology, clamping technology, diamond grinding wheel arc repair technology, bearing ring ultra-precision machining accuracy guarantee technology, experimental detection technology, etc. Compared with the traditional metallic bearing, the service life of the ceramic bearing is greatly improved, but the ceramic bearing still has a certain service life. In the service process, there are fatigue failures of rolling elements, inner and outer rings, cages, etc., such as surface wear, surface pitting and peeling off [43, 44]. The main reasons for the failure of the surface integrity of the bearing components are the defects of ceramic materials, the residual stress and micro cracks on the surface of the components after ultra-precision machining. The improvement of service life of ceramic bearing is mainly considered from two aspects: material preparation and component processing.

**Figure 7.** Fatigue failure of ceramic bearing components.

## 5. Development trend and Prospect of ceramic bearing

The development of ceramic bearing technology is closely related to the development needs and application fields of the industry. With the international hot research fields such as aerospace, naval industry and so on, the service behaviour of the bearing is constantly put forward higher use needs, which promotes the continuous progress of the bearing industry. The ceramic bearing officially develops rapidly under such special needs. The excellent performance of ceramic bearings has become an important basic guarantee for the development of high-end equipment [45].

In the future, in the face of extreme working conditions, the development trend of ceramic bearing technology mainly includes:

Self toughened high strength ceramic materials. In the sintering process of silicon nitride ceramic materials, the high purity and particle size distribution of the powder are selected, and the processes of deoxidization and decarbonization are adopted. Hot isostatic pressing sintering is adopted in the sintering process. The growth trend of silicon nitride grain length diameter ratio is controlled by microstructure engineering, the mechanism of surface crack growth is improved, and the controllability of crack growth is mastered.

Design standard for structural parameters of ceramic bearings. The properties and mechanical properties of ceramic materials are quite different from that of bearing steel. From the perspective of bearing structure design, the current bearing design theory is based on metallic materials. The related key technologies, such as bearing contact theory, thermodynamics theory, tribology theory and dynamic characteristics, need to be accurately analysed and designed in combination with the characteristics of ceramic materials.

Service state and long-term performance prediction of ceramic bearing under extreme working conditions. In view of the service requirements of ceramic bearing under extreme working conditions, the effects of environmental temperature, load and speed on the macro and micro morphology, contact lubrication state and friction and wear state of ceramic bearing components, as well as the law of heat accumulation and transfer inside the bearing are studied. Further develop the model and analysis method that combine the ceramic bearing test and simulation experiment research under real working condition, so as to realize the effective prediction of the long-term performance development or failure performance of ceramic bearing under real working condition.

Bearing testing machine for extreme conditions. For the extreme temperature, wide temperature range of ceramic bearing experimental equipment is the technical guarantee for large-scale application and mass production of ceramic bearings. At present, all kinds of bearing testing machines are designed to test bearing capacity, vibration, noise, temperature rise and service life of bearings under normal temperature and pressure. They cannot provide service parameters of bearings under extreme conditions, so they hinder the application of ceramic bearings in high-end equipment.

## 6. Conclusion

With the development of high-end equipment, high-performance ceramic bearing has been widely used in aero engine, high-end CNC machine tool spindle, marine military industry and other fields because of its advantages of light weight, wear resistance, high hardness and high performance. In this manuscript, the development status of ceramic bearings is reviewed, especially the excellent performance of high performance engineering ceramics under extreme working conditions. The necessity and feasibility of ceramic bearing used in liquid turbine pump bearing of rocket engine and large wind tunnel bearing in ultra-low temperature environment are mainly introduced. The present situation of ultra-precision machining technology and common bearing failure forms of engineering ceramic bearing components are summarized, and the development direction of ceramic bearing technology in the future is prospected. Compared with metallic bearing, ceramic bearing has excellent performance in dynamic characteristics, temperature rise, speed, life and other service performance, especially its self-lubricating characteristics can ensure the normal service of equipment in extreme temperature environment, and can greatly improve the application field of high-end equipment. At the same time, it can also promote the progress of engineering ceramic materials and bearing design and manufacturing.

## Acknowledgement

This work is supported by National Natural Science Foundation of China (No. 51975388).

**References**

[1] Wonil K, Jeonkook L and Yong-Bok L 2019 Theoretical and experimental approach to ball bearing frictional characteristics compared with cryogenic friction model and dry friction model *Mechanical Systems and Signal Processing* **124** 424-38
[2] Lin Y, Chen M and Zhong J 2009 Research on metadynamic recrystallization behavior of 42CrMo Steel *Transcation of Materials and Heat Treatement* 71-75
[3] Moazami Goudarzi M, Jenabali Jahromi S A and Nazarboland A 2009 Investigation of characteristics of tin-based white metals as bearing material *Mater. Des.* p 2283
[4] Bai X T, Wu Y H, Zhang K, Chen C Z and Yan H P 2017 Radiation noise of the bearing applied to the ceramic motorized spindle based on the sub-source decomposition method *Journal of Sound and Vibration* **410** 35-48
[5] Guo D M, Luo J B, Fang D N *et al.* 2017 Key Scientific issues on the design and construction of large wind tunnels (in Chinese) *Bull. Natl. Nat. Sci. Found. China* 420–27
[6] Liao D X, Huang Z L, Chen Z H and Tang G S 2014 Review on large-scale cryogenic wind tunnel and key technologies (in Chinese) *J. Exper. Fluid Mech.* **28** 1–6
[7] Gu L, Wang L Q, Zheng D Z *et al.* 2010 Study on the characteristics of transferred lubricating thin films on ceramic surfaces generated in liquid nitrogen *Tribol. Trans.* 161–68
[8] Nosaka M, Kikuchi M, Oike M, *et al.* 1999 Tribo-characteristics of cryogenic hybrid ceramic ball bearings for rocket turbopumps: Bearing wear and transfer film *Tribol. Trans.* 106–15
[9] Subramonian B and Basu B 2006 Development of a high-speed cryogenic tribometer: Design concept and experimental results *Mater. Sci. Eng.* A **415** 72–79
[10] Qiu M, Zhou Z S, Zhou D W, *et al.* 2018 Tribological properties of self-lubricating spherical plain bearings with PTFE/PPS fabric biners (in Chinese) *Tribology* 547–53
[11] Li W, Hu Z Q, Yang Y L, *et al.* 2016 Configuration of aerial self-lubricating spherical plain bearing testers (in Chinese) *China Mech. Eng.* **27** 742–47
[12] Babuska T F, Pitenis A A, Jones M R, *et al.* 2016 Temperature-dependent friction and wear behavior of PTFE and MoS2 *Tribol. Lett.* p 15
[13] Gu L, Wang L Q, Qi Y, *et al.* 2004 Study on the performance of super low temperature and high speed hybrid ceramic bearing *Journal of Harbin University of Technology* p 157
[14] Luo J B and Li J J 2010 Progress and future of Tribology *Lubrication Engineering* 1-12
[15] Ahmad M A, Kasolang S and Dwyer-Joyce R S 2014 Experimental study on the effects of oil groove location on temperature and pressure profiles in journal bearing lubrication *Tribology International* **74** 79-86
[16] Fengrong B, Kang S, Liu C, *et al.* 2015 Analysis of thermoelastohydrodynamic performance of journal misaligned engine main bearings *Chinese Journal of Mechanical Engineering* **28** 511-20
[17] Zhang H 2013 Finite element analysis of the rolling bearing clearance environmental field with ANSYS *AMM* **310** 249—52
[18] Lee J, Kim D H and Lee C M 2015 A study on the thermal characteristics and experiments of high-speed spindle for machine tools *Int. J. Precis. Eng. Man.* **14** 293-99
[19] Nagaraj K A 2016 Gigacycle rolling contact fatigue of bearing steels: A review *Int. J. Fatigue* **93** (part 2) 238-49
[20] Pinel S I, Signer H R and Zaretsky E V 2008 Design and Operating Characteristics of High-Speed, Small-Bore Cylindrical-Roller Bearings *Tribology Transactions* **44** 70-78
[21] Hertz H 1896 *On the Contact of Elastic Solids and on Hardness* Miscellaneous Papers (London: MacMillan)
[22] Farid A B and Moerlooze K D 2008 A model of the transient behavior of tractive rolling contacts *Advances in Tribology* 1-17
[23] Kalker J J 1977 A survey of the mechanics of contact between solids bodies *ZAMM* **57** 3-17
[24] Johnson K L 1985 *Contact Mechanics* (United Kingdom: Cambridge University Press)
[25] Zhang J, Fang B, Hong J and Zhu Y 2017 Effect of preload on ball-raceway contact state and

fatigue life of angular contact ball bearing *Tribiology International* **114** 365-72
[26] De Moerlooze K, Al-Bender F and Van Brussel H 2011 Modeling of the dynamic behavior of systems with rolling elements *Int. J. Nonlin. Mech.* **46** 222-33
[27] Rong Q, Lin T and Shao Y 2008 Finite element simulation of the motion process for deep-groove ball bearing *Journal of Advanced Manufacturing Systems* **7** 9–13
[28] Harsha S P, Sandeep K and Prakash R 2004 Non-linear dynamic behaviors of rolling element bearings due to surface waviness *Journal of Sound and Vibration* **272** 557-80
[29] Feng G, Zhang D, Chen K, Guo Y and Hao T 2017 Research on in-situ microscopic observation of dynamic contact and reciprocating sliding friction of GM-3 lining interface *Tribiology International* **115** 179-90
[30] Bakoglidis K D, Nedelcu I, Schmidt S, Greczynski G, Ehret P and Hultman L 2016 Rolling contact fatigue of bearing components coated with carbon nitride thin films *Tribiology International* **98** 100-07
[31] Khader I, Renz A and Kailer A 2017 A wear model for silicon nitride in dry sliding contact against a nickel-base alloy *Wear* **376-377** part A 376–77
[32] Cranmer D C 1985 Friction and wear properties of monolithic silicon-based ceramics *Journal of Materials Science* 2029-37
[33] Hu S, Brunetiere N, Huang W, Liu X and Wang Y 2017 The bi-Gaussian theory to understand sliding wear and friction *Tribiology International* **114** 186-91
[34] Wang W, Wen W, He N and Chen W 2018 Effect of load on tribological properties of silicon nitride/steel under rolling-sliding contact condition *Tribiology International* **125** 27-38
[35] El-Thalji I and Jantunen E 2015 Dynamic modelling of wear evolution in rolling bearings *Tribiology International* **84** 90-99
[36] Nazir M H, Khan Z A and Saeed A 2008 Experimental analysis and modelling of c-crack propagation in silicon nitride ball bearing element under rolling contact fatigue *Tribiology International* **126** 386-401
[37] Xu J, Luo J B, Liu S H, Xie G X and Ma L 2008 Tribological characteristics of aloe mucilage *Tribology - Materials, Surfaces & Interfaces* **2** 72-76
[38] Harris T A and Kotzalas M N 2006 *Rolling Bearing Analysis—Advanced Concepts of Bearing Technology* 275-83
[39] Bizarre L, Nonato F and Cavalca K L 2018 Formulation of five degrees of freedom ball bearing model accounting for the nonlinear stiffness and damping of elastohydrodynamic point contacts *Mechanism and Machine Theory* **124** 179-96
[40] Sopanen J and Mikkola A 2003 Dynamic model of a deep-groove ball bearing including localized and distributed defects Part 1: theory *Proceedings of the Institution of Mechanical Engineers, Part K: Journal of Multi-body Dynamics* **217** 201-11
[41] Sopanen J and Mikkola A 2003 Dynamic model of a deep-groove ball bearing including localized and distributed defects. Part 2: implementation and results *Proceedings of the Institution of Mechanical Engineers, Part K: Journal of Multi-body Dynamics* **217** 213-23
[42] Kerst S, Shyrokau B and Holweg E 2018 A semi-analytical bearing model considering outer race flexibility for model based bearing load monitoring Mechanical Systems and Signal Processing **104** 384-97
[43] Jang G H and Jeong S W 2003 Stability analysis of a rotating system due to the effect of ball bearing waviness *Journal of Tribology* **125** 91-101
[44] Jang G H and Jeong S W 2002 Nonlinear excitation model of ball bearing waviness in a rigid rotor supported by two or more ball bearings considering five degrees of freedom *Journal of Tribology* **124** 82-90
[45] Hwang Y K and Lee C M 2010 Development of a newly structured variable preload control device for a spindle rolling bearing by using an electromagnet *Int. J. Mach. Tool. Manu.* **50** 253-59

# Overall stability analysis of combined boom with telescopic main boom and fixed flying jib

**S M Liu[1], W L Qin[1], X Q Han[2] and L X Meng[1*]**

[1]School of Mechanical Engineering, Shenyang Jianzhu University, Shengyang, 110168, Liaoning, China
[2]Liaoning Academy of Safety Science, Shengyang, 110004, Liaoning, China

*E-mail: menglixia1983@163.com

**Abstract.** On the purpose of accurately analyze the overall stability of the combined boom, and it also provides calculation basis for the structural design of the crane combined boom and the revision of the Chinese *Design Rules for Cranes* GB3811-2008. In order to analyse the overall stability of the combined boom accurately, the deflection differential equations of combined boom is established based on the vertical and horizontal bending theory. What's more, combining the constraint and displacement boundary condition, the precise recursive expression of the buckling characteristic equations of the combined boom is derived. Taking the combined boom of 6-section telescopic main boom and fixed flying jib as an example, the proposed method and the ANSYS simulation method are used to calculate the critical bucking force of the combined boom, and the comparison results show that: the deduced instability characteristic equation of the combined boom is correct and accurate, which is necessary to the designing analysis in engineering.

## 1. Introduction

On the purpose of further improving the lifting height of the truck crane to meet the needs of high-rise and super high-rise building construction, the combined boom with telescopic main boom and fixed flying jib is used widely, as shown in figure 1. In order to effectively reduce the weight of the boom, its fixed flying jib is a variable cross-section lattice beam welded by many members. As the main load-bearing structure of the crane, stability analysis of the jib is one of the key issues in design calculation, and many researchers have conducted in-depth research [1].

In the past, the approximation method represented by the energy method is usually used to analyse the stability of the telescopic boom [4], and its accuracy is affected by the assumed deformation curve. The precise stability analysis of the telescopic boom mainly applies the differential equation method [5, 6], the instability characteristic equation established by this method can calculate the instability critical load of the structure accurately. The theoretical analysis methods for the stability of lattice members principally include equivalent length method, equivalent inertia moment method, differential equation method and finite element method, etc. The equivalent length method [8] and the equivalent moment of inertia method [9, 10] are the methods for the stability analysis of lattice members, both of these methods are equivalent lattice members into solid web members, and the analysis results have certain errors. *Standard for Design of Steel Structures* [11] and *Design Rules for Cranes* [12] in China, lattice members are equivalent to solid-web members. In [13], a tapered cantilever is equivalent to solid-web member of inertia moment vary quadratic, then the differential equation method is used to analyse the stability of the variable cross-section lattice structure. At present, the finite element is mainly used to calculate the stability of the complex combined boom structure in actual engineering, and the designer directly uses finite element software for modelling and analysis [14, 15]. However, the use of finite element software to build models is complicated, and each analysis case has little significance for the design and optimization of this type of structure, and lacks of in-depth theoretical analysis. In previous studies, researchers have conducted in-depth studies on the stability of telescopic booms or lattice booms, however, for the boom structure composed of telescopic boom and lattice boom that its precise recursive mathematical model for stability analysis has not yet been formed.

**Figure 1.** Truck crane.

In this paper, the combined boom with a telescopic main boom and a variable cross-section lattice fixed flying jib is used as the research object. The differential equation method is applied to establish deflection differential equations of the combined boom, and the precise recursive expression of the instability characteristic equation is derived. The theoretical derivation and quantitative description of the overall stability analysis of the combined boom structure are given, and providing calculation basis for structural design of truck crane combined boom.

## 2. Instability characteristic equation of combined boom

In the analysis of overall stability of the combined boom with telescopic main boom and fixed flying jib, the telescopic main boom is equivalent to a multi-stepped column, a variable cross-section lattice fixed flying jib is equivalent to a solid-web member with inertia moment vary quadratic [13], the equivalent mechanical model of the combined boom is shown in figure 2.

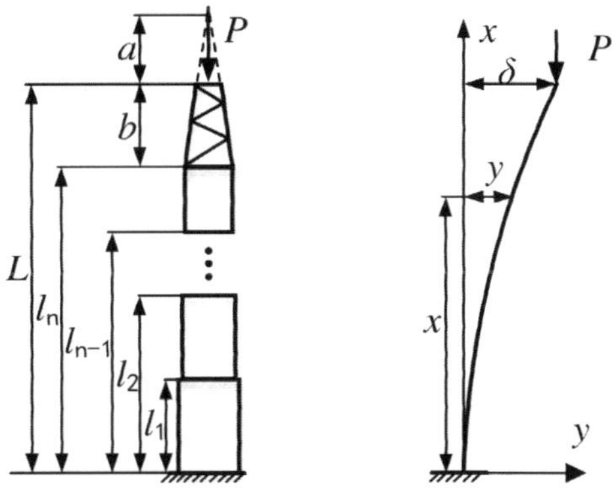


**Figure 2.** Schematic diagram of combined boom.

In figure 2, the total length of the telescopic boom is $l_n$, the length of the fixed jib is $b$, the total length of the combined boom is $L = l_n + b$. Suppose the moment of inertia of each telescopic boom is $I_i (i = 1,2,\ldots n)$, the elastic modulus is $E$, for variable cross-section lattice fixed jib, the moment of inertia of the large end section is $I_{g1}$, the moment of inertia of the small end section is $I_{g2}$, and then the sectional moment of inertia $I_g(x)$ at any section of the fixed jib can be expressed as:

$$I_g(x) = I_{g2}\left(\frac{\bar{L} - x}{a}\right)^2 \tag{1}$$

where, $\bar{L} = L + a$.

Based on the vertical and horizontal bending theory, the deflection differential equation of combined boom can be established as:

$$\begin{cases} EI_1 y_1'' = P(\delta - y_1) & (0 \le x \le l_1) \\ \vdots & \\ EI_n y_n'' = P(\delta - y_n) & (l_{n-1} \le x \le l_n) \\ EI_g(x) y_{n+1}'' = P(\delta - y_{n+1}) & (l_n \le x \le L) \end{cases} \tag{2}$$

Let $k_i = \sqrt{\dfrac{P}{EI_i}}$, the general solution of equation (2) is written as:

$$\begin{cases} y_i(x) = A_i \sin(k_i x) + B_i \cos(k_i x) + \delta & (i = 1,2,\ldots n) \\ y_{n+1}(x) = \sqrt{\dfrac{\bar{L} - x}{a}}\left(A_{n+1} \sin\left(\varepsilon \ln\left(\dfrac{\bar{L} - x}{a}\right)\right) + B_{n+1} \cos\left(\varepsilon \ln\left(\dfrac{\bar{L} - x}{a}\right)\right)\right) + \delta & \end{cases} \tag{3}$$

In equation (3), $\varepsilon$ is the axial force coefficient excluding the effect of shear deformation,

$$\varepsilon = \sqrt{\frac{Pa^2}{EI_{g2}} - \frac{1}{4}} \tag{4}$$

Applying boundary conditions at $x = 0,\ y = 0,\ y' = 0$, the integral constant is getted,

$$\begin{cases} A_1 = 0 \\ B_1 = -\delta \end{cases} \tag{5}$$

According to the displacement boundary conditions of each boom section of the telescopic boom at $x = l_i$, $y_i = y_{i+1}, y_i' = y_{i+1}'$, $(i = 1,2,\ldots n-1)$, the relationship between the integral constants can be obtained.

$$\begin{aligned} A_i \sin(k_i l_i) + B_i \cos(k_i l_i) &= A_{i+1} \sin(k_{i+1} l_i) + B_{i+1} \cos(k_{i+1} l_i) \\ A_i k_i \cos(k_i l_i) - B_i \sin(k_i l_i) &= A_{i+1} k_{i+1} \cos(k_{i+1} l_i) - B_{i+1} \sin(k_{i+1} l_i) \end{aligned} \tag{6}$$

Equation (6) is expressed as:

$$\begin{Bmatrix} A_{i+1} \\ B_{i+1} \end{Bmatrix} = \boldsymbol{Q}_i^{-1} \boldsymbol{U}_i \begin{Bmatrix} A_i \\ B_i \end{Bmatrix} \tag{7}$$

where

$$\boldsymbol{U}_i = \begin{bmatrix} \sin(k_i l_i) & \cos(k_i l_i) \\ k_i \cos(k_i l_i) & -k_i \sin(k_i l_i) \end{bmatrix} \tag{8}$$

$$\boldsymbol{Q}_i = \begin{bmatrix} \sin(k_{i+1}l_i) & \cos(k_{i+1}l_i) \\ k_{i+1}\cos(k_{i+1}l_i) & -k_{i+1}\sin(k_{i+1}l_i) \end{bmatrix} \tag{9}$$

Let

$$\boldsymbol{T}_i = \boldsymbol{Q}_i^{-1}\boldsymbol{U}_i \tag{10}$$

The recursive expressions of coefficients $A_n$ and $B_n$ can be derived.

$$\begin{Bmatrix} A_n \\ B_n \end{Bmatrix} = \boldsymbol{T}_{n-1}\boldsymbol{T}_{n-2}\cdots\boldsymbol{T}_1\begin{Bmatrix} A_1 \\ B_1 \end{Bmatrix} = \prod_{i=n-1}^{1}\boldsymbol{T}_i\begin{Bmatrix} A_1 \\ B_1 \end{Bmatrix} \tag{11}$$

According to the deformation compatibility condition between the telescopic main boom and the fixed jib, when $x = l_n$, $y_n = y_{n+1}, y'_n = y'_{n+1}$, then

$$\begin{cases} A_n\sin(k_nl_n) + B_n\cos(k_nl_n) = \sqrt{\lambda}\left(A_{n+1}s + B_{n+1}c\right) \\ A_nk_n\cos(k_nl_n) - B_nk_n\sin(k_nl_n) = -\dfrac{A_{n+1}}{a\sqrt{\lambda}}\left(\dfrac{s}{2}+\varepsilon c\right) - \dfrac{B_{n+1}}{a\sqrt{\lambda}}\left(\dfrac{c}{2}-\varepsilon s\right) \end{cases} \tag{12}$$

In equation (12),

$$\begin{cases} s = \sin(\varepsilon\ln\lambda) \\ c = \cos(\varepsilon\ln\lambda) \end{cases} \tag{13}$$

Where $\lambda$ is the taper coefficient of the variable cross-section beam, $\lambda = (a+b)/a = \sqrt{I_{g1}/I_{g2}}$.

Expressing the equation (12) in matrix form, then

$$\begin{Bmatrix} A_{n+1} \\ B_{n+1} \end{Bmatrix} = \boldsymbol{D}^{-1}\boldsymbol{U}_n\begin{Bmatrix} A_n \\ B_n \end{Bmatrix} \tag{14}$$

where

$$\boldsymbol{U}_n = \begin{bmatrix} \sin(k_nl_n) & \cos(k_nl_n) \\ k_n\cos(k_nl_n) & -k_n\sin(k_nl_n) \end{bmatrix} \tag{15}$$

$$\boldsymbol{D} = \begin{bmatrix} \sqrt{\lambda}s & \sqrt{\lambda}c \\ -\dfrac{1}{a\sqrt{\lambda}}\left(\dfrac{s}{2}+\varepsilon c\right) & -\dfrac{1}{a\sqrt{\lambda}}\left(\dfrac{c}{2}-\varepsilon s\right) \end{bmatrix} \tag{16}$$

Substituting (11) into (14) gives

$$\begin{Bmatrix} A_{n+1} \\ B_{n+1} \end{Bmatrix} = \boldsymbol{D}^{-1}\boldsymbol{U}_n\prod_{i=n-1}^{1}T_i\begin{Bmatrix} A_1 \\ B_1 \end{Bmatrix} \tag{17}$$

By boundary conditions at the top of the combined: $x = L$, $y_{n+1} = \delta$, the following equation is getted:

$$\begin{Bmatrix} 0 & 1 \end{Bmatrix}\begin{Bmatrix} A_{n+1} \\ B_{n+1} \end{Bmatrix} = 0 \tag{18}$$

By substituting (17) into (18), the recursive expression of instability characteristic equation of combined boom is obtained:

$$B_{n+1} = \begin{Bmatrix} 0 & 1 \end{Bmatrix}\boldsymbol{D}^{-1}\boldsymbol{U}_n\prod_{i=n-1}^{1}\boldsymbol{T}_i\begin{Bmatrix} A_1 \\ B_1 \end{Bmatrix} = 0 \tag{19}$$

Solving equation (19), the critical bucking force $P_{cr}$ is obtained when the combined boom structure is instability. In *Design Rules for Cranes* [12], the cross-section inertia moment of the first boom is usually used to represent the critical instability of the boom, so

$$P_{cr} = \frac{\pi^2 EI_1}{(\mu L)^2} \tag{20}$$

To facilitate analysis and comparison, the dimensionless effective length coefficient $\mu$ is often used to express the Euler's critical force of the structure. From equation (20), the effective length coefficient of the combined boom structure can be expressed as:

$$\mu = \frac{\pi}{L}\sqrt{\frac{EI_1}{P_{cr}}} \tag{21}$$

## 3. Numerical examples

Taking the combined boom of 6-section telescopic boom and fixed flying jib as an example, as shown in figure 3. Suppose the elastic modulus of the combined boom is $E = 2.05e11$, the moment of inertia of the telescopic boom's first section is $I_1 = 43.26\times10^{-3}\,\mathrm{m}^4$, the relationship of moment of inertia of each section boom are $\beta_2 = I_1 / I_2$, $\beta_3 = I_2 / I_3$, $\beta_4 = I_3 / I_4$, $\beta_5 = I_4 / I_5$, $\beta_6 = I_5 / I_6$, the total length of the combined boom is $L = 66\,\mathrm{m}$, the total length of telescopic boom is $l_6 = 60\,\mathrm{m}$, the length of each section boom is $l_1 = 0.2l_6, l_2 = 0.4l_6, l_3 = 0.6l_6, l_4 = 0.8l_6, l_5 = 0.8l_6$; the length of the fixed flying jib is $b = 0.1l_6$, and the moment of inertia of large end section is $I_{g1} = I_6$, the moment of inertia of small end section is $I_{g2}=I_{g1} / \lambda^2$, the moment of inertia of fixed jib at any cross section is $I_g(x) = I_{g2}(\frac{\bar{L}-x}{a})^2$.

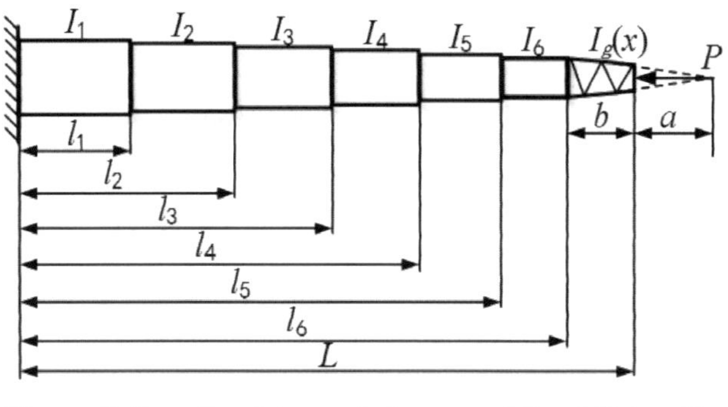


**Figure 3**. 6-section telescopic boom and fixed flying jib.

To verify the correctness of the expression results of the instability characteristic equation deduced in this paper, the simulation calculation results of the finite element software ANSYS are compared with the calculation results of equation (19). In ANSYS simulation, the combined boom is equivalent to a cantilever beam model, the telescopic main boom is equivalent to a multi-stage ladder column, the variable section lattice fixed flying jib is equivalent to a solid web structure with the moment of inertia changing twice along the cross-section, then the Beam44 beam element is used to simulate each boom section of the telescopic main boom and the fixed flying jib, where the section moment of inertia of the fixed flying jib changes twice along the section, but the section area remains unchanged, as shown in figure 4. In ANSYS, to ensure the accuracy of analysis, each boom section of the telescopic boom is divided into 10 units, and the fixed jib is divided into 20 units.

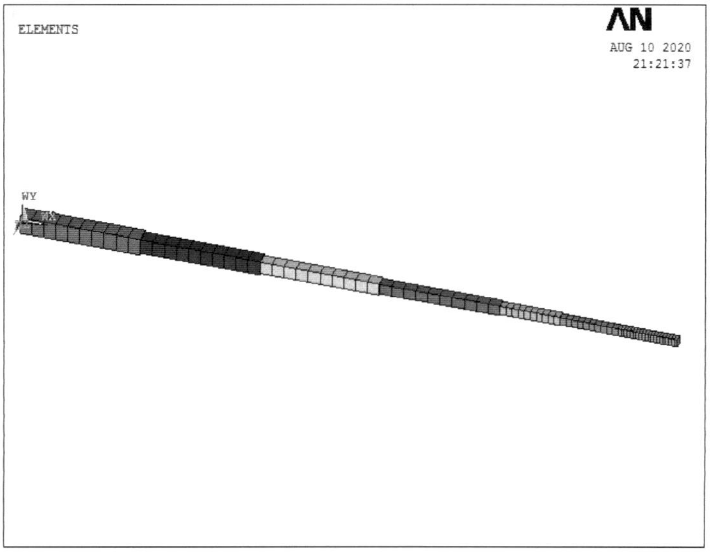


**Figure 4.** ANSYS simulation model.

In table 1, the instability critical force $P_{cr}$ of the combined boom is calculated under different structural parameters $\beta_i$ and $\lambda$ , where $\beta_i = \boldsymbol{I}_{i-1} / \boldsymbol{I}_i$ is the ratio of the moment of inertia of the adjacent boom section of the telescopic main boom, and $\lambda$ is the taper coefficient of the fixed flying jib; in order to facilitate the comparison of results, the dimensionless calculation length coefficient $\mu$ in equation (21) is used to express equation (19) and ANSYS simulation calculation results.

**Table 1.** Effective length coefficients under different sectional inertia moments ratio.

| $\beta_2 / \beta_3 / \beta_4 / \beta_5 / \beta_6$ | $\lambda$ | Simulation($\mu$) | Equation($\mu$) | Error% |
|---|---|---|---|---|
| 1.3/1.3/1.3/1.3/1.3 | 2 | 2.4098 | 2.4100 | 0.008 |
| | 4 | 2.4117 | 2.4119 | 0.008 |
| | 6 | 2.4128 | 2.4130 | 0.008 |
| | 8 | 2.4135 | 2.4137 | 0.008 |
| 1.6/1.6/1.6/1.6/1.6 | 2 | 2.9510 | 2.9512 | 0.007 |
| | 4 | 2.9569 | 2.9571 | 0.007 |
| | 6 | 2.9603 | 2.9605 | 0.007 |
| | 8 | 2.9626 | 2.9628 | 0.007 |
| 1.9/1.9/1.9/1.9/1.9 | 2 | 3.6525 | 3.6528 | 0.008 |
| | 4 | 3.6672 | 3.6674 | 0.005 |
| | 6 | 3.6758 | 3.6760 | 0.005 |
| | 8 | 3.6818 | 3.6819 | 0.003 |
| 2.5/2.5/2.5/2.5/2.5 | 2 | 5.6325 | 5.6328 | 0.005 |
| | 4 | 5.6871 | 5.6873 | 0.004 |
| | 6 | 5.7202 | 5.7202 | 0.000 |
| | 8 | 5.7437 | 5.7434 | -0.005 |

According to the analysis results of the length coefficients of combined boom in table 1, the maximum error between the solution in this paper and the simulation solution of ANSYS is 0.008%, which verifies the correctness and accuracy of the instability characteristic equation derived in this paper.

## 4. Conclusion

In this paper, the differential equation method is applied to study the overall stability of the combined boom with telescopic main boom and lattice fixed jib, the recursive expression of the instability characteristic equation of the combined boom is established, furthermore, the numerical solution of the recursive expression is compared with the ANSYS simulation solution. The results show that, the maximum error between the calculation results of the instability characteristic equation of the combined boom derived and the ANSYS simulation calculation results is 0.008%, it is proved that it is accurate and reasonable to calculate the instability critical force of the combined boom by applying the derived theoretical formula.It provides a new solution for the design calculation of the combined boom with a telescopic main boom and a lattice fixed jib, and provides theoretical data support for the Chinese crane design code *GB/T3811*, which has certain practical application value.

## References

[1] Fenglin Y, Wenjun M, Jie Z, Zhanjiao S, Guoshan S and Haibo L 2018 *J. of Mechanical Science and Technology* **32** (8) 3637-44
[2] Liming Z, Jinghao T, Weijie W, Erfei Z, Shuhui R, Qixun Z and Peng L 2019 *J. Advances in Mechanical Engineering* **11** (11) 1-12
[3] Li Q S 2001 *J. Journal of Constructional Steel Research* **57** (7) 753-77
[4] Timoshenko S P and Gere J M 1961 *Theory of Elastic Stability*(New York: Mc Graw-Hill) pp 49-63
[5] Nian-li L and Liang D 2015 *J. Engineering Mechanics* **32** (8) 217-22
[6] Xin W, Huaijun Y, Rixin Z and Hua X 2014 *J. China Mechanical Engineering* **25** 1744-47
[7] Li Q S 2008 *J. Journal of Constructional Steel Research* **64** (3) 367-76
[8] Han L and Nian-li L 2005 *J. Construction Mechanization* **9** 25-28
[9] Nian-li L and Jia W 2008 *J. Construction Machinery* **15** 79-84
[10] Xin W, Jing Z, Ru-min T and Shun-de G 2009 *J. 2nd Int. Symp. on Computational Intelligence and Design* vol 1 pp 213-17
[11] She S and Li-Jun W 2017 *GB50017-2017 Standard for Classification of Steel Structures* (Beijing: China Architecture & Building)
[12] Li W and Ge-ning X 2008 *GB/T3811-2008 Design Rules of Crane* (Beijing: Standards Press of China)
[13] Li-xia M, Nian-li L and Shi-ming L 2013 *J. Construction Machinery* **5** 77-81
[14] Kai W and Shen-ji Z 2005 *J. Machine Design and Research* **21** (5) 88-91
[15] Xin W, Ting-zheng A, Wei-fu L and Xiao-nan J 2015 *J. Machinery Design&Manufacture* **5** 210-14

# The simplified analysis method of nonlinear buckling and displacement of latticed structure

**S M Liu[1], W Yang[1], X Q Han[2] and L X Meng[1*]**

[1]School of Mechanical Engineering, Shenyang Jianzhu University, Shengyang, 110168,Liaoning,China
[2]Liaoning Academy of Safety Science, Shengyang ,110004, Liaoning,China

*E-mail: menglixia1983@163.com

**Abstract.** In order to simplify the modeling process of building complex lattice structures by finite element analysis software, a simplified method for nonlinear buckling and deformation analysis of lattice structures based on equivalent inertia moment method is proposed. By comparison with the displacements, the latticed structure is equivalent to a solid structure, and the expression of the equivalent inertia moment of a typical latticed structure is given. The finite element analysis software ANSYS is used to analyze the buckling problem and the deformation of latticed structure and the equivalent solid structure, furthermore, the load-displacement curves of the actual structure and equivalent model are obtained. The results show that the simplified analysis method of nonlinear buckling and deformation of latticed structure is feasible and can satisfy the needs of engineering design.

## 1. Introduction

Latticed structure is connected by a series of rods, which has the characteristics of light weight, reasonable stress and good wind passing. They are widely used in steel structure buildings, bridges, rocket launching towers, hoisting machinery, and so on. The buckling and nonlinear deformation of complex latticed structure are difficult in design and calculation, so it is necessary to find a simple, practical and effective simplified calculation method.

For structural stability, beams and rods are usually used as the research object to obtain the critical buckling load of structural instability. Xia Chaofan [1] established the relationship between the buckling loads of symmetric and asymmetric compression bars for the high-order modal analysis of the stability of slender compression bars. Humer [2] deduced the buckling load at the bifurcation of the equilibrium path, and used the elliptic integral to solve the post buckling. At present, the buckling and deformation of latticed structure are studied by analytical methods. Usually, latticed structure are abstracted as "beams", and then the deflection differential equations of beams are established for related research. Q.S.Li [3] used the second-order differential equations to give the buckling control equations for non-uniform members under concentrated axial compressive forces.

According to the flexural differential equation method, the overall stability of the crane's telescopic boom was studied by Simao [4]. Guo Guangjian [5] analyzed the stability of one-way compression and bending lattice member with two-limbed, and verified the stability of lattice members under three different cross-section arrangements. What's more, many researchers have adopted equivalent methods to study the buckling and deformation of latticed structure. By keeping the axial dimension of the component unchanged, Lu Nianli [6] equivalent latticed structure to solid structure, and gave the calculation method of its equivalent moment of inertia. [7] used the principle of equivalence to analyze the buckling of the variable cross-section boom structure. China's code for design of steel structures [8] and code for design of cranes [9] also use the idea of equivalence to express the critical load of the actual structure with the reduced slenderness ratio. The research of the above literature is mainly directed to the linear buckling and deformation of latticed structure, and its analysis object is a single member or a simple combined structure. However, For the nonlinear buckling and deformation analysis of complex structures composed of many members, the finite element method can only be used. Sun Boyong [10] conducted a buckling analysis on the lattice steel column, and the most favorable arrangement and applicable height range are obtained in his study. Xian-Rong Q [11] analyzed the tower crane's linear and nonlinear buckling process by finite elements method. Qin Y [12] obtained the maximum working range of crane and stability convergence load through eigenvalue buckling and nonlinear buckling analysis. The finite element method can obtain high-precision analysis results by simplifying the structure less, but the modeling is complicated and should be based on the preliminary design.

In order to simplify the modeling process and improve the analysis efficiency, the equivalent inertia moment method is adopted to equivalent the actual latticed structure into a solid structure, and the load-displacement curve tracking of the equivalent structure is established. Then, the finite element software ANSYS is used to establish the actual latticed structure for nonlinear analysis. Finally, the analysis results of the equivalent solid structure are compared with the analysis results of the actual latticed structure to analyze its nonlinear buckling and deformation, which provides a simple and practical calculation method for the nonlinear buckling and deformation analysis of the latticed structure.

## 2. Equivalent inertia moment method

In view of the stability of lattice structures, many literatures use the equivalent length method to solve the stability of latticed structure, but it is not feasible to perform an overall analysis of the combined structure, [13] uses the energy method to discuss the calculation of the equivalent moment of inertia of latticed structure. Through the internal force analysis of the two-limbed latticed structure, the formula for calculating the lateral displacement is obtained.

$$\Delta_1 = Q\frac{H^3}{3EI_1}\beta \tag{1}$$

where: $H$ is the height of the two-limbed latticed structure(m);

$Q$ is the horizontal force subjected by the latticed structure (N);

E is the elastic modulus of the material (N/m$^2$);

$I_l$ is the two-limbed latticed structure's moment of inertia of the limb to the neutral axis (m$^4$);

$\beta$ is the influence coefficient of the mesh of two-limbed latticed structure.

The mechanical model of the four-limbed latticed structure with root fixed is shown in figure 1. $a$ and $b$ are the distance between two limbs in different planes. When the horizontal force $F$ is applied, its overall lateral displacement $\Delta$ is the same as the lateral displacement $\Delta_l$ of the two-limbed member under the horizontal force of $F/2$. That is to say, $\Delta=\Delta_l$, then:

$$\Delta_1 = \frac{F}{2}\frac{H^3}{3EI_1}\beta \tag{2}$$

For a solid cantilever, let the height as $H$ and the moment of inertia as $I$, then the lateral displacement caused by the horizontal force $F$ at the end is written as:

$$\delta = \frac{H^3}{3EI}F \tag{3}$$

According to the equal displacement of the ends, the four-limbed latticed structure can be equivalent to a solid structure, let the equivalent moment of inertia is denoted as $I_{dx}$, then the displacement of four-limbed latticed structure can be expressed as:

$$\Delta = \frac{H^3}{3EI_{dx}}F \tag{4}$$

where: $I_{dx}$ is the equivalent moment of inertia of the four-limbed latticed structure ($m^4$).

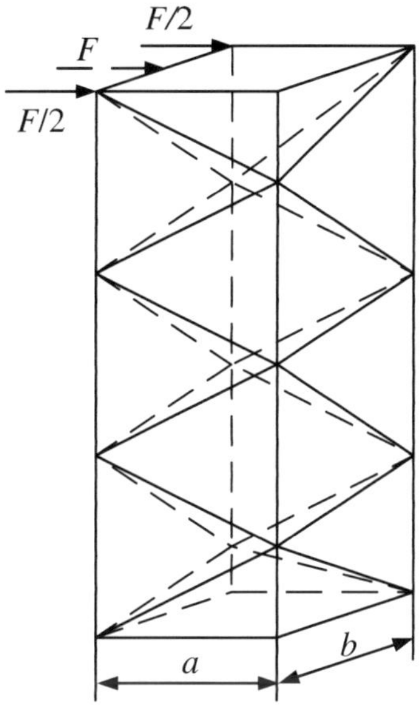


**Figure 1.** The load decomposition model of the four-limbed latticed structure.

According to $\Delta=\Delta_1$, can get

$$F\frac{H^3}{3EI_{dx}} = \frac{F}{2}\frac{H^3}{3EI_1}\beta \tag{5}$$

So, the expression of the equivalent moment of inertia is:

$$I_{dx} = \frac{2I_1}{\beta} = \frac{I_2}{\beta} \tag{6}$$

where: $I_2$ is the moment of inertia of the four limbs of the latticed structure to the neutral axis, $I_2 = 4A(a/2)^2$ or $I_2 = 4A(b/2)^2$.

The unfolded drawing of the four-limbed structure which commonly used in engineering is shown in figure 2. In order to facilitate the calculation of the equivalent moment of inertia of the latticed structure, the influence coefficient β of the four-limbed structure under different mesh layout is given as shown in table1.

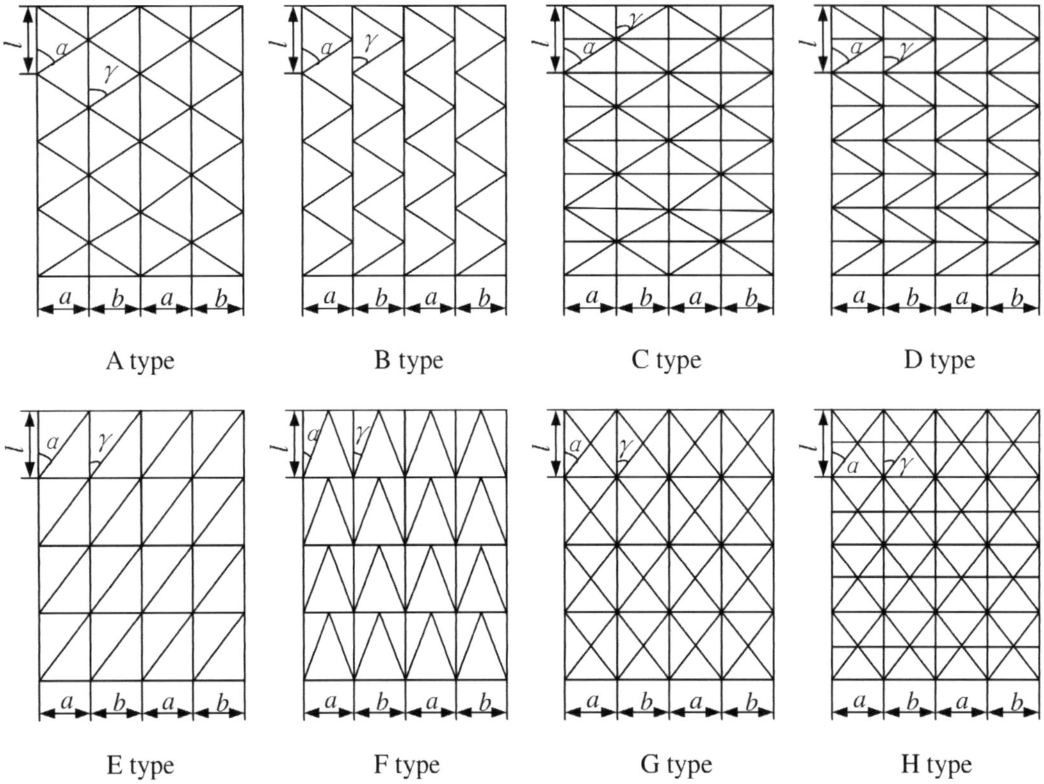


**Figure 2.** Different layout of mesh elements in common.

**Table 1.** Influence coefficient of mesh under horizontal load ($\beta$).

| Types of mesh | A, B | C | D | E, F |
|---|---|---|---|---|
| $\beta$ | $1+\frac{3}{2H^2}\frac{\lambda_2}{\lambda_1}$ | $1+\frac{3}{2H^2}\frac{\lambda_2+\lambda_3}{\lambda_1}$ | $1-\frac{3}{2n}+\frac{3}{4H^2}\frac{\lambda_2+\lambda_3}{\lambda_1}$ | $1+\frac{3}{4H^2}\frac{\lambda_2}{\lambda_1}$ |

where: $\lambda_1=\frac{1}{EA_x a^2}$, $\lambda_2=\frac{1}{EA_f \cos\alpha \sin^2\alpha}$, $\lambda_3=\frac{1}{EA_h}\tan\alpha$, $A_x$, $A_f$, $A_h$ are the cross-sectional area of chord, diagonal mesh, and transverse mesh, respectively.

## 3. Numerical examples

In this paper, an A-type latticed cantilever is used as the finite element model to do the analysis. The limb bars use 60 mm× 60 mm square steel; the meshs use 50 mm × 50 mm square steel; the height of a single layer is $l$ = 500mm; limb span is $a = b$ = 500mm; the total height is h; the elastic modulus of the material is E=2.05×10$^5$MPa, the finite element model of the actual latticed structure uses beam188 elements, and the finite element model of the equivalent solid structure uses beam44 elements. Take the slenderness ratio $\lambda = 2h/\sqrt{I/A}$ = 48, 64, 80, 96, 112, 128 as examples, based on the method of equivalent moment of inertia, the actual latticed structure is equivalent to a solid structure. The simplified schematic diagram of the actual latticed structure is shown in figure 3, and the inertia moment of the equivalent solid structure is shown in table 2.

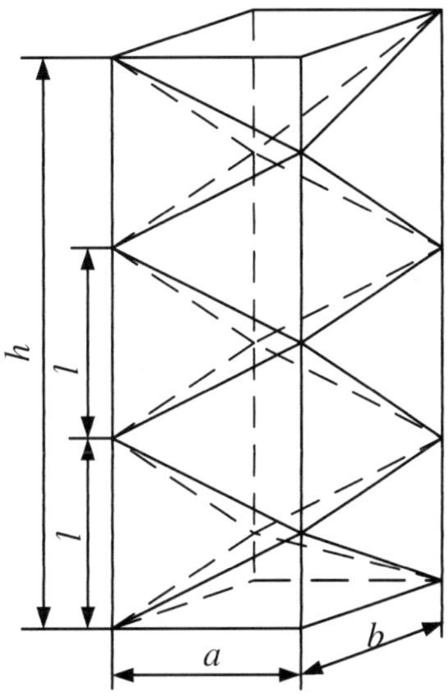


**Figure 3.** Simplified schematic diagram of the latticed structure.

**Table 2.** Equivalent moment of inertia of equivalent solid structure.

| Layer number | Slenderness ratio | $I_{dx}$ ($10^8$mm$^4$) |
|---|---|---|
| 12 | 48 | 8.6381 |
| 16 | 64 | 8.7927 |
| 20 | 80 | 8.8662 |
| 24 | 96 | 8.9067 |
| 28 | 112 | 8.9312 |
| 32 | 128 | 8.9481 |

*3.1.Linear buckling analysis*

Linear buckling is also called the first type of instability. In this type of instability, the critical buckling load $P_{cr}$ of the structure is obtained by solving the equilibrium equation (7),

$$\{[K]+\mu[K(\sigma_0)]\}\{\Delta u\}=0 \tag{7}$$

where: $[K]$ is the linear stiffness matrix;

$\mu$ is the eigenvalue value;

$K(\sigma_0)$ is the geometric stiffness matrix;

$\Delta u$ is the displacement increment matrix.

Buckling analysis was performed on the actual latticed structure and the relative equivalent solid structure, the comparison results of the critical buckling load are shown in figure 4.

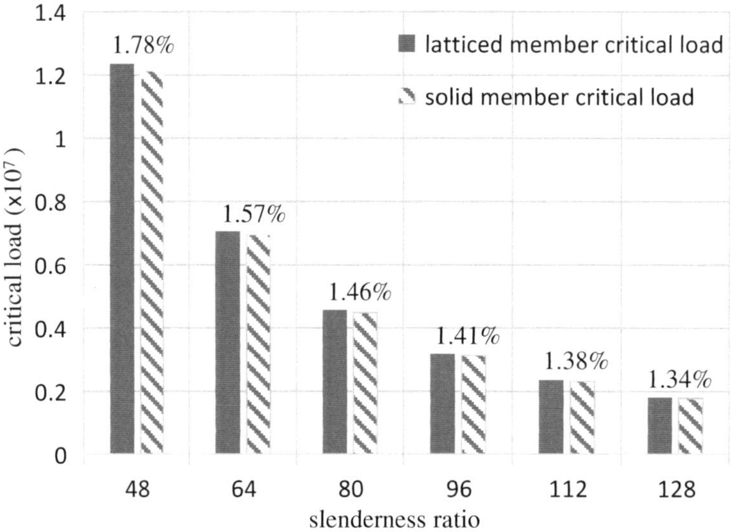


**Figure 4.** The comparison of critical load between latticed structure and the equivalent solid structure.

It can be seen from figure 4 that the errors between the critical load of the actual latticed structure and the equivalent solid structure are within 2%, and the error gradually decreases as the slenderness ratio increases. Therefore, the proposed equivalent analysis method can be used for the analysis of the eigenvalue buckling of the actual latticed structure in engineering.

*3.2.Geometrical nonlinear buckling analysis*

Econd type of instability occurs in most structures in actual engineering. In the second type of instability, the entire load-displacement path can be obtained by solving the nonlinear equation (8),

$$[K]\{u\}=\{P\} \tag{8}$$

where: [K] is the stiffness matrix;
{u} is the displacement matrix;
$\{P\}$ is the external force.

The nonlinear buckling analysis of the actual lattice structure and the equivalent solid mesh structure is carried out, and the arc length method is used to track the whole process of the load displacement curve of the structure. First, The load displacement curves of structures with different slenderness ratios are obtained, and the critical loads of structures are determined by the load deformation curves. Then, the displacement load curves of the equivalent solid structure is compared with that of the actual latticed structure. The load-displacement curves are shown in figure 5.

As can be seen from figure 5, the Y value at point A is the critical load for nonlinear buckling analysis of the actual latticed structure, the Y value at point B is the critical load for nonlinear buckling analysis of the equivalent solid structure, and the Y value at point C is the critical load for linear buckling analysis of the actual latticed structure, and the Y value at point D is the critical load of the linear buckling analysis of the equivalent solid structure. From the above calculation and analysis results, it can be seen that the linear analysis has a higher structural stability bearing capacity than the nonlinear analysis. When the slenderness ratio is the same, the trends of load-displacement curve of the actual latticed structure and the equivalent solid structure are basically the same. The structural deformation gradually increases as the load increases, and the structure maintains the load-bearing capacity before the first inflection point. The comparative analysis results indicate that the simplified analysis method based on the equivalent moment of inertia can be used for the nonlinear buckling analysis of latticed structure.

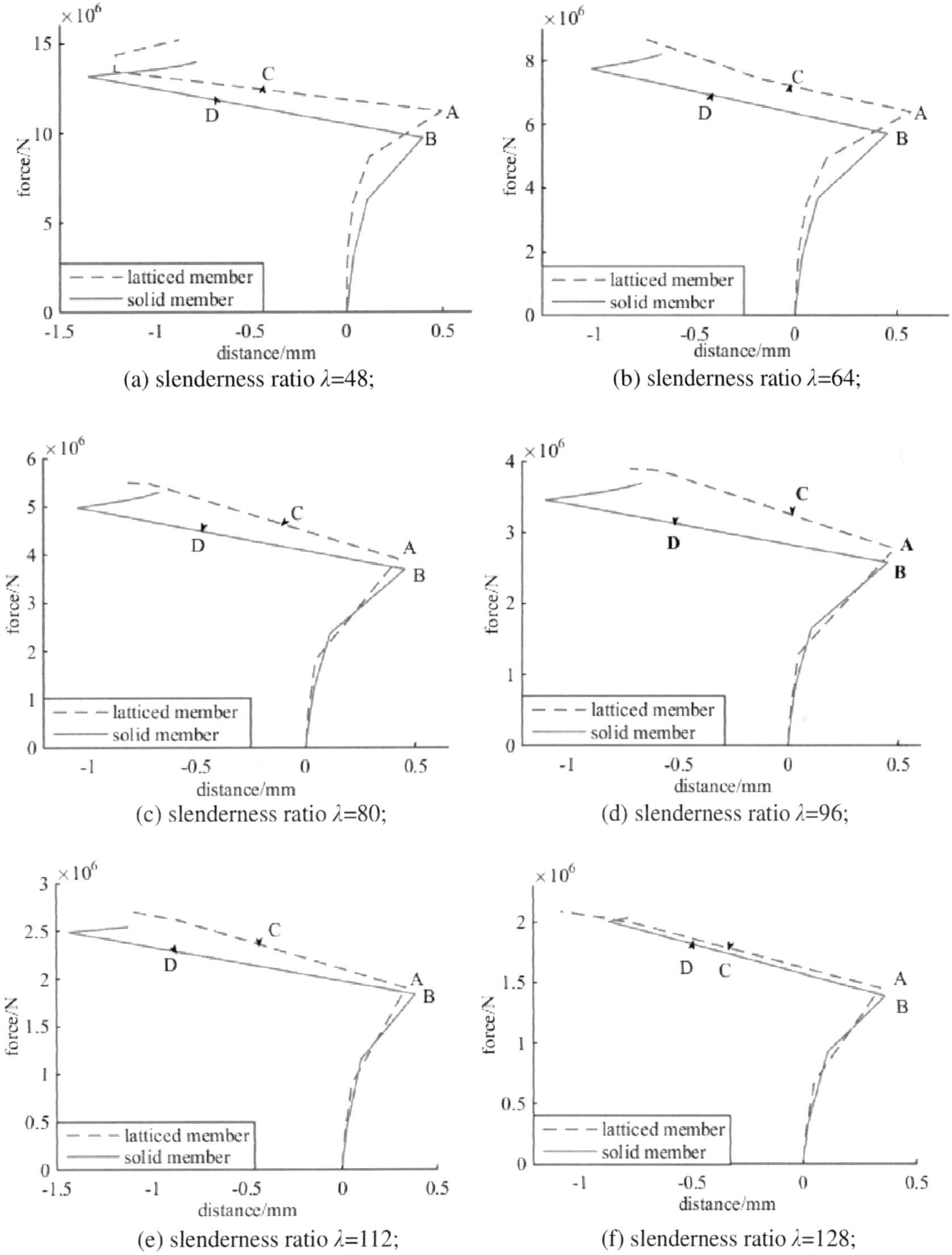


(a) slenderness ratio $\lambda$=48; (b) slenderness ratio $\lambda$=64; (c) slenderness ratio $\lambda$=80; (d) slenderness ratio $\lambda$=96; (e) slenderness ratio $\lambda$=112; (f) slenderness ratio $\lambda$=128;

**Figure 5.** Nonlinear buckling analysis of structural load-displacement curves.

*3.3.Geometrically nonlinear full process analysis*

The whole process of geometric nonlinearity analysis is carried out for the actual latticed structure and the equivalent solid structure. The structure is subjected to both lateral loads Q and axial loads P. The arc length method was used to obtain the load-displacement curves of the structure. Dimensionless coefficients are used in calculation, where $K = Qh^2 / EI$ , $m = P / P_{cr}$ . Figure 6 shows the load-displacement curves of actual latticed structure and equivalent solid latticed structure with different slenderness ratios at $m$ = 0.5, $K$ = 0.3, 0.6, 0.9, 1.2, and 1.5.

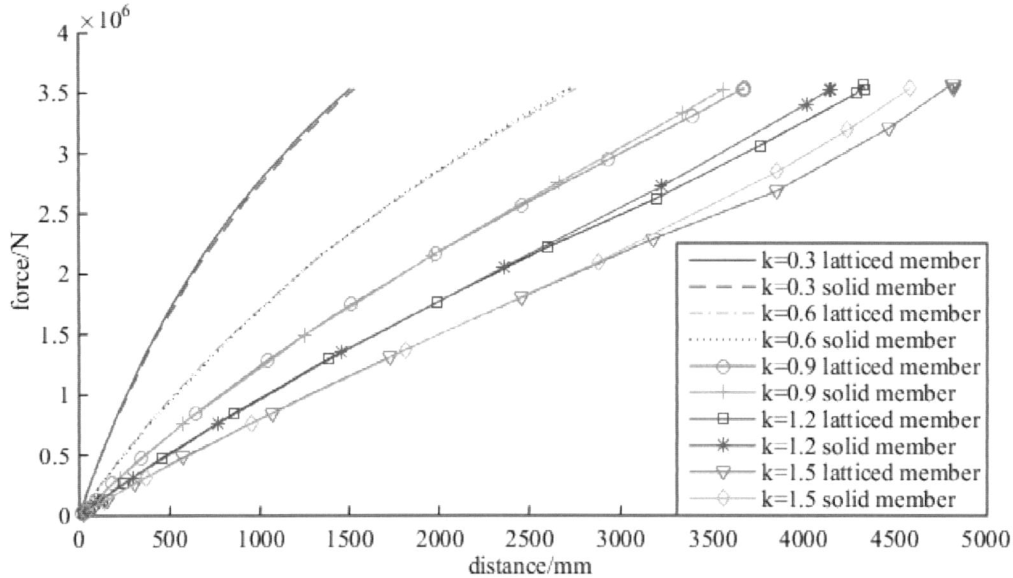


(a) slenderness ratio $\lambda$=64;

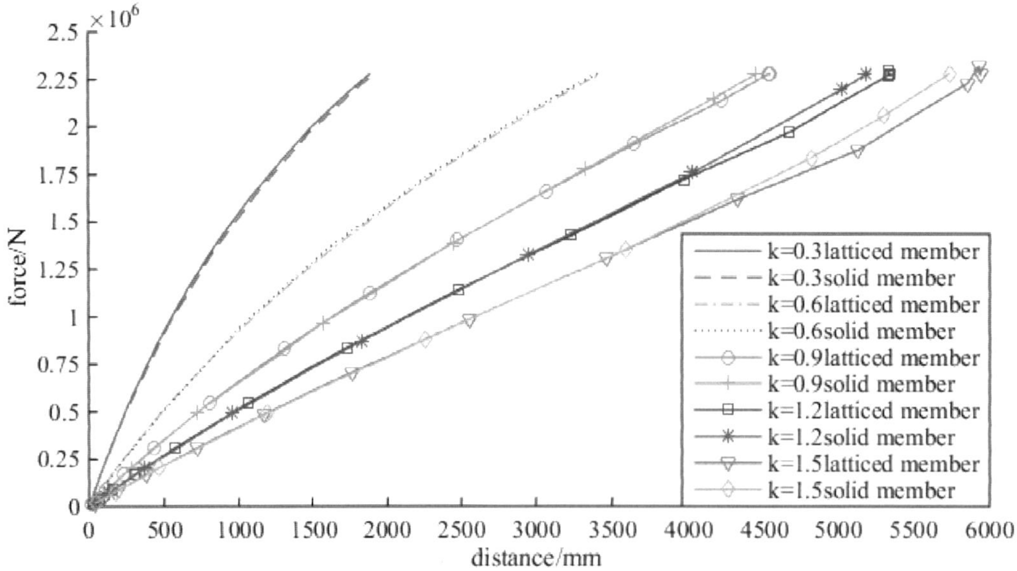


(b) slenderness ratio $\lambda$=80;

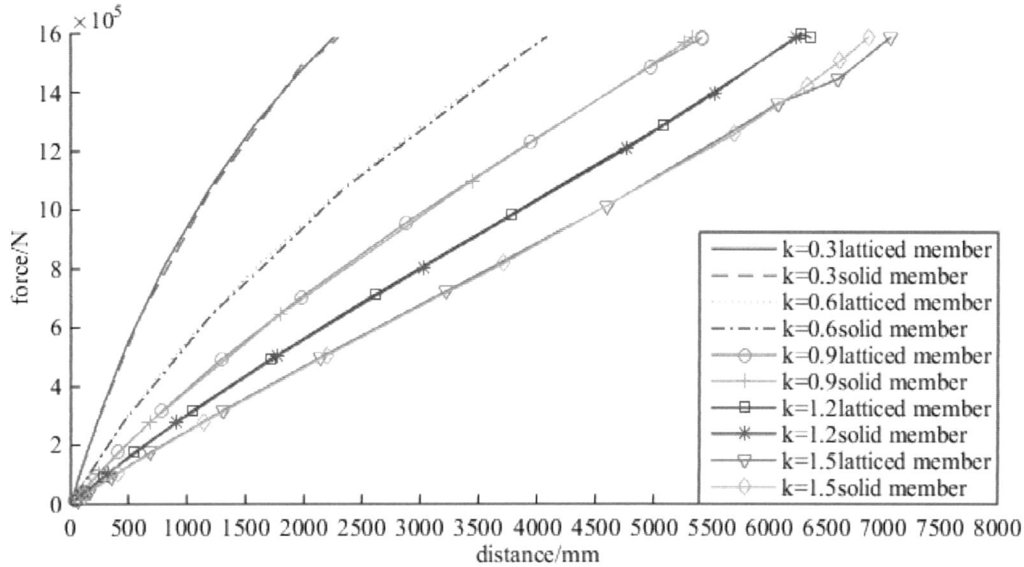


(c) slenderness ratio $\lambda$=96;

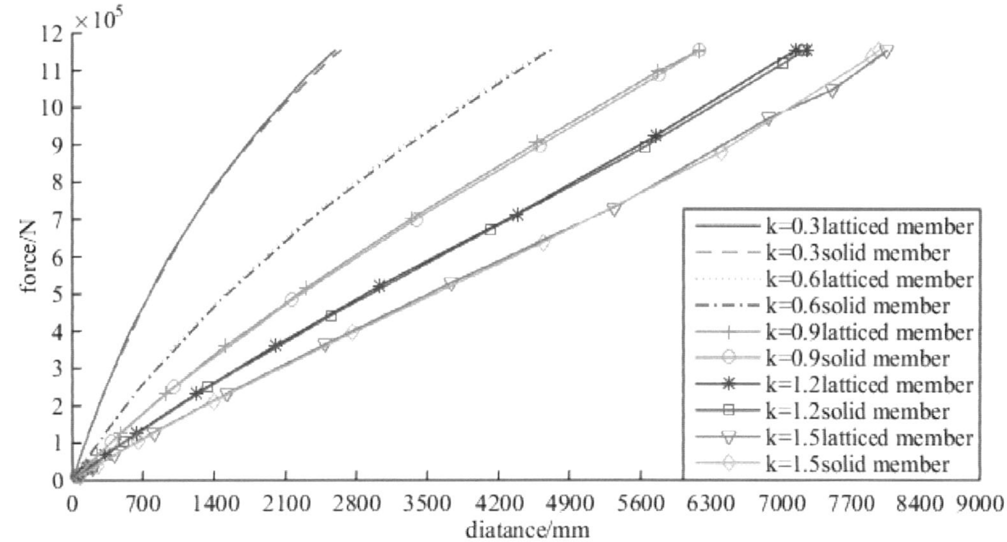


(d) slenderness ratio $\lambda$=112;

**Figure 6.** Nonlinear full-process analysis of structural load-displacement curves.

It can be seen from figure 6 that, when the slenderness ratio $\lambda$ and the coefficient k are unchanged, the trends of load-displacement curve of the actual latticed structure and the equivalent solid latticed structure are basically the same. Therefore, the comparative analysis shows that the proposed equivalent analysis method can also be adopted to analyze the whole process of geometric nonlinearity of the latticed structure in engineering.

## 4. Conclusion

In this paper, in order to simplify the modeling process of building complex lattice structures by finite element analysis software ,a simplified analysis method for nonlinear buckling and deformation of latticed structure based on the equivalent inertia moment method is proposed. The stability calculation of the actual lattice structure is transformed into the equivalent calculation of the solid mesh structure, and the theoretical formula of the equivalent moment of inertia is derived. The load-displacement curve of the actual latticed structure is compared with the load-displacement curve of the equivalent solid structure. The trends of load-displacement curve of the actual latticed structure and the equivalent solid latticed structure are basically the same and the errors between the critical load of the actual latticed structure and the equivalent solid structure are within 2%. The results show that the proposed equivalent analysis method can satisfy the analysis of nonlinear buckling and deformation of lattice structures and can satisfy the needs of engineering design.

**Reference**

[1] Chao-fan X, Yu-chen Z and Qing-yi X 2018 *J. Construction Quality* **36** 83-86
[2] Humer A 2013 *J. Acta Mechanica.* **224** 1493-1525
[3] Li. Q S 2003 *J. Engineering Structures* **25** 1289-99
[4] Simao, Pedro D, Girao C, Ana M, Bijlaard and Frans S K 2012 *J. Engineering Structures* **42** 51-82
[5] Guang-jian G and Jian-chao Z 2012 *J. Anhui Architecture* **19** 155-56
[6] Nian-li L, Jia W and Peng L 2008 *J. Construction Machinery* **15** 79-84
[7] Li-xia M 2008 MA. Eng. thesis Harbin Institute of Technology
[8] She S and Li-Jun W 2017 *Standard for Classification of Steel Structures* (Beijing: China Architecture & Building Press)
[9] Li W and Ge-ning X 2008 *Design Rules of Crane* (Beijing: Standards Press of China)
[10] Bo-yong S, Guang-hua Z and Feng-yuan W 2014 *D. Chongqing Architecture* **13** 50-52
[11] Xian-Rong Q , Florent C and Ying-Hua L 2013 *Chinese Journal of Construction Machinery* **11** 187-89 + 204
[12] Qin Y , Zhang L and Feng Z 2016 *International Journal of Science Technology and Society* **4** 57-62
[13] Jia W 2006 MA. Eng thesis Harbin Institute of Technology

# A brief review of numerical simulation in process machining

**C I Malea[1*], E L Nitu[1] and M D Iordache[1]**

[1]Manufacturing and Industrial Management Department, University of Pitesti, Faculty of Mechanics and Technology, Targul din Vale Street No.1, Pitesti, Romania

[*]E-mail: claudiu.malea@upit.ro

**Abstract**. Machining is a technological process that involves removing the material by generating chips, aiming to obtain flat, cylindrical, helical surfaces etc. This processing method is widespread in the industry and it is, therefore, necessary to define appropriate technical and economic solutions concerning the materials and geometry of the tools, the orientation and fixture of the devices, and the machine-tool used. It is also particularly important to use the most appropriate cutting parameters for modern industrial applications. Scientific research addresses these issues through analytical, semi-empirical, or numerical modeling. Numerical modeling and simulations can satisfy with very good precision the solutions of these aspects, with diverse methods and techniques used in the study of the cutting simulation. This paper presents an analysis of recent studies, briefly presenting the main approaches and techniques used in numerical modeling and simulation of machining. The input parameters, components, and output parameters of these numerical models are identified as well. Finally, the advantages and disadvantages of using these methods and techniques are summarized, as well as the problems to be developed in the area of numerical modeling and simulation of cutting processes.

## 1. Introduction

Cutting processes are still the main industrial process, applicable in all areas of manufacturing. Cutting procedures have a wide range of applications, the most common being turning, milling and drilling, figure 1. Understanding the mechanisms of chip formation continues to be a challenge, being also the main challenge for researchers as well as increasing the productivity of these types of processes and improving tool life.

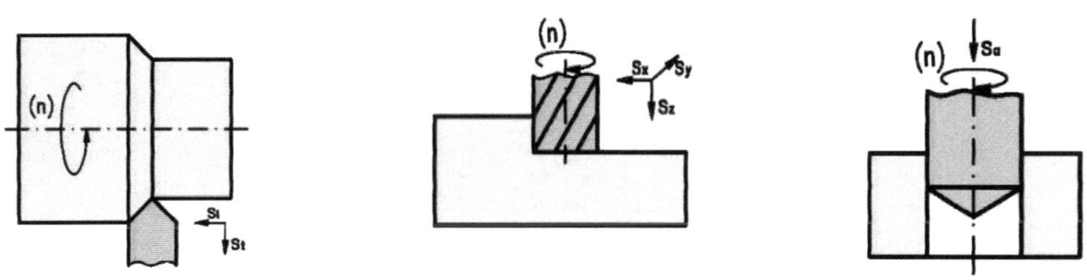


(a) (b) (c)

**Figure 1.** The main cutting processes: (a) turning; (b) milling; (c) drilling.

The objective of using numerical simulations for cutting processes is to generate, in the design phase of the technology, valid results compared to those obtained in real applications, such as cutting force, process temperatures, stresses and strains in the workpiece etc. This makes it possible, on one hand, to understand the phenomena and mechanisms associated with the cutting process and, on the other hand, to establish optimal values for the geometric parameters of the tools and for the parameters of the cutting speed, leading to the achievement of certain goals, such as: maximum productivity, minimum cost, etc.

Modern cutting materials and processes, such as high hardness alloys, high cutting speed machining, processing accompanied by certain ultrasonic techniques or liquid nitrogen cooling etc., support the need to develop appropriate numerical models to reduce the cost of processing and increase the quality of the parts.

In this paper, an analysis of recent studies is presented, briefly outlining the main methods and techniques used in numerical modeling and simulation of machining.

## 2. Specific elements for numerical modeling and simulation of machining

A numerical model describing a cutting process must have the ability to generate a chip of at least the same shape as that obtained experimentally. Therefore, it is necessary to learn the mechanisms of chip removal, being described several types of chips, which are continuous, lamellar, segmented and discontinuous. For most numerical models, continuous chip shapes are considered, with the focus on topical research and their validation by experiments, respectively, being oriented more towards the study of forces and temperatures, the strains and stresses, in the areas of tool- work piece contact. These aspects are due to the fact that current research on the modelling of cutting processes is not yet sufficiently developed to define the morphology corresponding to the formation of the chip, and the correct implementation of the contact conditions.

The chip is formed when the cutting body penetrates the material to be removed, advancing in the cutting direction. In other words, we can say that the chip forming process has three stages: plastic deformation, cutting initiation and forming.

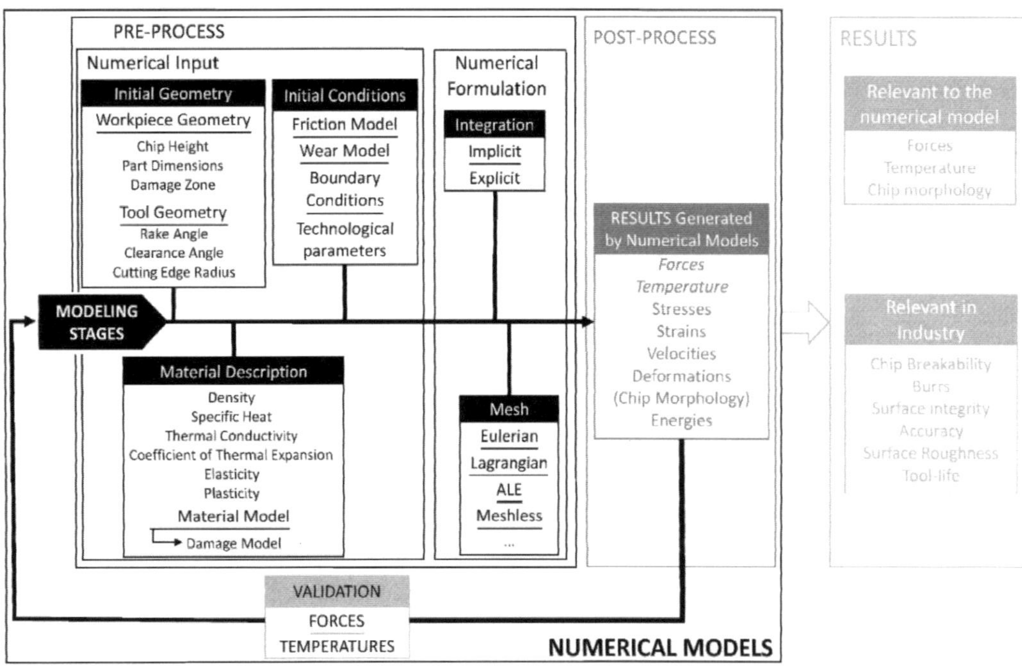


**Figure 2.** Numerical inputs, formulations and results.

To study the chip morphology, forces and temperatures occurring in the cutting process, respectively stresses and strains using numerical models that reproduce the experimental results as closely as possible, it is necessary to implement initial conditions related to geometry, material behavior etc., respectively appropriate approaches and formulations. They are presented schematically in figure 2 and detailed in the following sub-chapters. The relevance of the results obtained with such models is reflected in the correct choice of these parameters. In other words, the model should be updated/ improved until it generates adequate results compared to those obtained experimentally.

### *2.1. Numerical input*
This first stage consists in entering the data regarding the description of the geometries, the behavior of the material to be processed, the contact conditions, the failure criterion, the technological parameters etc.

*Model geometry*
For an orthogonal cutting model, the workpiece is described by: the thickness of the undeformed chip (h), the size of the failure area ($h_f$), where a model with such an area is used [1], as well as the dimensions of the part to be machined (l, L), and the cutting tool must be described by: the rake angle ($\alpha$), the flow angle ($\gamma$), the radius at the tip of the tool ($r_\varepsilon$), as shown in figure 3.

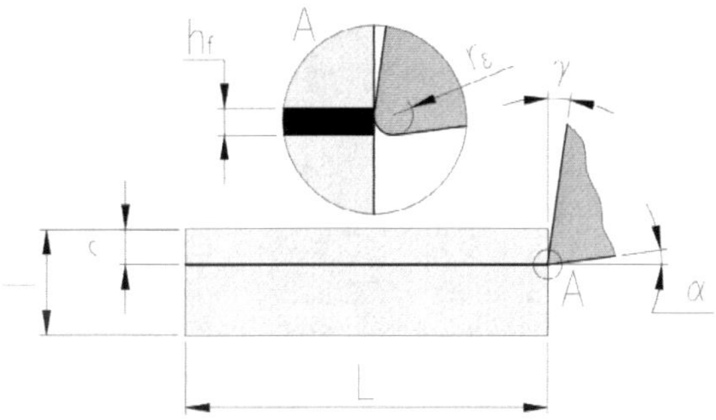


**Figure 3.** Model geometry.

*Material behavior*
The behavior of the alloy is defined by: density; thermal conductivity; elasticity: Young's modulus, Poisson's ratio; the fraction of transformation of mechanical deformation work into heat; thermal conductivity; specific heat; the model of plastic behavior.

The plastic model of material behavior is of the greatest interest for research in this field. Thus, paper [2] examines four models of the plastic behavior of the processed material: Johnson-Cook [3], Zerilli-Armstrong [4], Arrhenius [5] and Norton-Hoff [6], for different cutting speeds (150, 200 and 250 [m/min]). By comparing process force and temperature data obtained by numerical simulation with those obtained experimentally, it is concluded that Johnson-Cook and Zerilli-Armstrong plastic behavior models provide the best results compared to experimental ones.

The most commonly used model describing the plastic response of materials is the Johnson-Cook model [3], used in many studies [2, 7-22], presented in equation (1). It is a thermo-elasto-visco-plastic model, whose parameters are frequently identified by the shear compression specimen (SCS) [23], consisting of three types of tests: quasi-static at room temperature, quasi-static at various elevated temperatures and dynamic test at room temperature. Table 1 presents the values of the Johnson-Cook model for some materials identified in scientific literature.

$$\sigma_{JC} = (A+B\cdot(\overline{\varepsilon}^{pl})^{n})\left(1+C\cdot\ln\frac{\dot{\overline{\varepsilon}}^{pl}}{\dot{\overline{\varepsilon}}_{0}}\right)\left(1-\left(\frac{T-T_{t}}{T_{m}-T_{t}}\right)^{m}\right) \tag{1}$$

where:

- $\sigma_{JC}$ is the equivalent stress Johnson-Cook;
- A is the yield strength of the material;
- B, C, n and m are the parameters of the material model;
- $\overline{\varepsilon}^{pl}$ is the equivalent strain;
- $\dot{\overline{\varepsilon}}^{pl}$ is the equivalent strain rate normalized to a reference speed $\dot{\overline{\varepsilon}}_{0}$;
- T is the evolution of temperature;
- $T_t$ is the transition temperature;
- $T_m$ is the melting temperature of the material.

**Table 1.** Johnson-Cook model parameters for some materials identified in the literature.

| No. | ISO Material | A [Mpa] | B [Mpa] | n | C | m | Source |
|---|---|---|---|---|---|---|---|
| **1.** | *C45* | 50 | 176 | 0.517 | 0.105 | 0.56 | [24] |
| **2.** | *Ti6Al4V* | 875 | 793 | 0.386 | 0.033 | 0.71 | [8] |
| **3.** | *NiCr19Nb5Mo3* | 1262 | 1354 | 0.5 | 0.006 | 1.08 | [17] |
| **4.** | *X40CrMoV5-1* | 908.5 | 321.4 | 0.278 | 0.028 | 1.18 | [20] |
| **5.** | *36CrNiMo4* | 792 | 510 | 0.26 | 0.014 | 1.03 | [3] |
| **6.** | *50CrMoV13-15* | 1539 | 477 | 0.18 | 0.012 | 1 | [3] |
| **7.** | *42CrMo4* | 595 | 580 | 0.133 | 0.023 | 1.03 | [25] |
| **8.** | *Cu-OFE* | 90 | 292 | 0.31 | 0.025 | 1.09 | [3] |
| **9.** | *CuZn28* | 112 | 505 | 0.42 | 0.009 | 1.68 | [3] |
| **10.** | *AlCu4Mg1* | 265 | 426 | 0.34 | 0.015 | 1 | [3] |
| **11.** | *AlZn4Mg3* | 337 | 343 | 0.41 | 0.01 | 1 | [3] |
| **12.** | *AlZn5.5MgCu* | 527 | 575 | 0.72 | 0.017 | 1.61 | [26] |
| **13.** | *AlZn6MgCu* | 546 | 678 | 0.71 | 0.024 | 1.56 | [26] |
| **14.** | *AlMg4.5Mn* | 167 | 596 | 0.551 | 0.001 | 0.859 | [27] |

*Description of the element failure model*

Material removal and chip formation on the tool flow surface occurs due to shear forces, which lead to critical values in the plastic field of the curve describing the material behavior, eliminating the elements in the failure zone ($h_f$). In general, this failure of ductile materials can only be correctly interpreted in numerical models using FEM.

For the description of how the chip breaks, several models are presented in specialized studies, such as Cockroft-Latham [14, 21], modified Cockroft-Latham [12] and Johnson-Cook [7-9, 12, 13, 15-18, 20, 22] where the elimination of the elements occurs when the accumulated stresses in the failure zone reach a maximum value. Of these models, the most widely used is the Johnson-Cook model [3], from equation (2), which leads to the best results. Table 2 displays the parameters of the Johnson-Cook fracture model for some materials identified in scientific literature.

$$\overline{\varepsilon}_{f}^{pl} = (D_{1}+D_{2}+\varepsilon^{D_{3}\cdot\eta})\left(1+D_{4}\cdot\ln\frac{\dot{\overline{\varepsilon}}^{pl}}{\dot{\overline{\varepsilon}}_{0}}\right)\left(1-D_{5}\frac{T-T_{t}}{T_{m}-T_{t}}\right) \tag{2}$$

where:

- $D_1$, $D_2$, $D_3$, $D_4$, $D_5$ material parameters.

**Table 2.** Fracture values for the Johnson-Cook model, for certain alloys identified in the literature.

| No. | ISO Material | $D_1$ | $D_2$ | $D_3$ | $D_4$ | $D_5$ | Source |
|---|---|---|---|---|---|---|---|
| **1.** | *C45* | 0.04 | 1.52 | -6.9 | -0.023 | 1.3 | [24] |
| **2.** | *Ti6Al4V* | -0.09 | 0.25 | -0.5 | 0.014 | 3.87 | [8] |
| **3.** | *NiCr19Nb5Mo3* | 0.406 | 0.75 | 1.45 | 0.04 | 0.89 | [17] |
| **4.** | *X40CrMoV5-1* | -0.8 | 2.1 | -0.5 | 0.0002 | 0.61 | [20] |
| **5.** | *36CrNiMo4* | 0.05 | 3.44 | -2.12 | 0.002 | 0.61 | [3] |
| **6.** | *50CrMoV13-15* | 0 | 0.56 | -1.5 | 0 | 0 | [28] |
| **7.** | *42CrMo4* | 1.5 | 3.44 | -2.12 | 0.002 | 0.1 | [25] |
| **8.** | *Cu-OFE* | 0.54 | 4.89 | 3.03 | 0.014 | 1.12 | [3] |
| **9.** | *CuZn28* | 0 | 2.65 | -0.62 | 0.028 | 0 | [28] |
| **10.** | *AlCu4Mg1* | 0.13 | 0.13 | -1.5 | 0.011 | 0 | [28] |
| **11.** | *AlZn4Mg3* | 0.14 | 0.14 | -1.5 | 0.018 | 0 | [28] |
| **12.** | *AlZn5.5MgCu* | 0.11 | 0.572 | -3.446 | 0.016 | 1.099 | [26] |
| **13.** | *AlZn6MgCu* | -0.068 | 0.451 | -0.952 | 0.036 | 0.697 | [26] |
| **14.** | *AlMg4.5Mn* | 0.0261 | 0.263 | -0.349 | 0.247 | 16.8 | [27] |

*Modeling of contact conditions*

To obtain a valid numerical model, it is also needed to define the contact conditions that accurately reflect the interaction between tool-chip and chip-workpiece, as shown in figure 4. Most of the studies presented in the literature take contact conditions into account in the definition of models and, in most cases, use Coulomb type friction conditions [2, 12, 16, 17, 20, 21], which consider the coefficient of friction to be constant, equation (3), which contradicts the actual contact behavior.

$$\mu = \frac{F_n}{F_f} \tag{3}$$

where:

- $\mu$ is the friction coefficient;
- $F_n$ is the normal force;
- $F_f$ is the friction force.

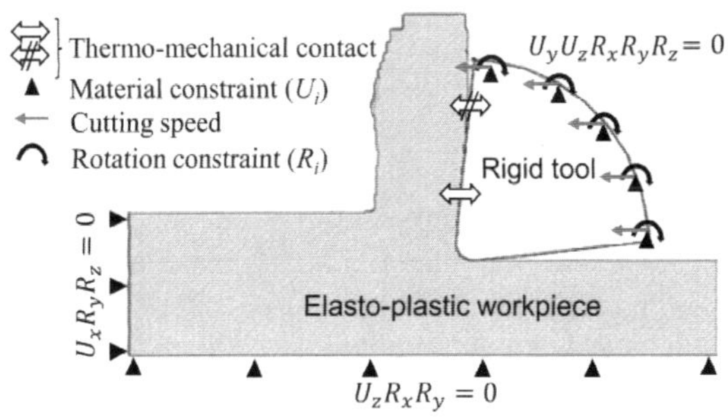


**Figure 4.** Contact conditions [22].

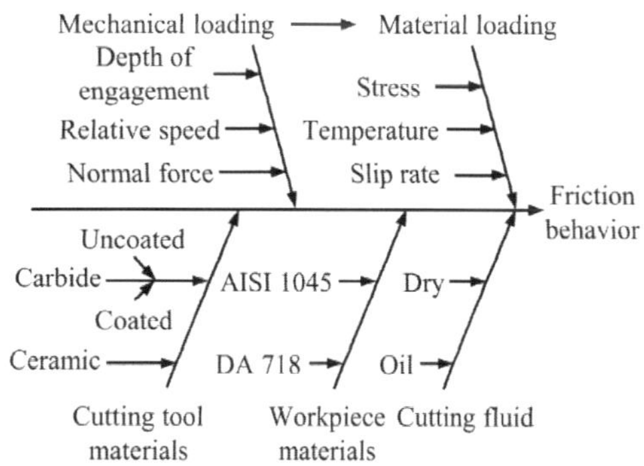


**Figure 5.** Cause-effect diagram for the interpretation of friction behavior [19].

In document [17], the coefficient of friction is considered to be temperature dependent and even an algorithm is proposed to optimize it. The results are favorable and the errors are less than 6%. In document [19], several models for contact conditions are compared and a new model is proposed, in which the friction coefficient depends on relative speed, temperature and contact pressure. A cause and effect diagram of figure 5 is proposed, in which a model interpreting the contact conditions has to take into account all the parameters introduced in a cutting process: from the external and internal mechanical stresses in the material being processed, to the types of material and the coolant lubricant used.

### *2.2. Numerical formulation*

*Discretization of space*

The most widely used method in the discretization of the numerical model is the finite element method, which analyses the deformations and mechanisms of the solid. For this method, the Lagrangian formulation [2, 10, 12-19, 21, 22, 29] is most often used from the mechanics of the continuous mediums. This is the analysis of the discretization, following the evolution of each point in time, or, in other words, the discretization follows the material. This type of formulation is used with remeshing and adaptive mesh [2, 8, 10, 16, 17, 22, 29]. The Eulerian formulation [16], in which the discretization remains fixed and the material evolves through it.

From these two classical types of formulations, other formulations have been developed. On one hand, the CEL formulation is used, which assumes that the material is defined using the Eulerian formulation, and the tool is featured with the Lagrangian formulation or vice versa. On the other hand, the ALE formulation merges the advantages of Lagrangian and Eulerian formulations and has the property that the discretization structure follows the evolution of the material, maintaining the same connections of the elements. Unlike the Lagrangian formulation, ALE does not need to regenerate the discrete structure, remaining the same structure in the contact area. This leads to much shorter calculation times.

In paper [30], most numerical simulation models are analyzed, describing non-discretization models and those based on particle-type methods, figure 6. These types of methods are extremely useful for describing failure models and for calculating very large deformations.

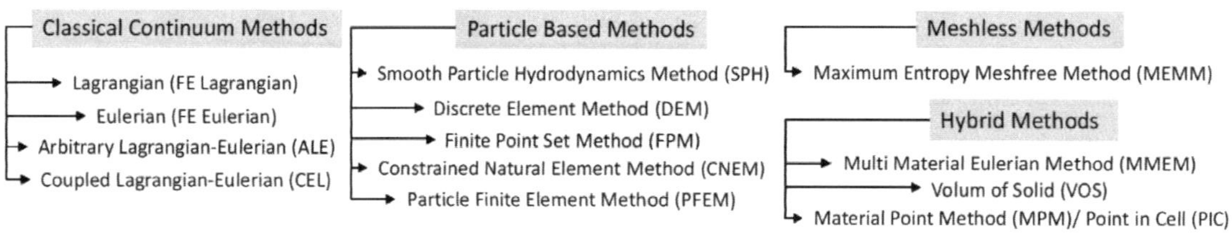


**Figure 6.** Types of numerical models used in cutting processes [30].

*Integration methods*

Computation time in models that simulate cutting processes is relatively long, due to the fact that they are complex dynamic problems. The implicit integration methods are mainly suitable for linear problems. Explicit methods, which are the most used in technical literature [2, 7-22], solve nonlinear and dynamic problems, as found in the cutting processes. Explicit methods are suitable for problems involving complex contacts between the tool-chip and chip-workpiece. No comparison is made in the literature between the two integration methods, in order to understand in which situations they are more appropriate in simulating cutting processes, although there are software products that use only one of the two methods.

## 3. Numerical results and models

*Results generated by the numerical models used for validation*

The process measurements that can be determined experimentally in cutting processes are rather limited in their scope. Thus displacements, displacement rates and temperatures in the contact zones or inside the tool can be measured by means of specialized devices [9, 12, 13, 18, 21, 22]. In order to determine other important parameters necessary for the understanding of cutting processes, such as tool and workpiece stresses, numerical models respond favorably to the interpretation of these variables, with the ability to generate relevant results, as opposed to conventional empirical tests.

One of the most important parameters in the validation of numerical models, but also the one most often used in the interpretation of the results, is the cutting force [2, 7-22]. Basically, a series of tests are performed on the basis of which the numerical models are calibrated, redefining the material and contact parameters, until they provide results compared to those obtained experimentally in relation to the cutting force. In papers, these calibrations are carried out until the models generate errors between a maximum of 5-10%.

The measured temperature [2, 7-9, 11, 14, 15, 17, 18, 22, 24] is another process value on the basis of which the numerical model is validated. This is measured experimentally using thermocouples inserted in tools [15, 20] or by means of thermal cameras [7, 8, 14]. Although in some cases [8], thermocouples may interfere with technological processing systems, they provide relevant information on the accuracy of temperature measurement. Instead, thermal cameras can record very high temperatures, which can occur in cutting processes.

*Results of numerical simulations relevant to research and industry*

Validated numerical models are then used in research and development studies of the cutting process, in order to improve/ optimize industrial applications, to process new materials or to process certain materials under particular conditions.

The multitude of studies in which the effect of the various *geometrical parameters of the tools* in machining is noted and thus leads to their geometrical optimization. These include:

- The influence of the flow angle [29] in the orthogonal cutting process on chip formation and interrupted chip generation is studied. The variation of the flow angle for different cutting thicknesses determines the points of stability and instability, figure 7, which leads to energy conservation in the cutting process;
- The effect of the flow angle [11] on the temperature and the chip is studied in the orthogonal cutting process. This study leads to the conclusion that with increasing the flow angle, temperature and the chip degree of segmentation decrease, for an undeformed chip thickness of 0.1 mm and a cutting velocity of 160 m/min, figure 8;

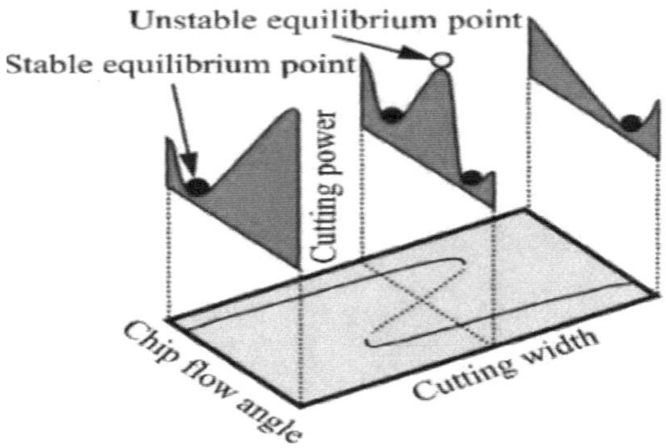


**Figure 7.** Stability equilibrium diagram [29].

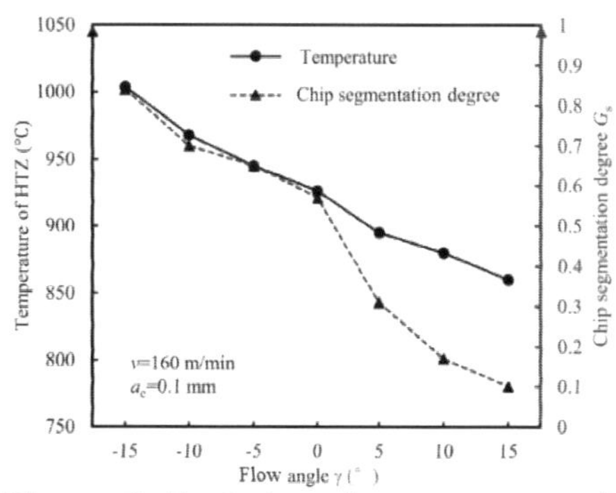


**Figure 8.** Evolution of temperature and chip segmentation degree for different flow angles [11].

- The tool tip radius influence [15] in the orthogonal cutting process on the cutting force, temperature and thickness of the chip is studied. The thickness of the chip is insignificantly influenced by the tool tip radius, while the cutting force and temperature are significantly influenced by it. It is observed that the cutting force and temperature are influenced by the technological parameters, and less by the material of the workpiece or the tool;

- The effect of different micro-milling on the inserts used in milling processes, figure 9 [7], and turning, figure 10 [21], on cutting force, temperature and chip formation is studied. These micro-milling operations improve cutting performance, conservation of cutting forces, reduction of workpiece and tool temperatures.

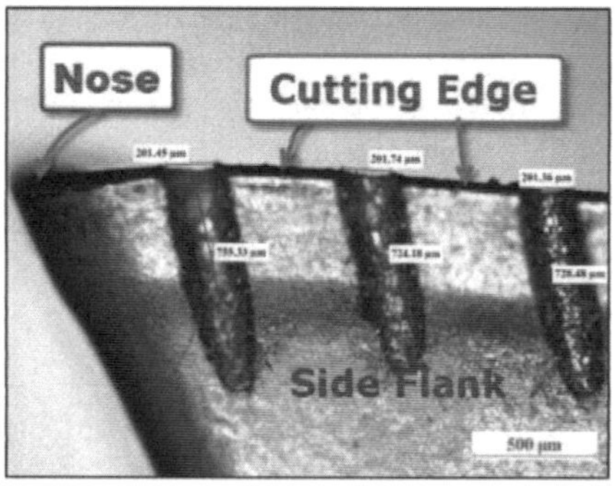


**Figure 9.**Micro-milling insert profile [7].

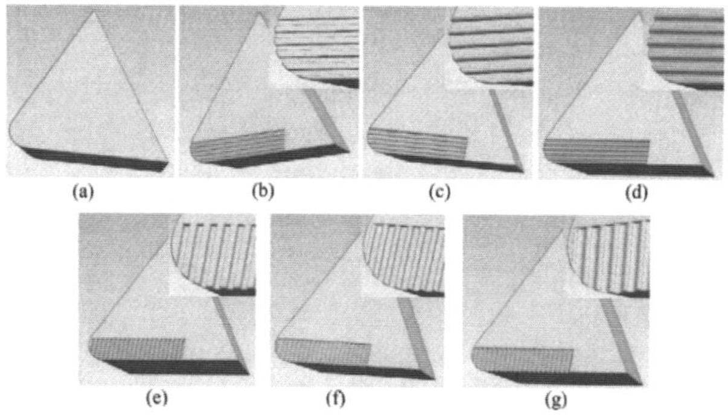


**Figure 10.** Types of inserts with micro-milling [21].

Another part of the studies is oriented towards the analysis of *the shapes of the chips* in the cutting processes. So:

- The optimal size of the chip thickness is studied [18], in the orthogonal cutting process, depending on the radius at the tip of the tool. It is found that the thickness of the chip must be at least a quarter of the size of the tool radius;
- The shape of the triangular [12, 15, 17], rectangular [8, 9, 16, 18, 19, 22] undeformed chip is studied in the orthogonal machining. The study of triangular chips, figure 11, is performed to imitate as accurately as possible the shape of the chips in the milling processes;
- The serrated chip formation is studied [8, 19], in orthogonal cutting processes, influenced by cutting speed and feed, figure 12. Following these analyzes, the morphology of the chips is studied by measuring the tips and depths of the serrated chips and comparison of these experimentally obtained dimensions with those generated by simulations.

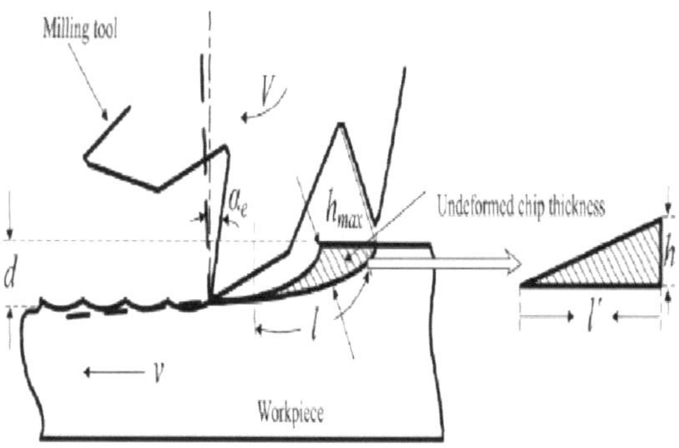


**Figure 11.** The triangular shape of the chip [10].

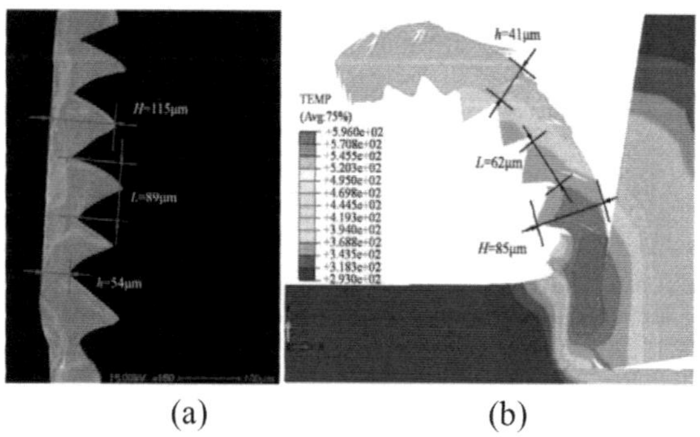


**Figure 12.** Deformed chip geometry (a) experimental (b) simulation [16].

At the same time, there are numerous studies based on the use of numerical simulation in which the influence of the different technological parameters is analyzed, on the dimensions of the processes (cutting forces, temperatures), the durability of the tool or the finished quality of surfaces (roughness, microdurability, residual stresses, etc.), the tool- workpiece contact length. These include:

- The effect of the cutting velocities of 160, 180 and 200 m/min is studied for different feed rates of 0.01, 0.055 and 0.1 mm/rev, in the turning process of high-hardness steel 42crmos4 [22], on the cutting force and contact length, figures 13 and 14. It is found that the simulation model provides results similar to those obtained experimentally, with errors of less than 10% in the case of the forces. The errors are greater for the tool- workpiece contact length, since the friction model is not the most suitable for this simulation;

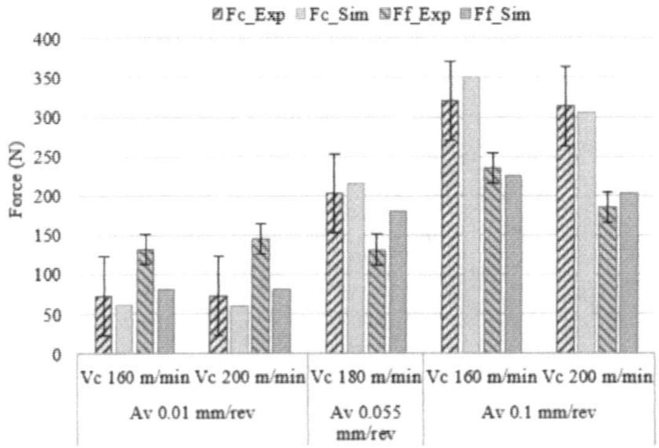


**Figure 13.** Comparison of cutting forces obtained experimentally and from simulations for different technological parameters [22].

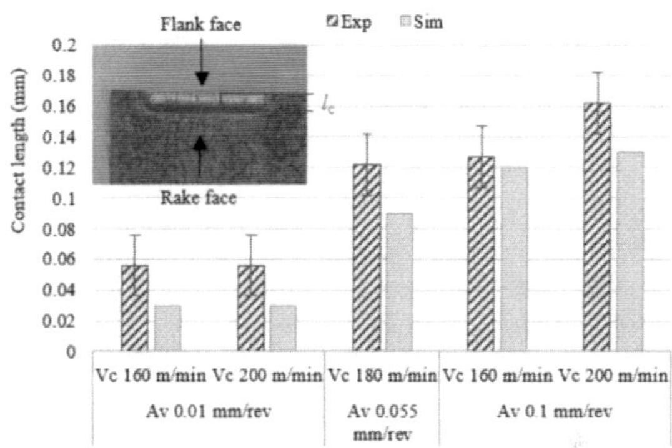


**Figure 14.** Comparison of tool-work piece contact length obtained experimentally and from simulation for different technological parameters [22].

- The effect of the cutting velocities of 25 and 40 m/min, for feed rates of 0.008 and 0.005 mm/rev, with an undeformed chip thickness of 2 mm, in the milling process with the high hardness steel Ti6Al4V [8], on the cutting force and temperature, figures 15 and 16. It is found that the proposed simulation model has errors of up to 6%.

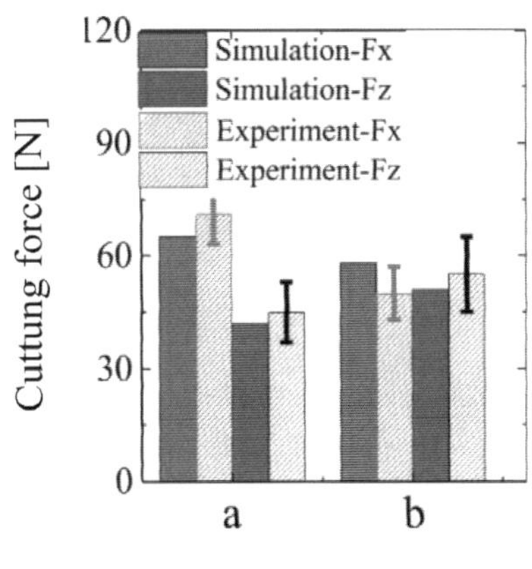


**Figure 15.** Comparison of cutting forces obtained experimentally and from simulations [8].
a. 25 m/min and 0.008 mm/rev
b. 40 m/min and0.005 mm/rev

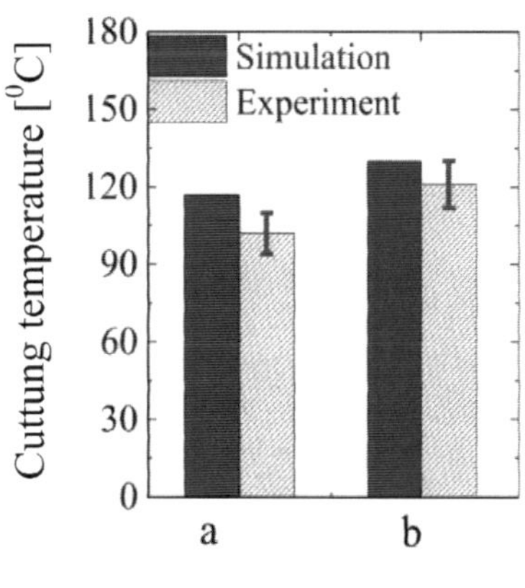


**Figure 16.** Comparison of temperatures measured [8] experimentally and in simulations.
a. 25 m/min and0.008 mm/rev
b. 40 m/min and0.005 mm/rev

## 4. Conclusions and perspectives

In the study of cutting processes, the use of numerical methods has become the main topic, providing more appropriate results for optimizing the technological parameters and geometry of cutting tools, compared to conventional analytical or empirical methods. For the description of the behavior of ductile materials, Johnson- Cook model, Lagrangian formulation and Johnson-Cook failure model are most often used. The contacts between the tool-chip and chip-workpiece must also be very well described, so that most often the classic Coulomb model of the friction coefficient is used. After the simulations, the most important step is the validation of the model comparing the results with those obtained experimentally. Usually, the numerical model is continuously improved, until the errors generated are decrease below 10% compared to those determined experimentally.

The main numerical models used to simulate cutting processes are those using the finite element method, but in recent years a large number of new models have emerged, mainly inspired by fluid mechanics, based on particle models and without meshing, which tend to provide even more relevant results.

The main conclusions regarding the limitations of the current numerical models used for the simulation of cutting processes, derived from the analyzed literature, are summarized as follows:

- the database on the specific parameters of the material models that have proved to be suitable (plastic behavior of the material, the way the chip breaks) is limited to certain alloys;
- the mechanisms of chip formation are still incomplete, although recent numerical models provide relevant information about them. This is due to the fact that they are not yet very well defined to meet all the conditions imposed on cutting processes. Among these are the modeling of how to break the chips, as well as the tool-chip and chip-workpiece contact requirements;
- in real cutting processes, almost always, coolants are used to reduce temperatures during cutting and thus improve the durability of the tool and increase the finished quality of surfaces. Numerical models are limited to predict possible behaviors that may occur from the use of such liquids.

Based on the conclusions summarized above, it follows that the main research directions for modelling and simulation of cutting processes are aimed at overcoming the limitations identified. For example, there is a tendency to move from two-dimensional orthogonal cutting processing studies to more complex three-dimensional calculations for turning, milling, drilling, etc., in order to obtain results that are as relevant as possible to the industry. Another main direction of the study is the development of robust numerical models that could meet today's industrial requirements, which simulate cutting processes with a short calculation time and the simplest use, also at the lowest cost.

This study will continue, on one hand, by increasing the number of bibliographical sources studied and, on the other hand, by focusing on specific cutting processes, such as turning and milling at high cutting speeds or the machining of hard alloys. Subsequently, it is intended to develop numerical models capable of simulating as accurately and with a reasonable calculation time real cutting processes, specific to current industry trends.

## References

[1] Shrot A and Bäker M 2012 Determination of Johnson-Cook parameters from machining simulations *Comput. Mater. Sci.* **52** 298

[2] Paturi U M R, Narala S K R and Kakustam S 2018 Investigations on the effects of different constitutive models in finite element simulation of machining *Materials Today: Proceedings* **5** 25295

[3] Johnson G R and Cook W H 1983 A Computational Constitutive Model and Data for Metals Subjected to Large Strain, High Strain Rates and High Pressures *The Seventh InternationalSymposium on Ballistics* **541**

[4] Armstrong R W and Zerilli F J 1987 Dislocation mechanics based constitutive relations for material dynamics calculations *J. Appl. Phys.* **61** 1816

[5] Arrhenius S 2017 Über die Dissociationswärme und den Einfluss der Temperatur auf den Dissociationsgrad der Elektrolyte *Zeitschrift für Physikalische Chemie* **4**

[6] Hoff N J 1954 Approximate analysis of structures in the presence of moderately large creep deformations *Q. Appl. Math.* **12** 49
[7] Muaz M and Choudhury S K 2020 A realistic 3D finite element model for simulating multiple rotations of modified milling inserts using coupled temperature-displacement analysis *Int. J. Adv. Manuf. Tech.* **107** 343
[8] Ying N, Feng J and Bo Z 2020 A novel 3D finite element simulation method for longitudinal torsional ultrasonic-assisted milling *Int. J. Adv. Manuf. Tech.* **106** 385
[9] Mohan E, Mamundi Azaath L, Natarajan U and Raghuraman S 2020 Comparative design analysis on turning of AISI 4340 using finite element model *Materials Today: Proceedings*
[10] Li J, Tao Z, Cai X, An Q and Chen M 2020 Experimental and finite element analysis of the formation mechanism of serrated chips of nickel-based superalloy Inconel 718 *Int. J. Adv. Manuf. Tech.* **107** 4969
[11] Anhai Li J Z and Jian Zang 2020 Effect of cutting parameters and tool rake angle on the chip formation and adiabatic shear characteristics in machining Ti-6Al-4V titanium alloy *Int. J. Adv. Manuf. Tech.* (7-8)
[12] Razanica S, Malakizadi A, Larsson R, Cedergren S and Josefson B L 2020 FE modeling and simulation of machining Alloy 718 based on ductile continuum damage *Int. J. Mech. Sci.* **171**
[13] Agmell M, Bushlya V, M'Saoubi R, Gutnichenko O, Zaporozhets O, Laakso S VA and Ståhl J E 2020 Investigation of mechanical and thermal loads in pcBN tooling during machining of Inconel 718 *Int. J. Adv. Manuf. Tech.* **107** 1451
[14] Mohruni A S, Zahir M, Yanis M, Sharif S and Yani I 2019 Investigation of Finite Element Modelling on Thin-Walled Machining of Ti6Al4V using DEFORM-3D *J. Phys. Conf. Ser.* 1167
[15] Javidikia M, Sadeghifar M, Songmene V and Jahazi M 2020 On the impacts of tool geometry and cutting conditions in straight turning of aluminum alloys 6061-T6: an experimentally validated numerical study *Int. J. Adv. Manuf. Tech* . **106** 4547
[16] Du M, Cheng Z and Wang S 2019 Finite element modeling of friction at the tool-chip workpiece interface in high speed machining of Ti6Al4V *Int. J. Mech. Sci.* **163**
[17] Klocke F, Döbbeler B, Peng B and Schneider S A M 2018 Tool-based inverse determination of material model of Direct aged Alloy 718 for FEM cutting simulation *Procedia CIRP* **77** 54
[18] Yao Y, Zhu H, Huang C, Wang J, Zhang P, Yao P and Wang X 2020 Determination of the minimum chip thickness and the effect of the plowing depth on the residual stress field in micro-cutting of 18 Ni maraging steel *Int. J. Adv. Manuf. Tech.* **106** 345
[19] Peng B, Bergs T, Schraknepper D, Smigielski T and Klocke F 2020 Development and validation of a new friction model for cutting processes *Int. J. Adv. Manuf. Tech.* **107** 4357
[20] Li B, Zhang S, Zhang Q and Li L 2019 Simulated and experimental analysis on serrated chip formation for hard milling process *J. Manuf. Process.* **44** 337
[21] Liu G ,Huang G, Su R, Özel T, Liu Y and Xu L 2019 3D FEM simulation of the turning process of stainless steel 17-4PH with differently texturized cutting tools *Int. J. Mech. Sci.* **155** 417
[22] Saez-de-Buruaga M, Gainza L, Aristimuno P, Soler D, Ortiz-de-Zarate G, Aizipuru O, Mielgo R and Arrazola P J 2019 FEM modeling of hard turning 42CrMoS4 steel *Procedia CIRP* **82** 77
[23] Dorogoy A and Rittel D 2009 Determination of the johnson-cook material parameters using the SCS specimen *Exp Mech.* **49** 881
[24] Murugesan M and Jung D W 2019 Johnson cook material and failure model parameters estimation of AISI-1045 medium carbon steel for metal forming applications *Materials* **12**
[25] Pantalé O, Bacaria J L, Dalverny O, Rakotomalala R and Caperaa S 2004 2D and 3D numerical models of metal cutting with damage effects *Comput. Method. Appl. M.* **193** p 4383
[26] Brar N S, Joshi V S and Harris B W 2009 Constitutive model constants for Al7075-T651 and Al 7075-T6 *AIP Conference Proceedings* **1195** 945

[27] Rashed A, Yazdani M, Babaluo A A and Hajizadeh Parvin P 2016 Investigation on high velocity impact performance of multi-layered alumina ceramic armors with polymeric interlayers *J. Compos. Mater.* **50** 3561
[28] Teng X 2005 *High velocity impact fracture* (Ph.D. Thesis)
[29] Zhu L X, Baoyi, Xiao Y M H and Wan X 2020 Theoretical modeling and experimental verification of chip flow angle catastrophe in double-edged cutting considering non-linear effect *Int. J. Mech. Sci.* **172**
[30] Rodríguez J M, Carbonell J M, and Jonsén P 2020 Numerical methods for the modelling of chip Formation *Arch. Comput. Methods Eng.* **27** 387

# CAD based design for the jewellery industry: A case study

**A Manavis[1], P Minaoglou[1], K Aidinli[1], N Efkolidis[1], P Kyratsis[1]***

[1] University of Western Macedonia, Dept. of Product and Systems Design Engineering, Kila Kozani, GR50100, Greece

*Email: pkyratsis@uowm.gr

**Abstract**. Jewel design is performed for many years and the designers and manufacturers follow a variety of inspirations. Traditionally the designs are generated based on the artistic sense of the creator, while at the same time a number of manufacturing restrictions apply. Nowadays, a series of advanced Computer Aided Design, CAD based tools are available in order to generate impressive concepts. Although, it is possible for designers to interface a CAD system and implement their design digitally, some prefer to use the programming capabilities offered by the CAD software developers. This is the way forward, in order to create unusual geometries for the jewellery industry and at the same time to develop the capability for generating a complete family of geometries that can satisfy even the most demanding customer. In the current paper, the latest approach is presented together with a case study.

## 1. Introduction

Traditionally the design activity is based on the use of a CAD system by the designer, who selects the appropriate commands in order to achieve the model of his perception. Although, this approach is used over the years, a number of disadvantages appeared, i.e the final design have limited amount of flexibility in changing the geometry after the CAD model is built, all geometries produced as alternative designs are very similar, the geometries achieved are relatively simple or unusual geometries demanded a great deal of effort from the designer point of view [1].

A step forward was the widespread use of the Application Programming Interface (API) of the commercially available CAD systems for producing general purpose geometries as the basis for product design. A series of standardized products and components are designed with CAD based applications that perform a variety of steps in a limited amount of time and increase the designers' productivity. The user is able to change a series of parameters from a user friendly computer interface and the rest is delivered automatically from the application [2]. Unfortunately, although this approach is very successful for prismatic geometries and mechanical parts, when it comes to jewellery the automatic design can become considerably harder and as a result computational design pieces of software have been developed and CAD based graphical algorithm editors have been in use [3].

## 2. Literature survey

A number of research teams dealt with the area of jewellery design from a variety point of view. Stamati and Fudos introduced ByzantineCAD, a parametric CAD system for designing pierced medieval jewellery. The design is implemented by selecting a series of parameter and constraints. The methodology is based on the use of voxel oriented feature based CAD principles. An additional characteristic of the proposed methodology is that the used algorithm, enlarges the designs without altering the size of the elementary structural elements [4]. Goel et al. worked towards the recreation of traditional Indian ornamental designs via computer aided geometric modelling. These patterns are very suitable for computer based applications. The main aim was to develop computer algorithms that would be able to design Indian jewellery automatically and thus to dramatically reduce the time needed for such a work in the traditional manner [5]. Parsricha and Greeninger applied zero-waste principles via 3D printing with an aim to achieve sustainability in the fashion industry. A number of CAD based applications were used together with 3D printing hardware and software in order to create their designs. Design thinking principles were used with biodegradable materials and both fashion and jewellery products were produced with success [6]. Thammachote et al. studied the development of a new gating system for achieving complete filling in casting of silver jewellery. The proposed shape and dimensioning were investigated using computational modelling and optimal solutions were achieved. Then a series of experimental work was performed in order to validate the simulations presented, thus casting defects were drastically reduced and at the same time the process was extremely productive [7]. Grobman et al. presented a case example for the efficiency of procedural shape modeling in for the mass customization of wedding rings. The current implementation of wedding rings implementation is integrated into the REx system which is based on the GML browser plugin [8]. The present paper aims in providing a systematic way in order to design jewels based on the principles of computational design. According to Grobman the idea of using early stage simulation for form generation, which inverse the traditional design process, tries to integrate numerous developments [9] in jewellery design practice. The idea is not only to provide an automatic tool for the design activities but an integrated tool that will design customized products for each customer, while at the same time a number of alternatives can become available. Those alternatives offer a family of products with minimum effort from the designer, while the geometry is significantly unusual.

## 3. Proposed framework for the jewellery industry

In order to be able to cope with the increasing demands of the jewellery industry for unusual geometrical forms and easy manufacturing processes, computational design principles are necessary together with the extensive use of 3D printing technologies. The inspiration was based on a technical issue such as the lost wax casting process. The geometry achieved reminds of the geometries appearing during lost wax casting i.e long geometries with relative constant section size and increased 3D network-like form.

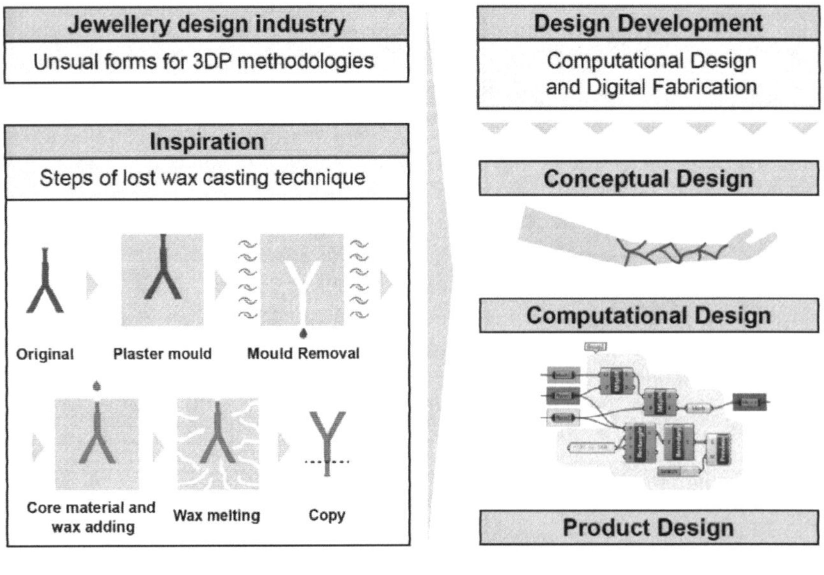


**Figure 1.** The proposed framework for the jewellery industry.

## 4. Case study: Bracelet

Figure 2 depicts the technical aspects of the inspiration used. It worth mentioning that this is an additional novelty of the presented work because a great deal of research effort is done towards discovering new ways and methodologies to advance the inspiration of the designers. In the present case study, a manufacturing process such the casting inspired the design team to perform the rest of the steps and finish with a completely unusual geometry for jewellery.

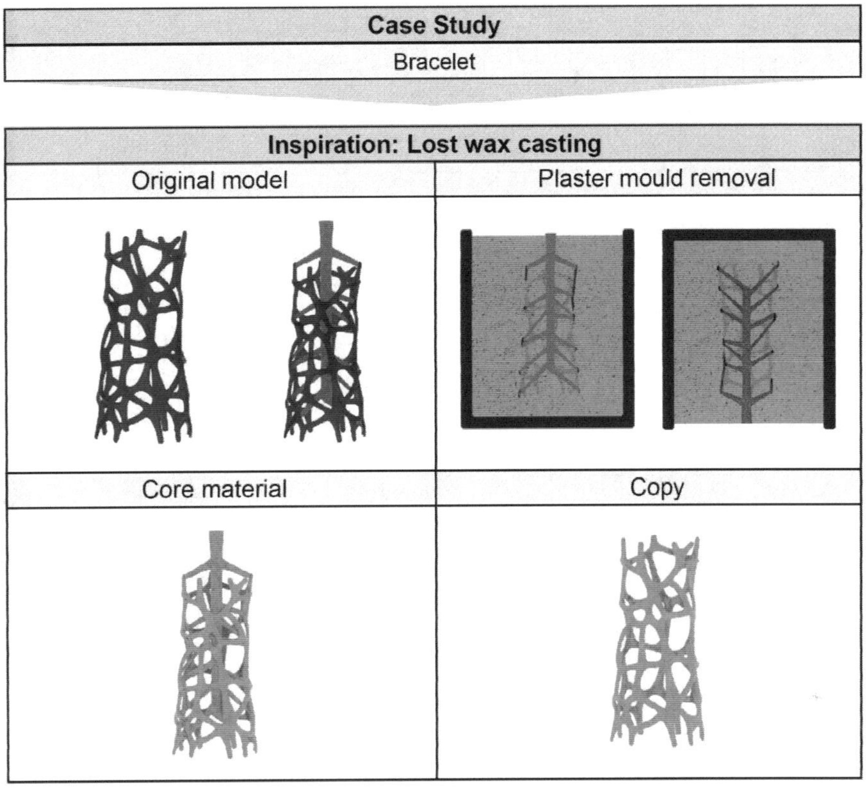


**Figure 2.** Conceptual stage for the design development.

From the technical aspects point of view, the CAD based use of graphical algorithm editor was performed via Grasshopper™. It is an editor integrated with Rhinoceros™ computer aided design platform. The idea was to use a .stl file of the customer's 3D scanned arm geometry (figure 3). Two planes were created in order to provide the start and the end of the geometry to be used in the automated design algorithm. This methodology emphasizes the customization of the proposed jewellery and increases its added value from the customer's perspective, because he gets a unique product that fits exactly on him. At the same time, the customer gets additional added value for his funds because he invests on his own 3D geometrical characteristics.

Grasshopper™ enables the development of parametric designs via visual programming. All the models generated are digital models within Rhinoceros™ and as a result the software's export capabilities can be used for downstream applications such as 3D printing. Another advantage of this approach is the opportunity to go beyond and use the RhinoScriptSyntax module, which incorporates capabilities performed within Rhinoceros™ operations. The library of resources available contains the basis for creating geometry, apply commands, manage document objects and develop application methods. All this functionality is using simple Python™ variables and makes the design process more productive.

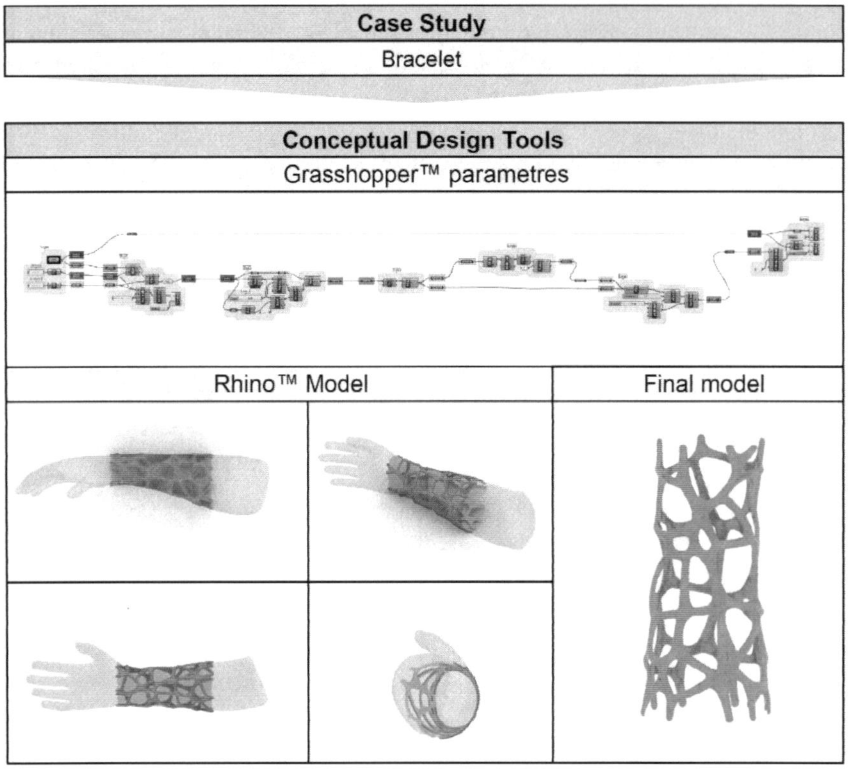


**Figure 3.** Computational product design.

## 5. Alternative forms based on computational design approach

Following the proposed approach, a number of parameters were incorporated within the model for increasing the size of the geometrical sections, the size of the gaps etc., while at the same time the basic principles for the customized jewel applies for each customer. As a result, different geometrical alternatives are designed and the customer is able to decide which alternative is more suitable for his taste. Although there is an increased demand for computation power, when the model is built with a large amount of variables and features, the output is designed automatically. The modelling approach offers the capability of a vast amount of alternatives that can satisfy even the most challenging demands from the customer's point of view.

If someone compares the time needed for such a complicate geometry to be designed from scratch every time there is a demand for change, the proposed methodology and tool offers a quick response and impressive results in a limited time period. When building this family of products that are suitable for customization, the designer is able to offer this flexibility without the need for repeating manually the appropriate steps needed for designing the final product. All these models can be very easily used for downstream applications that support i.e.:

- Fast prototyping with the use of 3D printing;
- High quality manufacturing with the combination of 3D scan data of the customer and the 3D CAD models produced;
- Virtual prototyping using rendering applications in order to promote both the product itself or customer wearing the product;
- Web-based exhibition of all the models for enterprise and designer promotion.

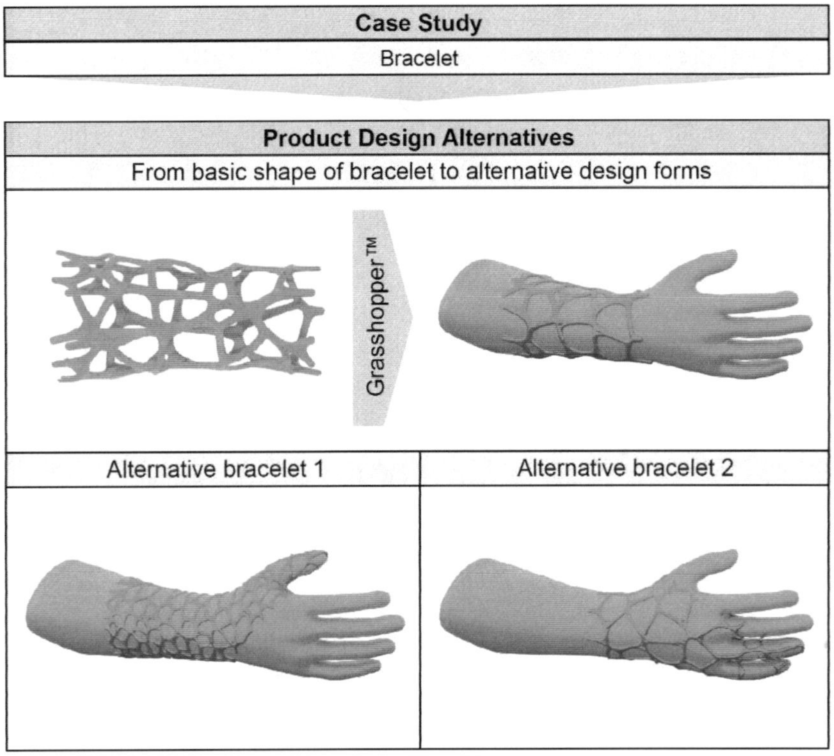


**Figure 4.** Alternative designs building families of customized products.

## 6. Conclusion

A complete framework for designing customized products for the jewellery industry was presented. The emphasis was given towards the design of a bracelet customized to a specific customer after his arm was 3D scanned and the data appropriately processed. In addition, the computational design tool offers the possibility of executing alternative designs for the same customer's geometrical data and thus offers a great deal of added value. The contribution of this paper aims to update the jewellery design point of view on digital tools usage (i.e. computational design). The proposed methodology and the presented case study depicts the advantages of a new era of CAD based product design that influences not only the automation of the design process but the customization of the products designed via the production of families of products.

## References

[1] Tzintzi V, Manavis A, Efkolidis N, Dimopoulos C, Kakoulis K and Kyratsis P 2017 *MATEC Web of Conferences* **112** 07025
[2] Kyratsis P, Gabis E, Tzotzis A, Tzetzis D and Kakoulis K 2019 *Int. J. Mod. Manuf. Technol.* **11(3)** 110
[3] Kyratsis P 2020 *Int. J. Mod. Manuf. Technol.* **12(1)** 82
[4] Stamati V and Fudos I 2005 *Comput. Aided Des.* **37** 431
[5] Goel V, Khanduja D, Garg TK and Tandom P 2015 *Comput. Aided Des. Appl* **12(4)** 457
[6] Pasricha A and Greeninger R 2018 *Fash Text* **5** 30
[7] Thammacht N, Dulyapraphant P and Bohez ELJ 2013 *Int. J. Adv. Manuf. Technol* **67** 797
[8] Berndt R, Schinko C, Krispel U, Settgast V, Havemann S, Eggeling E and Fellner D 2012 *Proc. Int. Conf. on Creative Content Technologies (Nizza, France)* (Wilmington, USA: Xpert Publishing Services) p 72
[9] Grobman Y and Ron R 2011 *Proc. of 29th eCAADe Confernce* (Ljubljana) (CumInCAD) p 107

IOP Conf. Series: Materials Science and Engineering 1009 (2021) 012037 doi:10.1088/1757-899X/1009/1/012037

# Computational design technologies for interior designers: a case study

**A Manavis[1], P Minaoglou[1], D Tzetzis[1], N Efkolidis[1], P Kyratsis[1]***

[1] University of Western Macedonia, Dept. of Product and Systems Design Engineering, Kila Kozani, GR50100, Greece

*E-mail: pkyratsis@uowm.gr

**Abstract**. Traditionally interior design is based on the inspiration generated by the designer who acts as an artist and engineer at the same time. Today, a number of digitally developed forms can be used with an aim to offer unusual geometries. Computational design technologies can be used in order to fully automate the generation of a design based on a series of parameters that can be altered within specific limits. In such a way, families of forms and proposals are made available and more opportunities for products dealing with interior design can be produced. These forms require the use of modern 3D printing technologies for their construction due to their geometrical complexity and their digital development. In the present paper, computational design will be presented together with a case study in the area of interior design.

## 1. Introduction

Computational design is a supportive methodology for letting designers further explore the design space in order to create innovative geometrical forms. It allows the creation of a geometry based on a number of parameters that can be altered within specific ranges. As a result, families of initial forms can be easily produced, thus pushing the design boundaries to new limits. Those limitations can be controlled by the way that the initial code is built and all the constrains set among the geometrical entities [1].

The commonly used CAD systems in product design, deal in details with the output of the design stage rather than at a conceptual level. Computational design deals with the initial form design and offers potential solutions to a significantly greater extent. The concept that needs to be expressed with an unusual geometry can be implemented and then further explore the concepts by changing the parameters and constrains used [2].

## 2. Literature survey

Gunduz et al. presented the implementation of computation design and design geometry courses within students and how a number of issues like cyclic process, feedback between geometric relations, material performance and joinery details were handled with an aim to achieve novel results [3]. Soleimani presented how the design process can be affected by progress in technology and computation. Computational methods were introduced to architecture schools and students were exposed in computational thinking and the fundamentals of generative analytical technologies. The key issues presented were the exploration of the computation capabilities potentials and the role that such a collection of tools and methodologies can affect the way of thinking about design [4]. These efforts provide a solid basis that computational design and its effects in industry have already been introduced systematically in education. According to Ovesen, students with a preference in mathematical thinking have gained a better understanding of form [5]. Menges dealt with how computational design enables the biomimetic design processes in architectural applications. Morphological exploration and form-finding get increasingly enriched by environmental performance feedback [6]. At the same time, he proves the need for a different approach in engineering disciplines. The paper deals with the difference between computer aided design (CAD) and computational design in architecture, the possibility of transferring principles from nature to computational design and issues such feature-based, constraint-based etc. [7]. The present paper offers a holistic approach for the interior designers who are willing to follow an alternative path when they face a more challenging inspiration environment. Further to the proposed methodology, a case study is presented with an aim to offer a paradigm of successful implementation of computational design.

## 3. Proposed methodology

Interior design faces a number of challenges from the inspiration and implementation point of view. Designers sometime feel that there is a lack of appropriate CAD tools for expressing their ideas or they limit their inspiration due to limitations for 3D representation of their idea. Figure 1 depicts the proposed methodology for interior designers to follow in the area of decoration, lighting, furniture etc. Following that approach, all the product development steps are implemented with CAD based tools i.e. the conceptual design sketches, the initial inspiration 3D forms and the finalized virtual products based on advanced render techniques.

With an aim to overcome the problems of 3D modeling unusual geometries, the use of computational design introduces extra freedom from the designer's point of view. Advanced tools offer the possibility to model every geometry the designer might have in mind. At the same time, through the use of 3D printing technologies, those design can be produced in a limited amount of time without the manufacturing restrictions of other traditional processes i.e. molding, casting.

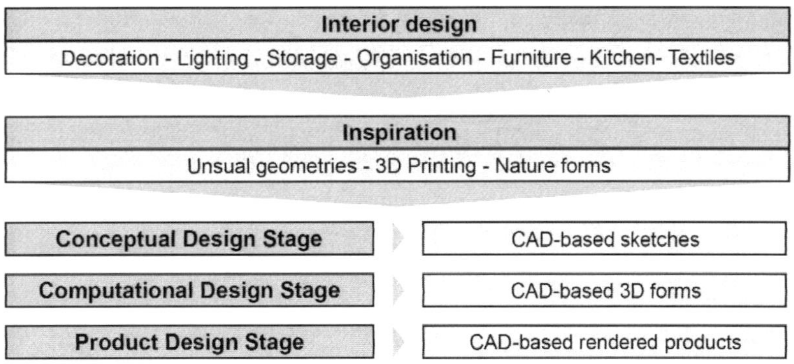


**Figure 1.** A holistic approach in interior product design.

## 4. Case study

The central idea was to model a room/office divider focused on simple geometries, such as boxes, based on the principles of computational design. The process starts with a number of nodes in space, that can be connected and implement an unusual but well-designed geometry. The basic aim is to create prototypes that will have the same external dimensions and with their appropriate assembly more alternative designs will be implemented. So their combination can offer the freedom to finally have an infinite amount of design combinations of the output presented. The components of the proposed structure will be kept constant with appropriate metallic connectors. The connectors will be placed towards all directions so positioning the component will be safe for use. Figure 2 provides an overview of the conceptual tools used in order to get the inspiration for the proposed design i.e. mid map, mood board, sketches.

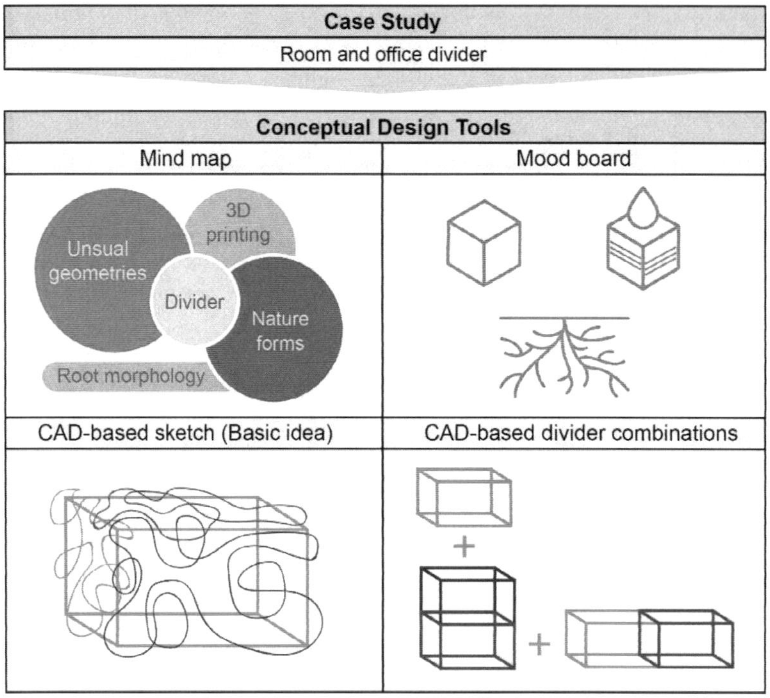


**Figure 2**. Conceptual design tools implementation.

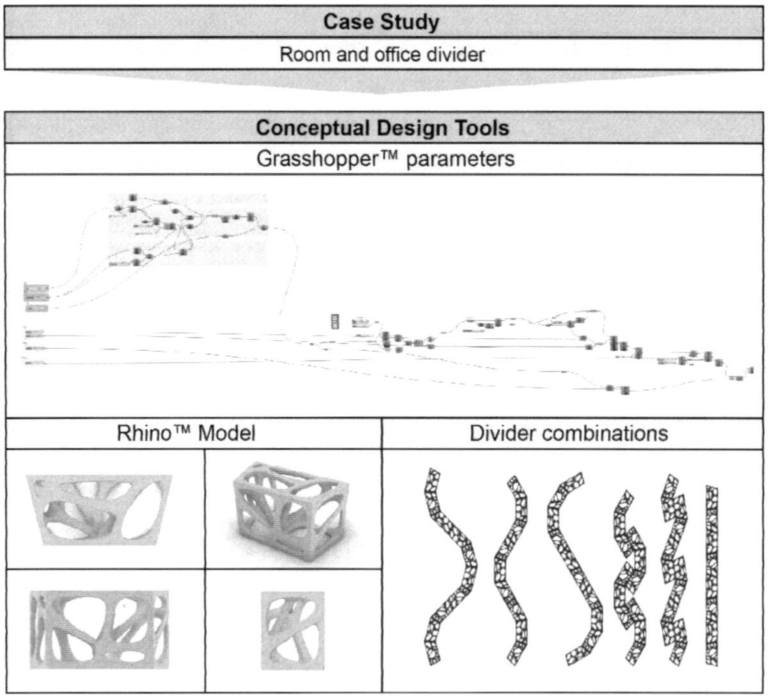


**Figure 3.** Applying computational design principles.

Grasshopper™ was used for creating the appropriate code. A series of geometrical parameters and constrains were introduced and the user was able to change them within a specific range of values. The Rhinoceros™ model created in each case, an angle was incorporated that permits the positioning of the divider in different directions, thus allowing the shape of the divider to change and provide a significant number of combinations (figure 3). Those configurations can create a straight line, a series of arches, U turns etc. In addition, some components can be subtracted in order to have the total geometry of the divider with a variety of alternative designs offered.

Due to the complicate geometry of the proposed box-like part, 3D printing should be used for its production. This offers additional advantages because more colors can be selected for the material used and at the same time specially manufactured wood like materials (powder of wood together with PLA) can be used. The result is to be able to offer a great deal of shapes, colors and textures when dividing a room or an office (figure 4).

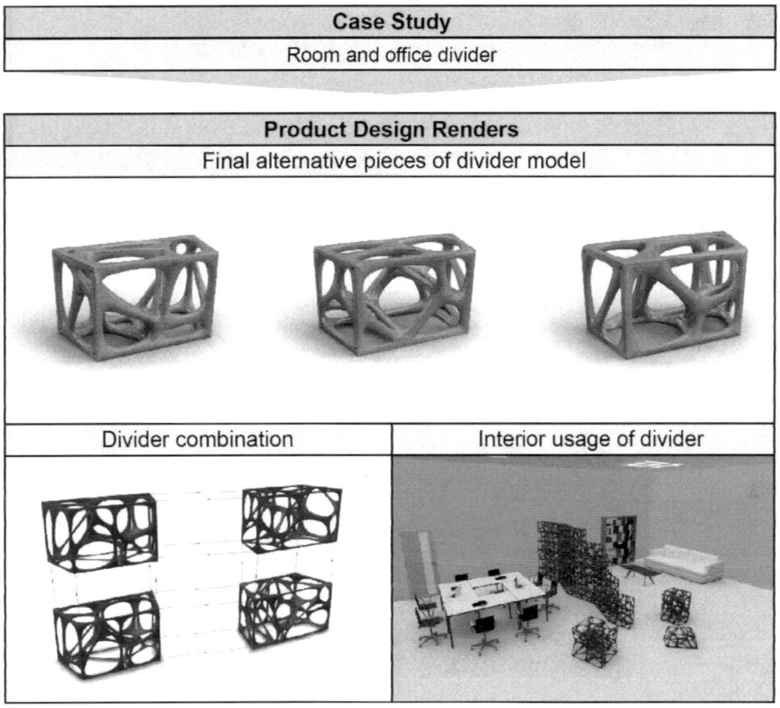


**Figure 4.** Alternative geometries of the divider.

## 5. Alternative forms and product family

Further to the different designs of the divider component, computational design offers additional advantages by exploring the use of the same principles in other components used in interior design products. The use of the same 3D printing technology with alternative materials and colours can be extended at components made for an office desk (figure 5). In the same way, the unusual forms and geometries produced can offer distinct advantage from the appearance point of view, when presenting a similar form with the previously described space divider units.

It offers a unique opportunity to link the traditional minimal design of an office desk with high tech based computational design of the desk's support with components that look more futuristic. In this way, one side of the desk looks similar with what the customer expects to see on a desk, while at the other side forms and geometries generated from computational design offer the sense of similarity with the divider previously described. The increase in productivity is significant because with similar computational code an extensive number of components can be produced in a limited time and manufactured following widely accepted 3D printing manufacturing processes.

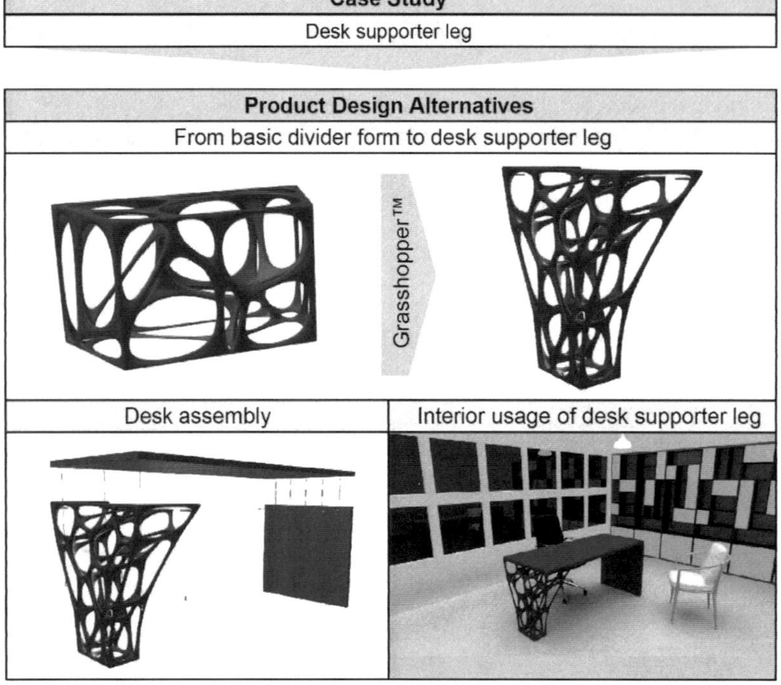


**Figure 5.** Family of products following the computational design approach.

## 6. Conclusion

A holistic approach in designing products was presented with an aim to use computational design together with a number of conceptual design tools and methodologies. When the proposed framework is used the designer can offer unique geometries for the customer, 3D printing capabilities for their manufacturing and a family of products that relate to each other and perform an eye catching feeling to the potential customer. The contribution of this paper aims to update the interior design point of view on digital tools usage (i.e. computational design). Although all these technologies are still under development and the design engineers try to find application for these tools, it seems that there is a shape shifting trend towards futuristic product design components to accompany ordinary traditionally presented products.

## References

[1] Kyratsis P 2020 *Int. J. Mod. Manuf. Technol.* **12(1)** 82
[2] Kyratsis P, Gabis E, Tzotzis A, Tzetzis D and Kakoulis K 2019 *Int. J. Mod. Manuf. Technol.* **11(3)** 110
[3] Gunduz G, Oral H, Yazar T 2018 *Proc. 36$^{th}$ eCAADe Conference - Computing for a better tomorrow* (*Łódź*) vol 2 p 439
[4] Soleimani A 2019 *Proc. of Reynolds Symposium'19: Digital Approaches to Design Education* (*Portland*)
[5] Oversen N 2014 *Proc. of Conference on Engineering and Product Design Education* (*Glasgow*) vol 139 (Design Society) p 714
[6] Holzer D 2016 *Automat. Constr.* **72** 3
[7] Menges A 2012 *Bioinsp. Biomim.* **7** 015003

IOP Conf. Series: Materials Science and Engineering 1009 (2021) 012038 doi:10.1088/1757-899X/1009/1/012038

# Progressive screw shaft manufacturing technology

**M Matejic[1]*, M V Dragoi[2], M Blagojevic[1], A C Filip[2] and I Miletic[1]**

[1] University of Kragujevac, Faculty of Engineering, Sestre Janjić 6, 34000 Kragujevac, Serbia

[2] Manufacturing Engineering Department, Transilvania University of Brașov, Eroilor 29, 500036 Brasov, Romania

*E-mail: mmatejic@kg.ac.rs

**Abstract.** The progressive screw shaft is one of the very challenging part for machining. It might have a decisive influence on the precision of the parts processed by the machine tool. In this paper is presented a new, innovative technology of its manufacturing. The first goal is to develop a technological process for CNC milling machine. Technological process is fully described and with all parameters given and commented. Technology for progressive screw shaft manufacturing is developed in Autodesk Inventor HSM software. The second goal of the paper is a comparative analysis of the here described process for milling the different profile shapes. In the paper are given the milling examples of rectangular and circular profiles. Paper aims to emphasize the advantages and disadvantages of the proposed method. Some conclusions and directions to next research steps are given at the end of the paper.

## 1. Introduction

The thread is defined by a profile which is wrapped around of cylinder in the form of helix. The threads can be internal, wrapped into internal cylindrical surface, or external, wrapped onto external cylindrical surface. In the machine practise there are many types of threads such as: metric, trapezoidal, various pipe threads etc. Threads can be used for static components, for assembling purpose, and rotary components. Threads in rotary components turn rotary motion into translation along the thread axis, and torque into axial force. The rotary components with thread are called power screw joints. There are numerous applications for power screw joints such as: car jacks, linear motion applications in tooling machines, usage in processing equipment, lifting applications etc. The described cases of thread usage have the constant thread pitch. In the other thread applications, such as transportation technique, pumps, positioners etc., thread can often have a variable pitch. Shaft with variable thread pitch is called progressive shaft. Progressive shafts can have two, three or more different pitches. Segments having different pitches are joined together to continue each other and form a single thread with variable pitch Manufacturing technology of progressive shafts is a very demanding one, and it can vary depending on size series in production. The progressive shafts are mostly produced on lathes with special tooling, CNC lathes, CNC machining centres and etc.

The progressive shafts, due to their specific geometry, is a very interesting topic for researchers worldwide from couple different aspects: application, manufacturing technology etc. R F Gillie considered variable pitch screw drive in vehicle stability control [1]. In that case variable pitch screw drives indicates to be very reliable solution. P D Lin and M F Lee did a research about NC data generation in order to produce variable pitch screws with four-axis milling machine [2]. They derived equations which was used for machine NC programming. J Y Liu and H S Yan performed geometry design of trapezoidal variable pitch lead screws [3]. Their paper was significant in this research field because it gave complete mathematical model of progressive screw geometry. Senanayake S and Clarke B did a design, manufacturing and testing of food extruder which consists of two meshed progressive screw shafts [4]. Slătineanu L et al. did a manufacturing of threads with variable pitch by using noncircular gears [5]. Their technology is based on turning. Li et al did a program for NC code generation depending of the progressive shaft input dimensions [6]. This research is interesting, but it is performed only for rectangular thread profile. This technology for progressive screw manufacturing is based, as well on turning. The group of authors performed study on the computer numerical control process of variable pitch, groove depth and groove width screw [7]. Gyenge et al. created new manufacturing technology for variable pitch and variable screw profile worms which is based on turning [8]. They were using ProE for doing a CNC machining and developing the technology. Nabeta et al. developed their own algorithm for turning of progressive screw shaft.

Apart from these, in this paper is described a problem with manufacturing progressive screw shat using three-axis CNC machine. For CNC programming Autodesk Inventor HSM was used. At the end of the paper conclusions are shown and comparison is made with other presented technologies for progressive shaft manufacturing.

## 2. Manufacturing problem description

The idea of solving the problem with progressive shaft manufacturing on three-axis CNC machining centre, for this particular paper, was developed within an industrial project of the authors. Namely, the authors held CNC machining training course in company which produces packing equipment. One of the training challenging part was introducing the company workers with machining technology of progressive shaft screw. Progressive shaft screw was used as positioner in the bottle filling machine. Examples of progressive shafts are shown in figure 1. Example in figure 1 (b) is done by proposed technology in this paper

**Figure 1.** Installed progressive shaft on the machine.

The progressive shafts, shown in figure 1 were made of Polyamide (PA6). The progressive shafts shown in figure 1 was used for different filling types of bottles. The type of the bottle, bottle geometry dictates the shape of progressive shaft profile. Most commonly used types of bottle are shown in figure 2.

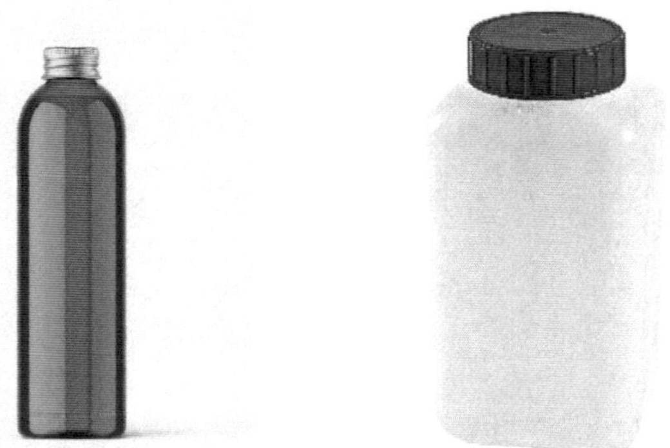

(a) (b)

**Figure 2.** Bottle types: (a) round bottle; (b) square bottle.

On the shaft beginning helix start with smallest pitch, while on the shaft end helix pitch is increased to wanted dimension for bottle distance, position one to another.

The company owns CNC lathes and three-axis CNC milling machine, both shown in figure 3. Both of the machines had the Siemens SINUMERIK 828D controlling unit.

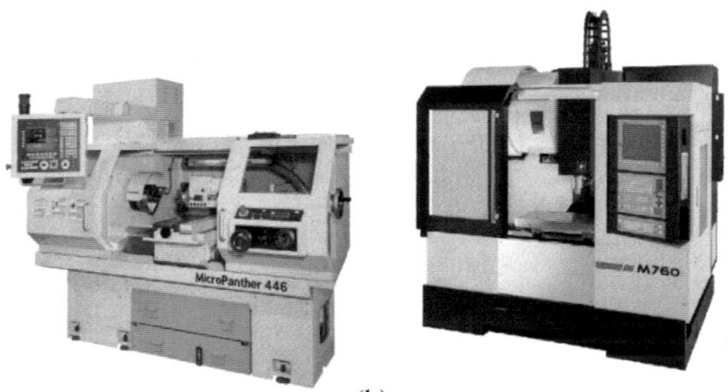


(a) (b)

**Figure 3.** CNC machines:
(a) CNC Lathe and (b) three-axis CNC milling machine.

The problem with machining by using CNC lathe was limited by purchased licenses in the controlling unit and as well with maximum thread pitches which can be achieved. Milling machine was limited because it was three-axis milling machine. The single way to solve this demanding task was introducing the fourth axis into milling machine table which can have the dividing head. The fourth axis was took from company conventional machine. The fourth axis apparatus is shown in figure 4.

**Figure 4.** The fourth axis introduction to three-axis milling machine table.

In the Siemens SINUMERIK 828D controlling unit the fourth axis is available for usage, but without appropriate equipment it cannot be used. Instead of that progressive shaft manufacturing was performed as 3D milling.

## 3. Manufacturing technology of progressive screw shafts

For developing of manufacturing technology the Autodesk Inventor HSM software was used. Autodesk Inventor HSM supports a very large amount of post processors for various types of CNC machines, which made it a logical choice for solving this problem. The both helix profiles was used: for square bottles, and for round bottles. The helix profile shapes are given in figure 5.

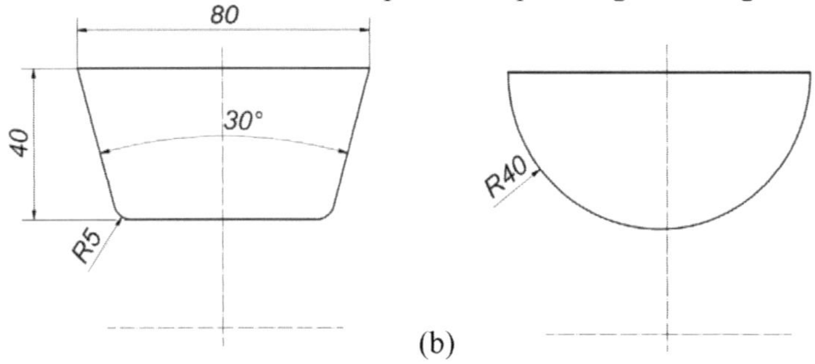


(a) (b)

**Figure 5.** Helix profile shapes:
(a) for rectangular bottles; (b) for round bottles.

The progressive shaft diameter is 120mm, while pitches are respectively, 90 mm, 120 mm and 150 mm. For both shafts the same diameter and same pitches were used. In figure 6 are sown shaft 3D models.

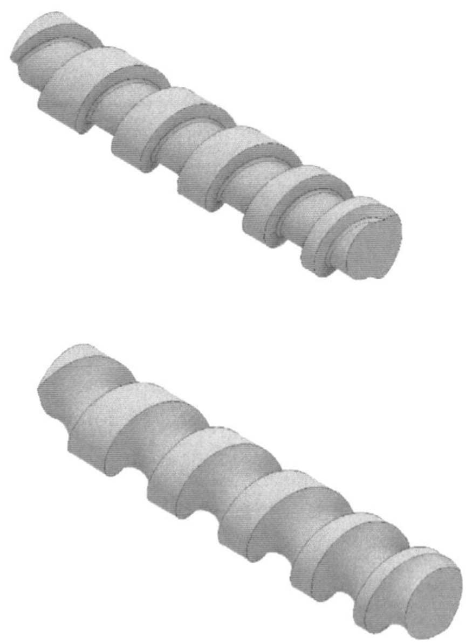

(a)

(b)

**Figure 6.** Shaft 3D models:
(a) for rectangular bottles; (b) for round bottles.

In order to easily do the CNC programming the corrections were done on the progressive shafts models at the joints of two neighbour pitches. Namely, the shaft models were divided into 6 different models states, which represents 6 various separate operations. The example of one model state per one operation is shown in figure 7.

IOP Conf. Series: Materials Science and Engineering 1009 (2021) 012038 doi:10.1088/1757-899X/1009/1/012038

**Figure 7.** Shaft model state example per operation.

Per each of those operations the roughing and finishing was performed. For roughing it was used bullnose mill with diameter of 10 mm. For finishing it was used ball mill with diameter of 10 mm as well. The machining parameters of cutting tools are given in figure 8. Machining parameters were adopted to be best for machining of PA6.

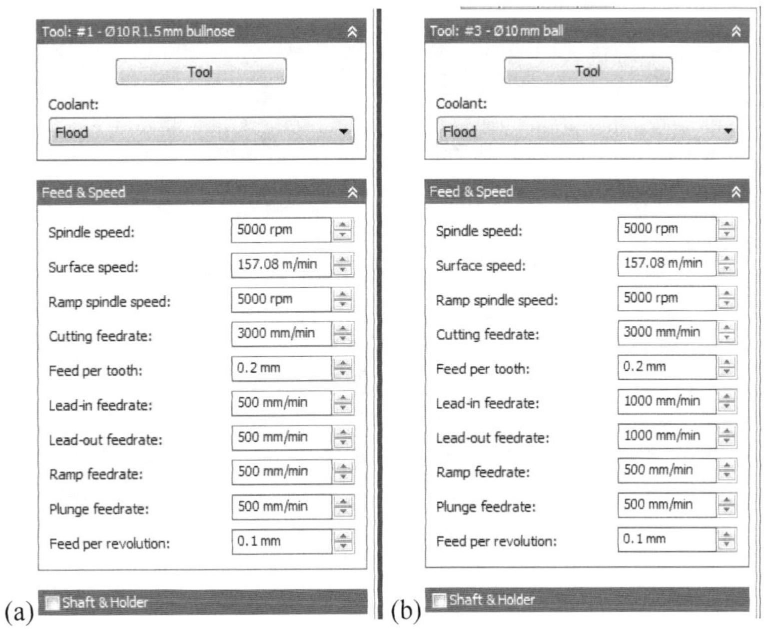


**Figure 8.** Cutting parameters: (a) bullnose mill; (b) ball mill.

The same parameters were used for machining of both shafts. Roughing phase was done for both shafts by usage of 3D pocket Inventor HSM option. In figure 9 is shown geometry for rouging cutting tool, and finishing cutting tool. These tools have to be set to work with cutting parameters shown in figure 8.

(a)
(b)
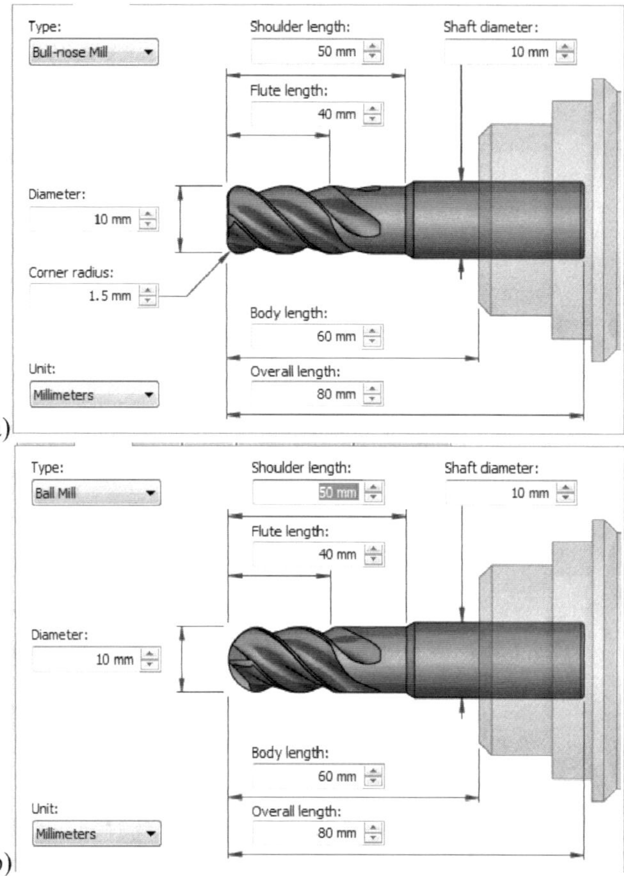


**Figure 9.** Cutting tools:
(a) bullnose mill-roughing tool;
(b) ball mill-finishing tool [10].

The roughing phase is shown in figure 10.

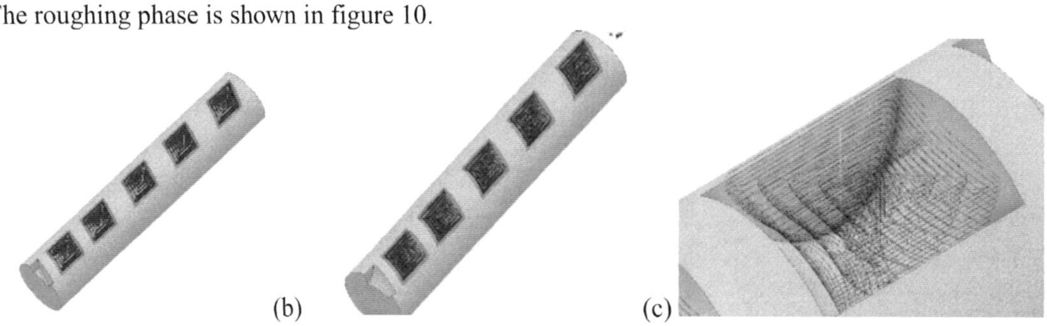

(a) (b) (c)

**Figure 10.** Roughing phase: (a) for rectangular bottles; (b) for round bottles; (c) path detail.

The finishing phase was performed by Parallel command in Inventor HSM. Doing the parallel command was performed by setting machining angle to be equal with helix angle, except on helix transitions. This action must be performed in order to get helix surface as smoothest as possible. On helix transitions it was used arithmetic mean of helix angles. Finishing phase is shown in figure 11.

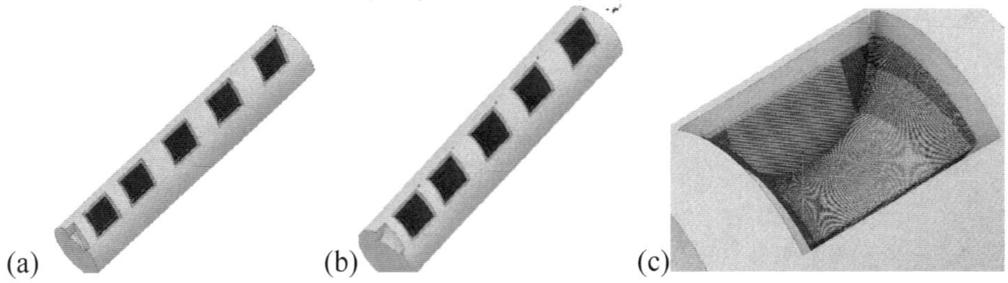

**Figure 11.** Finishing phase: (a) for rectangular bottles; (b) for round bottles; (c) path detail.

The machining time for square bottles progressive shaft per operation is 1 hour and 20 minutes, while for round bottles is around 1 hour and 11 minutes. This difference is caused of the larger section of square profile, so more material to be removed. Total machining time of progressive shaft for round bottles is about 11.25% shorter related to the machining time of progressive shaft for rectangular bottles.

## 4. Conclusions

In this paper is shown combined usage of old and new technologies in order to get optimal solution for progressive shaft manufacturing constrained with certain circumstances: those given by specific technological equipment endowment of a certain plant. This innovative solution allowed processing complex parts with no supplementary expenses for special technological equipment. The limitation in equipment was overridden by introducing an existing fourth axis to standard 3D CNC milling process. The fourth axis was fictive for the milling process. The progressive shaft manufacturing is bring down to making 3D pockets per certain angles. Advantages of used technology are:

- There were no need for additional special made apparatus;
- The both cases for square and round bottles progressive shat positioners was covered;
- Elix profile by this technology can be very customized;
- There were no need to use the complicated functions with massive NC code output;
- There were no need for customized cutting tools.
- Yet there are some disadvantages, as well:
- The shaft length is limited by CNC milling machine table; this can be overcome by joining together several parts end to end;
- Longer machining time related to other technologies;
- By $^{h}$ axis rotation possible errors on joints might occur;
- Profiles with negative edges cannot be performed without special cutting tools;
- Progressive shaft diameter is limited by dividing indexing head.

The presented machining process is very feasible and flexible. However, there is much space for improving the presented process, such as parameterizing the progressive shaft model, parameterizing the machining steps related to the helix profiles, exanimating various cutting tools to get process better etc. The most challenging one is obtaining a continuous increasing pitch, not one divided into segments with constant pitch. The future direction of the research will be mostly pointed to automate the CNC programming process.

**Acknowledgments**

This paper is a result of TR33015 project of the Ministry of education, science and technological development, in Serbia. The project is titled "Research and development of a Serbian zero-net energy house". We would like to thank the Ministry of education, science and technological development on their financial support during this research. Authors wishing to acknowledge assistance or encouragement from colleagues, special work by technical staff or financial support from organizations should do so in an unnumbered.

**References**

[1] Gillie R F 1978 Variable pitch screw drives *Mech. Mach. Theory* **13** 523
[2] Lin P D and Lee M F 1997 NC data generation for 4-axis machine tools equipped with rotary angle head attachments to produce variable pitch screws *Int. J. Mach. Tool. Manu.* **37** 341
[3] Liu J Y and Yan H S 1992 Geometry design of trapezoidal threaded variable pitch lead screws *Computers & Mathematics with Applications* **23** 65
[4] Senanayake S and Clarke B 1999 A simplified twin screw co-rotating food extruder: design, fabrication and testing *J. Food Eng.* **40** 129
[5] Slătineanu L, Dodun O, Coteaţă M, Coman I, Nagîţ G and Beşliu I 2016 Manufacture of threads with variable pitch by using noncircular gears. *In IOP Conference Series: Materials Science and Engineering* **147** 012009 IOP Publishing
[6] Li S, Han M and Deng L 2009 Research on CNC machining screw which is variable pitch & groove depth & groove width *In IEEE 10th International Conference on Computer-Aided Industrial Design & Conceptual Design* 710
[7] Han M, Li S and Deng L T 2011 Study on the computer numerical control process of variable pitch, groove depth and groove width screw *Advanced Materials Research* **201** 85
[8] Nabeta Y, Jain S, Beaucamp A and Takeuchi Y 2020 Development of a New Computer Aided Manufacturing System for Variable Pitch Screws Fabrication *of the Korean Society for Precision Engineering* **37** 195
[9] https://knowledge.autodesk.com/support/autodesk-hsm/learn explore/caas/CloudHelp/cloudhelp /2019/ENU/Inventor HSM/files/GUID-315BB66F-4FFF-4803-A36B-4599F71720DD-htm.html, accessed in May 2020

IOP Conf. Series: Materials Science and Engineering 1009 (2021) 012039 doi:10.1088/1757-899X/1009/1/012039

# Optimization of industrial management processes

**C Mărgineanu[1] and D Lixăndroiu[2]***

[1] Transilvania University of Brașov, Eroilor nr. 29,500036, Brașov, România
[2] Departament of Management and Economic Informatics, Faculty of Economics and Business Administration, Transilvania University of Brasov, Colina Universității nr. 1, Corpul A, etajul III, Brasov România

*E-mail: lixi.d@unitbv.ro

**Abstract**: The paper aims to present a technique for optimizing industrial management processes by mathematical modeling using linear inequality systems. An issue facing an industrial enterprise regarding production costs will be presented and solved, after which the results will be interpreted to formulate the optimal solution. The main phases of solving a problem with the help of linear programming are presented as well as the detailed analysis of the obtained results. The reduction of costs due to the processes presented in the analysis is presented alternatively with the existing costs and the paper details how to save 336 euros per day, as well as an increase in profit by 1935 euros. At the same time, the paper aims to present a software product to support economic decision modeling with which to perform calculations and present the optimal decision.

## 1. Introduction

Currently, along with the spread of the concept of globalization as well as against the background of changes in societies and the world economy, various production optimization techniques have been perfected to keep companies in step with the development of international trade. Against the background of the liberalization of world markets, the import/export policy as well as the investments that exceed the borders of a country, at the production level there is an increasing tendency to use specialized software to plan until the most small detail the use of the company's resources both material and time.

The accelerated development of technology and the rapid exchange of information impose new adapted and interdisciplinary requirements and approaches. These influences are felt both at the individual level and at the micro or macroeconomic level. Businesses are forced to be flexible and implement the changes generated by the development of technology in order to have sustainable growth and long-term success in the market. Well-defined strategies and well-developed procedures with long-term systematic approaches not only turn current dynamics into an internal advantage, but also have the ability to identify opportunities early [1].

The management of current industrial processes by enterprises has become increasingly complex, imposing new requirements for the implementation of effective decision support tools. Manual production planning has been replaced with integrated computer systems that can be accessed in real-time. This leads to a thorough control of the processes and a rigorous analysis of the information and feedback received [2].

In these circumstances, to maintain a high standard and to optimize processes, real-time communication is an absolute necessity that can only be met by a well-developed computer system geared to the needs of the undertaking concerned. Production process optimization is the technique of adjusting a process to perfect a specified set of parameters without violating certain constraints. The most common objectives are to minimize costs and maximize efficiency and/or efficiency.

The programs dedicated to solving production optimization problems are composed of tools, methods, processes, and mathematical models that guarantee a better approach to the technological process. The enterprise software development industry is growing at a rapid pace due to the fact that many organizations today use information technology to gain a competitive advantage [3].

The paper aims to present a technique for optimizing industrial management processes by mathematical modeling using linear inequality systems. The novelty proposed by the paper is the combination of a minimum function and a maximum function using linear programming techniques developed by a free software that can be used by an enterprise that does not have an integrated ERP system or software dedicated to optimizing technological processes.

## 2. Theoretical notions of linear programming

The steps to solving a mathematical problem require different calculations. Depending on their complexity, these calculations can be performed either manually or using a computer. This whole process generates an effort that is made in order to obtain the desired results and has led to the development of different problem-solving techniques [4].

Linear programming is a mathematical technique in which relationships with a high degree of complexity are represented by linear functions where after solving them the optimal points are generated. The applicability of this mathematical process is very common, even in everyday life, from optimizing production processes or optimizing the supply policy of goods to choosing a route between home and work where the goal is to choose the shortest route.

A linear programming problem is characterized by the fact that it represents through linear functions both the objective function and the constraints imposed by the enterprise. The solution to such a problem is based on the mathematical apparatus offered by linear algebra, the SIMPLEX algorithm, and the dual SIMPLEX algorithm [5].

The general form of a linear programming problem is given by the function:

$$Max\,[Min]F = c_1 \cdot x_1 + c_2 \cdot x_2 + \cdots + c_n \cdot x_n \quad (1)$$

The constraints imposed on this function are written as:

$$a_{11} \cdot x_1 + a_{12} \cdot x_2 + \cdots + a_{1n} \cdot x_n \leq b_1 \quad (2)$$
$$a_{21} \cdot x_1 + a_{22} \cdot x_2 + \cdots + a_{2n} \cdot x_n \leq b_2 \quad (3)$$

......................................................................

$$a_{m1} \cdot x_1 + a_{m2} \cdot x_2 + \cdots + a_{mn} \cdot x_n \leq b_m \quad (4)$$

and

$$x_1 > 0\ , x_2 > 0\ , \dots\ , x_n > 0 \quad (5)$$

where:

$x_i$ = unknown level of the studied product;
$c_i$ = p - d and represents the profit brought by a unit of product denoted by $x_i$, being calculated as the difference between the selling price (p) and the production cost (d);
$b_i$ = known level of resources;
$a_{ij}$ = quantity of resource i used to produce a unit of product j [5].

## 3. Presentation of a production optimization software

QM is an accessible software package for assisting managerial decisions on production/operations management issues. This software package is available for download for Windows operating systems, being easy to use even for those without computer training [6].

The graphical interface of this software is very intuitive and is similar to a program in the Microsoft Office package. A user familiar with a text editing program, spreadsheet program, or presentation package will easily be able to use this application. The program's interface includes the basic menu, toolbar, status bar, and help files that can be found in Windows programs.

To solve a linear programming problem the user will open the application and from the MODULE menu will select Linear Programming. After selecting the module, the user needs to press the New button, enter the number of variables and constraints that the problem has, and select whether we want to maximize or minimize a function. Besides that, the user will enter the function and the constraints in a specific table and press the solve button.

The optimal solution, if any, is presented together with the margins of the variables, the loss of resources, as well as together with graphs for a better understanding of the optimal decision presented.

The results obtained, the data, the graphs, and the solving steps can be easily exported or printed. At the same time, the software allows the comparison of the results and displays step by step the solution of the Simplex algorithm in case of linear programming problems. QM also assists with transport problems, forecast, inventory, breakeven point, etc.

## 4. Case study on production optimization conducted within Alpha enterprise

### *4.1. Maximizing profit*

Alpha enterprise, from Brasov, Romania produces six models of wooden kitchen furniture. To make a furniture product for sale it must go through six processing operations:

- O1 - sorting of raw materials,
- O2 - the processing of raw materials,
- O3 –assembling parts,
- O4 - applying the veneer,
- O5 - finishing and
- O6 - painting, varnishing, packaging.

The models and profits they bring are:

**Table 1.** Sales price, production cost and profit of each product in euros.

| Product → | EB7-P | EB7-MD | EB7-N | EM7-CZ | EM7-AB | EM7-L |
|---|---|---|---|---|---|---|
| **Sales price** | 294 | 322 | 308 | 273 | 316 | 303 |
| **Production cost** | 252 | 271 | 253 | 225 | 256 | 245 |
| **Profit** | 42 | 51 | 55 | 48 | 60 | 58 |

The time required for operations in minutes for each model and the time in hours for each operation are shown in table 2.

The marketing department, after an analysis of demand and supply, imposed the following restrictions:

- No product must exceed 30% of total production;
- Eb7-n and em7-l models to be produced at least 20% of the total production each;
- All products must be at least 10% of the total production each.

The company currently produces 95 EB7-P products, 113 EB7-MD products, 120 EB7-N products, 121 EM7-CZ products, 130 EM7-AB products, and 135 EB7-L products.

The production process generates a monthly profit of 37791 euros but does not comply with the second restriction imposed by the marketing department, the EB7-N model producing in a proportion of approximately 17% and the EM7-L model producing in a proportion of approximately 19 %.

**Table 2.** Production times required for each product for each operation.

| MODEL | O1-SRM | O2-PRM | O3-AP | O4-AV | O5-F | O6-PVP |
|---|---|---|---|---|---|---|
| EC7-P - marked a | 26 | 168 | 65 | 58 | 85 | 121 |
| EC7-MD - marked b | 24 | 154 | 69 | 65 | 87 | 130 |
| EC7-N - marked c | 28 | 198 | 63 | 62 | 69 | 97 |
| EC7-ES - marked d | 29 | 219 | 96 | 60 | 86 | 200 |
| EC7-CM5 - marked e | 25 | 205 | 71 | 61 | 98 | 185 |
| EC7-W6 - marked f | 37 | 200 | 85 | 54 | 101 | 195 |
| Availabletime (hours) | 1120 | 2324 | 1520 | 1423 | 1800 | 2000 |
| Available time (min) | 67200 | 139440 | 91200 | 85380 | 108000 | 120000 |

Objective function:

$$[\max]\ F\ (a;b;c;d;e;f) = 42a + 51b + 55c + 48d + 60e + 58f \quad (6)$$

Constraints:

O1: $26a + 24b + 28c + 29d + 25e + 37f \leq 67200$ (7)
O2: $168a + 154b + 198c + 219d + 205e + 200f \leq 139440$ (8)
O3: $65a + 69b + 63c + 96d + 71e + 85f \leq 91200$ (9)
O4: $58a + 65b + 62c + 60d + 61e + 54f \leq 85380$ (10)
O5: $85a + 87b + 69c + 86d + 98e + 101f \leq 108000$ (11)
O6: $121a + 130b + 97c + 200d + 185e + 195f \leq 120000$ (12)

Product criteria: all products must be equal to or less than 30% of total products:

C1: $a \leq 3/10\ (a + b + c + d + e + f) \rightarrow 7a - 3b - 3c - 3d - 3e - 3f \leq 0$ (13)
C2: $7b - 3a - 3c - 3d - 3e - 3f \leq 0$ (14)
C3: $7c - 3a - 3b - 3d - 3e - 3f \leq 0$ (15)
C4: $7d - 3a - 3b - 3c - 3e - 3f \leq 0$ (16)
C5: $7e - 3a - 3b - 3c - 3d - 3f \leq 0$ (17)
C6: $7f - 3a - 3b - 3c - 3d - 3e \leq 0$ (18)

Product criteria: product c and product f must be at least 20% of total products:

C7: $c \geq 20/100\ (a + b + c + d + e + f) \rightarrow 4c - a - b - d - e - f \geq 0$ (19)
C8: $f \geq 20/100\ (a + b + c + d + e + f) \rightarrow 4f - a - b - c - d - e \geq 0$ (20)

Product criteria: all products must be equal to or more than 10% of total products:

C9: $a \geq 10/100\ (a + b + c + d + e + f) \rightarrow 9a - b - c - d - e - f \geq 0$ (21)
C10: $9b - a - c - d - e - f \geq 0$
C11: $9c - a - b - d - e - f \geq 0$ (22)
C12: $9d - a - b - c - e - f \geq 0$ (23)
C13: $9e - a - b - c - d - f \geq 0$ (24)
C14: $9f - a - b - c - d - e \geq 0$ (25)
R1: $a, b, c, d, e, f, \geq 0$ (26)

After entering the data in specialized software (QM V5) the following results were obtained:

| | a | b | c | d | e | f | | RHS | Dual |
|---|---|---|---|---|---|---|---|---|---|
| 5 | 85 | 87 | 69 | 86 | 98 | 101 | <= | 108000 | 0 |
| 6 | 121 | 130 | 97 | 200 | 185 | 195 | <= | 120000 | 0 |
| 7 | 7 | -3 | -3 | -3 | -3 | -3 | <= | 0 | 0 |
| 8 | -3 | 7 | -3 | -3 | -3 | -3 | <= | 0 | 0 |
| 9 | -3 | -3 | 7 | -3 | -3 | -3 | <= | 0 | 0 |
| 10 | -3 | -3 | -3 | 7 | -3 | -3 | <= | 0 | 0 |
| 11 | -3 | -3 | -3 | -3 | 7 | -3 | <= | 0 | 0 |
| 12 | -3 | -3 | -3 | -3 | -3 | 7 | <= | 0 | 0 |
| 13 | -1 | -1 | 4 | -1 | -1 | -1 | >= | 0 | -1.72 |
| 14 | -1 | -1 | -1 | -1 | -1 | 4 | >= | 0 | -1.23 |
| 15 | 9 | -1 | -1 | -1 | -1 | -1 | >= | 0 | -1.3 |
| 16 | -1 | 9 | -1 | -1 | -1 | -1 | >= | 0 | 0 |
| 17 | -1 | -1 | 9 | -1 | -1 | -1 | >= | 0 | 0 |
| 18 | -1 | -1 | -1 | 9 | -1 | -1 | >= | 0 | -2.16 |
| 19 | -1 | -1 | -1 | -1 | 9 | -1 | >= | 0 | -.56 |
| 20 | -1 | -1 | -1 | -1 | -1 | 9 | >= | 0 | 0 |
| Solution | 75.37 | 226.12 | 150.75 | 75.37 | 75.37 | 150.75 | | 39872.3 | |

**Figure 1.** Display of optimal results.

**Table 3.** The comparative situation between the initial scenario and the optimally calculated one.

| MODEL | The current scenario | | | The optimal calculated scenario | | |
|---|---|---|---|---|---|---|
| | Nr. pieces produced | Profit per unit | Total profit per unit | Nr. optimal product pieces | Profit per unit | Total profit per unit |
| EB7-P | 95 | 42 | 3990 | 75 | 42 | 3150 |
| EB7-MD | 113 | 51 | 5763 | 226 | 51 | 11526 |
| EB7-N | 120 | 55 | 6600 | 150 | 55 | 8250 |
| EM7-CZ | 121 | 48 | 5808 | 75 | 48 | 3600 |
| EM7-AB | 130 | 60 | 7800 | 75 | 60 | 4500 |
| EM7-L | 135 | 58 | 7830 | 150 | 58 | 8700 |
| TOTAL PROFIT | | **37791** | | | **39726** | |

It is noted that the total profit can increase by 1935 euros to the value of 39726 euros compared to 37791 the current value which is an increase of about 5.13%. At the same time, this software presents the following situation:

| Variable | Value | Reduced ... | Original Val | Lower Bou... | Upper Bou... |
|---|---|---|---|---|---|
| a | 75.37 | 0 | 42 | -160.53 | 55.1 |
| b | 226.12 | 0 | 51 | 45.84 | Infinity |
| c | 150.75 | 0 | 55 | -46.26 | 64.01 |
| d | 75.37 | 0 | 48 | -154.53 | 70.37 |
| e | 75.37 | 0 | 60 | -187.48 | 65.74 |
| f | 150.75 | 0 | 58 | -43.26 | 64.48 |

**Figure 2.** Display of profit ranges.

The circled section shows the intervals in which the variables can be framed individually to obtain the same results. It follows that the product profit of the product marked with a can vary in the range (-160.53, 55.1) and so on. To obtain the same profit, however, only one value can be changed.

| | Dual Value | Slack/Surp... | Original Val | Lower Bou... | Upper Bou... |
|---|---|---|---|---|---|
| O1 | 0 | 45944.82 | 67200 | 21255.18 | Infinity |
| O2 | .29 | 0 | 139440 | 0 | 150000 |
| O3 | 0 | 35800.86 | 91200 | 55399.14 | Infinity |
| O4 | 0 | 39703.98 | 85380 | 45676.02 | Infinity |
| O5 | 0 | 42425.51 | 108000 | 65574.48 | Infinity |
| O6 | 0 | 8448 | 120000 | 111552 | Infinity |

**Figure 3.** Display of resource losses.

The circled section shows the time wasted time for each technological operation. Thus, it is observed that only the operation of raw materials processing (O2) is optimally standardized compared to other processes that have significant losses of time. For example, for sorting raw materials (O1) the optimal time to be allocated is 21255.18 minutes compared to 67200 minutes resulting in a loss of 45944.82 minutes.

### *4.2. Minimizing costs*

Within the same company, the production staff has a schedule from Monday to Friday between 8-17. Based on previous experience the necessary personnel for the supervision of the machines as well as for the direct relationship with the carriers and suppliers differs during the program as follows:

**Table 4.** Need for detailed staff by hours.

| Time interval | The need for staff |
|---|---|
| 8-9 | 5 |
| 9-10 | 6 |
| 10-11 | 5 |
| 11-12 | 6 |
| 12-13 | 8 |
| 13-14 | 9 |
| 14-15 | 7 |
| 15-16 | 8 |
| 16-17 | 5 |

Alpha enterprise can have two categories of employees:

1. Full-time employee (8-17), 5 days a week with a one-hour unpaid break between (12-13) or between (13-14), with a salary of 18 euros per hour;
2. The part-time employee who works 4 consecutive hours every day with a salary of 12 euros per hour. At the same time, the board of directors decided to limit part-time staff to a maximum of 5 employees.

The company currently employs 8 full-time and 6 part-time employees. Full-time employees are divided equally, four of them have a break from 12-13 and four have a break from 13-14. Part-time employees have a fixed schedule from 11 am to 3 pm. The production staff generates a cost of 1440 euros per day and an average hourly cost of 180 euros per hour.

To optimize the work schedule of employees and reduce costs, the following work schedules were taken into account.

**Table 5.** Distribution of staff

| Employees \ Time interval | 8-9 | 9-10 | 10-11 | 11-12 | 12-13 | 13-14 | 14-15 | 15-16 | 16-17 |
|---|---|---|---|---|---|---|---|---|---|
| Full-time lunch break at 13-14 (a) | ■ | ■ | ■ | | ■ | ■ | ■ | ■ | ■ |
| Full-time lunch break at 13-14 (b) | ■ | ■ | ■ | ■ | | ■ | ■ | ■ | ■ |
| Full-time lunch break at 13-14 (c) | ■ | ■ | ■ | ■ | ■ | | ■ | ■ | ■ |
| Part-time 8-12 (d) | ■ | ■ | ■ | ■ | | | | | |
| Part-time 9-13 (e) | | ■ | ■ | ■ | ■ | | | | |
| Part-time 10-14 (f) | | | ■ | ■ | ■ | ■ | | | |
| Part-time 11-15 (g) | | | | ■ | ■ | ■ | ■ | | |
| Part-time 12-16 (h) | | | | | ■ | ■ | ■ | ■ | |
| Part-time 13-17 (i) | | | | | | ■ | ■ | ■ | ■ |

Objective function:

$$\text{Min } F(a;b;c;d;e;f;g;h;i) = 144a + 144b + 144c + 48d + 48e + 48f + 48g + 48h + 48i \quad (27)$$

Constraints:

R1: $a + b + c + d \geq 5$ (28)

R2: $a + b + c + d + e \geq 6$ (29)

R3: $a + b + c + d + e + f \geq 5$ (30)

R4: $b + c + d + e + f + g \geq 6$ (31)

R5: $a + c + e + f + g + h \geq 8$ (32)

R6: $a + b + f + g + h + i \geq 9$ (33)

R7: $a + b + c + g + h + i \geq 7$ (34)

R8: $a + b + c + h + i \geq 8$ (35)

R9: $a + b + c + i \geq 5$ (36)

R10: $d + e + f + g + h + i \leq 5$ (37)

| | a | b | c | d | e | f | g | h | i | | RHS |
|---|---|---|---|---|---|---|---|---|---|---|---|
| Minimize | 144 | 144 | 144 | 48 | 48 | 48 | 48 | 48 | 48 | | |
| R 1 | 1 | 1 | 1 | 1 | 0 | 0 | 0 | 0 | 0 | >= | 5 |
| R 2 | 1 | 1 | 1 | 1 | 1 | 0 | 0 | 0 | 0 | >= | 6 |
| R 3 | 1 | 1 | 1 | 1 | 1 | 1 | 0 | 0 | 0 | >= | 5 |
| R 4 | 0 | 1 | 1 | 1 | 1 | 1 | 1 | 0 | 0 | >= | 6 |
| R 5 | 1 | 0 | 1 | 0 | 1 | 1 | 1 | 1 | 0 | >= | 8 |
| R 6 | 1 | 1 | 0 | 0 | 0 | 1 | 1 | 1 | 1 | >= | 9 |
| R 7 | 1 | 1 | 1 | 0 | 0 | 0 | 1 | 1 | 1 | >= | 7 |
| R 8 | 1 | 1 | 1 | 0 | 0 | 0 | 0 | 1 | 1 | >= | 8 |
| R 9 | 1 | 1 | 1 | 0 | 0 | 0 | 0 | 0 | 1 | >= | 5 |
| R 10 | 0 | 0 | 0 | 1 | 1 | 1 | 1 | 1 | 1 | <= | 5 |
| Variable type (click to set) | Integer | Integer | Integer | Integer | Integer | Integer | Integer | Integer | Integer | | |
| Solution-> | 2 | 2 | 2 | 0 | 0 | 3 | 0 | 2 | 0 | Optim... | 1104 |

**Figure 4.** The optimal management solution for productive personnel

**Table 6.** Current staff distribution compared to the optimal staff distribution.

| Time interval / Employees | Optimal number of employees | Number of employees currently | 8-9 | 9-10 | 10-11 | 11-12 | 12-13 | 13-14 | 14-15 | 15-16 | 16-17 |
|---|---|---|---|---|---|---|---|---|---|---|---|
| Full-time lunch break at 11-12 (a) | **2** | **4** | | | | | | | | | |
| Full-time lunch break at 12-13 (b) | **2** | **4** | | | | | | | | | |
| Full-time lunch break at 13-14 (c) | **2** | **0** | | | | | | | | | |
| Part-time 10-14(f) | **3** | **0** | | | | | | | | | |
| Part-time 11-15(g) | **0** | **6** | | | | | | | | | |
| Part-time 12-16(h) | **2** | **0** | | | | | | | | | |
| Need of staff by time interval | | | **5** | **6** | **5** | **6** | **8** | **9** | **7** | **8** | **5** |
| Current situation | | | **8** | **8** | **8** | **10** | **10** | **14** | **14** | **8** | **8** |
| Current scenario costs (euro/day) | | | | | | | **1440** | | | | |
| Optimal situation | | | **6** | **6** | **9** | **7** | **9** | **9** | **8** | **8** | **6** |
| Optimal scenario costs (euro/day) | | | | | | | **1104** | | | | |
| Saved difference (euro/day) | | | | | | | **336** | | | | |

After the analysis of the personnel distribution, it results that the company could save 336 euros per day in salary costs, which represents a 23.33% reduction from a cost of 1440 lei per day to a cost of 1104 euros per day. At an average of 21 working days per month, the company would save 7056 euros in salary costs. This saving is possible in the case of reorganizing the distribution of staff to 6 full-time employees, and 5 part-time employees with the schedule shown in table 5.

## 5. Conclusions

Nowadays, the decision-making process has grown exponentially in complexity, forcing managers to adapt, using dedicated software to make optimized, clear and precise decisions. Production organization techniques were automated by abandoning manual calculations.

With the simplification of international trade formalities as well as the expansion of online trade, a business to grow sustainably is obliged to optimize its production and technological processes. The globalization of the economy and the increasingly competitive markets in the industrial field force the managers to maintain the high standard, being assisted by well-developed information systems oriented to the needs of the enterprise in question.

Managerial decision support software products can become prohibitive due to the high monthly costs being considered a luxury for small businesses which forces managers to opt for free licensed software such as QM (the software used in this paper to perform the calculations).

By making a simple calculation, the company can maximize its profit by 1935 euros, which represents an increase of about 5.13% of the profits currently obtained. At the same time, the analysis through linear programming techniques shows that the company can reorganize the time allocated to each industrial process to avoid unnecessary loss of resources. The staff cost analysis also indicates that the company can minimize costs by about 23.33% per day, generating a monthly saving of about 7056 euros.

All these cost reductions and profit increases help the industrial enterprise to be more competitive in the market and to continue to have a sustainable growth of the turnover as well as it offers the possibility of expansion. at the same time, the presented analysis offers a perspective on the working times that are lost through a wrong standardization and presents the technological processes that consume a lot of working time of the productive employees.

## References

[1] Mehedintu G 2016 Optimizarea proceselor de lucrul prin intermediul managemetului schimbării (Optimizing work processes through management change) *The National Conference with International participation of the Faculty of Legal, Economic and Administrative Sciences, Brasov*

[2] Vieira M, Pinto-Varela T and Barbosa-Póvoa A 2019 A model-based decision support framework for the optimisation of production planning in the biopharmaceutical industry, *Computers & Industrial Engineering* **129** 354

[3] Gheisari M, Pham Q, Alazab M, Zhang X, Fernández-Campusano C and Srivastava G 2019 ECA: An Edge Computing Architecture for Privacy-Preserving in IoT-Based Smart City *Access IEEE* **7** 155779

[4] Eiselt H and Sandblom C L 2007 Linear Programming and its Applications *Ekonomski horizonti* **16** 77

[5] Lixăndroiu D 2014 *Modelarea deciziei economice (Modelling economic decision) (*Economica, București) p 29

[6] https://qm-for-windows.software.informer.com/5.2/, link accessed on 10.06.2020.

IOP Conf. Series: Materials Science and Engineering 1009 (2021) 012040 doi:10.1088/1757-899X/1009/1/012040

# A comparative assessment of reliability indicators and parameters of the bearings using different estimation methods

**C O Morariu**[1*] **and S M Zaharia**[1]

[1]Department of Manufacturing Engineering, Transilvania University of Brasov, 29 Eroilor, Brasov 500036, Romania

*E-mail: c.morariu@unitbv.ro

**Abstract**. The paper depicts the method of statistical processing of experimental results, obtained by testing bearings on the test stands, using the maximum likelihood estimators and/or Best Linear Invariant Estimator. To verify statistical assumptions regarding parameters and reliability indicators of bearings, as well as, to determine confidence intervals, the article contains two original algorithms that allow obtaining values of some specific random variables by Monte Carlo numerical simulation. The statistical inferences are made through these values. The paper also, contains two case studies using data from failures and time truncated tests of bearings.

## 1. Introduction

Bearings manufacturing is an industrial field where product testing is performed planned in specialized laboratories, using types of equipment to ensure identical testing conditions for all products analyzed.

After the purpose of the test, tests shall be carried out, in the bearings manufacture, to check the reliability indicators of the bearings. This test is also called a typing test and shall be repeated periodically at well-specified intervals.

After the test procedure, the following shall be used in the manufacture of the bearings [1, 2, 3]:

- Complete tests (type $n$ out of $n$). Complete tests are considered to be those tests carried out until all the products under test have failed. This type of test was practiced only at the beginning of the manufacture of bearings, later giving them up, taking into account the very long duration of the test;
- Censored tests or failures-truncated tests type $r$ out of $n$, are commonly used. A test is considered to be censored at the level $r$, if it is completed at the time of the occurrence of the $r$ failure (the number of damages being established before the beginning of the tests), figure 1;
- Time - truncated tests (with a fixed testing time, $t_c$). In this reliability test, the test is completed after a previously set duration $t_c$, figure 2.
- After the working mode, accelerated tests are used in the case of bearings reliability tests. To shorten the duration of the tests, the constraints shall be increased to the operating conditions [1, 4].
- The acceleration factor shall be determined with the relationship of the basic rating life for dynamically loaded ball bearings:

$$L_{10h} = \frac{1000000}{60 \cdot n} \cdot \left(\frac{C}{P}\right)^q \quad (1)$$

where:

$C$ = basic dynamic load rating $[N]$;
$P$ = equivalent load $[N]$;
$n$ = operating speed [rpm];
$q$ = exponent, having the values 3.0 for ball bearings and $10/3$ for roller bearings.

By volume of products, tests specific to the manufacture of bearings are tests carried out on small volume samples.

Depending on the nature of the products, the bearings are considered as irreparable elements.

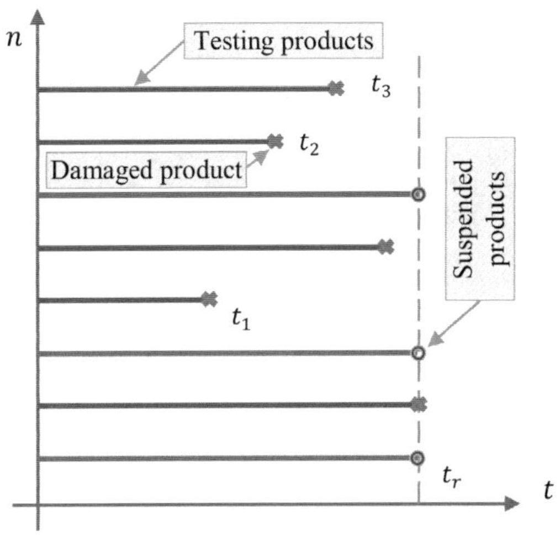


**Figure 1.** Censored tests.

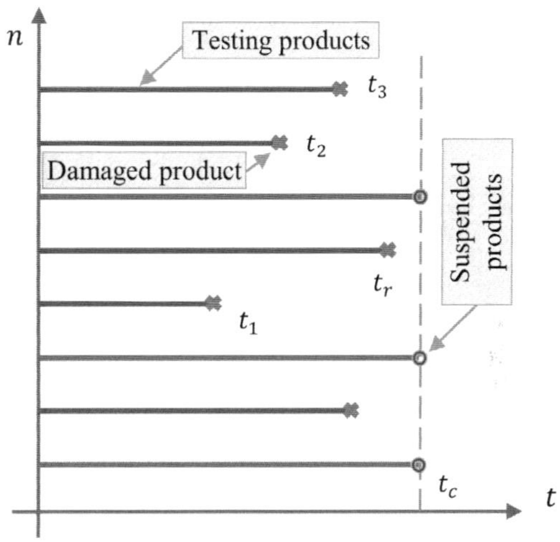


**Figure 2**. Time - truncated tests.

## 2. The reliability testing of bearings

The methodology of testing is a complex process (figure 3), which is carried out in several stages [5]:

- Sampling of the bearings lot;
- Visual appearance and measurement of external geometry (macro geometry);
- Measurement of the level of vibration;
- Measurement of internal geometry (microgeometry);
- Chemical-metallographic and physical-mechanical control;
- Reliability tests and statistical processing of experimental results;
- Post-mortem chemical-metallographic analysis of the damaged bearings;
- Witness bearings;
- Elaboration of the Experimental Report.

Of these steps, the most important is the reliability tests on the test stands, since the values of the reliability indicators obtained by statistical processing of the experimental results are extrapolated to the entire population.

For this reason, the parameters that characterize the test must be chosen very carefully, namely:

1. Sample volume ($n$) and number of items damaged during a test ($r$);
2. The operating condition of the bearings on the stands. There are two ways of accelerating bearings testing [68]:
   a. By increasing the test speed. It is recommended [1, 4]:

$$n \leq ( 0{,}4 \div 0{,}6 ) \cdot n_{lim} \quad (2)$$

where, $n_{lim}$, represents the catalog value of the limit speed, $[rpm]$.

The value of the speed of the test is limited by the dynamic effects that occur (changing the angle of the contact, a significant increase in centrifugal forces acting on rolling bodies, change in lubricant film sizes and load distribution in the case of contacts, uneven loading of rolling bodies, increasing the speed of the sliding of the rolling element and raceways, etc.), and which have not been considered in the calculation of the basic rating life.

b. By increasing the stresses applied to the tested bearings.
Equivalent dynamic load values that require bearings during the test are limited by the behavior of the material to contact fatigue [4].
The maximum recommended value [4], for Hertzian stress in the case of point contact is $\sigma_{max} = 3300\ MPa$. Thus [4], are recommended:
- For testing ball bearings: $C/P < 3.0$;
- For oscillating ball bearings: $C/P < 8.0$;
- For roller bearings: $C/P < 4.5$.

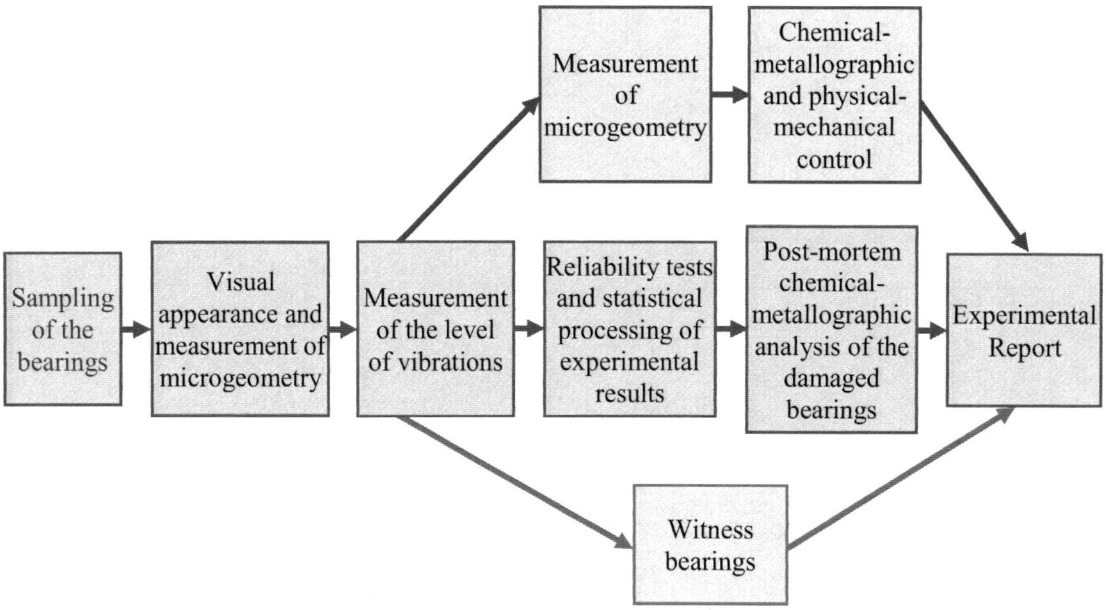


**Figure 3.** The methodology of reliability testing.

3. The estimation method used. For the statistical processing of the experimental results obtained from the reliability tests of the bearings and the estimation of the values of the reliability indicators are used:
   a. Graphical estimation methods [2, 3]. These are only recommended to provide consistent information on experimental data quickly;
   b. The least squares method [1, 2];
   c. Maximum likelihood method (MLE);
   d. Linear estimators, applied to censored type tests [6, 7].

## 3. Estimation methods used

The statistical model used for modelling reliability of bearings is the Weibull distribution. It is recommended to use the general case of the tri-parametric Weibull distribution, $\mathcal{W}(x, \beta, \eta, \gamma)$, for an accurate estimation, having the probability density function:

$$f(T) = \frac{\beta}{\eta} \cdot \left(\frac{T-\gamma}{\eta}\right)^{\beta-1} \cdot e^{-\left(\frac{T-\gamma}{\eta}\right)^{\beta}}, \qquad (T > \gamma;\ \beta, \eta > 0) \tag{3}$$

where: $\beta$ = shape parameter of Weibull distribution;
$\eta$ = scale parameter of Weibull distribution;
$\gamma$ = location parameter of Weibull distribution.

If we estimate the value of the scale parameter, $\gamma$, by a transformation of a random variable,

$$t = T - \gamma, \tag{4}$$

is obtained the model of the bi-parametric Weibull distribution, $\mathcal{W}(x, \beta, \eta)$, having the probability density function:

$$f(T) = \frac{\beta}{\eta} \cdot \left(\frac{t}{\eta}\right)^{\beta-1} \cdot e^{-\left(\frac{t}{\eta}\right)^{\beta}}, \text{ where } t > 0;\ \beta, \eta > 0 \tag{5}$$

The location parameter, $\gamma$, is estimated using the correlation coefficient ($\rho$) under conditions of a maximum value of this coefficient [5]:

$$\rho(T, C) = \frac{\sum_{i=1}^{n} ln(T_i - \gamma) \cdot c_i - \frac{\sum_{i=1}^{n} ln(T_i - \gamma) \cdot \sum_{i=1}^{n} c_i \cdot}{n}}{\left\{\left[\sum_{i=1}^{n} ln^2(T_i - \gamma) - \frac{\left[\sum_{i=1}^{n} ln(T_i - \gamma)\right]^2}{n}\right] \cdot \left[\sum_{i=1}^{n} c_i^2 - \frac{\left(\sum_{i=1}^{n} c_i\right)^2}{n}\right]\right\}^{1/2}} \tag{6}$$

In the previous equation, the values of the deterioration probabilities values, $c_i$, are calculated using the approximate computing relationship [2, 3]:

$$c_i = \frac{1}{1 + \frac{n - i + 1}{n} \cdot F_{0.50;2\cdot(n-i+1);2\cdot i}}, i = \overline{1, n} \tag{7}$$

where: $F_{p;v1;v2}$ = the $p$ quantile of Fisher-Snedecor's $F$-distribution, with $v1$ and $v2$ degrees of freedom.

Among the estimation methods, listed above, the maximum likelihood method, MLE, is usually used, due to the remarkable properties of these point estimators.

MLE is based on the likelihood function, $\mathcal{L}(t_i, \beta, \eta)$. For failures-truncated tests type $r$ out of $n$, the likelihood function is:

$$\mathcal{L}(t_i, \beta, \eta) = \frac{n!}{(n-r)!} \prod_{i=1}^{r} f(t_i, \beta, \eta) \cdot [1 - F(t_r, \beta, \eta)]^{n-r} \tag{8}$$

For the convenience of the calculation, it is usual to work with the logarithm of the probability function and the punctual estimates are obtained from the maximum condition of this function, as a solution of the system of equations:

$$\begin{cases} \frac{1}{\hat{\beta}} + \frac{1}{r} \cdot \sum_{i=1}^{r} lnt_i - \frac{\sum_{i=1}^{r} t_i^{\hat{\beta}} \cdot lnt_i + (n-r) \cdot t_r^{\hat{\beta}} \cdot lnt_r^{\hat{\beta}}}{\sum_{i=1}^{r} t_i^{\hat{\beta}} + (n-r) \cdot t_r^{\hat{\beta}} \cdot lnt_r^{\hat{\beta}}} = 0 \\ \hat{\eta}^{\hat{\beta}} = \frac{1}{r} \cdot \left[\sum_{i=1}^{r} t_i^{\hat{\beta}} + (n-r) \cdot t_r^{\hat{\beta}} \cdot lnt_r^{\hat{\beta}}\right] \end{cases} \tag{9}$$

The main indicator of the reliability of bearings, besides the Weibull distribution parameters, is the basic rating life, $L_{10}$, for a bearing or for a group of identical bearings which operate under the same conditions is durability associated with a reliability of 90%, at a conventional material quality manufacturing and operating conditions:

$$L_{10} = \gamma + \eta \cdot \left[ln\left(\frac{1}{1 - 0.10}\right)\right]^{\frac{1}{\beta}} \tag{10}$$

In other words, nominal durability is the 10th quantile of the functioning period, $t_{0.10}$, depending on the test procedure used in the manufacturing process of bearings.

Because the sample size is very small, the asymptotically properties of maximum likelihood estimators cannot be applied in this case, the estimations with confidence intervals were calculated based on three random variables, independent of the sample size ($n$) and the type of test [3, 7, 8]:

$$\begin{cases} v(r,n) = \dfrac{\hat{\beta}}{\beta} \\ k(r,n) = \hat{\beta} \cdot ln\left(\dfrac{\hat{\eta}}{\eta}\right) \\ u(r,n,p) = \hat{\beta} \cdot ln\left(\dfrac{\hat{t}_p}{t_p}\right) \end{cases} \quad (11)$$

The determination of these distributions is performed by the numerical simulation Monte Carlo. The original algorithm, developed in this sense, involves following steps presented in the figure 4.

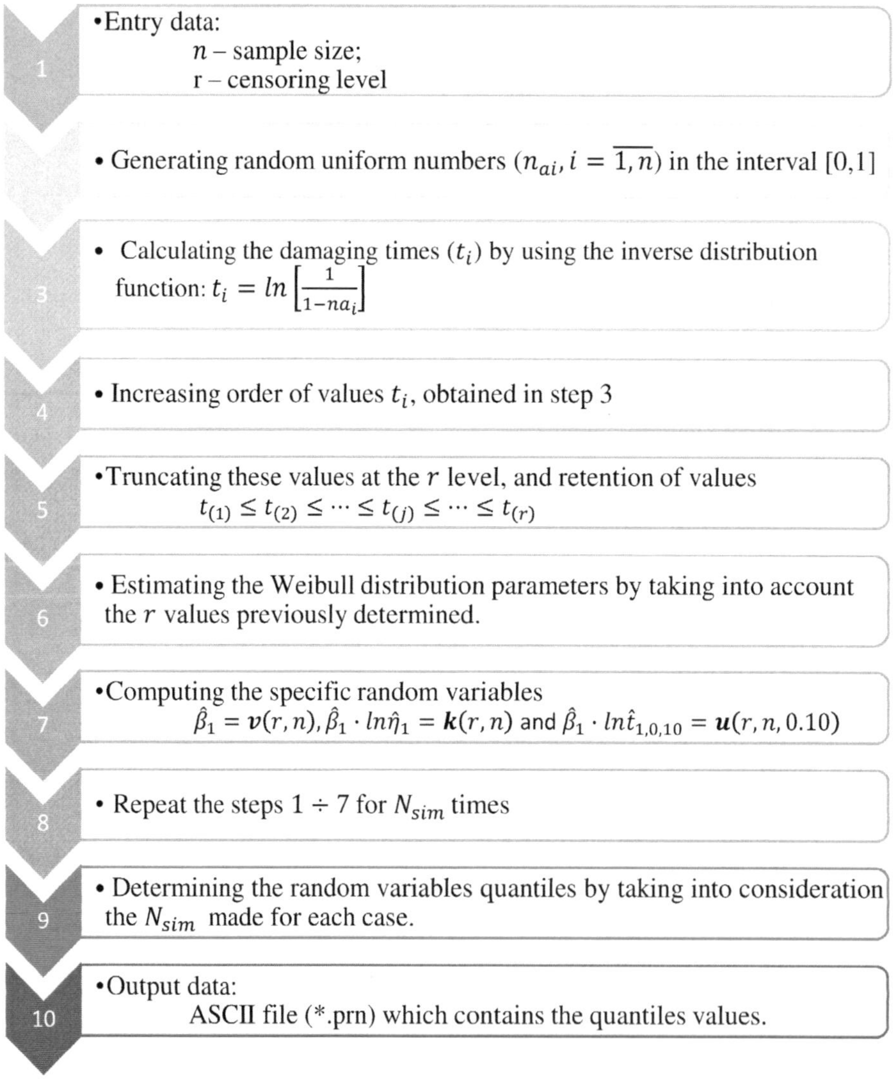


**Figure 4.** The algorithm used for calculating the values for $v(r,n), k(r,n)$ and $u(r,n,p)$.

Based on the original algorithm was carried out a Mathcad program for:

$$N_{sim} = 10000 \text{ simulations.}$$

For the shape parameter, the confidence interval (CI) corresponding to a confidence level $1-\alpha$, it is obtained as the solution of the equation of the probability:

$$Pr\left[v_{\alpha/2}(r,n) \le \frac{\hat{\beta}}{\beta} \le v_{1-\alpha/2}(r,n)\right] = 1-\alpha \tag{12}$$

and it has the form, [9]:

$$\frac{\hat{\beta}}{v_{1-\alpha/2}(r,n)} \le \beta \le \frac{\hat{\beta}}{v_{\alpha/2}(r,n)} \tag{13}$$

For the basic rating life, $L_{10}$, the corresponding CI is:

$$\hat{L}_{10} \cdot \left[e^{\frac{-u_{1-\alpha/2}(r,n,0.10)}{\hat{\beta}}}\right] \le L_{10} \le \hat{L}_{10} \cdot \left[e^{\frac{-u_{\alpha/2}(r,n,0.10)}{\hat{\beta}}}\right] \tag{14}$$

For Time - truncated tests with a fixed testing time, $t_c$, the likelihood function is:

$$\mathcal{L}(t_i,\beta,\eta) = \frac{n!}{(n-r)!}\prod_{i=1}^{r} f(t_i,\beta,\eta) \cdot [1-F(t_c,\beta,\eta)]^{n-r} \tag{15}$$

The maximum likelihood estimates are, therefore, the solutions of the following system of equations:

$$\begin{cases} \frac{1}{\hat{\beta}} + \frac{1}{r} \cdot \sum_{i=1}^{r} lnt_i - \frac{\sum_{i=1}^{r} t_i^{\hat{\beta}} \cdot lnt_i + (n-r) \cdot t_c^{\hat{\beta}} \cdot lnt_c^{\hat{\beta}}}{\sum_{i=1}^{r} t_i^{\hat{\beta}} + (n-r) \cdot t_r^{\hat{\beta}} \cdot lnt_r^{\hat{\beta}}} = 0 \\ \hat{\eta}^{\hat{\beta}} = \frac{1}{r} \cdot \left[\sum_{i=1}^{r} t_i^{\hat{\beta}} + (n-r) \cdot t_c^{\hat{\beta}} \cdot lnt_c^{\hat{\beta}}\right] \end{cases} \tag{16}$$

The estimations with confidence intervals were calculated based on two random variables, $v(r,n)$ and $k(r,n)$.

Thus, for the shape parameter the CI corresponding to a confidence level $1-\alpha$, has the form:

$$\frac{\hat{\beta}}{v_{1-\alpha/2}(r+1,n)} \le \beta \le \frac{\hat{\beta}}{v_{\alpha/2}(r,n)} \tag{17}$$

and for the basic rating life, $L_{10}$:

$$\hat{L}_{10} \cdot \left[e^{\frac{-u_{1-\alpha/2}(r,n,0.10)}{\hat{\beta}}}\right] \le L_{10} \le \hat{L}_{10} \cdot \left[e^{\frac{-u_{\alpha/2}(r+1,n,0.10)}{\hat{\beta}}}\right] \tag{18}$$

Another type of estimator, developed after the 1970's and which can be used to estimate the parameters and reliability indicators of bearings, are Best Linear Invariant Estimator (BLIE).

The point estimated values, based on the sampling data obtained, are determined by solving the system of equations [3, 6, 7]:

$$\begin{cases} \hat{\gamma}_{BLIE} = A(n,r,1) \cdot lnt_1 + A(n,r,2) \cdot lnt_2 + \cdots + A(n,r,r) \cdot lnt_r \\ \hat{\delta}_{BLIE} = C(n,r,1) \cdot lnt_1 + C(n,r,2) \cdot lnt_2 + \cdots + C(n,r,r) \cdot lnt_r \end{cases} \tag{19}$$

where, $A(n,r,i)$ *and* $C(n,r,i)$ are numerical coefficients.

The punctual estimations of Weibull parameters are:

$$\begin{cases} \hat{\beta}_{BLIE} = \dfrac{1}{\hat{\delta}_{BLIE}} \\ \hat{\eta}_{BLIE} = e^{\hat{\gamma}_{BLIE}} \end{cases} \tag{20}$$

In [7] it is shown that in the case of BLIE estimators, can be established the following random variables, independent of the population parameters:

$$\begin{cases} W(r,n) = \dfrac{\beta}{\hat{\beta}} \\ z(r,n) = \hat{\beta} \cdot ln\left(\dfrac{\hat{\eta}}{\eta}\right) \\ U(r,n,p) = \hat{\beta} \cdot ln\left(\dfrac{\hat{t}_p}{t_p}\right) \end{cases} \tag{21}$$

which allow the determination of confidence intervals, [9], for:

- Shape parameter:

$$\hat{\beta} \cdot W_{\alpha/2}(r,n) \leq \beta \leq \hat{\beta} \cdot W_{1-\alpha/2}(r,n) \tag{22}$$

- Basic rating life, $L_{10}$:

$$\hat{L}_{10} \cdot \left[ e^{\frac{-U_{1-\alpha/2}(r,n,0.10)}{\hat{\beta}}} \right] \leq L_{10} \leq \hat{L}_{10} \cdot \left[ e^{\frac{-U_{\alpha/2}(r,n,0.10)}{\hat{\beta}}} \right] \tag{23}$$

The determination of these distributions is performed by the numerical simulation Monte Carlo. The original algorithm, developed in this sense, involves following steps presented in the figure 5.

## 4. Case studies

Currently, it has been found that there is a trend in the procedure of testing radial ball bearings, to use complex devices that allow simultaneous testing, under characteristic conditions and on a large number of radial ball bearings. Through reliability tests, it can be obtained a series of information on reliability, durability, quality of materials, processing methods and technologies used, and manufacturing accuracy.

After testing on the test machines a lot of ball bearing ($n = 20$), type 6307, requested with radial load $Pr = 1160$ $[daN]$ at $n = 4000$ $[rpm]$, using failures-truncated tests [5], at level $r = 8$, we get the following failures times, $[h]$: 59, 98, 154, 172, 191, 232, 248, 300.

Based on the experimental results, the estimated value of the location parameter is:

$$\hat{\gamma} = \text{-}1.6868\ [h] \tag{24}$$

This value was determined under a correlation coefficient equal to:

$$\rho(\hat{\gamma}) = 0.9950 \tag{25}$$

which indicates a very good correlation with the Weibull two-parameter model.

The estimated values of the shape and scale parameters of the Weibull two-parameter using maximum likelihood method are:

$$\begin{cases} \hat{\beta} = 1.878 \\ \hat{\eta} = 430.006\ [h] \\ \hat{L}_{10} = 128.093\ [h] \end{cases} \tag{26}$$

The 90% confidence intervals for Weibull distribution parameters estimated based on experimental results are:

$$0.802 < \hat{\beta} < 2.730 \tag{27}$$

$$44.454 < \hat{L}_{10} < 188.025 \tag{28}$$

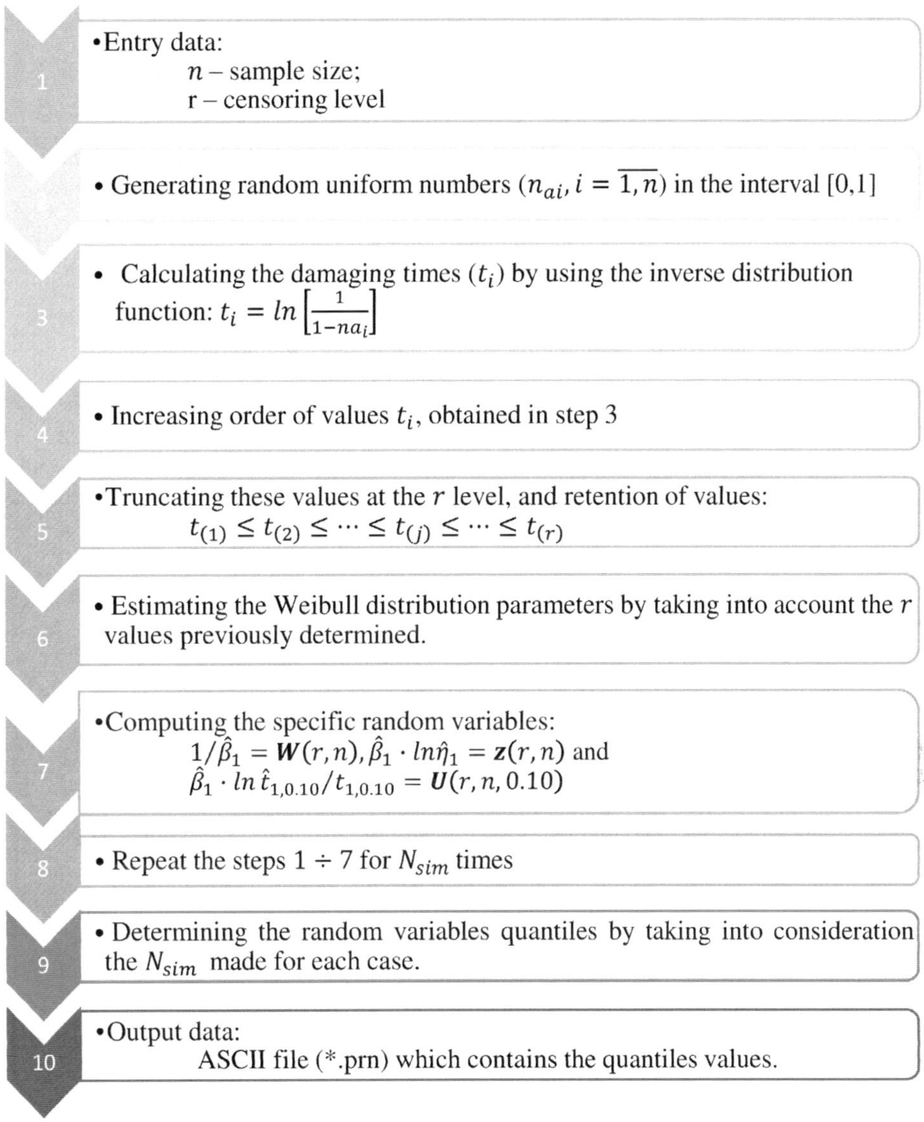


**Figure 5.** The algorithm used for calculating the values for $W(r,n), z(r,n)$ and $U(r,n,p)$.

The estimated values of the shape and scale parameters of the Weibull two-parameter using BLIE estimators are:

$$\begin{cases} \hat{\beta} = 1.863 \\ \hat{\eta} = 449.505\ [h] \\ \hat{L}_{10} = 128.093\ [h] \end{cases} \tag{29}$$

The 90% confidence intervals for Weibull distribution parameters estimated based on experimental results are:

$$0.797 < \hat{\beta} < 2.720 \tag{30}$$

$$44.126 < \hat{L}_{10} < 188.338 \tag{31}$$

Analyzing the results obtained, it is found that they are similar. BLIE type estimators representing a convenient alternative to MLE.

After testing on the stands a lot of ball bearing ($n = 20$), type 6007, requested with radial load $Pr = 550\ daN$[daN] at $n = 4000\ [rpm]$, using time-truncated tests [5], at $t_c = 250\ [h]$, we get the following failures times, $[h]$: 70, 92, 116, 120, 137, 152, 170, 172.

Based on the experimental results, the estimated value of the location parameter is:

$$\hat{\gamma} = 31.329\ [h] \tag{32}$$

This value was determined under a correlation coefficient equal to:

$$\rho(\hat{\gamma}) = 0.9957, \tag{33}$$

which indicates a very good correlation with the Weibull two-parameter model.

The estimated values of the shape and scale parameters of the Weibull two-parameter using maximum likelihood method are:

$$\begin{cases} \hat{\beta} = 1.330 \\ \hat{\eta} = 347.170\ [h] \\ \hat{L}_{10} = 65.325\ [h] \end{cases} \tag{34}$$

The 90% confidence intervals for Weibull distribution parameters estimated based on experimental results are:

$$0.622 < \hat{\beta} < 1.934 \tag{35}$$

$$23.775 < \hat{L}_{10} < 163.871 \tag{36}$$

The estimated values of the shape and scale parameters of the Weibull two-parameter using BLIE estimators are:

$$\begin{cases} \hat{\beta} = 2.484 \\ \hat{\eta} = 191.162[h] \\ \hat{L}_{10} = 108.589\ [h] \end{cases} \tag{37}$$

These values were obtained when the time interval $(172;\ 250]$ was ignored, thus obtaining a censored test. It is also found that the values obtained differ significantly from those obtained by the MLE.

## 5. Conclusion

From the analysis of the properties of the two types of estimators presented, as well as from the results obtained based on experimental data, it results:

- The maximum likelihood estimation is the only method that uses all the information obtained by tests of the failures and time truncated type;
- The Best Linear Invariant Estimator can be used only in the case of failures truncated tests;
- The punctual and/or confidence interval values of the parameters and reliability indicators for truncated tests are very similar;

- BLIE estimates can also be used for truncated attempts but in this case, a large amount of information is lost and results are very different;
- The maximum likelihood estimators have the great disadvantage that for certain combinations of $t_i$ values cannot solve equations using numerical methods;
- BLIE have the great disadvantage that for the estimation of parameters and reliability indicators of it is necessary to know the values of the coefficients $A(n,r,i)$ and $C(n,r,i)$.

From the analysis of the experimental data, presented in subchapter 4, no conclusions can be drawn regarding the superiority of an estimation method. To demonstrate this aspect, another study is needed, by Monte - Carlo numerical simulation. The accuracy of an estimation method can be assessed by analyzing the bias, consistency, mean square error and efficiency of the estimates obtained. These results will be the subject of future research.

**References**

[1] Changsen W 1991 *Analysis of Rolling Element Bearings* (London: Mechanical Engineering Publications LTD)
[2] Kececioglu D B 1991 *Reliability Engineering Handbook* (New Jersey: PTR Prentice-Hall, Inc. Englewood Cliffs)
[3] Kececioglu D B 1993 *Reliability & Life Testing Handbook* vol. 1 (New Jersey: PTR Prentice-Hall, Inc. Englewood Cliffs)
[4] Harris T A 1991 *Rolling Bearing Analysis*, 3-rd. Edition (John Wiley & Sons, Inc.)
[5] Morariu C O 1999 *Optimizing the reliability Tests of Bearings*, Phd thesis, Transilvania University of Brasov
[6] Kececioglu D B *Reliability & Life Testing Handbook* vol. 21 (New Jersey: PTR Prentice-Hall, Inc. Englewood Cliffs)
[7] Mann N R 1974 *Methods for Statistical Analysis of Reliability & Life Test Data* (John Wiley & Sons, Inc.)
[8] McCool J I 1974 *Inference on Weibull percentiles and Shape Parameter from Maximum Likelihood Estimates*, IEEE Transactions on Reliability **19** p 127
[9] McCool J I 2012 *Using the Weibull Distribution. Reliability. Modeling, and Inference* (John Wiley & Sons, Inc.)

IOP Conf. Series: Materials Science and Engineering 1009 (2021) 012041 doi:10.1088/1757-899X/1009/1/012041

# Influence of DC05 deep drawing steel real material properties on numerical simulation of incremental forming process

**D Nasulea[1]* and G Oancea[1]**

[1]Manufacturing Engineering Department, Transilvania University of Brașov, Eroilor nr. 29, 500036 Brasov, Romania

*E-mail: nasulea.marius.daniel@unitbv.ro

**Abstract.** The finite element method is one of the most useful virtual tools in industrial engineering, which allows quick and cheap predictions and verifications. As it is expected, the FEM simulations are also integrated for incremental forming processes to anticipate different aspects as forming forces, the material stress and strain, dimensional accuracy of parts, or simply to check if the desired part can be manufactured without material failure, before to start the effective manufacturing. The FEM predictions validity is influenced by many aspects of which the most significant are the sheet metal blank material properties. Each sheet blank material has its own mechanical properties, according with material standards, considered to be the theoretical properties. In most of the situations, those theoretical material properties are not matched with the real material properties. In this paper, it is presented the experimental method of determination of the real material properties for DC05 deep drawing steel, based on the tensile strength of the sheet specimens, and the true stress-strain curve which, in fact, represents the real material characteristic curve. A comparison is made between the FEM results, obtained using both, the theoretical and the determined real material properties, and the experimental trials for parts with frustum of a cone shapes.

## 1. Introduction

In modern industries, the efficiency and minimizing the manufacturing costs, keeping and even improving the products quality, are the successful keys for industrial companies or research organisations. For this purpose, one of the most useful tool is the finite elements method (FEM), being very cheap and easy to be implemented in products lifecycle development or as a preliminary check in the research stages, avoiding unwanted costs. FEM is a numerical simulation method that supposes, in the first stage, the product shape discretization, because, for complex shapes it is difficult to be analysed different aspects of the entire part. For a simplified calculation, the FEM method allows to divide the products into multiple simple sub-divisions, usually called finite elements, which are then individually analysed. The final result is a sum of the sub-divisions results. In the last years the FEM simulation methods are more and more performant and has gain a great confidence and good correlations of the provided results with the real word [1]. Taking into consideration FEM advantages, the researchers implemented this method also for incremental forming process simulations. There were used the most usual dedicated software such as ABAQUS, LS-DYNA or ANSYS, to analyse different process aspects as stress and strain, forming forces, thickness reduction, material behaviour, process parameters influence and tool path corrections, or even only as a preliminary check of the forming conditions or for material fracture identification[1-4].

However, the results provided by FEM simulations are not always in agreement with the reality. In order to ensure the confidence of the provided results, the right input parameters should be introduced in FEM model preparation stage. This paper presents how the material properties of DC05 deep drawing steel, used in a FEM simulation, influence the provided results, being the key element to obtain a great rate of confidence on real predictions on numerical simulation of incremental deformed parts. Two parts dimensional configurations will be simulated and validated by experiments.

## 2. Research description

One of the most important input parameters in FEM simulations, are the part material properties. The main topic of this research is to study the influence of the material properties in numerical simulation on single point incremental forming process (SPIF) for frustum of a cone shapes. The material used for parts simulation and manufacturing are sheet metal blanks of deep drawing steel DC05 material, with 1mm thickness. For SPIF process implementation, a technological system is required which suppose using a CNC milling machine, a blank fixing device and a forming tool. The setup configuration for the current research is composed from a CNC Victor Vcentre-55 milling machine, a fixing device which is able to clamp cone parts with the upper base diameter up to 100mm [5], and a head rounded tool, all these are presented in figure 1 [6]. A 12mm tool diameter were chosen because it is one of the most used values in the literature [7, 8]. In figure 1 there are presented also the frustum of a cone part dimensional parameters.

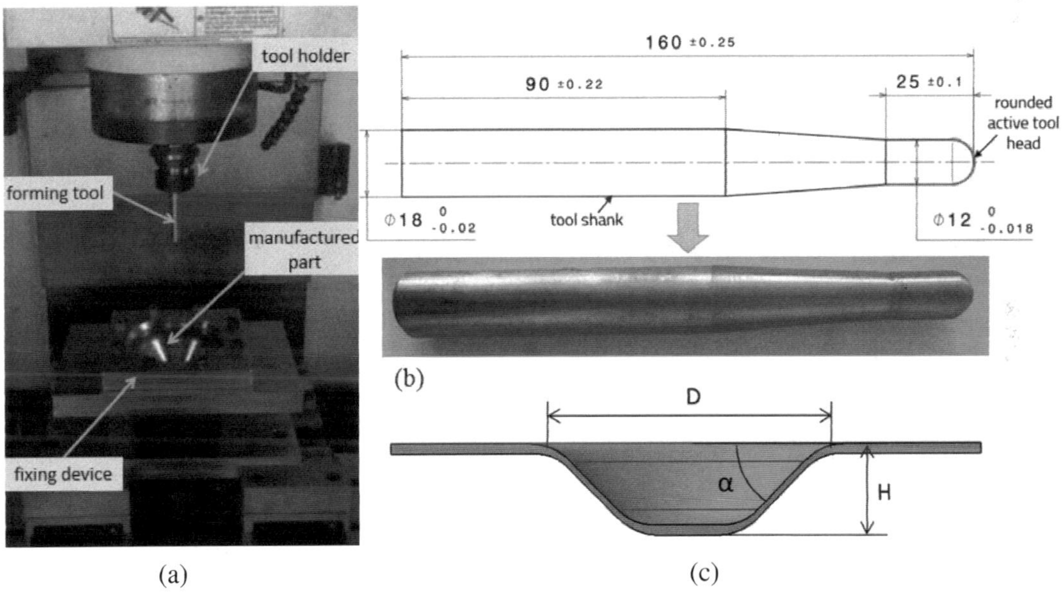


**Figure 1.** Technological system: (a) machine tool and blank fixing device, (b) forming tool dimensions, (c) parts parameters [6].

The fixing device presented in figure 1, was designed for clamping square sheet metal blanks. Because the FEM simulation for SPIF processes requires a considerable CPU time [7], necessary to analyse the forming process for one part, it was decided that, for FEM implementation, to be used round sheet metal blank. In this way the number of finite elements is decreased and the unnecessary runs for the blank corners are avoided, without influence upon the final provided results. The goal of the FEM simulations from the current research is to check if a certain part dimensional configuration can be deform without material fracture, the results being preceded by experimental validation.

IOP Conf. Series: Materials Science and Engineering 1009 (2021) 012041 doi:10.1088/1757-899X/1009/1/012041

## 3. Preliminary simulations

Two parts with different dimensional configurations were chosen for numerical simulation and experimental validation of the FEM results. The configurations, named C1 and C2, have the following dimensional parameters:

- C1: upper base diameter, D=85mm, part height, H=25mm and wall draw angle, $\alpha$=55°;
- C2: upper base diameter, D=95mm, part height, H=50mm and wall draw angle, $\alpha$=60°.

The first configuration is a small one and it is deformed using an incremental step depth $\Delta z$=1.5mm. The second one is bigger and with more difficult deformation conditions, being the reason of choosing an incremental step depth $\Delta z$=0.5mm. Both are simulated using for FEM model preparation the ANSYS parametric design language (APDL), and for process analyse, LS-Dyna explicit solver [6].The material properties introduced in FEM model preparation stage were according with DC05 standard SR EN 10130:2007, and there are presented in table 1 [6].

**Table 1.** Material properties for DC05 deep drawing steel according to SR EN 10130:2007 standard.

| DC05 deep drawing steel | | | | |
|---|---|---|---|---|
| Density [Kg/m$^3$] | Young modulus [MPa] | Poisson's ratio | Yield strength [MPa] | Tensile strength [MPa] |
| 7850 | 210000 | 0.3 | 180 | 270...330 |

Following all necessary steps for FEM implementation [1, 6], both part configurations were simulated and the post-processed results are presented in figure 2.

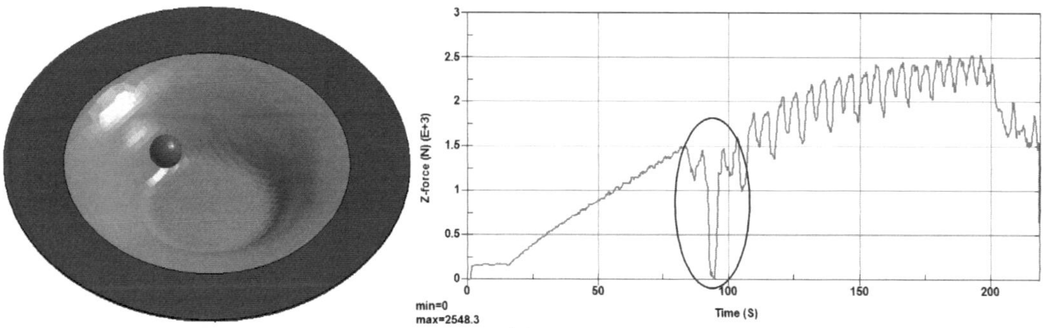


(a)

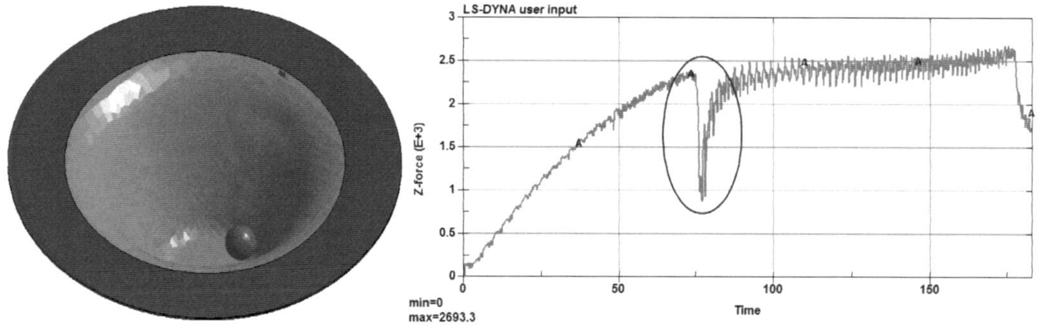


(b)

**Figure 2.** Simulation results for SPIF processing parts, FEM model and axial force plots: (a) C1, (b) C2.

According to figure 2, at a close look upon the FEM deformed models, it can be seen a series of shape undulation on the parts walls that can note the material fracture in those areas. This aspect is strongly visible also on both axial force plots, where, marked with blue colour, it can be seen that suddenly decrease of the forces on axial direction, even decreasing down to zero newton, which clearly suggests the deformation process failure because of material fracture. Therefore, it was decided to improve the FEM predictions by replacing the standard material properties with the real ones which are experimental determined for the specific used batch of DC05 material.

## 4. Experimental determination of the real mechanical properties for DC05 material

The real material properties are determined by applying the uniaxial tensile strength on 15 sheet metal specimens of which shape and dimensions are presented in figure 3.

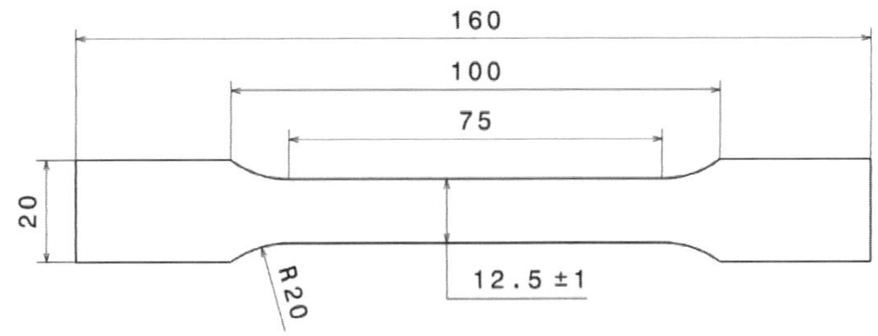


**Figure 3.** Parts manufactured without material [6].

Because the DC05 deep drawing steel sheet is a laminated blank, the material properties could be different depending on the directions in which they are measured. To avoid errors due to material anisotropy, it was decided to test five specimens cropped/machined along three direction, as follow (figure 4a) [6, 9].

- S1.1, S1.2,....., S1.5 - five specimen cropped along the material rolling direction;
- S2.1, S2.2,....., S2.5 - five specimen cropped on a 90° related to material rolling direction;
- S3.1, S3.2,....., S3.5 - five specimens cropped on a 45° related to material rolling direction.

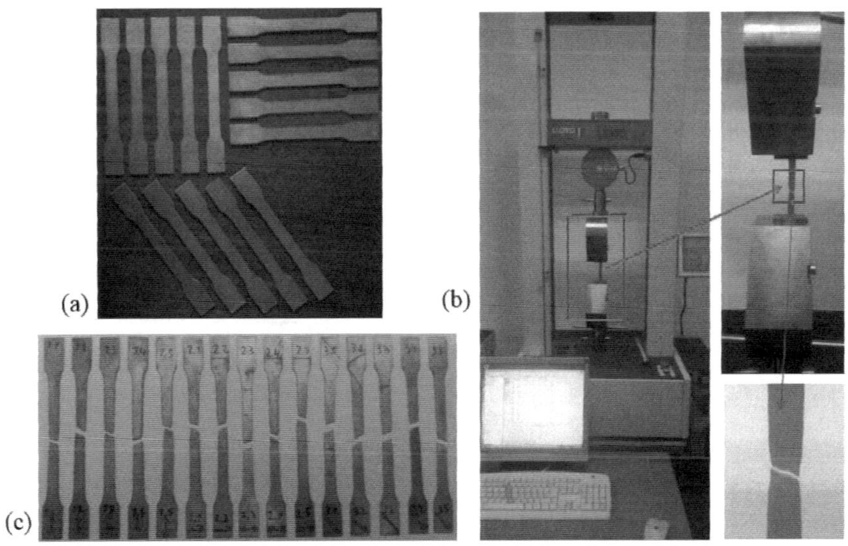


**Figure 4.** Tensile test setup: (a) 15 specimens, (b) tensile test machine, (c) tested specimens [6].

The specimens were manufactured using abrasive water jet cutting process on MAXIEM 1530 machine. The abrasive water jet technology was chosen because, compared to laser or plasma cutting, it does not affect material properties by heating the specimens, avoiding in this way obtaining erroneous results. Further, the specimens were tested using the uniaxial tensile test machine LLOYD LS100 Plus, together with the machine data management software NEXYGEN Plus (figure 4b) [6]. In figure 4c are presented all 15 specimens on which was applied the uniaxial tensile strength.

The result of the uniaxial tensile test is a set of values representing stress-strain values ($\sigma$, $\varepsilon$), based on which the material characteristics curve were represented for each tested specimen. Those curves represent the engineering stress – engineering strain graphics and they are presented in figure 5. It was selected for a further processing, the specimen S1.3, being the closest to the average values.

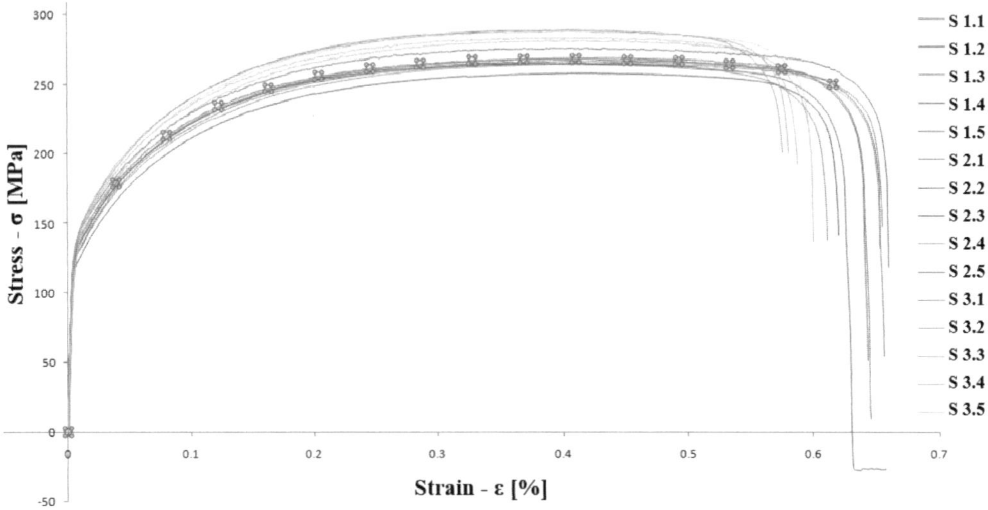


**Figure 5.** Engineering stress – engineering strain graphics for all 15 specimens [6].

The purpose of this further data processing of the engineering stress-strain values is to obtain the real material properties and, certainly, the true stress – true strain characteristic curve ($\sigma_{real}$, $\varepsilon_{real}$). The stress and strain values resulted direct from the machine tensile test are generated taking into consideration constant values for specimen gauge section area, $S_0$, and for gauge length $L_0$, and they are equal with the initial specimen dimensions. In fact, those two values are continuously changing during the tensile test, the real values for this two parameters being notated $S_i$ and $L_i$, the instantaneous values being different for each time increment. The difference between the initial specimen dimensions and the instantaneous values are presented in figure 6.

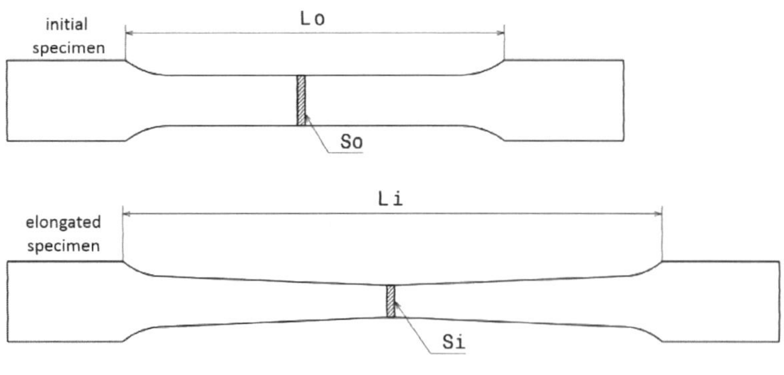


**Figure 6.** Difference between$S_0$ / $L_0$ and $S_i$ / $L_i$ during tensile test [6].

Taking into consideration those facts, it is necessary to follow the below presented steps, to obtain the true stress – true strain characteristic curve and also the real material properties [6]:

- From the law of constant volumes in material forming, results the following equation [6, 10]:

$$V_0 = V_i \Rightarrow S_0 L_0 = S_i L_i \quad (1)$$

- The instantaneous value for $S_i$ is [6]

$$S_i = \frac{S_0}{1+\frac{\Delta l}{L_0}} \quad (2)$$

- The instantaneous value for $L_i$ is [6, 11]:

$$L_i = L_0 + \Delta l \quad (3)$$

where $\Delta l$ is the specimen elongation;

- From the engineering stress relation, results the true stress relation is as follow [11]:

$$\sigma = \frac{F}{S_0} \Rightarrow \sigma_{real} = \frac{F}{S_i} \quad (4)$$

where $F$ is the traction force, also given by the tensile test machine;

- The real strain can be calculated using relation (5) [6]:

$$\varepsilon_{real} = \frac{\Delta l}{L_0} \quad (5)$$

- The true stress relation becomes [6]:

$$\sigma_{real} = \frac{F}{S_0}(1 + \varepsilon_{real}) \quad (6)$$

Using the mathematical relations above presented (1÷6), the initial set of values given by the tensile test machine was processed in an Excel worksheet, and the real values for true stress – true strain were obtained. Based on those processed values, the real characteristic curve was represented for specimen S1.3.

The Young modulus was also calculated using the relation (7), where α is the angle between the strain axis and the proportionality limit line from the real characteristic curve [6].

$$E = tg(\alpha). \quad (7)$$

Based on those processed data and on true stress – true strain characteristic curve ($\sigma_{real}$, $\varepsilon_{real}$), it was identified also the values for the real tensile strength and the real yield strength. In table 2 there are presented for comparison, the mechanical properties according to material standard and according to the real experimental determination.

**Table 2.** Material properties comparison –standard vs real experimental determination [6].

| DC05 deep drawing steel | | | | | |
|---|---|---|---|---|---|
| According to: | Density [Kg/m$^3$] | Young modulus [MPa] | Poisson's ratio | Yield strength [MPa] | Tensile strength [MPa] |
| SR EN 10130:2007: | 7850 | 210000 | 0.3 | 180 | 270...330 |
| Real processed data: | 7850 | 87952 | 0.3 | 190.59 | 388.67 |

Those new data are the true values of the material properties and they are specific for the used batch of DC05 material. Those are used further as input data for FEM numerical simulations, in order to obtain predictions as accurate as possible.

## 5. Second run of FEM simulations

Using those experimental determined values for DC05 material, the FEM simulations for C1 and C2 part configuration was solved once again. In addition to these real properties inputs, ANSYS APDL allows to specify also as input data, the real characteristics curve. The real characteristic curve is described in ANSYS as sets of points true stress – true strain ($\sigma_{real}$, $\varepsilon_{real}$), and this implementation can conduct to even more precise results. Thus, the simulations was repeated, obtaining much better results in comparison with the first trial, results which are presented in figure 7.

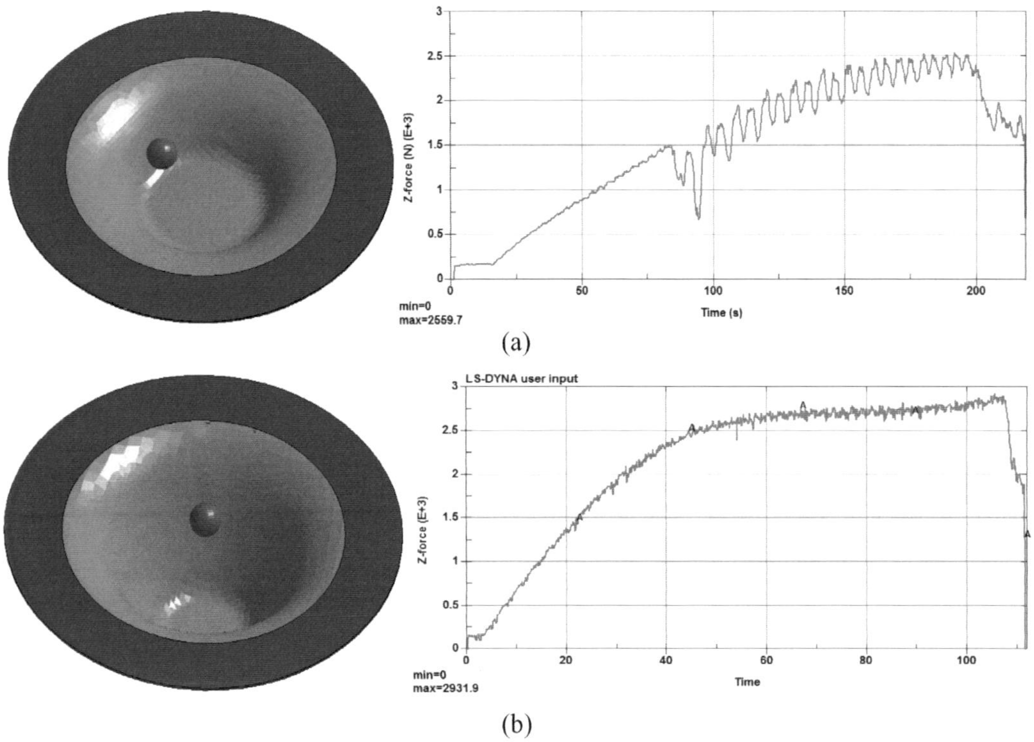


**Figure 7.** FEM simulation results using as input, the real material properties: (a) C1, (b) C2 [6].

For the new simulations, as can be seen in figure 7, the FEM deformed models has no shape undulation on its walls, and also on graphical representation of the axial loads, there are no sudden tendencies of force decreasing down to zero newton. Thus, it can be concluded that, according to FEM simulations using the real material properties, both parts can be manufacture without material fracture.

## 6. Experimental validation

In order to ensure that the FEM simulation provide confidence results, both parts configuration were manufactured using the same conditions implemented in the virtual analysis. For both, simulations and manufacturing, the used process parameters was a tool spindle speed of 200rpm, and a feed rate of 1500mm/min, following a spiral forming toolpath movement. In contrast with the first FEM results, both parts were deformed in good conditions, without material fracture, and they are presented in figure 8. Thus, there are no agreements between the manufacturing process and the FEM simulation results provided using, as mechanical properties, the theoretical ones from the material standard SR EN 10130:2007.

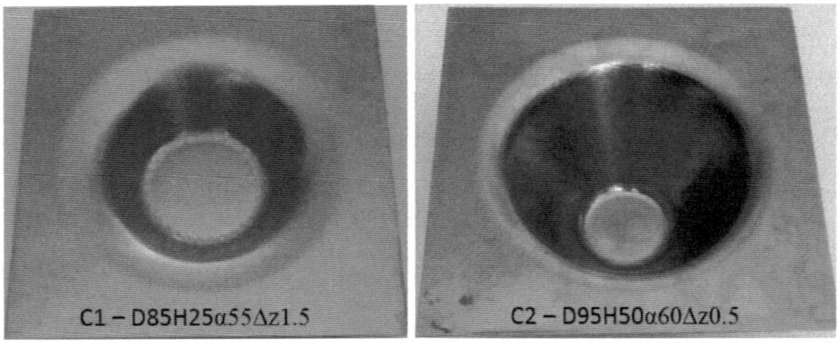


**Figure 8.** Parts manufactured without material fracture [6].

In comparison with the first simulations, for the second FEM trials, it can be said that the provided results are in a good agreement with the forming process. It can be noticed that, for real predictions, the right input data should be introduced in the FEM model preparation stage, with increased attention especially when refer on material properties. Therefore, because the material mechanical properties used as input data, has one of the major influence upon the numerical simulation results, defining the material behaviour during forming process, it is recommended to be determined the real material properties by experimental uniaxial tensile tests.

## 7. Conclusions

FEM simulation is a very advantageous engineering tool which can be used in product development and also in research. It is able to provide confidence results only if the proper input data are available.

The paper presented a research on how the material properties, as input data, can influence the FEM results and how the results are in connection with the practical experiments. It is presented the case of single point incremental forming, implemented for two frustum of a cone parts manufactured from DC05 deep drawing steel with 1mm thickness. These two parts were simulated, in a first stage, using the mechanical properties given by the material standard SR EN 10130:2007. Those results were erroneous because they were not in good agreements with the reality. It was also presented the method of real material properties determinations, based on uniaxial tensile tests and a series of relations available in the literature. Those real material properties, used in FEM simulation, conducted to improved results that match with the real forming process. It can be said that, if confidence results from FEM simulation are expected, the right input data should be provided, one of the most significant influence belonging to material properties. The material properties given by standards or suppliers are not always so accurate. The properties could be different from batch to batch and can be influenced by the date of material obtaining, storage conditions and other aspects.

## References

[1] Nasulea D and Oancea G 2018 *Stroj. Vestn-J. Mech. E* **64** 643
[2] Blaga A and Oleksik V 2013 *Adv. Mater. Sci. Eng.* **2013** 2
[3] Gatea S, Ou H and McCartney G 2016 *Int. J. Adv. Manuf. Tech.* **87** 479
[4] Fu Z, Mo J, Han F and Gong P 2013 *Int. J. Adv. Manuf. Tech .* **64** 1239
[5] Nasulea D and Oancea G 2018 *MATEC Web Conf.* **178** 02004
[6] Nasulea D 2019 *Research Regarding the Incremental Forming of DC05 Steel Sheet Parts (*PhD thesis,Transilvania University of Brasov, Romania)
[7] Nimbalkar D H and Nandedkar V M 2013 *Int. J. Eng. Res. Appl.* **3** 39
[8] Behera A K, De Sousa R A, Ingarao G and Oleksik V 2017 *J. Manuf. Process.* **27** 37
[9] Zein H and Al-Mufadi 2018 *Minia J. Eng. Technol.* **37** 152
[10] Bologa O 2014 *Cold plastic deformation processing (*"Lucian Blaga" University Press, Sibiu, Romania)
[11] Oleksik V S 2005 *Theoretical and Experimental Research on Incremental Forming of Indigenous Metal Sheets* (PhD thesis, Lucian Blaga University of Sibiu, Romania)

# Comparison of the roughness of concave cylindrical surfaces machined with spherical milling cutters and toroidal milling cutters

**A Osan[1], M Banica[1*], L Butnar[1] and V Nasui[1]**

[1]Technical University of Cluj-Napoca, North University Centre of Baia Mare, Dr. Victor Babes Street No. 62A, Romania

*E-mail: mihai.banica@cunbm.utcluj.ro

**Abstract.** The paper aims to process the concave cylindrical surfaces. This proposes a comparison between the processing of concave cylindrical surfaces with toroidal milling and the processing of concave cylindrical surfaces with spherical milling. The variable process parameters are the cutting speed, the feed per tooth and the inclination angle between the tool axis and the normal to surface. After processing, the surface with the best roughness is tracked according to the regimes and the tool used. A number of 27 concave cylindrical surfaces machined with toroidal milling cutter and 27 concave cylindrical surfaces machined with spherical head milling are machined. The aim is to measure the arithmetic roughness $R_a$ and the total roughness $R_t$, these being measured parallel and perpendicular to the direction of advance. In order to confirm the correctness of the measured values, microscopic images with the best surfaces will be captured.

## 1. Introduction

An important factor that influences the quality of the surface and the evolution of the roughness of the machined surface is the direction of inclination of the tool axis. In this case, it is the value of the angle at which the cutting edge is inclined compared to the normal surface. This factor requires the establishment of the optimal values of the angle of inclination of the tool axis, a value after which the roughness is the lowest. The values of the angle of inclination of the tool axis where high surface roughness's are determined are also avoided.

Han [1] investigated the influence of the tool tilt angle on the surface integrity, in particular the surface roughness and the residual effort in fast steel milling by milling experiments, including 8 cases of free-form surface milling. Finally, the optimum tilt angles of the tools, including the feed angle, the tilt angle and the milling method, were obtained for 5-axis milling with the ball mill.

The researcher Tonshoff, in the paper [2], indicates that the best results are obtained by inclining the semi-finished product by 15°. In this case the best results being from the point of view of the surface roughness, of the tool wear and of the productivity. Other works such as [3] indicate that the optimum value of the tilt angle is in the range 15°÷20°, while in the work [4] the optimal value indicated is 20°.

As a result of these analyses, it is necessary to investigate this process in order to identify both the optimum value of the inclination angle and the optimum inclination direction. Researches such as those of [5 - 9] analyse particular cases in order to identify the optimum value of the inclination angle.

IOP Conf. Series: Materials Science and Engineering 1009 (2021) 012042 doi:10.1088/1757-899X/1009/1/012042

## 2. The experimental part

This paper highlights the practical experiments performed on the concave cylindrical surface, 27 made with the toroidal milling presented by table 1 and 27 concave cylindrical surfaces made with the spherical milling as presented in table 2.

The organization has a decisive role in carrying out the experiments, so we noted with CCV-TR the concave cylindrical surfaces processed with toroidal milling cutter and with CCV-SF the concave cylindrical surfaces processed with spherical milling cutter.

The concave cylindrical surface has a cylindrical height along the plate with a diameter of Ø85mm and a maximum height of 5mm.

Referring to the cutting regimes, the three variables are presented in the tables, and constant throughout the experimentation is the cutting depth, which is 0.5mm ($a_p$=0.5mm) and the radial depth, having the value of 0.3mm ($a_e$=0.3mm). The experiments were performed in the presence of the coolant, the way of holding the specimens being in jaw vice.

It is necessary to emphasize the fact that during the whole processing of surfaces with both toroidal and spherical milling, the tool axis constantly maintains its inclination angle throughout the surface.

**Table 1.** Carrying out experiments on the processing of concave cylindrical surfaces with toroidal milling.

| No. | Cutting speed [m/min] | Tilt angle [°] | Advance on tooth [mm/tooth] | Surface CCV-TR |
|---|---|---|---|---|
| 1 | 80 | 15° | 0.11 | CCV-TR-1 |
| 2 | 80 | 15° | 0.15 | CCV-TR-2 |
| 3 | 80 | 15° | 0.19 | CCV-TR-3 |
| 4 | 80 | 35° | 0.11 | CCV-TR-4 |
| 5 | 80 | 35° | 0.15 | CCV-TR-5 |
| 6 | 80 | 35° | 0.19 | CCV-TR-6 |
| 7 | 80 | 55° | 0.11 | CCV-TR-7 |
| 8 | 80 | 55° | 0.15 | CCV-TR-8 |
| 9 | 80 | 55° | 0.19 | CCV-TR-9 |
| 10 | 170 | 15° | 0.11 | CCV-TR-10 |
| 11 | 170 | 15° | 0.15 | CCV-TR-11 |
| 12 | 170 | 15° | 0.19 | CCV-TR-12 |
| 13 | 170 | 35° | 0.11 | CCV-TR-13 |
| 14 | 170 | 35° | 0.15 | CCV-TR-14 |
| 15 | 170 | 35° | 0.19 | CCV-TR-15 |
| 16 | 170 | 55° | 0.11 | CCV-TR-16 |
| 17 | 170 | 55° | 0.15 | CCV-TR-17 |
| 18 | 170 | 55° | 0.19 | CCV-TR-18 |
| 19 | 210 | 15° | 0.11 | CCV-TR-19 |
| 20 | 210 | 15° | 0.15 | CCV-TR-20 |
| 21 | 210 | 15° | 0.19 | CCV-TR-21 |
| 22 | 210 | 35° | 0.11 | CCV-TR-22 |
| 23 | 210 | 35° | 0.15 | CCV-TR-23 |
| 24 | 210 | 35° | 0.19 | CCV-TR-24 |
| 25 | 210 | 55° | 0.11 | CCV-TR-25 |
| 26 | 210 | 55° | 0.15 | CCV-TR-26 |
| 27 | 210 | 55° | 0.19 | CCV-TR-27 |

**Table 2.** Carrying out experiments on the processing of concave cylindrical surfaces with spherical milling.

| No. | Cutting speed [m/min] | Tilt angle [°] | Advance on tooth [mm/tooth] | Surface CCV-SF |
|---|---|---|---|---|
| 1 | 280 | 15° | 0.05 | CCV-SF-1 |
| 2 | 280 | 15° | 0.09 | CCV-SF-2 |
| 3 | 280 | 15° | 0.13 | CCV-SF-3 |
| 4 | 280 | 35° | 0.05 | CCV-SF-4 |
| 5 | 280 | 35° | 0.09 | CCV-SF-5 |
| 6 | 280 | 35° | 0.13 | CCV-SF-6 |
| 7 | 280 | 55° | 0.05 | CCV-SF-7 |
| 8 | 280 | 55° | 0.09 | CCV-SF-8 |
| 9 | 280 | 55° | 0.13 | CCV-SF-9 |
| 10 | 370 | 15° | 0.05 | CCV-SF-10 |
| 11 | 370 | 15° | 0.09 | CCV-SF-11 |
| 12 | 370 | 15° | 0.13 | CCV-SF-12 |
| 13 | 370 | 35° | 0.05 | CCV-SF-13 |
| 14 | 370 | 35° | 0.09 | CCV-SF-14 |
| 15 | 370 | 35° | 0.13 | CCV-SF-15 |
| 16 | 370 | 55° | 0.05 | CCV-SF-16 |
| 17 | 370 | 55° | 0.09 | CCV-SF-17 |
| 18 | 370 | 55° | 0.13 | CCV-SF-18 |
| 19 | 430 | 15° | 0.05 | CCV-SF-19 |
| 20 | 430 | 15° | 0.09 | CCV-SF-20 |
| 21 | 430 | 15° | 0.13 | CCV-SF-21 |
| 22 | 430 | 35° | 0.05 | CCV-SF-22 |
| 23 | 430 | 35° | 0.09 | CCV-SF-23 |
| 24 | 430 | 35° | 0.13 | CCV-SF-24 |
| 25 | 430 | 55° | 0.05 | CCV-SF-25 |
| 26 | 430 | 55° | 0.09 | CCV-SF-26 |
| 27 | 430 | 55° | 0.13 | CCV-SF-27 |

For each test, the CAM programs will be made using the PowerMILL software, it gives the possibility to create the tool, to find the best strategy so that, for each experiment the processing will be done in one direction, the tool paths are parallel to the corresponding inputs and outputs.

Referring to the achievements of the program CAM machining the cylindrical surface of the concave shaping toroidal, if the angle of 15° was required for 147 lines in the case of inclination of 35° was required for 140 lines, and for 55°, 138 lines. In the case of the spherical milling machine, it worked the surface at 15° with 159 lines, at 35° with 162 lines, and for the 55° inclination, 142 lines were needed.

Figure 1 shows three images during the simulation of the toroidal tool path on the concave cylindrical surface, in the three cases of inclinations, and figure 2 consists of three images with the simulation of the route followed by the spherical milling on the concave cylindrical surface, in the three cases of inclination.

The image with the practical experiment itself carried out on the OKUMA MU-400VA numerical control centre is illustrated in figure 3; the figure contains three images with the toroidal mill that processes the concave cylindrical surface at the three types of inclinations.

In figure 4 there are three images with spherical milling during the processing of the concave cylindrical surface in the case of the three types of inclinations.

IOP Conf. Series: Materials Science and Engineering 1009 (2021) 012042 doi:10.1088/1757-899X/1009/1/012042

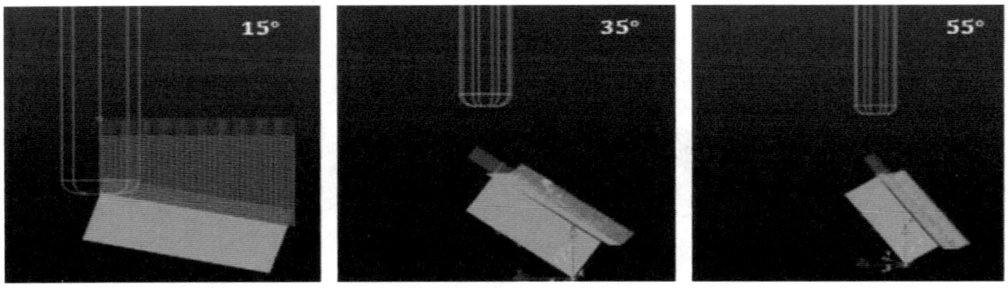


**Figure 1.** Images taken following the simulation of the toroidal mill trajectory in PowerMILL on CCV-TR.

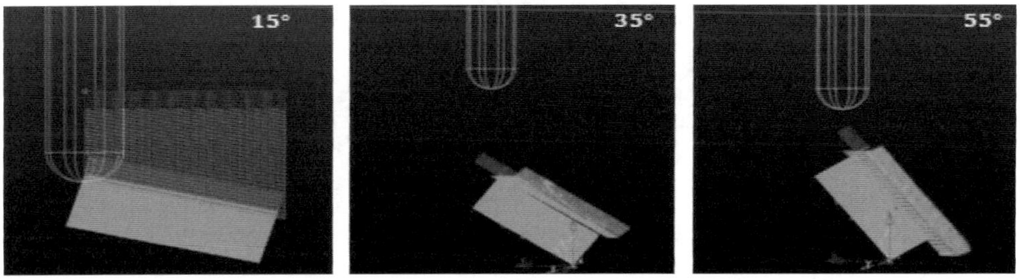


**Figure 2.** Images taken from the simulation of the toroidal milling trajectory in PowerMILL on CCV-SF.

**Figure 3.** The processing of the concave cylindrical surface with the toroidal mill in the case of the 3 types of inclinations.

**Figure 4.** Processing of the concave cylindrical surface with the spherical mill in the case of the 3 types of inclinations.

Thus, at the end of the experiments twenty-seven toroidal milling surfaces of concave cylindrical surfaces and twenty-seven other concave cylindrical surfaces are processed, and the roughness measurement will be carried out in the next chapters.

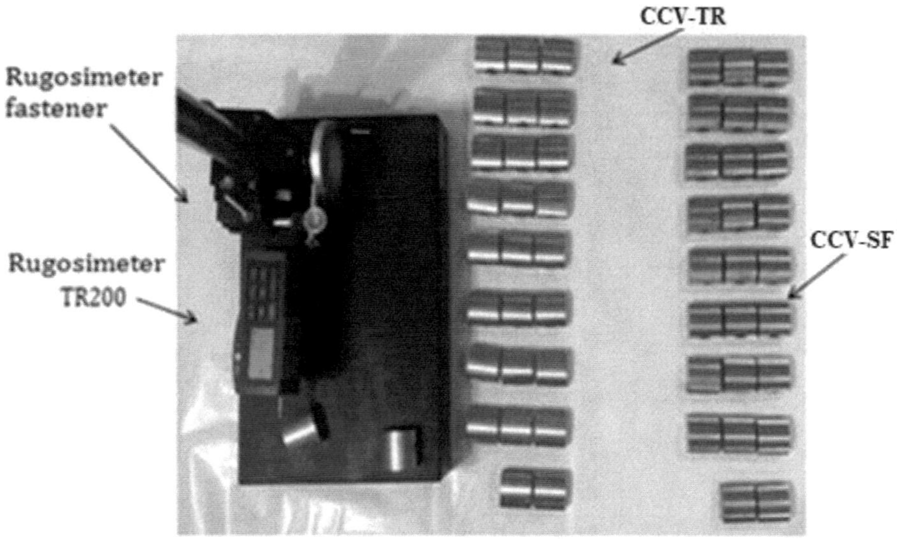


**Figure 5.** The concave cylindrical surfaces processed with the two types of milling cutters ready for measurement.

## 3. Certification of experimental results

In order to determine the quality of the surface, the principle of replication will be applied, thus three measurements will be made in the direction parallel to the advance direction, and the average of the three measurements will be made. Another three measurements will be made in the direction perpendicular to the advance, followed by calculating the average. Table 3 presents the results of machining the concave cylindrical surface with the spherical mill and table 4 presents the results of the processing of the concave cylindrical surface with the toroidal mill.

**Table 3.** The results related to the processing of the concave cylindrical surface with the ball mill.

| Surface type | Parallel roughness $R_a$ [μm] | Perpendicular roughness $R_a$ [μm] | Parallel roughness $R_t$ [μm] | Perpendicular roughness $R_t$ [μm] |
|---|---|---|---|---|
| CCV-SF-1 | 0.176 | 0.345 | 1.513 | 2.773 |
| CCV-SF-2 | 0.334 | 0.527 | 2.673 | 4.186 |
| CCV-SF-3 | 0.318 | 0.477 | 2.466 | 3.533 |
| CCV-SF-4 | 0.220 | 0.425 | 1.679 | 2.826 |
| CCV-SF-5 | 0.393 | 0.394 | 2.427 | 2.713 |
| CCV-SF-6 | 0.417 | 0.415 | 3.413 | 3.127 |
| CCV-SF-7 | 0.249 | 0.367 | 2.173 | 3.186 |
| CCV-SF-8 | 0.266 | 0.395 | 2.199 | 4.730 |
| CCV-SF-9 | 0.438 | 0.403 | 3.640 | 2.499 |
| CCV-SF-10 | 0.202 | 0.552 | 1.586 | 3.540 |
| CCV-SF-11 | 0.328 | 0.467 | 2.486 | 3.493 |
| CCV-SF-12 | 0.268 | 0.405 | 1.919 | 3.526 |
| CCV-SF-13 | 0.333 | 0.497 | 2.313 | 4.143 |
| CCV-SF-14 | 0.364 | 0.453 | 2.553 | 3.020 |
| CCV-SF-15 | 0.414 | 0.420 | 2.859 | 3.426 |
| CCV-SF-16 | 0.357 | 0.480 | 2.666 | 3.806 |
| CCV-SF-17 | 0.276 | 0.489 | 2.386 | 3.753 |
| CCV-SF-18 | 0.375 | 0.456 | 3.200 | 3.280 |
| CCV-SF-19 | 0.233 | 0.525 | 1.580 | 3.260 |
| CCV-SF-20 | 0.270 | 0.557 | 1.993 | 3.373 |

| | | | | |
|---|---|---|---|---|
| CCV-SF-21 | 0.306 | 0.557 | 2.686 | 4.739 |
| CCV-SF-22 | 0.261 | 0.394 | 1.826 | 3.093 |
| CCV-SF-23 | 0.399 | 0.505 | 2.917 | 3.546 |
| CCV-SF-24 | 0.286 | 0.462 | 2.280 | 3.486 |
| CCV-SF-25 | 0.295 | 0.531 | 2.186 | 4.300 |
| CCV-SF-26 | 0.374 | 0.484 | 2.546 | 4.520 |
| CCV-SF-27 | 0.422 | 0.459 | 2.666 | 3.380 |

By analysing the roughness values of the concave cylindrical surfaces processed with the spherical mill, the following deductions were reached.

According to figure 6, the lowest value obtained of the roughness $R_a$=0.176μm, is the case of the concave cylindrical surface with the number 1 (CCV-SF-1) with the cutting regimes $v_c$=280m/min, $f_z$=0.05mm/tooth and inclination of the tool axis of 15°, according to table 4.8.

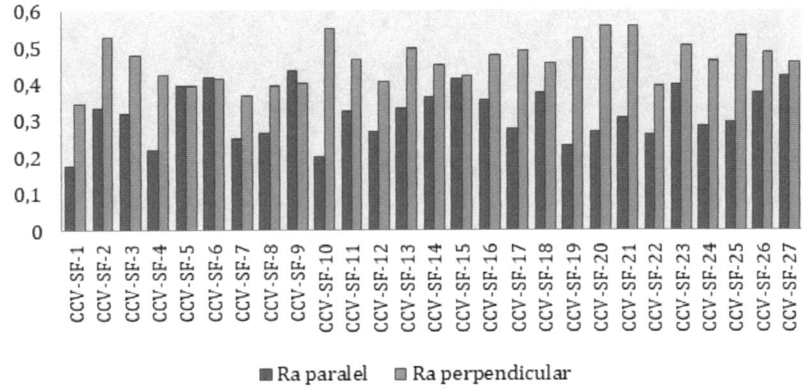


**Figure 6.** Arithmetic mean $R_a$ of the concave cylindrical surface processed with the spherical milling.

The highest roughness value $R_a$=0.552μm, was achieved on the concave cylindrical surface with the number 10 (CCV-SF-10) with the cutting speeds $v_c$=370m/min, $f_z$=0.05mm/tooth and the axis of the tool axis of 15°, according to table 3.

According to figure 7, the total minimum $R_t$ roughness is equal to 1.513μm on the concave cylindrical surface with the number 1 (CCV-SF-1), and the maximum $R_t$ is 4.739μm, generated on the concave cylindrical surface with the number 21 (CCV-SF-21) with the cutting speeds $v_c$=430m/min, $f_z$=0.13mm/tooth and the inclination of the tool axis by 15° according to table 3.

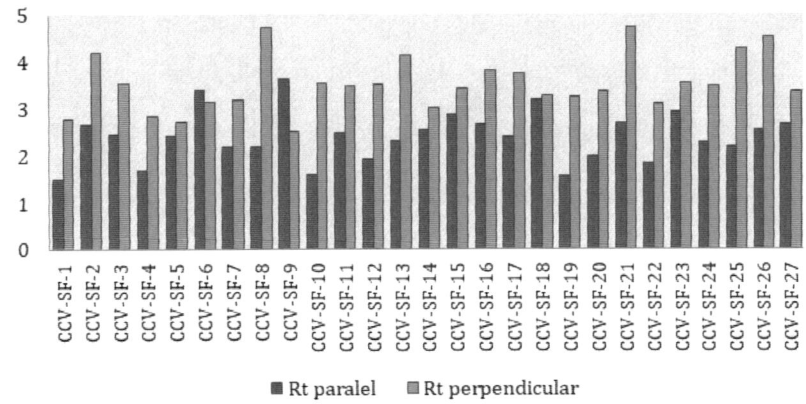


**Figure 7.** Arithmetic mean $R_t$ of the concave cylindrical surface processed with the spherical milling.

**Table 4.** Results related to the processing of the concave cylindrical surface with the toroidal mill.

| Surface type | Parallel roughness $R_a$ [μm] | Perpendicular roughness $R_a$ [μm] | Parallel roughness $R_t$ [μm] | Perpendicular roughness $R_t$ [μm] |
|---|---|---|---|---|
| CCV-TR-1 | 0.607 | 0.579 | 4.232 | 5.580 |
| CCV-TR-2 | 0.404 | 0.527 | 3.959 | 5.307 |
| CCV-TR-3 | 0.532 | 0.455 | 4.193 | 3.926 |
| CCV-TR-4 | 0.435 | 0.424 | 3.379 | 3.586 |
| CCV-TR-5 | 0.432 | 0.393 | 3.106 | 3.687 |
| CCV-TR-6 | 0.398 | 0.362 | 2.387 | 3.839 |
| CCV-TR-7 | 0.410 | 0.573 | 4.020 | 4.946 |
| CCV-TR-8 | 0.346 | 0.498 | 2.273 | 4.386 |
| CCV-TR-9 | 0.414 | 0.496 | 3.346 | 4.273 |
| CCV-TR-10 | 0.359 | 0.562 | 2.173 | 3.493 |
| CCV-TR-11 | 0.453 | 0.485 | 2.453 | 2.913 |
| CCV-TR-12 | 0.374 | 0.447 | 2.760 | 3.286 |
| CCV-TR-13 | 0.437 | 0.384 | 3.659 | 3.273 |
| CCV-TR-14 | 0.360 | 0.452 | 2.259 | 3.700 |
| CCV-TR-15 | 0.361 | 0.388 | 2.693 | 4.526 |
| CCV-TR-16 | 0.485 | 0.648 | 3.927 | 6.273 |
| CCV-TR-17 | 0.375 | 0.603 | 3.047 | 5.380 |
| CCV-TR-18 | 0.397 | 0.600 | 2.799 | 5.353 |
| CCV-TR-19 | 0.416 | 0.599 | 2.639 | 5.533 |
| CCV-TR-20 | 0.326 | 0.384 | 3.300 | 2.413 |
| CCV-TR-21 | 0.268 | 0.356 | 2.346 | 2.860 |
| CCV-TR-22 | 0.330 | 0.469 | 2.059 | 3.913 |
| CCV-TR-23 | 0.286 | 0.538 | 3.379 | 5.039 |
| CCV-TR-24 | 0.378 | 0.460 | 3.513 | 3.233 |
| CCV-TR-25 | 0.453 | 0.773 | 2.746 | 7.699 |
| CCV-TR-26 | 0.398 | 0.590 | 4.713 | 5.099 |
| CCV-TR-27 | 0.339 | 0.503 | 2.593 | 3.533 |

According to figure 8, the concave cylindrical surface processed with the toroidal milling saw the smallest value of $R_a$=0.268μm on the concave cylindrical surface with the number 21 (CCV-TR-21) with the cutting regimes $v_c$=210m/min, $f_z$=0.19mm/tooth and the inclination of the tool axis by 15°, according to table 4.

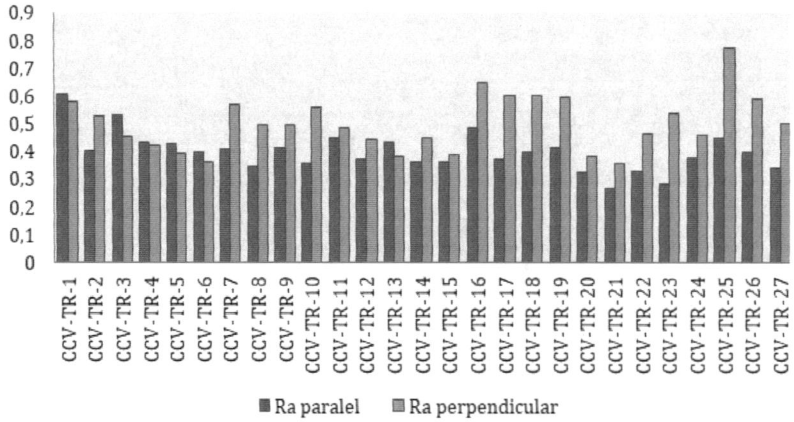


**Figure 8.** Arithmetic mean $R_a$ of the concave cylindrical surface processed with the toroidal milling

The highest $R_a$ value is 0.773μm, on the concave cylindrical surface with the number 25 (CCV-TR-25) with the cutting regimes $v_c$=210m/min, $f_z$=0.11mm/tooth and the inclination of the tool axis of 55°, according to table 4.

According to figure 9, the minimum $R_t$ obtained is 2.059μm, on the concave cylindrical surface with the number 22 (CCV-TR-22) with the cutting speeds $v_c$=210m/min, $f_z$=0.11mm/tooth and the inclination of the tool axis. 35° according to table 4.7, and the maximum value $R_t$ is 7.699μm, on the concave cylindrical surface with the number 25 (CCV-TR-25).

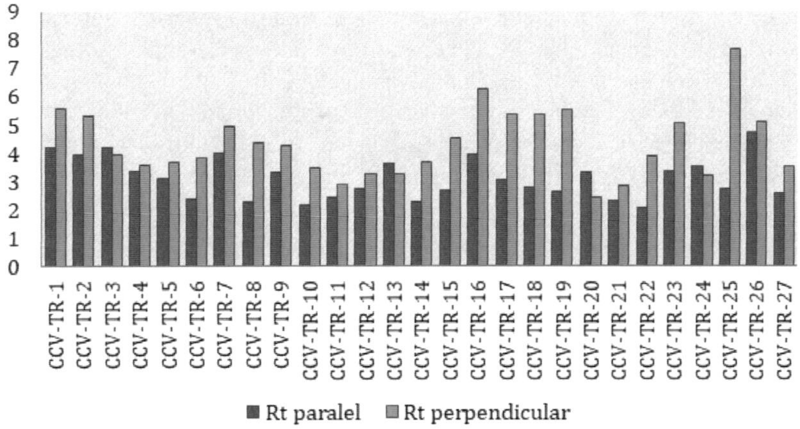


**Figure 9.** Arithmetic mean $R_t$ of the concave cylindrical surface processed with the toroidal milling.

Figure 10 shows microscopic images obtained with the IOR stereoscopic microscope and with the MM1-200 microscope of the best qualities of the concave cylindrical surfaces processed with spherical milling (CCV-SF) and toroidal (CCV-TR).

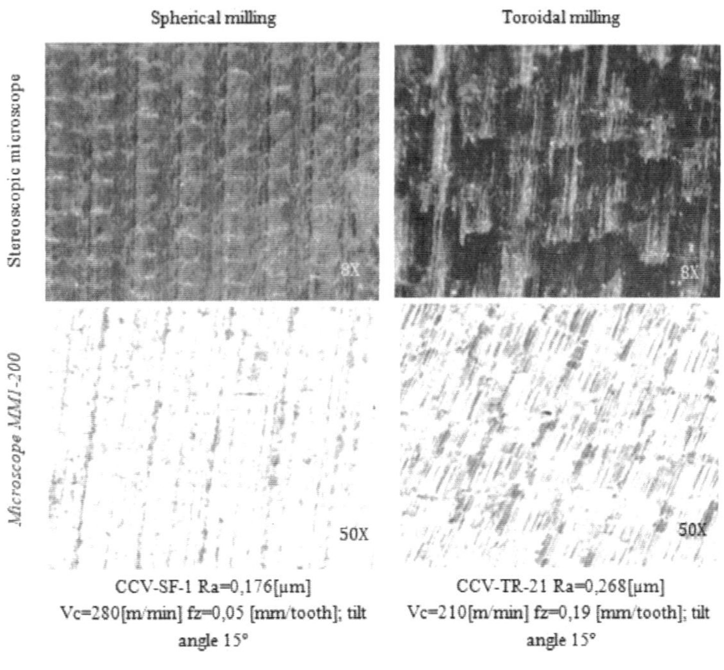


**Figure 10.** Microscope images with the best roughness of the concave cylindrical surfacesprocessed with the spherical mill and the toroidal mill.

## 4. Conclusions

The purpose of this work is to determine the surface quality which has a main role in the precision and lubrication of surfaces. It was chosen to compare the processed surfaces with the toroidal milling, respectively the spherical milling. The surface on which the roughness was investigated is the concave cylindrical surface.

As variable parameters, it was chosen to combine the cutting speed, the feed on the tooth and the inclination angle of the tool. During the practical experiments, twenty-seven surfaces were processed with toroidal milling and twenty-seven surfaces were processed with spherical milling.

Comparing the concave cylindrical surfaces processed with the spherical and the toroidal milling, the spherical milling has an advantage of 0.092μm compared to the toroidal milling, but also from the point of view of $R_t$ all the spherical milling is in advantage.

The concave cylindrical surface has the best roughness after processing with the spherical milling machine. In this case the spherical mill works the surface at an angle of inclination of 15°. In this case, the wear value of the spherical mill is slightly higher than in the case of the toroidal mill.

## References

[1] Han S G and Zhao J 2010 *Adv. Mat. Res.* **97** 2080
[2] Tonshoff H K and Hernandez C J 1998 *Mech. Work. Technol.* **20** 105
[3] Sadilek M and Cep R 2009 *World Acdmy of Sc, Eng and Technol.* **53** 661
[4] Shimzu K and Sasaki M Iwabe H 2006 *JSME Int. J.* **49** 28
[5] Paşca I 2012 *Proc. of the 16th Int. Conf. of Modern Technol. Qlty and Innovation.* **2** 709
[6] Osan A 2020 *Magazine of Hyd. Pn, Trib. Eco. Sens., Mech. Hidraulica* **1**
[7] Osan A 2019 *Magazine of Hyd. Pn, Trib. Eco. Sens., Mech. Hidraulica* **4**
[8] Osan A, Banica M and Nasui V 2019 *IOP Conf. Series: Mat. Sc. and Eng.***564** 012003
[9] Paşca I and Lobontiu M 2011 *The 16th Int.Conf. New Techn. and Prod. in Mach.* **18** 277

# Designing of an original extruding system for 3D printing of parts made of plastic material in powder-state form

**R Păcurar[1*], S Pascu[1], A Păcurar[1], D S Stan[2], E Teuţan[2], D I Băilă[3] and A Sadeh[4]**

[1] Department of Manufacturing Engineering, Technical University of Cluj-Napoca, B-dul Muncii 103-105, 400641, Romania
[2] Department of Mechatronics and Machine Dynamics, Technical University of Cluj-Napoca, B-dul Muncii 103-105, 400641, Romania
[3] Department of Manufacturing Engineering, Polytechnic University of Bucharest, Splaiul Independentei 313, 060032, Romania
[4] Department of Management of Technology, Holon Institute of Technology, GolombStreet, 5810201, Israel

*E-mail:razvan.pacurar@tcm.utcluj.ro

**Abstract.** Since their advent, various 3D printing technologies have been developed, out of which the Fused Filament Fabrication, FFF has completely revolutionized the industry in the last years. Classical FFF printers are using a thermoplastic filament, which is heated to its melting point and then extruded layer by layer to materialize a three dimensional part. One of the most important subassemblies of the 3D printer which is the most representative is the extrusion sub-assembly system and the thermal demands to which is subjected. In this article, one original constructive variant of the extruding sub-assembly system has been conceived and thermal analyses of the variant was performed by using the finite element method so as to be used not only for printing parts made of plastic materials in filament form, but also plastic materials in the powder state form at the Technical University of Cluj-Napoca, Romania.

## 1.Introduction

The manufacturing processes by using 3D printing methods are more and more widespread in many areas. These areas can be in the automotive, medical and aeronautical fields, etc [1-3]. These processes can be used for prototyping in order to avoid the high costs that are involved in these cases. Various 3D printing technologies have been developed, out of which the Fused Filament Fabrication (FFF) has completely revolutionized the industry in the last years [4]. FFF printers are using a thermoplastic filament, which is heated to its melting point and then extruded layer by layer to materialize a three dimensional part [5]. FFF printers are highly accurate machines, which due to their affordability have been included into various small and medium enterprises to be used for the production of functional prototyping and low volume production. One of the most important subassemblies of the 3D printer is the extrusion subassembly system and the thermal demands to which is subjected [6]. The extruder is usually adapted from the designing and manufacturing point of view according to the type of material that is used in 3D printing process, the accuracy, surface roughness and mechanical characteristics that are required for the realization of plastic parts [7-11]. The FFF technology uses the material in the filament form, which is melted within a specially designed head, the plastic filament being deposited afterwards in a form of a layer according to generated section data from a prepared 3D CAD model.

The accuracy and mechanical properties of plastic parts that are 3D printed are closed to the ones of plastic parts produced by injection moulding process [12, 13]. The main aim of research that has been presented in the current paper was to develop one new type of extruder that is suitable to print out parts made from plastic material in powder state form. Duraform Polyamide has been considered as raw material for the experiments. For the proposed research designing of a new extrusion mechanism for 3D printer Reprap Prusa i3 type has been considered, the validation of the constructive solution being made on the finite element analyses bases and experiments that were done for printing several samples that were subjected to tensile strength tests and surface roughness measurements that were made at the end.

## 2. Constructive design of the extrusion subassembly system

One of the most important subassemblies of the 3D printer which is the most representative is the extrusion subassembly system and the thermal demands to which is subjected. For the made research the 3D printer Reprap Prusa i3 type shown in figure1 has been considered.

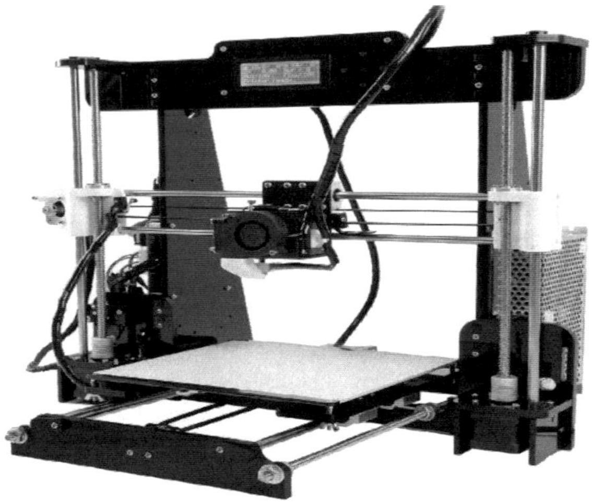

**Figure 1.** Reprap Prusa i3 [14].

**Table 1.** Technical characteristics of the 3D printing equipment [14].

| Brand Name | Anet; |
|---|---|
| Model Number | Anet A8 |
| Layer thickness | 0.1-0.3mm |
| Hotbed | 3mm Aluminium Heatbed |
| Structure | Metallic |
| Print speed | 100mm/s |
| Print materials | PLA,ABS,PVA,NYLON |
| File formats | STL/G-code |
| Max paper | 220x220x240mm |

In table 1 there are presented some specifications of the original 3D printer. In figure 2 is schematically presented the variant which was conceived and in figure 3 is presented the 3D model of this variant. The construction of this type of extruder is a simple one, based entirely on the classical extrusion concept. The rotation movement of the feeding screw combined with the gravitational force pushes the material towards the nozzle for extrusion. This aspect reduces the minimum required drilling length to achieve and providing optimum extrusion pressure. The melting of the material is realised by the resistance which is mounted in the crucible in the lower end of the extrusion subassembly. For good operating stability and high reliability the subassembly has been considered to be kept at a lower weight to avoid the high load of column, guiding systems, belt and motors. In order to reduce the amount of heat transferred to the upper part of the subassembly a weakly conductive thermal element such as ebonite has been taken into consideration. Ebonite can be machined by normal methods, but becomes flexible and shows rubber like elasticity when it is above its yield temperature which ranges between 80 °C to 280 °C. Taking into consideration the functionality aspects, two-hole crucible were designed and considered by mounting two 40W thermal resistors in order to reach the desired temperature faster, but also to provide a better thermal stability at the end.

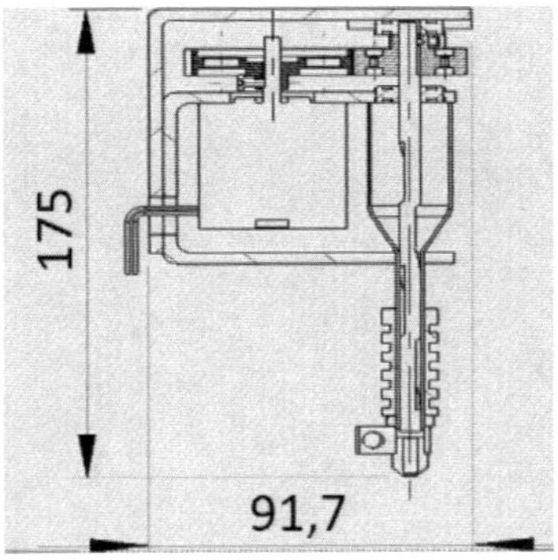


**Figure 2.** The 3D schematic of the extruder.

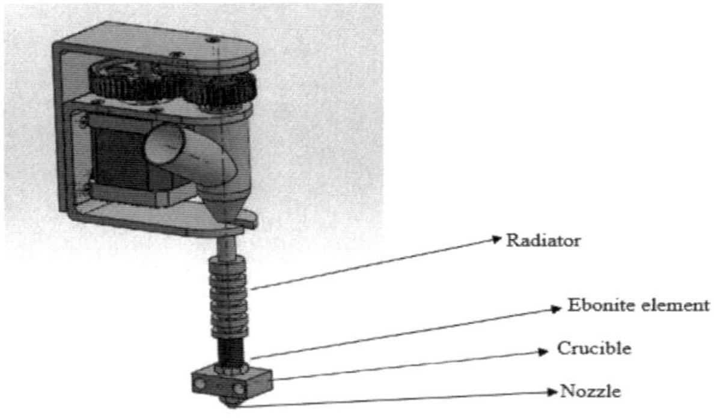


**Figure 3.** The 3D model of the extruder.

## 3. Finite element analysis of the new extrusion mechanism

Because the extrusion mechanism is subject to thermal loads, the simulating working conditions of the designed variants require the use of a program with the ability to perform the finite elements analysis of the extrusion mechanism. SolidWorks Simulation module that is compatible with the SolidWorks 2019 CAD program has been considered to simulate the thermal behaviour of extrusion mechanism. The SolidWorks Simulation module was preconfigured for stationary analyses. A total time for the nozzle was set to warm up the system (in this case 3600 seconds) and the time increment in which the temperature measurement was made was set from 15 to 15 seconds. At the initial data input stage it was considered the use of a 40 W resistance, the ambient thermal transfer coefficient being set at 15 $W/m^2 * K$ , the initial temperature of the components and the environment being pre-set at 20 °C (293,15 K). In figure 4 is shown how the 3D model was discretized within the program simulation. Solid mesh and curvature based mesh were used as main settings in the process of mesh generating. The total number of elements was 75082, and the number of nodes was 129301.The thermal and convection constraints to which the analysed model was subjected were introduced in this stage of the analysis as shown in figure 5. Taking into account the functioning principle of the extrusion subassembly a high attention must be focused on the thermal loads at which the sub-assembly mechanism is subjected, due to the thermal conduction phenomenon.

Mesh information

| | |
|---|---|
| Mesh type | Solid Mesh |
| Mesher Used: | Curvature-based mesh |
| Jacobian points | 4 Points |
| Maximum element size | 11.9497 mm |
| Minimum element size | 2.38994 mm |
| Mesh Quality | High |
| Remesh failed parts with incompatible mesh | On |

Mesh information - Details

| | |
|---|---|
| Total Nodes | 129301 |
| Total Elements | 75082 |
| Maximum Aspect Ratio | 40.035 |
| % of elements with Aspect Ratio < 3 | 63.7 |
| % of elements with Aspect Ratio > 10 | 5.3 |
| % of distorted elements(Jacobian) | 0 |
| Time to complete mesh(hh;mm;ss): | 00:00:16 |

**Figure 4**. Discretization of the 3D CAD model.

**Figure 5**. Boundary conditions.

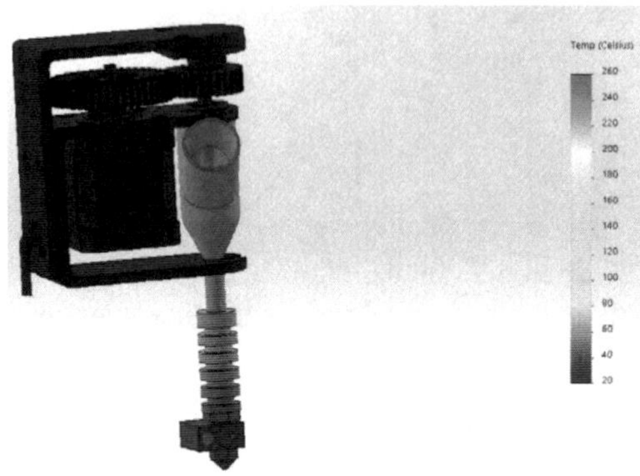

**Figure 6**. Temperature associated with the last time increment of transitory thermal process.

Taking into account the functioning principle of the extrusion mechanism, a high attention must be focused on the thermal loads at which it is subjected the sub-assembly mechanism, due to the thermal conduction phenomenon. In figure 6 is presented the temperature distribution associated with the last time increment of a transitory thermal process analysis. It can be observed that the heat distribution with temperatures is ranging from 20 °C to 260 °C. The maximum temperature was considered as being 260 °C and was reached by the crucible and the minimum temperature of 20°C was registered in the stepper motor area. This can be considered a positive result due to the fact that the main interest region is in the crucible area and not in the whole subassembly. Figure 8 shows the temperature over time, the temperature that was recorded at a certain time gradient during the transient-type analysis. This graph is valid only for the elements of the subassembly that are thermally loaded with a temperature higher than 20 °C according to figure 3 namely: nozzle, crucible and radiator. As one may notice in figure 8, the minimum temperature stabilizes at 100 °C after 500 seconds. The problem encountered in this variant after finite element analysis that was made is that, even after the 3600-second heating period, the nozzle tip shows variations of the temperature between 249 °C and 257 °C (figure 7), a difference of 11°C respectively 3 °C from the temperature at the resistance was set. This problem is due to the convection of temperature from the crucible to the rest of the subassembly. One of the most important conclusions of the analysis was that it is necessary to pre-heat the nozzle for 90 seconds, so the use of a thermostat for the rigorous control of the temperature in that area is an essential aspect to be taken into account for good operation of the extrusion mechanism. This variant of the nozzle has been considered for experiments that were made using the extruder adapted for printing plastic material like Duraform Polyamide in powder state form (the melting point of this material is 180 °C, which is below the temperature of heated nozzle) [15]. Temperature of 260 °C has been considered for the nozzle, since 3D printing extruder was tested for realization of parts using plastic materials in powder state form.

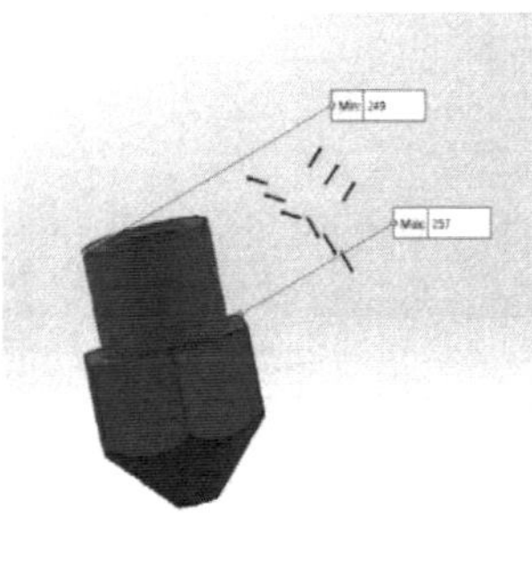

**Figure 7.** The temperature distribution of the nozzle.

| Timp | Tmin | Tmax |
|---|---|---|
| [s] | [°c] | [°c] |
| 0 | 20 | 20 |
| 15 | 20 | 260 |
| 30 | 20 | 260 |
| 90 | 23 | 260 |
| 120 | 36,9 | 260 |
| 150 | 58,8 | 260 |
| 210 | 81,5 | 260 |
| 300 | 93 | 260 |
| 405 | 96,8 | 260 |
| 615 | 98,4 | 260 |
| 1005 | 99,2 | 260 |
| 1500 | 99,4 | 260 |
| 3000 | 99,5 | 260 |

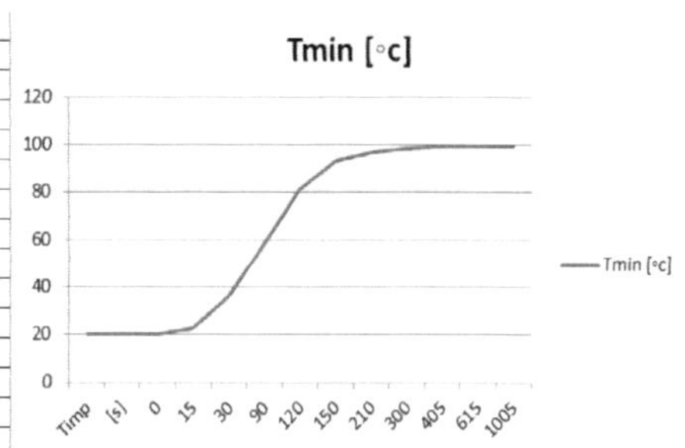


**Figure 8.** Temperature variation for the nozzle versus time.

## 4. The manufacturing and testing of the extrusion sub-assembly system

The 3D printing extruder was realized using several types of processes such as: welding, turning, bending, etc. Figure 9 shows the manufactured extruder subassembly of the 3D printer. Samples that were tested to determine the tensile strength were printed using this type of extruder that was mounted on Reprap Prusa i3 equipment. The samples were made of Duraform PA12 powder material, having a granulation of 50 microns, this type of material being normally used in the process of producing plastic parts by Selective Laser Sintering (SLS) process. The technological parameters that were used in the process of printing were the following ones: temperature of the nozzle 180 °C; bed temperature 100 °C; layer height 0.2mm; solid layers on the top 3, solid layer on the bottom 3, fill density 40%, fill pattern –line, fill angle 45°. Dimensions of samples that were printed are shown in figure 10, while an image with the printing process of one sample is emphasized in figure 11.

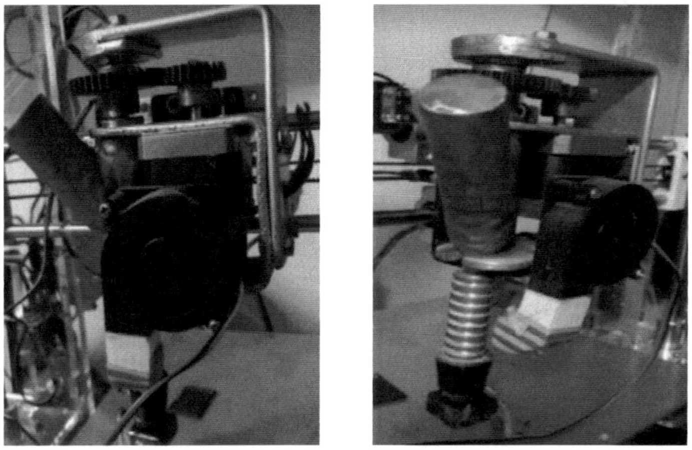

**Figure 9**. The manufactured extrusion subassembly.

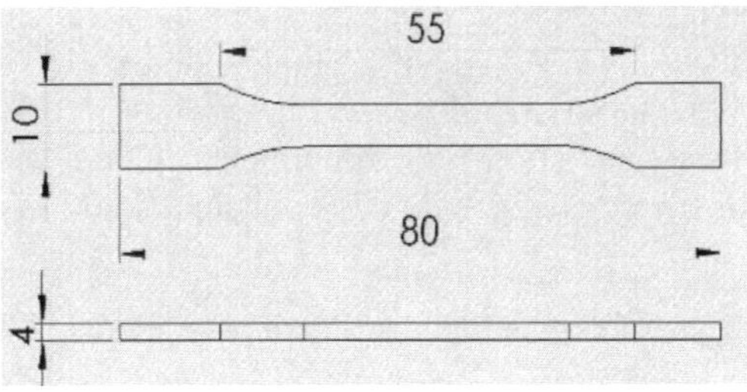


**Figure 10.** The dimensions of the designed sample.

**Figure 11.** 3D printed part made from Duraform PA 12 material by using the manufactured extrusion subassembly system.

Tests for determining the mechanical characteristics were carried out by using the Instron tensile-compression load cell Type 2511-308 at the Technical University from Cluj-Napoca (see figure 12). A comparison was made between the tensile strength of the samples which were printed using the designed extrusion mechanism and the tensile strength of samples that were made by Selective Laser Sintering (SLS) from Duraform PA 12 material. As shown in figure13, it was observed that the tensile strength in the case of the printed part using the designed extrusion mechanism had a value of 26.39 MPa, as compared to the one made by SLS, in which case the reached value was 39.6 MPa.

**Figure 12.** Instron tensile-compression testing equipment.

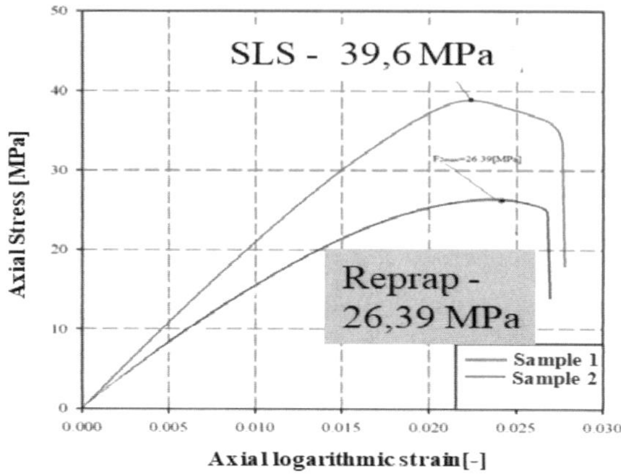


**Figure 13.** Stress-strain curves

Roughness measurements were made in the case of these samples using roughness tester Mitutoyo SJ-201 type. The reached results emphasized that the Ra values was significantly variable in the case of samples that were made using the extrusion mechanism mounted on Reprap Prusa i3 equipment. Values of Ra were between 7.39 μm and 20.17 μm due to the instability of the printing process. The best value of roughness was obtained in the case of the first deposited layers during the printing process, which were in contact with the heated bed, thus improving the roughness, while the highest value of roughness was obtained on the top surface, this phenomenon being caused mainly by to the high compaction of the plastic material while going through the heated nozzle. A solution for improving the roughness of the printed parts is to design and manufacture a sharpened head nozzle, reducing in this way the contact surface between the nozzle and the plastic material while going through the heated nozzle. Difficulties were encountered during the printing tests that were noticed also in terms of nozzle diameter, which are dependent by the layer height of the printed part. The solution that was found for improving the bounding of the successive deposited layers was to use a nozzle of Φ0.7 mm and a layer height of 0.25 mm. This correlation was important to avoid layer delamination. Before finding such a solution it was used a nozzle of Φ6 mm and a height of 0.5 mm for printing.

## 5. Conclusions

One original extrusion mechanism has been designed and tested in 3D printing process of parts made of Duraform PA12 material in powder state form as raw materials. Finite element analyses were made in order to analyse the transfer conduction and to reduce the amount of heat transferred from the nozzle to the upper part of the extrusion mechanism. Following the same analyses it was found that for a proper functioning of the printer it is necessary to pre-heat the nozzle for 90 seconds, so the use of a thermostat for the rigorous control of the temperature in that area is an essential aspect to be taken into account for good operation of the extrusion mechanism. Samples were produced using this type of extruding mechanism and tests were made in order to determine the tensile strength of parts made of Duraform PA material. The tensile strength had a value of 26.39 MPa, value that is comparable to the one reached in the case of samples made of Duraform PA material by SLS process (in this case the reached value was 39.6 MPa). Best value of roughness (7.39μm) was obtained in the case of the first deposited layers during the printing process. A solution for improving the roughness of the printed parts is to design and manufacture a sharpened head nozzle, reducing in this way the contact surface between the nozzle and the plastic material while going through the heated nozzle.

## References

[1] Heidari-Rarani M. et al. 2019 *Composites Part B: Engineering* **175** 107147
[2] Zidana A et al. 2019 *Int. J. of Pharmaceutics* **554** 292
[3] Shahrubudin N et al. 2019 *Procedia Manufacturing* **35** 1286
[4] Cheng C Y et al. 2020 *Chemical Engineering J.* **39615** 125242
[5] Wang J et al. 2020 *Composites Communications* **19** 142
[6] Baca D and Ahmad R 2020 *The Int. J. of Adv. Manufacturing Technology* **106** 4509
[7] Dinesh Y et al. 2020 *Materials Today: Proc.* **21(3)** 1592
[8] Yaoab T et al. 2020 *Composites Part B: Engineering* **188** 107894
[9] Pramanik D, Mandal A, Kuar A S 2020 *Materials Today: Proc. In press* 10.1016/j.matpr.2020.01.054
[10] Wang P, Zou B, Ding S 2019 *Applied Thermal Engineering* **161** 114064
[11] Gavali V C et al. *2020 Materials Today: Proc.* **22(4)** 1786
[12] Srinivasan R et al. *2020 Materials Today: Proc. In press*
[13] Sheoran A J and Kumar H 2020 *Materials Today: Proc.* **21(3)** 1659
[14] RepRap Prusa i3 3D printer https://reprap.org/wiki/Prusa_i3
[15] Brandrup J. Immergut E.H. Grulke E.A. 1999 *Polymer Handbook* 4th ed. (John Wiley & Sons: Hoboken, NJ, USA) p 2250.

# The methodology of economic recovery of commercial companies in crisis conditions

**O Panazan[1*], C Gheorghe[1] and G Calefariu[1]**

[1] Transilvania University of Brasov, 29 Eroilor, Brasov 500036, Romania

*E-mail: oana.panazan@unitbv.ro

**Abstract.** Even though commercial companies play a really important role in the economic functioning and development, now, in crisis conditions, the future for some of them is unknown. As a result of the release of crisis caused by „Coronavirus severe acute respiratory syndrome 2, abbreviated SARS-CoV-2", these companies are facing novel negative situations, not being able to manifest temporarily or normally the activity. From a systematic point of view, this crisis has the role of a really powerful disturbing factor. The creditors and investors are in a similar fragile situation. This paper identifies some scientific methods which can decrease the negative appeared problems, building models that take into account the limited resources that companies face and that they can lead to improving their economic situation. As an application, part of the essay presents the synthesis of the main assets and malfunctions of a company established in Romania. The elaboration of future strategies will be achieved through the systematic research of the admissible bases, aiming at maximizing the objective function. The type of equilibrium by reverse induction will be established, thus obtaining the set of the best cost variants. Linear programming will generate the optimal solution capable of exploiting the opportunities offered by the market with a minimal impact of disruptive factors.

## 1. Introduction

At the moment, organizations face intense national and international competition, with quick changes, caused by the emergence and evolution of the phenomena caused by COVID-19. Because of the pandemic, the legislative framework has frequently changed and the business environment has been in an extremely vulnerable situation. Social protection, the prohibition of some activities or their continuation under a modified form, the provision of some measures with impact on the business partners lead to the need for quick and impactful decisions, taken by the management of organizations. In all situations, in order to adapt to market conditions, the management decisions carry high risks for organizations. The paper aims to present some mathematical models related to technical, managerial and economic phenomena. These models can be useful tools that can be used permanently within organizations. Given the volatility that characterizes the economic situation due to COVID-19, both at local, European or global level, in the current context, these models become useful tools in finding new solutions.

## 2. Background to the study

As it is presented in literature, operational research gives scientifical methods and methodologies that leads to optimal decisions in management or the best solution in other domains. It makes up a dynamic domain of scientific research, in this moment are many complex optimization procedures and ways to be selected from, all of them having its own attributes and practical aim. It is uncomplicated because the unsuitable decision of optimization methods or starting values cannot lead to a global optimal solution [1]. Even if a challenger suggests similar approaches, he should not be capable of implementing them or should not get a considerable amount of cost cutting on development compared to the company which owns the same capabilities [2]. Operational research provides scientific methods and methodologies that lead to optimal management decisions or to the best solutions in other areas. It is a dynamic field of scientific research, both theoretical and applicative, which studies optimization problems resulting from the mathematical modelling of some phenomena and processes. Operational research involves the application of scientific methods, by a multidisciplinary team, the aim is to obtain results that best serve the interests of the organization as a whole [3].

The research uses operational research techniques and linear programming. Both belong to various disciplines, they make use of algorithms, computer techniques, mathematics, economics concepts, and obviously statistics [4, 5]. The most important problem is then reducing the lock of methodologies and case studies which makes use of operational research, linear programming, incorporating both artificial intelligence techniques, multicriteria decision theory [6], and uncertainty modeling [7 - 9]. Operational research is mainly the basic instrument for decision theory, prediction and improvement [10, 11]. Limiting time and costs will give the organization a competitive advantage over the competition. The company can establish an improved capacity in advance to the competitor replication, advantage in the competition can be sustained if it is known in do time [12, 13].

Essentially, an optimization problem is a problem of choice. It assumes as given a collection of entities generically called permissible solutions (or variants, scenarios). Solutions can be compared with each other and classified by means of an assessment criterion (performance). In this context, the question which arises is of finding the most appreciated solution, also called the optimal solution of the problem [3]. In the domain of business, operational research can bring an improvement in all sectors (resources, organization structure, demands), their analysis can lead to an important competition progress [14]. Currently, operational research training have started to incorporated computer strategies in automated learning by machine and using artificial intelligence, in this way can be improvingly applied in domains such as medicine [15] forecast and improvement [16]. Work by Daniel Kahneman and Amos Tversky presents a theory of how people make decisions in conditions of uncertainty called the theory of perspective [17].

Game theory is a method of examining economics that is based on the rational choice of variants when the results depend on the unknown strategies chosen by others. It is a mathematical discipline, which defines types of solutions built on methods and algorithms for solving. An example of game theory that analyzes the study of two-person play is provided by Antoine Cournot [18]. The mathematician Emile Borel suggested a formal game theory [19], he published several papers on game theory, the problem being approached in order to find out the opponent's movement in a game with imperfect information. He designed game theory for use in economic and military applications, and the ultimate goal was to determine if there was a "best" strategy for a particular game and to find that strategy. However, the development and popularization of game theory belongs to John Von Neumann, who published his first work on game theory in 1928 [20], later developing it with economist Oskar Morgenstern [21].

Over time it was acknowledged as a strong instrument in different domains (biology, logic, economics, psychology, political science). It focuses on their attractiveness and refinement, thus proposing a measure to improve the game, grounded on information about uncertainty in the results of the game [22]. Chess and Mah Jong variants are also the subject of the application of a logistic model [23]. Another model was built on analyzing the development of game strategies and was applied to time-limited games [24]. Merrill Flood conducted one of the best-known early experiments to investigate this question [25]. This paper introduced the convict's questioning considered to be the most exciting issues in the theory of games. Ken Binmore studied the role of game theory in the social sciences, while another work in this field is developed by two Nobel Prize winners (2002) D. Kahneman and A. Tversky [26]. They presented a theory of how people take decisions in conditions of uncertainty called perspective theory [27].

Complexity theory deals with the study of things that can be calculated when the resources we have at our disposal are limited. The fields of both theory of complexity and of proper algorithms refer to issues from opposite directions. A correct algorithm should be applied to find a solution to a problem and is at the same time an evidence of the correctness of the problem. Instead the theory of complexity the aim is to show that important issues are not to be resolved with limited means. Algorithms theory and theory of complexity try to estimate this. Therefore, the two fields of computer science, namely the theory of complexity, the design of correct algorithms and their analysis pass the boundaries between the choice of algorithm correctness implemented with accurate demands [28]. The complexity theory has identified some of the most fundamental and profound problems related to computing, has developed a powerful methodology for solving them and is generally considered one of the most challenging mathematical frontiers [36], deals with the study of calculable phenomena when resources are limited. From a certain point of view, the complexity theory is the opposite of the theory of algorithms. If this provides a solution to the problem within the limits imposed by resources, the complexity theory identifies the moment when resources are insufficient to solve the problem. Complexity is defined as a fabric of inseparable heterogeneous constituents [37].

The solving of algorithm issue is a basic topic in IT and a main part in programming [29]. Solving the problem and producing correct and optimal results constitute the main aims in computer training There are many difficulties in solving algorithmic problems, mainly in using abstract actions in addressing complex problems in giving options for correct outcome and for introducing it in programs [30 - 32]. As a result, obtaining good solutions is not common. Thus understanding the steps in solving a problem is necessary. If the probabilistic approach must first construct a statistic pattern and only after give its coordinates on figures a data string, in the domain of algorithms the modal is directly defined by these measurement. The data table can be organized as a set of ordinary and aleatory pieces [33]. The algorithmic method in search of the shortest program will generate the obtained sequence [34]. One such method is to describe the set of symbols "1" in a binary word in the programs prefix and to describe in a suffix a number of binary words with such a number of "1" [33, 35].

## 3. Research organization

The modelling of an economic process for the purpose of optimization involves two types of values: constant (they might change from one period to another, but over a period of time, they can be considered to be invariable) and variable (which will be defined in such a way as to satisfy a certain performance criterion, taking into account the carrying out, under limitative conditions, of the economic process). The correct identification of restrictions will have implications for the accuracy of the model and hence the value of the optimal solution. In developing the mathematical model for a practical optimization problem it is recommended to go through the following steps [3].

- Identifying variable values, they will become the decision variables of the model;
- Identification of limitative conditions that characterize the modelled process and their formalization in mathematical relationships-inequalities and / or equalities - called restrictions;
- Specifying the performance criterion and formalizing it in a function of the choice variants, called the function of the objective;
- Specifying the explicit conditions imposed on variable sizes, such as: to take only real non-negative values or only integer values, etc.

After finalizing the model, solving the optimization problem will lead to the determination of the values of the decision variables that will satisfy the restrictions and conditions imposed on the variables and will give the objective function the maximum or minimum value. The permissible solutions will be identified with those sets of numerical values granted to decision variables that satisfy the explicit restrictions and conditions, and the optimal solution will be that permissible solution that gives the objective function the highest or lowest value, as the case may be [3].

Here is an example of the management of company that assembles closing systems for logistics center, warehouses, shops or production halls that have an intense flow and require an optimal level of isolation. The management wants to maximize the profit of the company in relation to the interim management balances and to select the best option of costs in order to minimize them.

To maximize profit, the problem will be approached using operational research [38] and the solutions will be revealed by linear programming using Maple 2019 programs and C++ code from the CodeBlocks application. Then the game theory will be used to select the optimal consumption variant (figure 1).

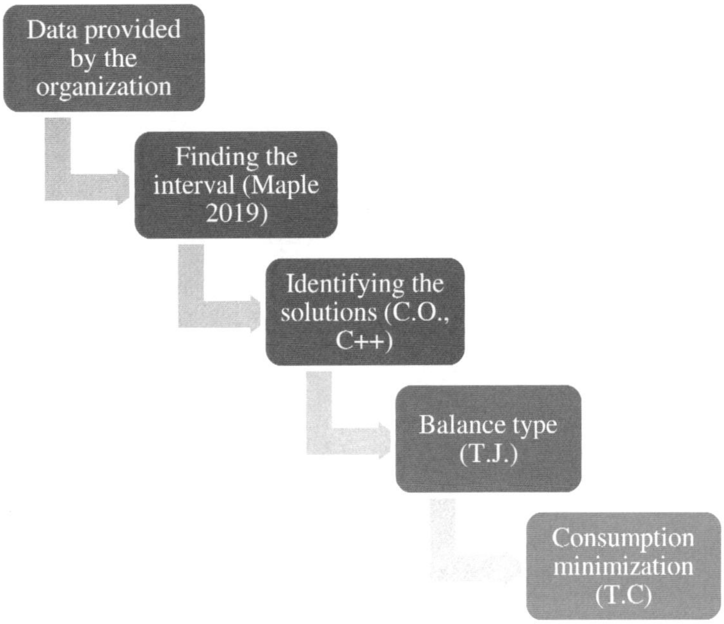


**Figure 1.** Scheme of resolution steps.

### *3.1. Data provided by organization*

The management of the organization seeks to achieve a profit share of 20 percent of the added value, the study of the indicator in dynamics revealed the following structure:

AV = 5 * CM+3 * PY-C where: AV- added value

CM - commercial margin

PY - production of the financial year

C - Third party consumption

The following variables will be used to perform the modelling: $x_1$(commercial margin), $x_2$(production of the financial year), $x_3$(third party consumption) and *fP* function $(x_1 x_2 x_3)$, profit of the organization dependent on the commercial margin, production of the financial year and consumption. The maximum of the function is obtained in relation (1), the limitative conditions (2) being the following:

$$\max f\text{B} = \frac{22*(5*x_1 + 3*x_2 - 2*x_3)}{100} \quad (1)$$

$$\begin{aligned} 5*x_1 + 3*x_2 &\geq 8\,295 \\ 5*x_1 - 2*x_3 &\leq 2\,752 \\ 3*x_2 - 2*x_3 &\leq -4\,986 \\ x_1 + x_2 + x_3 &= 10\,300 \\ x_1 \geq 0,\ x_2 \geq 0,\ x_3 &\geq 0 \end{aligned} \quad (2)$$

### *3.2. Finding the interval*

By entering the data into the Maple2019 application the program generates the following array of solutions (3):

>with (SolveTools[Inequality]):

>SolveTools[Inequality]:-LinearMultivariateSystem

({$5x_1+3x_2 \geq 8\,295$, $5x_1-2x_3 \leq 2\,752$, $3x_2-2x_3 \leq -4\,986$, $x_1+x_2+x_3=10\,300$, $x_1 \geq 0$, $x_2 \geq 0$, $x_3 \geq 0$,}, $[x_1, x_2, x_3]$)

$$\begin{gathered}\left[x_1=0,\ x_2=\frac{15\,614}{5},\ x_3=\frac{35\,886}{5}\right],\\ \left[x_1=\frac{85\,532}{31},\ x_2=\frac{62\,594}{31},\ x_3=\frac{171\,174}{31}\right],\\ \left[x_1 \in [0, 1\,659),\ x_2 \in \left[-\frac{5x_1}{3}+2\,765,\ \frac{15\,614}{5}-\frac{2x_1}{5}\right),\ x_3=10\,300-x_1-x_2\right],\\ \left[x_1 \in \left[1\,659, \frac{85\,532}{31}\right],\ x_2 \in \left[0, \frac{15\,614}{5}-\frac{2x_1}{5}\right),\ x_3=10\,300-x_1-x_2\right],\\ \left[x_1 \in \left(\frac{85\,532}{31}, 3\,336\right],\ x_2=11\,676-\frac{7x_1}{2},\ x_3=-1\,376+\frac{5x_1}{2}\right],\\ \left[x_1 \in \left(0, \frac{85\,532}{31}\right),\ x_2=\frac{15\,614}{5}-\frac{2x_1}{5},\ x_3=\frac{35\,886}{5}-\frac{3x_1}{5}\right],\\ \left[x_1 \in \left(\frac{85\,532}{31}, 3\,336\right),\ x_2 \in \left[0, 11\,676-\frac{7x_1}{2}\right),\ x_3=10\,300-x_1-x_2\right].\end{gathered} \tag{3}$$

### *3.3. Identifying the solutions*

In this way, the intervals that comply with all conditions were found. To find out the unknowns that lead to the maximum value, enter the data in a program such as C++ through the application CodeBlocks.

To start with, we will define 4 variables which will contain the maximum value of the maximized function (4) which is calculated by function B, which will be described below, and the 3 parameters (5) that led to this value.

```
int ma = 0                                  (4)
int x1, x2, x3                              (5)
```

The defined function called B calculates the value of the maximized function using the three given parameters which represent a set of values that is included in the array of solutions (6).

```
int B (int x1, int x2, int x3)
{
return (22*(5*x1+3*x2-2*x3)/100);           (6)
}
```

The returned valued will be stocked in a new variable (7) and compared with variable (4) and, if the value is greater in variable (7), then the used parameters in function B (6) will be stocked in the 3 variables described before (5). To illustrate, here is an example:

```
int p=B(0, 15614/5, 35886/5.0);
if(ma<p)
{
ma=p;
x1=0;                                       (7)
x2=15614/5;
x3=35886/5;
}
```

The same procedure will be applied for all intervals and unique solutions from the array of solutions which was given by Maple 2019. The maximum value of the function is 1 954 (m.u.), for $x_1$=2 752 (m.u.), $x_2$=2 044 (m.u.) and $x_3$=5 504 (m.u.)

### *3.4. Balance type*

This section aims to identify the Stackelberg balance type in order to minimize consumption. The company has three variants with the following cost matrix imposed by the demand curve. Game theory [39] consider a dyadic game of the three 3 variants, presented in the following matrix (8):

$$\begin{bmatrix} \begin{bmatrix} 52{,}33 & 52{,}80 \\ 52{,}21 & 52{,}15 \end{bmatrix} & \begin{bmatrix} 52{,}33 & 52{,}10 \\ 52{,}80 & 52{,}56 \end{bmatrix} & \begin{bmatrix} 51{,}70 & 51{,}75 \\ 51{,}76 & 51{,}66 \end{bmatrix} \\ \begin{bmatrix} 52{,}40 & 52{,}10 \\ 52{,}56 & 52{,}44 \end{bmatrix} & \begin{bmatrix} 51{,}72 & 51{,}72 \\ 51{,}82 & 52{,}21 \end{bmatrix} & \begin{bmatrix} 51{,}76 & 51{,}76 \\ 51{,}66 & 51{,}82 \end{bmatrix} \end{bmatrix} \tag{8}$$

V1, choosing strategy 1, chooses the first matrix line. V2, knowing that V1 chose line 1 of the matrix, chooses either the first strategy, or the second strategy. We assume it will choose strategy 1. In the end, V3 chooses its strategy, that is, the first or the second column. Suppose it chooses strategy 1. As a result, the next evolution of the game is obtained in the following matrix (9):

$$\begin{bmatrix} \begin{bmatrix} 52{,}33 & 52{,}80 \\ 52{,}21 & 52{,}15 \end{bmatrix} & \begin{bmatrix} 52{,}33 & 52{,}10 \\ 52{,}80 & 52{,}56 \end{bmatrix} & \begin{bmatrix} 51{,}70 & 51{,}75 \\ 51{,}76 & 51{,}66 \end{bmatrix} \\ [52{,}33 \quad 52{,}80] & [52{,}33 \quad 52{,}10] & [51{,}70 \quad 51{,}75] \\ 52{,}33 & 52{,}33 & 51{,}70 \end{bmatrix} \tag{9}$$

It highlights how to find the Stackelberg balance by reverse induction: V3 chooses minimum in each line from the matrices in column 3. Inductively V2, knowing how V3 will choose its strategy, chooses from the highlighted items the lowest. V1 chooses, knowing the choices V2 and V3. In the Stackelberg game V1 chooses the strategy and communicates it to V2, this knowing the choice of V1, selects its strategy and communicates to V3 both the strategy chosen and the strategy chosen by V1. V3 chooses its strategy knowing the strategies chosen by V1 and V2. Finally, after all the strategies have been chosen, each variant obtains the value corresponding to its cost function. So, the game has a single uncertain Stackelberg balance (2; 1; 2) with costs (52,10; 51,76; 51,72). Profile (2; 1; 2) is not Nash balance.

### *3.5. Consumption minimization*

The cost vector of size A = [52,33; 52,80; 52,21; 52,15; 52,40; 52,10; 52,56; 52,44; 52,33; 52,10; 52,80; 52,56; 51,76; 51,76; 51,82; 52,21; 51,70; 51,75; 51,76; 51,66; 51,76; 51,76; 51,66; 51,82], with distinct positive integer elements and a total cost: TC will be considered. Costs can be selected n times, and the program will generate the solution closest to the established criteria.

C++ code from the CodeBlocks app:

```
:
#include <iostream>
#include <fstream>
using namespace std;
ifstream fin ("date.in");
int main()
{
  double n;
  double a[50];
  fin>>n ;
  for(int i=0 ; i<24 ; i++)
    fin>>a[i];
  for(int i=0 ; i<23 ; i++)
    for(int j=i+1 ; j<24 ; j++)
```

```
        if(a[i]<a[j])
          swap(a[i],a[j]);
  int i=0;
  cout<<"For obtain total cost "<<n<<", we use:\n";
  while(n>0 && i<24)
  {
    int ap=0;
    while(n>a[i])
    {
      n-=a[i];
      ap++;
    }
    if(ap>0)
      cout<<"Cost "<<a[i]<<" de "<<ap<<" ori\n";
    i++;
  }
  if(n>0)
    cout<<"Cost "<<a[23]<<" de 1 time\n";
  return 0;
}
```

The solution that ensures a selection of a minimum number of costs is given by choosing the cost 52,80 by 208 times and the cost 51,66 once.

## 4. Results interpretation

It is important, especially at a time when societies are in a precarious balance due to the phenomena caused by COVID-19, to understand the methods by which organizations can grow. There are many studies that identify this possibility, but how they can grow and what are the factors that influence their evolution is a very wide field. Determining the best infrastructure of the organization is a crucial task. To this end, different approaches are applied which may lead to the achievement of the proposed objective. The study revealed a maximum profit of 1 954 (m.u.), in the case of a commercial margin of 2 752 (m.u.), production of the financial year in the amount of 2 044 (m.u.) and 5 504 (m.u.) consumptions. An uncertain Stackelberg balance was found, minimizing consumption can be achieved by selecting a minimum number of costs.

## 5. Conclusion

Based on the income statement, a series of value indicators can be determined regarding the volume and profitability of the organizations' activity, this fact being reflected in the analysis of the intermediate management balances. The added value is the surplus achieved by the company over the material consumption within the current activity. The increase of this indicator can be achieved taking into account the balance between the commercial margin and the production of the year on the one hand and the consumptions from third parties on the other hand. Each indicator is reflected in the final result, their analysis together or separately can lead to a strategy suitable for each organization. In the paper, we can notice the manner of solving the problem of operational research in Maple2019 and CodeBlocks applications, C++ code, accessing several optimization programs not being necessary, thus increasing the efficiency of solving the case. In most cases the answer to a simple question leads to a complex system of solutions. Game theory can predict how variants cost could be selected helping to design the economic system, but certainly research can continue in order to fully understand the purpose, applicability and usefulness of these theories.

**References**

[1] Writing Group of Operational Research 2000 *Operations Research* (Beijing: Tsinghua University Press)
[2] Bell P C 1989 Strategic Operational Research *J.O.R.* **49** 381
[3] Nica V T, 2011 *Cercetări operaționale* (București: Academia de Studii Economice)
[4] White J A 1991 An existence theorem for OR/MS *O.R* **39** 183
[5] Meisel S and Mattfeld D 2010 Synergies of operations research and data mining *E.J.O.R.* **206** 1
[6] Clemen R T and Reilly T 2013 *Making hard decision with Decision Tools* (U.S.A.: South-Western Cengage Learning)
[7] Howard R A 2012 *Dynamic probabilistic systems: Markov models* (U.S.A:, John Wiley&Son, INC)
[8] Abbas A E and Howard R A 2015 Foundations of decision analysis (England: Pearson Higher)
[9] Taha A H 2007 Operations research. An Introduction (UK: Pearson Prentice Hall)
[10] Anderson D, Sweenwy D J, Williams T A, et al 2018 *An Introduction to Management Science: Quantitative Approaches to Decision Making* (Cengage Learning EMEA)
[11] Albright S C and Winston W L *Practical Management Science* (Revised 3e South-Western Cengage Learning).
[12] Griffits G H and Finlay P N 2004 IS-enabled Sustainable Competitive Advantage in Financial Services and Manufacturing *S.I.S* **13** 29
[13] Flamholtz E and Hua W 2003 Searching for Competitive Advantage in the Black Box *E.M.J,* **21** 222
[14] Bell P C, Anderson C K and Kaider S P 2003 Strategic operations research and the edelman prize finalist applications 1989-1998 *O.R,* **51(1)** 17
[15] Li L et al 2004 Data mining techniques for cancer detection using serum proteomic profiling, *A.I. in medicine* **32(2)** 71
[16] Chui M, Manyika M, Miremadi M, Henke N, Chung R, Nel P and Malhotra S 2018 *Notes from AI frontier: Insights from hundreds of use cases* (McKinsey Global Institute)
[17] Kahneman D and Tversky A 1979 Prospect Theory: An Analysis of Decision under Risk *Econometrica* **47** 263
[18] Cournot A 1838 *Researches into the Mathematical Principles of the Theory of Wealth.*
[19] Borel E 1921 Game theory and left symmetric core integral equations
[20] Von Neumann J 1928 Theory of Parlor Games
[21] Von Neumann J and Morgenstern O 1944 *Theory of Games and Economic Behavior* (Princeton University Press)
[22] Iida H, Takeshita N and Yoshimura J 2003 *A metric for entertainment of board games: Its implication for evolution of chess variants* Entertainment Computing Technologies and Applications
[23] Iida H, Nahgashima J, Kajihara Y and Hashimoto Y 2004 An application of game refinement theory to Mah Jong *I.C.E.C*
[24] Sutiono A P, Purwarianti A, Iida H et al 2014 A Mathematical Model of Game Refinement *INTETAIN*
[25] Flood M M 1952 Some Experimental Games *Research Memorandum* RM-789
[26] Binmore K 1994 *Playing Fair* (MIT Press)
[27] Kahneman D and Tversky A 1979 Prospect Theory: An Analysis of Decision under Risk *Econometrica* **47**
[28] Wegener I 2003 *Complexity Theory. Exploring the Limits of Efficient Algorithms* (Springer-Verlag)
[29] Wing M J 2006 Computational thinking *Communications of ACM* **49**
[30] Ginat D 2001 Misleanding Intuition in Algorithmic Problem Solving *Proceedings of the thirty-seconds SIGCSE technical symposium on Computer Science Education*

[31] Haberman B, Averbuch H and Ginat D 2005 Is it really an algorithm?-The need for explicit discourse in *Proceedings of the 10th Annual Conf. of ITiCSE'05*
[32] Ye N and Salvendy G 1996 Expert-novice knowledge of computer programming at different levels of abstraction *Ergonomics* **39**
[33] Goppa V D 1995 *Introduction to the algebraic theory of information* (Moscow: Nauka)
[34] Kolmogorov A N 1965 Three approaches to the definition of the concept, quantity of information
[35] Melnyk S I and Melnik S S 2016 An algorithmic method of solving inverse problems of reconstruction of the macrostructure scattering media *9th International Kharkiv Symposium on Physics and Engineering of MSMW* (Kharkiv, Ukraine)
[36] Papadimitriou C H The Complexity of Knowledge Representation 0-8186-7386-9.
[37] Morin E 1991 *Introduction a la pensee complexe* (Paris: ESF)
[38] Dantzig G B 1960 Inductive Proof of the Simplex Method *IBM Journal of Research and Development* **4(5)** 505-506
[39] Woolridge M 2012 Does Game Theory Work? *IEEE Computer Society*

# Sustainable manufacturing systems – a simulation model

**A L Parv[1]*, B A Deaky[1] and R Stasiak-Betlejewska[2]**

[1]Manufacturing Engineering Department, Transilvania University of Brașov, Eroilor nr. 29, 500036 Brasov, Romania
[2]Institute of Production Engineering, Czestochowa University of Technology, Poland

* E-mail: luminita.parv@unitbv.ro

**Abstract**. Connecting companies to the paradigm of climate change leads to the creation of sustainable production systems. In this context, new approaches to processes allow the definition of sustainable manufacturing systems. The sustainability of manufacturing systems is an important aspect in order to insure the competitive power of enterprises. In this respect, the manufacturing systems must take into account economic factors, environmental concerns and social directions. Decisions regarding manufacturing development, optimization, or reorganization are driven by many factors and are often costly, with the benefits hard to justify before implementation. This paper describes an agent-based model simulation that analyses the behaviour of a sustainable manufacturing system.

## 1. Introduction

A sustainable manufacturing system (SMS) is a combination of the of waste reduction as Lean Manufacturing principles, reducing-recycling-reusing as Green Manufacturing principle, Total Quality Management, and Corporate Social Responsibility [1, 2].

This paper is aim at developing a simulation model that will be able to obtain an optimal configuration of SMS considering both economic criteria and environmental responsibilities, reflected in minimizing the total cost, total energy consumption and $CO_2$ emissions.

Hence, a feasible solution to the problem of SMS, its modelling, and implementation the model the Anylogic simulation environment, so that the application of the optimization can be done quickly [3].

The present paper is structured as follows: after a brief presentation of the theoretical aspects of a multimethod simulation model, we will discuss the development of an SMS, considering the following variables: suppliers s, the factory f and the warehouse w.

The model aims to determine the optimal solution taking into account the three objective functions, namely the investment cost required to configure the manufacturing system, the total energy consumption of the manufacturing system and the total CO2 emissions. The following chapter presents the solving methodology using the Anylogic environment [4]. The last chapter describes the conclusions related to the simulation model and possible further development.

## 2. The multimethod simulation model

Systemic methods are used to understand the organization of data. They present real-world objects in the form of entities. The relations between them are represented in the form of associations.

In order to implement the simulation it is necessary to do the modelling of the real-world entities. To make them into a single, efficient, and manageable multimethod model that represents the behaviour of the whole system, we should do a few more things. The simulation environments consists of three modelling methods [4, 11], respectively agent-based modelling dynamic systems, and discrete events.

Each method supposes a certain of level of abstraction. When the simulation requires a high degree of abstraction, the simulation environment can use dynamic systems.

As a research direction, the SMS model in this paper hereby described can be integrated into a supply chain, being thus connected to both the suppliers' network and the customers.

The conceptual model describes how a factory makes a product based on the resources provided by different suppliers. A factory can process raw materials supplied by different suppliers. The choice of suppliers is based on their production capacity and the ability to deliver current orders. This approach suggests the use of the model based on agents with a certain level of intelligence [4, 8].

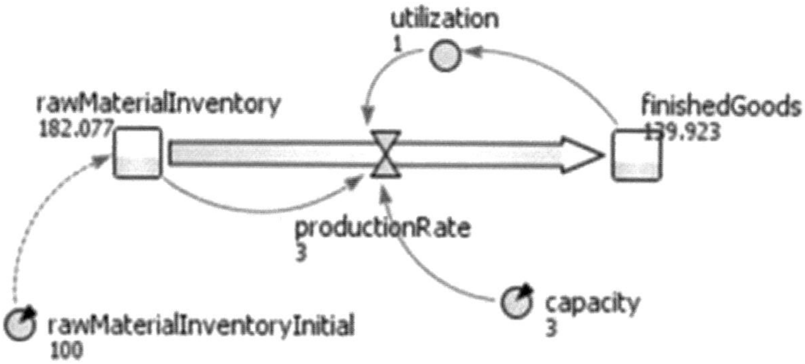


**Figure 1**. Conceptual model.

Customers request a product at the factory, and pick up the available items at the current time. The excess of demand over supply is lagging behind (figure 1).

Separate data sets for costs, energy consumption and carbon emissions will be associated to the simplified manufacturing system. The simulation results will be retrieved via Excel files.

**Figure 2**. Factory as an agent.

A factory is modelled as an agent (figure 2). They interact in an environment by sending messages to each other. There are two types in an environment when sending messages to each other.

From the perspective of an SMS, a product is made based on resources [5, 6], is processed on specific machines, and in the end other resources are released [7, 8].

Under the conditions of the designing of a sustainable manufacturing system, the solutions that can be obtained will have the following characteristics:

- Energy consumption and $CO_2$ emissions [9];
- Optimal number of machines [10];
- Optimal materials flow in the manufacturing system [11, 12].

Generally, the model uses a stationary policy. The cost, energy consumption and carbon emission parameters are associated with ordering, manufacturing, keeping products and having a backlog. The simulation output includes the structured average daily cost, energy, emission produced by all components of the supply chain.

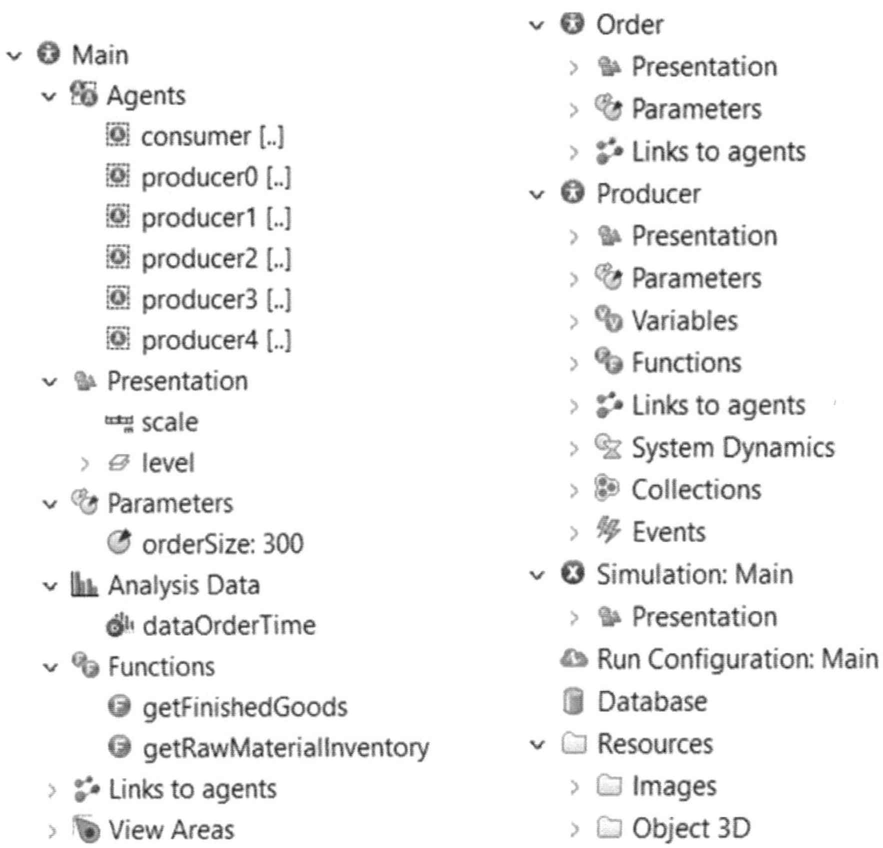


**Figure 3**. AB and SD design.

Inside the factory agent is represented a diagram of the dynamics of the system (figure 3). It represents the production process.

Therefore, this model is a multimethod one, based on Agent and System Dynamics [4].

There are two types of processes and corresponding messages: ordering and shipping. A black line between producers indicates that a producer is waiting for a supply from another producer (figure 4). The width of the line indicates the size of the command [3, 7].

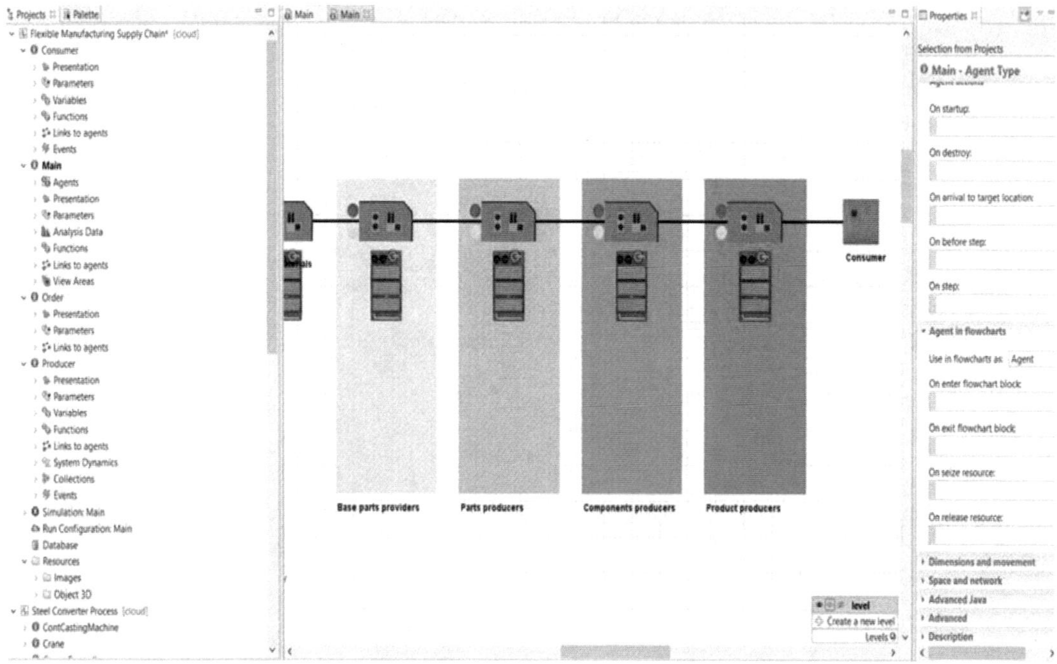


**Figure 4.** Simulation scene.

Parameters variation experiment performs several single model runs varying one or more parameters.

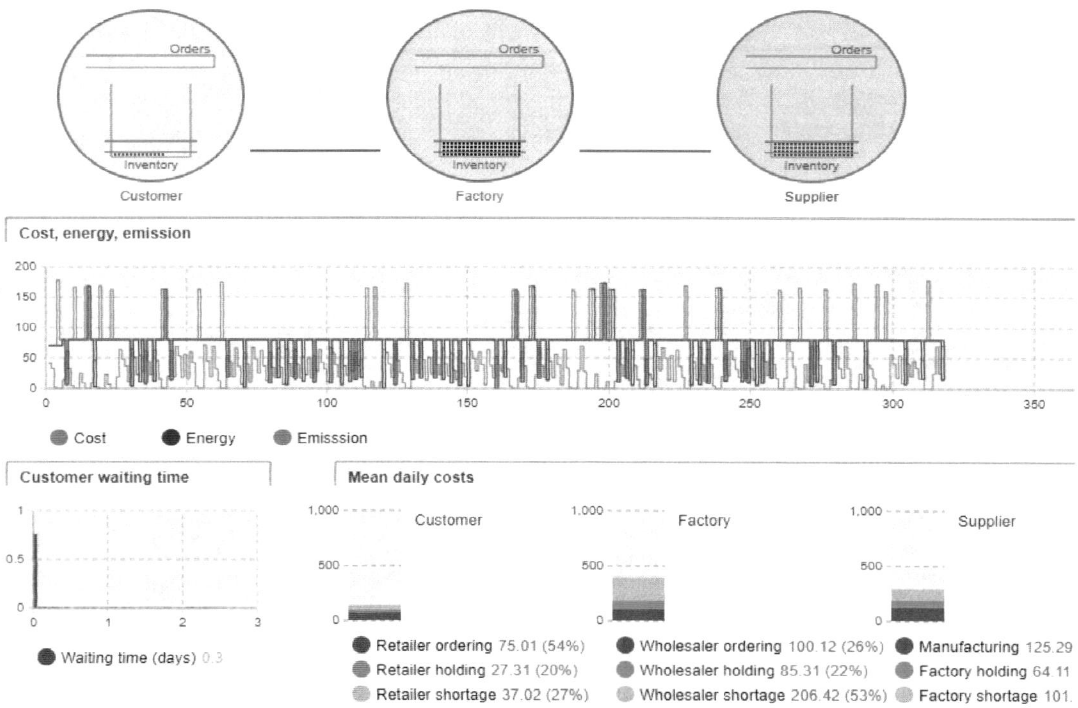


**Figure 5.** Running simulation.

Function defined for finished goods is described as follows:

```
if (factory.size()!=0) {
  double m=0;
  for (inti=0; i<factory.size()-1; i++)
    m+=((Factory)factory.get(i)).finishedGoods;
  return m;
}
else
    return 0;
```

Function defined for raw material inventory is described as follows:

```
if (factory.size()!=0) {
  double m=0;
  for (inti=0; i<factory.size()-1; i++)
    m+=((Factory)factory.get(i)).rawMaterialInventory;
  return m;
}
else
    return 0;
```

Using this experiment one can compare the behaviour of model with different cost, energy, and emissions values (figure 5). These parameters are associated with both waiting in the queue before the station, and with the processing. Each minute of waiting and each processing operation have a unit cost.

The system collects the values of cost, emissions and energy assigned to the parts that exit the system in a time series dataset.

## 3. Conclusion

Simulation, as an essential tool for the manufacturing system analysis and the impact of system changes evaluation, can assure informed decisions.

Specific processes and strategies can be modelled and simulated in manufacturing simulation software. This enables effective analysis, and provides an efficient way to experiment and reduce the costs of testing in the real world. Process simulation software, AnyLogic gives the power to get the most from your manufacturing [4].

Using process simulation software is key to the detailed analysis and effective operation of your organization.

By analysing the proposed optimization model, we can conclude that, this simulation model can determine the inventory level for the manufacturing system that result in minimum cost, energy consumption and carbon emission. The model includes a simulation experiment where ones can set up the parameters manually, and the automatic optimization experiment.

The mathematical and simulation model thus developed can be used to obtain an optimal configuration of SMS considering both economic criteria and environmental responsibilities, reflected in minimizing the total cost, total energy consumption and $CO_2$ emissions.

As a future direction of development, the simulation model will be extended with a detailed manufacturing module in the Anylogic environment [4], so that the application of the optimization can be done quickly.

The simulation model may be generalized at a complex manufacturing sustainable systems.

## References

[1] Nujoom R A O, Mohammed A M and Wang Q 2019 Drafting a cost-effective approach towards a sustainable manufacturing system design *Comput. Ind. Eng.* **133** 317
[2] Nujoom R A O, Wang Q and Mohammed A 2018 Optimisation of a sustainable manufacturing system design using the multi-objective approach *Int. J. Adv. Manuf. Tech.* **96** 2539
[3] Tilină D I, Zapciu M and Mohora C 2015 A system model to integrate the "Green Manufacturing" concept in Romanian manufacturing organisation *IOP Conf. Ser.: Mater.Sci. Eng.* **95** 012128
[4] www.anylogic.com . Accessed: april 2020.
[5] Mohammed A 2020 Towards 'gresilient' supply chain management: A quantitative study *Resour. Conserv. Recy.* **155** 104641
[6] Zindani D, Kumar K and Davim J P 2020 Sustainability Manufacturing Systems Design *Hashmi, S. and Choudhury, I. eds. Encyclopedia of Renewable and Sustainable Material Elsevier* **5** 512
[7] Ikumapayi O M E T, Akinlabi P, Onu S A, Akinlabi M and Agarana C 2019 A Generalized Model for Automation Cost Estimating Systems (ACES) for Sustainable Manufacturing *J. Phys.: Conf. Ser.* **1378** 032043
[8] Foit K, Banaś W, Gwiazda A and Hryniewiczet P 2017 The comparison of the use of holonic and agent based methods in modelling of manufacturing systems *IOP Conf. Ser.: Mater. Sci. Eng.* **227** 012046
[9] Tamás P, Illés B and Dobos P 2016 Waste reduction possibilities for manufacturing systems in the industry 4.0 *IOP Conf. Ser.: Mater. Sci. Eng.* **161** 012074
[10] Henao-Hernández I, Solano-Charris E L, Muñoz-Villamizar A, Santos J and Henríquez Machadoet R 2019 Control and monitoring for sustainable manufacturing in the Industry 4.0: A literature review *IFAC PapersOnLine* **52** 195
[11] Peter O and Mbohwaet C 2019 Industrial Energy Conservation Initiative and Prospect for Sustainable Manufacturing *Procedia Manufacturing* **35** 546
[12] Huang A, Badurdeen F and Jawahir I S 2018 Towards developing sustainable reconfigurable manufacturing systems *Procedia Manufacturing* **17** 1136

# Studies on structural MWCNT/epoxy nanocomposites for EMI shielding applications

**C Paun[1,3,*], C Obreja[1], F Comanescu[1], V Tucureanu[1], O Tutunaru[1], C Romanitan[1], O Ionescu[1], D E Gavrila[2], V Manescu Paltanea[3], V Stoica[4] and G Paltanea[3]**

[1] National Institute for Research and Development in Microtechnologies IMT-Bucharest, 126A Erou Iancu Nicolae Street, 077190, Bucharest, Romania
[2] Faculty of Applied Sciences, University POLITEHNICA of Bucharest, 313 Splaiul Independentei, 060042, Bucharest, Romania
[3] Faculty of Electrical Engineering, University POLITEHNICA of Bucharest, 313 Splaiul Independentei, 060042, Bucharest, Romania
[4] National Institute for Electrical Engineering ICPE-CA, 313 Splaiul Unirii, 030138, Bucharest, Romania

*E-mail: costel.paun@imt.ro

**Abstract.** This paper presents the preparation and characterization of MWCNT - Multi-Walled Carbon Nano-Tube/epoxy nanocomposites, as materials with a good yield for shielding electromagnetic interferences - EMI. Both the precursors used, and the nanocomposites obtained were characterized by different techniques: FTIR - Fourier transform infrared spectroscopy, Raman spectroscopy, SEM - Scanning Electron Microscope and XRD - X-ray diffraction. The electrical and mechanical properties of the nanocomposites were estimated compared to the epoxy resin without addition. Electromagnetic interference shielding tests were conducted in the 2.5 - 6.4 GHz frequency band. The results were found to be promising since for the MWCNT/epoxy sample it was obtained a 20 dB attenuation when compared with the ambient air.

## 1. Introduction

The replacement of carbon fibers in the structure of polymeric nanocomposites with carbon nanotubes was an important step in the field of nanotechnology [1]. The emergence of carbon nanotubes [CNT] has opened up important areas of applicability in the electronics industry, due to the thermal and strengthening properties at the nanometer scale [2], the electronic properties [3] and the superior chemical resistance [4, 5]. Due to the structure of single-cylinder or multi-cylinder graphene with a high aspect ratio and low density, MWCNT has special mechanical and electrical properties [6]. The special properties of CNT such as adhesion, chemical resistance and dimensionless stability contribute to obtaining epoxy nanocomposites with applicability in the automotive, aerospace, adhesive encapsulation, anticorrosive coatings, or the manufacture of materials with electromagnetic shielding properties. The hardening kinetics and chemical composition of epoxy resins influence the properties of nano-epoxy materials such as: Young's modulus, density, electrical and thermal conductivity. In the process of preparing CNT-epoxy nanocomposites, the liquid medium consisting of an epoxy or amine component, incorporates carbon nanotubes, leading to a good dispersion of them, then proceeds to the ultrasonic homogenization procedure [7], hardening of the resulting dispersion, by adding the resulting hardener composite in solid state. The amine hardener facilitates the opening of the epoxy resins ring and due to the reaction of the two components, the molecular opening of the epoxy resins ring and due to the reaction of the two components, the molecular mass increases, which implicitly leads to the hardening of the composite. In order to obtain certain properties of the CNT/epoxy nanocomposite, we use the functionalization processes [8] of nanotubes that control the rheological properties of the liquid composition or the Van der Waals bonds between the nanotubes and the liquid component [9]. Methods used in recent applications include: matrix use of carbon nanotubes with other nanoparticles [10, 11], nanotube surface functionalization [12], use of wetting surfactants [13], efficient methods for mixing and dispersing nanotubes in the liquid epoxy component [14], the use of mixed epoxy/copolymer nanocomposites [15] or fillers in combined form using nanotubes/graphene [16 - 20]. The use of a resin/hardener with low viscosity represents a major advantage in the process of preparing epoxy nanocomposites, which leads to the incorporation of a large amount of CNT, without modifying their rheological properties. The increase of the viscosity in case of a large specific surface leads to the modification of the structural properties of the nanocomposite, due to the inhomogenization of the dispersion and the incorporation of air particles. The following is a working method for obtaining the MWCNT dispersion in the epoxy system.

In this paper the electromagnetic waves shielding potential of some composite materials is evaluated. The analyzed composites show the efficiency of the electromagnetic waves in the 2.5 – 6.4 GHz band, which proves the utility of these materials in making some of the electromagnetic screens in environments in which the use of materials for classical screens (metals) is forbidden.

## 2. Materials and method

The epoxy component Epo-tek® 301 (Epoxy Technologies, Inc) is used as the polymer matrix, in liquid form, consisting of two components: the epoxy component (A) and the amine component (B). Carbon nanotubes (MWCNT –BPD30L520, NanoLab) having a length between 5-29 μm, purity > 95%, and an average diameter of 30 nm.

### *2.1. Preparation method of MWCNT / epoxy nanocomposites*

In order to characterize the materials (MWCNT/epoxy nanocomposites and epoxy polymer), specimens were made in the form of discs. The discs were made by casting in PVC (PolyVinyl Chloride) molds arranged on PET (Poly-Ethylene Terephthalate) lamination foils, supported by glass supports (figure 1), and for the stripping of the materials was used silicone Vaseline (High vacuum silicone grease – Dow Corning). Carbon nanotubes (MWCNT) are weighed, mixed, and homogenized in epoxy component (A), which with low viscosity (100-200 cP) makes it possible to easily incorporate materials with a large specific surface area.

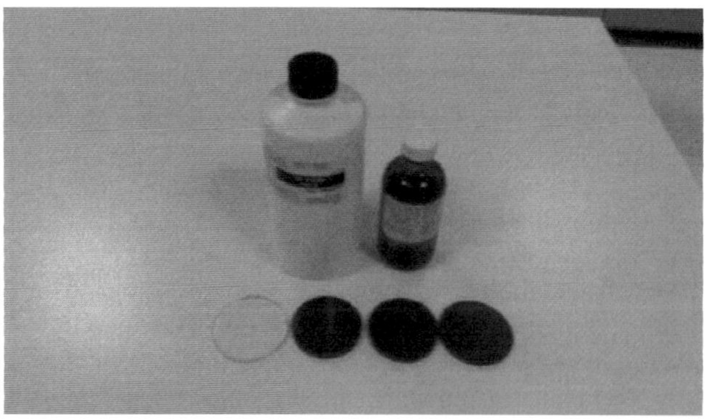

**Figure 1.** Precursors and samples of MWCNT/epoxy nanocomposites. From left to right: epoxy resin (A) (Epotek E301), amine hardener (B) (Epotek E301) and cured samples on nanocomposites.

A MWCNT/epoxy mixture is obtained and is subjected to the ultrasonic procedure for one hour in the bath (Elmasonic 300), using frequencies in the range of 25/45 kHz. The MWCNT/epoxy component thus obtained is mixed with the amine component (B) in a gravimetric mixing ratio of 4:1. The mixture is made at room temperature for 15 minutes, then 30 minutes for pre-reaction, stationary for 30 minutes. The resulting composition is poured into PVC discs in which it hardens for 24 hours. A gelation period of 2 hours was used, with a relative humidity of 50% and a temperature of 220°C. Then for 5 minutes the specimens are ultrasound in isopropyl alcohol (VLSi degrees, BASF), to remove the silicone grease and then follow the drying time. After stripping from the molds, 7 days of maturation follow, then the tests are performed on the respective specimens.

It was opted for concentrations of MWCNT of 0.3% and 0.6%, because MWCNT present large specific surface, which leads to the increase of the formed nano-composite viscosity. At high MWCNT concentrations is very difficult to homogenize the epoxy component (A) as well as hard to incorporate the amine component (B) for hardening and difficult to homogenize the formed (A + B) mixture. It was observed that at MWCNT concentrations lower than 1% in the nanocomposite, the percolation phenomenon exists and the unhardened mixture (A + B) can be easily poured in the molds without the occurrence of structural defects.

### *2.2. Methods for characterization*

All used materials, the starting substances, and the obtained nanocomposites, were characterized by different methods: X-ray diffraction, scanning electron microscopy (Nova NanoSEM 630), Raman spectroscopy, infrared spectroscopy. Mechanical tests, electrical resistance measurements and electromagnetic shielding tests were also performed.

Diffraction spectra were studied using the diffractometer (Rigaku Smartlab 9kW rotating anode).

IR analysis was performed on FTIR equipment (BRUCKER27). The FTIR Tensor 27 spectrometer (Bruker Optics), and the ATR (attenuated total reflectance) with Platinum accessory allowed the study of the structural transformation of the raw materials up to the composite stage. A scan speed of 64 scans was used and spectra were drawn over a range between 4000 and 400 $cm^{-1}$.

Scanning electron microscopy was used to morphologically characterize the obtained carbon nanotubes and nanocomposites. For analysis, a small amount of carbon nanotubes arranged on a conductive strip is used. The disks were cooled in liquid nitrogen and then fractured for morphological analysis of MWCNT/epoxy nanocomposites and highlighting their surface. Due to the dielectric properties of the epoxy material on the resulting small pieces a gold layer of 4 – 5 nm was deposited.

For Raman spectra a high-resolution equipment (LabRAM HR800 Horiba) was used.

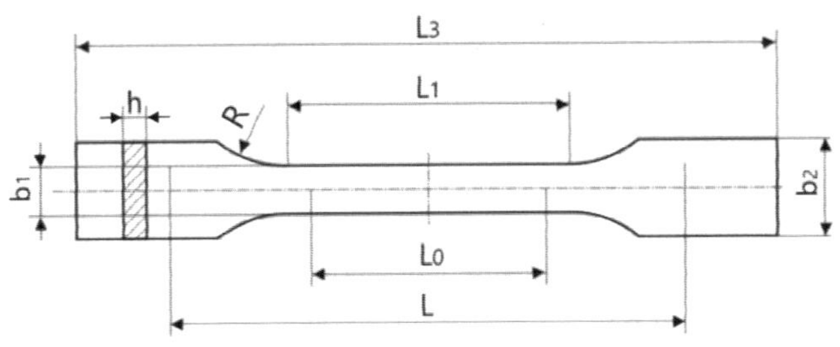


**Figure 2.** Dimensions of the tensile test specimen.

In figure 2 there are presented the dimensions of the tensile test specimen (type 1B) according to SR EN ISO 527-4: 2000 [21]: the total length ($L_3$ = 160 mm), length of narrow parallel part ($L_1$ = 60mm), radius (R = 60mm), width of extremities ($b_2$ = 20mm), width of the narrow part ($b_1$ = 10mm), thickness (h = 2.5mm), gauge length ($L_0$ = 50mm), initial distance between grips ($L_3$ = 114mm).The test system with which the mechanical measurements were performed, Mecmesin MultiTest 2,5i, with high performance load cells (ILC) of 2.5 kN, makes the obtained results to have very good accuracy and resolution. The precision mechanical system has a positioning error of ± 130 μm at a displacement of 300 mm. The Emperor software package provides a large programming capacity with many analysis options which leads to a very good characterization of the samples under analysis. The mechanical tests were performed on 3 material samples: MWCNT 0.3% / epoxy, MWCNT 0.6% / epoxy and epoxy resin (epoxy). The standards used are in accordance with the tested materials and to obtain comparative results in the three materials.

During the tensile tests performed, an axial force was applied to the tested specimen, until the material failed. The maximum traction force was 500 N and the gripping speed of the grips was 5 mm/minute. The test conditions for the 3 samples were identical. Calibration operations were performed on the measuring cell at the beginning of each test.

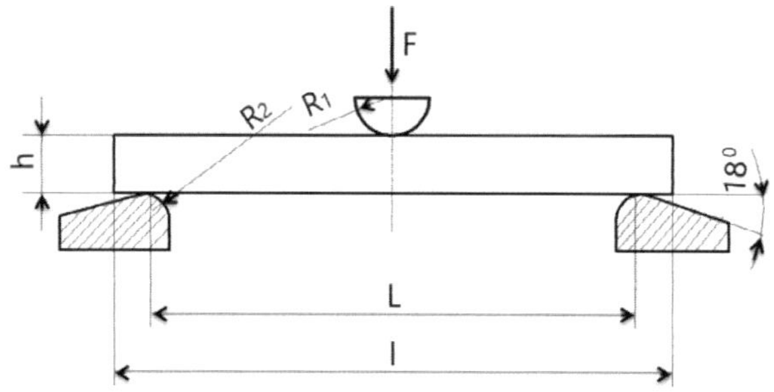


**Figure 3.** Dimensions of the flexure test specimen.

In figure 3 there are presented the dimensions of the flexure test specimen according to SR EN ISO 178:2019 [22]: length (l = 50 mm), support span (L = 40mm), width (b = 25 mm), thickness (h = 2.5 mm), radius of loading edge ($R_1$ = 5 mm), radius of supports ($R_2$ = 2 mm).

The specimens subjected to the flexure tests were subjected to forces perpendicular to the axis and to the half of the length of the specimen, registering the variations appeared, until the failing of the material.

The electrical resistance tests were performed using the Hioki 3532-50 LCR HITESTER device, with a frequency range of 42 Hz to 5 MHz, a response time of 5 ms, and ± 0.08% accuracy. The test samples had a length of 25 mm and a section of 12×5 mm$^2$. For each sample, tests were performed for three measurement frequencies.

EMI shielding of an electronic device is achieved by attenuating the passage of electromagnetic waves by reflection or by the absorption of the power of the incident ray. Due to the tendencies towards miniaturization of electronic devices, lightweight materials have been developed to replace metals with improved properties including corrosion resistance, flexibility, and speed in processing [18].

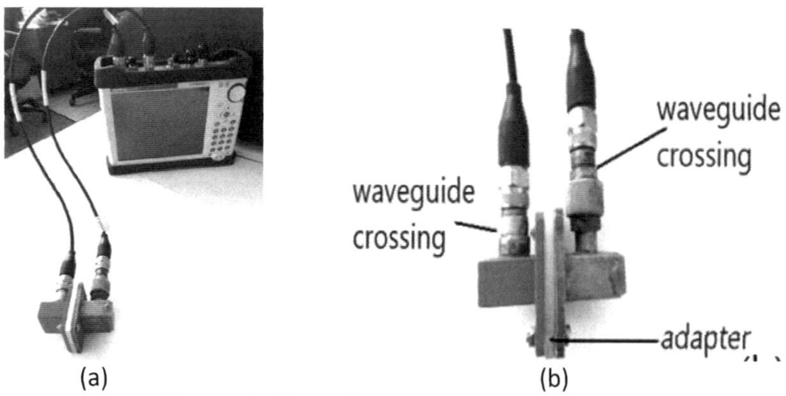


**Figure 4.** (a) Anritsu S 412 wave analyzer;
(b) Detail of waveguide switch adapter assembly.

A high-performance Anritsu S 412 vector network analyzer (figure 4a) with a 500 kHz to 6,400 MHz bandwidth was used in order to verify the attenuation introduced by the analyzed samples. Each sample was mounted in the adapter attached between the two coaxial cable-guide couplings type PVK 35X15, with a cut-off frequency of 2.5 GHz (figure 4b).

Tests were performed on three material samples, neat epoxy sample, 0.3% MWCNT/epoxy nanocomposite and 0.6% MWCNT/epoxy nanocomposite respectively, having the following dimensions: length = 35 mm, width = 15 mm, thickness = 3.5 mm. In the preparation process of samples upon mixing a sonication steps [19], debundling of MWCNT is effective without altering their length, leading to better EMI shielding efficiency [20].

## 3. Results and discussions

The absorption bands that are assigned to the C–H groups (figure 5) from the precursor components (aliphatic and aromatic), characterize the spectrum of the epoxy component (A).

The spectral bands 2970-2850 $cm^{-1}$ confirm the existence of methyl and methylene groups and the peak centered at 3055 $cm^{-1}$ confirms the C–H bonds in the epoxy and benzene ring. The C-O bonds in the range 1230 – 830 $cm^{-1}$ belong to epoxy precursor (A). The C–O–C epoxy group in precursor (A) is assigned the peak centered at 913 $cm^{-1}$ and is considered the imprint of this group. At approximately 3447 $cm^{-1}$ there is a low intensity band that shows the existence of O–H bonds which confirms the presence of dimers or compounds with high molecular weight or phenolic groups. The bands at 1607 and 1582 $cm^{-1}$, which belong to the C–C bonds in the benzene skeleton, confirm the benzene ring in the phenolic component. The peaks of the N–H bonds (3367, 3290 and 817 $cm^{-1}$) show the presence of the amine component (B). At 1013 $cm^{-1}$ is found the absorption peak that defines the C–N group. At 1460 and 1378 $cm^{-1}$ are found the peaks of the methyl, methylene (–CH3, –CH2) groups. In the spectrum of the sample there are also peaks that represent the vibration mode of the C–H, C–N and C–O bonds. In the spectrum that characterizes carbon nanotubes, there are no bands that show their functionalization. FTIR spectra for final products: hardened MWCNT/epoxy and epoxy (A + B) nanocomposites come to prove the respective reactions by the disappearance of the bands that would show the presence of C–O–C from epoxy (from 913 $cm^{-1}$) attributed to component (A) and N–H bonds attributed to component (B). In the area 4000-3000 $cm^{-1}$ there is a wide band that shows the formation of O–H bonds in the polymer and the crosslinked composite.

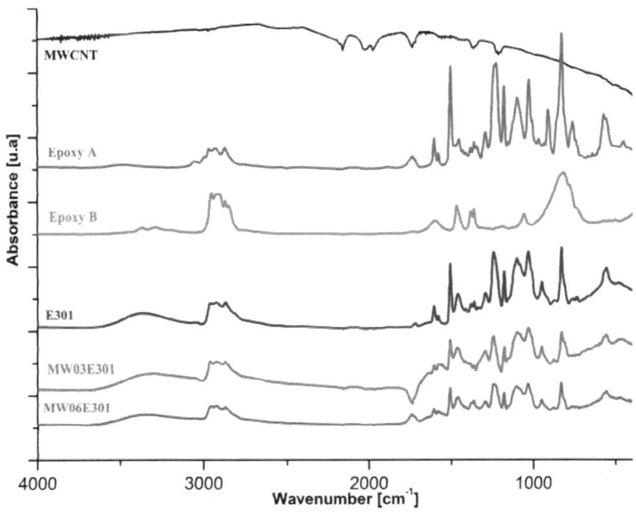


**Figure 5.** FTIR spectra for precursors consisting of solid cured epoxy and amine and MWCNT; hardened components made of epoxy (E301) and nanocomposites at concentrations of 0.3% (MW03E301) and 0.6% (MW06E301) respectively.

Magnifications of 70000×, 160000× and 500000×, respectively, of carbon nanotubes are shown in figure 6. A coiled filiform structure is observed, the fibers are randomly oriented and have a high aspct ratio.Carbon nanotubes have a diameter of 10-35 μm.

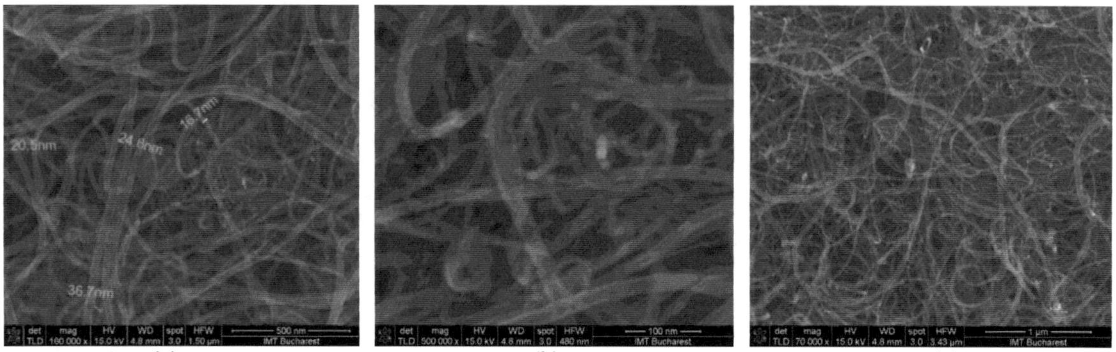

(a) (b) (c)

**Figure 6.** MWCNT at resolutions of 70000 x (c), 500000 x (b) and 160000x (a) respectively.

Figures 7a and figure 7b show the image of SEM MWCNT/epoxy nanocomposites, filler concentration of 0.3% for a magnification of 15000x, respectively 30000x. Figures 7c and figure 7d show the image of SEM MWCNT/epoxy nanocomposites, filler concentration of 0.6% for a magnification of 15000x. The fractured epoxy matrix shows a homogeneous distribution and a dense morphology. The appearance is slightly lamellar, and the carbon nanotubes show a uniform dispersion in the polymer matrix.

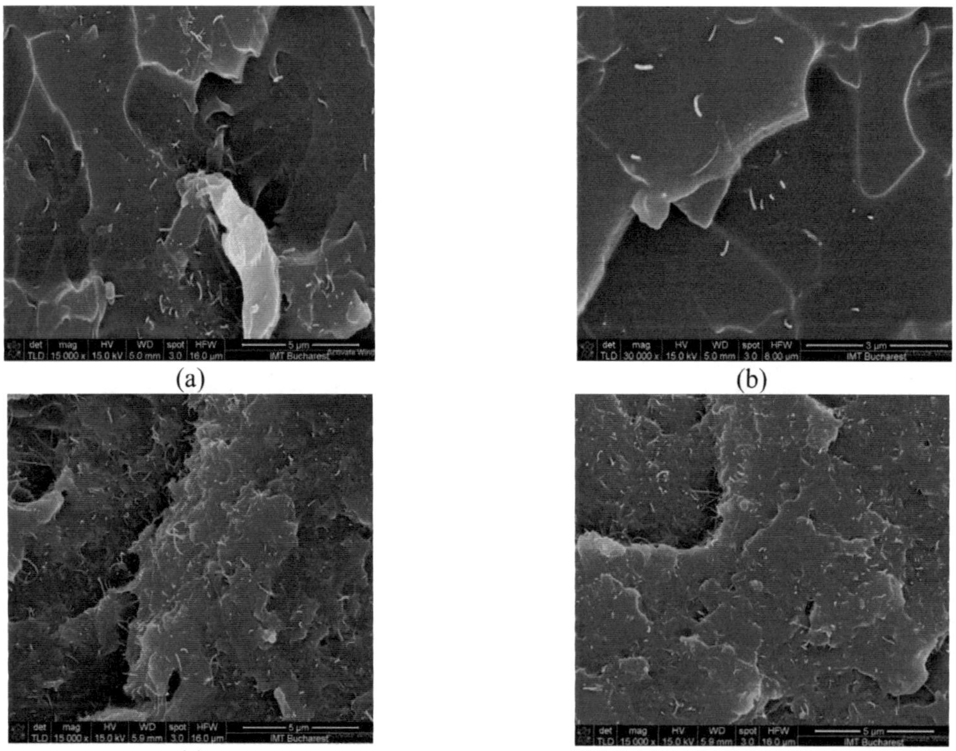

(a) (b)

(c) (d)

**Figure 7.** MWCNT/epoxy nanocomposites: (a, b) filler concentration of 0.3%;
(c, d) filler concentration of 0.6%.

The results of Raman spectrometry for carbon nanotubes (MWCNT) are shown in figure 8. The Raman spectrum in which the bands G and D located at 1582 $cm^{-1}$ (band G) confirm a graphical structure, the peak at 1327 $cm^{-1}$ (band D) shows the vibration of the C–C bonds that are activated in case of defects in the graphical system. There is a wide band having order 2 (G'), having low intensity, proportional to the increase of the density of the defects being at 2652 $cm^{-1}$.

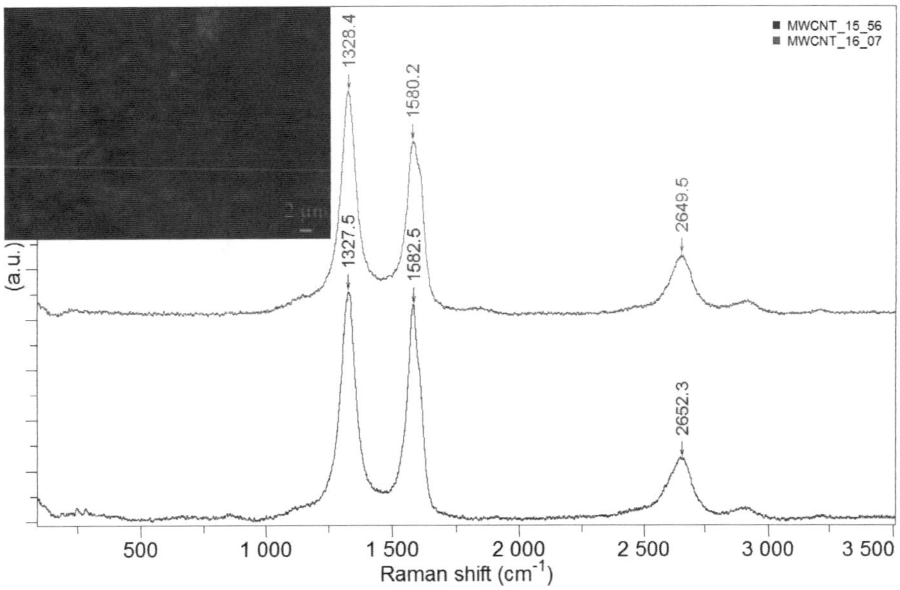


**Figure 8.** Raman spectrum for MWCNT before sonication. I nset picture: surface of analyzed sample.

The diffraction spectra resulting (figure 9) from the X-ray analysis, for the MWCNT/epoxy and MWCNT nanocomposites, show a wide diffraction peak due to orientations (101) and (100) in the range 43 – 450 and a peak that is at 26° having FWHM (Full width at half maximum) = 1.6°. Following the dispersion in the epoxy network, the MWCNT/epoxy nanocomposites, have two wide bands at 20° and 43°, which shows the existence of an amorphous structure. In the case of low concentration composites, the characteristic peak for MWCNT is not recorded in the analysis framework.

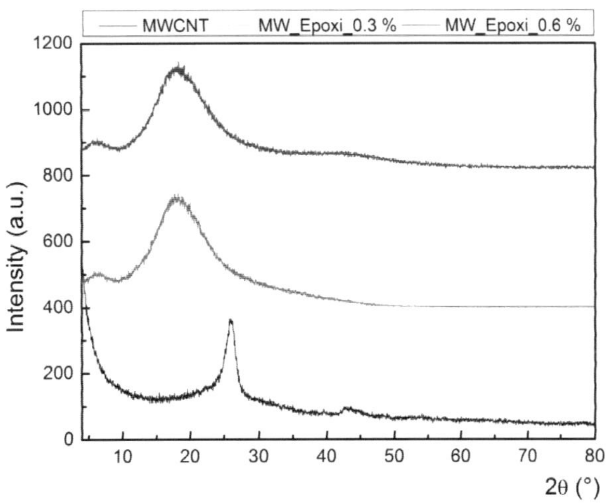


**Figure 9.** XRD spectra for MWCNT and epoxy / MWCNT nanocomposites.

The test piece made of epoxy resin, according to figure 10, went through the area of elastic deformation, following an approximately linear evolution, up to a force of 350 N. The material underwent an elongation of 3 mm, then entered the falling area (plastic deformation) and failed at an elongation of 3.3mm. In the case of the 0.3%/epoxy MWCNT sample, the elastic zone is increased to a force of 400 N, the material elongating by 2.8 mm, then the plastic deformation follows up to an elongation of 4.2 mm, following the rupture of the specimen. With the addition of 0.3% MWCNT can be seen an increase in tensile strength, the area of elastic and plastic deformation increases. For the MWCNT 0.6%/epoxy sample, an increase of the elastic deformation area can be observed. At an elongation of 7 mm the material fails. In conclusion, which makes it usable as an insulator, in elasticity and ductility of the respective material, which makes it usable as an insulator, in installations, electrical cables, in high voltage work areas. Tensile tests were performed according to SR EN ISO 527-4: 2000 [21].

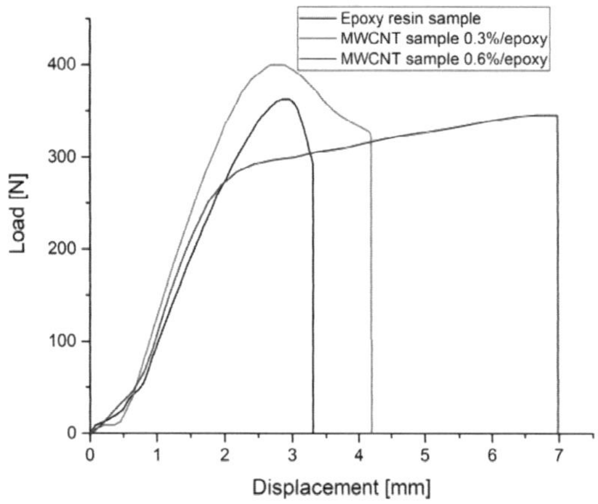


**Figure 10.** Tensile strength diagram for neat epoxy, 0.3% MWCNT composite and 0.6% MWCNT composite sample, respectively

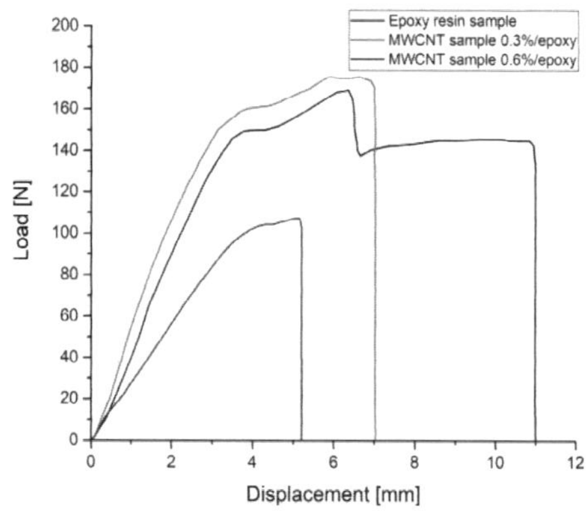


**Figure 11.** Flexure stress resistance diagram for epoxy resin sample, 0.3% MWCNT composite and 0.6% MWCNT composite, respectively.

From the obtained results according to figure 11, in the flexure of the specimens, it is found that the presence of MWCNT has an influence on the binding forces between the molecules of the respective polymer. As the amount of MWCNT increases, the intermolecular bonding forces decrease, the plastic deformation zone decreases. Flexure tests were performed according to SR EN ISO 178:2019 [22].

The results are presented in table 1 where it is observed that the increase of the MWCNT composition leads to an exponential decrease of the electrical resistivity.

**Table 1**. Measured values of electrical resistance for the three measurement samples.

| Frequency | Neat epoxy | 0.3% MWCNT /epoxy | 0.6% MWCNT /epoxy |
|---|---|---|---|
| f = 50 Hz | >1 GΩ | >1 GΩ | 0.35 MΩ |
| f = 800 Hz | 0.9 GΩ | 40 MΩ | 3 MΩ |
| f = 8 kHz | 373 MΩ | 10 MΩ | 0.9 MΩ |

The attenuation produced by the analyzed samples increases proportionally with the concentration of MWCNT present in the epoxy matrix. According to the results of figure 12 it was found that in the 2.5 GHz and 6.4 GHz band, the attenuation introduced by sample around the 5.5 GHz central frequency is approximately 20 dB between ambient air and 0.6% MWCNT/epoxy sample.

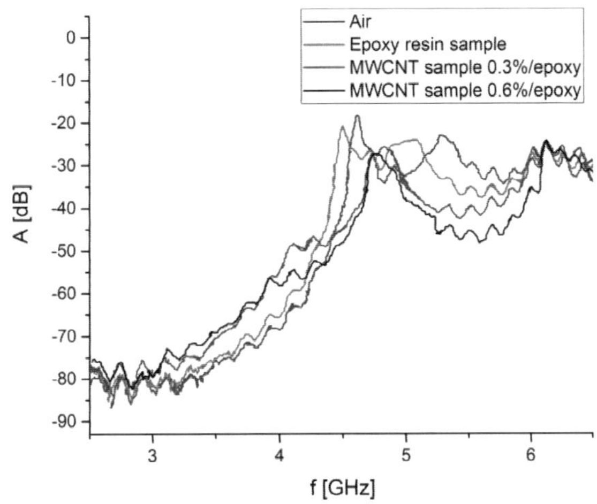


**Figure 12.** Attenuation produced by samples of neat epoxy, 0.3% MWCNT/epoxy and 0.6% MWCNT/epoxy, respectively.

## 4. Conclusions

Epoxy nanocomposites with carbon nanotubes were prepared using specific and easy methods. Two concentrations of 0.3% and 0.6% MWCNT embedded in epoxy resin were used. The characterization of the newly obtained materials reveals in content a homogeneous and flawless morphology. Mechanical tests show an increase in the tensile strength of newly formed nanocomposites compared to the epoxy material. In the case of the electrical resistance of nanocomposites, this decreases with the increase of the amount of MWCNT. In the case of EMI shielding tests, the attenuation was proportional to the amount of MWCNT present in the nanocomposite with a maximum value of 20 dB.

**Acknowledgements**

This work was supported by a grant of the Romanian Ministry of Research and Innovation, CCCDI–UEFISCDI, project number PN-III-P1-1.2-PCCDI-2017-0419/SENSIS – P3 – Infrared sensors for infrastructure security applications, within PNCDI III."

**References**

[1] Ijima S 1991 Helical microtubules of graphitic carbon *Nature* **354** 56
[2] Kumar A, Ghosh P, Yadav K L and Kumar K 2017 Thermo-Mechanical and Anti-Corrosive Properties of MWCNT/Epoxy Nanocomposite Fabricated by Innovative Technique *Composites Part B: Engineering* **113** 291
[3] Shena W, Fenga L, Liua X, Luo H, Liua Z, Tonga P and Zhang W 2016 Multiwall Carbon Nanotubes-Reinforced Epoxy Hybrid Coatings with High Electrical Conductivity and Corrosion Resistance Prepared via Electrostatic Spraying *Progress in Organic Coatings* **90** 139
[4] Kumar A, Kumar K, Gosh P K and Yadav K L 2018 KL MWCNT/$TiO_2$ Hybrid Nano Filler Toward High-Performance Epoxy Composite *Ultrasonics – Sonochemistry* **41** 37
[5] Khun N W, Rincon Troconis B C and Frankel G S 2014 Effects of Carbon Nanotube Content on Adhesion Strength and Wear and Corrosion Resistance of Epoxy Composite Coatings on AA2024-T3 *Progress in Organic Coatings* **77** 72

[6] Gojny F H, Wichmann M H G, Fiedler B and Schulte K 2005 Influence of Different Carbon Nanotubes on the Mechanical Properties of Epoxy Matrix Composites – A Comparative Study *Composite Science and Technology* **65** 2300
[7] Kausar A, Rafique I and Muhammad B 2016 Review of Applications of Polymer/Carbon Nanotubes and Epoxy/CNT Composites *Polymer-Plastics Technology and Engineering* **55 (11)** 1167
[8] Roy S, Petrova R S and Mitra S 2018 Effect of Carbon Nanotube (CNT) Functionalization in Epoxy-CNT Composites *Nanotechnol. Rev.* **7(6)** 475
[9] Hadavand B S, Javid K M and Gharagozlou M 2013 Mechanical Properties of Multi-Walled Carbon Nanotube/Epoxy Polysulfide Nanocomposite *Materials and Design* **50** 62
[10] Wang L, Qiu H, Liang C, Song P, Han Y, Han Y, Gu J, Kong J, Pan D and Guo Z 2019 Electromagnetic Interference Shielding MWCNT-$Fe_3O_4$@Ag/Epoxy Nanocomposites with Satisfactory Thermal Conductivity and High Thermal Stability *Carbon* **141** 506
[11] Rodriguez-Gonzalez J A, Rubio-González C, Jiménez-Mora M, Ramos-Galicia L and Velasco-Santos C 2018 Influence of the Hybrid Combination of Multiwalled Carbon Nanotubes and Graphene Oxide on Interlaminar Mechanical Properties of Carbon Fiber/Epoxy Laminates *Applied Composite Materials* **25(5)** 1115
[12] Q Zhang, X Zhao, G Sui and X Yang 2018 Surface Sizing Treated MWCNTs and Its Effect on the Wettability, Interfacial Interaction and Flexural Properties of MWCNT/Epoxy Nanocomposites *Nanomaterials* **8 9** p 680
[13] TY Chou, H-Y Tsai and MC Yip 2019 Preparation of CFRP With Modified MWCNT to Improve the Mechanical Properties and Torsional Fatigue of Epoxy/Polybenzoxazine Copolymer *Composites Part A* **118** p 30
[14] G Singer, P Siedlaczek, G Sinn, PH Kirner, R Schuller, R Wan-Wendner and HC Lichtenegger 2019 Vacuum Casting and Mechanical Characterization of Nanocomposites from Epoxy and Oxidized Multi-Walled Carbon Nanotubes *Molecules* **24 3** p 510
[15] H Duan, H Zhu, J Yang, J Gao, Y Yang, L Xu, G Zhao and Y Liu 2019 Effect of Carbon Nanofiller Dimension on Synergistic EMI Shielding Network of Epoxy/Metal Conductive Foams *Composites Part A: Applied Science and Manufacturing* **118** p 41
[16] E Wang, Y Dong, MD Zahidul Islam, L Yu, F Liu, S Chen, X Qi, Y Zhu, Y Fu, Z Xu and N Hu 2019 Effect of Graphene Oxide-Carbon Nanotube Hybrid Filler on the Mechanical Property and Thermal Response Speed of Shape Memory Epoxy Composites *Composite Science and Technology* **169** p 209
[17] L Zhang, S Bi and M Liu 2018 Lightweight Electromagnetic Interference Shielding Materials and their Mechanisms *Electromagnetic Materials and Devices* ed M G Han (Intech) pp 1-10.
[18] P Banerjee, Y Bhattacharjee and S Bose 2020 Lightweight Epoxy-Based Composites for EMI Shielding Applications *J. of Electronic Materials* **49** p 1702
[19] D Zhang, MG Villarreal, E Cabrera, A Benatar, LJ Lee and JM Castro 2019 Performance Study of Ultrasonic Assisted Processing of CNT Nanopaper/Solventless Epoxy Composite *Composites Part B: Applied Science and Manufacturing* **159** p 327
[20] BP Singh, K Saini, V Choudhary, S Teotia, S Pande, P Saini and RB Mathur 2014 Effect of Length of Carbon Nanotubes on Electromagnetic Interference Shielding and Mechanical Properties of Their Reinforced Epoxy Composites *J Nanopart. Res.* **16** p 2161
[21] SR EN ISO 527-4:2000: Plastics — Determination of tensile properties — Part 4: Test conditions for isotropic and orthotropic fiber-reinforced plastic composite.
[22] SR EN ISO 178:2019: Plastics — Determination of flexural properties.

# An approach regarding some performances of a FESTO pneumatic muscle actuator

**I M Petre**[1*]

[1] Department of Engineering and Industrial Management, Transilvania University of Brasov, 29 Eroilor, Brasov 500036, Romania

*E-mail: ioana.petre@unitbv.ro

**Abstract**. This paper presents some performances of a pneumatic actuated artificial muscle, manufactured by FESTO. Pneumatic muscles are novel actuators, constructed by an elastic membrane covered with a braided mesh, based on a contraction principle: inflated, it shorts its length with a certain dimension depending on the compressed air pressure. Those muscles are known for their wide application in industry and robotics, as actuators. The aim of this study is to find operational behaviour of a pneumatic muscle. Some experiments was made, in order to prove the maximum contraction of the muscle, the evolution of the contraction at different pressures and loads. The results obtained are graphical plotted. Based on the obtained results, it was made an analysis and identified the best fit equation for the data. The experimental research carried out highlighted that the pneumatic muscles offers many advantages and disadvantages in terms of their behaviour in operation.

## 1. Introduction

Pneumatic muscles are used in a wide range of applications. An optimal use of the pneumatic muscles is in assistive devices for disabled people or older ones. This type of actuator has a smooth movement and can obtain the desired stroke with minimal pain for the users. Others advantages over pneumatic cylinders are: higher power-to-weight ratio, smooth speed adjustment [1], and longer operating life (50 million cycles for FESTO pneumatic muscles [2].

The construction of the pneumatic muscles is simple: an inner tube covered by a braided mesh with fitting caps on both sides, which, under the action of pneumatic air, inflates and deflates, acting like a spring.

In this paper, it is presented some performances of a FESTO pneumatic muscle. The aim of the research is to obtain some real information about the maximum contraction of the muscle, in order to establish the possibility of using it as an actuator for a human leg assistive device.

The method used proposed the applying of different pressures to the pneumatic muscle and recording the contraction, in two situations: without load and with a 20 kg load. The aim is to establish the maximum possible contraction for the pneumatic muscle.

Some mathematical expressions can be developed for describing the correlation between contraction of the pneumatic muscle and the pressure.

## 2. Pneumatic muscle

Since their development, pneumatic muscles have been used in different applications. Nowadays, their applicability is mostly in industrial application as an alternative of the hydraulic or electric drives, as main motion power source. There are many studies focused on the applicability of the pneumatic muscle in industry: industrial manipulators [3-5], robotic arms [6, 7], assistive devices [8-10] and others.

Pneumatic muscle advantages are many, an important one to be specified is the low weight and the compliant behavior, assuring safe interactions with human.

### *2.1. Construction*

The pneumatic muscle is made by a rubber diaphragm with non-crimped fiber made of aramid yarns on the inside [2]. When internal pressure is applied, the muscle decreases its length, expanding its diameter, as it can be seen in figure 1.

Following the deformation of the tube due to the increase of the pressure, the actuator shortens by a certain amount, called stroke.

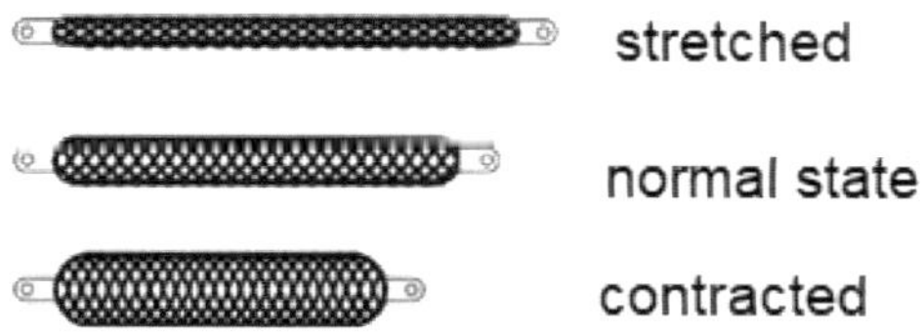


**Figure 1.** Working principle of a pneumatic muscle [11].

In the figure 2 it can be seen the construction of a pneumatic muscle: a rubber inner tube covered with a shell braided according to helical weaving. The braided sleeves act to constrain the expansion for maintaining the cylindrical shape.

In order to cover a wide range of applications, the pneumatic muscles can be constructed with different lengths and diameters. They are closed by two ends, one being the air input and the other the force attachment point.

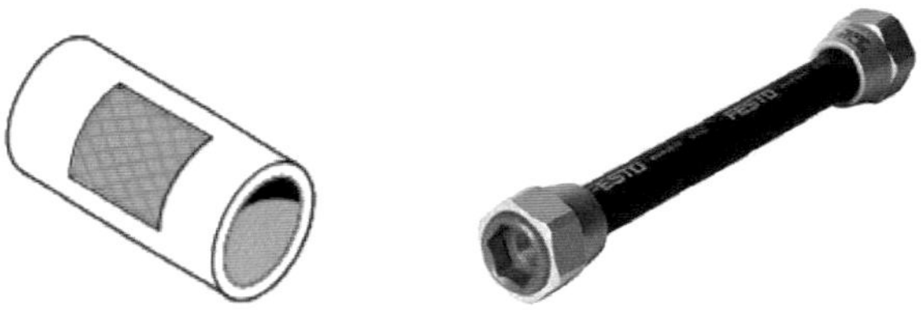

**Figure 2.** Construction of a pneumatic muscle [2].

### *2.2. Advantages and disadvantages*

The utilisation of the pneumatic muscle is preferred because of the advantages they have: a varying force-displacement relation at constant gas pressure, an adjustable compliance; the absence of friction and hysteresis; the ability to operate at a wide range of gas pressures, and thus to develop both very low and very high pulling forces; the possibility of direct connection to a robotic joint; cheaper to buy and install than other actuators and pneumatic cylinders; smooth and natural movement; fast -full contraction. Pneumatic muscles has many focal points for human-machine interface: low weight, high force to weight proportion, working principle same as human, safe energy [12].

The weak points of the pneumatic muscle are: the force which can be applied is only tensile in nature; its total displacement is only about 20% to 30% of its initial length; friction between the netting and the tube leads to a substantial hysteresis in the force-length characteristics; rubber is often needed to avoid the tube from bursting; rubber deformation will lower the force output of this type of muscle up to 60% [13].

## 3. Experimental setup

### *3.1. Components of the experimental stand*

The experimental setup consists of several components, as can be seen in figure 3. The pneumatic muscle used for the research is an FESTO origin pneumatic muscle of 750 mm length and 20 mm diameter. It is fixed at one side and different loads are applied at the other side. Under the pneumatic air influence, this muscle becomes pressurized and shortens with a certain stroke, of interest for the research.

Other components involved in experimental stand are: mechanical system with the support structure of the pneumatic muscle (1), pneumatic actuator system with compressor (2), compressed air preparing group, pneumatic distributors (3), transducers for determining the pressure and flow rate, interface for data acquisition (5) and computer equipment (4), loads of 20kg.

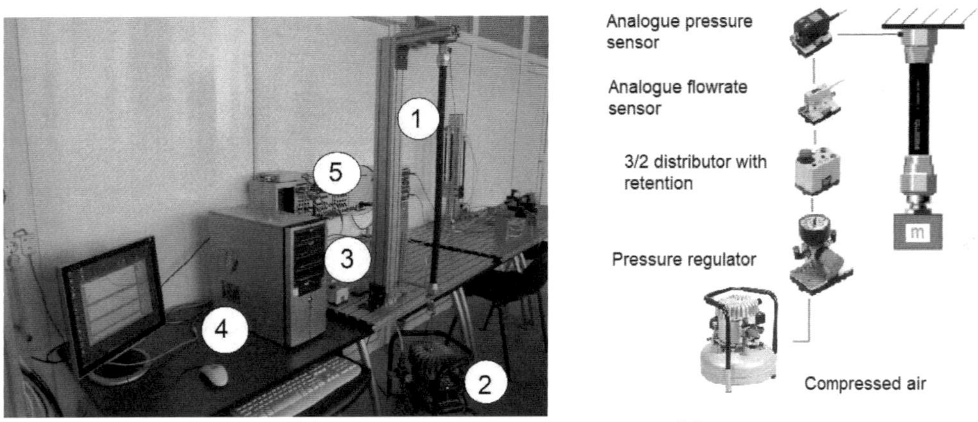


(a) (b)

**Figure 3.** Experimental setup: (a) photograph and (b) diagram.

The connection scheme is presented in figure 3 (b). The pressure of the compressed air delivered by the compressor is set from the pressure regulator valve with gauge, and the 3/2-way valve switch, normally closed is actuated for the pneumatic muscle feed. The value of the compressed air debit is read at the display of the flow rate sensor. The displacement of the muscle resulted is measured after each experiment.

### *3.2. Parameters and method*

The objective of the experiment was the study of the effects of loaded pressure on the length of the pneumatic muscle. The parameter used for this experiment is specified in table 1. The FESTO muscle used in the experiment has the nominal diameter of 20 mm and the length of 750 mm. The air pressure within the pneumatic muscle supplied by an air compressor was ranged between 0.0 bar and 6 bar (gage pressure). At the free end of the muscle, was attached a load of 20 kg.

**Table 1.** Experimental parameters.

| | Values |
|---|---|
| Muscle diameter (mm) | 20 |
| Muscle length (mm) | 750 |
| Load (kg) | 0, 20 |
| Pressure (bar) | 0, 1, 2, 3, 4, 5, 6 |
| Pneumatic supply | At one end |

The pneumatic muscle used is manually pressurized and the displacement resulted is measured. Next, the contraction is calculated, by reporting the measured displacement and the initial length.

The experiment is repeated using pressures between 0 and 6 bar, for 0 and 20 kg applied load at the free end of the muscle.

## 4. Results and discussion

The experimental results of the muscle contraction, for the considered parameters are presented in table 2, for the 0 kg load and in table 3, for the 20 kg load.

The contraction of the muscle, with no load applied (m = 0 kg) varies between 0 and 27.23%. The values of the contraction, with a 20 kg applied load, at the free end of the muscle varies between 0 and 25%.

**Table 2.** Experimental results of the contraction (%), *m = 0 kg*.

| Number of sample<br>Pressure [bar] | 1 | 2 | 3 | 4 | 5 | Average |
|---|---|---|---|---|---|---|
| 0 | 0 | 0 | 0 | 0 | 0 | 0 |
| 1 | 5.00 | 5.11 | 5.13 | 4.69 | 4.89 | 4.96533 |
| 2 | 15.04 | 15.27 | 15.53 | 14.79 | 14.99 | 15.1227 |
| 3 | 20.41 | 20.69 | 20.85 | 20.33 | 20.63 | 20.584 |
| 4 | 23.69 | 23.80 | 24.01 | 23.51 | 23.69 | 23.7413 |
| 5 | 25.93 | 25.84 | 26.01 | 25.36 | 25.52 | 25.7333 |
| 6 | 27.09 | 27.23 | 27.16 | 26.69 | 26.91 | 27.016 |

**Table 3.** Experimental results of the contraction (%), *m = 20 kg*.

| Number of sample<br>Pressure [bar] | 1 | 2 | 3 | 4 | 5 | Average |
|---|---|---|---|---|---|---|
| 0 | 0 | 0 | 0 | 0 | 0 | 0 |
| 1 | 2.15 | 1.93 | 2.36 | 2.36 | 2.44 | 2.248 |
| 2 | 8.12 | 7.71 | 8.60 | 8.68 | 9.01 | 8.424 |
| 3 | 15.69 | 15.21 | 15.76 | 16.05 | 16.21 | 15.7867 |
| 4 | 20.04 | 19.81 | 20.32 | 20.29 | 20.49 | 20.192 |
| 5 | 22.85 | 22.59 | 23.05 | 23.00 | 23.21 | 22.9413 |
| 6 | 24.83 | 24.64 | 25.00 | 24.87 | 24.96 | 24.8587 |

The aim of this research is to study how the contraction varies at different input pressures and at different applied loads. It can be observed that while the pressure increases, the contraction is also increasing. Also, it can be seen that, with no loads applied the values of the contraction are higher that the values obtained with a 20 kg considered load.

Figure 4 shows the average of contraction percentage obtained versus applied pressure with values between 0 and 6 bar. As seen in the graph, the maximum contraction of the pneumatic muscle is obtained in the situation of zero load (m = 0 kg) with the value of 27 %. This value deviates from the technical specifications of the pneumatic muscle analysed, providing a relative contraction of only 20%. The real possibility of pneumatic muscle to shorten more than the value from specifications is an advantage for obtaining larger strokes.

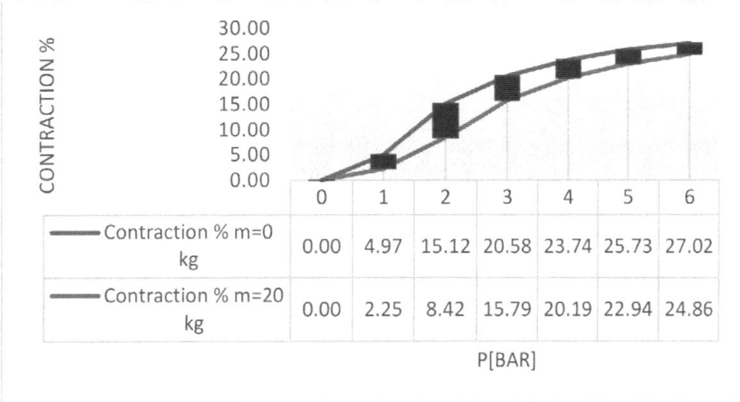


| | 0 | 1 | 2 | 3 | 4 | 5 | 6 |
|---|---|---|---|---|---|---|---|
| Contraction % m=0 kg | 0.00 | 4.97 | 15.12 | 20.58 | 23.74 | 25.73 | 27.02 |
| Contraction % m=20 kg | 0.00 | 2.25 | 8.42 | 15.79 | 20.19 | 22.94 | 24.86 |

**Figure 4.** Contraction percentage versus pressure.

In figure 5, using CurveExpert software, is presented the evolution of the pneumatic muscle contraction, for the case without load. The values of the contraction percentages obtained by measurement are represented by points, and the solid line shows the polynomial regression curve suggested by the considered software.

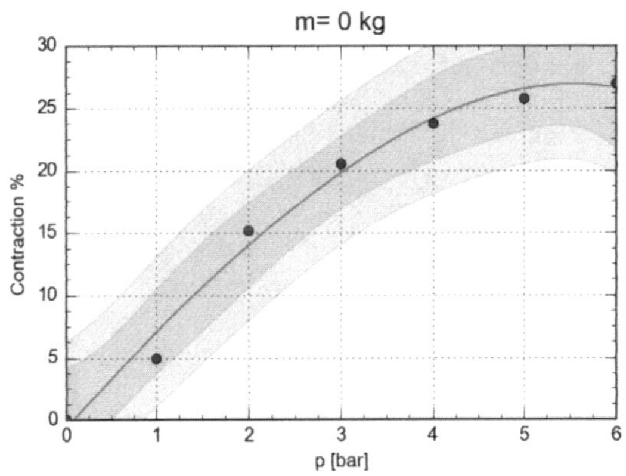


**Figure 5.** The evolution of the contraction percentage versus pressure.

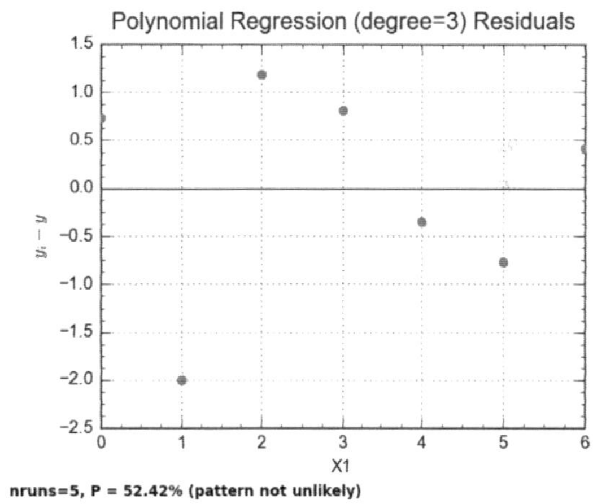


**Figure 6.** Residual values of the function for m=0 kg.

The function that most accurately describes the evolution of contractions in case of no load and for pressures between 0 ... 6 bar, is:

$$F(x) = -0.72 + 7.92 \cdot x - 0.16 \cdot x^2 - 0.06 \cdot x^3 \quad (1)$$

A function is suitable as a pattern if correlation coefficient r value is close to 1. The best model is considered the one with a high correlation coefficient, between 0.8 and 0.9. The determination coefficient $r^2$ offers a percentage interpretation of the obtained curve for the experimental dates.

For the considered function, the correlation coefficient (r) is 0,994 and the coefficient of determination ($r^2$) is 0,988. The obtained values show that the function is the best model for the analyzed situation.

Another method of validation of the function is by graphical form of the residual values. The residual values represent the difference between the experimental values and the ones obtained through mathematical calculation.

For the function determined to be validated residuals chart must follow the two conditions:

- Residual values are independent (there is no particular pattern of these, they are arranged at random);
- Residual values are normally distributed (Gaussian) with mean 0 (negative values are equal with positive values).

In this case, those conditions are respected and the method validates the function obtained, as can be seen in figure 7.

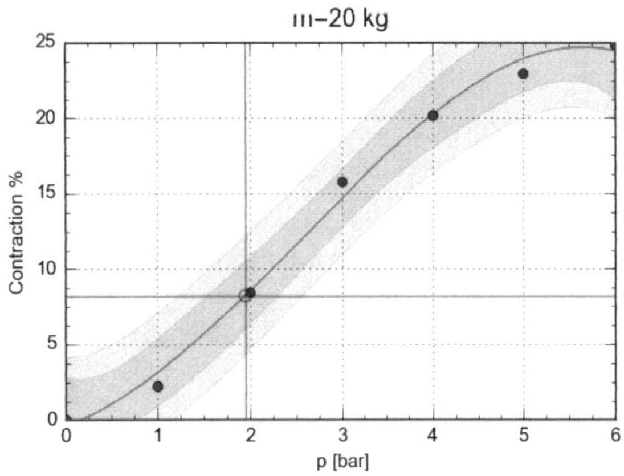


**Figure 7.** The evolution of the contraction percentage versus pressure.

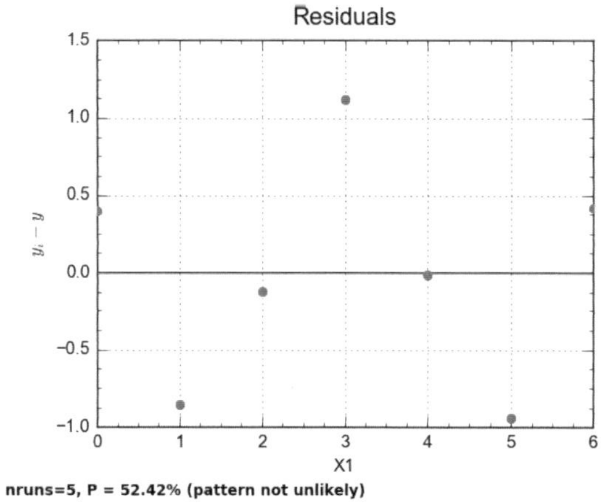


**Figure 8.** Residual values of the function for m=20 kg.

The function that most accurately describes the evolution of contractions in case of a 20 kg load and for pressures between 0 ... 6 bar (figure 7), is:

$$F(x) = -0.39 + 2.1 \cdot x + 1.6 \cdot x^2 - 0.21 \cdot x^3 \quad (2)$$

The correlation coefficient (r) is 0,997 and the coefficient of determination ($r^2$) is 0,994, meaning an accepted function. The graph from figure 8 validates the obtained function.

## 5. Conclusion

The aim of the experiments presented in this paper was to test the functioning of a FESTO origin pneumatic muscle under the action of pressures of 6, 5, 4, 3, 2, 1 and 0 bar, with different loads applied.

The contraction of the muscle increases as the different amount of loads and pressure applied increases. Based on the obtained results, it was made an analysis and identified the best fit equation for the data. The equations developed presents the correlation between the pressure applied and the contraction of the pneumatic muscle.

The results confirms the fact that, the considered pneumatic muscle, FESTO branded, has the possibility of performing higher contractions than those proposed in technical specifications. In practice, a contraction of 27% can be obtained. The obtained results gave a concrete and effective description to understand the real possibilities of the pneumatic muscle.

**References**

[1] Jouppila V, Gadsden S A and Elman A 2014 *Syst. Meas. and Con.* **136** (4)
[2] Fluidic muscle MAS available at: https://www.festo.com/cat/en-us_us/data/doc_enus /PDF/US/DMSP-MAS_ENUS.PDF
[3] Tondu B and Lopez P 2000 *IEEE Con. Syst. Mag* **20** (2) pp 15-38
[4] Deaconescu T and Deaconescu A 2009 *Proc. International MultiConference of Engineers and Computer Scientists Hong Kong 2009, International Association of Engineers (IAENG)* vol. 2 (Newswood Limited Hong Kong) pp 1823-1827
[5] Situm Z and Herceg S 2008 *16th Mediterranean Conf. on Control and Automation* doi: 10.1109/MED.2008.4602136 pp 926-931
[6] Jobbagy B, Simsik D, Karch J and Onofrejova D 2014 *Procedia Engineering* **96** pp 195-202
[7] Mohannad F et al 2018 *IOP Conf. Ser.: Mater. Sci. Eng.* **342** 012052
[8] Ferris D P, Czerniecki J M and Hannaford B 2005 *J.of Appl. Biomechanics* **21(2)** pp 189–197
[9] Sawicki G S and Ferris D P 2009 *J. of NeuroEngineering and Rehabilitation* **6(23)** pp 1–16
[10] Deaconescu A and Deaconescu T 2009 *World Congress on Engineering Londra 2009, International Association of Engineers (IAENG)* vol 2 ( Newswood Limited Hong Kong) pp 1516-1520
[11] Daerden F and Lefeber D 2002 *European J.of Mechanical and Environmental Engineering* **47(1)** pp 10-21
[12] Najmuddin W S W A and Mustaffa M T 2017 *IOP Conf. Series: J. of Physics: Conf. Series* **908** 012036
[13] Air Muscles 2008 available at: http://www.techalone.com

# Influence of process parameters on the AWJ cutting of the AL-EN AW 2017A (T4) aluminium alloy

**M C Radu[1*], C Schnakovszky[1], B A Chirita[1], N C Tampu[1], V A Ciubotariu[1] and E Herghelegiu[1]**

[1] Department of Industrial Systems Engineering and Management, „Vasile Alecsandri" University of Bacau, Calea Marasesti 157, 600115, Bacau, Romania

*E-mail: caxinte@ub.ro

**Abstract.** Abrasive water jet machining is an emerging non-conventional technology widely used in different industrial sectors due to its multiple economic and environmental benefits. The quality of processed surface depends on process parameters, abrasive material properties and processed material properties and is mainly expressed by surface roughness, shape and dimensional accuracy of cut, presence of striations and/or burr. The current paper aims to present the influence of water jet pressure and traverse speed on the quality of processed surfaces in the case of AL-EN AW 2017A - T4 aluminium alloy abrasive water jet cutting.

## 1. Introduction

Since 1987, when the World Commission on Environment and Development (the Brundtland Commission), convened to examine the degradation of the global environment, approved the report "Our Common Future", the stated goal of humanity, worldwide, is to promote sustainable development - a strategy to "meet the needs of the present without compromising the ability of future generations to meet their own needs" [1]. To achieve this goal, different initiatives have been adopted. Among them, promoting investment and innovation in clean technologies has been of crucial importance, as industry is increasingly integrated into global value chains.

Abrasive water jet machining (AWJM) is a relatively new, emerging non-conventional technology, widely used in different industrial sectors for processing a wide range of materials, metallic and non-metallic, soft and hard, from aluminium and mild steel [2] to titanium [3, 4], Inconel [5], composites [6, 7] etc. Its large applicability is due to the multiple economic and environmental benefits: it is eco-friendly (no hazardous substances used during processing but only water and non-harmful and recyclable abrasive material, no dust or hazardous emissions released), high productivity, good integrity of processed surfaces compared to other cutting technologies (no thermal affected zones), ability to cut complex shapes, high flexibility etc. [8].

In abrasive water jet cutting (figure 1), the material is removed due to an erosion process, created by the combined effect of the highly pressurised water jet and hard abrasive particles. The process can achieve high productivity, with material removal velocity up to 1000m/s when high water jet pressure, up to 600 [MPa], is used [9, 10]. The quality of processed surface depends on process parameters (water jet pressure, traverse speed, abrasive flow rate, standoff distance, focusing nozzle length and diameter, nozzle diameter), abrasive material properties (mesh size, hardness) and processed material properties (e.g. strength, hardness, modulus of elasticity) and is mainly expressed by surface roughness, shape and dimensional accuracy of cut, presence of striations and burr.

The current paper aims to present the influence of traverse speed and water jet pressure on the quality of processed surfaces in the case of AWJ cutting of an aluminium alloy, largely used in aeronautics industry.

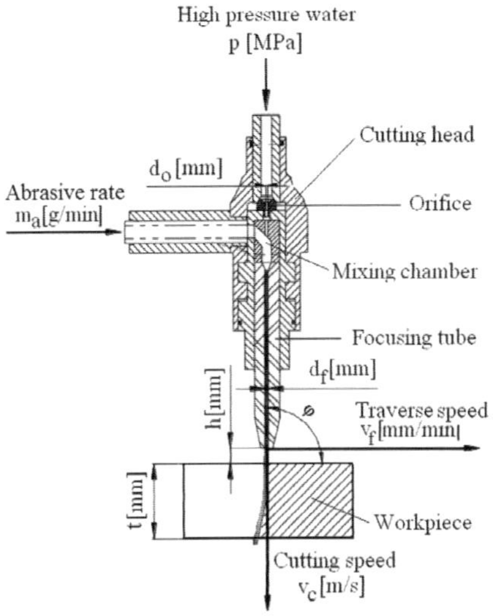


**Figure 1.** Abrasive water jet cutting [10].

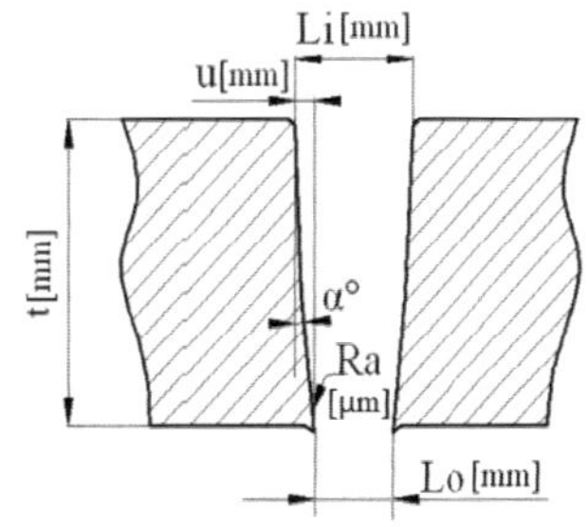


**Figure 2.** Quality parameters of AWJ cutting [11].

## 2. Experimental setup

The experimental tests were carried out on the AL-EN AW 2017A (T4), an aluminium alloy used in aviation, whose properties are presented in table 1. This alloy ensures high mechanical strength and it is suitable for aviation parts, where the development of technologies for the manufacture of light armor is desired. The thickness of the used sheet was 1.5 mm, a thickness imposed by the weight reduction of the developed products.

**Table 1.** Properties of the AL-EN AW 2017A (T4) aluminium alloy [12].

| Yield strength ($R_{p0,2}$), [MPa] | | Ultimate strength ($R_m$), [MPa] | | Elongation ($A_{5min}$), [%] | | | Hardness [HB] | | |
|---|---|---|---|---|---|---|---|---|---|
| 162*/145 | | 228*/225 | | 14 | | | 59*/55 | | |
| Chemical composition [max%] | | | | | | | | | |
| Cr | Cu | Fe | Mg | Mn | Ni | Si | Ti | Ti+Zr | Zn |
| 0.10 | 5 | 0.50 | 0.80 | 1.20 | 0.10 | 0.90 | 0.15 | 0.20 | 0.25 |

*values experimentally determined in the laboratories of the „Vasile Alecsandri" University of Bacau

Simple linear cuts through the material thickness were performed on a Hydro-Jet Eco 0615 machine, equipped with a piston pump capable of providing a maximum pressure of 150 MPa. The length of the cut was 40 mm. The stroke of the cutting head started 10mm outside the material to allow a proper homogenization of the water and abrasive material in the mixing chamber (figure 3).

Two process parameters were varied – water jet pressure and traverse speed, while the other parameters were kept constant (table 2). The minimum pressure was corelated with the transverse speed and was chosen to allow a proper cut throughout the material thickness. The maximum pressure was the highest available among the machine parameters.

Each trial was repeated three times to reduce the influence of the random errors. An average value of the results was calculated, and it is presented in this analysis.

In AWJ cutting, the quality parameters are defined by the ISO/WD/TC 44 N 1770 standard [11] (figure 2), and this paper analysis the following: top kerf width (Li), bottom kerf width (Lo), kerf taper angle (α) and surface roughness (Ra). The linear and angular dimensions were measured with a Leica MZ75 microscope while the surface roughness was measured with a Mitutoyo SJ 201 digital roughness device, at the middle of material thickness. Three measurements were performed for each cut and averaged for more accurate results.

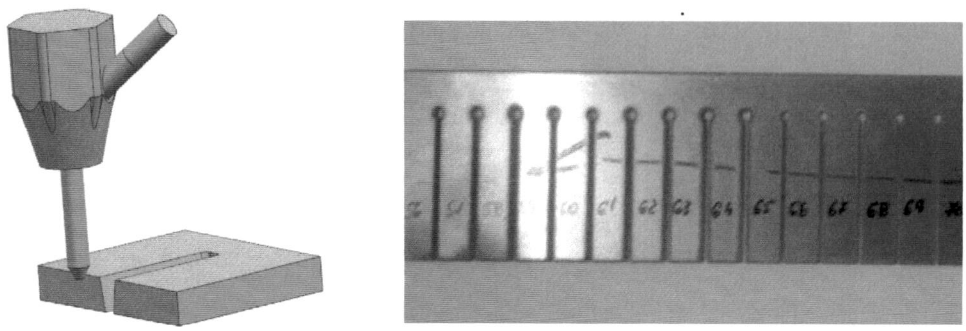

**Figure 3.** Abrasive water jet machining: cutting schema and processed part.

**Table 2.** Process parameters used for the experimental tests.

| Abrasive flow rate, ($Q$) [g/min] | Standoff distance, (h) [mm] | Nozzle diameter, ($d_0$) [mm] | Focusing nozzle length, ($L_f$) [mm] | Focusing nozzle diameter, ($L_f$) [mm] | Abrasive mesh [#] | Varied parameters | |
|---|---|---|---|---|---|---|---|
| | | | | | | Waterjet pressure, ($P$), [MPa] | Traverse speed ($Vf$), [mm/min] |
| | | | | | | 130 | 100 |
| 300 | 2 | 0.35 | 76 | 1.02 | 80 | 140 | 150 |
| | | | | | | 150 | 200 |

## 3. Results and discussion

The quality parameters for the water jet pressure of 130 [MPa] are presented in figure 4, as a function of traverse speed, while the surfaces macrostructure is presented in figure 5.

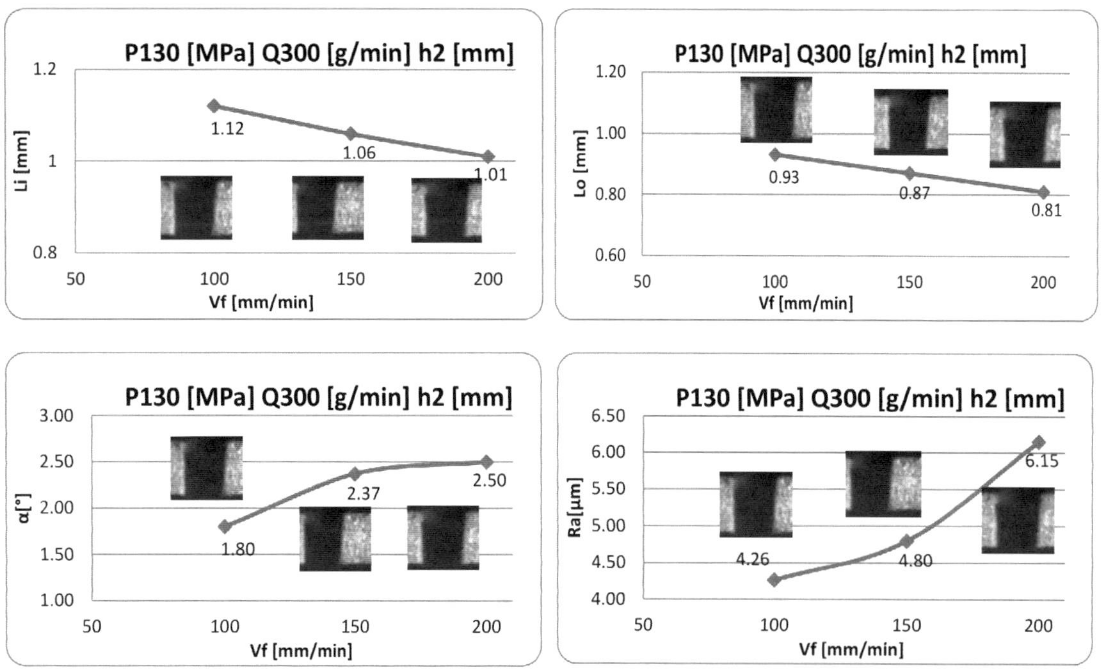


**Figure 4.** Influence of traverse speed on the quality parameters of cut when P = 130 [MPa].

As can be seen from the below graphs, the increase of water jet traverse speed leads to a narrower kerf, both, at the top and bottom (the Li and Lo parameters decreased with approximately 10% and 13%, respectively). This is because a lower value of the traverse speed allows the water jet to act for increased amount of time which leads to larger values of kerf widths. Meanwhile, the kerf taper increased with 0.7° (≈30%) with the increasing of water jet traverse speed. The surface roughness became coarser (with ≈40%), due to the small number of abrasive particles that collided the processed material.

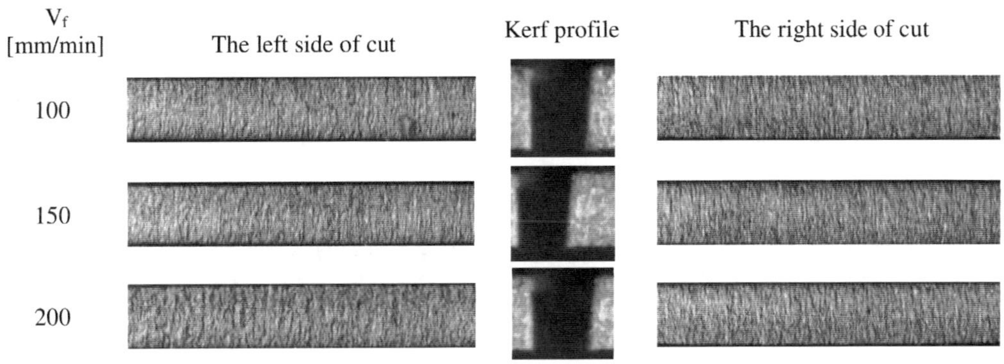


**Figure 5.** The surfaces macrostructure as a function of traverse speed when P = 130 [MPa].

The quality parameters for the water jet pressure of 140 [MPa] are presented in figure 6, as a function of traverse speed, while the surfaces macrostructure is presented in figure 7.

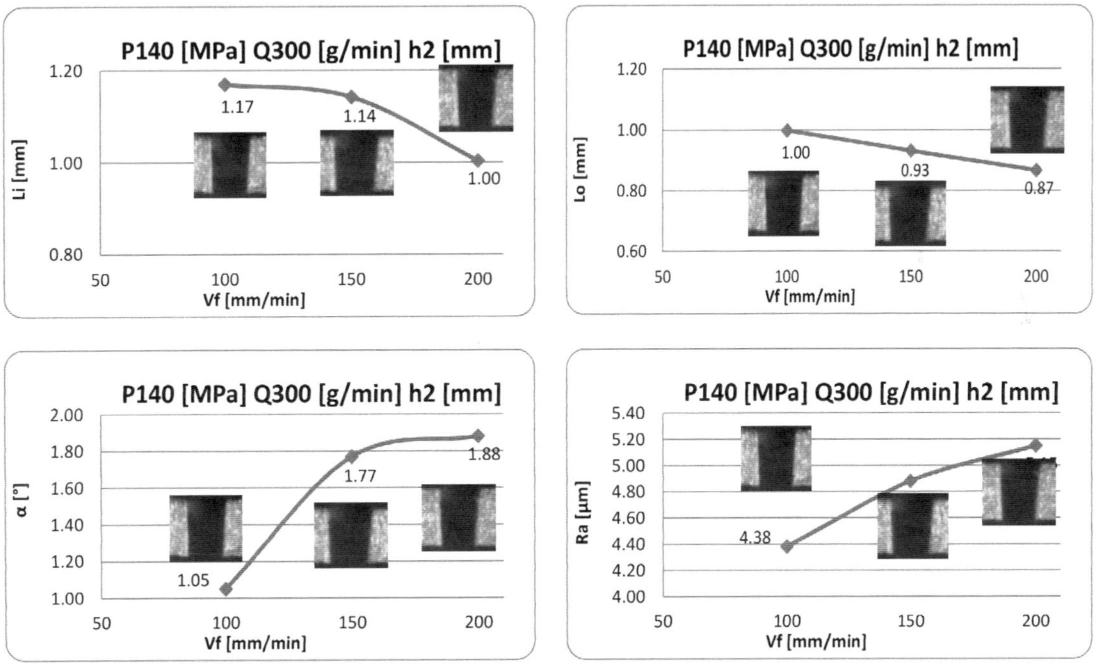


**Figure 6.** Influence of traverse speed on the quality parameters of cut when P = 140 [MPa].

In the case of using a higher water jet pressure, the graphs reveal the same tendency of variation for the quality parameters as in the previous case: the kerf width decreased with the increasing of transverse speed (the Li and Lo parameters decreased with 14.53% and 13%, respectively) while the kerf taper increased with ≈ 44% and the surface roughness increased with ≈ 15% with the increasing of transverse speed.

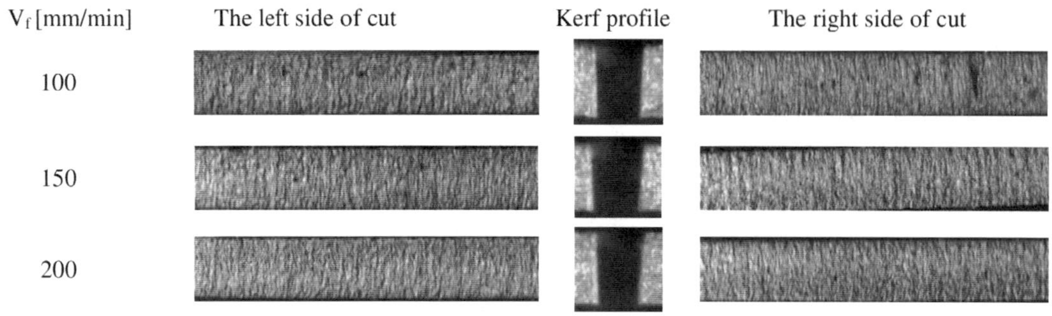


**Figure 7.** The surfaces macrostructure as a function of traverse speed when P = 140 [MPa].

The quality parameters for the water jet pressure of 150 [MPa] are presented in figure 8, as a function of traverse speed, while the surfaces macrostructure is presented in figure 9.

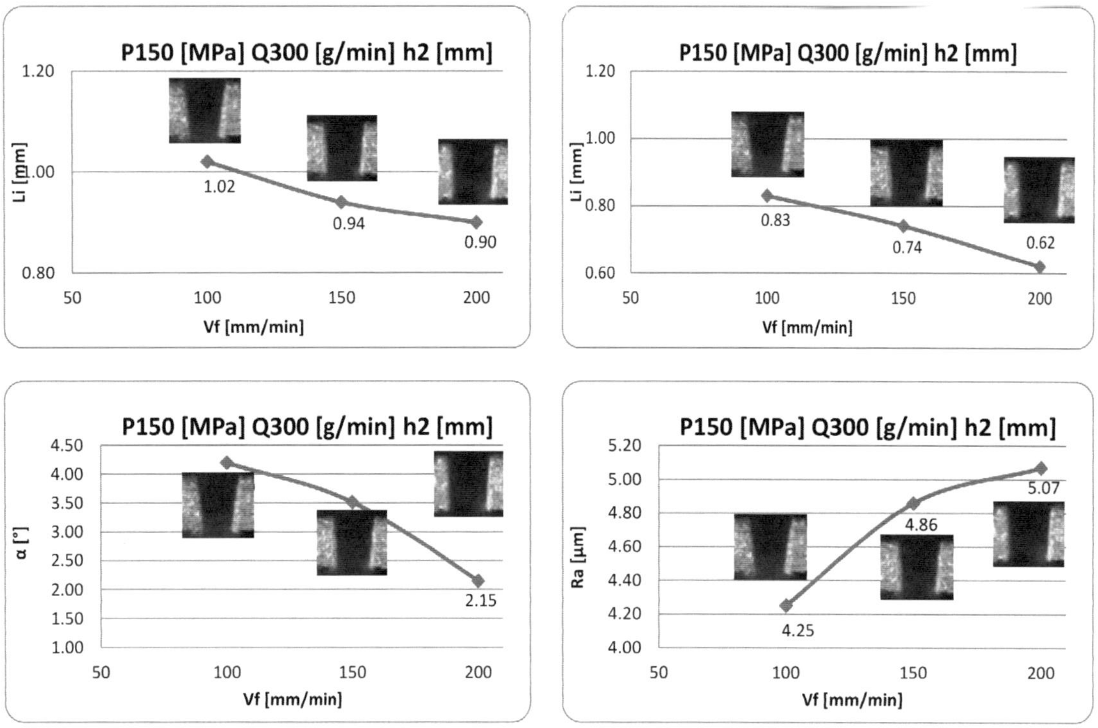


**Figure 8.** Influence of traverse speed on the quality parameters of cut when P = 150 [MPa].

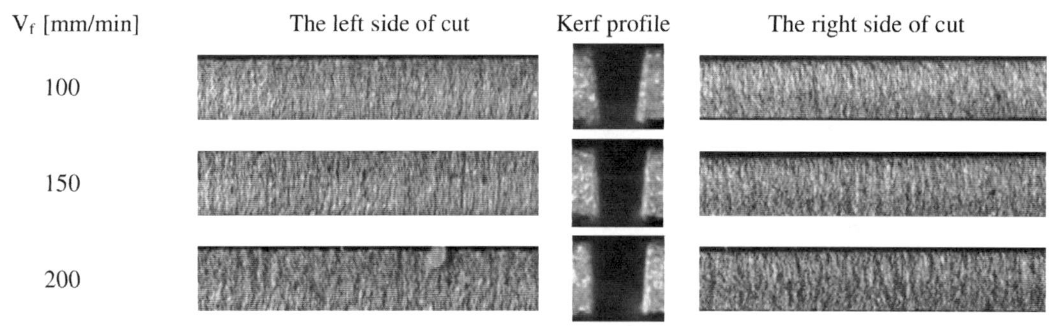


**Figure 9.** The surface macrostructure as a function of traverse speed when P = 150 [MPa].

For the water jet pressure of 150 [MPa], the increasing of the traverse speed determined a ≈ 12% decrease of kerf width at the top and ≈ 25% at the bottom. The taper angle of kerf increased with ≈ 49%, while the surface roughness increased with ≈ 16% with the increasing of the traverse speed. Higher water pressure gives an increased kinetic energy to the abrasive particles, which in turn are more capable to erode the material, thus leading to larger kerf widths.

## 4. Conclusions

The current paper presents a preliminary study in which the influence of water jet pressure and traverse speed on the quality of processed surfaces, expressed in terms of the ISO/WD/TC 44 N 1770 standard, in the case of AL-EN AW 2017A (T4) aluminium alloy abrasive water jet cutting is analysed. Further developments aim to extend the variety of materials, their thickness, the variation intervals for the control process parameters but also the process optimization using various methodologies such as design of experiment (DOE), neural networks, genetic algorithm, fuzzy logic etc.

In the case of AL-EN AW 2017A (T4) aluminium alloy AWJ cutting, within the investigated range of water jet pressure (130 - 150 MPa) and traverse speed (100 – 200 mm/min) and the other considered process parameters, the following aspects were highlighted by the obtained experimental results:

- The increasing of traverse speed led to a narrower kerf. The increasing of water jet pressure did not affect this tendency of variation but did affect the values of the kerf's width: the largest kerf was obtained for the water jet pressure of 140 MPa while the narrowest kerf resulted for the water jet pressure of 150 MPa.
- The kerf taper increased with increasing the traverse speed for water jet pressures of 130 MPa and 140 MPa; on the contrary, for waterjet pressure of 150 MPa, the kerf taper decreased for higher values of the traverse speed.
- The increasing of traverse speed led to a coarser roughness of the cuts. The water jet pressure has slightly affected the surface roughness, excepting the case when its value was minimum (130 MPa) and it was coupled with the highest traverse speed (200 mm/min) – in this case the coarsest surface roughness resulted.

Therefore, higher traverse speeds led to narrower but tapered kerfs and coarser surfaces. To improve the surface roughness, the tendency is to increase the abrasive flow rate since more abrasive particles would get in contact with the material and the erosion process would be intensified. However, the increasing of the abrasive flow rate would determine an agglomeration of the abrasive particles in the mixing chamber, their collisions would appear which would determine the reduction of their kinetic energy and, implicitly, the processing quality.

Higher traverse speeds coupled with higher water jet pressures (e.g. 150 MPa in this study) led to an improvement quality of the cut: narrower and less tapered kerf and better surface roughness.

Regarding the macrostructure of the processed surfaces, some small striations can be observed when higher values of the traverse speed were used.

## Acknowledgments

This work was supported by a grant of the Romanian Ministry of Research and Innovation, CCCDI – UEFISCDI, project number PN-III-P1-1.2-PCCDI-20170446 / 82PCCDI / 2018, within PNCDI III.

## References

[1] *** Report of the World Commission on Environment and Development: "Our Common Future", United Nations, 1987, https://sustainabledevelopment.un.org/content/documents /5987our-common-future.pdf, accessed on 06.06.2020
[2] Lemma E, Deam R and Chen L 2005 *J. Mater. Process. Technol.* **160** 188
[3] Mogul Y I, Nasir I and Myler P, Investigation and optimization for depth of cut and surface roughness for control depth milling in Titanium Ti6AL4V with abrasive water jet cutting, *Materials Today: Proceedings*, Article in Press. https://doi.org/10.1016/j.matpr.2019.12.229
[4] Lenin Raj S and Rajadurai A 2019 *Mater. Res. Express* **6(6)** 066532
[5] Bhandarkar V, Singh V and Gupta T V K 2020 *Materials Today: Proceedings* **23** 647
[6] Thakur R K, Singh K K and Ramkumar J 2020 *Materials Today: Proceedings* **21** 1551

[7] Deepak D and Davim P J 2019 *Procedia Manuf.* **35** 1211
[8] Natarajan Y, Murugesan P K, Mohan M and Ali Khan S A L 2020 *J. Manuf. Process.* **49** 271
[9] Radovanovic M and Herghelegiu E 2016 Nonconventional Technologies Review **20(2)** 39
[10] Mohamad W N F, Kasim M S, Norazlina M Y, Hafiz M S A, Izamshah, R and Mohamed S B 2020 *Results in Engineering* **6** 100101
[11] *** ISO/WD/TC 44 N 1770 standard
[12] ***https://www.euralliage.com/2017A.htm

# Mobile emergency notification apps: current state, barriers and future potential

**R Repanovici**[1*] **and A Nedelcu**[1]

[1]Department of Manufacturing Engineering, Transylvania University of Braşov, Mihai Viteazul 5, Braşov, Romania

*E-mail: mm.raluca@gmail.com

**Abstract**. The main objective of this research is to show that mobile emergency alert apps have the potential to support the Emergency Management system all over the world. Domain experts conducted the decision-making process by evaluating three methods for communicating in an emergency situation: Voice Call, SMS and Mobile Applications. To establish the relative weights of evaluation criteria, the Multi-Criteria Analysis was used. The findings suggest that the ideal solution for communicating during an emergency is alternative no. 3, the Mobile Application. Following the result, we discussed the current state, barriers and future potential of mobile emergency alert applications used for emergency notifications.

## 1. Introduction

Decision-making is the process of identifying alternatives and selecting the best solution for the defined objective. The decision-making process is based on different factors such as values, preferences and expectations of the decision makers. Multi-Criteria Analysis (MCA) method has been chosen to facilitate the decision-making process. MCA is useful for establishing the ranking order of identified solutions, and for which measurable criteria have been taken into account. The essential element of MCA is the decision-making process, where domain experts should establish the objectives and criteria, estimate relative importance weights and evaluate the contribution of each option to each performance criterion. For the alternatives' evaluation, the difficult point in decision-making process is the multiplicity and the diverseness of the criteria set, as more than two objectives should be considered. The criterion delimitates, defines properties and attributes of the analysed alternatives. MCA is able to work with a large number of criteria, but only the major criteria are considered. There are several appropriate solutions on the topic of means of access to emergency services, and the selection of the ideal solution from the optimal set is a complex and challenging process, as it depends on several criteria. In general, Multi-Criteria Analysis can be used to identify the most appropriate alternative, to rank the selected solutions, or simply to confirm solutions' usefulness. The fundamental steps of the MCA are to identify the alternatives, establish the decision-making criteria, assign criteria weights and rank the alternatives by assessing the performance levels [1, 2].

Currently, according to European Emergency Number Association (EENA), there are three major means of access to emergency services [3]:

- Voice calls: available worldwide;
- Emergency SMS (eSMS): available in some regions only;
- Mobile Emergency Applications: available in some regions only.

With the telecom market developing rapidly, mobile devices have the potential to improve citizens' safety and can be considered an important instrument for getting in touch with the emergency services and provide several opportunities for a more efficient help in emergencies by delivering more than just voice in these situations: accurate location, relevant information about the user, prediction of hazard and disasters etc.

Furthermore, the use of smartphone applications is increasing all over the world, and users expect to be able to contact emergency services with the latest technologies. Therefore, EENA concluded that mobile emergency applications are needed, and they should work in the same way as 112 call. According to EENA, over the last years, several mobile app development companies have contacted the European emergency services seeking for collaboration. The public-private partnership between the emergency authorities and the mobile application companies contributed to the successful development of mobile emergency applications. Some national public authorities together with EENA have tried to create a harmonised mobile emergency application but with no success. In consequence, several countries have developed country-specific mobile emergency applications which are only available in their region and can be used by their citizens [4].

## 2. MCA of Emergency Communications

The definition of "emergency" is broad and covers everything from heart attacks, crimes, road accidents, plane crashes, natural disasters, terrorist attacks and other similar situations. Regardless of emergency, the common denominator is the need to get in contact with the Public Safety Answering Point (PSAP) or the emergency dispatcher in order to get help and advice. In order to efficiently handle the case, the PSAP's call-takers are always searching for clear and verified information. To help the person in need, they can dispatch the right resources to the right place only if relevant information is provided [5, 6].

In order to understand what emergency communications means, we consulted the definitions recognised by the European Commission. The European Electronic Communications Code, Article 2 (37) defines emergency communication as "communication by means of interpersonal communications services between an end-user and the PSAP with the goal to request and receive emergency relief from emergency services". Furthermore, in the same document, Recital (255), the EU commission brings more clarity on emergency communication types, stating that "emergency communications are means of communication, that include not only voice communications but also SMS, messaging, video or other types of communications, that are enabled in a Member State to access emergency services" [7].

Currently, in many regions and countries of the European Union, the European emergency number 112 can only be accessed by voice communication. In 2017, EENA highlighted that voice calls between people in danger and PSAPs is not possible on all occasions due to permanent or temporary hearing or speech impairments or even to circumstances preventing or not advising voice communication (e.g. situations where generating any sound endanger the caller). In this context, it remains crucial to provide citizens and visitors with proper access to emergency services whenever they need, by offering the possibility to select optimal solution of communication, depending on their situation. Although the Universal Service Directive invited the European Union member states to guarantee equivalent access to emergency services to all users, including disabled citizens, this is still in progress, and even if some see SMS as the solution for making Emergency Services (ES) accessible, it cannot yet be considered an equivalent solution, although it can be suitable for citizens with hearing impairment to improve their ability to communicate with ES. Moreover, with the growing use of mobile applications for emergency communication, the potential usage of SMS for emergency access is potentially diminished [8].

### *2.1. Identifying the alternative means of access to emergency services*

Based on the telecommunications industry development, three alternative solutions for emergency communication are examined, and the features of each solution are outlined in this paper. The comparison of the three methods to transmit an alert during an emergency or disaster situation is reflected in table 1.

**Table 1.** Comparison between the means of access to emergency services [8, 9, 10].

| Features | | Voice Call | eSMS | Mobile App |
|---|---|---|---|---|
| Time to make the call | | 5 - 10 sec | > 10 sec | 5 - 10 sec |
| Time to answer the call | | ~ 10 sec | 0 sec | 0 sec |
| Time to gain address details | Automatically | 10 - 261 sec | - [b] | 0 sec |
| | Manually | ~ 30 sec [a] | 0 sec [c] | - |
| Location accuracy for automatically retrieved position | | 500 - 40000 m | - | < 100 m |
| Time to gain Event Description | | 30 - 278 sec | 0 sec | 0 sec |
| Total Call length (location and chief complaint) | | 30 - 278 sec | 0 sec | 0 sec |
| User Pre-registration | | Not needed | Required [d] | Required |
| Accessibility | All situations | Yes | Yes | Yes |
| | People with communication disabilities | No | Yes | Yes |
| | Silent emergency situations | No | Yes | Yes |

[a] Can take 3 minutes of extra questions for stressed/injured victims.
[b] Not available in most cases
[c] If provided by user in text message
[d] Required in some regions only

*2.1.1. Voice Calls.* Voice communication is at the present time, and must remain, a key method for communicating in emergency situations. But, while technology is changing the way we interact and communicate, in some EU countries, the European emergency number 112 is still not accessible by other means than voice calls. However, other types of communication such as SMS and/or mobile application could be reluctant and might confuse the users. On the other side, the lack of relevant information provided through a SMS and/or mobile app could impede the emergency responders to make fast decisions [8, 9, 10].

In 2019 EU reported that calls to "112" increased by 5% since the year before, reaching 141.141.731. Calls from mobile phones largely outweighed the number of calls from fixed phones. Based on the available data 73% of the calls were placed from a mobile phone. Meanwhile the total number of emergency calls dropped 2.5% to 293.510.378. Call abandons may be caused by network problems, call congestion, or people in difficult situations where noises can put them in a greater danger [10].

The call 112 process consists of the following steps before dispatching the resources: call reception - call's arrival to the dispatcher and assignment of the call to an operator, and call taking - retrieve caller's and incident location, incident identification and setting-up the case [11].

*2.1.2. eSMS.* Even if a lot has changed in mobile messaging, with an obvious transformation in mobile markets in favour of mobile applications, SMS remains a widely used in 2020. Most citizens send text messages as communication tool, and according to the latest marketing statistics, 23 billion text messages are sent each day worldwide [8, 12].

In many emergency circumstances, a voice call is the best alternative to warn emergency services, but Emergency SMS (eSMS) can be an option for people with communication disabilities, or in silent emergency situations. This way, access to 112 for people that are deaf, deafened, hard of hearing or with speech impairment, or for people that are in difficult situations where noises can put them in a greater danger (e.g. kidnapping), can be provided using eSMS [8].

EENA reported that many countries in Europe and all over the world are already using eSMS for communication with Emergency Services. In countries such as Estonia, Ireland, Latvia, Sweden and UK, registered users have the possibility to alert the emergency dispatcher via eSMS, whilst in some other regions such as Iceland and Lithuania, the eSMS to 112 service is open to everyone [8].

*2.1.3. Mobile Applications.* According to EENA, in present, mobile devices are the primary communication channel for many citizens, and account for over 70% of all emergency calls [13].

The progress in mobile technology and the strong adoption of smartphones can make the mobile applications a powerful tool to provide essential data to the emergency teams by providing instant identification of users and event information and, most importantly, accurate location of users. Over the last years, mobile emergency applications have been developed for a bi-directional communications flow (Authority-to-Citizen and Citizen-to-Authority) and by now, many European citizens are using them [14].

Initially, the primary objective behind mobile emergency solutions was to make use of the high-accuracy location techniques in smartphones and make this data available to emergency management systems, but it's obvious that important additional information can also be provided, such as caller identity, event description or relevant details about the emergency situation [15].

Similar to SMS, mobile apps can be an option for people with communication disabilities, or in silent emergency situations. Using an emergency mobile app has multiple benefits: allows all citizens to access emergency services, automatically sends relevant information authorities when the emergency occurs, provides caller's request and short description of the event, provides accurate location, can eliminate the language barrier [14].

### *2.2. Establishing the decision-making criteria*

The evaluation criteria of the three alternatives to transmit an alert can be conducted according to several aspects [1, 2]. In this study, we considered three aspect, as follows: social, economic and technological. Furthermore, the following eight (8) criteria were established in order to evaluate the alternative solutions and achieve a multi-criteria analysis:

A. Manufacturing and Maintenance Cost: refers to the cost of production, implementation and maintenance;
B. Utility: refers to the applicability of the considered solutions;
C. Quantities of Information: represents the amount of information provided during an emergency call;
D. Accuracy of Information: represents the accuracy of information provided during an emergency call;
E. Data Monitoring: represents the utility of information recorded during an emergency call;
F. User Perception: refers to the consumers' perception towards the three alternatives;
G. Usability: refers to the alternative solutions' capacity to provide a good user experience;
H. Time of Response: represents the time required by the Emergency Response teams.

### *2.3. Assessment of criteria weights*

For the evaluation of criteria weights, decision-making specialists established the corresponding importance of each criterion. As the criteria established do not carry the same importance, the experts determined the relationship between every pair of criteria by comparing each criterion against the others. Hence, the created matrix represents how these criteria relate to each other.

The weight factor of the MCA is assessed by calculating the weighting coefficients $\gamma i$ of each criterion. Within this scope, a quadratic table with the same number of rows and columns (equal to the number of considered criteria *Ncrt*, in this case *Ncrt* = *8*) is composed. In this table, each criterion on a row is compared to each criterion on a column.

As a criterion cannot be greater or smaller in importance than itself, the main diagonal of the table can only be filled with values of '0.5'. When a criterion on a row is compared to a criterion on a column, the weighting of the criteria is established, and a coefficient is assigned in relation to their qualitative importance. Thereby, while assessing the criteria weights, the experts faced three (3) possible situations, as follows:

- Greater in importance, the value '1' is assigned;
- Same importance, the value '0,5' is assigned;
- Smaller in importance, the value '0' is assigned.

On each row the score of each criterion is calculated and the criteria are ranked by their obtained scores. On the first position, the criterion with the highest score will be placed, and on the last level, the criterion with the highest lowest will be placed. If two criteria obtained the same score, the position will be the same: the level will have as value the arithmetic mean (average) of the positions corresponding to these criteria (e.g. in this study, criteria B and D obtained the same score and hold position 2 and 3 in the ranking order, so the assigned level of importance is 2.5).

After the relationship matrix is complete and the rank criteria is established, the weighting coefficients γi are set according to the FRISCO formula [1]:

$$\gamma i = \frac{p + \Delta p + m + 0{,}5}{-\Delta p' + \frac{Ncrt}{2}} \quad (1)$$

where: $\gamma i$ represents the weighting coefficients;

$p$ is the sum of the points gained by the respective criterion;

$\Delta p$ is the difference between the scores of the regarded criterion and the last ranked criterion;

$m$ represents the number of criteria outranked by the analysed element;

$N_{crt}$ represents the number of considered criteria;

$\Delta p'$ is the difference between the scores of the regarded criterion and the first ranked criterion.

The result of the analysis is determined by the objective and correct comparison of the relative importance of criteria. The weights γi of each criterion, calculated with FRISCO formula (1), are presented in table 2. The results show that Time of Response (H) is the most significant factor in evaluating the alternative solutions, succeeded by Utility (B) and Accuracy of Information (D).

**Table 2.** Criteria weights, calculated with FRISCO formula.

| | A | B | C | D | E | F | G | H | Scope (p) | Rank | $\gamma_i$ |
|---|---|---|---|---|---|---|---|---|---|---|---|
| **A** | 0.5 | 0 | 0 | 0 | 0 | 0 | 0.5 | 0 | 1 | 8 | 0.17 |
| **B** | 1 | 0.5 | 1 | 0.5 | 0.5 | 1 | 0.5 | 0.5 | 5.5 | 2.5 | 3.44 |
| **C** | 1 | 0 | 0.5 | 0 | 0.5 | 1 | 0 | 0 | 3 | 6 | 1.07 |
| **D** | 1 | 0.5 | 1 | 0.5 | 0.5 | 1 | 0.5 | 0.5 | 5.5 | 2.5 | 3.44 |
| **E** | 1 | 0.5 | 0.5 | 0.5 | 0.5 | 1 | 0 | 0 | 4 | 5 | 1.75 |
| **F** | 1 | 0 | 0 | 0 | 0 | 0.5 | 0.5 | 0 | 2 | 7 | 0.56 |
| **G** | 0.5 | 0.5 | 1 | 0.5 | 1 | 0.5 | 0.5 | 0.5 | 5 | 4 | 2.7 |
| **H** | 1 | 0.5 | 1 | 0.5 | 1 | 1 | 0.5 | 0.5 | 6 | 1 | 4.63 |

*2.4. Ranking alternatives*

The next step of the MCA is to analyse the three alternatives in relation to each criterion. Therefore, each alternative is awarded with importance grades (*Ni*) from 1 to 10. The performance value, or the importance grades (*Ni*), is given by the selected specialists to each alternative in relation to each criterion [1]. Table 3 summarise the multi-criteria ranking results of the emergency communications, namely the grades awarded to the considered emergency communication solutions, the performance index and the final ranking. In order to determine the final ranking of the three alternative solutions, the performance index ($F_i$) is calculated from the equation (2). $F_i$ is calculated for each alternative in relation to each criterion.

$$F_i = N_i \times \gamma_i \tag{2}$$

Based on the performance index values ($F_i$), the outcome is calculated for each of the studied alternatives and finally, the classification is established as presented in table 3. The top-ranked alternative according to the Multi-Criteria Analysis is considered to be the ideal solution. This study allows the following conclusions: First, the analysis of the considered alternatives revealed that the ideal solution for communicating during an emergency is alternative no. 3, the Mobile Application, with a final score of 172.75. According to the results, the SMS Text Service alternative is ranked very low, reflecting the need to innovate the Emergency Management (EM) system all over the world.

**Table 3.** Multi-criteria ranking results of the emergency communications.

| Criterion | $\gamma_i$ | Voice Calls | | SMS | | Mobile App | |
|---|---|---|---|---|---|---|---|
| | | $N_i$ | $F_i$ | $N_i$ | $F_i$ | $N_i$ | $F_i$ |
| A - Manufacturing & Maintenance Cost | 0.17 | 10 | 1.7 | 10 | 1.7 | 8 | 1.36 |
| B - Utility | 3.44 | 8 | 27.52 | 6 | 20.64 | 9 | 30.96 |
| C - Quantities of Information | 1.07 | 10 | 10.7 | 6 | 6.42 | 9 | 9.63 |
| D - Accuracy of Information | 3.44 | 10 | 34.4 | 7 | 24.08 | 10 | 34.4 |
| E - Data Monitoring | 1.75 | 10 | 17.5 | 9 | 15.75 | 10 | 17.5 |
| F - User Perception | 0.56 | 10 | 5.6 | 7 | 3.92 | 10 | 5.6 |
| G - Usability | 2.7 | 10 | 27 | 8 | 21.6 | 10 | 27 |
| H - Time of Response | 4.63 | 9 | 41.67 | 7 | 32.41 | 10 | 46.3 |
| Final Score | | - | 166.09 | - | 126.52 | - | 172.75 |
| Final Ranking | | II | | III | | I | |

The multi-criteria analysis of the alternative solutions was conducted with data given by the specialists from different decision-making groups. The evaluation method is based on collecting relevant evidence and classify the selected alternatives. The full results of our assessment indicate what to do next, in order to improve the Emergency Management system potential. In order to understands when and how it is better to implement mobile solutions in the Emergency Management field, the research continues by analysing the current stage, barriers and future potential of mobile emergency applications.

## 3. Mobile Emergency Applications Analysis

### *3.1. Current State*

The telecom market is developing rapidly, and the use of smartphones and mobile applications is increasing all over the world. Smartphone users install more and more mobile applications on their smartphones for daily usage. Recent studies conducted by EENA show that mobile devices can be a powerful tool to help in case of emergency, not only for those in need but also for the emergency helpers. In addition, citizens expect to be able to contact emergency services with mobile technologies and have clear expectations about the availability of 112 emergency mobile applications. As mobile telephony gives several opportunities to improve the help in emergencies, the researchers and involved actors need to investigate these possibilities [14].

Mobile applications can also support people in emergencies situations: road accidents, natural disasters, health problems or personal safety danger. The initial objective behind mobile emergency solutions was to allow emergency teams to make use of the high-accuracy location information provided by mobile devices. But before long, emergency authorities realized that additional information such as user's personal data, event details, or any other important information can be provided to the emergency dispatchers [15].

Over recent years there have been good results with mobile emergency applications and to date, there is a very diversified offer, which goes from providing escape guidelines during building fire, during natural disasters and nuclear accidents, or to help patients in medical emergencies. These applications provide more accurate information to emergency services [5].

The alerts sent through mobile emergency applications can reduce the total call length by shortening the time to make the call, the time to answer the call, the time to gain address details and the time to gain event description. In short, if the mobile alert is capable to provide emergency teams with relevant information for the two basic questions - what? and when? - , mobile emergency applications can enable a quicker dispatch of resources in case of emergencies [6].

However, in present, these applications only work in the region in which they were created, and this can confuse and discourage both citizens and emergency authorities. According to EENA, a mobile emergency application should be available all over the European Union (EU). This is the main reason EENA strongly believes that all the mobile emergency applications should work in a standardised way all over the EU and should be compliant with a pan-European standard [14].

*3.1.1. PEMEA Project [16].* Currently, the European Emergency Number Association is in the process of implementing a Pan-European Mobile Emergency Apps (PEMEA) policy. The PEMEA project's scope is to provide the European citizens the possibility to select and install a mobile emergency application that is user-friendly and meets customers' perception in terms of usability, functionality, utility and costs, with strong security and high privacy attributes, capable to provide in times of need instant identification of users, accurate location and event details to the nearest PSAP anywhere in Europe. A simplified call flow in PEMEA would be as follows:

1. The mobile emergency app sends the emergency data (telephone number, name, language, accurate location, location updates, communications, chat, etc.) to the PEMEA server.
2. The server sends the emergency data to the PSAP Service Provider (PSP) to which is connected. The PSP decides the next routing step in the PEMEA network, depending on the location of the user. The data might be sent to an Aggregating Service Provider (ASP) or it might go to other PSP or even directly to the PSAP in the region where the user is.
3. The emergency data sent by the App is received by the end PSAP (where the user's call is handled and where the application data is ultimately sent)
4. If the PSAP wants to invoke the establishment of a multimedia channel (i.e. data, voice, chat or video), then it can trigger the creation of the channel to the originating Application Provider (AP) and, subsequently, the AP would open it with the App. Several multimedia channels might be invoked in parallel by the PSAP and established during the emergency.
5. The PEMEA network nodes will manage the maintenance of the established channels till the end of the emergency.
6. All the PEMEA network nodes shall confirm that the rest of the entities to which they are connected are previously registered in the PEMEA Registry Authority (PRA). All the messages from non-registered entities shall be rejected.

The Network providers involved in the PEMEA project launched by EENA were: Deveryware for AP, PSP, ASP, Media Servers and Beta80 for AP, PSP and ASP.

The phase I of the project was conducted from June 2018 until the end of March 2019 and the findings revealed the capability of PEMEA project to provide:

- For citizens: access in roaming to emergency services all over the European Union through users' local mobile emergency application;

- For PSAPs: relevant information through a standardized application (e.g. accurate location, caller language, contacts, user's details, call tracking) so that emergency teams can provide the help as fast as possible.

In this phase of the project, eight participants were involved, as shown in table 4 [16]:

**Table 4.** List of participants in PEMEA project phase I.

| Participant (Software Provider) | Emergency Service Organization | Software Component | |
|---|---|---|---|
| | | Mobile App | PSAP System |
| Digia | 112 PSAP, Finland | 112 Suomi App | 112 Finland PSAP |
| Beta 80 | 112 PSAP, Italy | WhereAreU App | 122 Italy PSAP |
| University of Lujbiana | 112 PSAP, Slovenia | Snapring App | - |
| IPKom | | - | 112 Slovenia PSAP |
| Basque Country PSAP | 112 PSAP, Basque Country, Spain | SOS Dciak App | 112 Basque PSAP |
| Alterna | 112 PSAP, Murcia Region, Spain | 112 Murcia App | 112 Murcia PSAP |
| Omnitor | 112 PSAP, Sweden | 112 Omnitor App | - |
| Smart112 Mobile | 112 PSAP, Monaco | Smart112 App | - |

The scope of PEMEA project phase II, which was conducted from November 2019 until the end of May 2020, was to integrate and connect analysed applications with Public Safety Answering Points in roaming. The findings of this study revealed that the PEMEA network can provide the necessary infrastructure to allow emergency authorities (PSAPs) to connect with the citizens seeking for help, everywhere in Europe. The experiment showed that people in need were able to contact the closest PSAP connected to the PEMEA network using the local emergency application installed on the users' device. In this phase of the project, seven participants were involved, as shown in table 5 [16].

**Table 5.** List of participants in PEMEA project phase II.

| Participant (Software Provider) | Emergency Service Organization | Software Component | |
|---|---|---|---|
| | | Mobile App | PSAP System |
| University of Lujbiana | 112 PSAP, Slovenia | Snapring App | - |
| IPKom | | - | 112 Slovenia PSAP |
| Deveryware | PEMEA Consortium | DW Demo App | PSAP Interface Module |
| Telefonica | 112 PSAP, Catalonia Region, Spain | My112 App | 112 Catalonia PSAP |
| Cestel | 112 PSAP, Extremadura Region, Spain | 112 S-Visual App | 112 Extremadura PSAP |
| I-Urgence | 112 PSAP, SDIS-6 Region, France | I-Urgence App | - |
| Real Rider | PEMEA Consortium | MyNextbaseConnect App | - |

Thanks to PEMEA project and the available infrastructure, eight PSAPs in Europe and twelve mobile applications have been validated with the PEMEA standard. The PEMEA network is a live environment and ready to integrate other new mobile applications and/or emergency dispatchers in Europe, allowing the roaming of apps in different regions and countries of the European Union.

### *3.2. Barriers*

Regardless of the fast development of mobile applications, When it comes to mobile solutions in the emergency management field, significant barriers are holding back the improvement, including among others, Low Awareness of Mobile Emergency Apps, Lack of Regulation, Limited Evidence-Based Literature, Privacy and Security Concerns.

*3.2.1. Low Awareness of Mobile Emergency Apps.* Most users are unaware that mobile emergency applications are now available. Studies show that when citizens need to call 112 in an emergency situation, they tend to dial the emergency number on the keypad, instead of using a mobile application. This can be caused by any of the following:

- The user has forgotten that he installed an emergency app;
- The emergency app is not intuitive enough;
- It is too complicated to find the emergency app in a stressful situation;
- An emergency app is not installed.

That's why EENA recommends that emergency applications should preferably be added as basic applications in the smartphone itself, instead of a commercial application. [5]

*3.2.2. Lack of Regulation.* Another challenge that limits the potential of mobile emergency solutions is the lack of international regulation to ensure safety, accuracy, quality, and performance. All mobile manufacturers should use common standards, both for the caller, and the PSAP who needs to be able to receive and interpret the data.

Mobile technologies with emergency features are intended to improve emergency access and increase people's health and safety, but there are some cases where these emergency calls created chaos. In some PSAPs, the emergency alerts received via mobile applications have created difficult situations because no protocol has been previously defined, and the emergency responders were not prepared to handle these types of communications. Consequently, although citizens were in danger, emergency teams did not know how to manage these alerts and help those in need.

*3.2.3. Limited Evidence-Based Literature.* The formal literature for mobile emergency applications is limited. The most important evidence of research is conducted in mobile health applications, in areas as fitness and chronic diseases such as diabetes, cardiovascular health, mental health, and others. Until now, systematic reviews of mobile emergency applications are limited or poor. In addition, different studies show mixed results, and although some of them are demonstrating the efficacy of mobile emergency apps, the evidence base remains poor. In order to improve the emergency service sector, the main challenge of involved actors remains to develop at the speed of light a relevant process, evidence-based and with randomized controlled trial.

*3.2.4. Privacy and Security Concerns.* Mobile emergency applications utilization will increase the quantity of information collected and stored, in particular, users' personal data such as name, phone number, health information, location, etc. However, the need of privacy collides with the PSAP and first responders need to receive as much information as possible, as quick as possible and preferably as automated as possible. With the rapid development of mobile technology, laws and regulations are often outdated and not adapted to new circumstances. It is therefore important that regulatory authorities keep up the pace with technology so that regulations are updated constantly. For a secure evolution in mobile emergency applications, it is mandatory to guarantee data security and privacy for storage, transfer, encryption and authentication process [5].

### *3.3. Future Potential*

Despite of current barriers and limitations of mobile emergency solutions, the market is rapidly evolving, with the potential to make a dramatic impact into the emergency management system. As such, there is still more research needed to investigate the impact of mobile solutions in real life emergency situations, but the future evolution of these technologies include, among others, designing international standards and regulations for mobile emergency solutions, the integration of mobile solutions into the emergency system, promoting mobile emergency apps.

The PEMEA project conducted by EENA demonstrated that mobile emergency applications could provide citizens with fast and intuitive access to emergency services by providing improved communication channels (e.g. chat and audio-video), accurate location, users' information and call tracking at the time of emergency. In addition, a mobile emergency application will allow all citizens, including those with disabilities, to alert the emergency dispatcher when an emergency occurs and to automatically send crucial information to the nearest emergency dispatcher [14].

*3.3.1. Design international standards and regulations.* One of the major limitations for the mobile emergency market is the lack of international standards and regulations and this can be the one of the biggest reasons for the slow development of mobile emergency solutions. Providing access to emergency services without a strong cooperation between involved actors raise substantial doubts about the efficiency of such solutions.

Regulations differ between countries: while some countries have strict and complex regulation, others lack appropriate standards and regulations. That's why the EU and national regulators should do their part in making sure that the laws and regulations are up to date and in pace with the rapid development in the telecommunications market.

But changes in regulating mobile emergency market are about to happen, as in December 2019, EENA highlighted the need to increase the public-private partnership between the emergency authorities and the mobile application companies when developing such tools. EENA, together with signatories from all over the world, launched a call for tech companies to get involved in this consultation: "We all share the objective of keeping citizens safe and the conviction that technology can and should contribute to this goal. But this will only be achieved by combining the potential that technology can offer with the experience and expertise of emergency services" [17].

*3.3.2. Integrate mobile solutions into the emergency system.* Mobile solutions can improve the emergency management services due to their capability to provide accurate location and user's personal information in a short time. Mobile emergency applications are beneficial for both emergency professionals and citizens in danger by enabling professionals to save time and handle the case more efficiently.

For people with communication disabilities (e.g. speech or hearing disabilities) or people in silent emergency situations (e.g. kidnapping) the use of a mobile application can significantly improve their health and life. These citizens are often unable to contact the ES by dialling the emergency number on the keypad and therefore many communities are aiming to integrate mobile solutions.

At the time of writing, connectivity and communication of consumer mobile emergency applications is starting to increase.

*3.3.3. Promote mobile emergency Apps.* Due to a large number of mobile emergency applications available at present in major mobile apps stores, consumer, without guidance from public authorities, tend to choose from the most popular apps.

More studies substantiating improved outcomes are needed to encourage institutional stakeholders to adopt and integrate mobile solutions into the emergency system. Some countries have begun to develop mobile application and encourage their citizens to use the officially recognized applications in case of an emergency. Such a continued effort is expected in years to come, but further work is required to reach a maturity level.

## 4. Conclusion

To the present, the consumer mobile emergency market is unregulated. To develop the potential of mobile solutions for emergency management, multiple researches are conducted in this domain to accumulate data, establish evidence and expand data integration into the emergency system. To achieve great success in mobile emergency market, all the stakeholders including tech companies, professionals from emergency services and researchers must collaborate to overcome the barriers and the needs of this field.

EENA recognises the strong need of an emergency mobile application available in the market. Within the near future, EENA's goal is to promote the usage of mobile emergency apps that can provide accurate information in a secure way all over the European Union. EENA has taken firm steps in developing the specifications necessary to make to create a harmonized European mobile emergency application. Currently, EENA has officially recognized 28 mobile emergency applications in 13 countries [14]: Andorra, Australia, Denmark, Ecuador, Iceland, Ireland, Italy, Malaysia, Mexico, Norway, Spain, Switzerland, The Netherlands. According to the plan that several PSAPs have communicated to the PEMEA consortium, the following countries may have their connection to the PEMEA network ready before the end of 2020 [16]:

- Finland;
- Italy (several additional regions);
- Spain (several additional regions);
- France;
- Romania.

In the future, it is expected that all Europeans, including citizens with disabilities, will be able to use a harmonized mobile emergency application, regardless of where they are in European Union, and automatically send relevant information to the nearest PSAPs when the mobile alert is initiated.

**References**

[1] Department for Communities and Local Government 2009 *Multi-Criteria Analysis: A Manual* (London: Communities and Local Government Publications)
[2] Bobancu Ş 2015 *Creativitate şi Inventică* (Braşov: Editura Universităţii Transilvania)
[3] Vivier B 2019 Emergency Communications and the Eu Legislative Framework *EENA Operations Document*
[4] EENA 2014 112 Smartphones Apps *EENA Operations Document*
[5] Skoglund B 2017 Mobile Handset Requirements: Communication to Emergency Services *EENA Operations Document*
[6] Lanziani B and Biole M 2018 Recent advances in caller localisation for public safety answering point *Recent Trends in Control and Sensor Systems in Emergency Management. Advances in Intelligent Systems and Computing. Springer* **675** 40-57
[7] European Commission 2016 Proposal for a Directive of the European Parliament and of the Council Establishing the European Electronic Communications Code 2016/0288(COD)
[8] EENA 2015 SMS communication with PSAPs and EROs *EENA Operations Document*
[9] Lumbreras C and Machado G 2014 112 Caller location & GNSS *EENA Operations Document*
[10] European Commission 2020 Report on the Implementation of the European Emergency Number 112 COCOM pp 19-04
[11] EENA 2014 112 PSAPs Technology *EENA Operations Document*
[12] Dobrilova T 2020 35 Must-know SMS marketing statistics in 2020 *TechJury*
[13] Halliwell D and Lumbreras C 2018 Mobile Identity Platform for the Emergency Services *EENA Operations Document*
[14] Casse B, Gomez I and Winterbottom J 2015 112 Apps Strategy - Pan European Mobile Emergency *EENA Operations Document*
[15] Winterbottom J 2015 Pan-European Mobile Emergency Application (PEMEA) Requirements and Functional Architecture *EENA Operations Document*
[16] EENA 2020 Emergency Apps PEMEA - Final PEMEA Report *EENA Operations Document*
[17] EENA 2019 Emergency Services Call for Stronger Cooperation with Tech Companies *EENA Operations Document*

# Study of the deviation of shape for the parts obtained by additive manufacturing

**A D Rizea[1], D C Anghel[1] and D M Iordache[1*]**

[1] University of Pitesti, Manufacturing and Industrial Management Department, Târgul din Vale Street No.1, Romania

*E-mail: monica.iordache@upit.ro

**Abstract.** 3D printing or additive manufacturing is a parts manufacturing process using a digital model can be drawn using suitable computer aided design (CAD) They can be created using 3D printing products of almost any shape. This article presents research concerning the deviation in shape of the parts obtained by additive manufacturing (3D printing). The digital model was made with Catia V5 and was imported by the 3D machine software to obtain the prototyping codes for nine specimens. The specimens were made according to a Taguchi L9 plan. On the basis of the code obtained with the printer software, the parts have been prototyped, using the Z-ULTRAT material with a ZORTRAX M200 printer. The input data that were varied are: layer thickness, infill and orientation. For each part obtained by 3D printing were measured: circularity, flatness and dimensional accuracy. All parts were measured and a database was created and following the regression analysis performed with the Minitab19 software was establish the influence of the parameters on the deviations and dimensional accuracy.

## 1. Introduction

3D printing manufacturing has developed rapidly on a number of industrial sectors. 3D printing in the automotive industry is used for prototypes and functional parts. For example, Rolls-Royce is working for its customers to create fully customized vehicles using 3D printing [1]. Ford applies the technology of obtaining 3D printing for engine parts and prototypes, and in 2014 Local Motor manufactured the first car through 3D printing. [2]. BMW implemented FDM technology to build hand-tools for vehicle assembly and testing [3]. AUDI collaborates with SLM to produce spare parts using 3D printing manufacturing [4].

Also, 3D printing technology has penetrated in aerospace industry, electric and electronic, industry healthcare and medical industry, food industry, architecture and construction industry [2], [7]. This technology does not require tools or devices for the manufacture of a product or parts, which has an important advantage over conventional manufacturing.

It offers freedom of design and ecological advantages and parts with complex geometry can be obtained that would be difficult to obtain by another process. At the same time, complex models can be printed in 3D as a single piece, eliminating the need to assemble component parts. However, there are limitations in terms of 3D printing technology: high costs for large series production, parts with variable properties, low accuracy. Several study explores the characteristics and mechanical properties of composites developed through 3D printing techniques [5]. Dimensional accuracy and surface quality of parts manufactured by 3D printing are the concerns of users of these technologies. The geometric precision prediction of the parts manufactured by 3D printing technologies is analyzed based on experimental studies and analytical models. Experimental studies use simple specimens and experimental plans Taguchi, ANOVA, etc. in which different process parameters are varied. Thus, in the paper [6] were analyzed using the software Minitab and the Taguchi method, the orthogonal matrix L27, the circularity and roughness of the surfaces obtained by 3D printing. It turned out that the biggest influence has the bed temperature, followed by the number of loops, the temperature of the nozzle, the printing speed, the thickness of the layer and the filling. Another study shows that for a low-cost FDM machine, if the part is oriented vertically and its dimensions on Z are small, the best accuracy is obtained [7]. It is recommended that in order to obtain a deviation from the small shape, it is good to set the part so that those shapes sit on the machine bed, and the diameter of the nozzle influences the accuracy of the size and shape. [8].

This article shows the influence of manufacturing parameters (thickness of the layer, infill, orientation) on the dimensional accuracy, flatness and cilindricity of parts obtained by 3D printing. The results were interpreted in order to establish the influence on the deviation of shape for the parts obtained by additive manufacturing.

## 2. Experimental procedure

In order to establish the dimensional precision, a test specimen with approximate dimensions of 40 x 15 x 84 mm was designed. This part has three interior cylindrical surfaces. The diameters of the cylinders are in geometric progression. The ratio of the progression is equal to square root of 2. Thus, the test piece was designed using Catia software and exported as an STL file. ZORTRAX M200 printer machine is used for manufacturing the dimensional specimens.

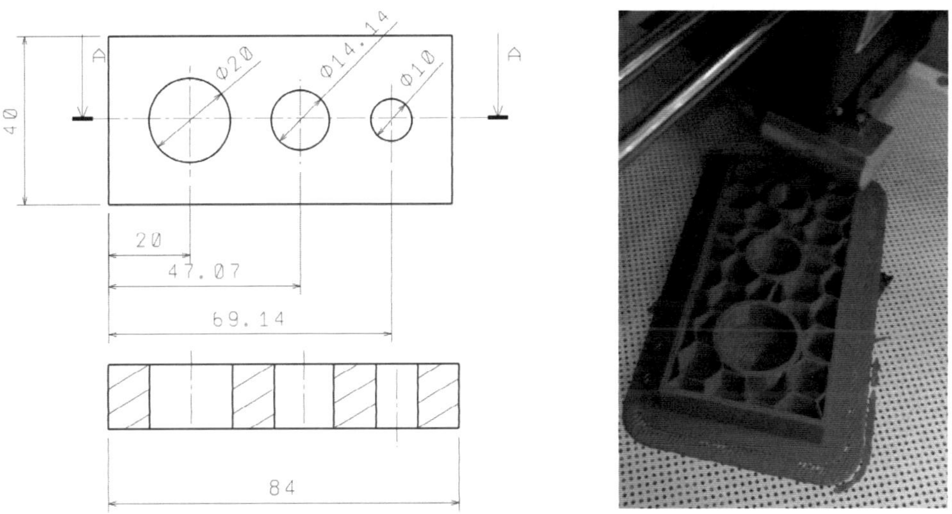


(a) (b)

**Figure 1.** Test specimen
(a) dimensions of the specimen; (b) manufacture of the specimen.

The work determines the accuracy of the specimens, taking into account three process parameters, namely: the thickness of the layer, the percentage of infill and the horizontal orientation of the part, figure 2. The process parameters were varied using the Taguchi method, L9 orthogonal matrix, the DOE experimental run generated using MINITAB 19.0. The diameters of the holes were chosen so as to be in geometric progression.

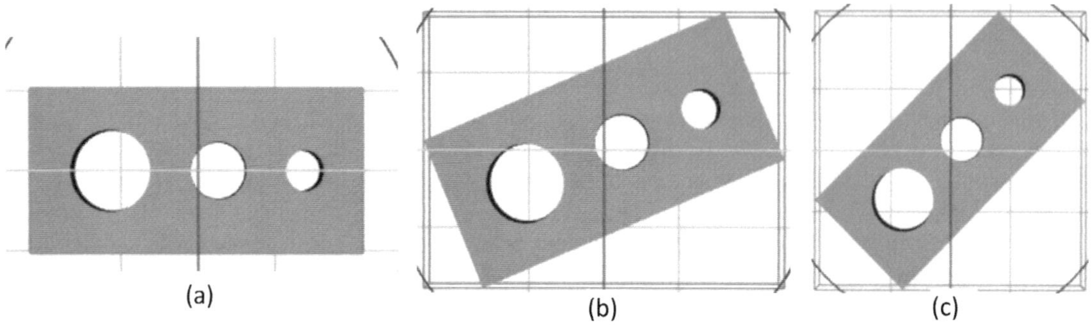


**Figure 2.** Orientation of the part.

Table 1 shows the input data, the symbol adopted for them and the 3-level values used in the experimental tests.

**Table 1.** Input data used for 3D printing of test specimens

| Symbols | Input data | Levels | | |
|---|---|---|---|---|
| | | 1 | 2 | 3 |
| T | Layer Thickness, mm | 0,09 | 0,14 | 0,19 |
| F | Infill, % | 10 | 50 | 90 |
| O | Orientation, degrees | 0 | 22,5 | 45 |

With these parameter values, 9 test specimens were manufactured using 3D printing. Each test piece was measured using a Zeiss Contura G3 coordinate measuring machine (with an accuracy of 1 μm) and CALIPSO software. Zeiss Contura enables tactile and optical measurement, figure 3. Thus, the diameters of the holes and the deviations from cylindricity and planarity were measured. At the same time, the diameters of the three holes were measured at three points on the depth.

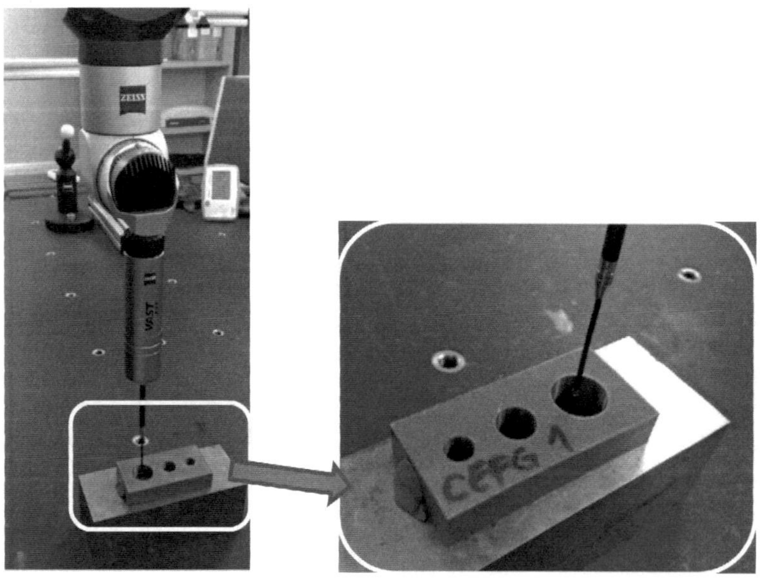

**Figure 3.** Cylindrical measurement with Zeiss Contura.

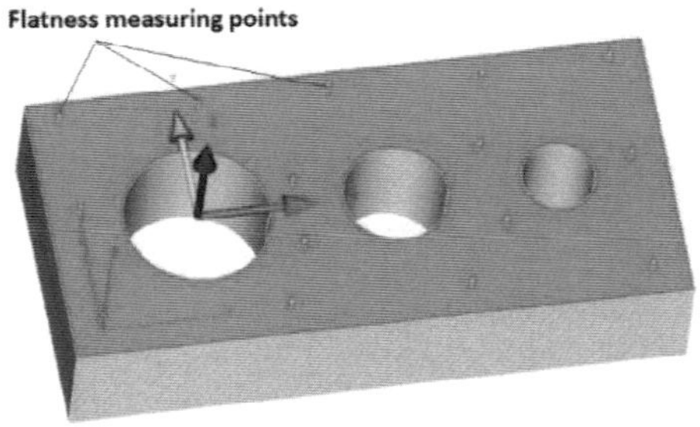


**Figure 4.** Flatness measuring points.

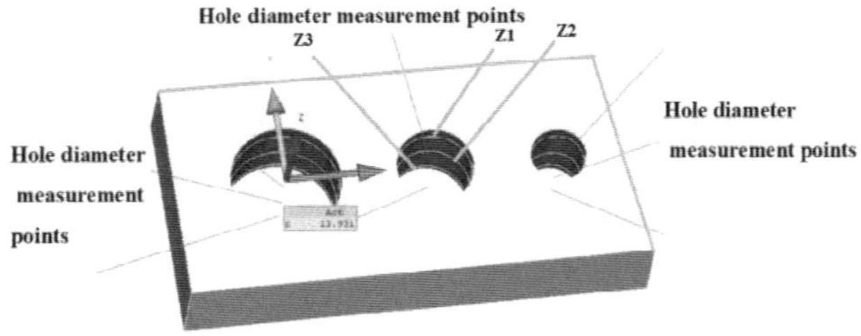


**Figure 5.** Hole diameter measurement points.

## 3. Results and discussion

The dimensions are measured in the points presented in figure 4 and figure 5. The results of measuring the cylindricality of the three holes and flatness are presented in table 2.

**Table 2.** The experimental plan and the results obtained.

| No | Layer Thickness, mm | Infill, % | Orientation, degrees | Cylindricity, mm | | | Flatness mm |
|---|---|---|---|---|---|---|---|
| | | | | Φ20 | Φ14.14 | Φ10 | |
| **1** | 0,09 | 10 | 0 | 0,012 | 0,019 | 0,007 | 0,081 |
| **2** | 0,09 | 50 | 22,5 | 0,025 | 0,014 | 0,008 | 0,115 |
| **3** | 0,09 | 90 | 45 | 0,027 | 0,020 | 0,001 | 0,031 |
| **4** | 0,14 | 10 | 22,5 | 0,009 | 0,011 | 0,009 | 0,133 |
| **5** | 0,14 | 50 | 45 | 0,035 | 0,000 | 0,002 | 0,062 |
| **6** | 0,14 | 90 | 0 | 0,011 | 0,014 | 0,008 | 0,039 |
| **7** | 0,19 | 10 | 45 | 0,039 | 0,011 | 0,004 | 0,106 |
| **8** | 0,19 | 50 | 0 | 0,029 | 0,024 | 0,013 | 0,076 |
| **9** | 0,19 | 90 | 22,5 | 0,006 | 0,010 | 0,020 | 0,049 |

Following the regression analysis performed with the Minitab19 software, the influence of the parameters on the flatness was obtained.

$$\text{Flatness} = 0.116397 + 0.0133333 \cdot T - 0.0008375 \cdot F + 2.22222e\text{-}005 \cdot O \quad (1)$$

Where: T is thickness of the layer, F is Infill and O represents Orientation,

The regression model presented above has a determination coefficient ($R^2$) of 0.67 and a p-value of 0.109. It is observed that the value of p is higher than the quantum of the confidence level (0.05), and the coefficient of the parameter O is very small. Thus, this influence is neglected and the regression analysis is repeated:

$$\text{Flatness} = 0.117 + 0.013 \cdot T - 0.000837 \cdot F \quad (2)$$

In this case p = 0.035. Taking into account the value of the linear regression coefficients, it can be concluded that the greater influence on the flatness has the thickness of the layer.

S represents the values of the standard distance data from the regression line, because the value is small it results that the equation predicts the response ($S_{regresion}$ = 0,0233), figure 6.

Because the standard error an of the infill factor coefficient is small (SE = 0.000238), it results that this coefficient is accurately estimated. p-value for infill is less than the α-value, as a result of which this parameter significantly influences the flatness. With the regression analysis, a relationship between the process parameters and the cylindricality could not be established (the correlation coefficient $R^2$ obtained is very small, 32%).

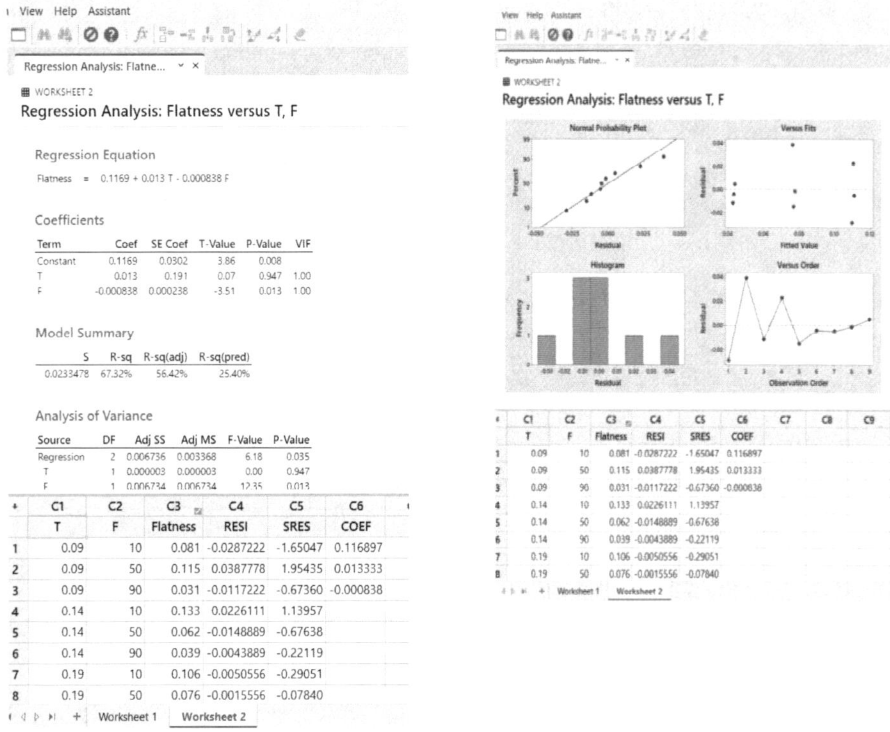


**Figure 6.** Regression by Minitab 19 for flatness.

Table 3 shows the diameters measured on the three circles (Z1, Z2, Z3), average and dimensional error E. Dimensional error (E) corresponds to the difference between the average dimensions for each direction of the three specimens and the true value.

**Table 3.** Experimental result (hole diameter).

| No | Φ20 | | | | Φ14.14 | | | | Φ10 | | | |
|---|---|---|---|---|---|---|---|---|---|---|---|---|
| | Z1 | Z2 | Z3 | E | Z1 | Z2 | Z3 | E | Z1 | Z2 | Z3 | E |
| **1** | 19,801 | 19,839 | 19,805 | 0,185 | 13,964 | 13,934 | 13,964 | 0,186 | 9,809 | 9,788 | 9,823 | 0,193 |
| **2** | 19,782 | 19,774 | 19,796 | 0,216 | 13,918 | 13,889 | 13,906 | 0,236 | 9,782 | 9,767 | 9,776 | 0,225 |
| **3** | 19,801 | 19,750 | 19,829 | 0,207 | 13,943 | 13,922 | 13,969 | 0,195 | 9,818 | 9,771 | 9,824 | 0,196 |
| **4** | 19,769 | 19,749 | 19,787 | 0,232 | 13,937 | 13,909 | 13,927 | 0,216 | 9,788 | 9,763 | 9,783 | 0,222 |
| **5** | 19,771 | 19,737 | 19,794 | 0,233 | 13,926 | 13,899 | 13,957 | 0,213 | 9,813 | 9,812 | 9,831 | 0,181 |
| **6** | 19,779 | 19,757 | 19,785 | 0,226 | 13,923 | 13,894 | 13,929 | 0,225 | 9,780 | 9,720 | 9,766 | 0,245 |
| **7** | 19,717 | 19,706 | 19,742 | 0,278 | 13,928 | 13,878 | 13,908 | 0,235 | 9,804 | 9,817 | 9,816 | 0,188 |
| **8** | 19,741 | 19,751 | 19,784 | 0,241 | 13,900 | 13,891 | 13,924 | 0,235 | 9,772 | 9,750 | 9,782 | 0,232 |
| **9** | 19,768 | 19,733 | 19,804 | 0,232 | 13,935 | 13,891 | 13,942 | 0,217 | 9,803 | 9,758 | 9,797 | 0,214 |

For each experiment the measured data were statistically analyzed. Thus, Mean (**M**), Range (**R**), Sample Standard Deviation (**SSD**) and Coefficient of Variation (**CV**) were calculated,(table 4).

**Table 4.** Statistical analysis for dimensional deviation.

| No | Φ20 | | | | Φ14.14 | | | | Φ10 | | | |
|---|---|---|---|---|---|---|---|---|---|---|---|---|
| | M | R | SSD | CV | M | R | SSD | CV | M | R | SSD | CV |
| **1** | 19,815 | 0,035 | 0,0178 | 0,0018 | 13,954 | 0,030 | 0,0171 | 0,0012 | 9,807 | 0,037 | 0,0205 | 0,0010 |
| **2** | 19,784 | 0,015 | 0,0075 | 0,0008 | 13,904 | 0,029 | 0,0144 | 0,0010 | 9,775 | 0,022 | 0,0109 | 0,0006 |
| **3** | 19,793 | 0,054 | 0,0294 | 0,0030 | 13,945 | 0,047 | 0,0237 | 0,0017 | 9,804 | 0,079 | 0,0402 | 0,0020 |
| **4** | 19,768 | 0,025 | 0,0134 | 0,0014 | 13,924 | 0,028 | 0,0144 | 0,0010 | 9,778 | 0,038 | 0,0192 | 0,0010 |
| **5** | 19,767 | 0,019 | 0,0106 | 0,0011 | 13,927 | 0,058 | 0,0290 | 0,0021 | 9,819 | 0,056 | 0,0284 | 0,0014 |
| **6** | 19,774 | 0,060 | 0,0311 | 0,0032 | 13,915 | 0,035 | 0,0188 | 0,0014 | 9,755 | 0,029 | 0,0151 | 0,0008 |
| **7** | 19,722 | 0,013 | 0,0072 | 0,0007 | 13,905 | 0,050 | 0,0250 | 0,0018 | 9,812 | 0,036 | 0,0185 | 0,0009 |
| **8** | 19,759 | 0,032 | 0,0163 | 0,0017 | 13,905 | 0,033 | 0,0171 | 0,0012 | 9,768 | 0,043 | 0,0224 | 0,0011 |
| **9** | 19,768 | 0,045 | 0,0243 | 0,0025 | 13,923 | 0,051 | 0,0278 | 0,0020 | 9,786 | 0,071 | 0,0357 | 0,0018 |

The coefficients of variation for an experiment have close values which shows that the diameter of the hole does not influence the final dimension. For a minimum and maximum layer thickness, the maximum coefficient of variation is obtained for the maximum infill.

Each piece was measured in several directions and at three heights from the base. Thus, it is observed that the diameter of the holes varies according to the height at which it is measured. Analyzing the results, we can say that the smallest diameter is obtained at the height of Z2 (approximately half the height of the piece) and it increases towards the extremities of part Z1 and Z3 (figure 7). This can be explained by taking into account different cooling times depending on height.

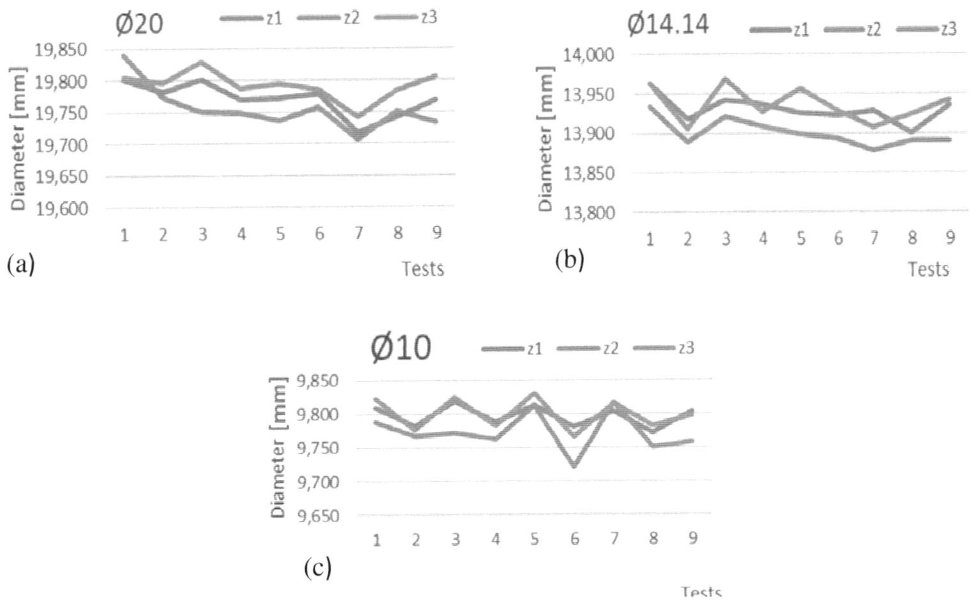


**Figure 7.** Influence of the height of the measured area on the value of the hole diameter (a) for diameter Φ20; (b) for diameter Φ14.14 c) for diameter Φ10.

Following the regression analysis performed with the Minitab19 software, the influence of the parameters on the dimensional error (for a diameter of 20 mm) was obtained.

$$E = 0.1562 + 0.479\ T - 0.000126\ F + 0.000480\ O \tag{3}$$

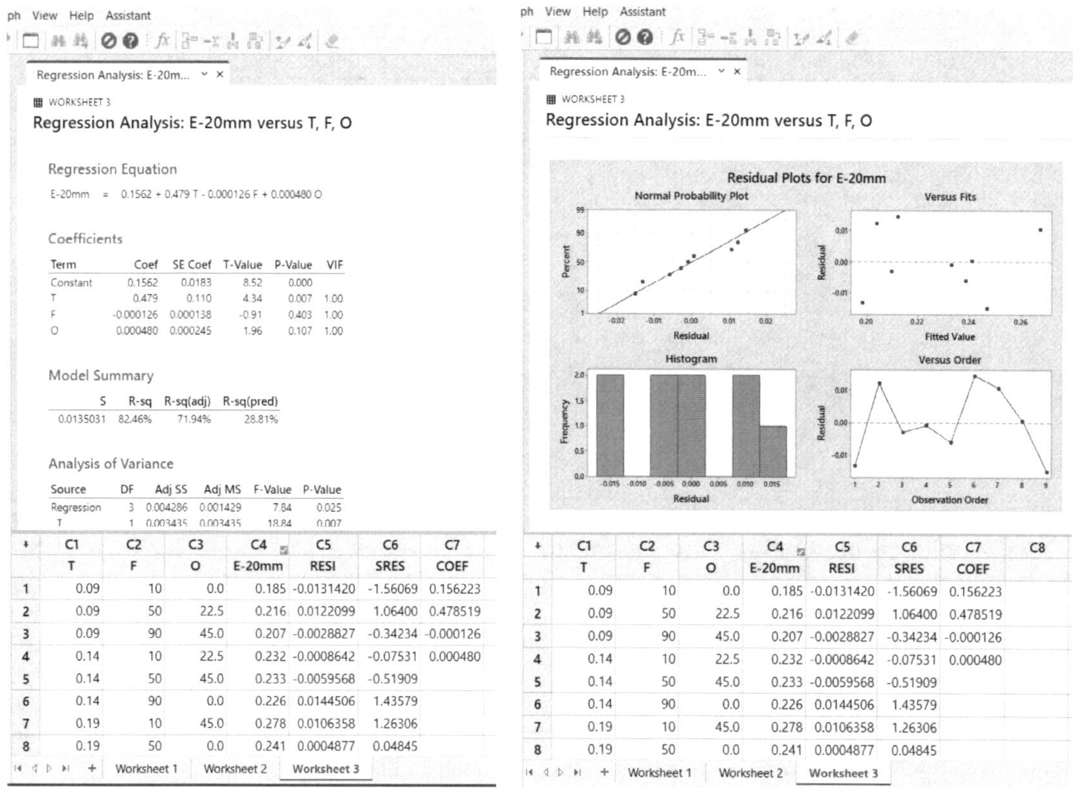


**Figure 8.** Regression by Minitab 19 for dimensional error.

The regression model has a determination coefficient of 0.82 and a p-value of 0.025, figure 8. S value is small it results that the equation predicts the response ($S_{regresion}$ = 0,0135). Analysing the Standard Error of the Coefficient shows that all coefficients are accurately estimated. The estimation of the coefficient is the most accurate in the case of the infill parameter, having associated the smallest standard error. Residues appear to follow a straight line, although values fall below this.

## 4. Conclusions

The paper presents an analysis on the influence of process parameters on the dimensional precision of the holes processed by additive fabrication. The influences on geometric precision are also presented by analysing the flatness of the surfaces and the cylindricality of the holes. This information is essential for the designer in the process of prescribing dimensional tolerances and geometric precision. The main conclusions are:

- For the deviation from the cylindricality was not able to obtain a relationship relevant from a quantitative point of view, but the geometry of the hole varies depending on the height of the piece. The diameters of the holes increase from half the height of the piece to the extremities;
- Dimensional deviation is primarily influenced by the infill F and thickness of the layer T. There is also a fixed coefficient that highlights the contraction of the piece;
- The main process parameter influencing the flatness is the thickness of the layer of material deposited, other parameters having very small influences.

**Acknowledgments**. This work was supported by a grant of the Romanian Ministry of Research and Innovation, CCCDI-UEFISCDI, project number PN-III-P1-1.2-PCCDI-2017-0224 / 77 PCDI/2018 within PNCDI III.

**References**

[1] Nichols M R 2019 How does the automotive industry benefit from 3D metal printing? *Met. Powder Rep.* doi:10.1016/j.mprp.2019.07.002

[2] Shahrubudin N, Lee TC and Ramlan R 2019 An Overview on 3D Printing Technology: Technological, Materials, and Applications *Procedia Manuf* **35** p 1286.

[3] Sreehitha V 2017 Impact of 3D printing in automotive industry, *Int J of Mech and Prod Eng* **5** 2 p 91.

[4] Petch M 2018 Audi gives update on use of SLM metal 3D printing for the automotive industry," 3D Printing Industry, [Online]. Available: https://3dprintingindustry.com/news/audi-gives-update-use-slm-metal-3d-printing-automotive-industry-129376/. [Accessed 2020].

[5] Singh B Kumar R and Singh Chohan J 2020 Polymer matrix composites in 3D printing: A state of art review *Mater. Today: Proc.* **28** p 1023

[6] Sajan N, John T D, Sivadasan M and Singh N K 2018 An investigation on circularity error of components processed on Fused Deposition Modeling (FDM). *Mater. Today: Proc.* **5**(**1**) p 1327.

[7] Mora S M, Gil J C and Camacho López A M 2019 Influence of manufacturing parameters in the dimensional characteristics of ABS parts obtained by FDM using reverse engineering techniques *Procedia Manuf* **41** p 968.

[8] Reyes-Rodríguez A, Dorado-Vicente R and Mayor-Vicario R 2017 Dimensional and form errors of PC parts printed via Fused Deposition Modelling *Procedia Manuf* **13** p 880

[9] Anghel D C, Rizea A D, Iordache D M and Beşliu-Gherghescu M L 2019 Studies on the influence of design parameters on the behaviour at shock of 3D-printed components fabricated by fused deposition modelling. *IOP Conference Series: Mater Sci Eng*, **564**

[10] Stănescu N D, Beşliu-Gherghescu M L and Rizea A D 2019 Determination of the shape of a beam obtained by fused deposition material with general loads *IOP Conference Series: Mater Sci Eng*, **564**

[11] Anghel D C, Beşliu-Gherghescu M L and Stănescu N D 2018 Optimized Prototyping for the Increasing of Performances of a Design Process *International Congress of Automotive and Transport Engineering*, (Springer, Cham) p 446.

[12] Anghel D C, Ene A and Belu N 2014 A Matlab neural network application for the study of working conditions *Adv Mat Res* **837** p 310.

IOP Conf. Series: Materials Science and Engineering 1009 (2021) 012051 doi:10.1088/1757-899X/1009/1/012051

# Making Industry 4.0 functionalities understandable based on the production of a model car

**S Roman[1]* and C Bunsen[1]***

[1]University of Applied Sciences Würzburg-Schweinfurt, Faculty of Mechanical Engineering, Laboratory for Product Development c-factory, Ignaz-Schön-Straße 11, 97421 Schweinfurt, Germany

*E-mail: sabina.roman@fhws.de, christoph.bunsen@fhws.de

**Abstract**. At the University of Applied Sciences Würzburg-Schweinfurt, Faculty of Mechanical Engineering, an Industry 4.0 use case was implemented with the c-factory, the so-called concept factory. This use case consists of building an individual model car. The focus here is not on the model car, but rather on mapping the entire product life cycle and the Industry 4.0 functionalities during the production. These are, as identified at the FHWS/FM: networking, the Internet of Things, flexible production, human-robot-collaboration, augmented reality, additive manufacturing and Big Data. The foundation stone was laid as part of a student project and, with its unique form of enabling didactic, serves not only to demonstrate and understand Industry 4.0, but also ensures that all participants take on the role of both learner and teacher. Through this, both sides support each other in expanding the use case and hereby broaden their own (educational) horizons.

## 1. Industry 4.0 at the University of Applied Sciences Würzburg-Schweinfurt

When people, machines and industrial processes are intelligently networked, we speak of Industry 4.0. Connecting production to state-of-the-art information and communication technology enables highly customized products on customers' requests instead of a fixed set of possible customizations. With the help of the Internet of Things, not only information is displayed, but also communication between devices and humans is made possible [1].

There is a clear trend towards individualization of products which requires flexible production processes. In the future, tasks will be completed jointly by humans collaborating with robots. Through additive manufacturing parts can be produced individually, for instance in geometric shapes that were previously impossible to create. Augmented reality joins the real world with the virtual world, e.g., to simulate new products in a familiar environment. Big Data is concerned with the management and interpretation of the data generated in each of these aspects.

With its i-factory, the University of Applied Sciences Würzburg-Schweinfurt (FHWS) wants to demonstrate an intelligent factory based on four pillars: digital innovation, digital intelligence for technical flexibility, digital integration and digital inclusion in the field of human resources. The concept-factory, the so-called c-factory, was established in the Faculty of Mechanical Engineering (FM) to show an Industry 4.0, i.e. the construction of a customized model car. The c-factory was developed in the context of a student project.

However, the focus here is not on the model car as such, but on mapping the entire product life cycle and the technical Industry 4.0 functionalities during production. These are (as mentioned above):

networking, the Internet of Things, flexible production, human-robot-collaboration, augmented reality, additive manufacturing and Big Data (figure 1).

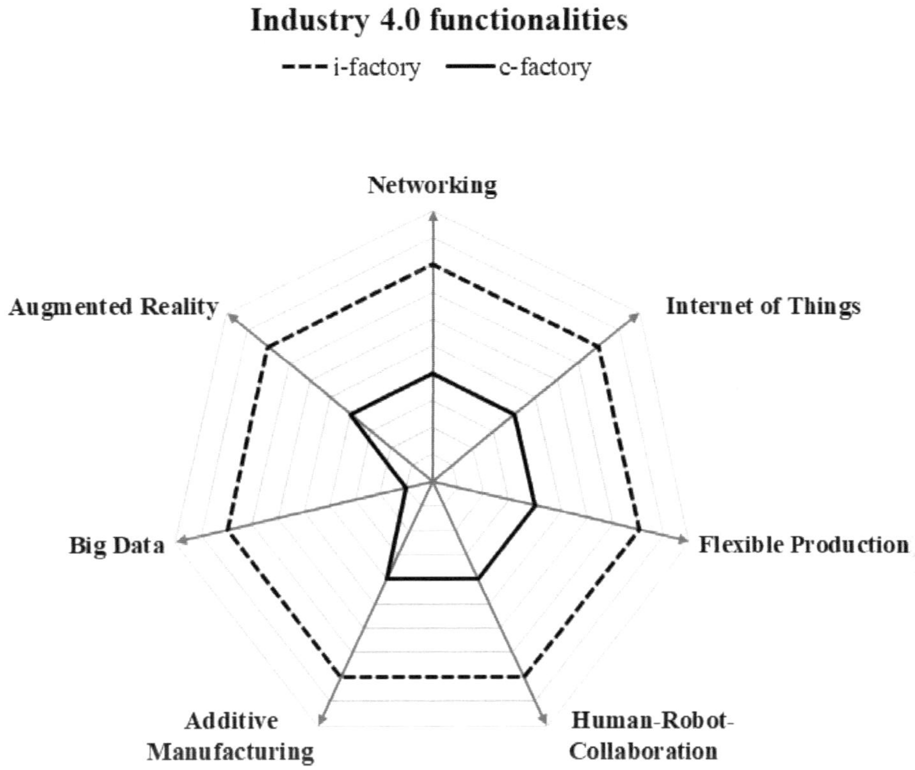


**Figure 1.** The technical dimensions of Industry 4.0, as identified at FHWS/FM.

The production of the model car demonstrates how these dimensions are applied in the c-factory. By personally carrying out these individual production steps, the user gains a much better understanding of the processes, their digitalization and implementation and their significance. In the conception of the case, special care was taken to include devices from different manufacturers in the process to meet the challenges of today's technology, as different standards and interfaces are still state-of-the-art today [2].

## 2. The use case

The c-factory realized a specific application to implement the technical functionalities of Industry 4.0 in a "hands-on" scenario by producing a small model car, a pick-up truck, which can be configured in almost any way, using various manufacturing processes. The truck has three components that are produced on site. The truck's cabin is injection molded, the chassis is milled individually and the loading area is printed three-dimensionally (3D). There are eight steps the user must follow in order to receive a finished product. These are: identification, configuration of the model car, sending the configuration to the cloud, using the transponder to start production, manufacturing and assembly of the vehicle, then doing the quality and functional checks (figure 2).

At the beginning, there is the identification at the system in order to configure a pick-up truck. There is a selection of the previously mentioned components: a small and a large cabin, a short and a long chassis and a single or double rear axle for the truck. Based on this selection, six possible standard versions can be created (figure 3).

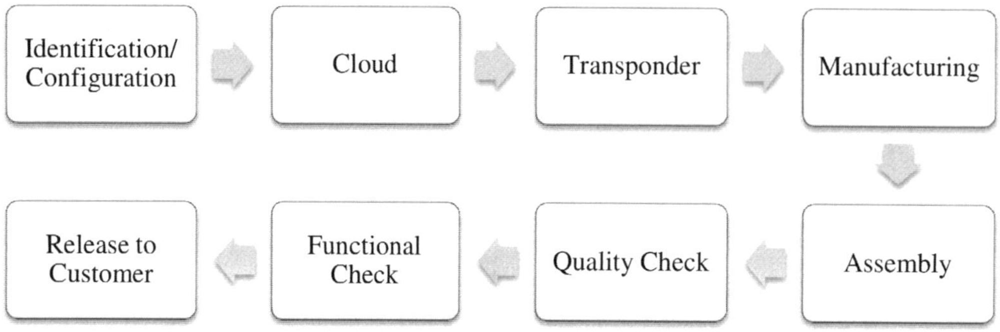


**Figure 2.** The eight steps involved in the production of a c-factory truck.

**Figure 3.** The six possible standard versions of the c-factory trucks.

However, the customer is given the opportunity to further individualize his self-configured vehicle by designing an individual loading area with the help of computer-aided design (CAD) software. Once this process is completed, the created configuration is given its own identification number (ID), which is made available as a quick response (QR) code. The configuration is stored in the cloud and linked to a transponder. The transponder is later needed to trigger the production of the components.

Only the selected components during the configuration are produced on site, whereas additional elements, that are required to complete the vehicle, are purchased parts such as axles and screws. Each of these components is produced differently. The chassis is milled, the cabin is injection molded and the loading area is printed in 3D. Once all required components have been produced, the vehicle can be assembled by the user at the designated station. This step is guided virtually with the help of augmented reality glasses. Light signals are used to point where the needed parts are located, while the glasses show an animation to demonstrate the assembly procedure. This is followed by a quality and then a functional check. Once the vehicle passes both checks it may be released to the customer.

## 3. The process in detail

As already mentioned, a transponder is used to request the required data for each production step. Therefore, the customer can carry out the production of the truck independently and without the support of trained personnel.

### *3.1. Preparatory steps*

As mentioned above, everything starts with the user's identification on the system and the vehicle is given a name. Afterwards the configuration process starts, where the user can select a cabin, a chassis and a loading area for the pick-up truck. The cabin can be short or long, the chassis short, long or long with a double rear axle. With these variations, either a suitable standard loading area is suggested, or an individual one can be chosen. Such a loading area is designed in an extra step and must be printed separately in 3D. After all components have been selected, the configuration is saved in the cloud and is also assigned its own identification number (ID), which is immediately displayed on the monitor in the form of a QR code that leads to the vehicle's own website.

A truck was created to better understand every production step in the c-factory. In this example, the tuck was given a king-size cabin ("Kabine" in German), a deluxe chassis ("Chassis" in German), a double-rear axle ("Achse tandem" in German) and an individual loading area ("Ladefläche individuell" in German).

The configured pick-up truck was named "CoSME2020". Before finally saving the configuration, an overview of all selected elements was displayed for the user. By pressing the "zurück" button (German for "back"), the user could re-start the configuration process. By pressing the "speichern" button (German for "save"), a communication with the cloud was established (and thus to the database) and the entered data was saved (figure 4).

**Figure 4.** Screenshot of the configured vehicle with its selected components (king-size cabin, deluxe chassis, double rear axle and individual truck bed).

The vehicle was then assigned its unique chassis number, which was displayed on the screen as a QR code (figure 5).

**Figure 5.** Screenshot of the created QR code for the configured vehicle.

When scanned, the code guides the user to the vehicle's own website (figure 6). Since nothing has been produced yet, only the "configuration" button is highlighted and the selected components, the chosen name and a time stamp are listed.

The highlighting of the buttons serves as a visual indication of which steps have already been carried out. An orange button, as the "configuration" button in figure 6, indicates that the step has been completed, whereas grey buttons stand for not yet completed production steps.

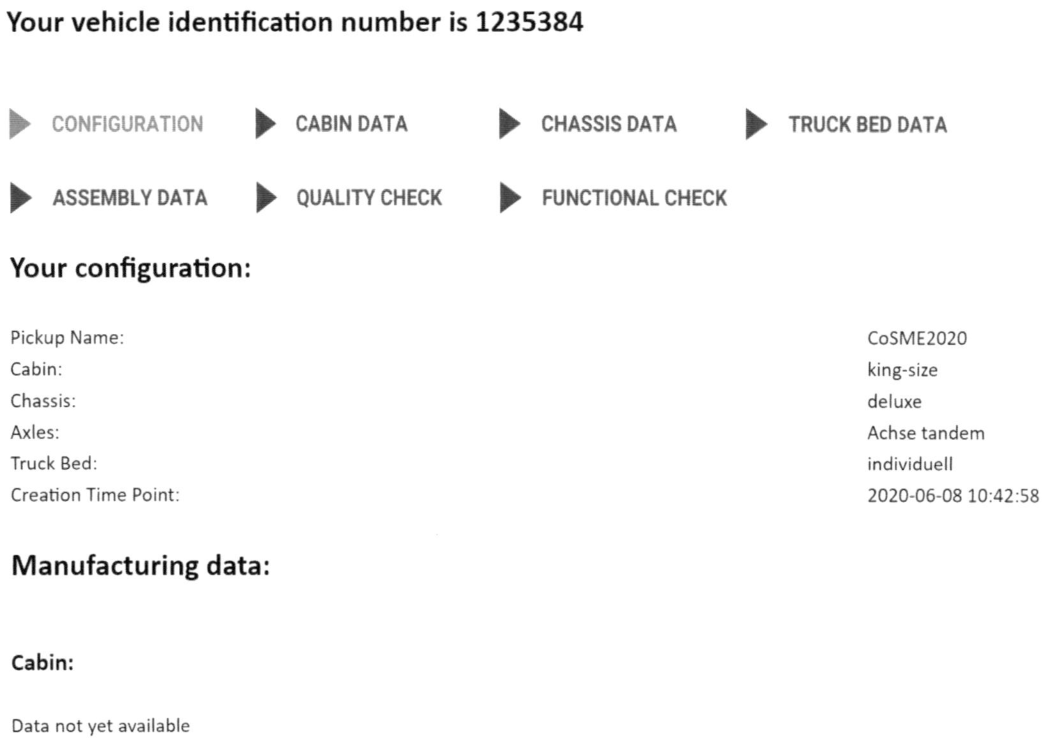


**Figure 6.** Screenshot of the model car's website immediately after its configuration.

This step creates an understanding of flexible production and networking. A gradual introduction to the topic, which illustrates the variety of products with a simple example, is well-suited to introduce both students and visitors to the topic of Industry 4.0.

Scanning the QR code with your own smartphone also demonstrates that devices are indeed networked with each other and that the configuration is sent to the cloud.

*3.2. Production of the model car*

Once the configuration has been made, production can begin. Now the transponder is needed in order to start manufacturing the chassis, cabin and loading area. First the chassis is produced. As mentioned before, there are three different versions possible: short, long and long with double axle.

Depending on the version, the drill holes for the axle brackets and screws must be positioned differently. The vehicle's ID and name are also engraved. All data are automatically transmitted to the milling machine with the help of the transponder. Communication with the database is established, and the required information for the milling process is transferred and entered into the G-code program (programming language to control automated machine tools like milling machines). The handling is done by a robot arm, which also needs to know whether to pick a long or a short chassis that is to be inserted into the milling machine (figure 7).

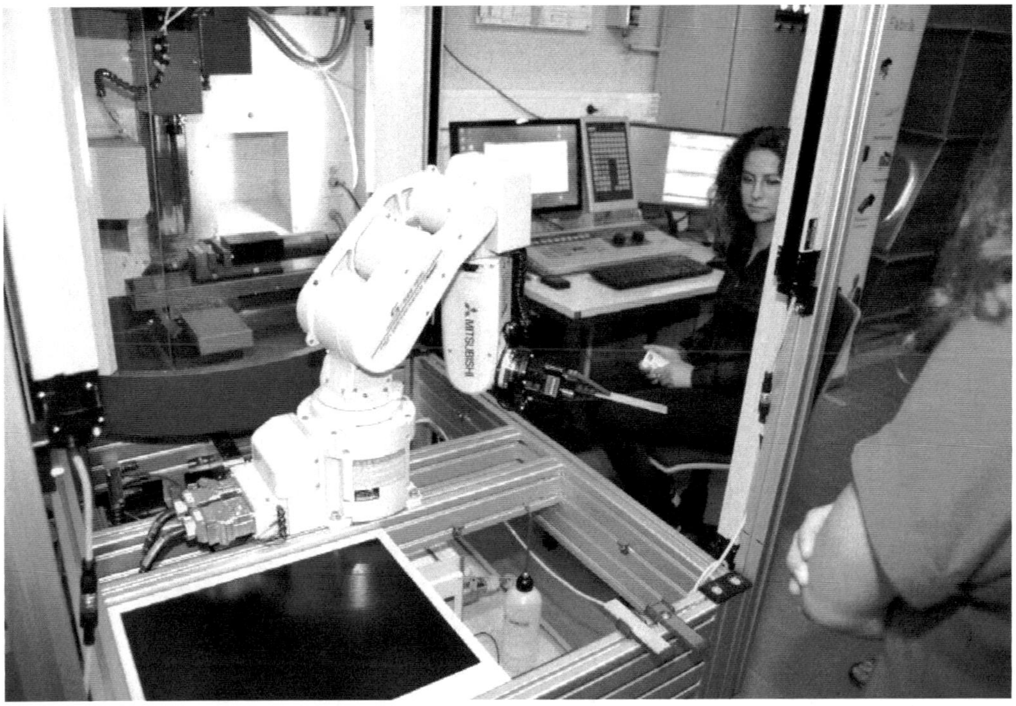


**Figure 7.** Student observing the handling robot while feeding a chassis into the milling machine. Photo credit FHWS.

As mentioned above, there is constant communication between the computer that controls the milling machine and the robot. It must be ensured that the correct chassis is milled according to the configuration. It also must be ensured that after the front side is finished, the door of the milling machine is opened so that the robot can turn over the part. When this operation is done (and the robot is outside of the machine), the door must be closed again, and the machining of the back side must be started. When this operation is finished, the door must be opened again so that the robot can remove the chassis and present it to the user for a visual quality check. Then the user can decide whether it is a good part or a reject. Afterwards, the resulting data is saved and made visible on the website (figure 8).

**Chassis:**

| | |
|---|---|
| Milling Time Front Side: | 12.96 min |
| Milling Time Back Side: | 2.24 min |
| Cutting Speed: | 7 m/s |
| Feed: | 7 mm/s |
| Quality Check: | passed |

**Figure 8.** Excerpt from the vehicle's website, showing the production data of the milling machine.

The cabin is the next required element. Analogous to the milling machine, communication to the database is established to obtain the necessary information. In this case the cabin size is transmitted, i.e. small or large cabin, the vehicle's name and ID. After the cabin has been injection molded, a QR code and the vehicle's name are written by laser on the cabin's roof. Afterwards, the cabin is dispensed via a conveyor belt (figure 9) and shown on the website (figure 10).

**Figure 9.** Handling robot placing an injection molded cabin on a conveyor belt.

**Cabin:**

| | |
|---|---|
| Injection Pressure: | 750 Pa |
| Residual Cooling Time: | 65 s |
| Cycle Period: | 120 s |
| Quality Check: | passed |

**Figure 10.** Excerpt from the vehicle's website, showing the production data of the injection molding machine.

Finally, the loading area is produced. Standard loading areas are produced in stock and can be found at the assembly site, whereas individual loading areas must be designed first. The designing process can also be started by using the transponder. Again, communication with the database is established to ensure that the designed element will fit and can also be mounted on the corresponding chassis (cut-outs and mounting holes must match). A CAD module is started, where the corresponding standard loading area is loaded. Based on this loading area, individual changes such as length, width and height can be made. Furthermore, the user can add labels and change the shape of the rear lights (figure 11).

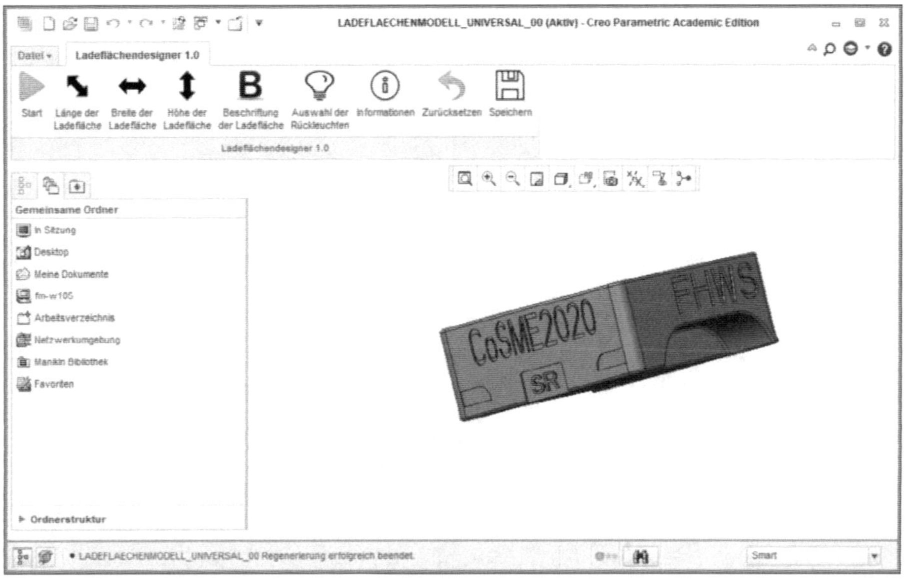


**Figure 11.** CAD module for the individualization of loading areas, the corresponding standard loading area can be seen.

A few modifications have been made for this example. Firstly, the loading area is narrower than a standard one. Figure 12 shows that while the loading area is transformed, the wheel arch hasn't been changed. This is due to the logic that ensures that important elements must not be altered.

Additionally, the shape of the rear lights has been changed, they are round in the final version. Furthermore, a label has been added, namely "Brasov".

After saving this individual loading area, all data (length, width, height, label and shape of the rear lights) was transferred to the database, as they not only serve to complete the data set but are also required later during quality inspection.

Both individual and standard loading areas are 3D printed.

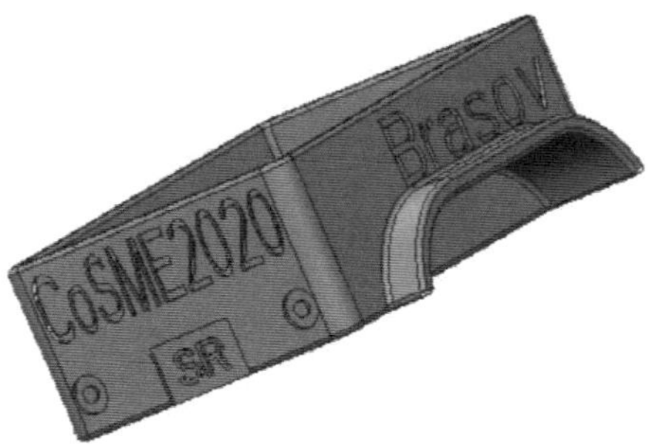


**Figure 12.** The finalised loading area. It is narrower than standard ones, and a label "Brasov" has been added to one side.

**Figure 13.** The designed loading area during 3D printing. An NFC chip was inserted.

However, the printing process is briefly interrupted to allow near-field communication (NFC) chips to be inserted (figure 13).

After the printing process is finished, all production data such as duration, used printer, material, color, time stamp and batch are stored in the cloud and assigned to the NFC chip. Therefore, the production data of each loading area can be tracked as well.

As their production time varies, cabins, chassis and loading areas must be stored, so they are neither mixed up nor lost. Their storage and transport is carried out by robots (figure 14). An autonomous robot brings the designated elements to the storage station and from the storage station to the assembly site.

**Figure 14.** Students observing an autonomous robot transporting parts to the storage location. Photo credit FHWS.

Once again, the transponder is needed for the identification at the system. Based on the ID, the user can store or retrieve a cabin, chassis, loading area or a pick-up truck. Every transaction is handled by a small robot.

This step demonstrates modern production conditions to students and/or visitors. The worker does not have to create each order individually, as these are automatically transmitted to the production unit via the network. Handling robots support humans by taking over dangerous or difficult tasks. The human is not replaced, but supported and assumes the role of the supervisor and decision maker (good part/rejects).

### *3.3. The quality and functional checks*

After all components have been produced and made available, the user can assemble the pick-up truck. Light signals as well as an augmented reality-based assembly instruction help with the completion. The light signals show where purchased parts or the self-produced components are located. The assembly aid is made possible with the Microsoft Hololens, an augmented-reality headset. The user is not only informed where the required parts and tools are located, but every single action is demonstrated (figure 15). The picture shows an extract from the assembly guide. The user is informed by colored rectangles where the axle should be positioned, whereas the small circle indicates where the user is looking.

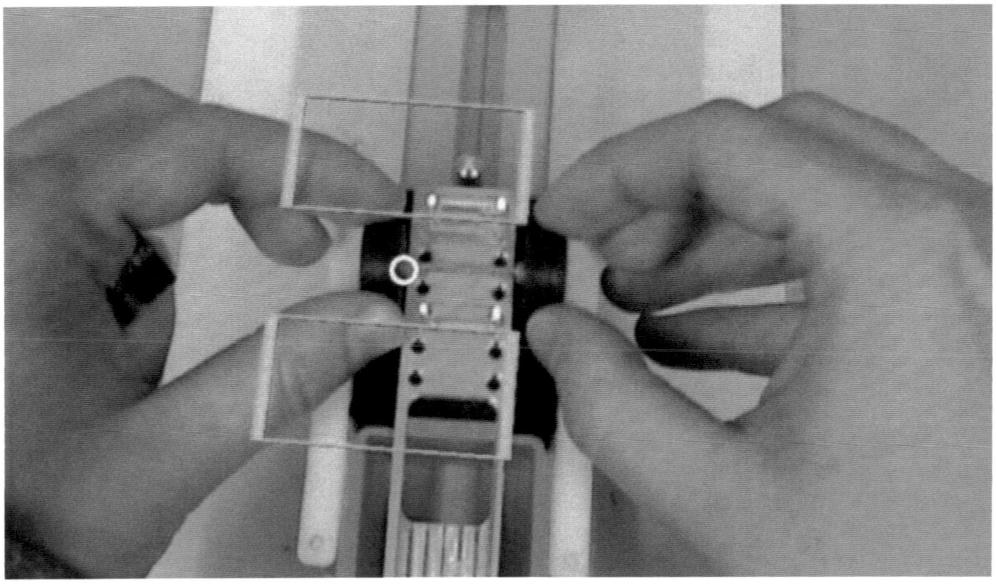

**Figure 15.** Screenshot from a guided assembly instruction with augmented reality. Photo credit: FHWS.

Now the quality inspection can be carried out. This is the first step in the production chain, where the transponder isn't needed, as the machine can interpret the QR code on the cabin's roof.

A database query provides the required information for the inspection, such as the vehicle's dimensions based on the selected elements. This allows the machine to carry out an automated target/actual comparison. As mentioned above, the dimensions of the individual loading area are also transmitted to the database.

Thus, the quality checks of vehicles with individualized loading areas can also be carried out automatically, as the target data is made available. The result of the test is automatically transferred to the cloud including the measured values (figure 16) and the taken images (figure 17).

**Quality Check:**

| | |
|---|---|
| Check Time: | 08-06-2020 17:21:11 Uhr |
| Assembly: | Correct |
| Errors: | None |
| **Measured Values:** | |
| Length Vehicle: | 125.08 mm |
| Length Truck Bed: | 525.70 mm |
| Width Truck Bed: | 36.01 mm |
| Height Truck Bed: | 17.00 mm |
| Length Cabin: | 28.65 mm |
| Number of Axles: | 3 |

**Figure 16.** Excerpt from the website showing the result of the quality check and the measured values.

**Figure 17.** Excerpt from the website, showing the pictures that were taken and used for the quality check.

The last step is a functional test on the inclined plane. The vehicle's speed is measured and a photo is taken. The data, consisting of measured values (figure 18) and image taken during the test drive (figure 19), are stored in the cloud and can be accessed via the vehicle's own website. As already mentioned, both testing machines are able to interpret the printed QR code in order to generate queries to the database, which return the target data. Consequently, the test results are also saved and made available in the cloud.

**Functional Check:**

| | |
|---|---|
| Check Time: | 08-06-2020 17:20:54 Uhr |
| Check: | OK |

**Test Values:**

| | |
|---|---|
| Average Speed: | 0.74 m/s |
| Maximum Speed: | 0.83 m/s |

**Figure 18.** Excerpt from the website showing the result of the functional check.

**Figure 19.** Picture of the vehicle, taken during the functional check.

This step demonstrates modern testing methods, which enable several quality feature checks within a short period of time, as they are suitable for both serial and flexible production. Students are taught how important coding skills have become nowadays. This simple example shows the following: a QR code is read in and a database query is generated. In return, the database must provide the corresponding target data to the test machine. The machine records the actual data and compares it to the provided target data. Then, a decision must be made on good part or reject, according to preset tolerance ranges. And this must take place within few seconds. In order to do so, the machine must be programmed to carry out all of these steps.

### *3.4. Monitoring the production*

All production progress can be tracked on the vehicle's own website, as the machines report their status back to the cloud/database. In addition to the vehicle's configuration data, all production data that has been generated during the manufacturing of the cabin, chassis and loading area are stored and displayed. Likewise, not only the results of the quality and function tests are displayed, but all created images as well. The user can not only read the facts but also see how the checks have been performed based on the provided images.

This step demonstrates a transparent production. The user can remotely access the production data and status of the self-created and configured vehicle at any time. Furthermore, the user can also follow the production progress and also review the generated data. Thus, the vehicle does not only transport a storyline, but also a small piece of traceable and understandable Industry 4.0.

## 4. Conclusion

The c-factory uses the production of a model car to better illustrate Industry 4.0 functionalities. How were these implemented? About networking: all machines are connected to the cloud and there is a bidirectional data flow, as data can be both written and read. The Internet of Things was realized in the vehicle's own website. Process information and progress can be viewed by accessing the corresponding website. Since the model car can be customized in many ways, a flexible production is guaranteed. The loading area of the vehicle is 3D printed and thus the additive manufacturing was implemented. The production is supported by different handling robots, they provide both safety (risk of injury) and assistance by transporting and storing the components. At the assembly station, the user is guided with the help of augmented reality glasses so that the car can be assembled without difficulty. All process and production data are collected and evaluated, yet given the relatively small amount of data, one cannot quite speak of Big Data.

The c-factory is not only reserved for students and lecturers, but also gives visitors from industry and trade the opportunity to reflect in workshops on the necessary skills for professional activities in the field of Industry 4.0.

Students were responsible for the technical implementation of the c-factory, from the creation of the digital twin to the commissioning of the machines and writing the necessary software. The c-factory is constantly being further developed in student projects across faculties in order to add more functionalities, such as business processes. In doing so, the c-factory acts as an inclusive knowledge incubator for both lecturers and students, for supervisors and tutors. The lecturers take on the role of coaches and by expanding the project, both sides benefit from the added content and their expanded knowledge horizon.

The c-factory's didactic concept won the prize "Best Machine House 2019" of The Mechanical Engineering Industry Association (VDMA). The jury was impressed by the teacher's willingness to take on a completely new didactic approach and thus question the previous teaching, while the students work in a realistic industrial environment to physically and digitally map the lifecycle of a technical product [3].

Furthermore, the c-factory was selected as a development and application environment for small and medium-sized companies (SMEs) by the so called "I4KMU" project, which is founded by the German Federal Ministry of Education and Research (BMBF). SMEs face challenges in the implementation of new concepts and ideas in the field of Industry 4.0, as they lack the structure of larger companies. They do not have the test environment and not enough personnel for the extensive application process to obtain funding. Thus, the c-factory has successfully achieved the classification as an Industry 4.0 test environment and can provide professional support to SMEs [4].

**References**

[1] Chaudhuri A 2019 *Internet of Things, for Things, and by Things* (New York: Auerbach Publications) p 5

[2] Becker M 2019 *Standards und Schnittstellen als Arbeitsgegenstand in Industrie 4.0 Kontexten Industrie 4.0 Risiken und Chancen für die Berufsbildung* (Berufsbildung, Arbeit und Innovation vol 5 ed Spöttl G and Windelband L (Bielefeld: wbv Publikation) p 109

[3] Zeitler M 2019 VDMA: *Würzburg-Schweinfurt University of Applied Sciences is "Best Machine House 2019",* The Mechanical Engineering Association, last accessed 31st of July 2020, https://www.vdma.org/en/v2viewer/-/v2article/render/43586571

[4] I4KMU, 12.12.2018,*FHWS – Entwicklungs- und Anwendungsumgebungen I4.0 // "c-factory und Starterkits I4.0*, last accessed 31st of July 2020, https://i4kmu.de/fhws-entwicklungs-und-anwendungsumgebungen-i4-0-c-factory-und-starterkits-i4-0/

# Multi-operative modular tool for cutting and mechanical hardening

**D M Rosca[1] and M Folea[1*]**

[1]Department of Manufacturing Engineering, Transilvania University of Brasov, 29 Eroilor, Brasov 500036, Romania

*E-mail: m.folea@unitbv.ro

**Abstract.** It is well known the important role of the tools for mechanical processing in manufacturing within the present context of the rush for profitability and technical optimization for an environmentally friendly industrial production. Modularization of tools is a way to ensure their versatility being at the same time economically advantageous. In this paper, two milling cutters are adapted in order to be used also for surface hardening. The third tool described here is a newly designed modular tool that, according to needs, can be used to perform rough, finish milling or mechanical hardening.

## 1. Introduction

The processing of plane/flat surfaces by various technological processes is very common in the manufacturing industry. It stands out particularly the milling process through which can be generated surfaces in a wide dimensional range.

Due to the permanent need to ensure the competitiveness of manufactured components, it is continuously necessary to design and manufacture more efficient technological equipment able to reduce the processing time without affecting the quality of the products. In this respect, a great interest was demonstrated in recent years for reconfigurable manufacturing systems (RMS) easily adaptable to frequent requirement changes [1–7]. Convertibility, defined as "the capability of transforming the functionality of existing systems and machines to fit new production requirements" is one of the core characteristics of RMS [3] along with scalability, diagnosability, customization, modularity and integrability. For Wiendahl et al. [6] the essential features of RMS are modularity and integrability. Integrating multiple processing technologies in a single machine tool could lead to higher productivity with lower costs. Brecher et al. [7] emphasise the importance of reducing process complexity by performing different operations in one workplace and, if possible by a single clamping of the tool. Designing multipurpose tools as are those presented by the authors ensures convertibility, integrability and modularity of milling and mechanical hardening tools.

Three cases of constructive flexibility of the active part of milling cutters are described in this paper. The design of the first tool, a Romascon type milling cutter, is modified to improve milling performances and to allow an easy transformation in a press hardening tool. The second tool is a commercially available robust face milling cutter for which it is provided a solution that allows reconfiguration into a hardening tool. The third solution presented is a new concept for a tool with two subassemblies that can be interchanged with little effort, one subassembly for rough milling and the other one for press hardening.

## 2. Reconfiguration of existing milling cutters

### *2.1. Redesign of Romascon milling cutter*

The newly designed tool is a front milling cutter (Romascon type) provided with blade-shaped teeth as shown in figure 1. The Romascon milling cutters were patented in Romania more than half a century ago by professor Vitalie Belous and are still used and appreciated for being robust, easy to manufacture and having brazed carbide inserts suitable for re-sharpening [8, 9].

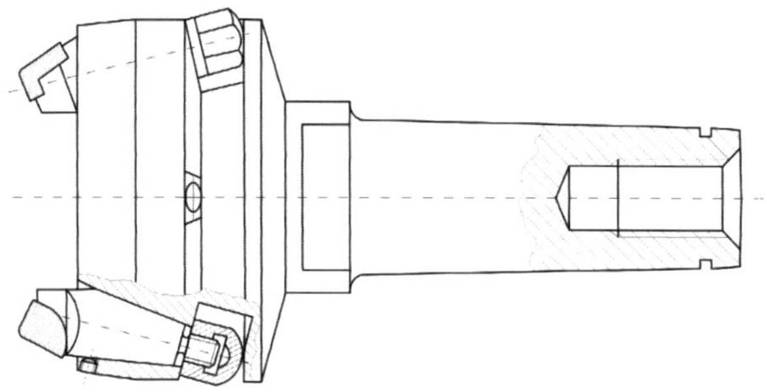

**Figure 1.** General appearance of Romascon milling cutter.

In the initial phase of operation, the tips of all teeth are arranged in the same frontal plane (AB), on the diameter $D_1$ (figure 2). To use a general purpose milling cutter for removing a thick layer of material in a single pass, working with higher values of the cutting depth $a_p$, a new design is proposed as follows.

A milling cutter is generally considered a tool with an even number of teeth/blades. Half of them will be replaced with longer blades, their tips (in the mounted state) reaching the level of point C. It is important that the replacement must be done alternatively: a tooth remains with the tip in the initial position (so in the frontal plane AB) and the tooth immediately following will have the tip at the level of the CD points (figure 2). As a result of this repositioning, it results that the second milling plane generated by the longer teeth/blades will appear; their tips will be located on a milling diameter $D_2 < D_1$. This new active area will allow the removal of a layer of BD thickness in the workpiece material. A total cutting depth will be obtained on the entire active area of this constructive configuration:

$$a_{p\ total} = a_{p1} + a_{p2} \text{ (mm)} \tag{1}$$

where $a_{p1}$ – axial cutting depth related to milling with the initial teeth positioning of the cutter;

$a_{p2}$ – axial cutting depth corresponding to milling with longer teeth ($a_{p2}$ = BD).

The cutting process will take place in better conditions if the cutting depth $a_{p1}$ is equal to approximately two-thirds of the total cutting depth is adopted.

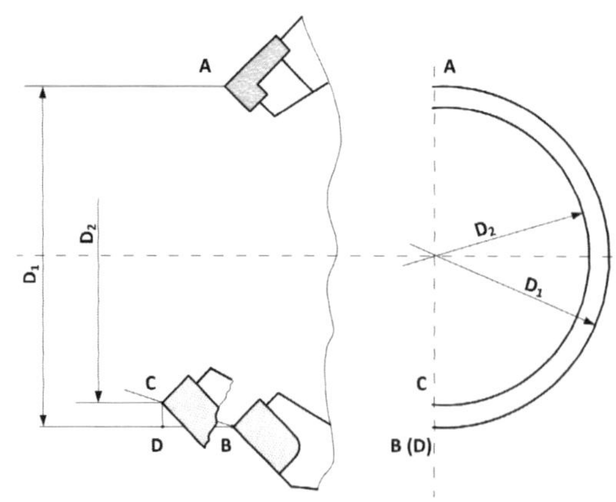


**Figure 2.** Romascon milling cutter reconfiguration.

Through constructive flexibility, several advantages are obtained, such as:

- Chip fragmentation on width, therefore better conditions for the formation and removal of chips;
- Reduced specific cutting force due to the fact that chip thickness is doubled in comparison with the one related to the work with the unmodified milling cutter;
- Lower values of torque, so lower specific energy consumption.

*2.2. Reconfiguration of face milling cutter*

Increasing the degree of use by modularizing the active part of the tools can be applied to other milling cutters, e.g. a face milling cutter with each tooth/blade fixed with two screws (figure 3).

The modularization of this tool consists of completing its design with a prismatic mark of thickness *a* having the shape of a plate (pos.1). This modification can only be done on a face mill with an even number of teeth because the new mark is to be placed every two teeth as in Romascon type milling cutter. In consequence, half of the total number of teeth of the milling cutter, those radially repositioned, will have the tips located on a diameter $D_1 > D_2$, being the first to come into contact with the workpiece, then after the distance *a* (in the feed direction), the other teeth enter the cutting. Those teeth, located on the diameter $D_2$ must have the tips at a distance *b* on the axial direction from those of the teeth arranged on the diameter $D_1$. From the point of view of the cutting parameters, *b* has the significance of the cutting depth for the teeth with the tips on the diameter $D_2$.

Besides the advantages of modularization, this modified milling cutter (figure 3) with large diameter and numerous teeth has other benefits: more uniform cutting forces and more favourable cutting conditions and a better surface roughness.

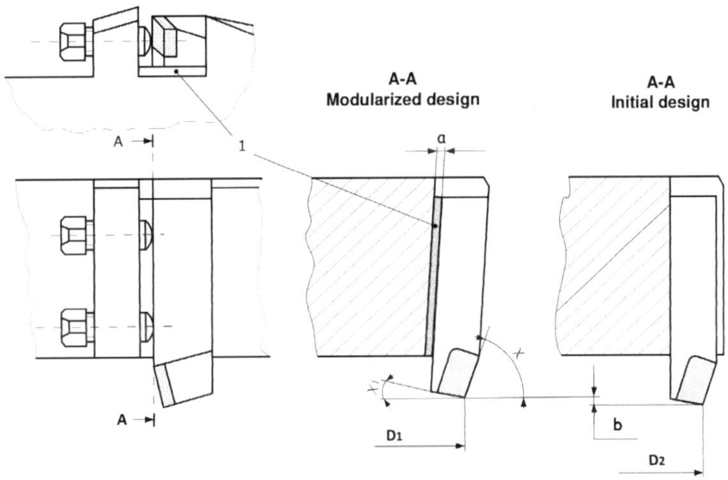


**Figure 3**. Modularization of face milling cutter.

Another major advantage of the previously presented milling cutter is that it can be reconfigured to a press hardening tool. This can be obtained if the repositioned teeth are replaced with a subassembly that contains a constructive element (e.g. a ball) that allows the pressing of the flat surface of the workpiece. It will be therefore a specific subset of the finishing process by pressing [10] designed to increase surface quality parameters: surface roughness, hardness/microhardness of the surface layer, dimensional accuracy, etc. Also, by applying plastic deformation at the surface, wear resistance and corrosion resistance of the workpiece are ameliorated.

*2.3. Multi-operative original tool design*

In this section is presented an original design of a combined tool for cutting and press hardening.

According to figure 4, this equipment is made of two different parts with distinct functions: one part for milling (pos.1) and the other part for press hardening (pos.2).

In the central zone, there is a cylindrical bore with a diameter *d* similar to that of commercially available face milling cutters, so clamping the combined tool on a traditional machine tool or on a numerically controlled one is possible using a standard tool holder.

Milling is performed with the two inserts 3 which are mechanically fixed in the milling cutter 1, and the plastic pressing/deformation occurs under the action of a spherical deforming part (ball) 4. As can be observed in figure 4, the initial positioning of the support-piece 5 is done with the spacer 6 and the fine adjustment of the *z* dimension is carried out using the pressing subassembly 2. The optimal value of positional dimension *z* is influenced by many factors like surface roughness before press hardening, shape and size of the deforming part, press hardening process parameters, physical and mechanical properties of the workpiece, etc.

In literature [11], the following relation has been established for the calculus of *z* taking into account surface roughness parameter $R_z$:

$$z = 1.35\left(R_z^{initial} - R_z^{final}\right) * 10^{-3}\,(\text{mm}) \tag{2}$$

where $R_z^{initial}$ and $R_z^{final}$ are values of measured $R_z$ before and, respectively, desired $R_z$ after press hardening.

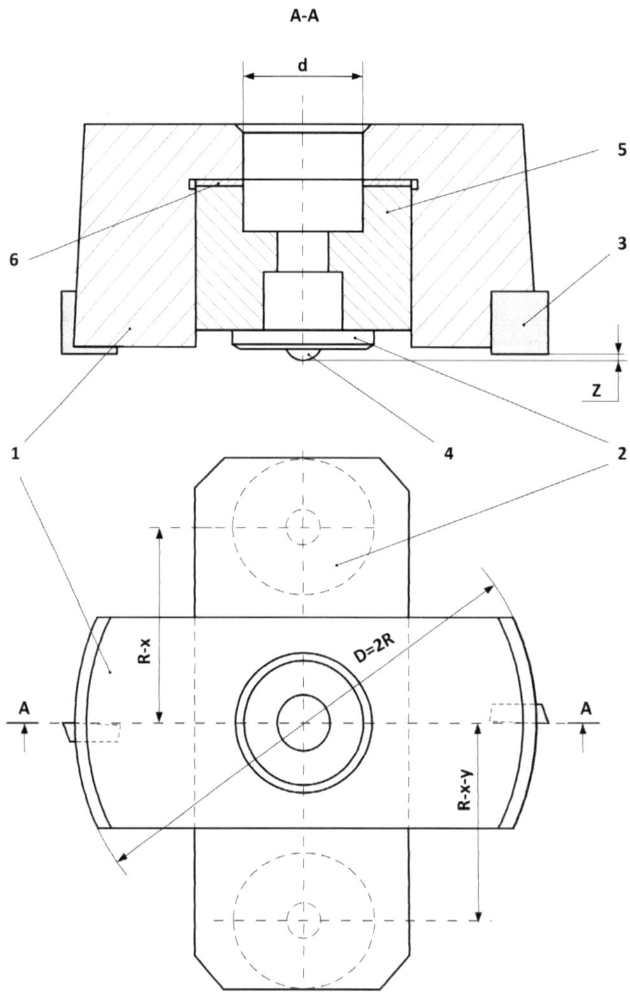


**Figure 4.** Modular milling and press hardening tool: milling cutter (1), pressing subassembly (2), inserts (3), ball (4), support-piece (5) and spacer (6).

For the purpose of obtaining very smooth surfaces with superior accuracy, the two deforming subassemblies were positioned differently: one subassembly is at a distance R-x, in relation to the axis of symmetry of the combined tool and the other at a smaller distance, respectively R-x-y. Therefore a first pressing action will be performed by the more distant subassembly (R-x), and then a second pressing action will be performed by the ball of the subassembly which is closer to the axis of symmetry of the combined tool.

## 3. Conclusion

The authors provided three solutions for modularity and convertibility of milling cutters that could be used on a regular milling machine. Two of them are adaptations of existing tools increasing their performance and making them easily convertible into press hardening tools. The third multipurpose tool is an original concept that in the future is intended to be manufactured and tested.

By implementing in the processes of mechanical processing the technological equipment proposed in this paper, technical-economic effects superior to those based on current technologies could be obtained.

## References

[1] Schmid A, Löwer M, Katzwinkel T, Schmidt W, Siebrecht J, Feldhusen J, Brecher C, Bois-Reymond F du and Rey J 2016 Life Cycle of Multi Technology Machine Tools – Modularization and Integral Design *Procedia CIRP* **50** p 719

[2] Schmid A, Katzwinkel T, Schmidt W, Siebrecht J and Löwer M 2016 *Reconfigurability and modularization for integrated machine tools based on functional analysis: a systematic approach International Design Conference - Design 2016* (Dubrovnik, Croatia) p 367

[3] Koren Y, Gu X and Guo W 2018 Reconfigurable manufacturing systems: Principles, design, and future trends *Front. Mech. Eng.* **13** 121

[4] Singh A, Gupta S, Asjad M and Gupta P 2017 Reconfigurable manufacturing systems: journey and the road ahead *Int. J. Syst. Assur. Eng. Manag.* **8** 1849

[5] Andersen A-L, Brunoe T D, Nielsen K and Rösiö C 2017 Towards a generic design method for reconfigurable manufacturing systems: Analysis and synthesis of current design methods and evaluation of supportive tools *J. Manuf. Syst.* **42** 179

[6] Wiendahl H-P, ElMaraghy H A, Nyhuis P, Zäh M F, Wiendahl H-H, Duffie N and Brieke M 2007 Changeable Manufacturing - Classification, Design and Operation *CIRP Ann.* **56** 783

[7] Brecher C, du Bois-Reymond F, Nittinger J, Breitbach T, Do-Khac D, Fey M and Schmidt S 2016 Qualifying multi-technology machine tools for complex machining processes *CIRP J. Manuf. Sci. Technol.* **13** 1

[8] Belous V 1980 *Sinteza sculelor aschietoare (Synthesis of cutting tools)* (Iasi, Romania: Editura Junimea)

[9] Urdea G-B, Florea V-A and Dumitrescu I 2019 The study of the milling heads ROMASCON by 3D modeling *Ann. Univ. Petroşani Mech. Eng.* **21** 105

[10] Buzatu C, Nedelcu A, Piukovici I and Badea L 1998 *Prelucrari de netezire a suprafetelor in constructia de masini (Surface finishing in manufacturing)* (Brasov, Romania: Editura Lux Libris)

[11] Mares G and Rosca D M 2002 *Tehnologii de suprafatare (Surfacing technologies)* (Brasov, Romania: Universitatea Transilvania)

# Analysis of scheduling activities in the processes of improving surface characteristics of parts

**M M Roşu[1], L Vlădăreanu[2] and M Iliescu[2]***

[1] University Politehnica of Bucharest, Department of Manufacturing Engineering, Splaiul Independentei, 313, sector 6, Bucharest, Romania
[2] Institute of Solid Mechanics, Romanian Academy, Constantin Mille 15, Bucharest, Romania

*E-mail: mihaiela.iliescu@imsar.ro

**Abstract**. In the framework of planning and scheduling process for improving surface characteristics the laser cladding process is one of the best solutions, due to the fact that it involves deposition of high-quality material layers by laser radiation. This process is efficient whenever there is the need of improving the surface characteristics of metallic parts, component of products that usually require high cost maintenance because of their severe wear while functioning. This paper presents a time- efficiency analysis of two types of surface properties improving techniques. One is laser cladding and the other is thermal spraying and there are envisaged the work in process time and the availability of capacity in each of the required manufacturing systems. In addition to the efficiency obtained by scheduling activities, the laser cladding process with nano powders ensures a reduction in the use of expensive materials, about 80% compared to traditionally manufacturing, higher productivity and lower energy consumption, about 20% reduction, compared to metallizing. So, compared to various existing solutions, laser nano-cladding technologies are versatile and applicable to various industrial sectors as energy, automotive, mining, food, etc.

## 1. Introduction

In order to reduce the cost of mechanical components with high surface properties, such as corrosion resistance, wear resistance and hardness, there is the issue of optimizing the production process through optimal planning and programming of production capacities [1] but also by using surface treatment advanced technologies to improve surface properties of metal components, to the detriment of the use of special alloys that are very expensive.

One of the efficient techniques for improving the surface properties of materials is laser processing, which by using the laser radiation modifies the structure and surface characteristics of materials. Laser surface treatments can be divided into direct processes, which need only the heat generated for hardening and melting (annealing, melting, solidification and hardening of the material in well-defined areas of the part) and thermo-chemical processes that need addition material for alloying and deposition [2, 3].

The objective of the research presented by this paper is to analyze the performance of the laser cladding process, compared to other classical process - thermal spraying, for improving the surfaces properties of metal parts (with intensive wearing in exploitation), in the context of optimization of production capacity. From a technological point of view, the analysis of researches conducted by specialists in the field, the laser cladding process enables a local deposition, on well-defined areas, of different types of materials (alloy steel powders, ceramic powders) and thus ensures improved mechanical characteristics (compared with the base material/substrate) such as: wear resistance, lubrication properties, corrosion resistance, etc.

## 2. Characteristics of the studied deposition techniques

One of the most important characteristics/properties of LC (Laser Cladding), according to the specialized literature, is the *wear resistance*.

In [4, 5] is presented an example of the use of the laser cladding process for the repairing and/or maintenance of parts, such as disc and wheels in the railway industry.

In [6] are presented aspects of determining the wear resistance of layers of Ni alloys, obtained from powders deposited by LC. The abrasion tests were performed with a specific equipment, according to the Brinell-Haworth scheme, where silicate granules with a particle size of 200-600 μm were introduced in the friction zone.

Another example of LC application is the laser deposition for laser and metal powder reconditioning of punches used in Abkant type bending presses [7]. The application in this case was made on a robotic cell, digitally controlled by a controller, and the laser beam is transmitted via fiber optics. The powder was distributed with a precise dosing device, with the possibility of drying and preheating the filler material.

Another important characteristic of the LC is the *hardness*. The determination of the hardness characteristics, for different types of high chromium alloy steel powders, deposited on 316L steel substrate, by the laser cladding process, with coaxial spray head, is presented in [8]. Also, in [7] is highlighted the way to determine the microhardness of the layers deposited by laser cladding, LC. The mechanical characteristics of the deposited layers were determined by analysing the microhardness in the fusion zone and in the heat affected zone.

According to [9], thermal spraying is a generic term for a group of metallic and non-metallic layers deposition processes. There are two basic types of thermal spray processes. The first type, called the "combustion process", is based on burning a mixture of hydrocarbons (acetylene, propane, kerosene) and oxygen (or air) to create the enthalpy and kinetic energy needed to plasticize and propel particles to the surface. In the second process type, the electric power (electric arc or plasma) is the energy source needed to spray the particles.

The essential feature of the layers deposited by thermal spraying is their wide applicability due, on the one hand, to the composition of the deposited material, which is almost unlimited, and on the other hand, to the temperature of the parts, which is kept low during the spraying process 100÷260°C, which considerably reduces the danger of deformation, oxidation and phase transformations.

## 3. Research methodology

In any production process for obtaining parts with high characteristics, optimization results of the planning and scheduling working flow activities of each of the processes refer to decreases of production costs and of the costs related to the exploitation life of the product, by high corrosion resistance, wear resistance and hardness. Thus, the purpose of this paper is to study the way of optimizing the production process by using adequate technologies to improve the surface characteristics of parts, the results pointing toward application of high performance techniques like laser cladding, over the classical one, like thermal spraying (metallizing).

In table 1 are presented the process parameters, according [10], for two type of surface improving techniques: thermal spraying and laser cladding.

In order to establish the performance of laser cladding process, there will be analysed the process flow of both type of techniques (laser cladding and thermal spraying), only for a type of part. In table 2 are presented the flow process and time for work in process (WIP) for both types of the analysed techniques.

**Table 1.** Process parameters [10].

| Type of process | Thermal Spray | Laser Cladding |
|---|---|---|
| Heat source | Electric arc | Laser beam |
| Coating thickness | 0.5 -5 mm | 0.2 -2 mm |
| Typical deposition rate | ≤10 kg/hour | ≤5 kg/hour |
| Dilution | 5 – 15% | ≤5% |
| Type of boding | Metallurgical | Metallurgical |
| Bond strength | ≤800 MPa | ≤800 MPa |
| Heat input | High | Low - medium |
| Porosity | ≤0.1% | ≤0.1% |
| Comparative capital cost | Low | High |
| Comparative running cost | Medium - low | Low |

**Table 2.** Time for work in process and process flow.

| Thermal Spraying | | | | Laser cladding | | | |
|---|---|---|---|---|---|---|---|
| Process flow of thermal spaying | Processing Equipment | Machine code | WIP [min] | Process flow of cladding process | Processing Equipment | Machine code | WIP [min] |
| Surface cleaning (degreasing) | Manual / special enceinte | M1 | 3 | Surface cleaning (degreasing) | Manual/ special enceinte | M1 | 3 |
| Surface machining by turning | CNC center | M2 | 1.5 | Cladding | MetcoClad system | M5 | 1.2 |
| Surface roughening – by abrasive jet / metco method | Grit blasting equipment | M3 | 2 | Post processing – seldom required- grinding | CNC center | M2 | 2 |
| Metaling Spraying | METCO – Oerlikon system | M4 | 1.5 | | | | |
| Post processing by turning | CNC center | M2 | 2.5 | | | | |

**Table 3.** Time for work in process and process flow for the Thermal Spraying technique.

| Sequences activities, time [min] | Surface cleaning (degreasing) | Surface machining by turning | Surface roughening – by abrasive jet/METCO method | Metaling Spraying | Post processing by turning | Total time per each step of WIP [min] | Forecast production rate |
|---|---|---|---|---|---|---|---|
| Working station | M1 | M2 | M3 | M4 | M2 | | |
| Setup time | 0.3 | 0.15 | 0.2 | 0.15 | 0.25 | 1.05 | 0.5 |
| Process time | 2.4 | 1.2 | 1.6 | 1.2 | 2 | 8.4 | |
| Cleaning and removing time | 0.3 | 0.15 | 0.2 | 0.15 | 0.25 | 1.05 | |
| Total | 3 | 1.5 | 2 | 1.5 | 2.5 | 10.5 | |

The time for each sequenced WIP is share per each step of WIP. The main processing steps for both studied techniques are: setup time, process time and cleaning/removing time. In tables 3 and 4 are presented the time per each step and calculated total time per each step and each working station.

**Table 4.** Time for work in process and process flow for the Laser Cladding technique.

| Sequences activities, time [min] | Surface cleaning (degreasing) | Cladding | Post processing – seldom required-grinding | Total time per each step of WIP[min] | Forecast production rate |
|---|---|---|---|---|---|
| Working station | M1 | M5 | M2 | | |
| Setup time | 0.3 | 0.1 | 0.4 | 0.8 | 0.5 |
| Process time | 2.4 | 1 | 2.4 | 5.8 | |
| Cleaning and removing time | 0.3 | 0.1 | 0.8 | 1.2 | |
| Total | 3 | 1.2 | 2 | 6.2 | |

For a rate of product planning in process by 0.5, the used capacity is presented in table 5, for each analysed process. The used capacity is calculated reporting the total planned time of each working station (resource) calculate considering the forecast of production rate at total available capacity of resources.

The used capacity for analysed case is presented in figure 1. If is necessary to do the planning of production process for different kind of parts, the used capacity will be higher and, in this case, it is necessary to do other analyses regarding the allocation of resources.

**Table 5.** Used capacity of resources for one type of part.

| Working station | Production capacity | Thermal spraying | Capacity using for thermal spraying process | Cladding Process | Capacity using for cladding process | Working hours per day | Forecast production rate |
|---|---|---|---|---|---|---|---|
| M1 | 1 | 1.5 | 18.75% | 1.5 | 18.75% | 8 | 0.5 |
| M2 | 1 | 2 | 25.00% | 1 | 12.50% | | |
| M3 | 1 | 1 | 12.50% | - | - | | |
| M4 | 1 | 0.75 | 9.38% | - | - | | |
| M5 | 1 | - | - | 0.6 | 7.50% | | |
| Operator | 1 | 1.05 | 13.13% | 1 | 15.50% | | |

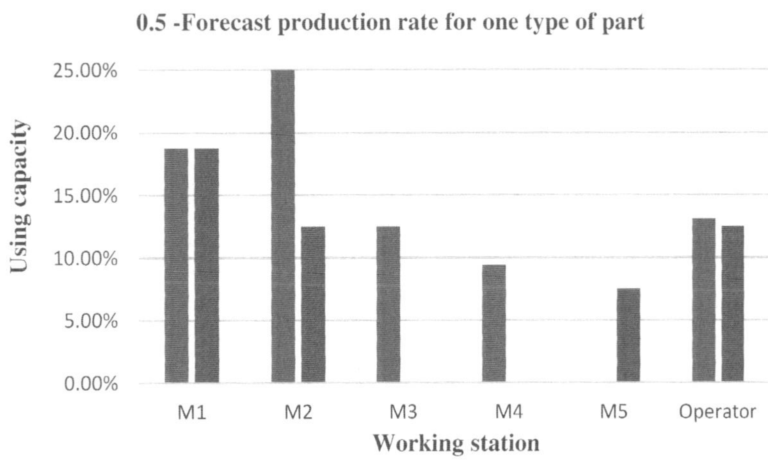


**Figure 1.** Used capacity.

Doing the simulation for different production rate forecast and for different type of parts, resulted in the values presented in table 6.

**Table 6.** Used capacity of resources for different type of parts.

| Working station | Total time thermal spraying | Part 1 0.5 | Part 2 0.35 | Part 3 0.3 | Part 4 0.25 | Part 5 0.43 | Part 6 0.47 | Total using capacity |
|---|---|---|---|---|---|---|---|---|
| M1 | 3 | 18.75% | 13.13% | 11.25% | 9.38% | 16.13% | 17.63% | 86.25% |
| M2 | 4 | 25.00% | 17.50% | 15.00% | 12.50% | 21.50% | 23.50% | **115.00%** |
| M3 | 2 | 12.50% | 8.75% | 7.50% | 6.25% | 10.75% | 11.75% | 57.50% |
| M4 | 1.5 | 9.38% | 6.56% | 5.63% | 4.69% | 8.06% | 8.81% | 43.13% |
| Operator | 2.1 | 13.13% | 9.19% | 7.88% | 6.56% | 11.29% | 12.34% | 60.38% |
| Working station | Total time laser cladding | Part 1 0.5 | Part 2 0.35 | Part 3 0.3 | Part 4 0.25 | Part 5 0.43 | Part 6 0.47 | Total using capacity |
| M1 | 3 | 18.75% | 13.13% | 11.25% | 9.38% | 16.13% | 17.63% | 86.25% |
| M2 | 2 | 12.50% | 8.75% | 7.50% | 6.25% | 10.75% | 11.75% | 57.50% |
| M5 | 1.2 | 7.50% | 5.25% | 4.50% | 3.75% | 6.45% | 7.05% | 34.50% |
| Operator | 2 | 12.50% | 8.75% | 7.50% | 6.25% | 10.75% | 11.75% | 57.50% |

As is shown in table 6, the used capacity for working station M2 is overall located; in this case it is necessary to have available two resources for thermal spaying process. This scenario involves additional cost with fixed asset, and additional space in production workshop.

## 4. Results

In the framework of these complex processes for improving surface characteristics, the process specific parameters (power, speed, quantity of powders etc.) must to be correlated with planning and scheduling parameters for obtaining quality mechanical components at the client ordering.

As can be seen in the simulation performed to determine the capacity utilization in for two studied techniques, the LC ensures a lower capacity utilization comparing with the case of thermal spaying, which leads to the conclusion that in terms of planning and scheduling the production process, the laser cladding process is more efficient, ensuring shorter processing times and less production capacity utilization. Also, considering the process parameters, the LC process is characterized by a low heating/cooling time of molten metal baths.

## 5. Conclusion

The new laser cladding technologies will ensure a reduction in the use of expensive materials (additional material is deposited accurately where is necessary) higher productivity and lower energy consumption (about 20% reduction, compared to metallizing).

There can be used a very wide range of materials and additional material completely merge with the base material. The material is not excessively heated, which reduces the distortion of the substrate that would require further corrective processing by machining, the process can be fully automated and integrated into production lines.

Compared to existing solutions, LC technologies are versatile and applicable to several industrial sectors (energy, automotive, mining, food, etc.) also, these improve the durability and mechanical and chemical resistance of industrial parts (due to special nano-powders mixes).

One disadvantage of the LC technique could stand in the relatively large initial investment that must be made to purchase the equipment and training of the staff.

**References**

[1] Rosu M M 2009 *Contributions to the development of operational management models of production projects* (PhD Thesis, University Politehnica of Bucharest)
[2] Schuocker D 1998 *Handbook of the Eurolaser Academy* (Ed. Chapman-Hall)
[3] Steen W M 2003 Laser Material processing *Springer*
[4] Lewis S R, Lewis R and Fletcher D I 2015 Assessment of Laser Cladding as an Option for Repairing/ Enhancing Rails *WEAR Journal* **330** 581
[5] Niederhause S 2006 *Laser Cladded Steel Microstructures and Mechanical Properties Relevance for Railway Applications* (Doctoral Thesis, Chalmers University of Technology, Goteborg, Sweden )
[6] Birukov V P, Tatarkin D Yu, Chriptovish E V and Fichkov A A 2017 Influence of the modes Of laser cladding on bond strength and wear resistance of coatings *IOP Conf. Series*: *Jour.of Phys.* **941** 012035
[7] Staicu A R 2013 *Reconditioning of press brake tools by laser cladding* (PhD Thesis, Transilvania University of Brasov)
[8] Abouda E, Dal M, Aubry P, Tarfa T N, DemirciI, Gorny C and Malot T 2016 Effect of Laser Cladding Parameters on the Microstructure and Properties of High Chromium Hardfacing Alloys *Phys. Procedia* **83** 684
[9] Sahoo P 1993 High-Performance Wear Coatings - the Quest Continues *Powder Metall. Int.* **2** 73
[10] https://www.metallisation.com/applications/metallisation-launches-new-laser-cladding-system

IOP Conf. Series: Materials Science and Engineering 1009 (2021) 012054 doi:10.1088/1757-899X/1009/1/012054

# Comparison between aluminum alloys behavior in incremental sheet metal forming process of frustum pyramid shaped parts

**G P Rusu [1]*, A Bârsan[1]*, M O Popp[1] and A Maroşan[1]**

[1]Lucian Blaga University of Sibiu, Department of Industrial Machinery and Equipment, Emil Cioran Street No. 4, Romania

*E-mail: gabriela.rusu@ulbsibiu.ro; alexandru.barsan@ulbsibiu.ro

**Abstract**. Single point incremental forming is a new emerging technology, used mainly for prototype development or small series production, with good prospects in industrial implementation. However, the parts manufactured through this process often presents low accuracy, thus the industrial implementation is yet limited. Incremental forming process can be done with the help of different technological equipment, such as industrial robots and numerically controlled machines. Since the process is still in development phase, it is a good practice to use finite element analysis software packages in order to reduce time and costs to produce accurate parts. The aim of this paper is to investigate the behavior of different aluminum alloys deformed through SPIF regarding forming forces, thickness distribution and springback effect. Based on numerical simulations performed, the material which produces the most accurate parts for the frustum cone can be chosen for further experimental research.

## 1. Introduction

The usual sheet metal forming processes involve high costs due to the high complexity of the molds and the equipment used to carry out the processes, as in the case of deep drawing and stamping of the sheets. Due to these costs, their use in small series and prototype production is avoided and cheaper solutions are sought to produce these parts. In order to produce these prototype parts or small batches, researchers have been searching for new methods to produce them, and one cheap way to deform the sheet metal is known as incremental forming.

This innovative manufacturing process is based on a hemispherical punch which follows a well-defined trajectory by pressing on the sheet and thus deforming it in the desired shape. Figure 1 shows the principle of incremental forming process, which involves apart from the sheet metal and the forming tool, a fixing element divided in two parts, a clamping plate, and a clamping ring. Some researchers also make use of a supporting element, but this is not mandatory. This process appears for the first time in the scientific literature in 1967 [1].

This process does not require a dedicated die nor a special tool, thus reducing the costs by using cheap elements in comparison with other processes used in automotive industry, but sometimes it can take a large amount of time to produce only one part. In addition, prototype parts can be produced also for aerospace and medical industries, in which an increase of the number of parts has been recognized in recent years [2]. The implementation of this process at a large scale can be done with the help of numerically controlled machines or industrial robots [3]. Industrial robots used in this process must perform a continuous movement by following the desired toolpath. For this, CAM software packages dedicated to milling process adapted for robots with serial structure are used, a fact that must be taken into account when choosing the right robot, as well as the right dimension of the robot in order to withstand the loads and forces that appear during the forming of the material [4]. Khan et. al. used CATIA software in paper [5] to generate a toolpath for a frustrum cone shaped part. Nasulea used a special CAM software called TMPG to generate frustrum pyramid shape [6].

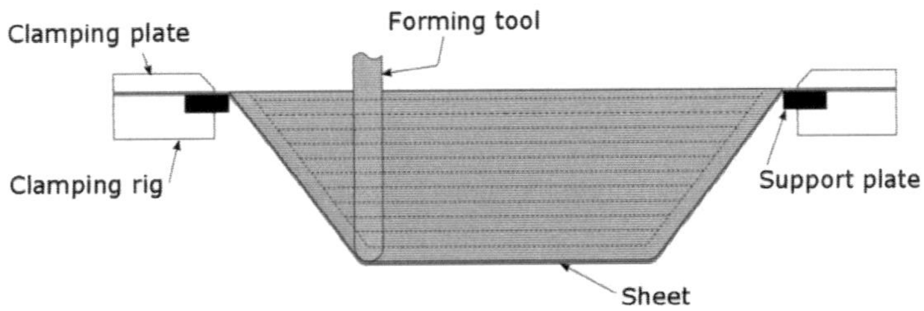


**Figure 1.** Single point incremental forming process principle [7].

The process being still in the development phase, several researches are carried out in order to improve it and one of the most important aspects that concerns the researchers in this field is the dimensional and shape accuracy of the parts. Based on the literature study on the precision obtained of the parts produced by SPIF, there is no clear distinction between dimensional and shape accuracy, but some researchers try to evaluate the degree of process implementation in industry through different methods (such as Analytic Hierarchy Process known as AHP) [8]. In many cases, the profile of the finished part is analyzed, which can be obtained by sectioning it with a plane containing one of the symmetry axis. After reconstruction of the real profile of the part, this is compared with the theoretical profile (CAD model) and thus some drawbacks can be evaluated and quantified such as the spring back effect and thinning of the material in different areas.

Research in this direction was carried out by Ambrogio et al. who in the paper [9] compared the profile obtained in the analysis through finite elements method with the real profile of a frustrum pyramid shaped part of 240 x 240 mm dimension, made of an aluminum alloy AA 1050-O with 1 mm sheet thickness, 30 mm depth and a constant wall angle of 55° as in figure 2.

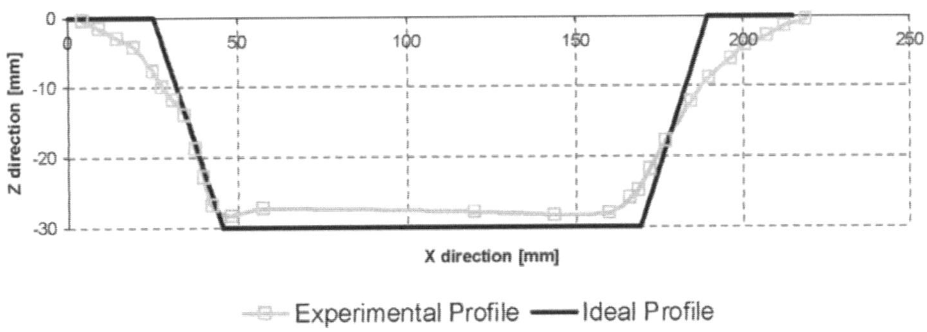


**Figure 2.** Comparison between theoretical and experimental profile of a frustum pyramid [9].

Some researchers have been concerned with simplifying the finite element analysis models by implementing a new type of element used to mesh the sheet metal FEA program, ABAQUS, as can be seen in the work of Ben Ayes [10].

## 2. Theoretical aspects

As mentioned above, the parts manufactured through single point incremental forming have several drawbacks caused by the behavior of the material. The most common disadvantages of this process are the material thinning and springback. Material thinning is caused mainly due to the combined phenomena that take place: two direction stretching and its bending. Excessive thinning can eventually produce the appearance of material failure, which is not acceptable in any industry. Extensive research was carried out by Oleksik in paper [11] where the influences of different geometric parameters on the thinning of the material was investigated. Other research in this field focused on influence of the toolpath over the material thinning and it was observed in papers [12] and [13] that a circular level toolpath has the highest influence. This drawback was also studied for sheets made of polyamide and polyethylene, and from the comparison between them in paper [14] it was concluded that polyamide sheets presented a better result than the case of polypropylene sheets.

One other big drawback of the sheets deformed through single point incremental forming process is the springback of the sheet which can be observed in two cases: during the process when the tool is in contact with the sheet, known as local springback and after the load of the tool is removed, which is known as global springback. Reducing the springback is one of the interests of the researchers because this leads to obtaining parts with higher accuracy and thus the process closer to implementation in industry. Some researchers used a method based on the usage of direct current both during and after the process in order to reduce the springback [15]. Other methods of reducing the springback are the use of laser beam [16] or a partial die located under the sheet [17]. Other defects that may occur during the process of incremental forming is the appearance of waves located on the wall of the frustrum shaped parts as shown in figure 3:

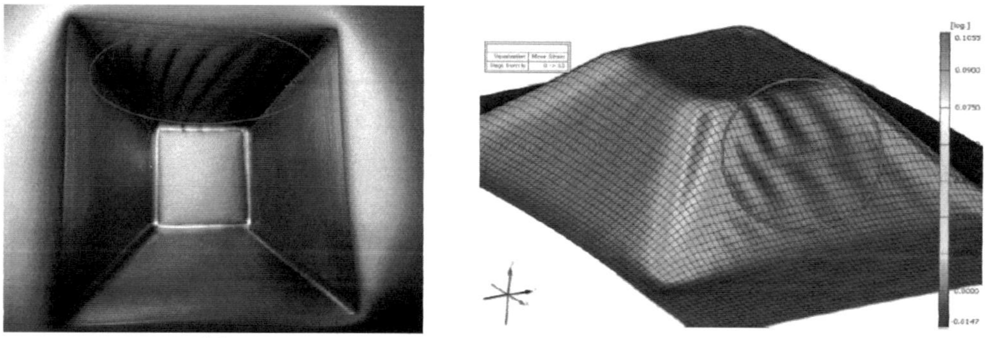

(a) (b)

**Figure 3.** Waves defects (a) observed experimentally;
(b) observed into numerical investigation [18].

This process requires a machine capable of applying the necessary forces to deform the part in the desired shape. Therefore, it is very important to know the maximum force required to deform a piece into its shape in order to choose the right equipment. The force prediction can be done with the help of finite element method as presented in the papers [19] and [20], or with the help of MATLAB software packages as in the paper [21].

## 3. Results and FEM model

The aim of this paper is to investigate the material behavior of three different aluminum alloys sheet metal. Finite element analysis was performed in ABAQUS/EXPLICIT to achieve the purpose of this paper. The desired shape to be obtained through single point incremental forming process is a frustrum pyramid with the large base of 100 x 100 mm, 30 mm depth and a constant wall angle of 45°. The

FEM model of the process contains the following assembly components: two retaining rings modeled as discrete rigid, a tool in shape of a hemispherical punch that is analytical rigid and the deformable body of the sheet metal. The initial shape of the blank was 200 x 200 mm with a thickness of 0.8 mm. The different sheet metal used consists of three types of aluminum alloys: AA 2007, EN AW 5754 and EN AW 6060 with the mechanical properties according to the standard EN 573 – 1 which are showed in table 1. These aluminum alloys were chosen due to the fact that in the Center for Studies and Research for Plastic Deformations from University Lucian Blaga of Sibiu, the material base contains all the necessary material curves and coefficients in order to perform finite element analysis.

**Table 1.** Mechanical properties.

| Material | Yield stress Rp0.2 [MPa] | Tensile strength Rm [MPa] | Elongation A [%] |
|---|---|---|---|
| **AA2007** | 220 | 340 | 8 |
| **EN AW 6060** | 160 | 215 | 12 |
| **EN AW 5754** | 80 | 180 | 16 |

The mesh of the deformable body and discrete rigid components from the assembly of the process is presented in table 2.

**Table 2.** Part mesh.

| **Piece** | **Element type** | **Number of elements** |
|---|---|---|
| Two retaining rings | R3D4 | 7680 |
| Sheet metal | S4R | 17689 |

After running all the simulations, a comparison is done between the theoretical profile of the parts and the analytical profiles of the frustrum pyramid obtained through single point incremental forming process where can be seen the different springback behavior of the materials. This comparison can be observed in figure 4 where the AA 2007 alloy with a yield stress of 220 MPa showed the highest springback. No clear difference can be observed between the other two alloys used in terms of the profile obtained.

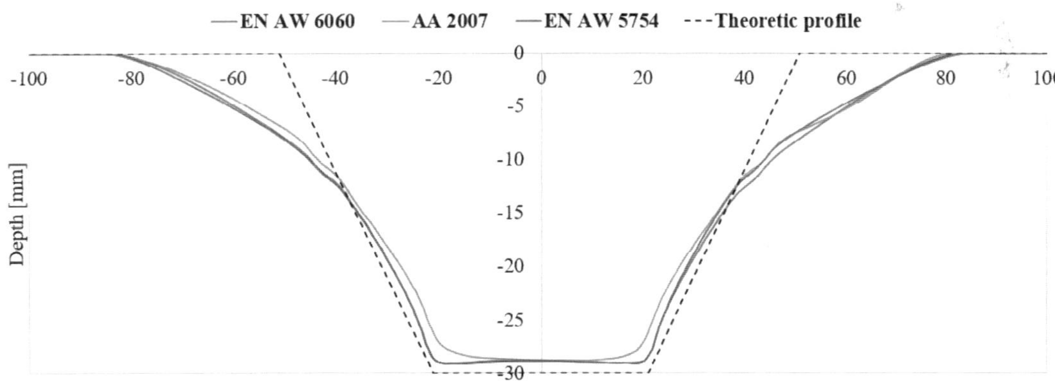


**Figure 4.** Comparison between analytical profile and theoretical profile.

Table 3 shows the error given by the springback of the sheets measured in the lowest point of the parts.

**Table 3.** Springback, thickness reduction and maximum Fz.

| Material | Springback [mm] | Thickness reduction [%] | Maximum value of the force measured on vertical axis Z [kN] |
|---|---|---|---|
| **AA 2007** | 1.19 | 52 | 1.35 |
| **EN AW 6060** | 0.91 | 53 | 0.99 |
| **EN AW 5754** | 0.91 | 53 | 0.89 |

The maximum thinning of the sheets used is shown in the previous table and the thickness distribution over the entire part surface for EN AW 6060 is shown in figure 5.

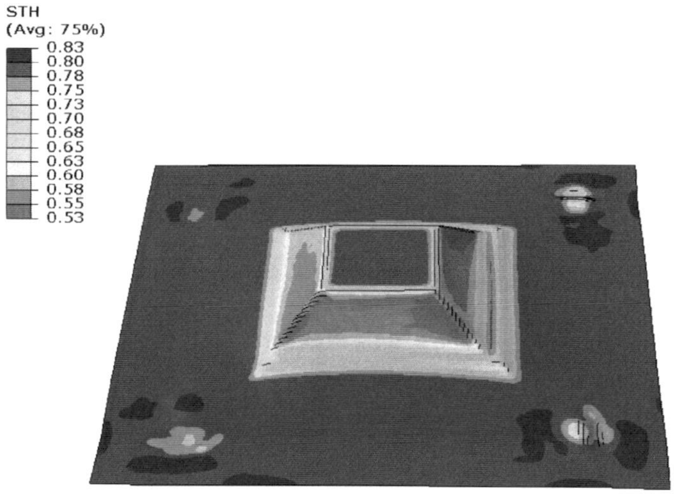


**Figure 5.** Thickness distribution for EN AW 6060 alloy.

The forces distribution over depth is presented in figure 6 and it can be observed that there is a dependence relationship between the maxim force on the Z axis and the yield stress of the alloys.

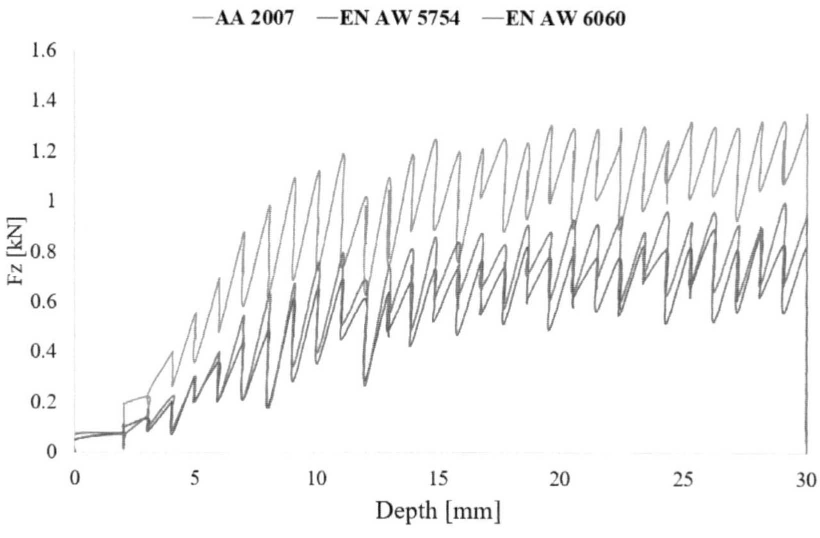


**Figure 6.** $F_z$ distribution over depth of frustrum pyramid.

Another defect of accuracy can be seen in the case of AA 2007 alloy, which is the most stiffest out of the ones investigated in this paper, in form of waves effect at the base of the frustrum pyramid similar with the results presented in paper [15] and it can be observed in figure 7 which shows the displacement in the Z direction.

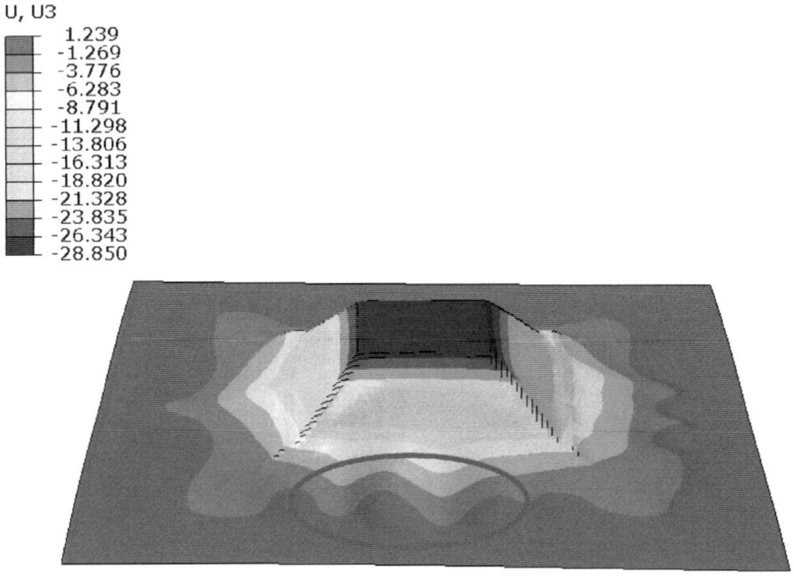


**Figure 7.** AA 2007 frustrum pyramid with waves effect.

## 4. Conclusions and further research

As observed from the analysis performed with the help of finite element method over three different types of aluminum alloys in single point incremental forming process of frustrum pyramid shaped parts, AA 2007 alloy presented the highest springback which conducted to the final part having the lowest accuracy. However, the other two alloys showed a better behavior regarding springbackand also the forces needed to deform the sheet metals were as much as 27% lower than in the case of AA 2007. The thickness reduction is similar in all three cases, thus not playing an important role in differentiating the materials.

In conclusion, it is fair to say that from an analytical point of view that all three types of aluminum alloys could be used to produce frustrum pyramid shaped parts through single point incremental forming process, but it would be better if either of EN AW 6060 or EN AW 5754 alloys would be chosen for this process because they have similar behavior and parts with higher accuracy can be achieved, which can lead to a step further for this process of implementation in industry.

Further research in this area is proposed by the authors to investigate harder types of materials such as steel and titanium and to experimentally validate the results obtained through the analytical models presented in this and future papers.

**Acknowledgments**

This research presented here was partially financed by a grant of the Romanian Ministry of Research and Innovation CCCDI-UEFISCDI, project number PN-III-P1-1.2-PCCDI-2017-0446/ nr. 82 PCCDI/2018, within PNCDI III, project title: "Smart manufacturing technologies for advanced production of parts from automotive and aeronautics industries".

**References**

[1] Leszak E and Halli E 1967 *Apparatus and process for incremental dieless forming* US Patent
[2] Racz S G, Breaz R E, Tera M, Gîrjob C, Biriş C, Chicea A L and Bologa O 2018 Incremental Forming of Titanium Ti6Al4V Alloy for Cranioplasty Plates—Decision-Making Process and Technological Approaches *Metals* **103**
[3] Racz S G, Oleksik V S and Breaz R E 2019 Incremental forming – CAE/CAN approaches and

results *IOP Conference Series Materials Science and Engineering* **591**
[4] Breaz R E, Bologa O and Racz S G 2017 Selecting industrial robots for milling applications using AHP *Procedia computer science* **122** 46
[5] Khan S and Pradhan S K 2020 Experimentation and FE simulation of single point incremental forming *Materials Today: Proceedings* **27** 2334
[6] Nasulea D and Oancea G 2019 Research on manufacturing of pyramidal frustum parts using single point incremental forming process *IOP Conference Series: Materials Science and Engineering* **564**
[7] Duflou J R, Habraken A M, Cao J, Malhotra R, Bambach M, Adams D, Vanhove H, Mohammadi A and Jeswiet J 2017 Single point incremental forming: state-of-the-art and prospects *Int. J. Mater. Form.* **11** 743
[8] Breaz R E, Bologa O and Racz S G 2018 Using the Analytic Hierarchy Process (AHP) and fuzzy logic to evaluate the possibility of introducing single point incremental forming on industrial scale *Procedia computer science* **139** 408
[9] Ambrogio G, Filice L and De Napoli L 2005 A simple approach for reducing profile diverting in a single point incremental forming process *Proceedings of the Institution of Mechanical Engineers* **219** 823
[10] Ben Ayed L, Robert C, Delamézièrea A, Nouari M and Batoz J L 2014 Simplified numerical approach for incremental sheet metal forming process *Eng. Struct* **62** 75
[11] Oleksik V 2014 Influence of geometrical parameters, wall angle and part shape on thickness reduction of single point incremental forming *Procedia Eng.* **81** 2280
[12] Girjob C, Racz S G, Bologa O and Biris C 2015 FEM Simulation of Laminated Lightweight Materials Processed through Single Point Incremental Forming *Appl. Mech. Mater.*p 38
[13] Girjob C, Racz S G, Bologa O and Biris C 2015 Study of the Formability of Light Metallic Materials *Appl. Mech. Mater.* **809** 289
[14] Roşca N A and Oleksik M 2019 Simulation of the Single Point Incremental Forming of Polyamide and Polyethylene Sheets *MATEC Web Conf.* **290**
[15] Ruszkiewicz B J, Scriva C., Reese Z C, Nikhare C P, Roth J T and Ragai I 2015 Direct Electric Current Spot Treatment's Effect on Springback of 90 Degree Bent 2024-T3 Aluminum *ASME 2015 International Manufacturing Science and Engineering Conference* **1**
[16] Duflou J R, Callebaut B, Verbert J and De Baerdemaeker H 2007 Laser assisted incrementalforming: formability and accuracy improvement *CIRP Annals* **56** 273
[17] Li J, Bai T and Zhou Z 2014 Numerical simulation and experimental investigation of incremental sheet forming with an elastic support *Int. J. Adv. Manuf. Tech.* **74** 1649
[18] Oleksik M 2018 Comparative study about different experimental layouts used on single point incremental forming process *Acta Universitatis Cibiensis. Technical Series* **70**
[19] Oleksik V, Bologa O, Breaz R and Racz G 2008 Comparison between the numericalsimulations of incremental sheet forming and conventional stretch forming process *Int. J. Mater. Form.* **1** 1187
[20] Oleksik V, Pascu A, Gavrus A and Oleksik M 2010 Experimental studies regarding the single point incremental forming process *Academic Journal of Manufacturing Engineering* **8** 51
[21] Crenganis M and Csiszar A 2019 A Dynamic Model for KUKA KR6 in SPIF *Mater. Sci. Forum.* **957** 156

# Research on reliable running time of motorized spindle based on health indicator

**Z Si[1], L Fan[1*], Z Jia[1] and J Liu[1]**

[1]School of Mechanical Engineering, Shenyang Jianzhu University, Shenyang 110168

*E-mail: fanliting@sjzu.edu.cn

**Abstract.** Motorized spindle is one of the key components of CNC machine tools, and its degradation can be monitored in real time to predict the future reliable running time of motorized spindle, It is of great significance to reduce maintenance support cost and improve equipment stability to provide support for maintenance decision .In this paper, based on the comprehensive health index of motorized spindle and reliable running time forecast method, first of all, through the health indicator as a measure of motorized spindle health status characteristic parameters and characteristic parameters for single and multiple health indicator formula respectively, many characteristic parameters by using the method of sequence relations act and the entropy weight method to achieve the unity of the subjective and objective. Then, after the health factor modification, the comprehensive health indicator formula is obtained. Finally, an exponential model based on time series analysis is established to fit the variation trend of comprehensive health indicator of motorized spindle with time, and the expression of health indicator model is obtained. An example is given to verify the rationality and feasibility of the proposed method, and a comparative analysis is made with the reliability of references. The final results show that the proposed method in this paper is effective and reasonable. Can achieve the purpose of reliable running time estimation of motorized spindle.

## 1. Introduction

Reliable running time estimation, as an important core technology for fault prediction and health management, is also the most difficult and most challenging comprehensive technology. It aims to predict the future reliable working time of the equipment, provide support for maintenance decisions. It is of great significance to reduce maintenance support costs and improve the stability of the equipment. However, as one of the core components of the machine tool, the motorized spindle has the characteristics of high accuracy, long life and high reliability, and it is difficult to obtain the motorized spindle failure life data through reliability experiments in a short time [1, 2]. The traditional reliability method needs to obtain the failure life data of the product to infer the overall reliability [3]. When the product life is longer and the reliability higher, it is difficult to obtain its failure life data in a short time. Therefore, for the motorized spindle with high reliability and long life, this paper adopts the degradation data based on state detection, through the change of some characteristic parameters in the degraded data to achieve the purpose of detecting the safe operation of motorized spindle equipment, excavates and abstracts the characteristic information hidden in a large number of condition monitoring data and failure models to reflect the healthy state of the equipment. Rely on real data and reliable methods to evaluate the health status of the equipment itself, establish the health file for the whole life cycle of the equipment, and provide the basis for the decision of equipment maintenance and support. So as to carry out ideal maintenance on the basis of the state monitoring. The health index [4, 5] is a quantitative value used to measure and characterize the health status of the equipment. The current status of the equipment can be used to judge the current status of the equipment. Then, combined with the historical test data and real-time performance information, the changing trend of the health index with time is fitted, and the distribution of the health index with time at a certain time in the future can be predicted. Estimate the reliable running time of the motorized spindle.

At present, many experts and scholars have done a lot of research on the reliability and remaining useful life of the motorized spindle, but it is almost blank to estimate the reliable running time by combining multiple characteristic parameters in the field of motorized spindle. Previous reliability evaluations of motorized spindle included both fault-free data and those based on the performance degradation data of motorized spindle and failure data. But most of them are based on a certain key parameter as the research point to evaluate the reliability and remaining useful life of the motorized spindle. Lack of comprehensive comprehensive health assessment. With no fault data: Jiang Xi [6] and others based on the Bayes method combined with the virtual augmented sample method to study the reliability of the motorized spindle. Finally, the reliability evaluation results of the motorized spindle were obtained and the rationality was verified. References [7, 8, 9] also made an in-depth study of the Bayes method and its application. In terms of performance degradation data of motorized spindle: Qiu Ronghua [10] et al. established a degradation model of the motorized spindle based on the degradation data of the radial jump of the motorized spindle, and obtained its service life. Chi Yulun [11] et al. used the acoustic emission signal as the performance measurement of the machine tool spindle, established a degradation model, obtained the failure life, and combined the virtual augmented sample method to evaluate the machine tool spindle reliability. When there is failure data of the motorized spindle failure: Zhao Jinping [12] and other statistics and analysis of the failure data, forming observation samples, and fitting the probability density distribution function and cumulative distribution function curve of the equipment failure interval time, so as to infer the distribution law obeys the Weibull distribution. Finally, the reliability evaluation indexes of the equipment were calculated based on the statistical results. Yang Bin [13] established a reliability covariant regression model considering multiple performance degradation to realize the reliability evaluation of the motorized spindle. The proposed method overcomes the problem that it is difficult to carry out analysis without failure data in the traditional reliability assessment work and provides a reference for the reliability assessment work under accelerated test. Foreign scholars Jayaram [14] and others put forward an analysis model of product performance degradation data, and analysed and predicted the real-time reliability level of the spindle products using the maximum likelihood method.

But the research on health index mostly focuses on the remaining battery life and circuit equipment. The remaining battery life research: Lu Zhaoquan [15], etc. proposed the estimation of lithium battery power based on health factors. The experimental data and Matlab simulation proved that the method has high feasibility. Pang Jingyue [16] and others proposed a method framework for constructing a health factor for predicting the remaining life of lithium-ion batteries by using charge-discharge detection parameters of lithium-ion batteries, and a method system for predicting the remaining life of lithium-ion batteries online. In terms of circuit equipment: Zhang Fengxia [17] and others proposed the life prediction method of track circuit equipment based on health index. The feasibility is verified through an example and it provides a theoretical basis for the maintenance of the equipment by the relevant maintenance department. Shi Changkai [18] and others made quantitative assessment of distribution network risk based on the real-time health index of equipment. The examples show that the proposed method can obtain the failure rate, failure time and load risk of various load points, and prove the effectiveness of a quantitative assessment method for distribution network risks based on real-time health index. Chunbo Yang [19] and others proposed the evaluation of equipment health status based on the comprehensive health index. An example is used to analyze the effectiveness and rationality of the method.

This paper establishes a complete model for evaluating the health status of motorized spindle. It considers both single feature degradation and multiple feature degradation, and comprehensively considers the balance between feature parameters to obtain a comprehensive health index formula.

Finally, the change trend of the comprehensive health index over time based on the time series is established, so that the purpose of predicting the reliable running time of the motorized spindle in the future is achieved, the equipment can be repaired in time and the safety of the equipment is improved.

## 2. Reliable running time prediction

As shown in figure 1, firstly, appropriate sensors are selected for the motorized spindle system to monitor the real-time status of the motorized spindle. The obtained historical status monitoring data are extracted for characteristic parameters, and representative and convenient characteristic parameters are selected. According to the formula of comprehensive health index, the obtained characteristic parameters were weighted for the first time according to the order relation method. Order relation analysis is an analytical method combining qualitative analysis with quantitative analysis. However, because the order relation method has strong subjectivity, in order to exclude the influence of subjective factors, the entropy weight method is used for the second weighting to achieve the unification of main and objective. Then the comprehensive health index is obtained according to the formula of comprehensive health index. After that, the curve fitting of the historical comprehensive health index was carried out to obtain the curve of the comprehensive health index over time, and the fault prediction was carried out to predict the future reliable running time of the motorized spindle, so as to carry out maintenance in advance and provide maintenance decisions.

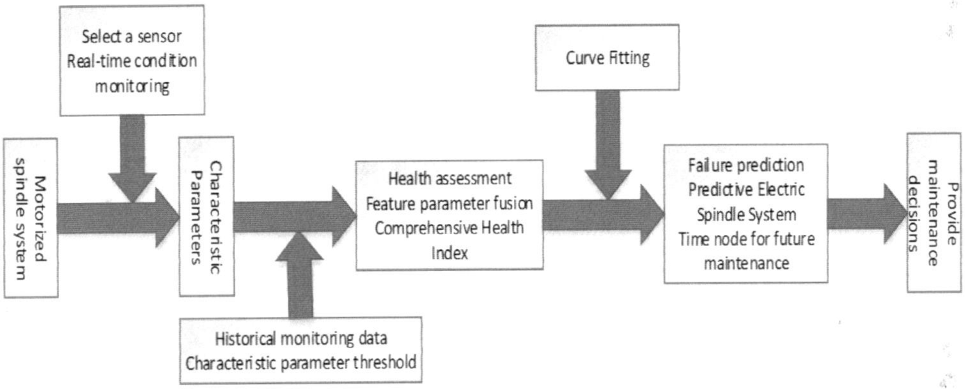


**Figure 1.** Predictive framework of reliable running time of motorized spindle based on comprehensive health indicator.

## 3. Health indicator of motorized spindle

### *3.1. Health index with single feature parameter*

The Health Index of motorized spindle is usually denoted as HI, and the value range of HI is defined as [0, 1]. The closer the *HI* is to 1, the better the Health state of the equipment is; the closer it is to 0, the worse the Health state is 1 means best condition, 0 means complete failure.

During the operation of motorized spindle, the standard value of each index changes continuously with time. Through the design parameters of the equipment and certain historical statistical data, the state degradation quantities $x_{i,j}(t)$ of this parameter and the minimum limit value $x_{min}$ and maximum limit value $x_{max}$ of this indicator parameter can be obtained. When the actual measured value exceeds the threshold range, the health index is 0, and the machine should be stopped immediately for maintenance. When the actual measured value coincides with the standard value, the health index is 1, which is the best condition of the equipment. When the actual measured value is between the threshold range, the value between [0, 1] can be obtained to quantify the equipment state. The normalization is calculated as follows:

$$h_{i,j}(t) = \frac{x_{max} - x_{i,j}(t)}{x_{max} - x_{min}} \tag{1}$$

where $x_{i,j}(t)$ is the health index based on the characteristic parameter $x_j$ of device $i$; $x_{i,j}(t)$ is the measured value of characteristic parameter $x_j$ at time $t$; $x_{max}$ is the maximum value and $x_{min}$ is the minimum value.

### *3.2. Health index with multiple feature parameters*

During the operation of the spindle, there will be a lot of degraded data, such as vibration, current, voltage, temperature, noise, etc. Due to the complexity of the equipment operating environment, it is often difficult to evaluate the whole equipment operating state by the change of the characteristic parameters of the single state of the equipment. In this paper, the comprehensive weighting method is adopted to evaluate the health state of multiple feature parameters of the equipment. The weight value reflects the importance of each feature parameter on the operation state of the equipment, and reasonable weight distribution is the basis for accurate evaluation of the operation state of the equipment. Order relation analysis is an analytical method combining qualitative analysis with quantitative analysis. First, the mutual importance of feature parameters is compared in pairs, and the importance of feature parameters is sorted according to the expert experience: $y_1>y_2>y_3>y_4...>y_j$. Then determine the relative importance between the two adjacent indicators $y_{j+1}$ and $y_j$, and obtain the weight coefficients of each characteristic parameter as follows:

$$p_{j+1} = \left(1 + \sum_{j=1}^{n-1} \prod_{k=j}^{n-1} \frac{y_j}{y_{j+1}}\right)^{-1} \tag{2}$$

$$p_j = (\frac{y_j}{y_{j+1}})^{-1} p_{j+1,\ \ j=1,2,\ldots n-1} \tag{3}$$

In the equation, $p_j$ is the weight value of the j-th characteristic parameter obtained by the order relationship method.

Due to the influence of subjective arbitrariness on weight allocation in order relation method, the results depend on expert experience knowledge. In order to eliminate the influence of subjective factors, an entropy weight method based on objective data relationship is used to redistribute weight. Entropy weight method is a method to obtain the entropy weight of each characteristic parameter through information entropy based on the variation degree of the characteristic parameter, and then obtain the weight of each characteristic parameter. The greater the difference of an index, the smaller the entropy weight, indicating that the index provides more information, plays a greater role in the evaluation, and the greater the weight. The calculation equation is:

$$y'_{i,j} = \frac{y_{i,j}}{\sum_{i=1}^{m} y_{i,j}} \tag{4}$$

$$e_j = -\frac{1}{\ln n} \sum_{i=1}^{m} y_{i,j} \ln y_{i,j} \tag{5}$$

$$q_j = \frac{1 - \mathrm{e_j}}{\mathrm{n} - \sum_{j=1}^{n} e_j} \tag{6}$$

where: $y_{i,j}$ is the sample data; $y'_{i,j}$ is the proportion of the i-th sample data under the j-th feature parameter; $m$ is the number of sample data; $n$ is the number of feature parameters; $e_j$ is the entropy weight of the j-th feature parameter; $q_j$ is the weight value of the j-th feature parameter obtained by the entropy weight method. In order to make the combination weights as close to $p$ and $q$ as possible to achieve subjective and objective unity, an optimization model of the least square method is established to obtain the combination weights.

$$minH(\omega) = \begin{Bmatrix} \sum_{j=1}^{n}[(\omega_j - p_j)^2 + (\omega_j - q_j)^2] \\ s.t. \sum_{j=1}^{n}\omega_j = 1, \omega_j \geq 0, j = 1,2,3 \ldots . n \end{Bmatrix} \quad (7)$$

The expression of equipment health index based on multiple characteristic parameters is as follows:

$$h_j(t) = \sum_{j=1}^{m} \omega_j \, h_{i,j}(t) \quad (8)$$

Among them, $h_i(t)$ is the health index of the device; $m$ is the number of characteristic parameters; $\omega_j$ is the combined weight of the j-th characteristic parameter, which reflects the influence of this parameter on the state of the device ($0 \leq \omega_j \leq 1$ and $\sum_{j=1}^{m} \omega_j = 1$).

### 3.3. Health indicator model

The relationship between the health of power equipment and time can be characterized as:

$$H = H_0 \cdot e^{B(t-t_0)} \quad (9)$$

This equation is derived from an empirical formula of the British EA company [20], where H represents the health status of the power equipment, which is continuously increasing with the use of time. The larger the value, the worse the health status, and the lower the value, the better the health status. The health index required by this article is continuously decreasing with the continuous use time. Therefore, in order to be applicable to the remaining reliable running time evaluation model of the motorized spindle of this article, it is improved and the improvement is as follows:

$$H_i(t) = 1 - (1 - H_0) \cdot e^{B(t-t_0)} \quad (10)$$

In the equation: $H_0$ is the health index at the time $t_0$ of the device, taking 0.95, quoted from the literature [20]. $H_i(t)$ is the health index of the device at time t. B is the aging coefficient. The $t_0$ is the time when the equipment is initially put into operation. The $t$ is time when the equipment is evaluated.

In order to obtain its value, the expected operating life $T_d$ and aging coefficient $B$ of the motorized spindle must be obtained first. The two calculation equations are:

$$T_d = \frac{T_D}{f_1 \cdot f_2} \quad (11)$$

$$B = \frac{\ln(1 - HI_{ty}) - \ln(1 - H_0)}{T_d} \quad (12)$$

where: $T_D$ is the design life of the equipment; $f_1$ is the load correction coefficient, which is about 1.05; $f_2$ is the environmental correction coefficient, which is about 1.05, which is quoted from the literature [19]; $HI_{ty}$ is the health indicator when the equipment is decommissioned, which is generally taken as 0.2, which is quoted from the literature [20].

Taking the aging of equipment into consideration, defining a variable health factor is defined as:

$$\mu = \frac{Hi(t)}{H_0} \quad (13)$$

As the motorized spindle runs for a long time, its degradation generally follows an exponential form. Therefore, the comprehensive health index of the motorized spindle after the weighted and variable health factor correction is:

$$HI(t) = \mu \cdot hi(t) \quad (14)$$

*HI(t)* is a quantified expression of the comprehensive operating state. With the continuous operation of the motorized spindle, its changing trend also presents complex changing characteristics. In order to be able to correctly predict the future health status of the motorized spindle and the time nodes that need to be maintained, a health index sequence that can reflect the time-varying health status of the motorized spindle *HI (t) = {HI (t1), HI (t2), ..., HI (tn)}*. And build a suitable prediction model to describe this sequence. By collecting the historical monitoring data of the motorized spindle, a comprehensive health index is obtained, and these data are fitted using the Matlab curve fitting toolbox to obtain a function image in the form of an index of health index and time to establish a life cycle health curve. The exponential model constructed is as follows:

$$HI(t) = a \cdot e^{kt} \tag{15}$$

## 4. Case for verification

In this experiment, experimental data from reference [21] were selected and there was only one characteristic parameter, so its weight was 1. Some test data are shown in table 1:

**Table 1.** The corresponding radial jump at different times.

| Runout (mm) | Time (h) | | | | | | | | | |
|---|---|---|---|---|---|---|---|---|---|---|
| | 12 | 24 | 36 | 48 | 60 | 72 | 84 | 96 | 108 | 120 |
| Sample 1 | 6.4 | 6.8 | 6.9 | 7.0 | 6.4 | 6.2 | 6.8 | 7.1 | 7.1 | 7.0 |
| Sample 2 | 5.6 | 5.6 | 6.3 | 6.4 | 5.9 | 6.5 | 6.6 | 6.7 | 6.1 | 6.1 |
| | Time (h) | | | | | | | | | |
| | 132 | 144 | 156 | 168 | 180 | 192 | 204 | 216 | 228 | 240 |
| Sample 1 | 7.3 | 7.4 | 7.9 | 7.6 | 7.1 | 7.4 | 7.3 | 7.6 | 7.2 | 7.5 |
| Sample 2 | 5.8 | 5.8 | 5.9 | 6.2 | 6.7 | 6.2 | 6.7 | 6.7 | 6.3 | 6.1 |
| | Time (h) | | | | | | | | | |
| | 252 | 264 | 276 | 288 | 300 | 312 | 324 | 336 | 348 | 360 |
| Sample 1 | 8.2 | 8.3 | 8.2 | 8.4 | 8.7 | 8.5 | 8.8 | 8.4 | 8.5 | 8.5 |
| Sample 2 | 6.8 | 7.2 | 6.8 | 6.7 | 6.6 | 7.1 | 6.7 | 6.5 | 6.9 | 7.3 |

The motorized spindle has been running for 360h, and its normal value should be between 5.0-20um. The expected operating life of *TD* is 2500h and 12000h respectively. The expected operating life *Td* is 2268h and 10884h, respectively. The calculated comprehensive health index is shown in table 2:

**Table 2.** Corresponding health index at different times.

| Runout (mm) | Time (h) | | | | | | | | | |
|---|---|---|---|---|---|---|---|---|---|---|
| | 12 | 24 | 36 | 48 | 60 | 72 | 84 | 96 | 108 | 120 |
| Sample 1 | 0.94 | 0.93 | 0.93 | 0.92 | 0.92 | 0.91 | 0.91 | 0.90 | 0.90 | 0.90 |
| Sample 2 | 0.97 | 0.97 | 0.96 | 0.97 | 0.96 | 0.96 | 0.96 | 0.96 | 0.96 | 0.96 |
| | Time (h) | | | | | | | | | |
| | 132 | 144 | 156 | 168 | 180 | 192 | 204 | 216 | 228 | 240 |
| Sample 1 | 0.89 | 0.89 | 0.88 | 0.88 | 0.87 | 0.87 | 0.87 | 0.86 | 0.86 | 0.85 |
| Sample 2 | 0.96 | 0.96 | 0.95 | 0.95 | 0.95 | 0.95 | 0.95 | 0.95 | 0.95 | 0.95 |
| | Time (h) | | | | | | | | | |
| | 252 | 264 | 276 | 288 | 300 | 312 | 324 | 336 | 348 | 360 |
| Sample 1 | 0.85 | 0.85 | 0.84 | 0.84 | 0.83 | 0.83 | 0.83 | 0.82 | 0.82 | 0.81 |
| Sample 2 | 0.95 | 0.95 | 0.94 | 0.94 | 0.95 | 0.94 | 0.94 | 0.94 | 0.94 | 0.94 |

According to expert experience, the health status is divided into the following five levels [19], and the mapping relationship is shown in table 3:

**Table 3.** Health Index Rating.

| Grade | Health index value range | Health level | Health description |
|---|---|---|---|
| First level | [0.8-1] | Health | Very good health and safe equipment |
| Secondary | [0.6-0.8] | Healthier | Good health and safe equipment |
| Third grade | [0.4-0.6] | Sub-health | The equipment is not safe, and there is a slight abnormality. Monitoring and troubleshooting should be strengthened |
| Fourth grade | [0.2-0.4] | Malfunction | The equipment is very unsafe and serious abnormalities should occur, and maintenance should be arranged as soon as possible |
| Fifth grade | [0-0.2] | Serious failure | Device does not work |

These two axes have been running for 360 hours, with health indices of 0.81 and 0.94. Still in good health. Now, the first 20 groups of health indexes calculated will be fitted using MATLAB toolbox, and the obtained fitting curves are shown in figures 1 and 2 respectively:

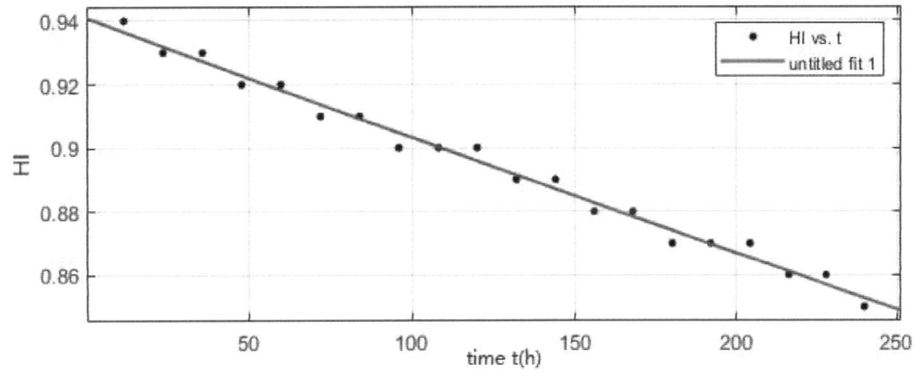


**Figure 2.** Curve fit of the first 20 sets of data for sample 1.

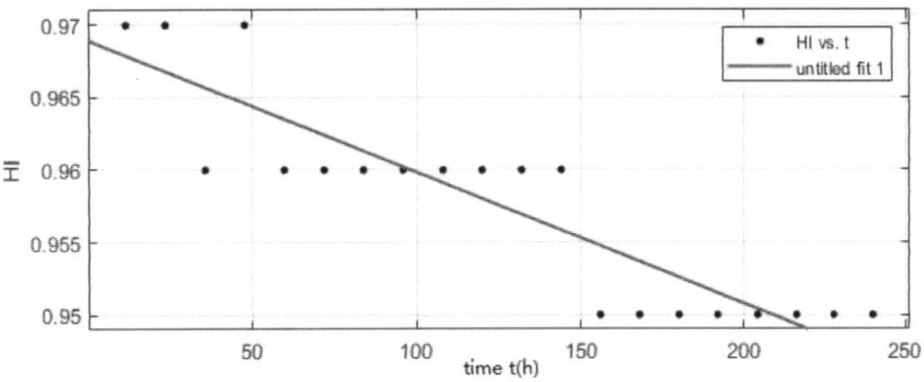


**Figure 3.** Curve fit of the first 20 sets of data for sample 2.

The obtained expressions about the comprehensive health index of the motorized spindle are:

$$HI_1(t) = 0.9411 \cdot \exp(-4.109 \times 10^{-4} \cdot t) \tag{16}$$

$$HI_2(t) = 0.969 \cdot \exp(-9.495 \times 10^{-5} \cdot t) \tag{17}$$

According to the expressions of the two comprehensive health index, the health index of group 30 are predicted to be 0.81 and 0.94, respectively. As before, it is proved that the expression of the comprehensive health index function obtained by the fitting is reliable.

According to the comprehensive health index expression, the change curves of the health index over time are shown in figure 3 and figure 4.

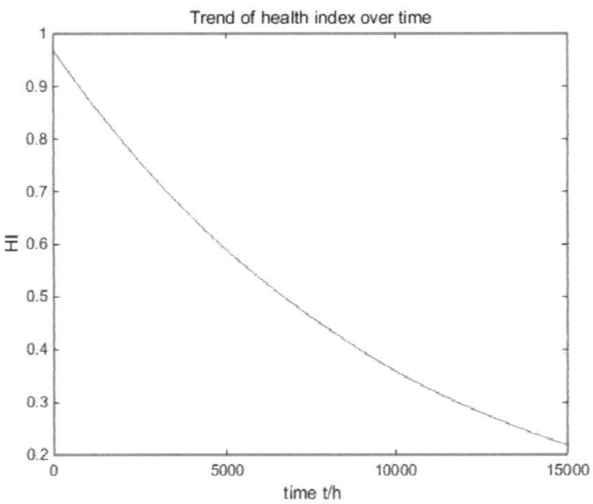


**Figure 4.** Health curve of sample 1.

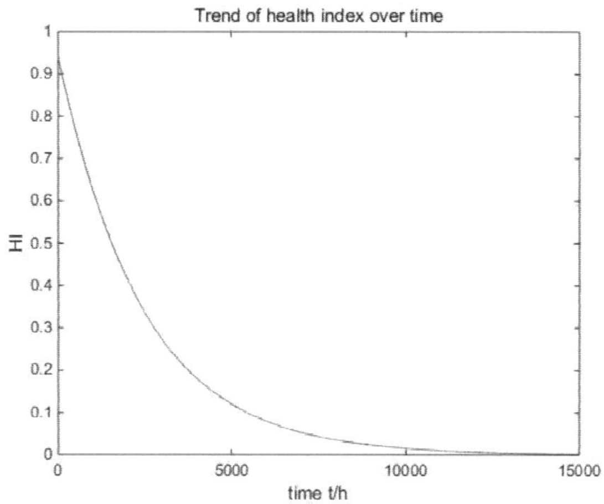


**Figure 5.** Health curve of sample 2.

The graph obtained by putting two curves on one graph is shown in figure 5.

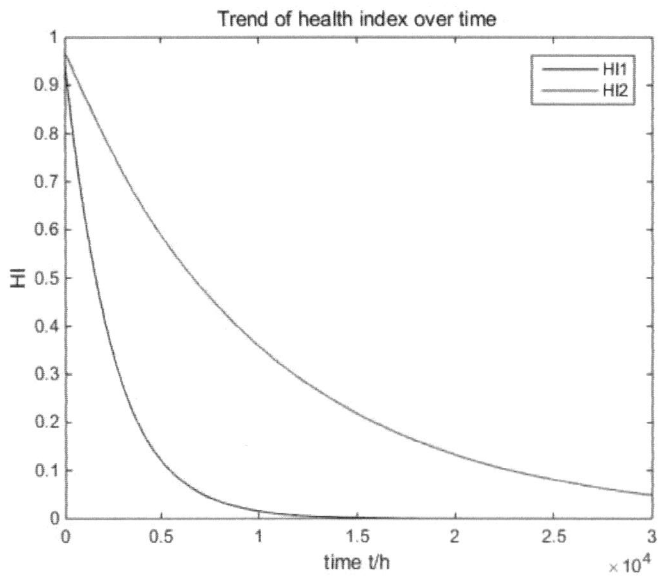


**Figure 6.** Health curve of sample 1 and 2.

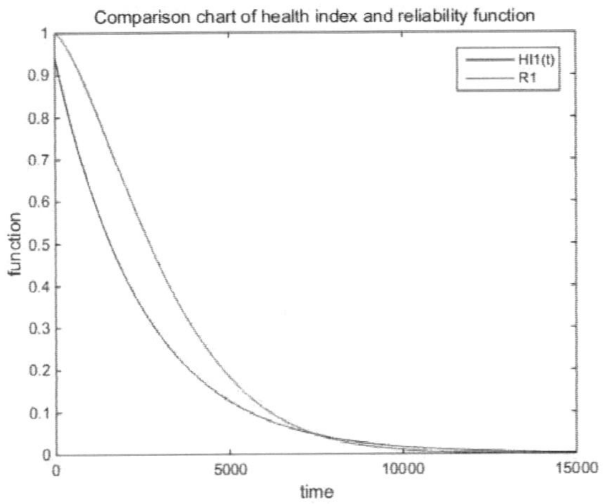


**Figure 7.** Sample 1 health index and reliability function comparison chart.

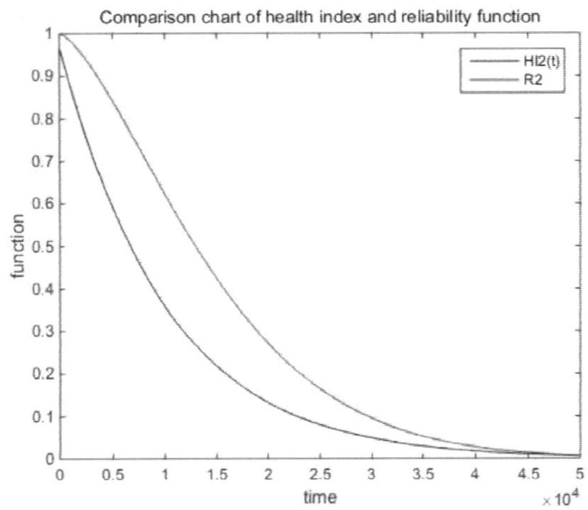


**Figure 8.** Sample 2 health index and reliability function comparison chart.

Respectively from the figure 4 and figure 5 sample 1 and sample 2 health indicator image can be seen that the electric spindle started to run as a bit faster to deterioration of the health index, such as running a period of time, motorized spindle after running-in, a little bit slow to deterioration of the health index, wait for a long time after the operation, the healthy index to converge to zero until failure.

According to table 3, when the health level reaches level 3, that is, $HI$=0.6, $t$ is equal to 1120 and 5048, respectively. Therefore, when the motorized spindle runs for 760h and 4688h again, it will enter level 3, so monitoring and troubleshooting should be strengthened. When the health level reaches level 4, i.e. $HI$=0.4, the equipment is very unsafe and seriously abnormal. Maintenance should be arranged as soon as possible, and $t$ is equal to 2082 and 9318 respectively. There is little difference with the experimental results of 2500 and 12000 in literature [21], and the error rate is only 0.16 and 0.18. In addition, it can be seen from the comparison of the health index and the reference reliability of samples 1 and 2 in figure 7 and figure 8 that there is a slight difference between them in the middle period of operation, but the change trend is the same, which conforms to the trend of the reliable working time of electric spindle changing with time. Therefore, the proposed method is proved to be feasible to some extent. The future reliable working time of the two motorized spindles is 1722h and 8958h respectively.

## 5. Conclusion

Aiming at the problem that reliability analysis method based on life data is difficult to evaluate the reliability of motorized spindle, this paper adopts the method of reliable working time estimation of motorized spindle based on comprehensive health index to solve the problem of reliable working time estimation. The main research work is:

(1) Through the health index as a measure of motorized spindle health status characteristic parameters and characteristic parameters for single and multiple health index formula established respectively, and the more characteristic parameters using the method of order relation with entropy weight method weighted method for data fusion, to achieve the subjective and objective unifies, and then through the health factor correction, get comprehensive health index formula.

(2) An exponential model based on time series analysis was established to fit the variation trend of comprehensive health index of motorized spindle over time, and the model expression of health index was obtained to solve the problem of reliable working time estimation of motorized spindle in the future.

(3) Through an example to verify the rationality and feasibility of the proposed method, the obtained curve in the form of exponential function also conforms to the performance degradation law of

the motorized spindle, and finally the two motorized spindles will enter the third-level sub-health state after running for 760h and 4688h respectively, so the monitoring and fault elimination should be strengthened. And compared with the remaining life of the reference, the error rate is only 0.16 and 0.18.The health index curve is similar to the reliability curve. The results show that the proposed method is effective and reasonable. Can achieve the purpose of reliable working time estimation of motorized spindle.

The evaluation results show that the health index of the motorized spindle changes with time in line with the performance degradation law of mechanical products. In addition, this paper applies the concept of health index to the motorized spindle system for the first time, which has never been done before. Moreover, it can be combined with a variety of monitoring data to make the reliable working time of motorized spindle more comprehensive, which has an advantage over the literature [21]. However, there are still some shortcomings. Since there are only experimental data of degradation quantity, only a single characteristic quantity is used in this paper to measure the degradation characteristics of electric spindle. If there are more other degradation data, it will be better to verify.

**References**

[1] Liao L and Lee J 2009 A novel method for machine performance degradation assessment based on fixed cycle features test *J. Sound and Vib.* **326** (3-5) 894-908

[2] Tang J Y, He P, Liang H Q *et al.* 2013 Comprehensive reliability evaluation of correlation failure of high-long-life products with multiple failure modes *Journal of Mechanical Engineering* 66-73

[3] Zhao J Y 2005 *Reliability Modeling and Application Research Based on Performance Degradation* Data National University of Defense Technology

[4] Shi C K , Ning X, Sun Z T *et al.* 2018 Quantitative risk assessment of distribution network based on equipment real-time health index *High Voltage Technology* **44** (2) 534-40

[5] Zhou L M, Ma Z and Sheng W X 2016 Recent research progress on the health index theory of modern distribution networks *Power Supply and Supply* **33** (1) 3-7

[6] Xiang J, Liu H Z, Jiao J *et al.* 2015 Research on the reliability of the smallest sub-samples of the motorized spindle based on the Bayes method *Journal of Vibration and Shock* **34** (4) 122-7

[7] Liu Y, Li Y F, Huang H Z, Zuo M J and Sun Z 2010 Optimal preventive maintenance policy under fuzzy Bayesian reliability assessment environments *IIE Transactions* **42** (10) 734-45

[8] Sarhan A 2003 Non-parametric empirical Bayes procedure *Reliab. Eng. Syst. Safe.* **80** (2) 115-22

[9] Taheri S M and Zarei R 2011 Bayesian system reliability assessment under the vague environment Appl. Soft Comput. **11** (2) 1614-22

[10] Qiu R H, Ju K L, Dong Y G *et al.* 2016 Reliability test study of a small sample motorized spindle based on performance degradation *China Mechanical Engineering* **27** (20)

[11] Chi Y L, Zhong X L, Wu C P and Li H L 2017 Research on reliability of spindle performance degradation based on acoustic emission signal monitoring *Mechanical Strength* **39** (05) 1086-91

[12] Zhao J P, Xiong J X and Liu J S 2017 Analysis and evaluation of equipment operation reliability based on fault data *High Technology Letters* **27** (04) 359-63

[13] Yang B 2018 *Reliability Evaluation of Motorized Spindle Based on Performance Degradation* Jilin University

[14] Jayaram J S R and Girish T 2005 Reliability Prediction through degradation data modeling usinga quasi-likelihood approach *J. Chem. Technol. Biot.* **83** (5) 410-15

[15] Lu Z Q, Tao J F and Wang W 2018 Estimation of lithium battery capacity based on health factors *Power Technology* **42** (11) 1615-17

[16] Pang J Y, Ma Y T, Liu D T and Peng Y 2014 Indirect prediction method for remaining life of lithium ion batteries *China Science and Technology Paper* **9** (01) 28-36

[17] Zhang F X and Mi G S 2015 Research on life prediction method of track circuit equipment based on health index *Journal of the China Railway Society* **37** (12) 61-66

[18] Shi C K, Ning X, Sun Z T, Xi C Q, Du Y and Ma G M 2018 Quantitative evaluation of distribution network risk based on real-time health index of equipment *High Voltage Technology* **44** (2) 534-40
[19] Yang C B, Tao Q, Zhang J, Cheng Z Y and Fan Y Z 2019 Equipment condition assessment based on comprehensive health index *Power System Protection and Control* **47** (10) 104-09
[20] Cai H F 2012 *Comprehensive Assessment of Remaining Technical Life of Power Transformers* Zhengzhou University
[21] Jiang X, Liu H Z, Tong J J, Yuan D N, Liu L L 2014 Comparison of reliability of motorized spindle based on pseudo life distribution and bayes method J. *Mech. Sci. Technol.* **33** (11) 1694-99

# Robotized application of assembly and soldering – case study

**I Staretu[1]***

[1]Transilvania University of Brasov, Romania, Technical Sciences Academy of Romania, Bucharest, Romania

*E-mail: istaretu@yahoo.ro

**Abstract.** Robotic assembly is increasingly encountered in various technological processes. This can also be followed by robotic operations such as soldering some components. This paper presents the robotics of two operations to which an electric contactor is subjected. The first operation is the robotic fitting of a rubber bandage. The second robotic operation is the soldering of two electrical contacts. For the transfer of the pieces, a poly-gripper is used and it consists of six grippers with jaws, each of which grasp an electric contactor. This solution was chosen due to the large number of contactors to increase the productivity of the process. The technical projects and 3D models of the gripper and the poly-gripper are shown. For each robotic operation, the structure of the robotic cell is presented and the simulation of the transfer operations of the electrical contactors from the initial positions to the workstations and then to the storage places points of view is shown too.

## 1. Introduction

Currently, industrial robotics is highly evolving. Robotic applications are multiplied at the level of different technological processes. These processes can be complex or simpler. There are situations when a unit of average complexity requires operations that cannot be achieved at the desired productivity and efficiency parameters otherwise than with robotic equipment. This paper presents the case of a part to which a rubber band has to be added and after completion with specific electrical components, two double-wired electric contacts have to be soldered.

Given a relatively large number of parts, a solution of simultaneous execution was chosen on six parts, with two robotized operations: adding the rubber band and then soldering the electrical contacts. Robotized soldering is widely applied to electronic circuits [1-4]. Other significant data on robotic soldering applications are given in the papers [5-10]. However, it can be specified that the papers dealing with the problem of robotic soldering are relatively few, and those that present concrete solutions are very rare, which significantly raises the importance of this paper, which is absolutely original, in addition to the literature on robotic soldering applications.

## 2. Description of the working assembly

The assembly envisaged is an electric contactor consisting of a base part on which several components are mounted. In a first phase, a rubber band 1 according to figure 1 a, b has to be added, after which two electrical components (2 and 3) are added and then two stitches are necessary (4 and 5, figure 1c) of the wire ends coming from these electrical components. Until the soldering of the electrical contacts, a plastic ring 6 was added to the central housing.

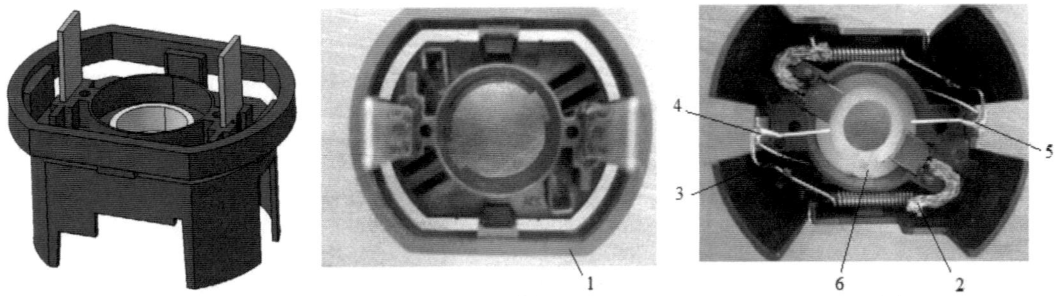


(a) (b) (c)

**Figure 1.** The electric contactor (a) without electric components (b) and with the electric components (c).

## 3. Designing the gripper and poly-gripper

In order to grip the basic part of the given assembly, a original gripper (figure 2) was proposed, whose jaws have stepped tips [11-16]. The step corresponding to the large diameter is required for gripping and displacement in order to add the rubber band when the central housing with a diameter of 20 mm is free, and the second step is required for gripping to ease handling for the soldering operation after adding a plastic ring in the central slot.

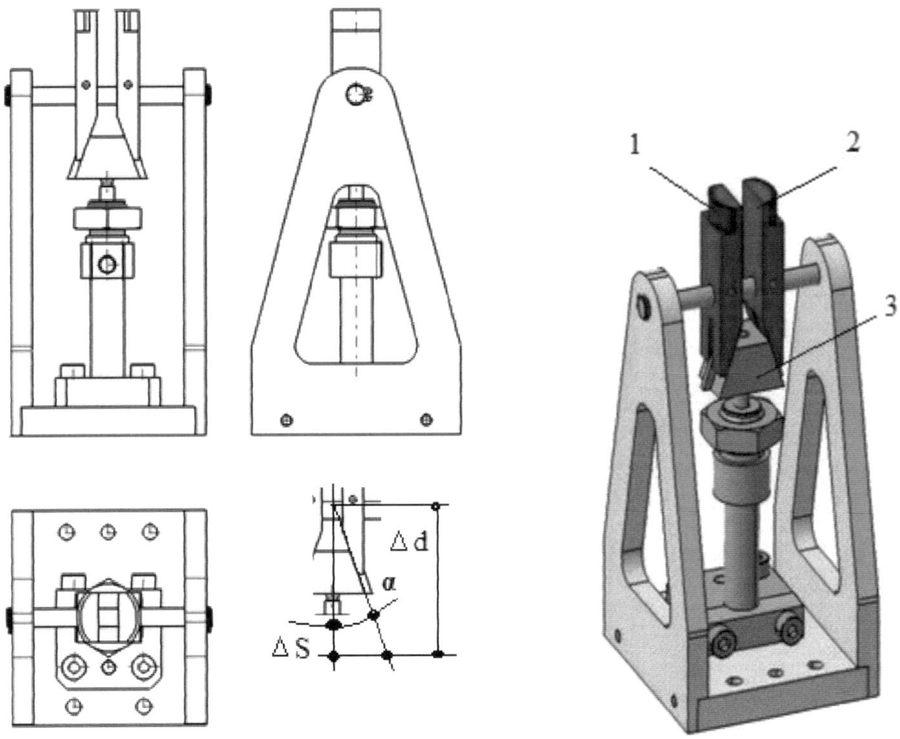


(a) (b)

**Figure 2.** Technical drawing (a) and 3D model of the gripper (b).

Jaws 1 and 2 (figure 2b) are displaced by means of a part, 3 mounted on the axle of a Festo linear pneumatic motor. The horizontal positioning of the jaws is $\Delta S = \Delta d * tg\alpha$ (figure 2a). Since the number of assemblies is estimated to be large to very large (10,000 per month), for higher efficiency the solution of designing a poly-gripper (figure 3) was adopted, which contains 6 grippers arranged on a support plate, which can be attached to the final part of the robot arm. In more detail, if we used a single gripper with jaws, the time of making a part (only applying the rubber band and then soldering the electrical contacts} was estimated at 240 seconds, a relatively long time, which is why it was proposed to use a poly-gripper consisting of 6 grippers, which can interact simultaneously with 6 contactors, so that the time for completing a single contactor part will be 40 seconds, which significantly increases the efficiency of the solution used.

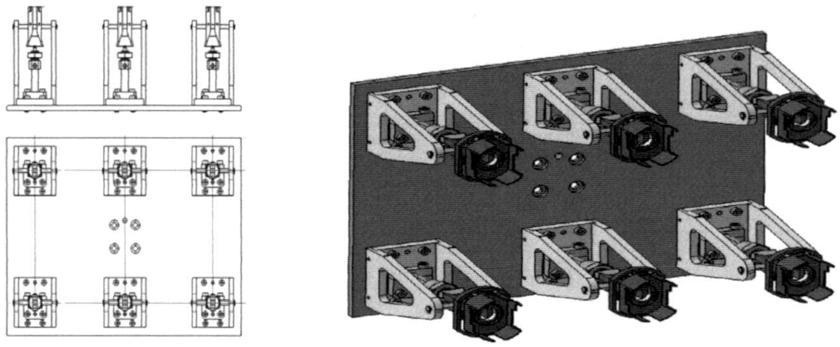

(a) (b)

**Figure 3.** Technical drawing (a) and 3D model of the poly-gripper (b).

## 4. Structure and functional simulation of the robotic cell

The robotic cell was designed in two variants. A first CI variant is for the operation of adding the rubber band, and the second CII variant is intended for the operation of soldering the electrical contacts. The structure of the robotic cell is original and was finalized after a careful analysis of the problem to be solved and of several known solutions [1, 4, 5]. To test the proposed solution, a simulation of the stages of making an electrical contactor is performed, which is described below. The simulation is useful for the validation of the proposed solution, but also for its optimization, being possible the necessary modifications in the CAD design phase, after which the solution is put into practice, thus the costs necessary for the modifications of the solution already made, generally important costs being eliminated.

### *4.1. The robotic CI cell for adding the rubber band*

*4.1.1. Structure of the robotic CI cell.* Figure 4 shows the structure of the robotic cell CI for adding the rubber band. It consists of the robot 1 equipped with the poly-gripper 2, the table 3 on which the plate 3.1 is found with the parts before the rubber band is added. Then, there is table 4 on which the special plate 4.1 is placed where the rubber band is inserted, then the gripped parts are transferred to the second plate 3.2 on table 3. Command unit of the robot is noted with 5.

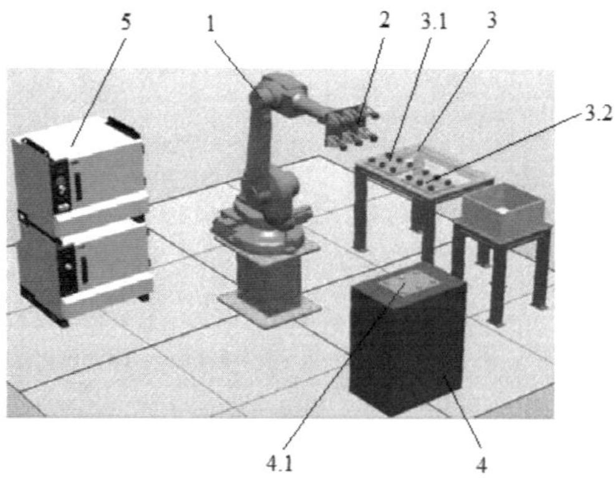


**Figure 4.** Structure of the robotic CI cell.

*4.1.2. Functional simulation of the robotic CI cell.* The robotized operation for adding the rubber band has the following main sequences (figure 5): moving the robot 1 from the initial position to table 3 (figure 5a); gripping the 6 parts on the support plate 3.1 (figure 5b); transferring the parts to the rubber banding table (figure 5c); placing the parts on the special worktop to add the rubber band (figure 5d); gripping the parts (figure 5e) and placing them on the support plate 3.2 on table 3 (figure 5f).

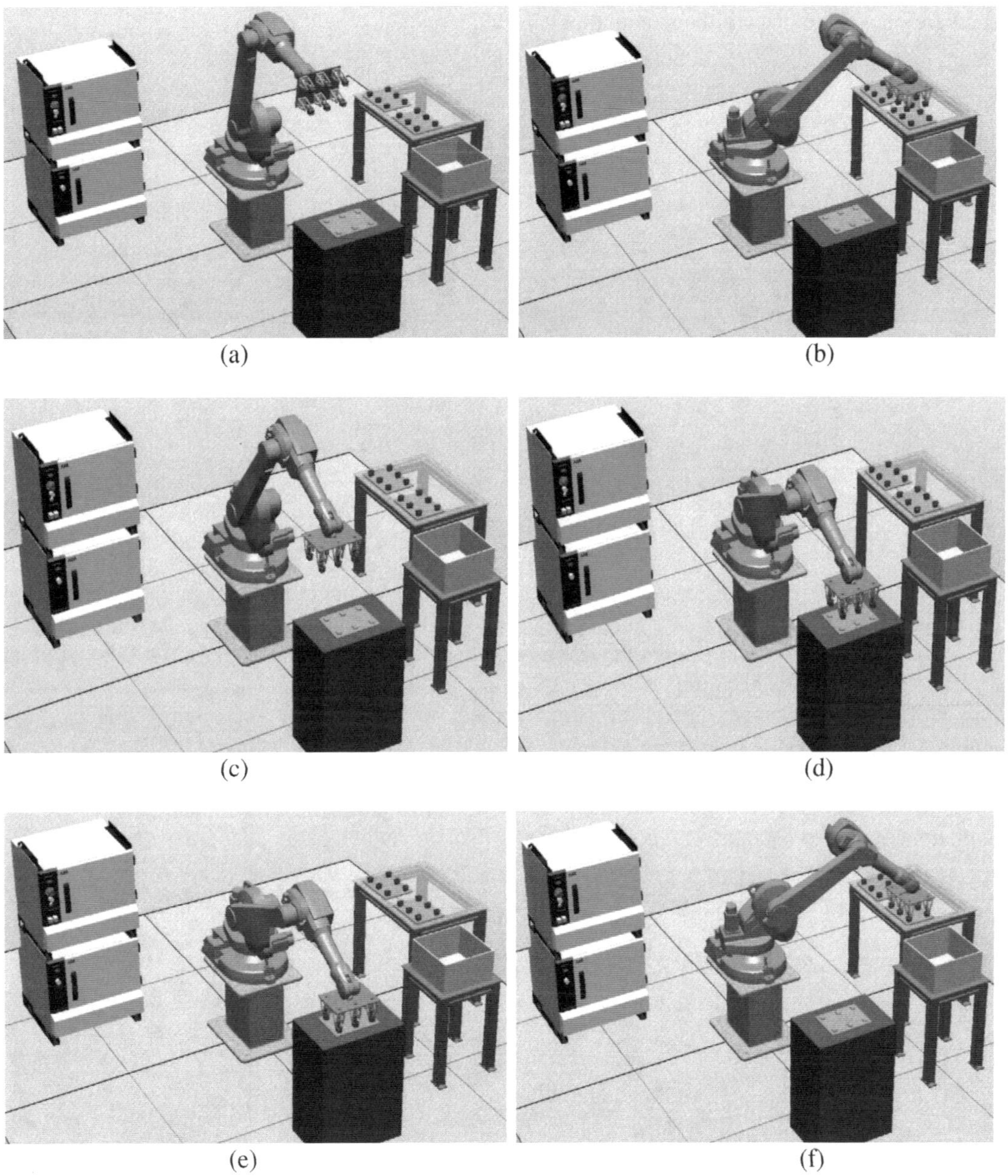


**Figure 5.** Functional simulation of the robotic CI cell.

Subsequently, the parts go into other operations to add electrical components.

### *4.2. CII robotic cell for soldering electrical contacts*

*4.2.1. Structure of the robotized cell CII.* The robotized cell for soldering electrical contacts consists of (figure 6): the robot 1 equipped with the poly-gripper 2, the support table 3 on which there are the support plates for the assemblies prepared for the purpose of soldering the electrical contacts, the table 4 on which there is the soldering robot 5 of UNIX-DF404S type [17] and table 6 on which box 7 is located where the assemblies will be stored after soldering.

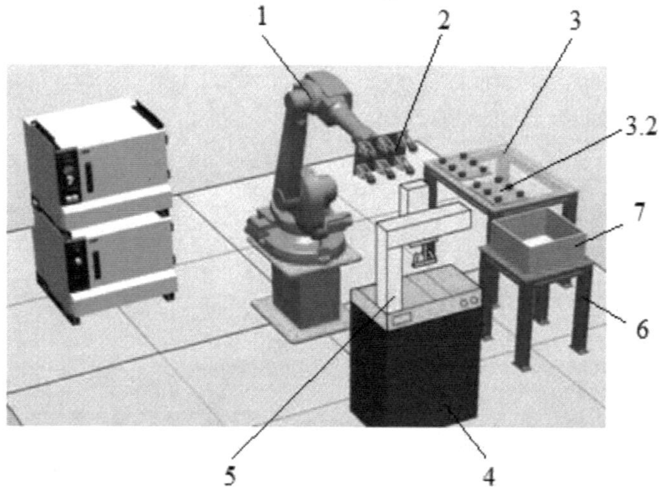


**Figure 6.** Structure of the robotized cell CII.

*4.2.2. Functional simulation of robotic cell CII.* The robotized operation for soldering electrical contacts has the following sequences (figure 6): moving robot 1 from the initial position to the support table 3, where the support plates containing the assemblies are located before soldering the electrical contacts; the six assemblies gripping on the first support plate 3.2; moving the robot to the soldering robot table; placing the 6 assemblies on the support plate on the soldering robot table; soldering the electrical contacts by the soldering robot using a special soldering unit; gripping the six units after soldering with the poly-gripper 2; displacement of the robot with the poly-gripper containing the six assemblies to the storage box (figure 7a) and release from the grippers of the 6 contactors with the electrical contacts soldered in the storage box 7 (figure 7b).

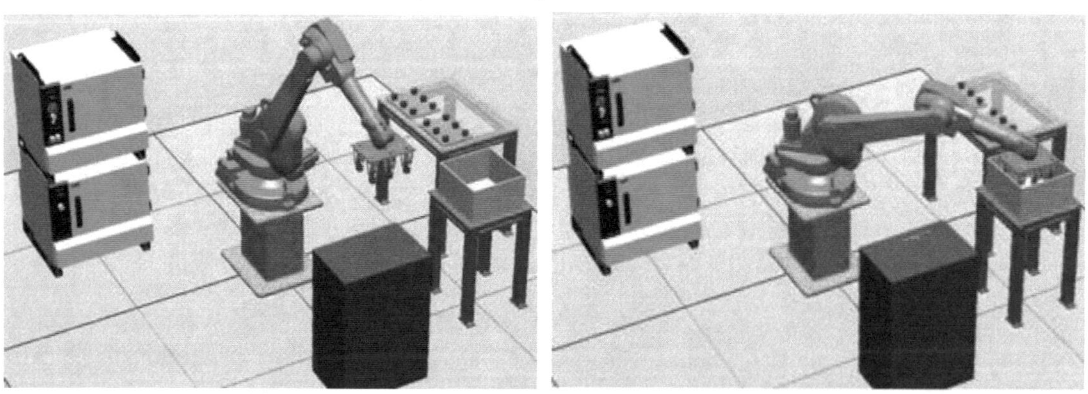

(a) (b)

**Figure 7.** Release from the grippers of the 6 contactors with the electrical contacts soldered in the storage box 7 (without the soldering robot 5-figure 6).

Obviously, the functional simulation described can be diversified to simulate operation in different situations such as the use of a single gripper for a contactor to accurately estimate the required execution times.

## 5. Conclusions

On the basis of the information presented in this paper, the following conclusions can be drawn:

- In order to obtain an average complexity unit, some phases of the technological flow can be robotized;
- In order to increase efficiency when conditions allow it, poly-grippers can be used that can grip and ensure the transfer of a larger number of units simultaneously;
- Depending on the construction features, appropriate adaptations will be used to reduce the robotic components needed, such as: stepping jaws to allow the grippers to be used for different configurations of the assembly involved in the robotized operation;
- The robotic cell used can be reconfigurable so that it can perform several types of operations in that technological flow;
- In order to verify the correct operation of the robotic cell and to optimize the structure and operation by modifications without additional costs, it is useful to simulate the CAD operation of the proposed robotic solution.

## References

[1] Ruggeri S, Fontana G, Basile V, Valori M and Fassi I 2017 *Procedia Manufacturing* **11** 441
[2] Zhang S *et al.* 2017 *Opt. Express* **25** 28838
[3] Xiang Li, Xing Su and Yunhui Liu 2017 *Proc. 18th International Conference on Advanced Robotics (ICAR)* (Hong Kong) 216
[4] Li X, Su X and Liu Y 2017 *Proc. IEEE/RSJ International Conference on Intelligent Robots and Systems* (Vancouver) 1651
[5] Chia K L, Choo B L, Lee S J, Luang C T S, Tan M S and Truman T 2018 *Pan Pacific Microelectronics Symposium (Pan Pacific)* Waimea HI 227
[6] Norman R A 2015 *Perma Works LLC, Grant DE-FG36-08GO10185*
[7] Santiago S B, Oliveira Almeida E L de and Dias J O 2019 *Proc. of the International Conference on Industrial Engineering and Operations Management* (Toronto) 195
[8] Draghiciu N, Nicolae, Burca A and Galasel I T 2017 *J. Comput. Sc. Control Syst.* **10** 11
[9] Ribes- Pleguezuelo P, Koechlin C, Burkhardt T, Hornaff M, Kamm A, Gramens S, Beckert E, Fiault G, Eberhardt R and Tünnermann A 2016 *Proc. SPIE 10562, International Conference on Space Optics — ICSO* (Biarritz) **10562** 105620Q-1
[10] Neubert J and Lipson H 2016 *Autonomous Robots* **40** 139
[11] Staretu I 2011 *Gripping systems* (Tewksbury, Massachusetts: Derc Publishing House)
[12] Staretu I 2010 *Sisteme de prehensiune (Grip systems)* (Brasov: Lux Libris)
[13] Monkman G J, Hesse S, Steinmann R and Schunk H 2007 *Robot Grippers* (Weinheim: Weley VCH Verlag GmbH & Co. KGaA)
[14] Staretu A I 2015 *Military Technical Courier* **LXIII** p 30
[15] Bolboe M, Staretu I and Alexandru P 2014 *Journal of Mechanics Engineering and Automation* **4** 234
[16] Staretu I and Ionescu M 2005 *Proc. of the* $9^{th}$ *IFToMM International Symposium on Theory of Machines and Mechanism* (Bucharest) 815
[17] Information on: https://www.japanunix.com/en/products/automation/410series/410r/

# Generation of precessional gear teeth by plastic deformation

**N Trifan**[1*], **R Ciobanu**[1] **and O Ciobanu**[1]

[1]Technical University of Moldova, Department Fundamentals of Machinery Design, Chisinau, Republic of Moldova

*E-mail: nicolae.trifan@bpm.utm.md

**Abstract**. The correct choice of the dimensions of the workpiece product is one of the main problems, the solution of which depends on the reduction of the consumption of materials and energy, as well as the quality of the wheels obtained by plastic deformation. Referring to the manufacture of conical wheels with convex-concave profile of the teeth by knurling, the height of the teeth is determined by the condition of the equal volume of metal that moves from the gaps between the teeth to their tip during the plastic deformation by rolling. The use of methods for the manufacture of teeth without chipping will allow the increase of the utilization coefficient of expensive materials.

## 1. Introduction

Gears are machine parts that occupy an essential place in the operation of various mechanical constructions. The execution of the gears at a high-quality level and at a low cost, puts in front of the specialist multiple problems.

From the design stage, the specialist must take certain measures to achieve a high reliability of the designed gears. Shapes of teeth profiles that can be obtained by plastic deformation [1, 2] for use in precessional multipliers are presented. Starting from the correct choice of materials and dimensions of the workpiece product, the designer must adopt constructive solutions of the gears that can be made in the most economical conditions. Previous research by the author has been presented in [3, 4, 5].

## 2. Determination of the displacement value of the deformation node in the direction of the axial feed

Based on the computerized model of the toothed wheel plastic deformation device, the working strokes (vertical table feed), the number of revolutions of the main machine tool shaft, were analyzed to ensure the deformation speed prescribed by the technical literature.

Determination of the vertical feed of the table is necessary for the complete teeth formation [3, 4, 6].

Determining the total stroke of the plastic deformation rollers in order to obtain the full height of the teeth requires the determination of the height of the teeth, which is calculated from:

$$h_{th} = R_{out} \cdot tg2\theta \tag{1}$$

where: $R_{out}$ is the outer conical radius of the blank, mm;

$\theta$ is the angle of nutation, degrees.

The height increase of the tooth according to the vertical advance of the plastic deformation node, in which the plastic deformation rollers are fixed, was determined according to the relation:

$$\Delta h_{pc} = s_{pc} \cdot cos(\delta + \beta + \theta) \quad (2)$$

where: δ is the angle of the conical axoid, degrees;

$s_{pc}$ is the advance of the plastic deformation node to a precession cycle, mm;

β is angle of conicity of the rollers, degrees.

The value of the stroke of the plastic deformation node in which the plastic deformation rollers are fixed:

$$s_t = \frac{h_{th}}{cos(\delta + \beta + \theta)} \quad (3)$$

According to the literature recommendations for the total teeth profile formation [5], the summary advance was divided into steps, measured by the digital indicator. The results of calculating the penetration advance according to tooth height are presented (see table 1).

**Table 1.** The advance depending on the height of the teeth.

| Step No | The angle of nutation θ, degrees | Angle of the conical axoid δ, degrees | Angle of conicity of the rollers β, degrees | Height of the teeth $h_i$, mm | The advance, $S_{pci}$, mm |
|---|---|---|---|---|---|
| 1. | | | | 13.72 | 13.73 |
| 2. | | | | 14.22 | 14.23 |
| 3. | 3.0 | 0 | 3.5 | 14.72 | 14.73 |
| 4. | | | | 15.22 | 15.23 |
| 5. | | | | 15.52 | 15.53 |

## 3. Forces determination for plastic deformation with precession tool

Knowledge of the forces necessary for the formation of the teeth is an important factor for the calculation of the plastic deformation of toothed wheels. It should be noted that in all of the conceptual developed diagrams the plastic deformation tool performs sphero-spatial motion, which reduces by about 60% the value of the forces required for tooth deformation due to the reduction of the instant contact surface of the tool with the blank. The progressive plastic deformation aims to reduce the deformation pressure by dividing the deformation width into several sectors. The plastic deformation of the teeth with a precession tool and deformation with i rolls is further examined. In figure 1 (a) the action of the deformation rollers on the blank is replaced by an equivalent roll. The torque transmission (forces) from the main shaft is carried out with arms help $F_{hi}=f(e)$ figure 1 (b), measured in the planes passing through the application points of the $F_{rhi}$, levers $L_1$ şi $L_2$: $F_{r_{hi}} \cdot L_1 = R_m \cdot L_2$. The deforming forces acting on the workpiece, when the deformation process is already stabilized, are shown in figure 1 (c). The deformation rolls act with the normal force $F_n$ and the tangential force $F_t$, and $R_m$ is their result.

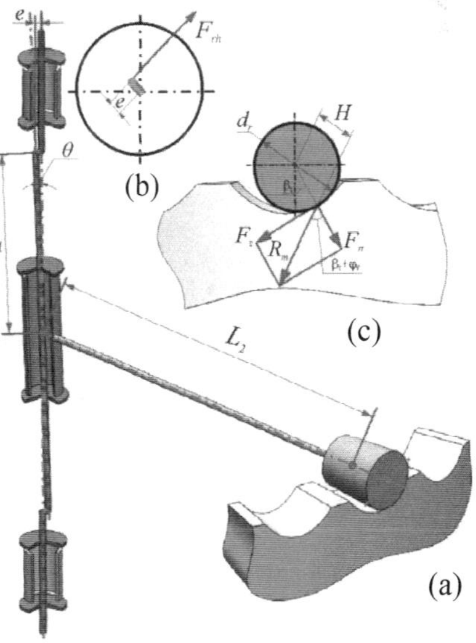


**Figure 1.** Scheme for determining the applied torque to the main shaft.

The normal force, being the sum of the elementary forces relative to the surface unit of the deformation zone, results from the relation: $F_n = AP_m$ where A is the contact surface in the deformation zone, and $P_m$ is the mean pressure in the deformation zone. From figure 1 (c) results:

$$R_m = \frac{F_n}{cos(\beta_1 + \varphi_f)} \quad (4)$$

The friction $\varphi_f$ and pressure $\beta_1$ angles, with small values, can be approximated: $R_m=F_n$, resulting in: $R_m = AP_m$. The average $P_m$, pressure, in the case of hot deformation, as indicated in the literature, varies within very wide limits: $P_m = (0{,}5\ldots9)\tau_r$ in which $\tau_r$ is the tensile strength of the material of the workpiece subjected to deformation.

The torque T required to deform the material is determined from the relationship:

$$T = R_m H = AP_m \frac{d_r}{2} sin(\beta_1 + \varphi_f) \quad (5)$$

The torque applied to the crank shaft is determined from the relationships shown in figure 1.

The $F_{rh}$ force is determined from the relationships:

$$F_{rh}L_1 = R_m\, cos(\beta_1 + \varphi_f)L_2$$
$$F_{rh} = \frac{R_m\, cos(\beta_1 + \varphi_f)L_2}{L_1} \quad (6)$$

The torque applied to the crank shaft is determined from the relationships, which from figure 1 (b) results: $T_h = F_{rh}e$. Replacing the expression for $F_{rh}$, also $L_2 = R_{m2}$ (the average conic length of the manufactured wheel) and $L_1 = e/tg\theta$ obtain the relation: $T_h = R_m\, cos(\beta_1 + \varphi_f)R_{m2}tg\theta \approx AP_m R_{m2} tg\theta$.

In the obtained relationship, $R_m$ represents the equivalent resultant force, with which it acts on the equivalent of the roll of the semi-fabric teeth.

In fact, thanks to the large number of rollers, which simultaneously participate in the plastic deformation process of the teeth of the blank $(Z_2\text{-}1)/2$, the load, which returns to a deformation roll, is smaller.

At the basis of the elaboration and design of the experimental devices of plastic deformation were taken in account the following technical solutions.

In order to simplify the construction of the device (figure 2), a construction that differs from the wheel machining device has been developed in the absence of the mechanism for correcting the profiles of the teeth [1].

(a) (b)

**Figure 2.** Device construction: (a) measuring the speed of the crankshaft; (b) the plastic deformation node.

IOP Conf. Series: Materials Science and Engineering 1009 (2021) 012057 doi:10.1088/1757-899X/1009/1/012057

Also, the device does not have the mechanism of transformation of the movement, which requires the machining of the blank by teeth of the linear profile.

According to the researches carried out, the maximum load, which corresponds to a deformation roll, does not exceed 20-30% from the total load, taking into account the unevenness of the distribution of the deformation force between the rollers simultaneously participating in the plastic deformation process of the teeth, characterized by possible manufacturing errors and deformation of the technological equipment parts [5].

Figure 3 (a) – (d) shows the evolution of the plastic deformation of the teeth of the central wheels of the precessional gear with convex-concave profile of the teeth.

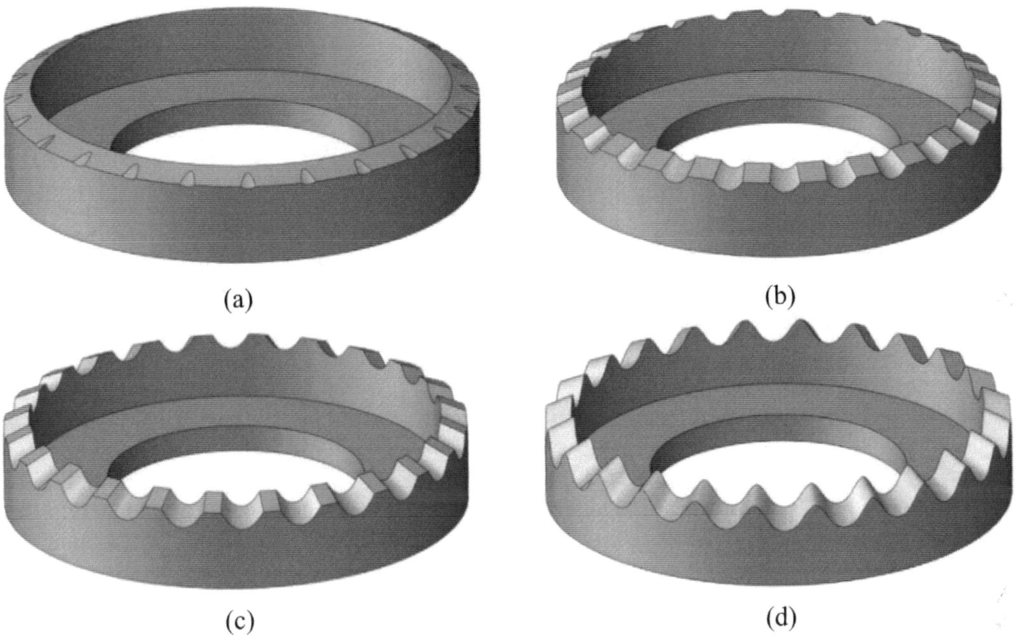


**Figure 3.** The evolution of the plastic deformation of the teeth from the precessional gear: (a) the beginning of the deformation; (b) 25% tooth height; (c) 75% tooth height; (d) 100% tooth height.

## 4. Arguing the choice of the teeth profile of the central wheel of the precessional multiplier

In figure 4 there are presented teeth profiles selected from the set of profiles obtained by the authors in order to have some geometric parameters depending on the pre-existing angle: the angle of the conical axoid $\delta = 0 \ldots 30^{\circ}$; nutation angle $\theta = 1.5^{\circ} \ldots 3^{\circ}$; the conicity angle of the rollers $\beta = 4^{\circ} \ldots 7^{\circ}$; number of teeth $z_1$ and rollers $z_2$, $z_1 = 10 \ldots 60$; $Z$, mm the height of teeth; $Y$, mm the width at the foot of the teeth; the relationship between the number of central and crown wheels of the satellite, $z_1 = z_2 \pm 1$.

The analysis of the forces acting in contact with the teeth of the central wheel and the rollers of the satellite crown showed that the minimization of power losses in the multiplier gear is achieved at high gearing angles (low pressure angles) [7]. At the same time, at low gear angles, self-braking effects occur during the operation of the transmission in multiplication mode [5, 7, 8].

The nutation angle θ (inclination of the crank) must be as large as possible to ensure a higher load-bearing capacity (a torque applied to the leading element) and, at the same time, as small as possible to ensure minimum height and extended profile of the teeth.

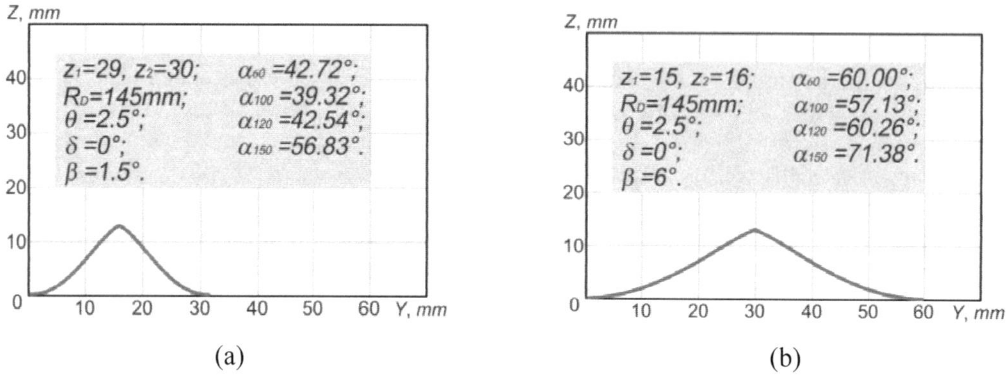


**Figure 4.** Tooth profiles: (a) samples of tooth profiles, which are not recommended for operation in multiplication mode; (b) tooth profiles which are recommended.

## 5. Conclusion and future research

The identification of the profiles of the teeth that can be obtained by plastic deformation will allow the establishment of the manufacturing technology still at the design stage. For future research, it is proposed to study more deeply the processes that take place in the plastic deformation with precessional tool of the teeth with convex-concave profile, including at micro - and macro - level, with the prediction of the mechanical properties of the surface layer.

**Acknowledgments**

This paper was elaborated under Moldova State Project 160-PS from 03.01.2020 with title *Increasing the competitiveness of precessional transmission by developing and capitalizing „conform" toothed gear and expanding their range of application.*

## References

[1] Behrens B A and Odening D 2009 Process and tool design for precision forging of geared components *Int. J. of Mater. Form.* **2** 125-28

[2] Irani M, Amini Najafabadi R and Dallali Isfahani T 2014 A study of precision forging of steel spur gears: upper bound model and experiments *International Journal of ISSI* **11** (1) 18-23

[3] Trifan N 2014 *Contributions to the Generation of Precessional Gear Teeth by Plastic Deformation* PhD Thesis UTM p 174

[4] Bostan I, Dulgheru V and Trifan N 2018 Generation of the teeth precessional gears by plastic deformation *IOP Conf Series: Materials Science and Engineering - The 8th Int. Conf. on Advanced Concepts in Mechanical Engineering - ACME 2018 JUNE 07 - 08* (Iași: Romania) **444** 022001

[5] Bostan I 2019 *Precessional Transmissions: Contact Geometry, Surface Generation and Applications* (Ch.: Ed. Bons Offices SRL) p 639

[6] Bostan I, Dulgheru V and Trifan N 2016 Characteristics of materials and thermal treatments applied to gearwheels obtained by plastic deformation *IOP Conf. Series: Materials Science and Engineering 7th Int. Conf. on Advanced Concepts in Mechanical Engineering 9–10 June 2016* (Iasi: Romania) **147** p 29 – 33

[7] Ciobanu R 2014 *Contributions to Research Planetary Precessional Multipliers Gear Box* PhD Thesis UTM p 145

[8] Bostan I 2019 *Precessional Transmissions: Synthesis, Kinematics and Elements of Calculation* (Ch.: Ed. Bons Offices SRL) p 477

# Investigation of an axial hydrokinetic turbine with the CFD program Flow Simulation

**K Tuzharov[1], S Iliev[1]*, V Vilag[2] and J Vilag[2]**

[1] University of Ruse, Faculty of Industrial Agriculture, Faculty of Transport, 7017 Ruse, Bulgaria,
[2] COMOTI Romanian Research and Development Institute for Gas Turbines, Department "Aviation and industrial turbines. Gas turbine assembly", D Iuliu Maniu Bd., sector 6, cod 061126, OP 76, CP174

*E-mail:spi@uni-ruse.bg

**Abstract**. In the work, the characteristics of a hydrokinetic axial turbine are obtained by numerical CFD modelling. Using the new research tool, Fluid Computational Dynamics, arises the question as to the accuracy of the solution. That's why for solving the problem, has been used new numerical method "Sliding mesh", which is included in the "Flow Simulation" soon. Besides, the numerical results for frequency characteristics are compared with the characteristics obtained by modified analytical model. It was found that for both solutions - for the output powersof the turbine, the differ is (5 ... 15) % and for the torques of the working shaft the differ is (4...18) %. Because it's accepted that the analytical solution is more accurate as it is done by a modified classical method, the compliance of the shape and energy performance characteristics is considered to be a numerical model verification obtained by simulation with the Flow Simulation CFD.

## 1. Introduction

The Interest of hydro kinetic turbines began in 1973 due to the worsening energy and ecological crises on the planet. Recently, the new research tool, Fluid Computational Mechanics (CFD), has been used for their research. In this approach, the flow area is calculated by means of different forms of the Navi-Stocks' private differential equations, including closing turbulent equations known as turbulent models, with which the differentiate system is reduced into a solvable state. The fluid area (computing area) is divided into multiple volumes (cells) for each of which the flow parameters are calculated. This method does not make too many simplifying prerequisites. The question arises as to the accuracy of the solution as compared to the analytical methods of research.

## 2. Exposition

The object of the study is a model of the hydrokinetic turbine (figure 1), consisting of a grid 1, a rotor 2 which acts as a casing of the electric generator, the flow channel in the housing 3, the impeller 4 consisting of four blades, a hub, the diffuser 5 and the carrier 6 for connecting the hydro unit to the turbine housing. The impeller is with an outside diameter $d_2$=0.935 m , internal diameter $d_1$=0.180 m, the aerodynamic profile of the blades is Espero with relative thickness$\delta$=0.2, with chords at external and internal diameters, respectively $c_2$=0.160 m, and $c_1$=0.135 m. The flow of the working fluid through the hydrokinetic turbine actuates the impeller 4, achieving the equality between torque created by the turbine and the resistor torque.

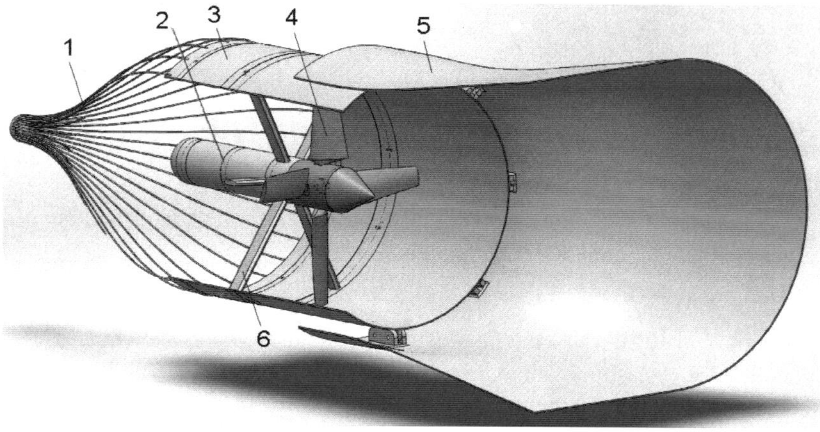


**Figure 1.** Hydrokinetics turbine. 1-protecting grid; 2- electric generator housing; 3-turbine housing; 4-wheel drive (blades, hub); 5-diffuser; 6-struts for connecting the hydro unit to the turbine housing.

*The purpose* of the report is to obtain the frequency characteristics of the hydrokinetic turbine which are dependent on the rotational force and shaft power from the impeller speed at constant velocity v=2m/s of the flow. To accomplish the goal, the following tasks were performed: a physical model of the Solid Works turbine and a numerical model (figure 2a) was created with the CFD Flow Simulation program using the Wizard option (built-in project management program). The type of analysis is internal, three-dimensional, with the computing area before and after the turbine being extended (not shown in figure 2a) in order to achieve convergence and higher accuracy of the solution.

Prerequisites for the study are that the working fluid is pure water, the flow is turbulent with a laminar sublayer on the walls, the walls are thermoinsulation (the flow is adiabatic). Taking into account that there are sealing cavities in the geometric model to be excluded from the pattern, the *Exclude cavities without flow conditions* options are activated.

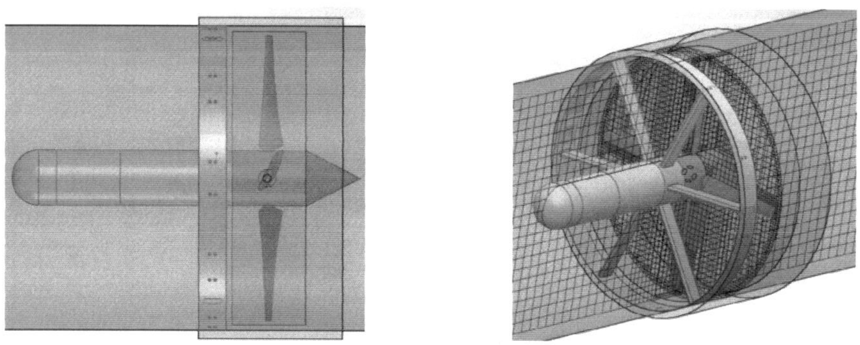

(a) (b)

**Figure 2.** CFD model and computing mesh in isometry.

The general project settings are shown in table 1. The table shows that the default conditions are left by default by activating the Pressure potential option. In Flow Simulation, turbulent fluid flows are modeled with the Navier-Stokes equations [1, 2] which describe the laws of storage of matter, impulse and energy. Flow parameters are averaged over time, according to Reynolds. As a result, the equations have additional members - Reynolds tensions. To close the system of equations in Flow Simulation, the equations for the transfer of turbulent kinetic energy and its dissipation within the turbulent model $k-\varepsilon$ [1, 2] are used.

**Table 1.** The general project settings.

| | |
|---|---|
| Physical features | *Rotation* |
| Wall Conditions | Default smooth walls |
| Initial and Ambient Conditions | Thermodynamic parameters: Temperature=293 K; Pressure: 101325 Pa; Turbulence parameters: turbulence intensity 5%, turbulence length 0,01 m; turbulence energy k=1J/kg and decipation $\varepsilon$ = 1 kW/kg. |
| Result and Geometry Resolution | Level of Result resolution equal to 5; Minimum gap size = 0.01 m, other options according to default. |

An auxiliary body was created to simulate the rotation, which covered the turbine with a reserve. The body diameter is 1050 mm, the width is 250 mm. These dimensions are determined by the condition that they exceed the outside diameter and the width of the impeller. For the rotating area, the current values are set to the stepped angular velocity $\omega = 0..20\, s^{-1}$ with step $\Delta\omega = 5s^{-1}$and the Disable attribute is assigned, which turns the body into a transparent for the flowing environment.

The boundary conditions are set by closing the edges of the extended computing area with lids. The velocity of the flow (v=2m/s) and the outlet pressure (p=101325 Pa) and ambient temperature (T=273K) are set on the surface of the inlet. The roughness of the walls is neglected.

To set up the home network, the Automatic settings mode is turned off, and the number of cells of the base network by coordinates, the criterion for partitioning the dividing network with the liquid medium, narrowing the narrow channels to 3 cells is determined, provided the system can a refining level is achieved 6.

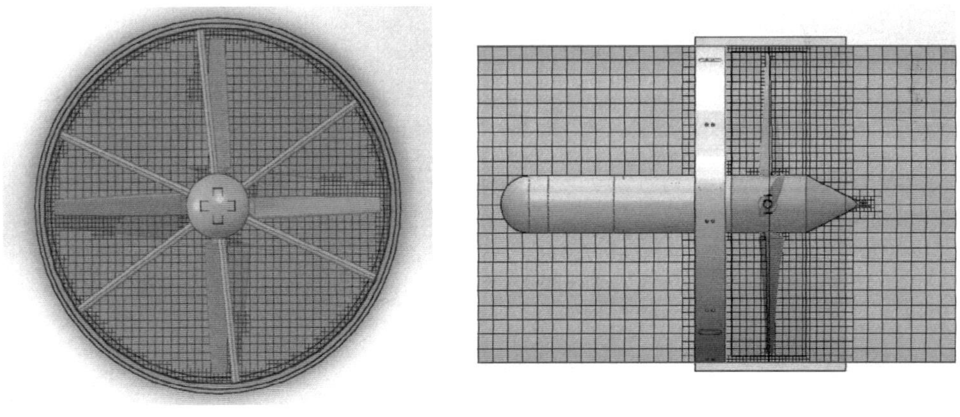

(a) (b)

**Figure 3**. Computing meshes about planes x-y and z-y.

To optimize the network around the impeller, a functionalized *Local Initial Mesh* is used. A body has been created in advance that covers the area. The *Tolerance Refinement Criterion* in the Solid / Fluid Interface option has been increased to 6 by setting the crushing level of the cells. Split to level 2 all the cells that fall in the Refining cells option and increase to level 4 the desired number of cells across the narrow channels in the Narrow channels option.

Fragments of the resultant mesh in isometry are shown in figure 2b, and in transverse x-y and longitudinal z-y sections in figure 3. It yields 142,745 liquid and 41,587 partial cells. As far as the result is concerned, the mesh is sufficiently rational - along the blades there are 30..40 cells and in the width of 12..20 cells, the mesh thickens at the blades and the hub walls, as well as in the gap between the rotor and the flow tube.

A way to compare the theoretical currents obtained by different methods is by comparing the results for control points, control lines or control loops. Figure 4a shows two control loops (a circle with a radius of 350 mm and a rectangle with sides 700x1350) and control lines in the direction of the blades with a length of 350 mm.

For the purposes of the project, the determination of the pressure and flow distribution in the flow, the average values of pressure and velocity in the inlet and outlet sections, the reactive torque in relation to the z axis, which the flow creates on the rotor, the inlet power of the flow, the output mechanical power over turbine shaft, torque and power ratios, and turbine speed.

The pressure and flow velocities in the flow are obtained as a result of the solution of the Navier-Stokes private differential equations. The torque $M_z$ and the force $F_z$ values for different rotation angles at a certain pressure distribution are modeled using the built-in universal functions:

$$M_z = \int_{r1}^{r2} dM_z(r)\ , F_z = \int_{r1}^{r2} dF_z(r) \tag{1}$$

where$r_1$,$r_2$ are the inner and outer radii of the impeller; r - the current radius along the blades of the impeller; $dM_z$and $dF_z$- the current values of the elementary torque generated by the pressure distribution under and above the blades and the elementary axle force generated by the difference between the pressures before and after the impeller.

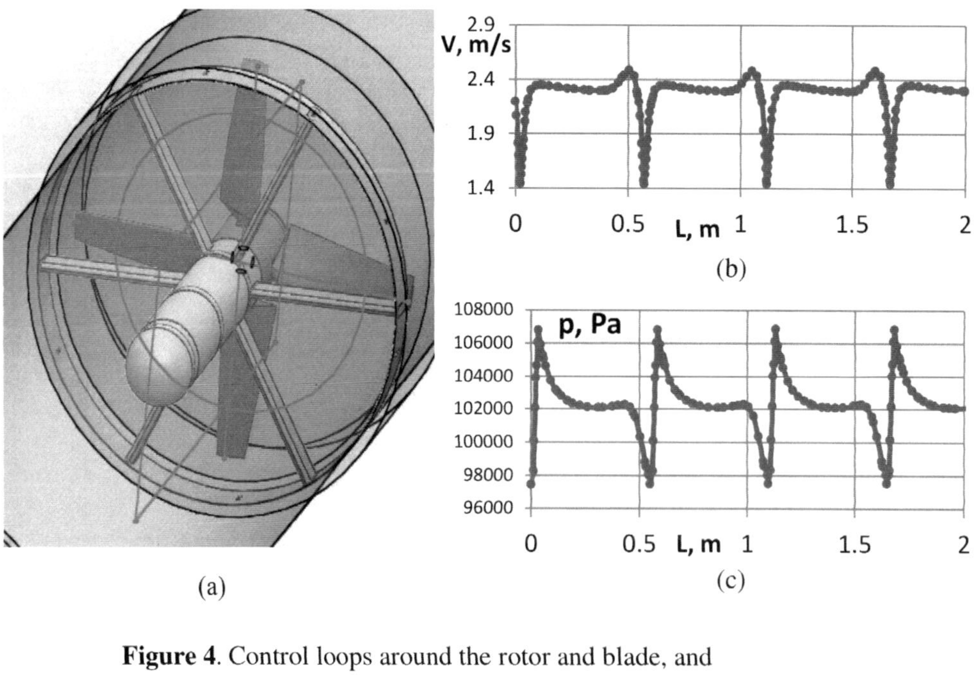


**Figure 4**. Control loops around the rotor and blade, and velocity and pressure distribution along a circle with a radius of 350 mm.

The average torque values and the front force are calculated on the basis of the data for the instantaneous values with the dependencies: $M=\frac{\sum_1^N M_z}{N}$ и $F=\frac{\sum_1^N F_z}{N}$, where N is the number of steps for one complete rotation of the rotor.

The powers and the dimensionless turbine coefficients are calculated by inserting dependencies for the input power with the Insert *Equation Goals option:*

$$P_1=\frac{\pi}{8}\rho D^2 \upsilon^3 \quad (2)$$

- The output power

$$P_2=M\omega \quad (3)$$

- The speed of rotation

$$n=\frac{30}{\pi}\omega \quad (4)$$

- The speed ratio

$$\lambda=\frac{\pi}{30}\frac{R}{\upsilon}n \quad (5)$$

- The torque ratio

$$C_M=\frac{8M}{\rho\pi D^2\upsilon^3} \quad (6)$$

- The power factor

$$C_P=C_M\lambda \quad (7)$$

## 3. Results and analysis

As a whole, in the course of the solution, the task exhibits stability and rapid convergence especially for the power indicators - torque and axle force, as well as for all project objectives – the power in the input and output of the rotor, the torque and power coefficients. Figure 5 shows that the task reaches at the 150th iteration.

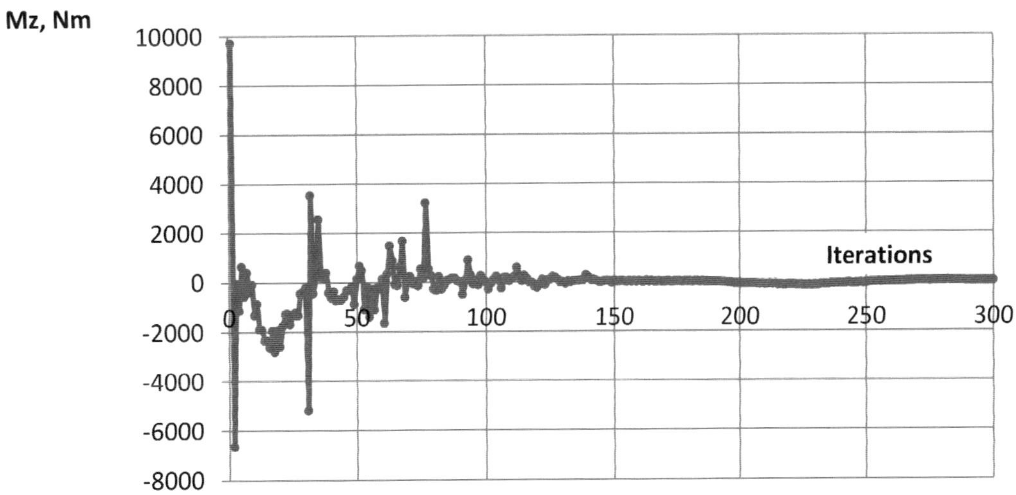


**Figure 5.** Convergence of the torque solution depending on the number of iterations.

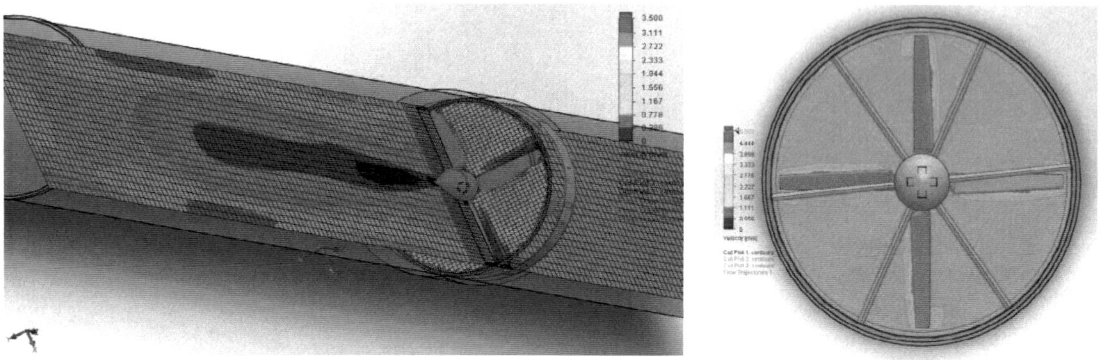

(a) (b)

**Figure 6.** Longitudinal and cross section of a velocity field (isotache) at rotational speed $n=76,5\ min^{-1}$.

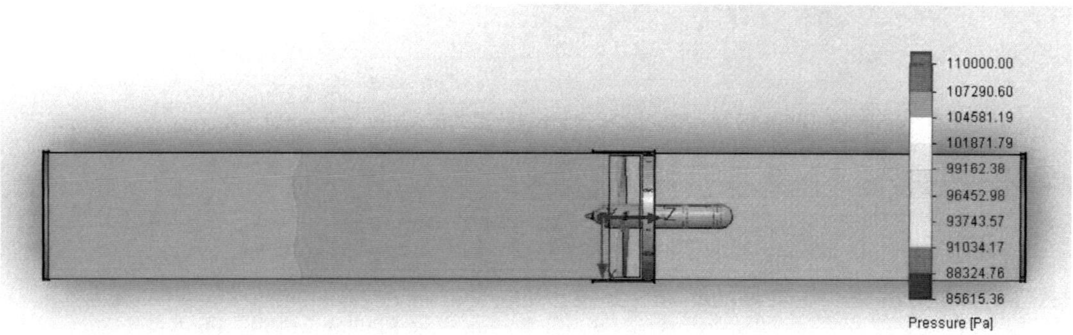


**Figure 7.** Lengthways section of static pressure field.

As stated in the introduction with the Reynolds equation, the distribution of velocities and pressure through the turbine flow channel are determined.

The results for the velocity field in the longitudinal and cross section of the flow channel at rotor speeds n=76.5 $min^{-1}$ are shown in figure 6, and in figure 7 for the static pressure field in longitudinal section. It can be seen that the flow before the rotor is steady, uniform with an average constant velocity 2 m/s (figure 6a). The carriers 6 (figure 1) negligibly disturb the flow, but the impeller exchanges energy with the water flow, creating a strong redistribution of speed. Its highest values are on the front edge and back of the blades, and the lowest on the low part of the blades caused by the circulating velocity around them. Because of the contraction, the velocity through the rotor and beyond it has higher values 2.7 m/s - on n= 76.5 $min^{-1}$and 3.3 m/s at n=214 $min^{-1}$.Separation of the flow after the impeller creates a highly swirling zone with low speeds 0 ..0.5 m/s which, as the rotational speed of the rotor increases, moves away from the plane in which it is located.

The velocity field forms the pressure field, which has three clearly defined zones - one before the rotor, and two after the rotor. With average pressure values, respectively, $p_1$=104600 Pa, $p_2$=97500 Pa, and $p_3$=93600 Pa.

The pressure field forms the force interaction of the flow with the rotor, respectively the power indicators - the torque and the force from the front pressure.

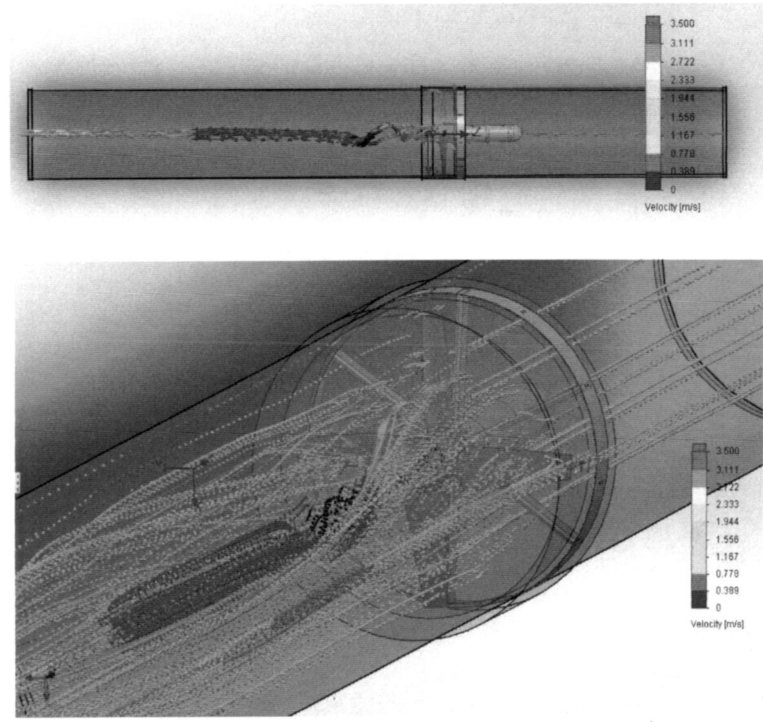


a

b

**Figure 8**. Trajectories and isotaxs at n=214 min$^{-1}$.

Another characteristic phenomenon that occurs after the rotor is the rotation of the flow. It complicates the flow and is illustrated with the current lines shown in figure 8. In front of the rotor they are parallel and then rotate in the direction of rotation of the blades with an angular velocity depending on the rotor angular velocity. Current lines are combined with the isotahs (lines at constant speed). The absolute speed after the impeller increases to 3.5 m/s due to the contraction of the flow path and the flow rotation.

The distribution of velocity and pressure on closed control loops (figure 4b, c) around the impeller, for example, on a 700 mm diameter circle, as mentioned above, can serve as a benchmark for different test methods.

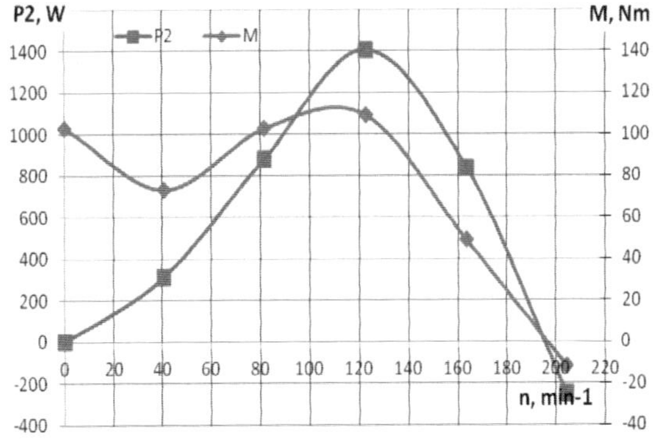


**Figure 9.** Frequency characteristics obtained by CFD simulation.

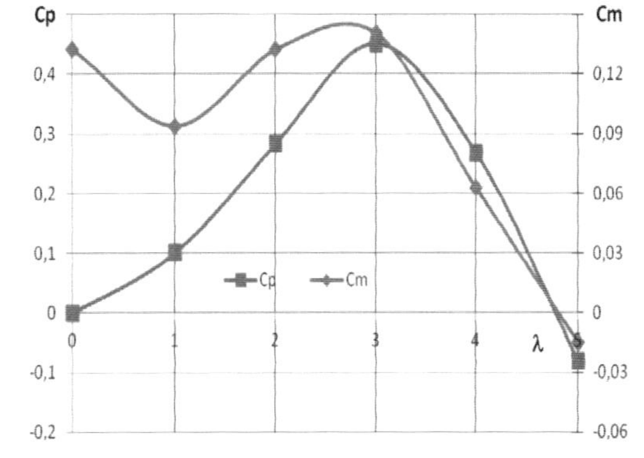


**Figure 10.** Aerodynamic characteristics $Cm = f(\lambda)$ и $Cp = f(\lambda)$ obtained by CFD simulation.

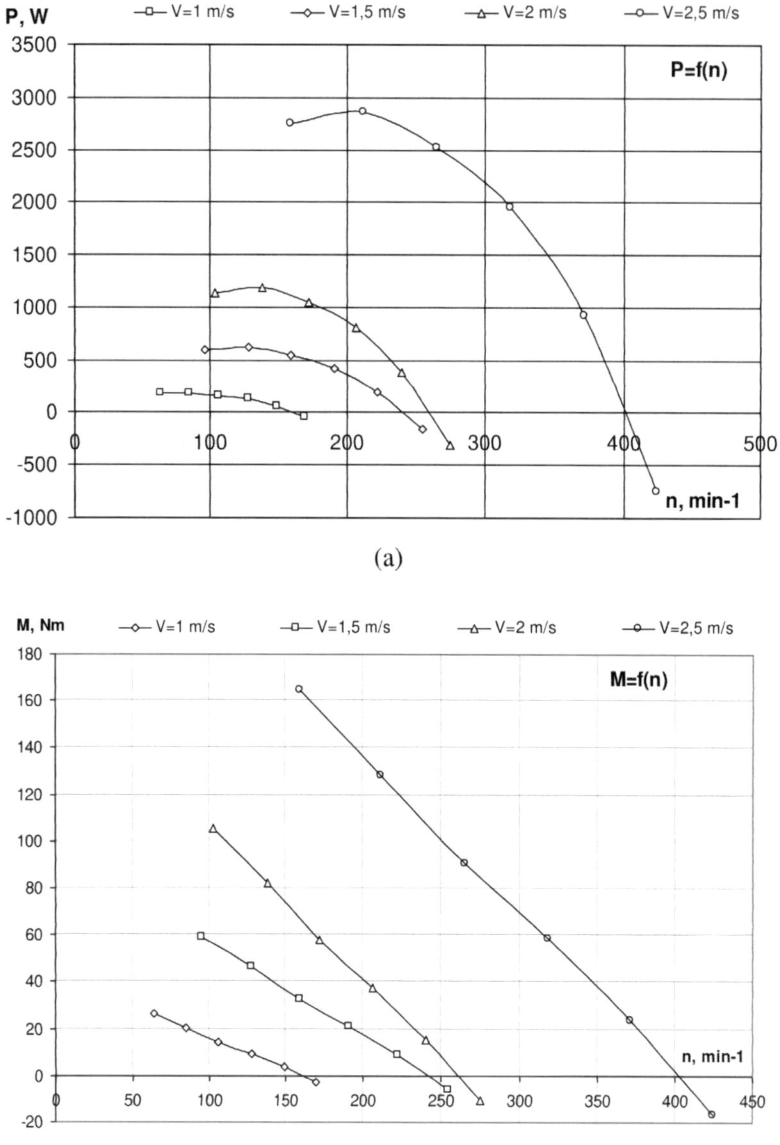


**Figure 11.** Frequency power and torque characteristics obtained by analytical method.

The results of the calculations in the form of torque M frequency characteristics and shaft power $P_2$are shown in figure 9. The value of the starting torque at zero rotor speed is 100 Nm, at speed n=50 $min^{-1}$ decreases to value of 75 Nm at the nominal speed n=125 $min^{-1}$, rises to a nominal value $M_{nom}$=110 Nm, then decreases smoothly and at zero synchronous rotation speed n=190 $min^{-1}$reaches zero. The power curve is parabola with the maximum at the rated rotation speed, which value is Pmax= 1400 W.

After undemensioning of the frequency characteristics the aerodynamic characteristics, which are dependencies between the torque coefficients $C_m$ and power coefficient from the speed ratio$\lambda$, are obtained (figure 10).In figure 11 are given the frequency characteristics of the shaft power and torque of the hydrokinetic turbine, obtained by the analytical method at different water velocities [4]. The comparison of the results is based on characteristics at water velocity v=2 m/s (lines with triangular marks). The match between the characteristics is very good at rotations of 100....150 $min^{-1}$, that is, in the field of nominal operating modes.

## 4. Conclusions

From the frequency characteristics for the torque and power (figure 9) and the hydrodynamic characteristics between the torque coefficients and the power versus speed ratio (figure 10) it is apparent that they correspond qualitatively and quantitatively to the experimentally obtained characteristics presented in the literature.

In the comparison of the frequency characteristics of figure 9 with the analytically obtained frequency characteristics [4] shown in figure 11, it was found that for both solutions at the rotational speeds $n = 100..150\,min^{-1}$, the powers differ by (5 ... 15)% and the torques with (4..18)%. It is accepted that the analytical solution is more accurate as it is done by a modified classical method [3, 4].

Compliance of the shape and energy performance characteristics is considered to be a numerical model verification obtained by simulation with the Flow Simulation CFD.

## Acknowledgments

The present paper has been written with the Project No 2020-RU-03's financial assistance.

## References

[1] Terziev A 2014 Specifics in numerical modeling of flow past a square-cylinder *Science Conference of Ruse University* **53** 143

[2] Tuzharov K, Popov G, Zheleva I, Klimentov K, Nikolaev I, Kostov B and Ahmedov A 2015 Theoretical study of tachometric flowmeter with CFD product Flow Simulation. *Science Conference of Ruse University* **46**

[3] Fateev F M 1958 *Wind turbines and wind installations* (Agryculture edition, Moskva)

[4] Tuzharov K, Popov K G, Klimentov K, Nikolaev I, Kostov B 2017 Characteristics of Modeling Hydrokinetic Turbine *56th Science Conference of Ruse University* Ruse

# Stress and vibration analysis of a drone

**M Urdea**[1*]

[1]Department of Manufacturing Engineering, Transilvania University of Brasov, 29 Eroilor, Brasov 500036, Romania

*E-mail: urdeam@unitbv.ro

**Abstract.** This paper refers to drones. The development of drones has been extremely fast in the last few years. They can be used in many different areas of military or of civil life, for industrial and agricultural purposes. In some fields this unmanned aerial vehicles may open up new horizons. These vehicles can solve a range of problems, in law enforcement contexts. Their two basic functions are flight and navigation. The purpose of this paper is to present this device, to develop a CAD model of it using SolidWorks and to perform FE analysis study. The drone frame was simulated and analysed in various conditions such as lift and vibration.

## 1. Introduction

A drone is an unmanned aircraft. It can be named a flying robot that can be controlled during the flight by a special software which works in conjunction with on-board sensors and GPS. On the other hand it can be associated with unmanned aerial vehicles (UAVs).

They were used initially only in the army, but now, the drones, are also used in the civil life, like traffic monitoring, civil surveillance, weather monitoring, search and rescue, surveillance of volcanoes, of fires, sensational footage in nature, aerial photos and movies, construction, utilities and telecom businesses, etc.

In the history of the army it is assumed that the first drones appeared in 1935 as unmanned aircraft. They have also evolved in this direction in the military, having been noted in civil plan since 2006. The use of drones on a large scale began in 2007, as a border monitoring between the U.S. and Mexico. In the last years the purpose of these devices is very important, their technology is improving, namely the altitude, the flight duration, the weight of payload, the speed, etc. In 2013 Amazon Company developed a project for delivering packages using drones and after that other companies have followed it [1].

The two basic functions of the drones are flight and navigation. The drones have, normally, a support like a frame, this is typically made of lightweight, composite materials. The goal is to reduce weight and to increase manoeuvrability in the flight. Drones require a flight controller, to launch, navigate and land the drone. Every controllers communicate with the drone using radio waves 2.4 MHz [2].

Other important components are: Electronic Speed Controllers (ESC - transmit voltage to motors), an electronic circuit that controls the motor speed and direction, GPS module (for some drone), Flight Controller (FC), video transmitter (VTX), 4 batteries, antenna, receiver (receives signals from the station), propellers, cameras, sensors, accelerometer, for measuring the speed and altimeter, for measuring altitude.

Most of the frame drones are manufactories using various materials such as aluminium, carbon fibre, plastic, ABS, ABS-PC. The manufacturing of the frame and the arms can be processed by laser cutting, by milling, by using 3D printing technology, etc.

Figures 1 and figure 2 present racing drones with complete equipment. The frame of the second drone will be a model for studying the future device.

**Figure 1.** Racing drone.

**Figure 2.** Drone used in analysing process.

## 2. The drone frame

It starts from an existing frame, ready for a future racing drone (figure 3). Figure 4 presents the frame modelled in SolidWorks using for dimensions a real racing drone frame. The materiel of the components for the studied frame is aluminium 6061 Alloy. Aluminium is ideal for drones because is a light and strong material, resistant to corrosion.

The frame is designed in SolidWorks with four arms, motors and propellers as in figure 5.

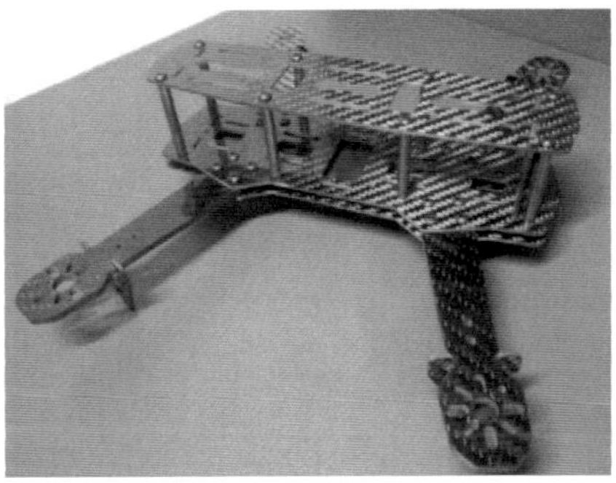

**Figure 3.** The drone frame.

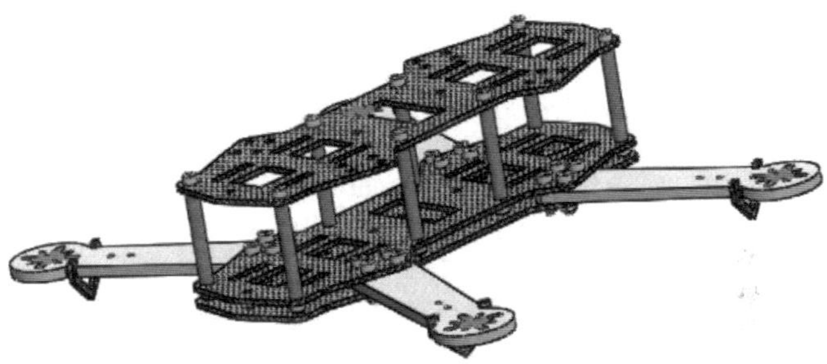

**Figure 4.** The drone frame modelled in SolidWorks.

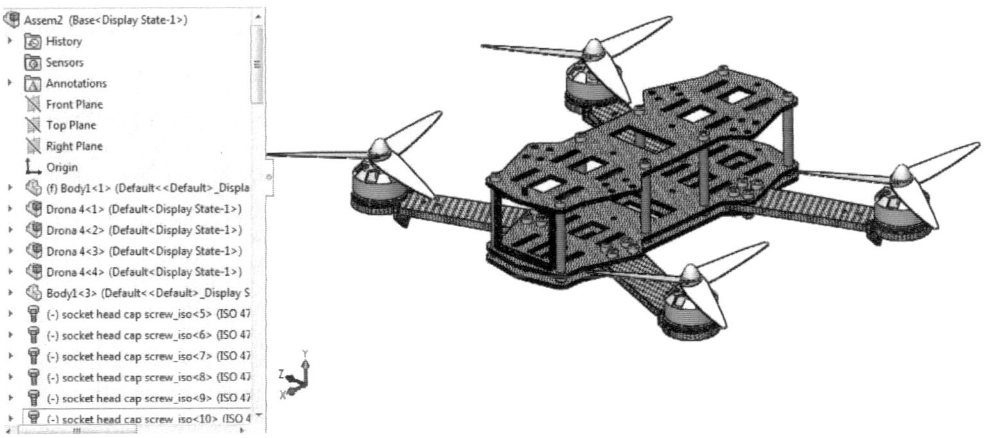


**Figure 5.** The modelled drone frame.

## 3. Static analysis case for the frame

### *3.1. Preparing the frame for simulation*

For performing FE analysis it is important to define all the forces acting on the frame. The weight of the frame and all components (ESP, GPS module, flight controller, motors-weight, battery, antenna, propellers and camera) is normal to the ground. Static loading involves a static study for calculating the response of bodies on which loads are applied statically or slowly.

For simulation, it is very important to create customized SolidWorks coordinate system, for remote loads locations. Figure 6 presents the coordinate systems for all the loads, this will be used for defining the remote loads in the static analysis. For theses loads the Remote Loads function will be used, figure 7 presents the Remote Loads defined as external loads. Gravity is attached like an external load [3, 4].

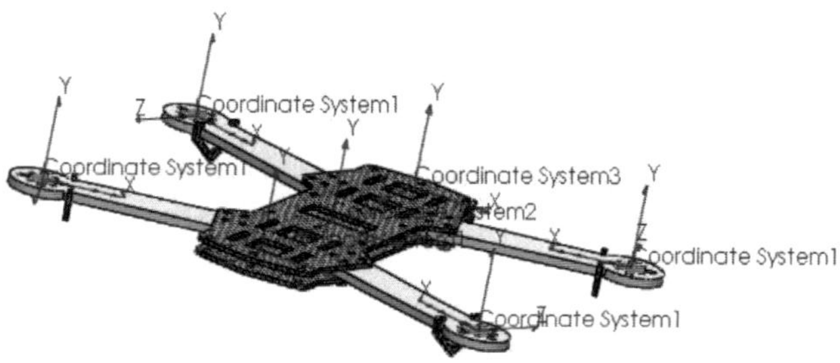


**Figure 6.** The drone frame with coordinate system.

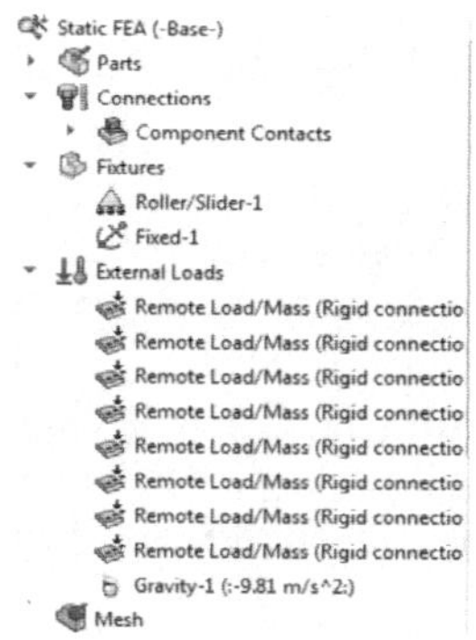


**Figure 7.** Remote Loads.

The steps for activating the remote loads are presented in the following figures. For each load a mass is assigned in the Remote Loads/Mass window, also the location and the suitable coordinate system are selected. There results some parameters visible for motor positioned on the drone arm in the cylindrical area, (figure 8) and for all electronics components positioned in the frame centre (figure 9). The weight of the components is estimated as follows: battery 0.165kg, motor 0.04kg, VTX video transmitter 0.01, ESC 0.03kg, FC 0.01, receiver 0.01, other components 0.02 and camera positioned in front of the frame, 0.015kg.

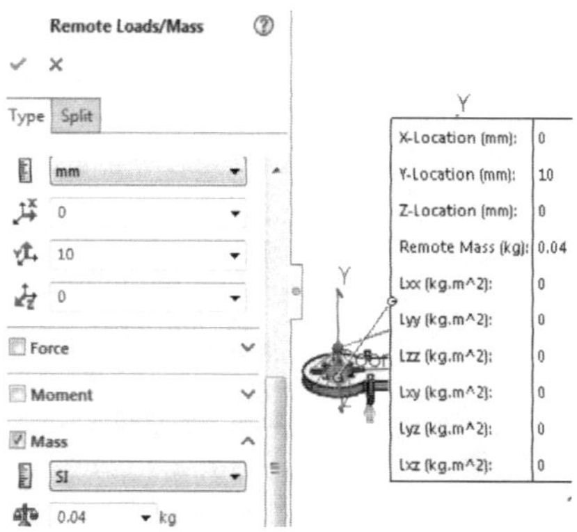


**Figure 8.** Remote Loads for the motor.

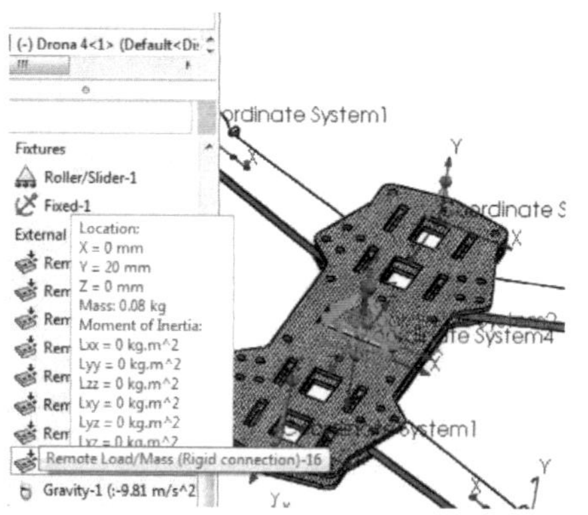


**Figure 9.** Remote Loads/Mass window.

The forces acting on the frame. The traction force of the motors was multiplied by a safety factor and divided for each arm. For a static analysis the fixtures ware attached**.** Namely Fixed Geometry to an arm leg and Roller Slider to the other three arm legs [3].

IOP Conf. Series: Materials Science and Engineering 1009 (2021) 012059 doi:10.1088/1757-899X/1009/1/012059

In this step the frame is prepared for the static simulation. Figure 10 presents the external loads with the remote loads, gravity and the fixture.

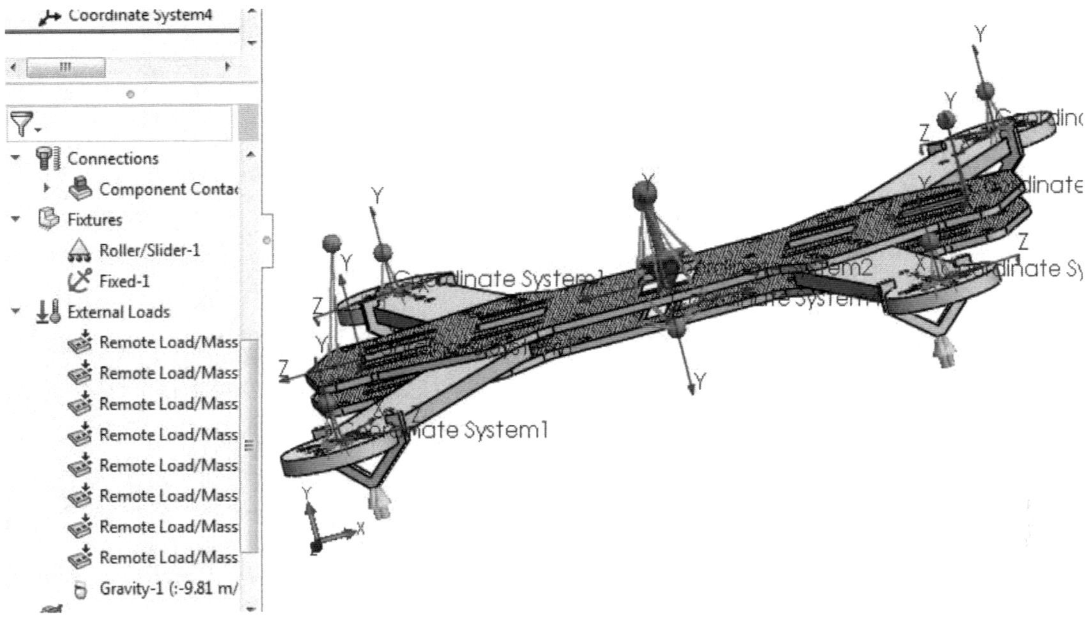


**Figure 10.** The drone frame prepared for static analysis.

### 3.2. *The analysis results*

The next step is named post-processing. In this part, after meshing, images are generated corresponding to analysis results, specific to static analysis cases, namely Von Mises Stress (figure 11), used to measure the level of stress and Displacement (figure 12) used for measuring the maximum deflection. The most deformed area is the centre of the drone.

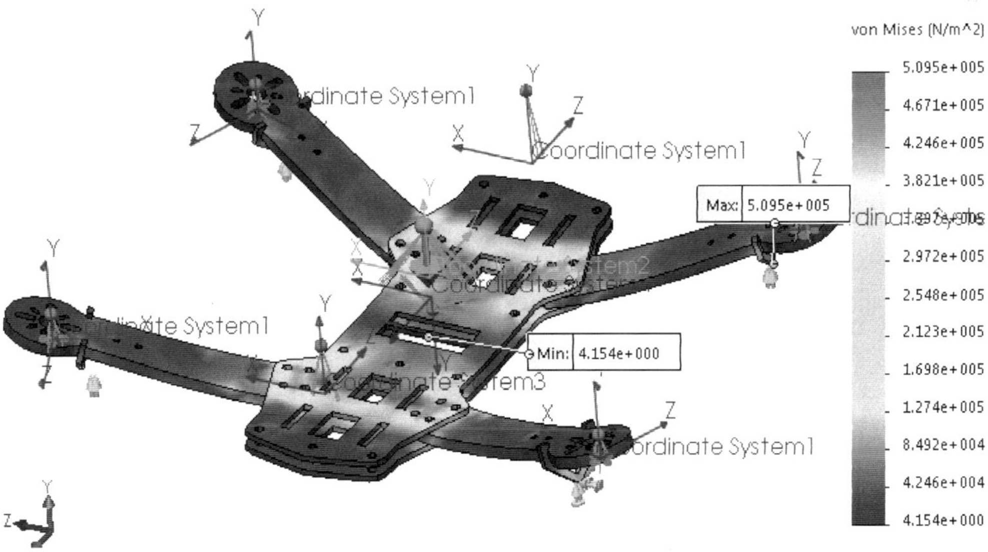


**Figure 11.** Von Mises for static analysis.

IOP Conf. Series: Materials Science and Engineering 1009 (2021) 012059 doi:10.1088/1757-899X/1009/1/012059

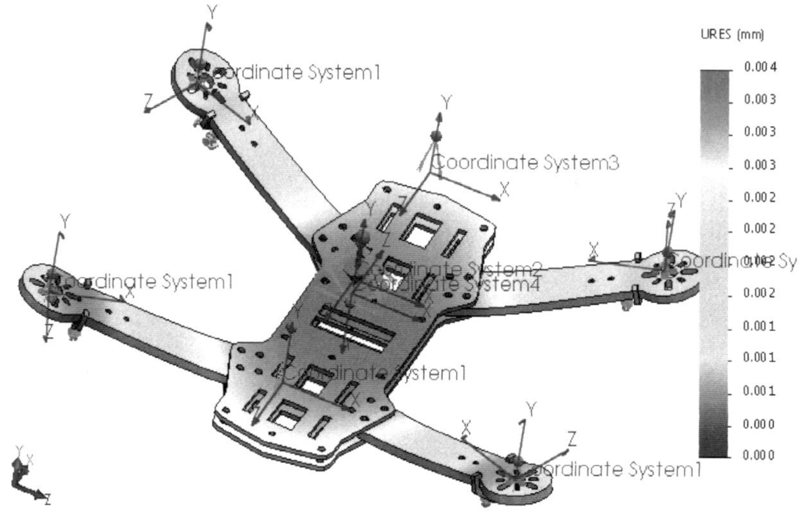


**Figure 12.** Displacements for static analysis.

## 4. Linear dynamic study for an arm

The simulation for the arm of drone can require a Dynamic study, namely a Linear Dynamic with the Harmonic (Harmonic-fatigue of sinusoidal loading) option. Dynamic analysis is important when an applied load, due to the vibration, isn't constant and isn't applied slowly. In linear dynamic analysis, the applied loads are time-dependent [4].

Run Frequency is activated from Dynamic pull down menu and Remote Loads are defined for the arm according to the previous chapters. List Resonant Frequency open the List Mode window for dynamic study with Frequency values. Figure 13 presents the frequency (Hertz) default value, and figure 14 presents the frequency curve defined using the new period for every frequency. It is activated from Force window using Curve option namely Edit button. In this situation, for this analysis, 1N on the arm, is defined [3].

The selection Curve could be used for specifying the own variation of load versus frequency. The frequency column is the same from the List Modes, a new curve has been edited using other period values [3, 4].

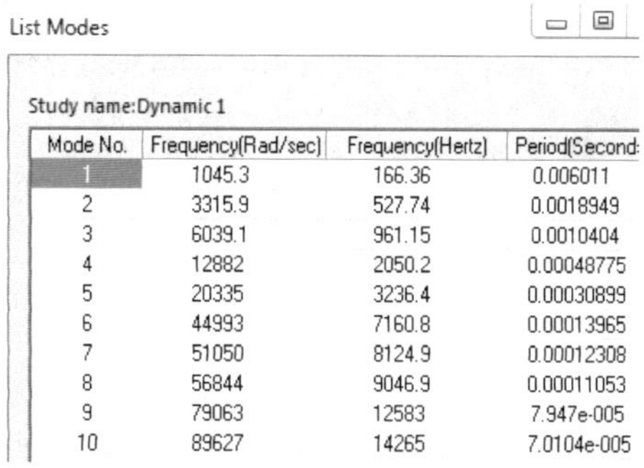

List Modes

Study name:Dynamic 1

| Mode No. | Frequency(Rad/sec) | Frequency(Hertz) | Period(Second |
|---|---|---|---|
| 1 | 1045.3 | 166.36 | 0.006011 |
| 2 | 3315.9 | 527.74 | 0.0018949 |
| 3 | 6039.1 | 961.15 | 0.0010404 |
| 4 | 12882 | 2050.2 | 0.00048775 |
| 5 | 20335 | 3236.4 | 0.00030899 |
| 6 | 44993 | 7160.8 | 0.00013965 |
| 7 | 51050 | 8124.9 | 0.00012308 |
| 8 | 56844 | 9046.9 | 0.00011053 |
| 9 | 79063 | 12583 | 7.947e-005 |
| 10 | 89627 | 14265 | 7.0104e-005 |

**Figure 13.** Frequency - default values.

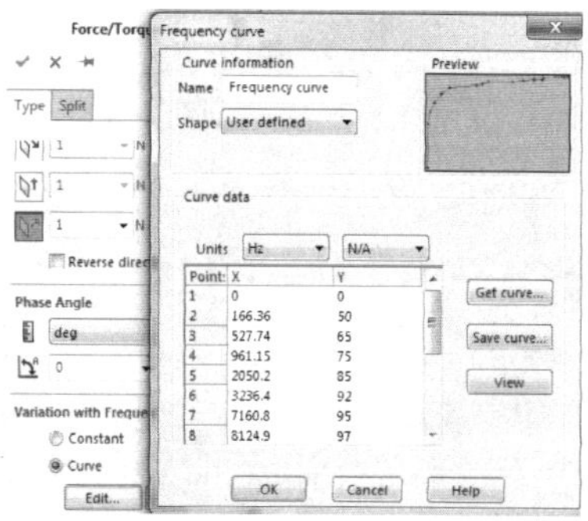


**Figure 14.** New Frequency curve.

With the Dynamic Properties one can open the Harmonic window and Solver/Direct sparse solver can be selected as in figure 15. The Direct Sparse solver is available for frequency, linear dynamic, and nonlinear studies [4]. The following values are defined in this window: Number of frequencies 10, by Frequencies Options, Frequencies upper limits 12583 by Harmonic option [5] (figure 16).

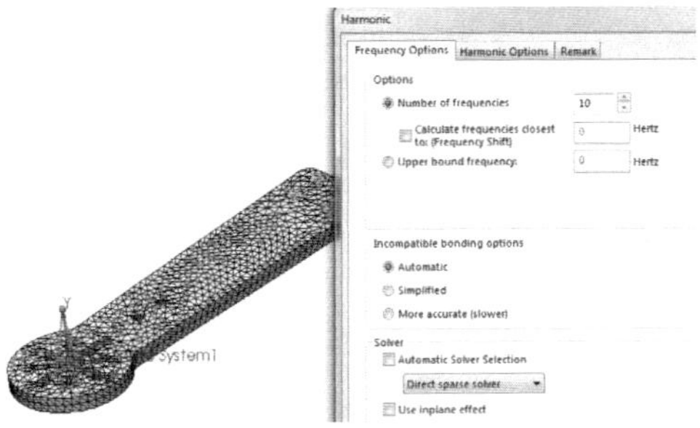


Figure 15. Direct Sparse solver.

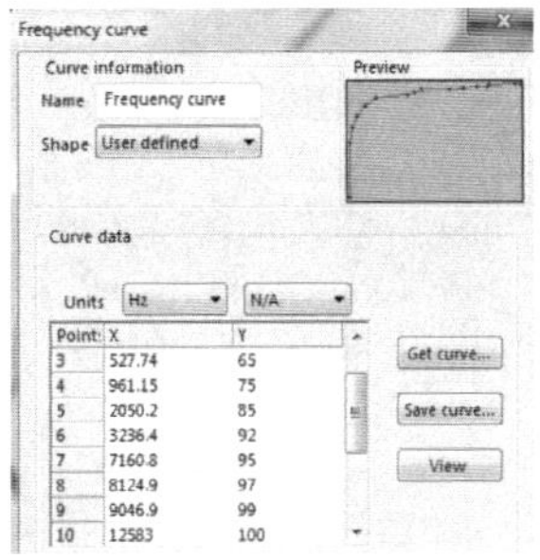


Figure 16. Frequencies upper limits.

The final results of the linear dynamic study from an arm are the Von Mises stress distribution, the most requested area is the bolt area and the Displacements, the most deformed area is on the end of the arm, there is small displacement (figure 17 and figure 18).

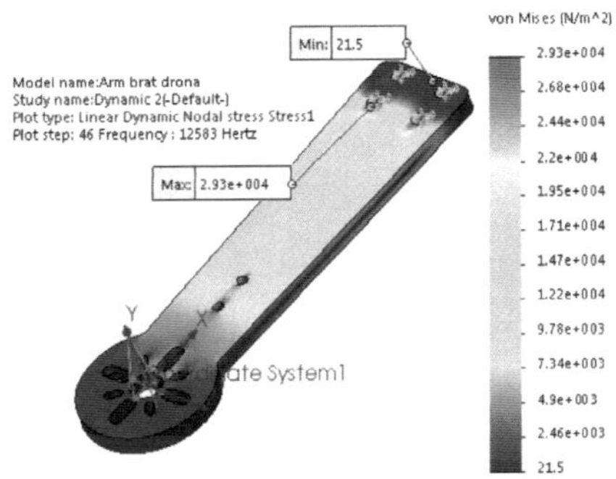


Figure 17. Von Mises.

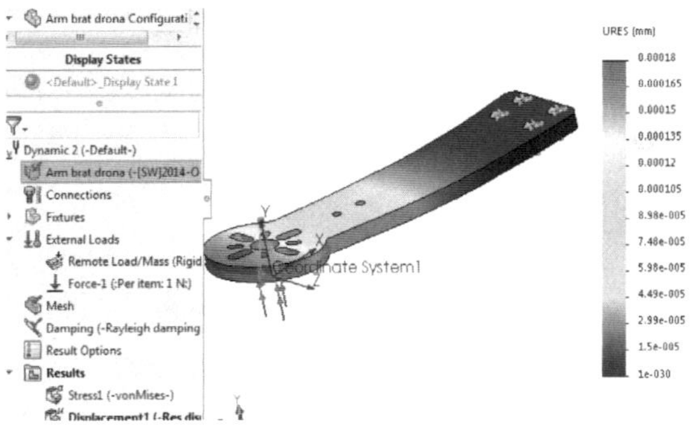


Figure 18. Displacement.

## 5. Natural frequency study for the drone frame

Each structure tends to vibrate with frequencies. A natural frequency study calculates the natural frequencies, adding also the body's mode shapes. For this study one can use the same External Loads structure and the FFEPlus Solver, the solver used for calculating the resonant frequencies. This option is viewed in the Properties/Frequency window, also the number of frequency [6, 7] (figure 19).

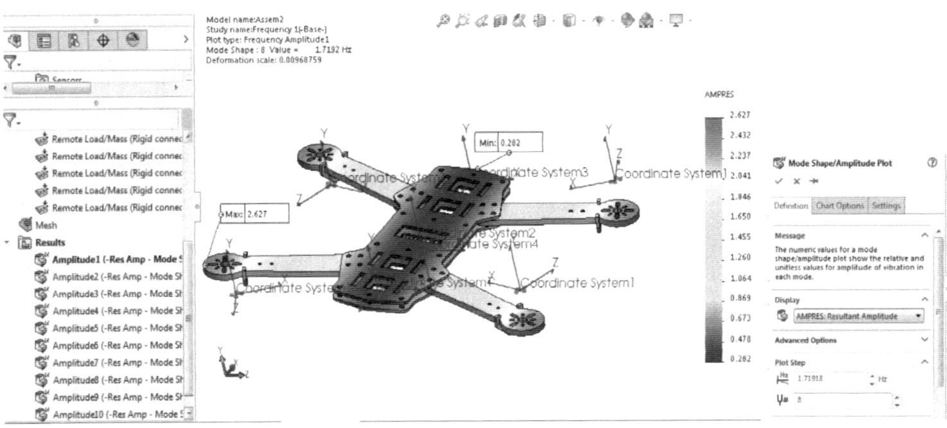


Figure 19. Amplitudes.

After running a frequency study, the amplitude of vibration results in each mode (figure 19). Here the Mode Shape/Amplitude Plot can allow to manage the mode shapes for frequency. Figure 19 presents one of the resulted amplitude values.

There can be multiple results using double click an icon in a Results folder. For example, List Resonant Frequencies gives the list with requested number of resonant frequencies. The most requested area is the extremity area of the arms.

## 6. Conclusion

Several finite element analysis studies can be performed on a drone using SolidWorks. For instance: Assembly Static Analysis, Nonlinear Static Analysis, Nonlinear Dynamic Analysis, Structural Analysis, Vibration Analysis, Natural frequency study, and so on. These analyses can also be performed on an arm or even on a sectioned arm being symmetrical. SolidWorks is able to perform these analyses.

In this paper the results are based only on the use of FEA analysis in SolidWorks. In these analysis it was considered that the centre of gravity of the body coincides with the origin of the axes.

This paper uses SolidWorks modelling, supports the importance of SolidWorks study and aims to demonstrate the ease of use of this program and finally the many possibilities for developing finite element analysis. For the demonstration, a drone was chosen, being approved by inexperienced people. The material chosen was aluminium, lightweight, corrosion resistant. Attempts were also made to use AS34 carbon fibre, but the material library did not fully define this material.

The work can be continued with other studies by using carbon fibre in the arms or in the whole frame, in that case in which ANSYS will be used.

## References

[1] TechTarget https://internetofthingsagenda.techtarget.com/definition/drone
[2] Sai M P and Zheng L 2018 Design and analysis of 3d printed quadrotor frame *Int. Adv. Research J. in Science, Eng and Tech* IARJSET **5** 66
[3] Stress Analysis of a Quadcopter https://www.youtube.com/watch?v=XEyCI54WgwM
[4] SolidWorks Help
[5] http://help.solidworks.com/2017/english/solidworks/cworks/c_frequency_analysis.htm
[6] Vibration Analysis of a Quadcopter https://www.youtube.com/watch?v=UMCrHFmEjI0
[7] Brijesh P, Sukhija R P and Sai Prassana Kumar J V 2018 Structural analysis of arm of multicopter with various loads *Int. J. of Emerging Tech. and Adv. Eng.* IJETAE **8** 215

# Static and dynamic characteristics analysis of self-balancing motorized spindle

**Z Wang[1*], W Z He[1], H M Zhou[1] and S Y Du[1]**

[1]School of Mechanical Engineering, Shenyang Jianzhu University, Shenyang 110168, China

*E-mail: juven1126@163.com

**Abstract**. In order to effectively suppress the vibration of motorized spindle caused by unbalance fault, realize online dynamic balance control of motorized spindle unit. A self-balancing motorized spindle with dynamic balancing device is designed and the finite element analysis of self-balancing motorized spindle is carried out. The static analysis is carried out, and the static deformation diagram of spindle is obtained after constraint and load are applied. Then modal analysis and harmonic response analysis are carried out, and the first six natural frequencies, mode shapes and displacement frequency curves of spindle are obtained. The results show that: The deformation of motorized spindle is small and rigidity of spindle meets design requirements: Within the working speed range, the spindle will not resonate and work stably and reliably. The rationality and reliability of structure design of self-balancing motorized spindle are verified.

## 1. Introduction

As an executive part, motorized spindle unite is a very important part. Its main function is to drive the cutter to rotate at high speed to complete the cutting task. In the cutting process, the spindle unit mainly bears the cutting force and the driving force from the machine tool. Modern CNC machine tools have high power, fast cutting speed, high-speed spindle rotation, and little human intervention in the process of work, only by the reliability of the machine itself to ensure the quality. So static and dynamic characteristics of the spindle must be good. It depends on the reasonable structure design, the application of high-performance materials and the continuous optimization and improvement of spindle. Cutter is installed on the spindle for cutting, so the static and dynamic characteristics of spindle affect processing quality through the cutter. In addition, the speed characteristics of spindle and the service life of the spindle unit parts are closely related to dynamic characteristics of spindle.

Therefore, in the past few years a large number of scholars have studied static and dynamic characteristics and dynamic balance of spindle, and achieved a series of achievements. Dongju Chen et al. have studied some properties of spindle supported by hydrostatic bearings, and analyzed the influence of parameter changes on transshape of spindle and eccentric vibration response. The analysis results show that the position and rigidity of bearings have a high influence on the machining accuracy [1]. Jorgensen et al. built a model of spindle system based on Timoshenko beam theory, taking into account factors affecting the dynamic performance of the spindle such as high-speed centrifugal force, cutting force and non-linear bearing rigidity. Then they analyzed the dynamics performance of the motorized spindle system. The analysis results indicate that the cutting force parameters, speed, etc. all affect the natural frequency of the spindle unit [2]. Shin et al. considered the thermal deformation of the bearing, established the bearing load-deformation model, and combined the model with the discrete dynamic model of spindle to obtain some more accurate dynamic parameters [3]. Alfares neglected the centrifugal force and damping of the system by adopting the parametric discussion method, studied the influence of axial pre-tension force of angular contact bearing of spindle, and established a five-degree-of-freedom grinder spindle mathematical model [4]. Rodrigues put forward a method of dynamic balance regulation and control based on sphere as dynamic balance actuator [5]. Chen et al. studied a real-time dynamic balancing regulation of rotor using liquid balancing ring actuator [6]. Zhang et al. studied the improvement of power transfer function in unbalanced regulation and control based on influence coefficient method and analyzed the correlation between power transfer function and influence coefficient [7]. Yue et al. studied the multi-order multi-plane transient balancing method for high-speed rotor system and completed the correction of the third order unbalance [8]. Sarhan A A D et al. established three-dimensional model of FUSM milling machine by using software ANSYS Workbench. Natural frequency and vibration mode were obtained by finite element analysis. The experimental modal analysis of milling machine tool was carried out by using vibration testing system. By comparing the two analysis results, the rationality of model establishment and analysis of machine tool was verified [9]. Matsubara A used spring damping in radial direction and moment direction to simulate rolling bearings and established the mathematical model of lathe spindle system. The experimental confirmed that damping coefficient of the system is determined by damping moment and the natural frequency is determined by radial stiffness value [10]. Although a series of research achievements have been made on dynamic characteristics and dynamic balance problems, few research have been done on the dynamic and static analysis of the special structure of self-balancing s motorized spindle. So the special spindle which can install the dynamic balance head is analyzed dynamically and statically in this paper.

## 2. The establishment of the finite element model of self-balance motorized spindle

The self-balancing motorized spindle unit model has been established in SolidWorks. Considering the practical operability, after simplifying it, the three-dimensional model to be imported into ANSYS Workbench for finite element analysis is shown in figure 1.

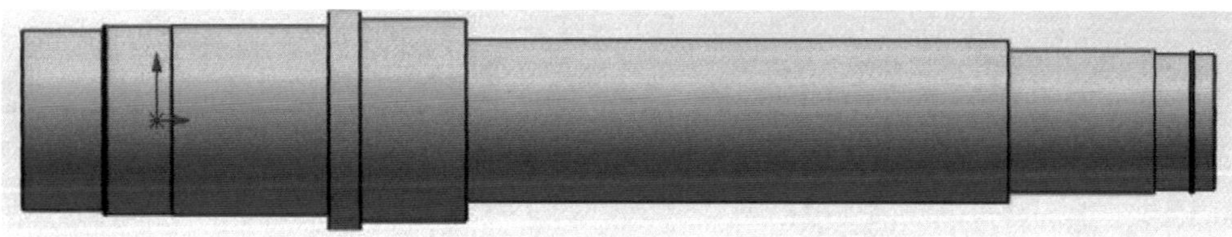

**Figure 1.** Three-dimensional spindle model.

The material properties of the spindle need to be set in finite element analysis of the spindle. According to experience, the material properties of the spindle can be selected as 40Cr. Poisson's ratio, elastic modulus and density need to be set. Material properties are shown in table 1.

**Table 1.** Spindle material properties.

| Material Properties | Density($kg/m^3$) | Young's Modulus(Pa) | Poisson's Ratio |
|---|---|---|---|
| 40Cr | 7820 | 215000 | 0.3 |

For finite element analysis of spindle, grid partition is the key step. Grid partition includes selection of grid type, size of grid unit and number of node units. The quality of grid partition directly determines the reliability of calculation accuracy. Sparse partition easily leads to the calculation accuracy cannot meet the requirements, and too close will affect the calculation speed. In this paper, the method of combining automatic partition and size control is used for the mesh partition of the spindle. Most of the meshes are hexahedron units and element size is controlled at 6mm. Figure 2 shows the finite element mesh model of spindle.

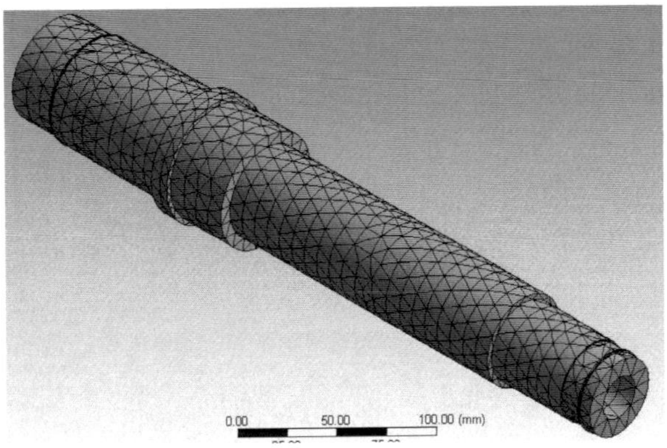


**Figure 2.** Finite element meshing of spindle model.

## 3. Static characteristic analysis of spindle

### *3.1. Apply constraints and loads*

In practical engineering, the bearing system will have inelastic deformation, which will affect the machining quality of the machine tool .In this paper, the elastic constraint in Ansys Workbench is equivalent to the elastic support, and the radial stiffness of the bearing is taken as the constraint stiffness value.

First, cylindrical constraint is added to the front and rear bearing segment surfaces of the motorized spindle. Cylindrical surface is restrained and two cylindrical surfaces are set free in radial and axial directions, with tangential restraint. Fixing constraints are added to the front bearing set of the spindle and the rear end face of the cylindrical surface of the spindle to which the front bearing set is installed is selected. Finally, elastic constraints are added to the simplified bearing set.

To apply a load is to apply a concentrated force to the front load acting surface of the spindle. Specific operation, select Loads>Force, then select the milling force action surface at the front of the shaft, set the load type to Components, Z-axis negative direction, the size of 200N. The applied restraints and loads are shown in figure 3.

**Figure 3.** Restraints and loads.

### *3.2. Solution of Static Analysis*

The main manifestation of static analysis is the static deformation of the model under external load. The main shaft is solved under the static analysis module. Figure 4 and figure 5 show the static total deformation displacement diagram and the displacement nebulae in three directions respectively.

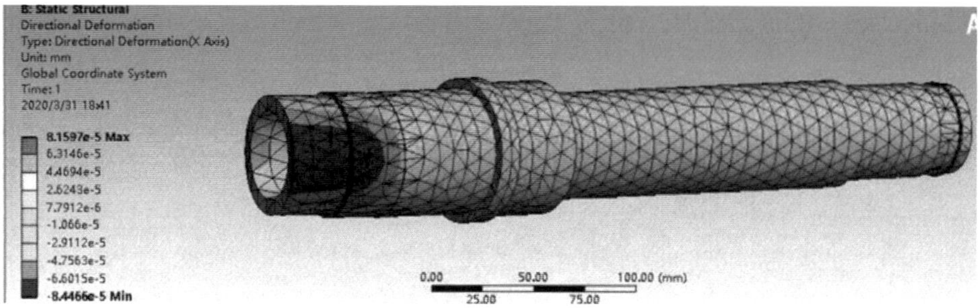


(a)

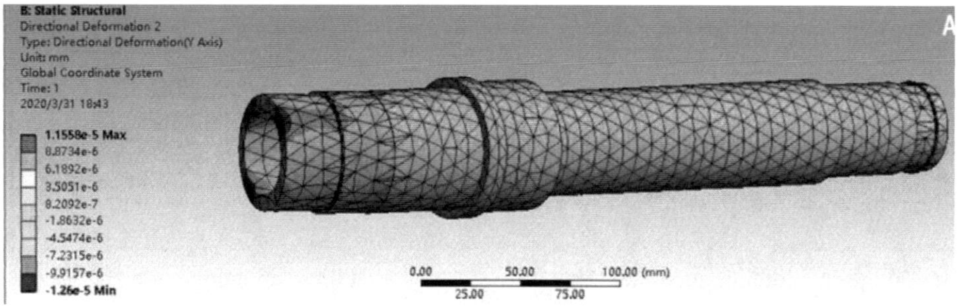


(b)

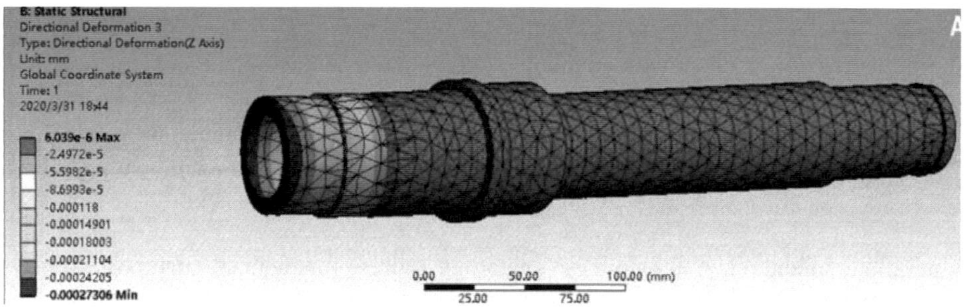


(c)

**Figure 4.** Static deformations diagram of the spindle: (a) X-axis to the static deformation maps; (b) Y-axis to the static deformation maps; (c) Z-axis to the static deformation maps.

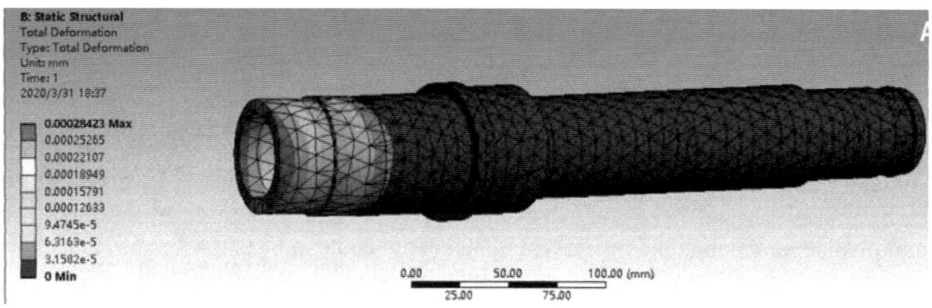


**Figure 5.** Stress change diagram of the high speed shop spindle.

From the static total deformation diagram of the spindle, it can be seen that the deformation mainly is in the first hard of the spindle, while the deformation in other parts is small. Since the radial force 500N is applied in the Z-axis direction, the deformation of X-axis and Y-axis is small and that of Z-axis is large. The maximum value at 500N is expressed as $\delta=2.8x10^{-7}$ $m=0.28\mu m$. Other deformations are smaller than this value and can be ignored. From the static stiffness calculation formula of the motorized spindle, it is clear that the stiffness value of the motorized spindle is reasonable under this small amount of deformation.

## 4. Modal analysis of spindle

The main requirement of high-speed machining center is high accuracy and speed. During the development process, it is inevitable to encounter deformation problems caused by vibration, which affects the machining quality and precision of high-speed machining center. Therefore, it is very important to comprehend the dynamic performance of the core components of the self-balancing spindle to improve the working ability of the whole motorized spindle. Since the dynamic performance of the mechanical model is mainly related to the natural frequency and the external excitation to the system. The dynamic performance analysis of self-balancing spindle starts with the modal analysis and the harmonic response analysis.

Modeling, material selection, meshing and constraints of modal analysis are the same as those of static analysis and are no longer analyzed. It is judged empirically that when studying the modal characteristics of the motorized spindle structure system, its low-order modal values have more influence on the vibration. Therefore, the first six modes are taken in this paper. The results of the modal analysis are shown in figure 6 and figure 7.

| | Mode | ✓ Frequency [Hz] |
|---|---|---|
| 1 | 1. | 3884.6 |
| 2 | 2. | 3886. |
| 3 | 3. | 7773.9 |
| 4 | 4. | 8177. |
| 5 | 5. | 8177.9 |
| 6 | 6. | 8957.4 |

**Figure 6.** Natural frequencies of the spindle.

These results indicate that results that the overall vibration mode of spindle is relatively stable and reasonable, and the first-order frequency of spindle is 3884.6Hz, which is enough to meet the requirements of the dynamic and static stiffness of the spindle. Converting the natural frequency of the spindle to the first critical speed is: $n=60x3884.6=2.3x10^5 r/min$. Obviously, it is greater than the maximum speed. Therefore, the resonance will not occur within the normal working range of the spindle.

## 5. Harmonic response analysis of spindle

It is based on the modal analysis of spindle. When the spindle rotates to drive the cutting tool to process the workpiece, the workpiece will transfer a periodic excitation force to the spindle. When frequency of excitation force coincides with natural frequency of spindle, it will cause the resonance of the spindle and seriously affect the quality of the machining. Therefore, the resonance phenomenon must be avoided. Harmonic response analysis is very important part. The displacement and strain response curves of the important points on spindle at different frequencies can be obtained by it. As modeling process is consistent with modal analysis and the boundary constraints are also consistent with that of harmonic response analysis, the finite element model for modal analysis can be directly selected.

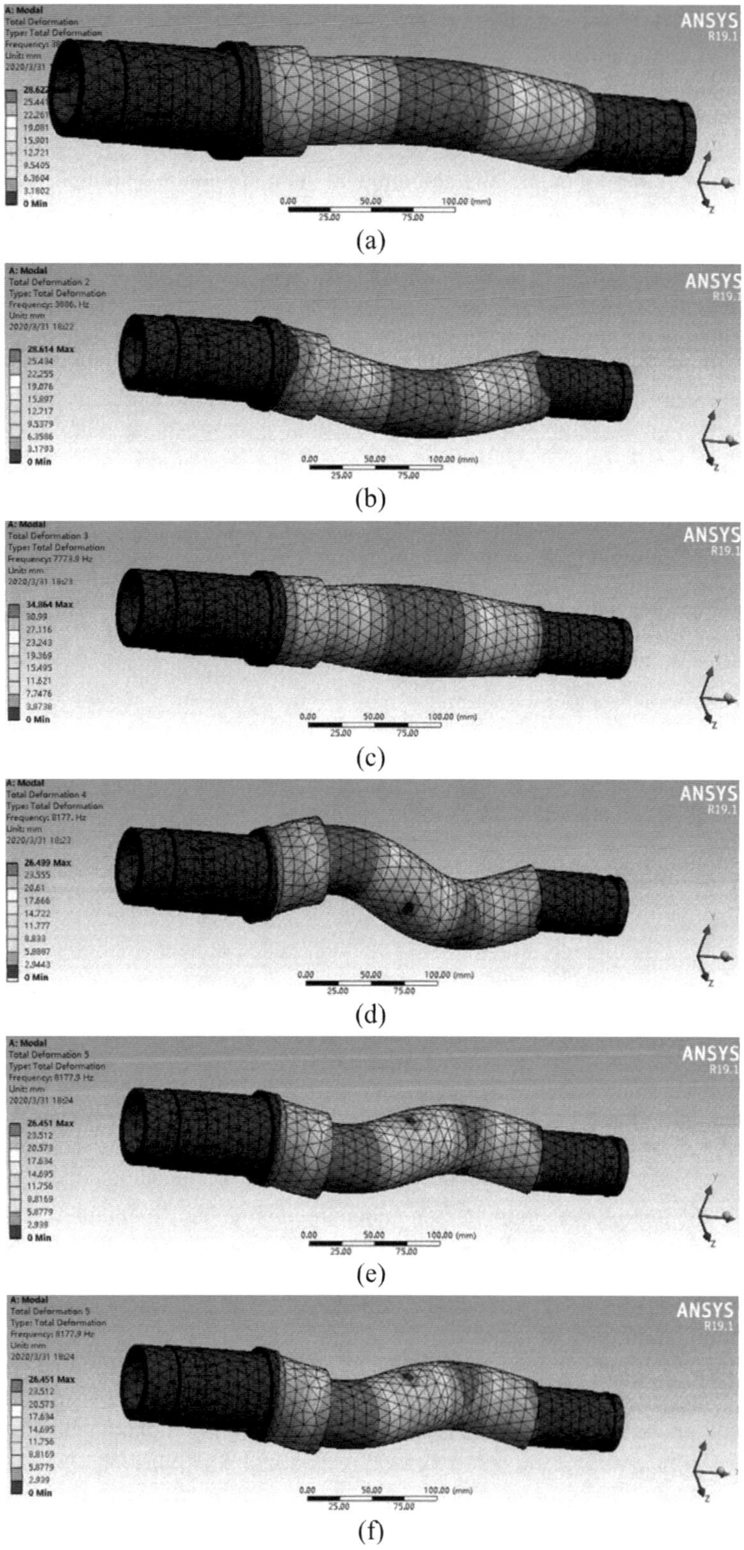


**Figure 7.** Sixth-order mode diagrams of the spindle: (a) first order; (b) second order; (c) 3$^{th}$ order; (d) 4$^{th}$ order; (e) 5$^{th}$ order; (f) 6$^{th}$ order.

The magnitude of the excitation force loaded in this harmonic response analysis is 200N in Z direction. As a result of natural frequency of the sixth-order mode is 8957.4Hz in modal analysis. Therefore, harmonic response is studied by taking the excitation frequency of 0-90000Hz, as shown in figure 8.

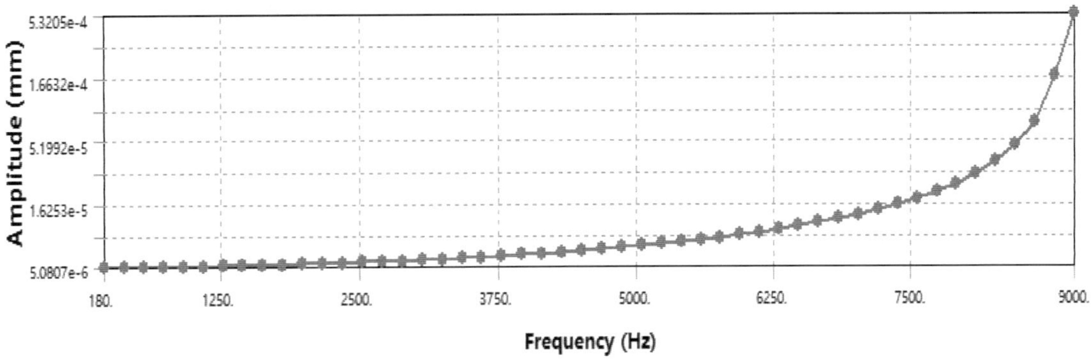


**Figure 8.** Front axle end face 0HZ-9000HZ frequency amplitude curve.

It can be seen from the above diagram that the resonance frequency of the electro-spindle occurs near 9000Hz. So the image range is narrowed down to 8800-9100Hz, as shown in figure 9.

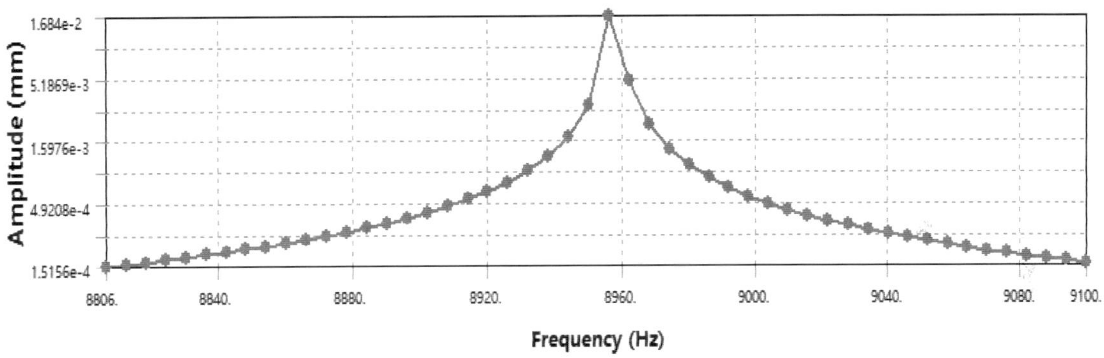


**Figure 9.** Front axle end face 8800HZ-9100HZ frequency amplitude curve.

The results show that the first peak value of the curve appears at the frequency 8956HZ of the external excitation load, i.e. the first resonance region, which is very close to the sixth-order natural frequency in modal analysis results of spindle. From the above analysis results of harmonic response, working speed of spindle is less than the speed at which resonance occurs. Spindle will not resonate and works stably and reliably.

## 6. Conclusion

Through static analysis of motorized spindle, static deformation diagram is obtained. The deformation of the motorized spindle is small under the load, and the structure meets the rigidity requirements. The modal analysis of the motorized spindle is done, and the first six order natural frequency and mode diagram are obtained. Finally, the harmonic response is analyzed and the displacement frequency curve is obtained. The results show that spindle will not have resonance, and the work is stable and reliable. Next, dynamic model of self-balancing motorized spindle can be established from the point of view of numerical analysis to study the influence of gyroscopic moment, bearing stiffness et al. on system dynamic performance.

**Acknowledgements**
This work was supported by National Science Foundation of China: 51-805337, Key Laboratory of Vi bration and Control of Aero-Propulsion System, Ministry of Education, Northeastern University: VCA ME202008 and National Science Foundation of China: 51675353.

**References**
[1] Chen D J, Fan J W and Zhang F H 2012 Dynamic and static characteristics of a hydrostatic spindle for machine tools *J. Manuf. Syst.* **31** 26
[2] Jorgensen B R and Shin Y C 1998 Dynamics of spindle-bearing systems at high speeds including cutting load effects *J. Manuf. Sci. Eng.-Trans. ASME* **120** 387
[3] Jorgensen B R and Shin Y C 1997 Dynamics of machine tool spindle/bearing systems under thermal growth *J. Tribol.-Trans. ASME* **119** 875
[4] Alfares M A and Elsharkawy A A 2003 Effects of axial preloading of angular contact ball bearings on the dynamics of a grinding machine spindle system *J. Mater. Process. Technol.* **136** 48
[5] Rodrigues D J, Champneys A R and Friswell M I 2010 Experimental investigation of a single-plane automatic balancing mechanism for a rigid rotor *J. Sound. Vib.* **330** 385
[6] Chen H W, Zhang Q J and Fan S Y 2011 Study on steady-state response of a vertical axis automatic washing machine with a hydraulic balancer using a new approach and a method for getting a smaller deflection angle *J. Sound. Vib.* **330** 2017
[7] Zhang Y, Mei X and Shao M 2013 An improved holospectrum-based balancing method for rotor systems with anisotropic stiffness *Proc. Inst. Mech. Eng. Part C-J. Eng. Mech. Eng. Sci.* **227** 246
[8] Yue C, Ren X and Yang Y 2015 Unbalance identification of speed-variant rotary machinery without phase angle measurement *Shock Vib.* **2015** 1
[9] Sarhan A A D and Matsubara A 2015 Investigate about the characterization of machine tool spindle stiffness for intelligent CNC end milling *Pergamon Press.* **34** 133
[10] Matsubara A, Tsujimoto S and Kono D 2015 Evaluation of dynamic stiffness of machine tool spindle by non-contact excitation tests *CIRP Ann-Manuf. Technol.* **64** 365

# Continuous improvement of the quality management system within an automotion company in Romania

**M V Zerbes[1]*, L G Popescu[1], M Olteanu[2] and A Olteanu[3]**

[1] "Lucian Blaga" University of Sibiu, Engineering Faculty, 4 Emil Cioran Street, Sibiu, 550025, Romania
[2] Kuka Automatizare Romania S.R.L.
[3] ThyssenkruppBilstein S.A.

*E-mail: mihai.zerbes@ulbsibiu.ro

**Abstract**. In this article, the authors present a case study on the methodology used in the process of continuous improvement of the performance of the quality management system, carried out in a company belonging to the field of production of devices and automation systems. To start this process, the internal audit subsystem and the analysis process performed by the management were first streamlined. In order to improve the non-compliant processes, within the quality management system an improvement strategy was established based on the stages of the 6 sigma methodology, namely the DMAIC model. Following the application of the methodology proposed by the authors, improvements of the processes between 44 and 96% were obtained depending on their type.

## 1. Introduction

Quality management is a set of activities aimed at achieving objectives, through the optimal use of financial, human, informational and physical resources. This set includes planning, coordination, organization, control and quality assurance activities.

The organization proposes a series of strategic objectives in the economic, social, technical, commercial field, which are reached through operational objectives. This type of objectives can be: obtaining quality products and services corresponding to the requirements, in the requested quantity, within the agreed term and which are available at the desired place or market, all these, under the conditions of minimum costs [1].

As provided by the ISO 9001 standard, the efficient operation of the quality management process can be reach if we resort to its systemic approach. In consequence, the quality management system (QMS) is a systematic set of operating procedures, guidelines and goals. This are documented, maintained and controlled by the company in order to achieve its quality goals and the continuous improvement of the QMS [2].

A QMS is simply a way for the organization to demonstrate that it understands the requirements of its customers and to produce and deliver products and / or services that meet these requirements, at an ever-increasing level.

The quality management system allows an organization to achieve the goals and objectives set in its policy and strategy. At the same time, it offers coherence and satisfaction in terms of methods, materials and equipment.

In order to be effective, the Quality Management System needs to interact with all the activities of the organization, starting from the identification of the client's requirements and aiming at obtaining the customer's satisfaction [3].

In this paper, the authors analyse the process of improving the performance of a QMS implemented according to SR EN ISO 9001: 2015 in an organization in the field of production of automation devices and systems.

The process of improving the performance of a QMS, applied in practice by the authors, have as the foundation the cycle summarized in figure 1.

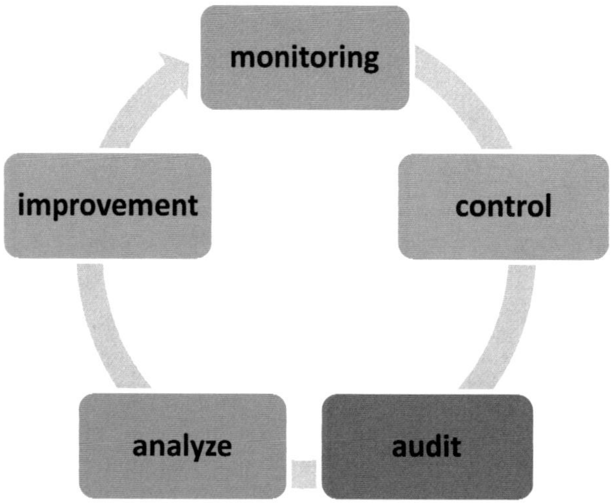


**Figure 1.** The performance improvement cycle of a QMS.

In this paper, because of the lack of space and to highlight relevant contributions, the authors focus on the audit and correct threads. Thus, the analysis of data obtained from an audit is approached by the authors, as a hybrid process between the classic audit conducted according to ISO 19011 and the specific approach to the process of measuring the reliability of technical systems (of course adapted to new requirements).

Regarding the continuous improvement, the sub-process is managed using a strategy based on the DMAIC model specific to the 6-sigma methodology, of course also simplified and adapted to the needs of the organization.

## 2. Reliability and performance of the QMS within the organizational framework

In the organizational framework, a QMS is reliable and efficient if there are not found major non-conformities following the audit process.

At the time of QMS development, the authors together with the company's representatives set among the main objectives a threshold of 0.90, within which the system is considered reliable and a threshold of 90% after which the processes are considered efficient.

Taking into account the reliability function, we calculated the level of reliability resulting from the internal system audit performed, as follows:

$$R_{(t)}=N_{(t)}/N_0=172/202=0,85 \quad (1)$$

where:

- $N_{(t)}$ = 172 represents the number of requirements identified as compliant following the audit;
- $N_0$ = 202 represents the total number of audited requirements.

In order to, better understand how these values were obtained, we present the centralized data in table 1 and figure 2.

**Table 1**. Audited requirements of standard and non-conformities/observations identified by the audit system.

| Organizational entity - Audited process / activity | Representatives of the auditee | Elements ISO 9001 | $N_0$ | CN | $N_{(t)}$ | $N_{(t)}/N_0$ |
|---|---|---|---|---|---|---|
| Management | General manager<br>CEO<br>Quality Manager (QMS Manager) | 4.1; 4.2; 4.3; 4.4;<br>5.1; 5.2; 5.3;<br>6.1; 6.2; 6.3;<br>7.1; 7.2; 7.3; 7.4; 7.5;<br>8.1; 8.2; 8.4; 8.5(missing 8.5.3); 8.6; 8.7;<br>9.1; 9.2; 9.3;<br>10.1; 10.2; 10.3 | 27 | 1 | 26 | **0,96** |
| Human resources | Human resources manager | 4.4;<br>6.3;<br>7.1; 7.2; 7.3; 7.4; 7.5;<br>8.1; 8.4(missing 8.5.6); 8.7;<br>9.1; 9.3;<br>10.1; 10.2; 10.3 | 15 | 0 | 15 | **1** |
| Logistics | Logistics Manager | 4.2; 4.4;<br>6.1; 6.2; 6.3;<br>7.1; 7.2; 7.3; 7.4; 7.5;<br>8.1; 8.2; 8.4; 8.5(missing 8.5.1, 8.5.5, 8.5.6); 8.6; 8.7;<br>9.1; 9.2;<br>10.1; 10.2; 10.3 | 21 | 4 | 17 | **0,81** |
| Quality assurance | Quality Manager<br>QMS Consultant | 4.2; 4.4;<br>6.1; 6.2; 6.3;<br>7.1; 7.2; 7.3; 7.4; 7.5;<br>8.1; 8.2; 8.4; 8.5; 8.6; 8.7;<br>9.1; 9.2; 10.1; 10.2; 10.3 | 21 | 6 | 15 | **0,71** |
| 3D measurement | Quality Manager<br>QMS Consultant | 4.2; 4.4;<br>6.1; 6.2; 6.3;<br>7.1; 7.2; 7.3; 7.4; 7.5;<br>8.1; 8.2; 8.4; 8.5; 8.6; 8.7;<br>9.1; 9.2;<br>10.1; 10.2; 10.3 | 21 | 6 | 15 | **0,71** |
| Acquisitions | Acquisitions manager | 4.4; 6.1; 6.3;<br>7.1; 7.2; 7.3; 7.4; 7.5;<br>8.1; 8.2; 8.4; 8.5(missing 8.5.3 and 8.5.5); 8.6; 8.7;<br>9.1; 9.3; 10.1; 10.2; 10.3 | 19 | 2 | 17 | **0,89** |
| Assembly | Assembly Manager | 4.2; 4.4; 6.1; 6.2; 6.3;<br>7.1; 7.2; 7.3; 7.4; 7.5;<br>8.1; 8.2; 8.4; 8.5; 8.6; 8.7;<br>9.1; 9.2; 10.1; 10.3 | 20 | 3 | 17 | **0,85** |
| Production | Production manager | 4.2; 4.4;<br>6.1; 6.2; 6.3;<br>7.1; 7.2; 7.3; 7.4; 7.5;<br>8.1; 8.2; 8.4(missing 8.4.3);<br>8.5(missing 8.5.3 and 8.5.5);<br>8.6; 8.7;9.1; 9.3;10.1; 10.2;<br>10.3 | 21 | 5 | 16 | **0,76** |

| Organizational entity - Audited process / activity | Representatives of the auditee | Elements ISO 9001 | $N_0$ | CN | $N_{(t)}$ | $N_{(t)}/N_0$ |
|---|---|---|---|---|---|---|
| Production Launch | Responsible for Production Launch | 4.4; 6.1; 6.3; 7.1; 7.2; 7.3; 7.4; 7.5; 8.1; 8.2; 8.4; 8.5(missing 8.5.3 and 8.5.5); 8.6; 8.7; 9.1; 9.3; 10.1; 10.2; 10.3 | 19 | 3 | 16 | **0,84** |
| Marketing | Financial Manage | 4.4; 6.3; 7.1; 7.2; 7.3; 7.4; 7.5; 8.1; 8.2; 8.4; 8.5(missing 8.5.3 and 8.5.5); 8.6; 8.7; 9.1; 9.3;10.1; 10.2; 10.3 | 18 | 0 | 18 | **1** |
| | **Total** | | **202** | **30** | **172** | **0,85** |

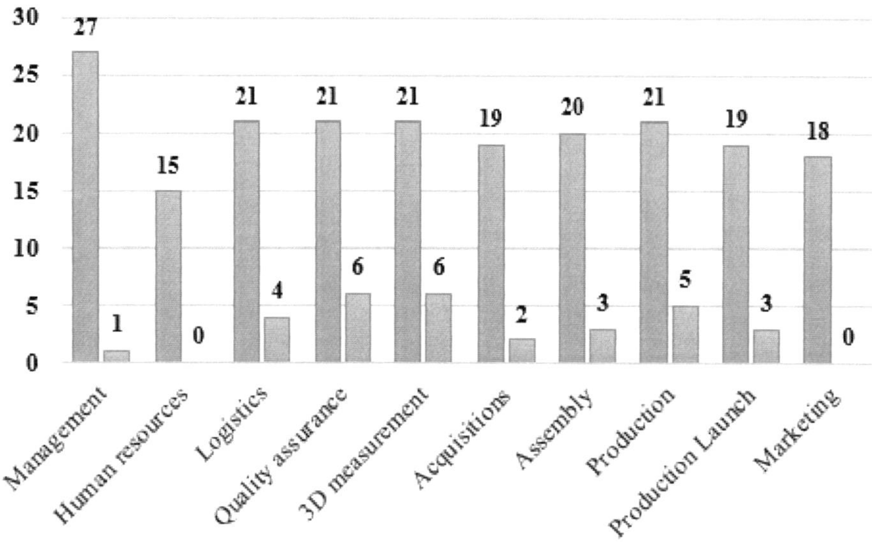


**Figure 2**. Number of non-conformities by departments.

As can be seen from the above:

- The department with the most audited requirements is the Department of Management, having a single non-conformity resulting from the audit;
- The Human Resources and Marketing Departments did not obtain any non-conformity following the audit;
- The departments with the most non-conformities, six in number, are Quality Assurance and 3D Measurement.

Figure 3 summarise these data.

Following the analysis of the audit report, a ranking was made of the reliability levels related to each department in the company.

From the presented diagram, it appears that only three of the departments (Management, Human Resources and Marketing) have reached the level of reliability established by the organization, the remaining seven do not fall within the acceptance level of 0.90.

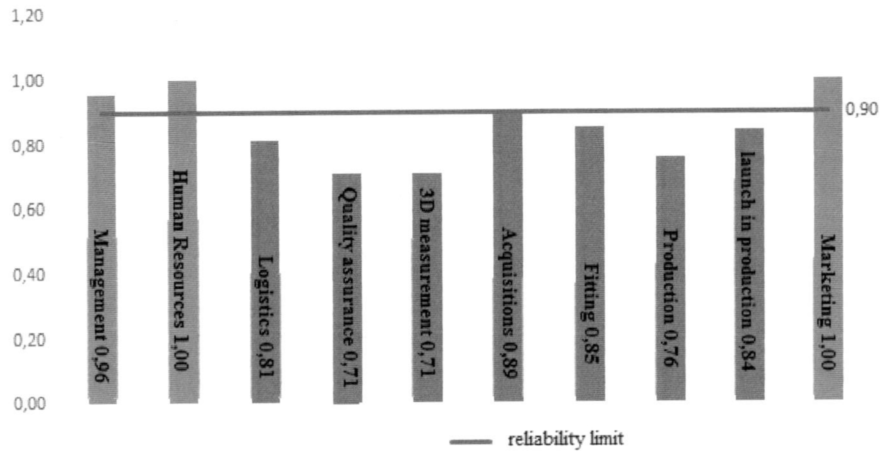


**Figure 3.** Level of system reliability by departments.

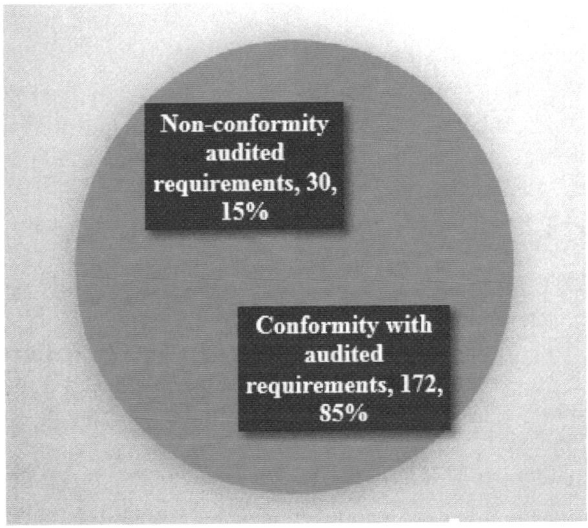


**Figure 4.** The share of non-conformities of the Management System.

It can be seen that out of the total audit requirements, in number of 202, 172 (85%) of them conform with the standard, and 30 (15%) are not conforming with the audited standard requirements.

Following the system audit performed it appears that the organization has not reached the established reliability level of 0.90, the non-conformities identified are minor and can be corrected by the next follow-up audit in 2020.

The organization does not demonstrate the efficient implementation and maintenance / improvement of the management system.

The organization did not establish or maintain its management system in accordance with the requirements of the standard and did not demonstrate the system's ability to meet requirements for products and / or services within the scope, policies and objectives of the organization.

During the audit process, the management system did not fully demonstrate overall conformity with the requirements of the audit standard.

In conclusion, we can say that the organization's QMS is functional but not reliable for the moment.

In order to improve the non-conformities identified following the audit, a strategy based on the DMAIC model was adopted.

In order to exemplify the way in which this model is applied, we present, briefly, the way of evaluating the performance of a process within the organization.

**Table 2.** Audited requirements of the standard and non-conformities / observations identified following the process audit.

| Organizational entity – Audited process / activity | Representatives of the auditee | Elements ISO 9001 | PT | NI | PT/(NI*0,1) |
|---|---|---|---|---|---|
| Production/ CNC 5 | Workstation operator | 9,1 | 59 | 8 | **73,75%** |

The level of process performance is calculated from process audits as follows:

$$PP=PT/(NI*0,1)=59/(8*0,1)=73,75\% \quad (2)$$

where:

- PP = represents the performance of the process;
- PT = 59 represents the total score obtained in the questionnaire after the audit;
- NI = 8 represents the number of questions addressed in the questionnaire.

Following the audit process, it was found that not all reach the level of performance set by the organization.

An audit process following, established that the performance level was not reached is presented above, for the 5-axis CNC production process, workstation 06.

Thus, following the audit process performed at the CNC 5 axis 06 workstation, the performance level was not reached because it does not fall within the acceptance margin of 90%, the degree of fulfilment being 73.75%, being found a non-conformity minor and two observations.

In order to improve the non-compliant processes, we established as a strategy to improve the non-conformities identified on the processes, the application of the new strategy as presented below.

## 3. Process improvement using the DMAIC model

### *3.1. Project definition (D)*

The following criteria were used to select the improvement project:

- Importance - After the completion of the project, obvious results must appear to justify the effort;
- Risk - If there are known or potential risks, the project is likely to last, or even fail to achieve the expected results;
- Chronicity - The project needs to correct a frequently occurring issue, not a recent one;
- Impact - Impact must be measurable;
- Quick fix - Projects must be less than one-year long.

The following problems was identified within the target company (table 3).

For the evaluation of the problems, scores will be given on a scale from 1 to 5 where:

- 1 - Unimportant problem;
- 5 - Very important problem.

The selection criteria are weighted according to importance.

**Table 3.** Evaluation matrix.

| No. | Problem | Importance 0.3 | Risk 0.2 | Chronicity 0.1 | Impact 0.2 | Quick fix 0.2 | Total |
|---|---|---|---|---|---|---|---|
| **1** | **Internal complaints for non-compliant processing** | **5** | **4** | **4** | **5** | **4** | **4,5** |
| 2 | Internal complaints for non-compliant processing | 4 | 3 | 4 | 3 | 3 | 3,4 |
| 3 | Dead time between operations | 3 | 3 | 3 | 2 | 3 | 2,8 |
| 4 | Delays in delivery | 3 | 3 | 4 | 3 | 5 | 3,4 |
| 5 | Price 10% higher for purchased tools | 1 | 2 | 2 | 1 | 1 | 1,3 |
| 6 | Processing times 5% longer than standard times | 1 | 1 | 2 | 1 | 1 | 1,1 |
| 7 | Expensive maintenance for old machines | 1 | 1 | 2 | 2 | 1 | 1,3 |

*3.1.1. Identifying the priority problem.* The problem that had the highest score is: Internal complaints for non-compliant processing.

*3.1.2. Defining the problem.* In the last year, the percentage of complaints for inward processing is 5% due to non-compliant processing. Thus, the loading capacity of the CNCs is affected by increasing the reprocessing times by 2%, the timely delivery of the parts being affected by 1.5%.

*3.1.3. Defining the mission.* We aim to reduce the percentage of complaints for landmarks identified as non-compliant internally to no more than 2.5% by the end of January 2020.

Rework times will reach at least 1%.

Delivery delays will be at least 1%.

*3.1.4. Team selection.* The following team was selected to solve the problem:

**Table 4.** Team selection.

| Role in the team | No personnel |
|---|---|
| Foreman | 1 |
| CNC supervisor | 1 |
| CNC operator | 4 |
| Quality controller | 1 |
| Logistics operator | 1 |
| Production launch technician | 1 |
| Technologist | 1 |

*3.2. Data measurement and analysis (MA)*

From 01.01.2019 to 31.12.2019, 9400 domestically produced parts were checked.

Following the checks carried out by the quality controllers, 470 parts were declared non-compliant and complaints were made following the internal processing.

Delivery time has been exceeded for 141 of the parts.

*3.2.1. Identifying the cause.* It was used as a cause-and-effect tool to identify the causes.

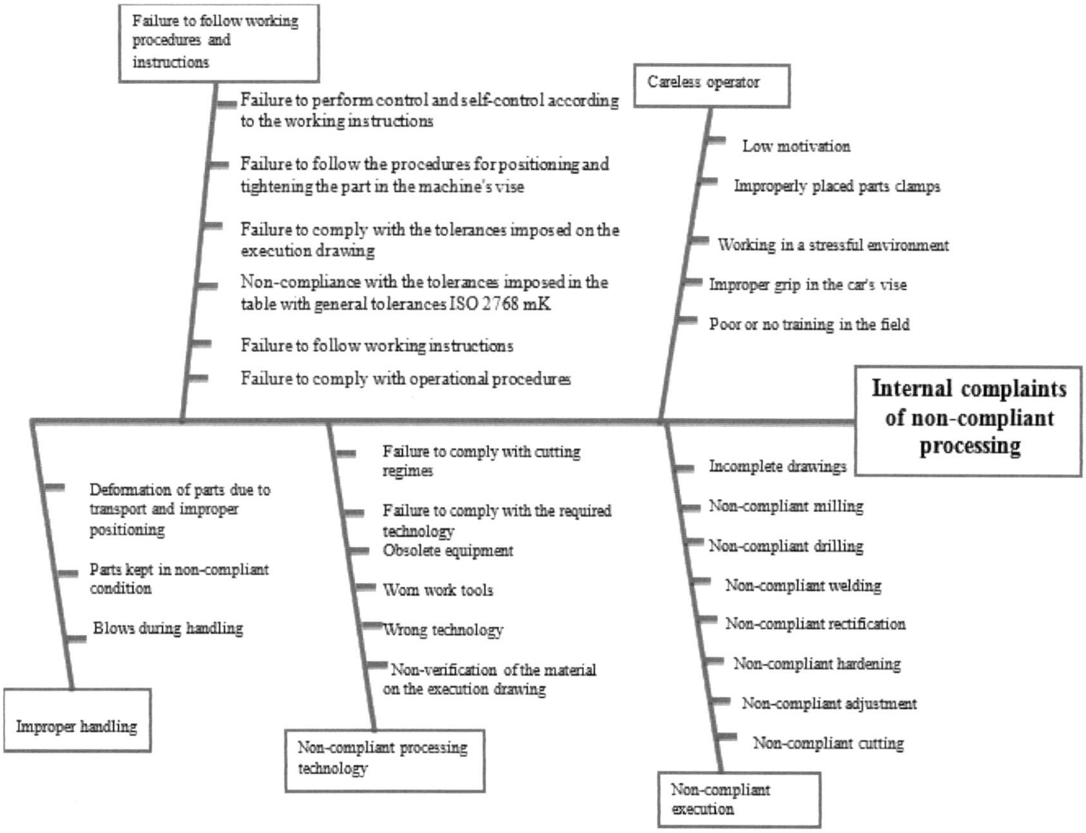


**Figure 5**. Cause-effect diagram.

*3.1.4 Weighting of causes.* Factors for which it was decided to claim landmarks are summarized in table 5.

**Table 5.** Factors for which it was decided to claim the landmarks.

| Cause | Share |
|---|---|
| Careless operator | 53 pieces – 11.28% |
| Failure to follow working procedures and instructions | 20 pieces – 4.25% |
| Improper handling | 2 pieces – 0.43% |
| Non-compliant processing technology | 75 pieces – 15.95% |
| Non-compliant execution | 320pieces – 68.09% |

Based on these weights, a Pareto diagram was developed to rank the causes.

**Table 6.** Weighting of causes.

| Cause | Frequency | Cumulative frequency | Percent |
|---|---|---|---|
| Non-compliant execution | 320 | 320 | 68,09% |
| Non-compliant processing technology | 75 | 395 | 84,04% |
| Careless operator | 53 | 448 | 95,32% |
| Failure to follow working procedures and instructions | 20 | 468 | 99,57% |
| Improper handling | 2 | 470 | 100,00% |

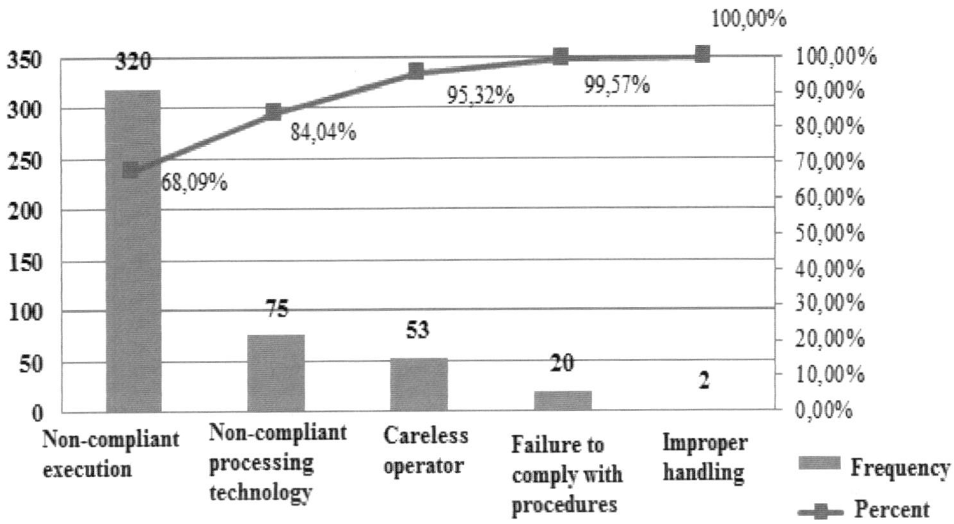


**Figure 6.** Pareto diagram.

### 3.3. *Improvement*

To solve the main causes, three improvement alternatives have been established for each of them.

**Table 7.** Improving the main causes.

| Cause | Alternative 1 | Alternative 2 | Alternative 3 |
|---|---|---|---|
| Non-compliant execution | Check the drawing before starting processing | Checking devices and tools before processing | Performing self-control at the end of the landmark |
| Non-compliant processing technology | Checking the material on the drawing and the stock of material | Standardization of processing capabilities of equipment from within and from suppliers | Training of operators in order to know the cutting regimes |
| Careless operator | Operator training | Checking the clamps after positioning the part on the machine | Performing self-control at the end of the landmark |

To select the best solution, for each case the alternatives were evaluate according to the following criteria:

- costs;
- implementation difficulty;
- necessary human resources.

These were evaluated with grades from 1 to 5 as follows:

- 1 is the least important criterion;
- 5 is the most important criterion.

Following the evaluation of the first case, the most efficient solution was decided: "Checking devices and tools before processing".

**Table 8.** Cause assessment 1.

**Cause 1:** Non-compliant execution

| **Selection criteria** | **Alternative 1** | **Alternative 2** | **Alternative 3** |
|---|---|---|---|
| Costs | 4 | 5 | 3 |
| Difficulty implementing | 2 | 4 | 2 |
| Human resources required | 2 | 4 | 1 |
| **Score obtained** | **8** | **13** | **6** |

**Solution:** Check devices and tools before processing

Following the evaluation of the second case, the most efficient solution was decided: "Standardization of processing capabilities of equipment from within and from suppliers."

**Table 9.** Cause assessment 2.

**Cause 2:** Non-compliant processing technology

| **Selection criteria** | **Alternative 1** | **Alternative 2** | **Alternative 3** |
|---|---|---|---|
| Costs | 3 | 5 | 4 |
| Difficulty implementing | 3 | 4 | 2 |
| Human resources required | 2 | 5 | 4 |
| **Score obtained** | **8** | **14** | **10** |

**Solution:** Standardization of processing capabilities of equipment from within and from suppliers

Following the evaluation of the third case, the most efficient solution was decided to be: "Training of operators".

**Table 10.** Cause assessment 3.

**Cause 3:** Careless operator

| **Selection criteria** | **Alternative 1** | **Alternative 2** | **Alternative 3** |
|---|---|---|---|
| Costs | 5 | 4 | 2 |
| Difficulty implementing | 5 | 4 | 3 |
| Human resources required | 5 | 2 | 3 |
| **Score obtained** | **15** | **10** | **8** |

**Solution:** Training of operators

Once the most convenient alternatives have been established, corrective actions will be taken in the cases, the necessary resources will be established, a person in charge will have to be appointed in order to respect the given term and the stage of accomplishing the proposed action; all this will be followed periodically.

**Table 11.** Corrective actions.

| No | Causes (with the largest share) | Corrective action | Necessary resources | Responsible | Deadline | Achievement stage% (effectiveness) / date of evaluation |
|---|---|---|---|---|---|---|
| 1 | Non-compliant execution | Checking devices and tools before processing | Time allotted for checking tools and devices before machining | CNC operator | At the beginning of each processing for an unlimited period of time starting with March 1st | 100% / 28.01.2020 |
| 2 | Non-compliant processing technology | Standardization of processing capabilities of equipment from within and from suppliers | Manual of internal machinery and equipment suppliers, 2 employees | foreman | 27.01.2020 | 100% / 28.01.2020 |
| 3 | Careless operator | Standardization of processing capabilities of equipment from within and from suppliers | Training course, budget of 5000 RON allocated for the course | foreman | 27.01.2020 | 100% / 28.01.2020 |

## 4. Conclusions

Because of the application of the strategy based on the DMAIC model, the improvement process had a visible positive trend. Therefore, the implementation of the corrective measures established and planned until January 28 2020 generated the following results:

- 5300 domestically produced parts were checked in just 3 weeks compared to 9400 in the previous year;
- Following the verifications performed by the quality controllers, 79 (1.5%) parts were reported as non-compliant due to the internal processing compared to 470 (5%) last year;
- The delivery term was exceeded for 5 (0.09%) benchmarks compared to 141 (1.5%) in 2019.

In other words:

- The verification rate increased from 181 to 1325 landmarks per week, is over 7.3 times;
- The number of non-compliant parts decreased 3.3 times;
- Delivery time has improved 0.6 percent in just 4 weeks compared to last year.

Which led to the validation of the proposed improvement methodology.

This model of calculation based on adapting the reliability equation (as an element of originality) can be successfully applied in any organization (regardless of field of activity) that has implemented a management system in accordance with a benchmark adopted or required by the market or imposed by legal regulations. It should be mentioned that, in addition to the standards already mentioned, it must be taken into account other international, European, national or even domestic standards such as: ISO 22000, EMAS, ISO 45001, VDA, ISO / IEC 27001, ISO / IEC 17025, ISO 15189 etc. or the Code of internal managerial control of public entities.

IOP Conf. Series: Materials Science and Engineering 1009 (2021) 012061 doi:10.1088/1757-899X/1009/1/012061

**References**

[1] Grecu V 2017 *Sustainable management through data analysis and interpretation (* ”Lucian Blaga” University of Sibiu) p 13
[2] Usman M, Shafi M, Savino M, Rashid Z, Yehya M I, Naqvi A A, Iqbal S and Mariluisa M M 2019 Investigating the Role of QMS implementation on customers’ satisfaction: A Case Study of SMEs **52** 2032
[2] SSC 2012 What is a quality management system? Smart Staf Consulting SRL http://smartstaff.ro/sistem-de-management-al-calitatii/
[3] ASRO 2015aSR EN ISO 9000 2015 Quality management systems. Fundamental principles and vocabulary
[4] ASRO 2015bSR EN ISO 9001 2015 Quality management systems. Requirements
[5] ASRO 2018SR EN ISO 19011 2018 Guidelines for auditing management systems

# Study on vibration characteristics of motorized spindle based on multi-physics fields

**K Zhang[1], Z N Wang[1], Z Wang[1], Q Y Wang[1] and J Ma[1*]**

[1] School of Mechanical Engineering, Shenyang Jianzhu University, Shenyang, 110168, China

*E-mail: 591672678@qq.com

**Abstract.** Ceramic motorized spindle plays an important role on CNC machine tool. Its dynamic characteristic determines the machining quality of spindle. This paper established a dynamics model of ceramic bearing, and explored the relationship of bearing stiffness and vibration response. The temperature field model of ceramic motorized spindle was established to analyze the temperature field. The motorized spindle air gap influence on unbalanced magnetic pull was studied. The dynamic model of ceramic motorized spindle was established through considering the multiple physical fields influence factors, the changing law of dynamic characteristics was studied. It has been verified by experiments. The average error of ceramic spindle decreased from 5.1% to 2.1%. The establishment of multi-fields dynamic model provided an important guidance for spindle design.

## 1. Introduction

The ceramic motorized spindle has a complex dynamic - thermal - magnetic coupling relationship. The thermal field and magnetic field are directly related to the vibration of the spindle. The stable dynamic output performance of ceramic motorized spindle is the important factor to ensure the machining quality. Ceramic motorized spindle has been widely used in aerospace, petrochemical and other fields due to its characteristics of small deformation, high rigidity and reverse magnetism. At present, scholars at home and abroad have done relevant research on temperature field [1], magnetic field [2], vibration [3, 4] and other aspects.

Aydin Gunduz [5] established a dynamic model through a five dimensional stiffness matrix to analyze the axial bearing preloads effect on the vibration response. Zhao Wanhua [6] proposed a new analytical modeling to calculate contact stiffness and clamping force, structure parameters based on the rotor system axial, and radial deformation. I.S. Barmanov [7] revealed radial and axial displacements of interference influence on ball bearing. Van-The Than [8] presented high-speed spindle related to preload which predict nonlinear thermal characteristics to find the thermal effects on bearing. Dexing Zheng [9] proposed angular contact ball bearings thermal performances model considering oil-air lubrication to analyze the bearing heat generation and transfer. Chen Xiaoan [10] established the electromagnetism and mechanical system to study the unbalanced force response on dynamic performance under different preload. Zhang Ke [11] established a magnetic model to calculate magnetic radial pull which have effect on spindle system dynamic characteristic.

However, the coupling of multiple physical fields on spindle system vibration has not been deeply studied. In the paper, the bearing dynamic characteristic, spindle magnetic and thermal distribution models are established. The relationship between the bearing stiffness, thermal displacement, unbalanced magnetic pull and vibration are analyzed. The ceramic spindle system dynamic mechanism is revealed through multiple physical fields models. The dynamic model provides technical support for motorized spindle design.

## 2. High speed ceramic spindle model

### *2.1. Ceramic bearing stiffness model*

The bearing ball will appear centrifugal force when are in high speed rotating. There are obvious axial and radial thermal displacement in the some parts of bearing. Under the combined action of dynamic and thermal fields, the contact angle and load of the bearing ball was variational. The bearing stiffness model is a nonlinear equation composed of the deformation equation and centrifugal force equation through quasi-statics. The relationship of bearing center position is shown in figure 1.Before bearing loading, BD line is the curvature center of bearing, which is affected by ball load. The curvature center changes on different lines. The bearing line position is shown in equation (1).

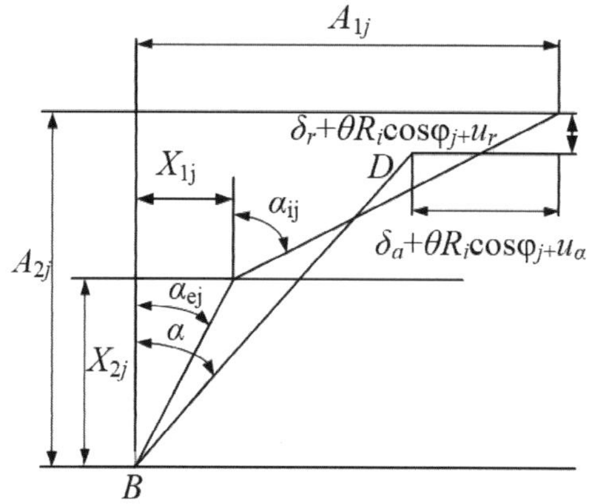


**Figure 1.** Center position relationship of bearing ball and channel curvature.

$$
\begin{aligned}
A_{1j} &= BD\sin\alpha_o + \delta_a + \theta R_i \cos\varphi_j + u_a \\
A_{2j} &= BD\cos\alpha_o + \delta_r \cos\varphi_j + u_r \\
BD &= (f_o + f_i - 1)D_b
\end{aligned} \tag{1}
$$

where: $\delta_a$ and $\delta_r$ are axial and radial displacement of bearing respectively, $R_i$ is radius of curvature center, $u_a$ and $u_r$ are axial and radial thermal displacement of bearing.

The complex angle relationship between bearing ring raceway and ball is shown in equation (2).

$$
\begin{aligned}
\cos a_{oj} &= \frac{X_{2j}}{(f_o - 0.5)D_b + \delta_{oj}} \\
\sin a_{oj} &= \frac{X_{1j}}{(f_o - 0.5)D_b + \delta_{oj}} \\
\cos a_{ij} &= \frac{A_{2j} - X_{2j}}{(f_i - 0.5)D_b + \delta_{ij}} \\
\sin a_{ij} &= \frac{A_{1j} - X_{1j}}{(f_i - 0.5)D_b + \delta_{ij}}
\end{aligned} \tag{2}
$$

### 2.2. *Ceramic bearing magnetization and thermal model*

There is a complex magnetic relationship inside the motorized spindle. According to the Maxwell principle, the spindle motor will generate the radial unbalanced magnetic pull. The motor magnetic pull have a direct correlation with magnetic field density, which is shown in equation (3) [11],

$$F_r(\theta,t) = \frac{B^2(\theta,t)}{2\mu_0} \tag{3}$$

where $B(\theta,t)$ is air gap magnetic density.

The spindle owns cooling system, however, there are still complex heat phenomenon in the spindle. Based on the Comsol software and thermal model, the temperature field diagram of the spindle rotor system in figure 2.

**Figure 2.** The temperature field of the motorized spindle rotor system.

From figure 2, the relationship between the temperature rise of the rotor and the spindle speed is non-increasing based on Finite Element software. The temperature of rotor is the highest in the spindle due to the motor loss.

## 3. Analysis and verification of motorized spindle model

### 3.1. *The analysis of multi-physical fields model*

According to the multiple physical field coupling spindle model, through the calculation of bearing model, the distribution of bearing stiffness is analyzed in figure 3. The stiffness of metal bearing is smaller. The operation mechanism of the metal bearing stiffness appear to change at 18000rpm.

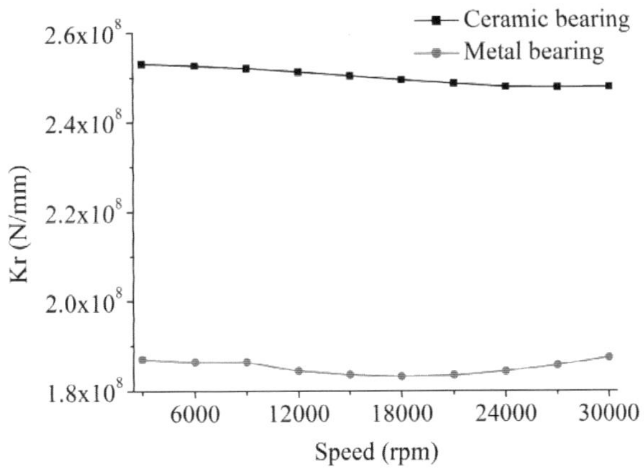


**Figure 3.** Bearing stiffness.

Through the ANSYS Maxwell software and the magnetic model, figure 4 presented the magnetic radial force distribution of the spindles. The maximum magnetic radial force of the metal spindle can exceed 150N is larger than that of ceramic spindle.

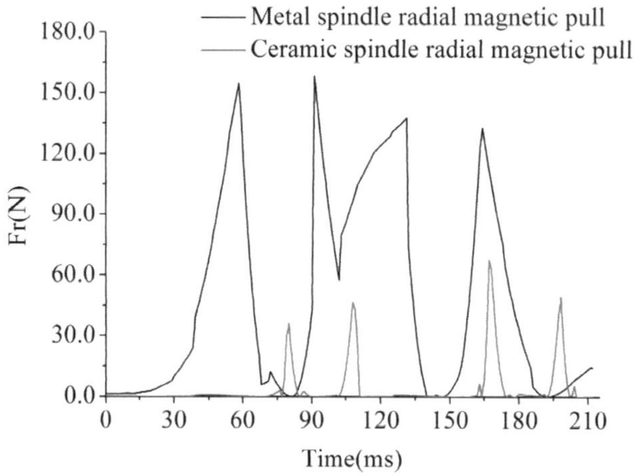


**Figure 4.** The calculation of magnetic pull.

### 3.2. Experimental verification

Spindle bearing lubrication adopted oil and gas lubrication under the experiment conditions. The spindle is cooled by cooling water system. The preload force supports 400 N. The spindle end vibration is measured through laser vibration instrument and data acquisition instrument.The measurement of the vibration response of the ceramic motorized spindle is shown in figure 5.

**Figure 5.** The measurement of the spindle end vibration velocity.

The relationship between spindle vibration velocity and speed is presented in figure 6. There is a direct correlation between vibration and rotational speed. The vibration speed is exceed 18mm/s under 18000rpm. Considering the multiple physical fields, the simulation results have a better accuracy comparison with the experiment.

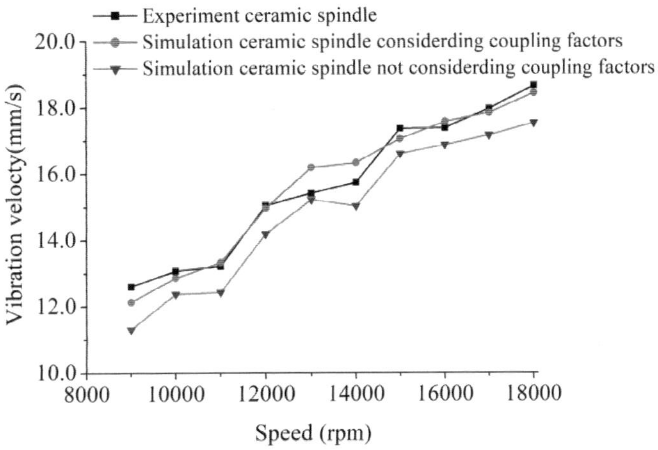


**Figure 6.** The relationship between spindle vibration velocity and speed.

## 4. Conclusion

Considering the dynamic-thermal-magnetic multiply fields coupling, the temperature and vibration of spindle are studied and analyzed based on the spindle model. The conclusions are as follows:

- Due to the ceramic material, the stiffness of ceramic bearing is different with that of the metal through the bearing model; The temperature of rotor is the highest in the spindle due to the motor loss;
- The maximum magnetic radial force of the metal spindle is larger than ceramic spindle based on the electromagnetic model, the magnetic radial force, the stiffness and thermal field have a great influence on the motorized spindle vibration.The average error of ceramic spindle is 2.1%;
- The multi-physical fields dynamic model is helpful to the design and development of ceramic spindle.

**Acknowledgments**

This research was funded by the National Natural Science Foundation of China [51675353 and 51805337], the guiding programme of Liaoning Nature Foundation [Grant number 2019-ZD-0653].

**References**

[1] Teng L, Weiguo G, Dawei Z, Yifan Z, Wenfen C, Cunman L and Yanling T 2017 Analytical modeling for thermal errors of motorized spindle unit *Int. J. Mach. Tool. Manu.* **112** 53-70

[2] Weidi H, Chunbiao G and Shixi Y 2016 Dynamic modeling and vibration response analysis of high speed motorized spindle *Journal of Zhejiang University (Engineering Science)* **11** 2198-206

[3] Shi H T and Bai X T 2020 Model-based uneven loading condition monitoring of full ceramic ball bearings in starved lubrication. *Mechanical Systems and Signal Processing* **139** 106583

[4] Abele E, Altintas Y and Brecher C 2010 Machine tool spindle units *CIRP Annals-Manuf. Tech.* **59** 781–802

[5] Aydin G, Jason T and Rajendra S 2012 Effect of bearing preloads on the modal characteristics of a shaft-bearing assembly: experiments on double row angular contact ball bearings *Mechanical Systems and Signal Processing* **31** 176-95

[6] Wanhua Z, Chao D, Jun Z and Haitao L 2013 Analytical Modeling Method of Dynamics for the Spindle Rotor System *J. Mech. Eng.* **6** 44-51

[7] Barmanov I S and Ortikov M N 2017 Ball Bearing Dynamics at the Interference fit on Balls *Procedia Eng.* **176** 19-24

[8] Van-The T and Jin H H 2016 Nonlinear thermal effects on high-speed spindle bearings subjected to preload *Tribology International* **96** pp 361-72
[9] Dexing Z and Weifang C 2017 Thermal performances on angular contact ball bearing of high-speedspindle considering structural constraints under oil-air lubrication *Tribology International* **109** pp 593-601
[10] Xiaoan C, Wenqu C, Huimin K and Ye H 2012 Dynamic analysis of high speed motorized spindles with eccentrics *Journal of Chongqing University* **3** pp 26-32
[11] Ke Z, Zinan W, Huaitao S, Xiaotian B and Zhan W 2019 Research on vibration characteristics of a ceramic spindle based on the reverse magnetic effect *Shock and vibration* **2019** pp 1-15

IOP Conf. Series: Materials Science and Engineering 1009 (2021) 012063 doi:10.1088/1757-899X/1009/1/012063

# Moving object detection of assembly components based on improved background subtraction algorithm

**K Zhang[1], S Tong[1*], H Shi[1], G Yue[1] and J Zhao[1]**

[1]College of Mechanical Engineering, Shenyang Jianzhu University, China.

*Email: tongshenghao@sjzu.edu.cn

**Abstract.** In the automatic assembly of mechanical components, it is difficult to guarantee the quality of assembly image acquisition caused by the change of assembly environment. Firstly, the color space is compared and analysed, and histogram equalization is used to preprocess the image to reduce the impact of image brightness on subsequent processing. Then, based on the texture features of the image, an image reconstruction idea from coarse to fine is created, and an image representation method from macro to micro is established to suppress the pseudo texture and noise texture in image processing. On this basis, an adaptive threshold algorithm based on the target distance is proposed to improve the detection accuracy of the assembly target. Experimental results show that the proposed algorithm can significantly improve the detection accuracy of moving objects in assembly video frames.

## 1. Introduction

Moving object detection has been widely used in automatic positioning, security monitoring and other scenes. It is particularly important for the development of computer vision to correctly detect moving objects from video [1, 2]. Background modelling is the core part of fixed camera video monitoring, which is used to segment moving objects from the background [3]. Therefore, background modelling in video surveillance is also called moving object detection. In this technology, the moving object is called foreground object, because its pixels are not suitable for background model, and can be extracted by "background subtraction" [4, 5].

In recent years, scholars have been studying the moving object detection algorithm using image features and color parameters for feature extraction [6]. In order to segment the moving foreground in real-time and accurately in complex scenes such as illumination change and dynamic background interference, Jin [7] proposed a moving object detection method based on Region Histogram modelling in color attribute space, aiming at the shortcomings of traditional methods based on color features and pixels. Firstly, the RGB color space is mapped to a more robust low-dimensional color attribute space, and the histogram is established in the local range of the pixel with the color attribute as the feature. At the same time, the spatial information of the pixel in each partition of the histogram is recorded, and the $K$ spatial histograms are used to form the background model of each pixel, and each histogram is giving different weights according to its matching degree. However, it is difficult to set the weight parameters in the actual operation of this method, and the results are greatly influenced by the weight parameters. Yin [8] proposes an anomaly detection algorithm based on the background of regular changes. Firstly, the region segmentation algorithm based on three frame difference method is used to segment the image into static background region and dynamic background region including moving objects. At the same time of segmenting static background area and dynamic background area,

Surendra algorithm based on three frame difference is used to extract static background area information and establish background model. Ding [9] proposed vibe background modelling with color and edge features to solve the problem of single feature in traditional background modelling methods. The problem of noise, breakpoint and internal hole in moving object detection results in three frame difference method was solved, and the results of image processing were compensated based on morphological processing method. But he did not solve the problem of image noise caused by the movement of the target in the video image. Although traditional background subtraction algorithms have been widely used in different environments, they still face various challenges [10]. One of them is that the foreground object is often incomplete with a lot of noise, which means that it contains holes or scattered fragments [11, 12]. These defects can cause false detection and affect the efficiency of advanced applications such as target location and appearance recognition [13, 14, 15]. To solve this problem, a more stable background modeling method is needed to extract the complete and clear foreground objects.

In this paper, from the perspective of improving the image threshold, the research of moving object detection in component assembly location video image is carried out. In the assembly location video of components, different image frames will produce different sizes and positions of targets, and the size and position parameters of targets will have different effects on the extraction of target features. Therefore, the corner points are set on the edge of the target feature extracted from the previous frame image, and the distance norm model of the pixels and corner points in different regions of the image is established. Based on this, an adaptive threshold algorithm based on the target distance is proposed to suppress the noise and improve the detection accuracy of the moving target.

## 2. Fundamental principles

Image threshold processing is an essential step in many visual tasks. The purpose of threshold processing is to classify pixels as "dark" or "bright". Image thresholding the segmentation of a digital image based on specific features of the pixel, such as intensity values. To create a binary image, categorizing each pixel into one of two categories. Integral image is a tool. When there is a mapping function from pixel to value and calculate the sum of function values in a rectangular area, integral image is a very efficient tool. In the case of no integral image, it need to calculate the sum of the mapping function of each pixel in each rectangular window by calculating the function value of each pixel separately and finally stacking. When computing the sum of multiple overlapping rectangular windows, the use of integral images can significantly reduce the amount of computation. In order to calculate the integral image, it should store the sum of all the items of the left and upper pixels in each position. For each pixel, calculate integration image using the following equation [4].

$$I(x,y) = f(x,y) + I(x-1,y) + I(x,y-1) - I(x-1,y-1) \tag{1}$$

The sum of the functions of any rectangle with upper left and lower right corners would be calculated as [6].

$$\sum_{x-x_1}^{x_2} \sum_{y-y_1}^{y_2} f(x,y) = I(x_2,y_2) - I(x_2,y_1-1) - I(x_1-1,y_2) + I(x_1-1,y_1-1) \tag{2}$$

The job of this method is to compare with nearby pixels. In addition, because the neighborhood samples are not evenly distributed in all directions, it is not a good representation to use the average step size in the whole image. This kind of problem is solved by using integral image, and the same output is produced for different processing methods. This is not to calculate the average value of the last s pixels, but to calculate the average value of the sxs pixel window centered on each pixel. This is a better average because it takes neighboring pixels on all sides into account. The time of average calculation using integral image is linear. The integral image is calculated when the image is traversed for the first time. In the second pass, use the integral image to calculate the s×s average value of each pixel, and using the time as a constant, then compare. Black or white will be set after the pixel value is compared.

## 3. Algorithm for moving object detection

*3.1. Feature extraction based on background subtraction*
For a perfect assembly location environment, the gray value of each pixel in the image is in line with the random probability distribution when there is no redundant moving target, the illumination does not change and other environmental factors do not change. Because of the actual engineering environment and in the process of image acquisition, it is inevitable to introduce noise artificially. These gray values can be kept in a certain gray value, and random excitation within a certain range. This kind of scene is called "background". Setting up background model and updating background are the most important problems in background subtraction.

Create a background image frame B, note that the current image frame is A, and the gray values at the pixel points of the current image frame and the background image frame are respectively recorded as $B(i, j)$ and $A(i, j)$. According to equation (1), the difference image $C_n$ is shown as:

$$C_n(i,j) = |A(i,j) - B(i,j)| \tag{3}$$

After the background model is established successfully, the next step is target detection. Set threshold $\sigma$, according to equation (2), binary image is obtained by binary processing of the difference image pixels $T_n$. And, the image T with complete moving target can be obtained.

$$T_n(i,j) = \begin{cases} 1, & C_n(i,j) > \sigma \\ 0, & \text{else} \end{cases} \tag{4}$$

The system needs to update the background model in real time when the background is slightly changed due to light changes. In order to quickly adapt to changes in the scene, the following formula is carried out.

$$B_{n+1} = B_n - \frac{B_n}{n+1} + \frac{\alpha_{i+1}}{n+1} \tag{5}$$

The operation process is shown in equation (3). '$\alpha$' represents the update weight coefficient. The general rule of background update is that if the current point is determined as a background point, the background value of the point is updated, otherwise it is unchanged. This can save the update time and speed.

*3.2. Adaptive threshold modeling*
High threshold will lead to fragmentation of foreground objects, while low threshold will lead to noise of pseudo foreground objects. To solve this problem, we propose an adaptive threshold algorithm based on target distance to improve the accuracy of target feature extraction. Different image frames will produce different sizes and positions of targets, and these parameters will have different effects on the extraction of target features. As shown in figure 1, we first divide the image into different blocks, the corner points on the edge of the target feature of the previous frame must be set, the distance norm model between pixels and corner points of different blocks of the image must be established, and the adaptive threshold algorithm must be established to complete the target extraction of the image.

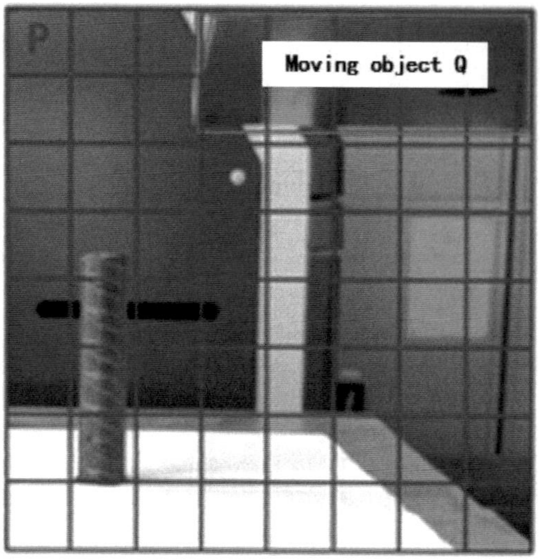


**Figure 1.** Block subdivision.

In order to ensure real-time, the distance norm between each pixel in the image and the corner of the target edge of the previous frame cannot be calculated. Therefore, use the method of dividing the image into different blocks to calculate the distance norm between the center pixel of each block and the edge corner of the previous image, which can not only improve the accuracy of feature extraction, but also reduce the amount of calculation. Here, it use P to represent the set of image block pixel center points, and Q to represent the set of target edge corner points. Distance model *D(P,Q)* is defined as follows:

$$D(P,Q) = d(P,Q) + d(P,Q) \tag{6}$$

$$d(P,Q) = \min(q \in Q), (p \in P) \|p - q\| \tag{7}$$

$$d(Q,P) = \max(q \in Q), (p \in P) \|q - p\| \tag{8}$$

$\|\bullet\|$ represents the distance norm between two sets of points. On this basis, we define the adaptive threshold as follows:

$$K_p = \frac{\alpha_{n+1}}{D(P,Q_n)} \bar{K} . \tag{9}$$

$\bar{K}$ represents setting a threshold. Then the feature extraction can be expressed as follows:

$$b_{i,j} = \begin{cases} 0, & c_{i,j} < K_p \\ 1 & \text{or} \end{cases} . \tag{10}$$

Among them, $\alpha$ represents the updated weight coefficient, $b_{i,j}$ represents the extracted image feature value, and $c_{i,j}$ represents the differential image feature value.

## 4. Experiment

In order to verify the performance of the proposed algorithm, an assembly component positioning experimental platform is built in the laboratory, as shown in figure 2. On the acquired assembly component location video data set, the algorithm proposed in this paper is compared with the traditional background subtraction algorithm, literature [7] and literature [9], and the results are shown in figure 3.

**Figure 2.** Assembly platform.

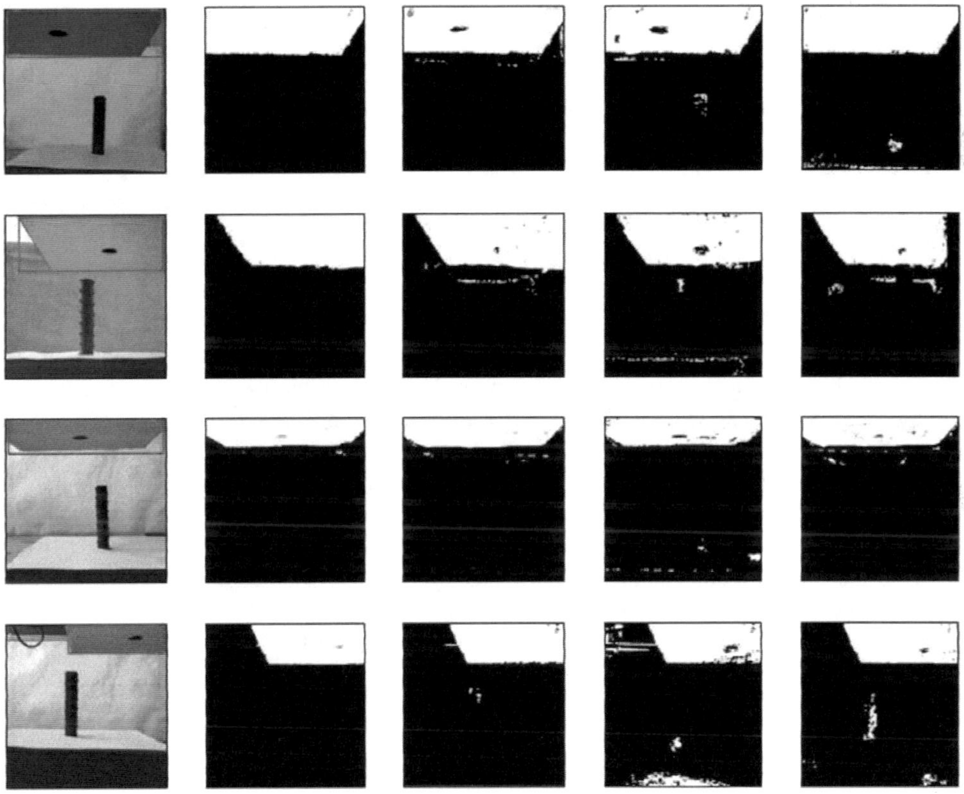

(a) (b) (c) (d) (e)

**Figure 3.** Comparison results: (a) input image; (b) ours; (c) literature [7]; (d) literature [9]; (e) traditional.

In column (a), the red rectangle box indicates the target to be extracted. From column (b), it can be seen that the target edge extracted by this method is clear, the internal defects are small, the external noise is small, and the detection results are obviously better than other detection methods. As column (c), it shows that there is external noise near the target extracted by literature [7], and the extraction accuracy is lower than that of this method. As column (d), it shows that the internal defects and external noise of the target extracted by the algorithm in literature [9] are more than other methods, and the extraction accuracy is lower. From the detection results of (e) column, it can be seen that the target extracted by traditional background subtraction algorithm has more external noise, and there will be incomplete extraction of the target, which shows that the method has poor results in external interference, and the accuracy of target feature extraction is low.

## 5. Conclusion

In this paper, we aiming at the problem of low target detection accuracy of background subtraction algorithm based on fixed threshold, the assembly location video image has been studied. This paper introduces the principle of background subtraction algorithm, carries out background modeling for assembly target, sets corner points on the edge of target feature extracted from the previous frame image, establishes the distance norm model of pixels and corner points in different regions of this frame image, and proposes an adaptive threshold algorithm. In the laboratory, the assembly positioning experiment platform is built, the video image data set is made, and the assembly positioning images at different times are selected for comparative experiments. The results show that the algorithm in this paper greatly improves the target detection performance of the assembly video image, and has a strong ability to suppress image noise.

## Acknowledgments

This work was supported by the National Key R&D Plan of China, grant number 2017YFC0703903, 2017YFC0704003 and the Shenyang Science and Technology Plan Project (Y19-1-004).

## References

[1] He Y 2016 *Moving Object Detection Based on Improved Frame Difference Method and Background Difference Method* Yanshan University
[2] Liu T 2016 *Research on Moving Object Detection Algorithm Based on Background Modelling* Zhengzhou University
[3] Liu Z, He S and Hu W 2017 Moving object detection of video sequence based on background subtraction *Computer Application* **37** (6) 1777-81
[4] Hu Z, Xi Y and Tong J 2018 Adaptive moving target detection in the background of uneven clutter *Command Information System and Technology* **9** 28-32
[5] Zhang Y and Liu Y 2017 Moving object detection based on frame difference method and background difference method *Computer Technology and Development* **27** (2) 25-28
[6] Li Y, Xu X and Chen J 2017 Application of background update of dynamic feature block matching in motion detection *Journal of Instrumentation* **38** (2) 445-53
[7] Jin J, Dang J and Wang Y 2019 Spatial histogram background modeling of regional color attributes *Chinese Journal of Image Graphics* 714-23
[8] Yin R, Yang J and Lu W 2019 Research on anomaly detection algorithm based on the background of regular change *Control Engineering* **26** (8)
[9] Ding Z and Lu W 2019 Moving object detection algorithm based on vibe background modeling. *Computer System Application* **28** (04) 187-91
[10] Di C 2017 *Research on Video Detection and Tracking Methods of Moving Objects Based on Features* University of Chinese Academy of Sciences
[11] Yin H, Chen B, Chai Y *et al.* 2016 Overview of vision based object detection and tracking *Journal of Automation* **42** (10) 1466-89
[12] Wang D, Fan Y and Wang X 2017 A motion detection algorithm based on gradient and vibe algorithm *Television Technology* **41** (3) 100-04
[13] Zhan Z, An Y and Cui W 2017 Image threshold segmentation algorithm and comparative study *Information Communication* (4) 86-89
[14] Chen J 2017 *Research on Threshold Segmentation of Non-uniform Illumination Image* Hunan University of Technology
[15] Hu Z, Fan W and Yu G 2017 Adaptive threshold image foreground extraction based on OpenCV *Photoelectric Technology Application* **32** (1) 56-59

# AUTHOR INDEX